AIRCRAFT DESIGN

Aerospace Series List

Introduction to UAV Systems 4e	Fahlstrom and Gleason	August 2012
Theory of Lift: Introductory Computational Aerodynamics with MATLAB®/ Octave	McBain	August 2012
Sense and Avoid in UAS: Research and Applications	Angelov	April 2012
Morphing Aerospace Vehicles and Structures	Valasek	April 2012
Gas Turbine Propulsion Systems	MacIsaac and Langton	July 2011
Basic Helicopter Aerodynamics, 3rd Edition	Seddon and Newman	July 2011
Advanced Control of Aircraft, Spacecraft and Rockets	Tewari	July 2011
Cooperative Path Planning of Unmanned Aerial Vehicles	Tsourdos et al	November 2010
Principles of Flight for Pilots	Swatton	October 2010
Air Travel and Health: A Systems Perspective	Seabridge et al	September 2010
Design and Analysis of Composite Structures: With applications to aerospace Structures	Kassapoglou	September 2010
Unmanned Aircraft Systems: UAVS Design, Development and Deployment	Austin	April 2010
Introduction to Antenna Placement & Installations	Macnamara	April 2010
Principles of Flight Simulation	Allerton	October 2009
Aircraft Fuel Systems	Langton et al	May 2009
The Global Airline Industry	Belobaba	April 2009
Computational Modelling and Simulation of Aircraft and the Environment: Volume 1 – Platform Kinematics and Synthetic Environment	Diston	April 2009
Handbook of Space Technology	Ley, Wittmann Hallmann	April 2009
Aircraft Performance Theory and Practice for Pilots	Swatton	August 2008
Surrogate Modelling in Engineering Design: A Practical Guide	Forrester, Sobester, Keane	August 2008
Aircraft Systems, 3rd Edition	Moir & Seabridge	March 2008
Introduction to Aircraft Aeroelasticity And Loads	Wright & Cooper	December 2007
Stability and Control of Aircraft Systems	Langton	September 2006
Military Avionics Systems	Moir & Seabridge	February 2006
Design and Development of Aircraft Systems	Moir & Seabridge	June 2004
Aircraft Loading and Structural Layout	Howe	May 2004
Aircraft Display Systems	Jukes	December 2003
Civil Avionics Systems	Moir & Seabridge	December 2002

AIRCRAFT DESIGN

A Systems Engineering Approach

Mohammad H. Sadraey

Daniel Webster College, New Hampshire, USA

WILEY

A John Wiley & Sons, Ltd., Publication

This edition first published 2013

Registered office
John Wiley & Sons Ltd, The Atrium, Southern Gate, Chichester, West Sussex, PO19 8SQ, United Kingdom

For details of our global editorial offices, for customer services and for information about how to apply for permission to reuse the copyright material in this book please see our website at www.wiley.com.

Library of Congress Cataloging-in-Publication Data

Sadraey, Mohammad H.
Aircraft design : a systems engineering approach / Mohammad H. Sadraey.
pages cm
Includes bibliographical references and index.
ISBN 978-1-119-95340-1 (hardback)
1. Airplanes–Design and construction. I. Title.
TL671.2.S3136 2012
629.134′1–dc23

2012009907

A catalogue record for this book is available from the British Library.

Print ISBN: 9781119953401

Set in 10/12pt Times by Laserwords Private Limited, Chennai, India.

Printed and bound in Malaysia by Vivar Printing Sdn Bhd

1 2012

To Fatemeh Zafarani, Ahmad, and Atieh
For all their love and understanding

Contents

Preface

Objectives

The objective of this book is to provide a basic text for courses in the design of heavier-than-air vehicles at both the upper division undergraduate and beginning graduate levels. Aircraft design is a special topic in the aeronautical/aerospace engineering discipline. The academic major of aeronautical/aerospace engineering traditionally tends to have four main areas of expertise: aerodynamics, flight dynamics, propulsion, and structure. A qualified aircraft designer employs all these four scientific concepts and principles and integrates them using special design techniques to design a coordinated unique system; an *aircraft*. Design is a combination of science, art, and techniques. A designer not only must have sufficient level of knowledge in these four areas, but also needs to employ mathematics, skills, experiences, creativity, art, and system design techniques. It is true that aircraft design is not completely teachable in classrooms, but combining class lectures with a semester-long aircraft design project provides the best opportunity for students to learn and experience aircraft design.

Every aeronautical engineering discipline offers at least one course in aircraft design or aerospace system design. The lack of an aircraft design textbook with academic features – such as full coverage of all aspects of an air vehicle, aeronautical concepts, design methods, design flowcharts, design examples, and end-of-chapter problems – combined with the newly developed systems engineering techniques was the main motivation to write this book.

In the past several years, I have talked to various aircraft design instructors and students at conferences and AIAA Design/Build/Fly design competitions. I came to the conclusion that the great design books published by such pioneers as Roskam, Torenbeek, Nicolai, Stinton, and Raymer need more development and expansion. This is to meet the ever-increasing need of universities and colleges for aircraft design education, and of industries for design implementation. The new text should possess significant features such as systems engineering approaches, design procedures, solved examples, and end-of-chapter problems. This book was written with the aim of filling the gap for aeronautical/aerospace engineering students and also for practicing engineers.

Approach

The process of air vehicle design is a complex combination of numerous disciplines which have to be blended together to yield the optimum design to meet a given set of requirements. The systems engineering approach is defined as an interdisciplinary approach encompassing the entire technical effort to evolve and verify an integrated and lifecycle-balanced set of system people, products, and process solutions that satisfy customer needs. Multi-discipline system engineering design involves the application of a systems engineering process and requires engineers with substantive knowledge of design across multiple technical areas and improved tools and methods for doing it. Complex aircraft systems, due to the high cost and the risks associated with their development, become a prime candidate for the adoption of systems engineering methodologies. The systems engineering technique has been applied in the development of many manned airplanes. An aircraft is a system composed of a set of interrelated components working together toward some common objective or purpose. Primary objectives include safe flight achieved at a low cost. Every system is made up of components or subsystems, and any subsystem can be broken down into smaller components. For example, in an air transportation system, the aircraft, terminal, ground support equipment, and controls are all subsystems.

Throughout the text, the systems engineering approach is examined and implemented. The book has been arranged to facilitate the student's gradual understanding of design techniques. Statement proofs are provided whenever they contribute to the understanding of the subject matter presented. Special effort has been made to provide example problems so that the reader will have a clear understanding of the topic discussed. The reader is encouraged to study all such solved problems carefully; this will allow the interested reader to obtain a deeper understanding of the materials and tools.

Features

Some of the unique features of this textbook are as follows. It:

- follows a systems engineering approach;
- is organized based on components design (e.g., wing design, tail design, and fuselage design);
- provides design steps and procedures in each chapter;
- derives a number of design equations that are unique to the book;
- provides several fully solved design examples at the component level;
- has many end-of-chapter problems for readers to practice;
- includes a lot of aircraft figures/images to emphasize the application of the concepts;
- describes some real design stories that stress the significance of safety in aircraft design;
- provides various aircraft configurations, geometries, and weights data to demonstrate real-world applications and examples;
- covers a variety of design techniques/processes so that the designer has freedom and flexibility to satisfy the design requirements in several ways;
- encourages and promotes the creativity of the reader.

For these reasons, as aeronautical/aerospace engineering students transit to practicing engineers, they will find that this text is indispensable as a reference text. Some materials, such as "design optimization" and "design of control surfaces," may be taught at the graduate level. The reader is expected to have a basic knowledge of the fundamentals and concepts of aerodynamics, propulsion, aero-structure, aircraft performance, and flight dynamics (stability and control) at aeronautical/aerospace engineering senior level.

The following is a true statement: "design techniques are not understood unless practiced." Therefore, the reader is strongly encouraged to experience the design techniques and concepts through applied projects. Instructors are also encouraged to define an open-ended semester/year-long aircraft design project to help the students to practice and learn through application and experiencing the iterative nature of the design technique. It is my sincere wish that this book will help aspiring students and design engineers to learn and create more efficient and safer aircraft.

Outline

The text consists of 12 chapters and is organized in a standard fashion according to the systems engineering discipline: conceptual design, preliminary design, and detail design. In summary, Chapter 3 presents the aircraft conceptual design; Chapter 4 introduces the aircraft preliminary design; and Chapters 5–12 cover the aircraft detail design. The outline of this book is as follows.

Chapter 1 is an introduction to design fundamentals and covers such topics as engineering design principles, design project planning, decision-making processes, feasibility analysis, and tort of negligence. Design standards and requirements such as Federal Aviation Regulations (FARs) and Military Standards are reviewed in this chapter, and addressed further throughout the text.

Chapter 2 deals with the systems engineering approach. Major design phases according to systems engineering are introduced: conceptual system design, preliminary system design, and detail system design. In this chapter, several concepts and fundamental definitions such as technical performance measures, functional analysis, system trade-off analysis, design review, and design requirements are reviewed. Implementations of systems engineering into aircraft design via aircraft design phases, aircraft design flowcharts, aircraft design groups, and design evaluation and feedback loops are explained. At the end of the chapter, the overall aircraft design procedure in terms of design steps is outlined.

Chapter 3 covers aircraft conceptual design, and examines the aircraft configuration selection. The primary function of each aircraft component such as wing, fuselage, tail, landing gear, and engine is introduced. Furthermore, various configuration alternatives for each component are reviewed. In addition, the aircraft classification and design constraints are addressed. In this chapter the design optimization and its mathematical tools are briefly reviewed. The chapter ends with a configuration selection process and methodology, and also a trade-off analysis technique.

Chapter 4 discusses the topic of aircraft preliminary design. In this chapter, the technique to determine three aircraft fundamental parameters is presented. These parameters are: maximum take-off weight, wing area, and engine thrust/power. The weight build-up technique is examined for estimation of the aircraft maximum take-off weight. The matching plot technique is utilized in the calculation of wing area, and engine thrust/power. These three parameters are computed based on the aircraft performance requirements such as range, endurance, maximum speed, take-off run, rate-of-climb, and ceiling. Two fully solved examples illustrate the application of the two techniques.

Chapters 5–9 and 12 present detail design of the aircraft components of wing, tail, fuselage, propulsion system, landing gear, and control surfaces respectively. In these chapters, the techniques to calculate all aircraft components parameters such as wing/tail span, chord, airfoil, incidence, sweep angle, tail arm, tail area, landing gear height, wheel base, wheel track, fuselage diameter, fuselage length, cabin design, cockpit design, number of engines, and engine selection are examined. Furthermore, the features of various component configurations and their relationship with the design requirements (e.g., performance, stability, control, and cost) are addressed. Chapter 12 introduces the detail design of the conventional control surfaces of aileron, elevator, and rudder. In each chapter, the design flowchart and design step for each component is also presented. Each chapter is accompanied by several examples, including a fully solved chapter example to demonstrate the applications of design techniques and methods.

Chapter 10 introduces the technique to calculate the weight of the aircraft components, equipment, and subsystems. The technique is derived mainly based on past aircraft weight data and statistics.

Chapter 11 addresses the topic of aircraft weight distribution, and weight and balance. The aircraft center of gravity (cg) calculation, aircraft most aft and most forward cg, and cg range are also covered in this chapter. In addition, the technique to determine the aircraft mass moment of inertia about three axes (i.e., x, y, and z) is examined.

Unit Systems

In this text, the emphasis is on SI units or the metric system, which employs the meter (m) as the unit of length, the kilogram (kg) as the unit of mass, and the second (s) as the unit of time. It is true that metric units are more universal and technically consistent than British units. However, currently, many FARs are published in British units, where the foot (ft) is the unit of length, the slug is the unit of mass, the pound (lb) is the unit of force (weight), and the second (s) is the unit of time. In FARs, the pound is used as the unit for force and weight, the knot for airspeed, and the foot for altitude. Thus, in various locations, the knot is mainly used as the unit of airspeed, the pound for weight and force, and the foot for altitude. Therefore, in this text, a combination of SI and British unit systems is utilized. A common mistake in the literature (even in the *Jane's* publications) is the application of kg for the unit of aircraft weight. Throughout the text, whenever the unit of kg is used, the term "aircraft mass" is employed. Some texts have created the pound-mass (lbm) as the unit of mass, and the pound-force (lbf) as the unit of weight. This initiative may generate some confusion; so in this text, only one pound (lb) is employed as the unit of weight and force.

Series Preface

The field of aerospace is wide ranging and multi-disciplinary, covering a large variety of products, disciplines and domains, not merely in engineering but in many related supporting activities. These combine to enable the aerospace industry to produce exciting and technologically advanced vehicles. The wealth of knowledge and experience that has been gained by expert practitionersin the various aerospace fields needs to be passed onto others working in the industry, including those just entering from University.

The *Aerospace Series* aims to be a practical and topical series of books aimed at engineering professionals, operators, users and allied professions such as commercial and legal executives in the aerospace industry. The range of topics is intended to be wide ranging, covering design and development, manufacture, operation and support of aircraft as well as topics such as infrastructure operations and developments in research and technology. The intention is to provide a source of relevant information that will be of interest and benefit to all those people working in aerospace.

Aircraft design brings together the key aeronautical engineering disciplines: aerodynamics, flight dynamics, propulsion and structures, which must be combined to produce designs that meet today's stringent performance, economic and environmental demands. As such, aircraft designis a key component of all undergraduate aerospace engineering courses, and all aerospace students usually tackle some form of aircraft design project.

This book, *Aircraft Design: A Systems Engineering Approach*, extends the classical aircraft design approaches through the implementation of systems engineering techniques for the conceptual, preliminary and detailed design of heavier-than-air vehicles. As a very readable and informative text reference, with plenty of examples from a wide range of contemporary aircraft designs, and solved examples at the end of each chapter, it is a worthy addition to the Wiley Aerospace Series.

Peter Belobaba, Jonathan Cooper, Roy Langton and Allan Seabridge

Series Preface

Acknowledgments

I am enormously grateful to the Almighty for the opportunity to serve the aerospace community by writing this text. The author would like to acknowledge the many contributors and photographers who have contributed to this text. I am especially grateful to those who provided great aircraft photographs: Anne Deus (Germany); Jenney Coffey (UK); Anthony Osborne (UK); A J Best (UK); Vlamidir Mikitarenko (Germany); Rainer Bexten (Germany); Hideki Nakamura (Japan); Akira Uekawa (Japan); Luis David Sanchez (Puerto Rico); Tom Houquet (Belgium); Toshi Aoki (Japan); Miloslav Storoska (Slovakia); Tom Otley (Panacea Publishing International, UK); Jonas Lövgren (SAAB, Sweden); Jeff Miller (Gulfstream Aerospace Corporation, USA); Michael de Boer (Netherland); Konstantin von Wedelstaedt (Germany); Augusto G. Gomez R. (Mexico); Randy Crew (Singapore); Robert Domandl; Serghei Podlesnii (Moldova); Orlando J. Junior (Brazil); Balázs Farkas (Hungary); and Christopher Huber and www.airliners.net. In addition, the efforts of the author were helped immeasurably by the many insights and constructive suggestions provided by students and instructors over the past 16 years. Unattributed figures are held in the public domain and are from either the US Government Departments and Agencies, or Wikipedia.

Putting a book together requires the talents of many people, and talented people abound at John Wiley & Sons, Inc. My sincere gratitude goes to Paul Petralia, commissioning editor, for coordinating the whole publication process; Clarissa Lim for coordinating the production project; Sarah Lewis for editing the manuscript; Jayashree Saishankar for typesetting; and Sandra Grayson for helping in the copyright process. I am particularly grateful to my editors, Liz Wingett and Sophia Travis, for their comments and guidance. My special thanks go to the outstanding copyeditors and proofreaders who are essential in creating an error-free text. I especially owe a large debt of gratitude to the reviewers of this text. Their ideas, suggestions, and criticisms have helped me to write more clearly and accurately and have influenced the evolution of this book markedly.

Symbols and Acronyms

Symbols

Symbol	Name	Unit
a	Speed of sound	m/s, ft/s
a	Acceleration	m/s^2, ft/s^2
A	Area	m^2, ft^2
AR	Aspect ratio	–
b	Lifting surface/control surface span	m, ft
B	Wheel base	m, ft
C	Specific fuel consumption	N/h kW, lb/h hp
$\overline{C}$	Mean aerodynamic chord	m, ft
C_D, C_L, C_y	Drag, lift, and side-force coefficients	–
C_l, C_m, C_n	Rolling, pitching, and yawing moment coefficients	–
C_h	Hinge moment coefficient	–
C_{D_y}	Aircraft side drag coefficient	–
C_{D_β}	Rate of change of drag coefficient w.r.t. sideslip angle; $\partial C_D / \partial \beta$	1/rad
$C_{m_{ac\_wf}}$	Wing/fuselage pitching moment coefficient (about the wing/fuselage aerodynamic center)	–
C_{m_α}	Rate of change of pitching moment w.r.t. angle of attack	1/rad
C_{m_q}	Rate of change of pitch rate w.r.t. angle of attack	1/rad
$C_{L_{\delta_E}}$	$\partial C_L / \partial \delta_E$	1/rad
$C_{m_{\delta_E}}$	$\partial C_m / \partial \delta_E$	1/rad
$C_{l_{\delta A}}$	$\partial C_l / \partial \delta_A$	1/rad
$C_{n_{\delta R}}$	$\partial C_n / \partial \delta_R$	1/rad
C_{n_β}	Rate of change of yawing moment coefficient w.r.t. sideslip angle	1/rad
C_{n_r}	Rate of change of yawing moment coefficient w.r.t. yaw rate	1/rad

Symbol	Name	Unit
C_{D_o}	Zero-lift drag coefficient	–
C_{D_i}	Induced drag coefficient	–
C_f	Skin friction coefficient	–
$C_{L\alpha}$	Wing/tail/aircraft (3D) lift curve slope	1/rad
$C_{l\alpha}$	Airfoil (2D) lift curve slope	1/rad
$C_{L_{max}}$	Maximum lift coefficient	–
D	Drag force, drag	N, lb
D	Diameter	m, ft
d_c	Distance between the aircraft cg and center of the projected side area	m, ft
E	Endurance	h, s
E	Modulus of elasticity	N/m^2, Pa, lb/in^2, psi
e	Oswald span efficiency factor	–
F	Force, friction force	N, lb
F_C	Centrifugal force	N, lb
FOM	Figure of merit	–
g	Gravity constant	$9.81\ m/s^2$, $32.17\ ft/s^2$
G	Fuel weight fraction	–
GR	Gearbox ratio	–
G_C	Ratio between the linear/angular movement of the stick/wheel to deflection of the control surface	deg/m, deg/ft, deg/deg
H	Altitude	m, ft
h, h_o	Non-dimensional distance between cg (h) or ac (h_o) and a reference line	–
H	Height, wheel height	m, ft
H	Control surface hinge moment	Nm, lb ft
i_h	Tail incidence	deg, rad
i_w	Wing incidence	deg, rad
l	Length, tail arm	m, ft
I	Mass moment of inertia	kg m^2, slug ft^2
I	Second moment of area	m^4, ft^4
I	Index (e.g., design, performance)	–
K	Induced drag factor	–
L, L_A	Rolling moment	Nm, lb ft
L	Length	m, ft
L	Lift force, lift	N, lb
$(L/D)_{max}$	Maximum lift-to-drag ratio	–
M	Mach number	–
M, M_A	Pitching moment	Nm, lb ft
m	mass	kg, slug
$\dot{m}$	Engine air mass flow rate	kg/s, lb/s
MTOW	Maximum take-off weight	N, lb
MAC	Mean aerodynamic chord	m, ft
n	Load factor	–
n	Number of rows in cabin	–

Symbol	Name	Unit
n	Rotational speed	rpm, rad/s
N	Normal force	N, lb
N	Number of an item	–
N, N_A	Yawing moment	Nm, lb ft
P	Pressure	N/m^2, Pa, lb/in^2, psi
P	Power	W, kW, hp, lb ft/s
P_s	Seat pitch	m, ft
P_{req}	Required power	W, kW, hp, lb ft/s
P_{av}	Available power	W, kW, hp, lb ft/s
P_{exc}	Excess power	W, kW, hp, lb ft/s
P, p	Roll rate	rad/s, deg/s
q, $\overline{q}$	Dynamic pressure	N/m^2, Pa, lb/in^2, psi
Q, q	Pitch rate	rad/s, deg/s
Q	Fuel flow rate	kg/s, lb/s
R	Range	m, km, ft, mile, mi, nmi
R	Air gas constant	287.26 J/kg K
R	Radius	m, ft
R	Rank	–
Re	Reynolds number	–
ROC	Rate of climb	m/s, ft/min, fpm
R, r	Yaw rate	rad/s, deg/s
s	Semispan ($b/2$)	m, ft
S	Planform area of lifting/control surface	m^2, ft^2
S_A	Airborne section of the take-off run	m, ft
S_G	Ground roll	m, ft
S_{TO}	Take-off run	m, ft
SFC	Specific fuel consumption	N/h/kW, lb/h/hp, 1/s, 1/ft
SM	Static margin	–
t	Time	s, min, h
T	Engine thrust	N, lb
T	Temperature	°C, °R, K
T	Wheel track	m, ft
T, t	Thickness	m, ft
t/c	Airfoil thickness-to-chord ratio	–
T/W	Thrust-to-weight ratio	–
U	Forward airspeed	m/s, ft/min, km/h, mi/h, knot
V	Velocity, speed, airspeed	m/s, ft/min, km/h, mi/h, knot
V	Volume	m^3, ft^3
V_{max}	Maximum speed	m/s, ft/min, km/h, mi/h, knot
V_{mc}	Minimum controllable speed	m/s, ft/min, km/h, mi/h, knot
V_{min_D}	Minimum drag speed	m/s, ft/min, km/h, mi/h, knot
V_{min_P}	Minimum power speed	m/s, ft/min, km/h, mi/h, knot
V_R	Rotation speed	m/s, ft/min, km/h, mi/h, knot
$V_{ROC_{max}}$	Maximum rate of climb speed	m/s, ft/min, km/h, mi/h, knot
V_s	Stall speed	m/s, ft/min, km/h, mi/h, knot
V_T	True speed	m/s, ft/min, km/h, mi/h, knot

Symbol	Name	Unit
V_{TO}	Take-off speed	m/s, ft/min, km/h, mi/h, knot
V_W	Wind speed	m/s, ft/min, km/h, mi/h, knot
$\overline{V}_H$, $\overline{V}_V$	Horizontal/vertical tail volume coefficient	–
W	Weight	N, lb
W	Width	m, ft
W_f	Fuel weight	N, lb
W_{TO}	Maximum take-off weight	N, lb
W/P	Power loading	N/W, lb/hp
W/S	Wing loading	N/m^2, lb/ft^2
x, y, z	Displacement in x, y, and z direction	m, ft
Y	Side force	N, lb
y	Beam deflection	m, ft

Greek symbols

Symbol	Name	Unit
α	Angle of attack	deg, rad
β	Sideslip angle	deg, rad
γ	Climb angle	deg, rad
θ	Pitch angle, pitch attitude	deg, rad
λ	Taper ratio	–
ϕ	Bank angle	deg, rad
δ	Pressure ratio	–
δ	Control surface deflection	deg, rad
σ	Air density ratio	–
σ	Sidewash angle	deg, rad
ρ	Air density, materials density	kg/m^3, $slug/ft^3$
μ	Dynamic viscosity	kg/m s, lb s/ft^2
μ	Friction coefficient	–
μ	Mach angle	rad, deg
η	Efficiency, dynamic pressure ratio	–
Λ	Sweep angle	deg, rad
ω	Angular velocity	rad/s, deg/s
ω_n	Natural frequency	rad/s, deg/s
ω	Frequency	rad/s, deg/s
ψ	Yaw angle, heading angle	deg, rad
π	3.14	–
Ω	Spin rate	rad/s, deg/s, rpm
τ	Control surface angle of attack effectiveness	–
Γ	Dihedral angle	deg, rad
ε	Downwash angle	degr, rad
$\partial\varepsilon/\partial\alpha$	Downwash slope	–
$\partial\sigma/\partial\beta$	Sidewash slope	–
$\ddot{\theta}$	Take-off rotation angular acceleration	deg/s^2, rad/s^2
$\Delta\overline{x}_{cg}$	Non-dimensional range of center of gravity	–

Subscripts

Note	AR, S, b, λ, Λ, Γ, and C without a subscript indicate a wing property
0, o	Zero-lift, sea level, about aerodynamic center
0.25	Quarter chord
1	Steady-state value
a, A	Aileron
aft	The most aft location
A	Aerodynamic
ac	Aerodynamic center
avg	Average
a	Aircraft
b	Baggage
c/4	Relative to the quarter chord
c/2	Relative to the 50% of the chord
cs	Control surface
cross	Cross-section
C	Crew, ceiling, cruise, cabin
d	Design
D	Drag
e, E	Elevator, equivalent, empty, exit
eff	Effective
E	Engine
f	Fuel, fuselage, flap, friction
for	The most forward location
GL	Glide
h	Horizontal tail
i	Item number, inboard, ideal, initial, inlet
ISA	International Standard Atmosphere
L	Lift, left, landing
LG	Landing gear
max	Maximum
min	Minimum
m	Pitching moment
mg	Main gear
mat	Materials
o	Outboard
opt	Optimum
ot	Overturn
p	Propeller
PL	Payload
r, R	Rudder
R	Rotation
r	Root
ref	Reference
s	Stall, stick
ss	Steady-state
SL	Sea level
S	Side
SR	Spin recovery

t	Tip, tab, twist, horizontal tail
T	True
TO	Take-off
tot	Total
ult	Ultimate
v, V	Vertical tail
VT	Vertical tail
w, W	Wing, wind
wet	Wetted
wf	Wing/fuselage
x, y, or z	In the x, y, or z direction
xx, yy, or zz	About the x-, y-, or z-axis

Acronyms

ac or AC	Aerodynamic center
ca	Center of area, center of action
cg or CG	Center of gravity
APU	Auxiliary power unit
CAD	Computer-aided design
CAM	Computer-aided manufacturing
CDR	Conceptual design review
CFD	Computational fluid dynamics
cp	Center of pressure
DOF	Degrees of freedom
DOD	Department of Defense
EASA	European Aviation Safety Agency
ETR	Evaluation and test review
FDR	Final (critical) design review
FAA	Federal Aviation Administration
FAR	Federal Aviation Regulations
FBW	Fly-by-wire
GA	General aviation
HALE	High-altitude long-endurance
HLD	High-lift device
IATA	International Air Transport Association
ISA	International Standard Atmosphere
JAR	Joint aviation requirements
KTAS	Knot true air speed
KEAS	Knot equivalent air speed
LG	Landing gear
LE	Leading edge
MAC	Mean aerodynamic chord
MDO	Multidisciplinary design optimization
MIL-STD	Military Standards
NACA	National Advisory Committee for Aeronautics
NASA	National Aeronautics and Space Administration
NTSB	National Transportation Safety Board

np or NP	Neutral point
OEI	One engine inoperative
PDR	Preliminary design review
rpm	Revolutions per minute
rad	Radian
RCS	Radar cross-section
STOL	Short take-off and landing
TE	Trailing edge
Turboprop	Turbopropeller
VTOL	Vertical take off and landing
WWII	World War II

Conversion Factors

Length, Altitude, Range

1 in = 2.54 cm = 25.4 mm
1 ft = 0.3048 m = 12 in
1 statue mile (mi) = 5280 ft = 1.609 km
1 nautical mile (nmi) = 6076 ft = 1.852 km
1 km = 0.6214 mi = 3280.8 ft
1 m = 3.281 ft = 39.37 in

Area

1 m^2 = 10.764 ft^2
1 ft^2 = 0.093 m^2

Volume

1 l = 0.001 m^3 = 1000 cm^3 = 0.0353 ft^3 = 0.264 US gal
1 ft^3 = 0.0283 m^3 = 7.481 US gal
1 US gal = 0.1337 ft^3 = 3.785 l
1 m^3 = 1000 l = 264.17 US gal = 35.315 ft^3

Speed, Airspeed, Rate of Climb

1 knot = 0.514 m/s = 1.151 mi/h = 1.852 km/h = 1.688 ft/s = 101.27 ft/min
1 mi/h = 1.609 km/h = 1.467 ft/s = 0.447 m/s = 0.869 knot = 88 ft/min
1 km/h = 0.6214 mi/h = 0.2778 m/s = 0.9113 ft/s = 0.54 knot = 54.68 ft/min
1 ft/min = 0.01 knot = 0.011 mi/h = 0.018 km/h = 0.0051 m/s = 0.017 ft/s

Mass

1 slug = 14.59 kg
1 kg = 1000 g = 0.0685 slug

Force, Weight, Thrust

1 N = 0.225 lb
1 lb = 4.448 N

Mass and Weight

1 N (weight) = 1 kg m/s^2 (weight) → 0.102 kg (mass)
1 lb (weight) = 1 slug ft/s^2 (weight) = 4.448 N (weight) → 0.454 kg (mass)
1 kg (mass) → 9.807 N (weight) = 2.205 lb (weight)

Work, Energy

1 J = 0.7376 ft lb
1 BTU = 1055 J = 778.17 ft lb
1 cal = 4.187 J = 3.09 ft lb

Power

1 hp = 550 ft lb/s = 745.7 W = 33 000 lb ft/min
1 kW = 737.56 ft lb/s = 1.341 hp

Mass Moment of Inertia

1 kg m^2 = 0.738 slug ft^2
1 slug ft^2 = 1.356 kg m^2

Pressure

1 Pa = 1 N/m^2 = 0.00015 lb/in^2 = 0.00015 psi
1 atm = 101 325 Pa= 1.013 bar = 14.7 lb/in^2 = 14.7 psi
1 psi = 6895 Pa = 0.068 atm

Time, Endurance

1 day = 24 h = 1440 min = 86 400 s
1 h = 60 min = 3600 s

Angle

1 rad = 180/π deg = 57.3 deg
1 deg = π/180 rad = 0.01745 rad

1

Aircraft Design Fundamentals

1.1 Introduction to Design

Aircraft design is essentially a branch of engineering design. Design is primarily an analytical process which is usually accompanied by drawing/drafting. Design contains its own body of knowledge, independent of the science-based analysis tools usually coupled with it. Design is a more advanced version of a problem-solving technique that many people use routinely. Design is exciting, challenging, satisfying, and rewarding. The general procedure for solving a mathematical problem is straightforward. Design is much more subjective, there is rarely a single "correct" answer. The world of design involves many challenges, uncertainties, ambiguities, and inconsistencies. This chapter is intended to familiarize the reader with the basic fundamentals and overall process of design. This book has been written primarily to provide the basic tools and concepts required to create an optimum/efficient aircraft design that will meet the necessary design requirements.

A very basic and simplified model of a design process is shown schematically in Figure 1.1. In general, a design process includes three major operations: analysis, synthesis, and evaluation. Analysis is the process of predicting the performance or behavior of a design candidate. Evaluation is the process of performance calculation and comparing the predicted performance of each feasible design candidate to determine the deficiencies. The noun synthesis refers to a combination of two or more entities that together form something new. In this text, synthesis is employed interchangeably with design. Hence, synthesis is defined as the creative process of putting known things together into new and more useful combinations. Synthesis is the vehicle of the design, with evaluation being its compass. The candidate designs that fail to satisfy (partially or completely) the requirements are reiterated. That is new values, features, characteristics, or parameters are determined during synthesis operation. The redesigned candidate is reanalyzed again for compliance with the design requirements. This iterative process is continued until the design requirements are met. A design process requires both integration and iteration, invoking a process that coordinates synthesis, analysis, and evaluation. These three operations must be integrated and applied iteratively and continuously throughout the lifecycle of the design.

Aircraft Design: A Systems Engineering Approach, First Edition. Mohammad H. Sadraey.

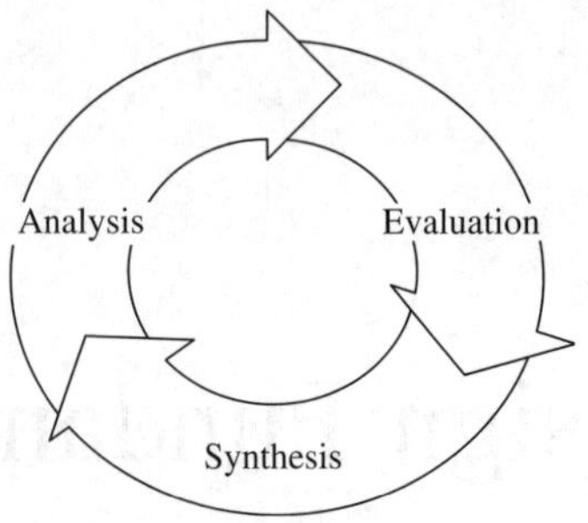

Figure 1.1 Interrelationship between synthesis, analysis, and evaluation

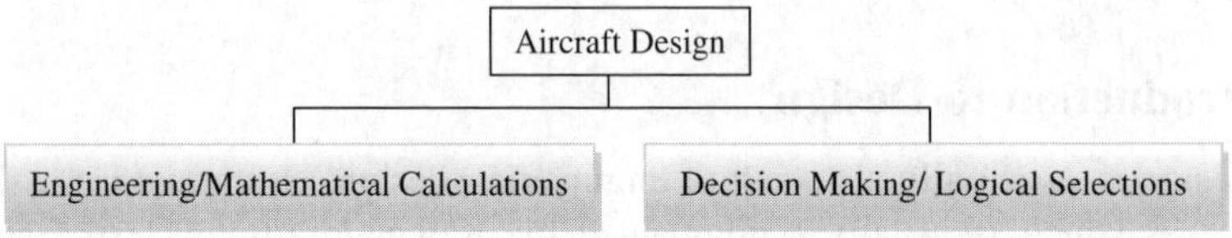

Figure 1.2 Two main groups of design activities in aircraft design

A design operation often involves two activities: (i) problem solving through mathematical calculations and (ii) choosing a preferred one among alternatives (Figure 1.2). The first activity is performed in Chapters 4–12 in designing various aircraft components. The second design activity is in general a decision-making process. The fundamentals of decision making are reviewed in Section 1.4; and employed entirely in aircraft conceptual design (Chapter 3). In addition, there are various decision-making processes in aircraft components design (e.g., wing design, tail design, and propulsion system design), as will be discussed in several chapters. The major components that comprise a conventional aircraft are wing, fuselage, horizontal tail, vertical tail, engine, landing gear, and equipment. The decision-making process plays a significant role in the configuration design of these primary components.

The traditional engineering education is structured to emphasize mathematics, physical sciences, and engineering sciences. The problem is the lack of sufficient concentration on design and creativity. Creative thinking and its attitudes are essential to design success. Producing a new design requires an ability to be creative and overcome strong barriers. To address this significant issue a new organization, CDIO,[1] was established in the late 1990s. The CDIO initiative is defined to be an innovative educational framework for producing the next generation of engineers. The framework provides students with an education stressing engineering fundamentals set within the context of conceiving/designing/implementing/operating real-world systems and products. This textbook has been written with a strong emphasis on creativity, and the freedom of the designer to go beyond current aircraft designs.

[1] www.cdio.org.

Throughout this text, various techniques for generating creative design alternatives are introduced. An effective approach in creative design as a source of new ideas is *brainstorming*. Brainstorming is a structured group-oriented technique for conceiving design alternatives. It consists of a group of individuals letting their imaginations run wild, but in accordance with central procedural rules. The ultimate goal is that the group members will inspire and support each other. The outcome is that the group will be able to conceptualize design alternatives that are more elegant than those the individuals could have achieved independently. In order to encourage members to describe their ideas, even totally impractical ones, a crucial brainstorming rule is that no criticism of individuals or ideas is permitted. The emphasis is on generating as many ideas and concepts as possible, without worrying about their validity. Rectifying, organizing, and combining the ideas suggested in a brainstorming session is performed out of the group meeting. The brainstorming technique is mainly applicable at the conceptual design phase (see Chapters 2 and 3).

In general, aircraft design requires the participation of six (Figure 1.3) fundamental disciplines: (i) flight dynamics, (ii) aerodynamics, (iii) propulsion, (iv) aero-structure, (v) management skills, and (vi) engineering design. The first four items are primary expertise areas of aeronautical engineering. This text has no particular chapters on any of these four topics; so the reader is expected to be familiar with the fundamentals, concepts, technical terms, and engineering techniques in such areas. Management is defined [1] as coordinating work activities so that they are completed efficiently and effectively with and through other people. An aircraft designer needs to be equipped with managerial skills and act as a manager throughout the design process. This topic is not covered in this text; however, a few aspects of management – such as project planning and decision making – are reviewed in this chapter (Sections 1.3 and 1.4).

Finally, engineering design [2–4] is at the heart of the design process and is assumed as the sixth discipline necessary for design of an air vehicle. Section 1.2 briefly examines various aspects of engineering design. It must be noted that aircraft engineering design has its own science, concepts, fundamentals, technical terms, and techniques. Chapters 3–12 all address various aspects of designing aircraft components as well as introducing aircraft design procedures.

This chapter will first examine the engineering design profession. Next, design project planning is addressed and tools such as Gantt charts are introduced. Then the principle of decision making, a very significant section of any design process, is presented. Feasibility study is also discussed in Section 1.5. Finally, the tort of negligence will be described to warn aircraft design engineers to take the utmost care in order to prevent liability.

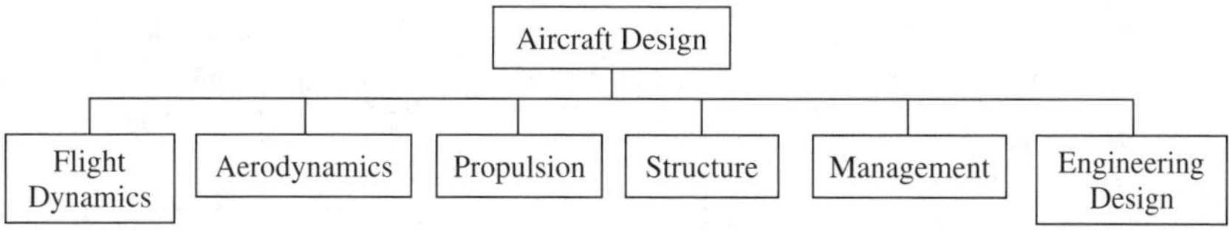

Figure 1.3 Aircraft design required tools and expertise

1.2 Engineering Design

Aircraft design is essentially a branch of engineering design. Design is the culmination of all engineering activities, embodying engineering operations and analysis as tools to achieve design objectives. Many engineering professors find it more difficult to teach design than to teach traditional engineering science-based analytical topics. Every undergraduate engineering curriculum has a design component, although the extent and structure of that component may vary widely. Engineering design fundamentals are common to all engineering disciplines – aeronautical, mechanical, electrical, civil, and computer. Engineering design is a methodical approach to dealing with a particular class of large and complex projects. Engineering design provides the design engineer with a realistic design process. Design is the central activity of the engineering profession, and it is concerned with approaches and management as well as design techniques and tools. In this section, the fundamentals of engineering design as well as the definitions of a few technical terms are presented.

There is a clear distinction between classical mathematics and science problem-solving techniques, and design operation. There is inherently a beauty embedded in the design process which is usually felt after the design output is created. The mathematics and science problems have three main features: (i) the problems are well-posed in a compact form, (ii) the solutions to each problem are unique and compact, and (iii) the problems have an identifiable closure. However, a real-world engineering design problem does not share these characteristics. In fact, engineering design problems are usually poorly posed, do not have a unique solution, and are also open-ended. The Accreditation Board of Engineering and Technology (ABET) [5] defines engineering design as follows:

> Engineering design is the process of devising a system, component, or process to meet desired needs. It is a decision making process (often iterative), in which the basic sciences and mathematics and engineering sciences are applied to convert resources optimally to meet a stated objectives. Among the fundamental elements of the design process are the establishment of objectives and criteria, synthesis, analysis, construction, testing, and evaluation.

Just as the ABET statement is only one of many definitions of engineering design, there are several approaches to describing how design is done. This text formalizes the ABET description into a simplified step-by-step model of the design process based on a systems engineering approach [6]. A very basic block diagram of the design process is shown in Figure 1.4. It represents the road from customer need to design output, including feedback based on evaluation. The problem formulation is discussed in this section, and project

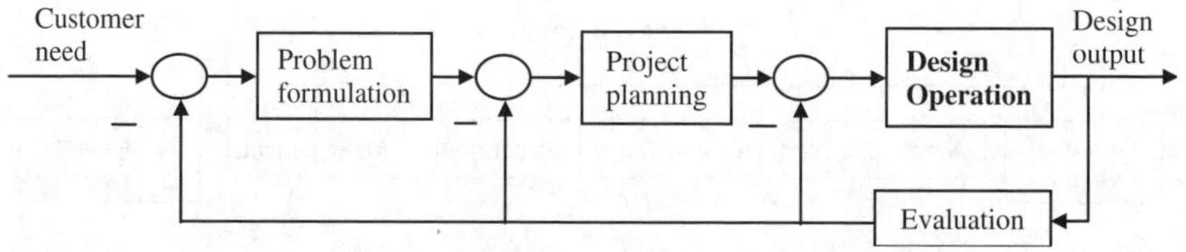

Figure 1.4 Engineering design block diagram

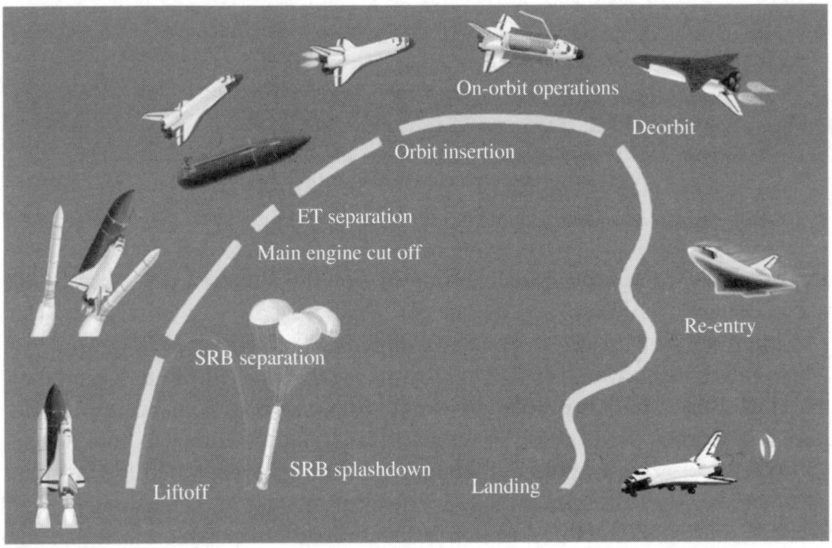

Figure 1.5 The original Space Shuttle concept and mission profile. Reproduced from permission of NASA

planning is examined in Section 1.4. A large part of this text is on design operations, including Chapters 3–12.

The evaluation not only influences the design operation, but most of the time may affect problem formulation and project planning. A clear current example is the Space Shuttle, which started in 1981 but retired in 2011. After more than 30 years of successful operations (135 space missions), the National Aeronautics and Space Administration (NASA) figured out that the current design concept is not viable. Besides economic factors, two reasons that forced NASA to re-engineer the Space Shuttle (Figure 1.5) are the disasters that happened in 1986 and 2003. On January 28, 1986 Space Shuttle Challenger broke apart, just 73 seconds into its flight, leading to the deaths of its seven crew members. On February 1, 2003, shortly before it was scheduled to conclude its 28th mission, Space Shuttle Columbia disintegrated over Texas during re-entry into the Earth's atmosphere, resulting in the death of all seven crew members. Until another US launch vehicle is ready, crews will travel to and from the International Space Station aboard Russian Soyuz spacecraft or possibly a future American commercial spacecraft.

After the need is clearly defined, the designer has to turn his/her attention to describing how he/she envisions meeting the need. This fundamental step requires achieving a delicate balance between establishing the general scope of the design efforts, and avoiding being so specific that opportunities are unnecessarily narrowed for creative design solutions. Problem formulation includes recognizing the need, identifying the customer, market assessment, defining the problem, functional analysis, and establishing design requirements. A problem statement needs to be constructed in such a way that it consists of three components: goal, objectives, and constraints (Figure 1.6).

A **goal** statement is a brief, general, and ideal response to the need statement. The need describes the current, unsatisfactory situation, while the goal describes the ideal future

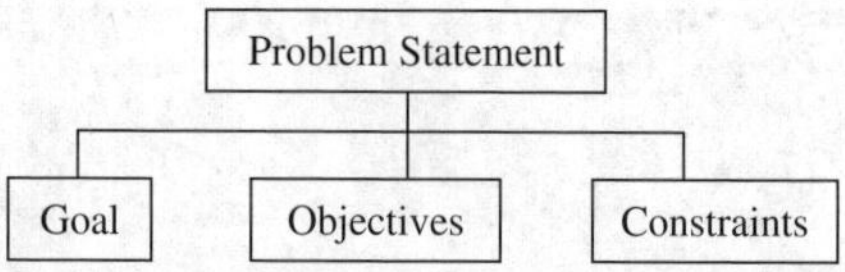

Figure 1.6 Three elements of a problem statement

condition to which we aspire in order to improve on the situation described by the need. The goal is defined by describing the current situation that is unsatisfactory. Hence the goal is to improve the current situation to a higher level. The goal is generally so ideal that it could never be accomplished. The goal is usually revised through a process called benchmarking. **Benchmarking** involves explicitly comparing your design to that of the competitor which does the best job in terms of satisfying customer requirements.

The **objectives** are quantifiable expectations of performance which identify those performance characteristics of a design that are of most interest to the customer. In addition, the objectives must include a description of conditions under which a design must perform. In the lifecycle, the objective is to specify the *whats* and not the *hows*; that is, *what* needs to be accomplished versus *how* it is to be done. When the operating conditions are specified, the designer is able to evaluate the performance of different design options under comparable conditions. Each of the objectives must be defined using words that convey the desirable aspect of performance. The term "performance specification" is often a synonym for objectives. However, the term "design specification" refers to the detailed description of the completed design, including all dimensions, material properties, weight, and fabrication instructions.

Restrictions of function or form are called **constraints**; they limit our freedom to design. Constraints define the permissible conditions of design features and the permissible range of the design and performance parameters. They are features that all design must have in order to be eligible for consideration. Most engineering design projects essentially include a variety of realistic constraints, such as economic factors, safety, reliability, aesthetics, ethics, and social impacts. For instance, the height of the new system cannot exceed 1.4 m; or its mass may not exceed 3.6 kg; or it must operate year-round during cold and hot days.

The value-free descriptors associated with each objective are referred to as **criteria**. For instance, an objective for a design is that it must be "inexpensive." The criterion associated with this objective is "cost." The criteria are quantified using the same bases for measurement and the same unit as their corresponding objectives. In other words, the criteria are more compact ways of identifying objectives. Table 1.1 demonstrates a number of typical design objectives and related criteria to design a vehicle.

Fundamentally, design products are developed and created to satisfy needs and wants and provide utility to the customer. The customer's needs have to be translated into **design requirements** through goal and objectives. Design requirements mainly include customer requirements plus engineering requirements. The customer requirements refer to objectives as articulated by the customer or client. The engineering requirements refer to the design and performance parameters that can contribute to achieving the customer requirements.

Table 1.1 Typical design objectives and related criteria for a vehicle design project

No.	Objective	Basis for measurement	Criterion	Units
1	Inexpensive in market	Unit manufacturing cost	Manufacturing cost	Dollar
2	Inexpensive in operation	Fuel consumption per kilometer	Operating cost	l/km
3	Light	Total weight	Weight	N
4	Small size	Geometry	Dimensions	m
5	Fast	Speed of operation	Performance	km/h
6	Maintainable	Man-hours to maintain	Maintainability	Man-hour
7	Producible	Required technology for manufacturing	Manufacturability	–
8	Recyclable	Amount of hazardous or non-recyclable materials	Disposability	kg
9	Maneuverable	Turn radius	Maneuverability	m
10	Comfortable	Ergonomic standards	Human factor	–
11	Airworthiness	Safety standards	Safety	–
12	No human casualty in operation	Level of injury to passengers in a mishap	Crashworthiness	–

Figure 1.7 illustrates conceptually the status of various design features during the design process. It indicates that there will be a large commitment in terms of configuration, manufacturing technology, and maintenance techniques at the early stages of a design program. In addition, it is at this point that major decisions are made and product-specific knowledge is limited. Moreover, it is estimated that about 70% of the projected lifecycle cost for a given product can be committed based on engineering design and management decisions during the early stages of design. As the design progresses, changes to the design get harder and harder. Therefore, the impact of a decision at the early stages of a design program is more profound than a decision at the later stages. Hence, it is crucial to be highly confident about any decision a designer makes at the conceptual design phase.

The cost of aircraft design is about 1% of the total lifecycle cost; however, this 1% determines the other 99%. Furthermore, the design cost is about 20% of the production (acquisition) cost. Thus, any necessary investment in design team members is worth it. Most aircraft manufacturers do not make any profit in the first couple of years of production, in the hope that in the future, they will make money. The large aircraft manufacturers get back their money after about 10 years; after that, they will make a profit. In the past, there were a few examples where aircraft manufacturers were bankrupted and only resurrected by government through long-term loans.

Wind-tunnel testing costs from 200 US$/hour for GA (General Aviation) small aircraft to 5000 US$/hour for large transport aircraft. The design and fabrication of some

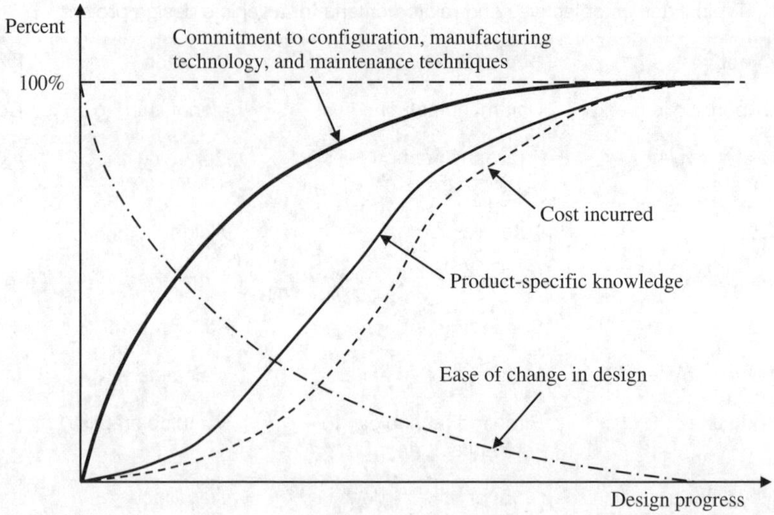

Figure 1.7 Status of various design features during the design process

aircraft – such as supersonic transport aircraft Aerospatiale-BAC Concorde (Figures 7.24 and 11.15) – was a great achievement, but when the international market does not purchase it, the production has to be stopped.

1.3 Design Project Planning

In order for a design project schedule to be effective, it is necessary to have some procedure for monitoring progress; and in a broader sense for encouraging personnel to progress. An effective general form of project management control device is the Gantt chart. It presents a project overview which is almost immediately understandable to non-systems personnel; hence it has great value as a means of informing management of project status. A Gantt chart has three main features:

1. It informs the manager and chief designer of what tasks are assigned and who has been assigned them.
2. It indicates the estimated dates on which tasks are assumed to start and end, and represents graphically the estimated ration of the task.
3. It indicates the actual dates on which tasks were started and completed and pictures this information.

Like many other planning/management tools, Gantt charts provide the manager/chief designer with an early warning if some jobs will not be completed on schedule and/or if others are ahead of schedule. Gantt charts are also helpful in that they present graphically immediate feedback regarding estimates of personnel skill and job complexity. Table 1.2 illustrates a typical Gantt chart for the design of a light single-seat aircraft in the form of a combined bar/milestone chart. Such a chart provides the chief designer with

Table 1.2 A typical Gantt chart for the design of a light single-seat aircraft

Job/task		Week/month/year					
Job	Task	January 2013	February 2013	March 2013	April 2013	May 2013	June 2013
Problem definition	Identification of design requirements						
	Feasibility analysis						
Conceptual design	Configuration design						
	Conceptual design review						
Preliminary design	Calculations						
	Preliminary design review						
Detail design	Wing design						
	Tail design						
	Fuselage design						
	Propulsion system						
	Landing gear						
	Equipment/ subsystems						
	Integration						
	Wind-tunnel testing						
	Weight distribution						
	Performance/stability analysis						
	Control surfaces design						
	Evaluation and test review						
Flight testing	Production of prototype						
	Flight tests						
Critical design review	Critical design review						
Certification	Certification						

Figure 1.8 Airbus A-380, the newest Airbus production. Reproduced from permission of Anne Deus

a scheduling method and enables him/her to rapidly track and assess the design activities on a weekly/monthly basis. An aircraft project such as Airbus A-380 (Figure 1.8) will not be successful without design project planning.

A preferred method of scheduling is through the use of program networks [2] such as the program evaluation and review technique (PERT) and the critical path method (CPM). The application of network scheduling is appropriate for both small- and large-scale design projects and is of particular value for a system development where there are several interdependencies. The definitions of new terms in Table 1.2, such as preliminary design and critical design review, and their associated techniques are addressed in Chapter 2.

1.4 Decision Making

First and foremost, it must be emphasized that any engineering selection must be supported by logical and scientific reasoning and analysis. The designer is not expected to select a configuration just because he/she likes it. There must be sufficient evidence and reasons which prove that the current selection is the best.

The main challenge in decision making is that there are usually multiple criteria along with a risk associated with each one. In this section, a few techniques and tools for aiding decision making under complex conditions are introduced. However, in most design projects there are stages where there are several acceptable design alternatives and the designer has to select only one of them. In such cases, there are no straightforward governing equations to be solved mathematically. Thus, the only way to reach the solution is to choose from a list of design options. There are frequently many circumstances in which there are multiple solutions for a design problem but one option does not clearly dominate the others in all areas of comparison.

A simple example is a transportation design problem where a designer is required to design a vehicle to transfer one person from one city to another. It is assumed that the two cities are both seaports and located at a distance of 300 km. The design solution alternatives are bicycle, motorbike, automobile, train, bus, ship, and aircraft. A traveler may select to travel using any of these vehicles. Three common criteria in most engineering design projects are: (i) cost, (ii) performance, and (iii) safety (and reliability). Table 1.3 shows a typical comparison of these design options and the ranking of each alternative.

Table 1.3 A typical multi-criteria decision-making problem (1 is the most desirable)

No.	Design option (vehicle)	Criteria		
		Cost (of operation)	Safety	Performance (maximum speed)
1	Bicycle	1	1	7
2	Motorbike	2	7	3
3	Automobile	5	6	4
4	Bus	3	5	5
5	Train	4	3	2
6	Ship	6	4	6
7	Aircraft	7	2	1

As the ranking illustrates, no one option clearly ranks first with respect to all three criteria to dominate the other six alternatives.

If the designer cares only for the cost of operation and safety, he/she has to select the bicycle, but if the only criterion was travel speed, the aircraft would be chosen as the vehicle. The bicycle is often the slowest vehicle; however it is the cheapest way to travel. In contrast, the aircraft does the best job in terms of speed (fastest to travel), but it is usually the most expensive option. It is evident that, for a typical traveler and designer, all the criteria matter. Thus, the question is how to come up with the best decision and the optimum vehicle. This example (Table 1.3) represents a typical multi-criteria decision-making problem that a design engineer frequently faces in a typical engineering design project. After the type of vehicle is selected, the calculations begin to determine geometry and other engineering characteristics.

A designer must recognize the importance of making the best decision and the adverse consequences of making a poor decision. In the majority of design cases, the best decision is the right decision, and a poor decision is the wrong one. The right decision implies design success, while a wrong decision results in a failure of the design. As the level of design problem complexity and sophistication increases in a particular situation, a more sophisticated approach is needed.

The approach for making the best decision to select/determine the best alternative is to take five steps, as follows.

- **Step 1**. Specify all the alternatives to be included in the exercise. Try to generate as many design concepts as possible using the brainstorming technique. However, given the resources required to include and consider all alternatives, you need to give considerable thought to reducing the alternatives to a manageable number.
- **Step 2**. The second step in selecting the best design is to identify and establish the criteria (e.g., Table 1.1). These criteria serve later as the guidelines for developing the options. Some design references employ the term "figures of merit" instead of criteria.

- **Step 3**. The next step is to define the metrics. The metrics are defined as a shorthand way of referring to the criteria performance measures and their units. Metrics are the tool to overcome a non-comparable complex situation (e.g., comparing apples and oranges) by establishing a common evaluation scale and mapping each criterion's metric onto this scale. A simple evaluation scale is to map each criterion as either excellent, adequate, or poor. So, each design option may be rated with respect to each criterion using this common scale. A better and more quantifiable scale is a numerical scale, as demonstrated in Table 1.4. Typical metrics for measuring performance of an aircraft are maximum speed, take-off run, rate-of-climb, range, endurance, turn radius, turn rate, and ceiling.
- **Step 4**. The fourth step is to deal with criteria that have unequal significance. A designer should not frequently treat all criteria as being equally important. The designer must try to ascertain how important each requirement (i.e., criterion) is to the customer. The simplest approach is to assign numerical weights to each criterion (or even at a metrics level) to indicate its importance relative to other criteria. These weights ideally reflect the designer's judgment of relative importance. Judgment as to whether one design alternative is superior to another may be highly dependent on the values and preferences of the evaluator. In some cases, the designer has no way other than relying on personal "feelings" and "judgments" for the basis of the numerical weights. As a starting point, you may pair up each criterion with every other criterion one at a time and judge which of the items in each pair is more important than the other. The weights may later be normalized (i.e., mathematically convert each number to a fraction of 1) in order to make them easier to compare.

  A prerequisite to identifying the weight of each criterion is prioritization. Table 3.6 demonstrates the priorities of various aircraft designers against 10 design criteria. When the number of criteria is small, this task is straightforward. For large and complex systems, a systems engineering approach must be employed (Chapter 2). A cookbook method is no substitute for experience and sound professional judgment in what is inherently a subjective process. Reference [2] describes a higher-level approach which is referred to as the analytical hierarchy process (AHP) method; it is worth considering for sophisticated systems.
- **Step 5**. Select the alternative which gains the highest numerical value. It is expected that the output of the decision-making process will yield the most desirable result.

  The designer may conduct the decision-making process by developing a software package to minimize or maximize a specific index. In case there are uncertainties in evaluating criteria, a sophisticated robust decision rule should attempt to incorporate the uncertainties into the decision-making process. One of the difficulties of dealing with uncertainties is coming up with the probabilities of the uncertain parameters and factors. This is best performed in a process referred to as "sensitivity analysis."

1.5 Feasibility Analysis

In the early stages of design and by employing brainstorming, a few promising concepts are suggested which seem consistent with the scheduling and available resources. Prior to committing resources and personnel to the detail design phase, an important design activity – feasibility analysis – must be performed. There are a number of phases through

Table 1.4 Common scale and criteria metrics and three examples

No.	Common scale		Criteria metrics		
	Preferred level	Value	Example 1: length (m)	Example 2: maximum speed (km/h)	Example 3: mass (kg)
1	Perfect	10	35	60	500
2	Excellent	9	29.1	52	550
3	Very good	8	25.7	41	620
4	Good	7	21.4	32	680
5	Satisfactory	6	18.4	27	740
6	Adequate	5	16.6	21	790
7	Tolerable	4	12.7	17	830
8	Poor	3	8.4	17	910
9	Very poor	2	6.7	14	960
10	Inadequate	1	4.3	10	1020
11	Useless	0	2.5	7	1100

which the system design and development process must invariably pass. Foremost among them is the identification of the customer-related need and, from that, the determination of what the system is to do. This is followed by a feasibility study to discover potential technical solutions, and the determination of system requirements.

It is at this early stage in the lifecycle that major decisions are made relative to adapting a specific design approach and technology application, which have a great impact on the lifecycle cost of a product. At this phase, the designer addresses the fundamental question of whether to proceed with the selected concept. It is evident that there is no benefit or future in spending any more time and resources attempting to achieve an unrealistic objective. Some revolutionary concepts initially seem attractive but when it comes to the reality, they are found to be too imaginary. Feasibility study distinguishes between a creative design concept and an imaginary idea. Feasibility evaluation determines the degree to which each concept alternative satisfies the design criteria.

In the feasibility analysis, the answers to the following two questions are sought:

1. Are the goals achievable, are the objectives realistic, or can the design requirements be met?
2. Is the current design concept feasible?

If the answer to the first question is no, the design goal and objectives, and hence the design requirements, must be changed. Then, no matter what the source of the design requirements – either direct customer order or market analysis – they must be changed (Figure 1.9). When the answer to the second question is negative, a new concept must be selected. Finding the answers to these questions is not always easy. To determine the answers other professionals beside design engineers – such as financial experts or manufacturing engineers – must often be involved in the feasibility study. The feasibility

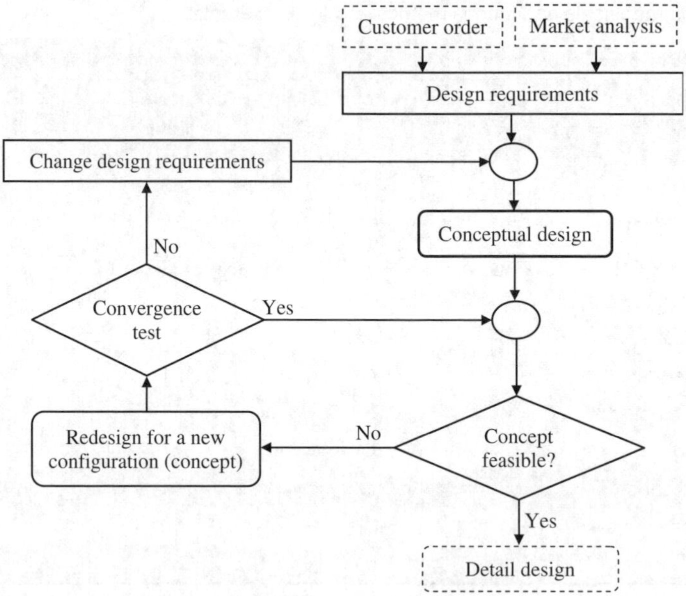

Figure 1.9 Feasibility analysis process

analysis will refine the design requirements and narrow down the initial promising design concepts to a few feasible ones. It is at this stage that uncertainties are identified.

When several concepts are analyzed and the convergency test illustrates that none of the promising concepts are feasible, the customer is informed that the objectives are not achievable within the current limits of science and technology. At this time, it is recommended that the customer reduces the level of his/her expectations. In contrast, the results of a feasibility study will significantly impact the operational characteristics of the product and its design for producibility, supportability, disposability, and detectability. The selection and application of a given technology or given materials has reliability and maintainability implications, will influence manufacturing operations, and will affect the product operating cost.

For instance, Boeing 787 Dreamliner (Figure 1.10) is the first commercial transport aircraft with full composite structure. The composite materials may have reduced the aircraft

Figure 1.10 Boeing 787 Dreamliner. Reproduced from permission of A J Best

weight, but will certainly influence the reliability, maintenance, and entire lifecycle. All these considerations should be dealt with during the feasibility study before a commitment is made to pursue extensive design activities. The systems engineering approach has a systematic view of feasibility analysis. Thus, a primary objective of systems engineering is to ensure the proper coordination and timely integration of all systems elements (and the activities associated with each) from the beginning. The systems engineering approach is introduced in Chapter 2.

1.6 Tort of Negligence

The issue of legal liability is crucial to an aircraft design engineer. Liability is basically part of the system of civil law. In civil law, the issue is not one of innocence or guilt; it is a question of who is at fault in a dispute, or who violated an agreement, or who failed to fulfill obligations. Liability law belongs to that branch of civil law known as torts. The area of tort law known as negligence involves harm caused by carelessness, not intentional harm. Negligence is a failure to exercise the care that a reasonably prudent person would exercise in like circumstances. Designers and manufacturers who sell their products to the public face many uncertainties regarding the legal ramifications of their actions. Design engineers and manufacturers are responsible and liable for harm done by their product or design to a customer or third party. Thus a designer has the responsibility to act in a careful and prudent manner. The negligence is applied to a designer when the product was defective or a design created a concealed danger.

Thousands of disasters have occurred throughout aviation history, for a great number of which the designers (not the pilots) have been responsible. Disasters include aircraft crashes, mishaps, and accidents. In all of these cases, harm (bodily or financially) has been done to a customer or to the public. The primary source of such incidents is the designer's carelessness in design, error in calculations, or lack of prediction of the future. In the area of accident prediction, Murphy's Law applies which states:

> If any event can happen, it will happen; or anything that can go wrong will go wrong.

For instance, one application of this law relates to liquid containers. The direct application of the law is as follows: every system in an aircraft which carries a liquid will **leak**. An aircraft with an air-breathing engine carries fuel and a passenger aircraft carries water. Thus, the aircraft designer must avoid installing electrical wiring and avionic systems in the belly, below the toilet or liquid container or fuel tank. Reference [7] describes a number of war stories based on actual events that happened in the design and development of aircraft programs. For instance, one story relates how the unacceptable field performance of the first F-18 fighter was traced to an error in the calculation of aerodynamic forces in the ground effect.

Another war story describes the Fowler flaps crunching in the first flight of the General Dynamics strike aircraft F-111A, when the pilot engaged the wing sweep system to sweep the wing aft after landing. The accident was clearly the designer's fault, in not expecting such an event. The solution was to employ an interlocking device to prevent a pilot from sweeping the wings with the flap down. One of the continuing functions of a design engineer is to compile development and operations "lessons learned" documents and ensure their integration into future systems development activities. Lessons learned files from previous projects are especially valuable in risk identification and characterization, and must be employed in feasibility studies.

The following three aircraft-related cases arose out of tragic accidents occurring at different times, and where the relatives of the victims brought a wrongful death case to court. In all three cases the court found the company (i.e., the designer) *negligent* and *liable*. Once a judgment has been made in favor of the plaintiff in a liability case, a monetary award is made. However, in more serious cases, punitive damages may also be awarded. In the area of astronautics, most satellite mishaps stem from engineering mistakes. To prevent the same errors from being repeated, some references have compiled lessons that the space community should heed.

- **Case 1**: United States versus "*Weber Aircraft Corp*." in 1984. When the engine of an Air Force aircraft failed in flight, the pilot was severely injured when he ejected from the plane. After Air Force collateral and safety investigations of the incident had been completed, the pilot filed a damages action against respondents as the entities responsible for the design and manufacture of the plane's ejection equipment.
- **Case 2**: Jack King and 69 European plaintiffs versus "*Cessna Aircraft Company*" in a tragic plane crash that occurred at Linate Airport in Milan, Italy, on October 8, 2001. On that foggy morning, a private Cessna jet operated by Air Evex, a German charter company, made a wrong turn and taxied toward an active runway, causing it to collide with Scandinavian Airlines Flight 686, which was just taking flight. One hundred and eighteen people died, including everyone on board both planes and four people on the ground, and others on the ground were injured.
- **Case 3**: Starting in 1991, a number of accidents and incidents involving the Boeing 737 were the result of the airplanes' unexpected rudder movement. One incident occurred on September 8, 1994 when a Boeing 737-300 of USAIR Flight 427 crashed near Pittsburgh, PA, killing 132 people. Another incident was when the Boeing 737 Flight 185 of SilkAiron plunged from 35 000 ft into a muddy river in Indonesia on December 19, 1997, killing all 104 people aboard. The Los Angeles Superior Court jury decided defects in the rudder control system caused the crash and Parker Hannifin Corp., the world's largest maker of hydraulic equipment, was told to pay US$43.6 million to the families of three people killed. On the contrary, the US National Transportation Safety Board (NTSB) concluded that there were no mechanical defects and the pilot intentionally caused the crash. The Federal Aviation Administration (FAA) ultimately ordered an upgrade of all Boeing 737 rudder control systems by November 12, 2002.
- **Case 4**: A Continental Airlines Boeing 737 went off the runway during takeoff from Denver International Airport in Colorado, plunging into a ravine and shearing off its landing gear and left engine. At least 58 people were injured in the crash that happened on December 20, 2008. The entire right side of the plane was burned, and melted plastic from overhead compartments dripped onto the seats. Note that the plane's left engine was ripped away along with all the landing gear. NTSB published that the probable cause of this accident was the captain's error (cessation of right rudder input).

Figure 1.11(a) shows a Tupolev Tu-154 which crashed while attempting to land in poor weather conditions on September 14, 1991 in Mexico City. Luckily all 112 occupants survived. Figure 1.11(b) illustrates the transport aircraft Ilyushin Il-76 freighter,

(a) (b)

Figure 1.11 Two aircraft in tragic accidents: (a) Tupolev Tu-154 crashed due to poor weather conditions; (b) An Ilyushin Il-76 freighter which caught fire on the ground. Reproduced from permission of (a) Augusto G. Gomez; (b) Serghei Podlesnii Part (a) reproduced from permission of Augusto G. Gomez

which caught fire on the ground while it was being loaded in preparation for a flight to Brazzaville, Congo on May 10, 2007.

The threat of liability law suits must spur on designers and manufacturers to be more sensitive to safety issues and to address them in more creative and innovative ways. The liability threat should not have a stifling effect on creative design and technological innovation. For this reason, the employment of safety factors is highly recommended. Federal Aviation Regulations have addressed this issue in many ways, but it does not suffice; aircraft designers and all involved engineers must be prudent and careful in the design process. A prudent design strategy is to employ the utmost care; to anticipate relevant wrongful events; and to incorporate some features into products to make them more robust.

There is a famous 10^9 rule in aircraft design which is acceptable within society. This rule states that one death in 1 000 000 000 aircraft travelers is accepted. Even one human death is a great disaster to a community, but stupidity and negligence can sometimes lead to a deadly crash. In terms of statistics, about 300 people are killed every year in aviation-related accidents in the USA while about 45 000 are killed in car accidents. Therefore, the aircraft is much safer than the car, and air travel is 150 times safer than road travel. About one-third of aviation accidents are because of CFIT (controlled flight into terrain). When a pilot makes a mistake and hits a mountain, a designer has almost no influence on this incident. Not every pilot mistake has a solution by the aircraft designer; some mistakes may be avoided by design, but not all. Reference [7] describes several stories about pilot mistakes as well as designer mistakes. All stories are beneficial to aircraft designers and have lessens to be learned.

References

[1] Robbins, S.P. and Coulter, M. (2008) *Management*, 10th edn, Pearson Prentice Hall.
[2] Dieter, G. and Schmidt, L. (2008) *Engineering Design*, 4th edn, McGraw-Hill.
[3] Hyman, B. (2003) *Fundamentals of Engineering Design*, 2nd edn, Prentice Hall.
[4] Eggert, R.J. (2005) *Engineering Design*, Pearson Prentice Hall.
[5] ABET Constitution, Accreditation Board for Engineering and Technology (2012), www.abet.org.
[6] Blanchard, B.S. and Fabrycky, W.J. (2006) *Systems Engineering and Analysis*, 4th edn, Prentice Hall.
[7] Roskam, J. (2006) *Roskam's Airplanes War Stories*, DAR Corporation.

2

Systems Engineering Approach

2.1 Introduction

The systems engineering (SE) discipline was originally developed to help understand and manage complexity. The scale of complexity found in modern aircraft systems necessitates an approach different from that applied traditionally. The formal instruction in modern SE principles is cited as beginning more than 40 years ago. The applications of systems engineering began during the late 1950s, when the early Department of Defense (DOD) view of SE was documented. This was due to the race to space [1] and the development of the nuclear missiles program that were essential for US survival. The first attempt to formalize systems engineering within an engineering curriculum occurred at MIT in 1950. In the 1960s, systems engineering gained widespread acceptance within the DOD as the preferred approach to engineer military systems. The systems engineering approach was revitalized in the mid-1990s.

Increased participation using systems engineering processes and practices during the system development and demonstration phase is seen as key to implementing this approach. The industry, academia, and government revitalization efforts include publishing systems engineering processes, methods, and templates to guide people to implement systems engineering. In 2003 and 2004, the DOD [2, 3] issued a number of policies that placed renewed emphasis on the application of systems engineering, stating that it is essential to the Department's ability to meet the challenge of developing and maintaining needed capability. It was noted that this is especially true as systems become more complex in a family-of-systems, or system-of-systems. In addition, NASA [4] developed and published a systems engineering handbook in 2007.

In general, systems may be classified as either natural or human-made. Human-made or technical systems (e.g., aircraft) come into being by human intervention in the natural order utilizing pervasive technologies. System is an assemblage or combination of elements, members, components, and parts forming a complex or unitary whole. A random group of items in a room would not qualify as a system because of the absence of unity, functional relationship, and useful purpose. Systems are composed of components, attributes, and relationships. The purposeful action performed by a system is its function.

Aircraft Design: A Systems Engineering Approach, First Edition. Mohammad H. Sadraey.

A system view is only one way of understanding complexity. The systems engineering approach is defined as "an interdisciplinary approach encompassing the entire technical effort to evolve and verify an integrated and lifecycle balanced set of system people, product, and process solutions that satisfy customer needs." Multi-discipline SE design involves the application of a systems engineering process and requires engineers with substantive knowledge of design across multiple technical areas and improved tools and methods for doing it.

Industry, government, and academia share responsibility for the development of the future engineers needed to keep aerospace products and capabilities at the leading edge of technology. One of the enablers is fundamental knowledge of systems engineering and its practical application to systems that involve multiple disciplines. Nonetheless, engineering education programs continue to focus on the traditional educational product – highly qualified but single-discipline engineers and technologists. Meeting the demand for multidisciplinary systems engineering designers requires teaching something different than is found in current textbooks.

Education in systems engineering [5–7] is often seen as an extension of the regular engineering programs. The formal instruction in SE principles is cited as beginning more than 40 years ago. The applications of systems engineering began during the late 1950s because of the race to space and the development of the nuclear missiles program that were essential for national survival. There are a limited number of undergraduate university programs in systems engineering. The International Council on Systems Engineering (INCOSE) maintains a directory of systems engineering [8] academic programs worldwide. As per the 2006 INCOSE directory, there are about 75 institutions in the United States offering a total of about 130 undergraduate and graduate programs in systems engineering. Education in systems engineering can be broken down into two basic categories: systems engineering-centric or domain-centric. SE-centric programs treat systems engineering as a separate discipline, focusing their courses on systems engineering practice and techniques. In 2006, there were 31 institutions offering 48 degree programs in the systems engineering-centric category and 48 institutions offering 82 domain-centric SE degree programs across a number of engineering domains.

This chapter is devoted to briefly introducing the fundamentals of the systems engineering discipline, principles, design phases, design flowcharts, design evaluation, and systems engineering application to the aircraft design process. The systems engineering approach is employed throughout this book to present aircraft design techniques. This chapter is organized as follows. In Section 2.2, the fundamentals of systems engineering are presented. Sections 2.3–2.5 provide the features of the conceptual design phase, the preliminary design phase, and the detail design phase, respectively. In Section 2.6, design for operational feasibility is introduced which covers topics such as maintainability, producibility, detectability, usability, supportability, affordability, recyclability, and disposability. The design review, evaluation, and feedback are important steps in the systems engineering technique and are covered in Section 2.7. Finally, the application of systems engineering for the design of an air vehicle or systems engineering approach in aircraft design is established in Section 2.8.

2.2 Fundamentals of Systems Engineering

An aircraft is a system composed of a set of interrelated components working together toward some common aerial objective or purpose. Primary objectives include safe flight achieved at a low cost. Aircraft are extremely complex products comprising many subsystems, components, and parts. They are but one system operating within a global air transportation or defense "system-of-systems." The conception, design, production, operation, and maintenance of aircraft are influenced by many factors including technical, economic, political, organizational, financial, and regulatory. The engineering of an aircraft as a system requires methods, tools, and processes which can successfully address these many complexities. Aircraft systems, due to the high cost and the risks associated with their development, are a major user of systems engineering methodologies.

Systems engineering is the fundamental discipline embodying these methods, tools, and processes. It also addresses the overall strategy for developing system-level requirements which meet users' needs, meet investors' expectations, incorporate knowledge from past experience, and satisfy regulatory and other constraints. Systems engineering, on the other hand, is the process used to develop integrated human, hardware, and/or software components such that the resulting system or product meets the system-level requirements. To apply the systems engineering technique, one must decide "what constitutes the system." To provide a framework, the following levels are defined where (in this book) the level 2 or system level is addressed.

- **Level 1, System-of-systems level**. The air transportation/defense system which includes aircraft, missiles, satellites, ground stations, airports, air traffic management, etc.
- **Level 2, System level**. The aircraft and/or related systems which include aircraft, users, operators, trainers, manufacturing plants, maintenance shops, etc.
- **Level 3, Subsystem level**. Major aircraft subsystems which include the flight control mechanism, hydraulic, electric, avionic, powerplant, fuel, air conditioning, structure, seat, etc.
- **Level 4, Component level**. Components which include the wing, fuselage, tail, landing gear, radar, pumps, nacelles, control surfaces, auxiliary power unit (APU), etc.
- **Level 5, Part level**. Parts which include fittings, fasteners, blades, propeller, screws, nuts, ribs, spars, frame, stiffener, skin, shaft, wires, pipes, etc.

To ensure economic competitiveness, engineering must become more closely associated with economics and economic feasibility, which is best accomplished through a lifecycle approach to engineering. The system lifecycle includes design, development, production, operation, support, and disposal. The design process is divided into three major phases: (i) conceptual design phase, (ii) preliminary design phase, and (iii) detail design phase. These are artificial categories that, along with test and evaluation, make up the four basic phases of system design.

The summation of conceptual design, preliminary design, detail design, and production and/or construction is referred to as the acquisition phase (Figure 2.1), while the summation of product use, support, phase-out, and disposal is called the utilization phase. It is essential that aircraft designers be sensitive to utilization outcomes during the early stages of the design and development process. They also need to conduct lifecycle engineering studies as early as possible in the design process. Figure 2.2 illustrates the relationship among four major design activities in a systems engineering approach. The design process primarily starts with the conceptual design phase, based on design requirements. The details of the conceptual design phase are presented in Section 2.3. The preliminary design begins right after the conceptual design phase and employs the output of this phase.

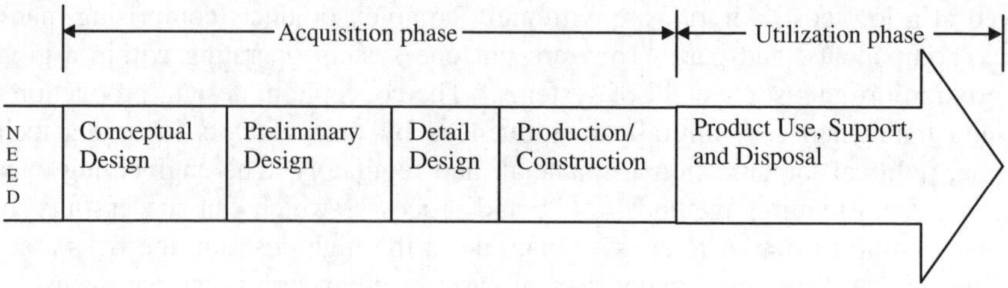

Figure 2.1 The system life cycle

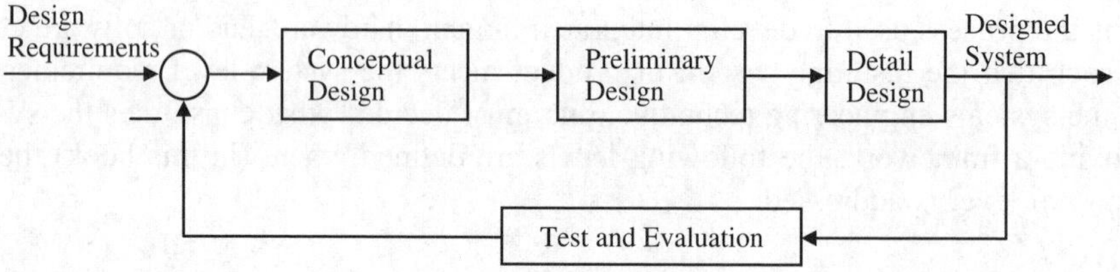

Figure 2.2 Relationship among four major design activities

Section 2.4 is devoted to the preliminary design phase. The detail design phase begins right after the preliminary design phase and utilizes the output of this phase. Section 2.5 reviews the detail design phase.

After each round of design, a test and evaluation is conducted to compare the characteristics of the designed system with the design requirements. If the system does not meet the requirements in any way, feedback is sent to the design groups to make the necessary corrections. This iteration is continued until all design requirements are satisfied. Figure 2.2 models the entire design process as a feedback control system where an error signal is produced if there is any difference between input (design requirements) and output (features of the designed system). The test and evaluation is introduced in Section 2.7.

Although the overall design phases are generally accepted, a particular design process (such as the waterfall model, spiral model, or "Vee" model) must be tailored to a specific program need. The interested reader is encouraged to refer to Refs [1, 9, 10] for more details on various models. The systems engineering and aerospace engineering influences on design are illustrated in Figure 2.3. Both influences reverse as the design process progresses; the influence of systems engineering decreases, while the area influence of aerospace engineering increases. Thus, there is a need to ensure that the aerospace engineering techniques are properly integrated.

From the perspective of systems engineering, the design of aircraft should not only transform a need into an air vehicle, but also ensure the aircraft's compatibility with related physical and functional requirements. Therefore, it should consider operational outcomes expressed as safety, producibility, affordability, reliability, maintainability, usability, supportability, serviceability, detectability, disposability, as well as the requirements on performance, stability, control, and effectiveness. A major objective of systems engineering

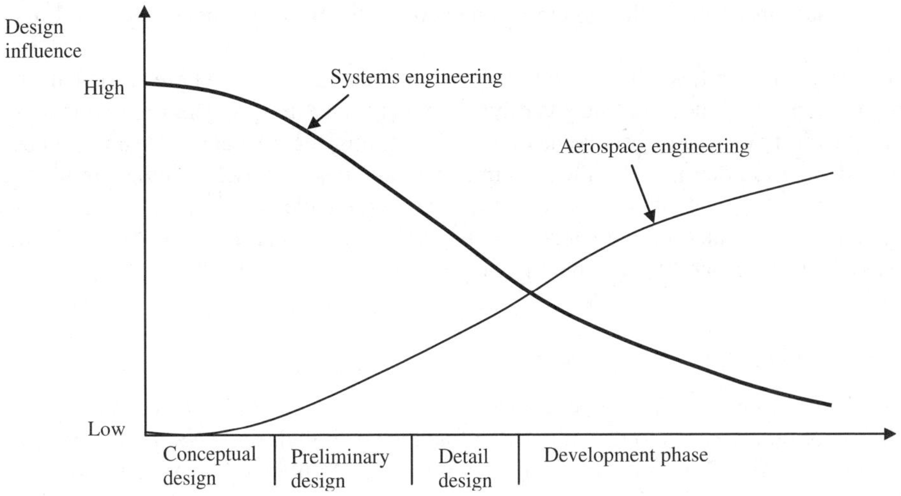

Figure 2.3 Systems engineering and aerospace engineering influence on design

is to develop a good set of requirements in order to define a single baseline from which all lower-level requirements may evolve.

Published works on systems engineering that go into intricate detail regarding its implementation are vast, with a great number of well-written papers and texts on the subject. Reference [11] proposes the metrics of systems engineering along with a systems engineering diagnostic method. Published works on systems engineering are plentiful and cover a variety of areas from need validation to retirement of assets/capabilities. References [12–15] address various aspects of the applications of systems engineering techniques in aerospace engineering. Multi-disciplinary multi-university design of a high-altitude inflatable-wing aircraft with systems engineering is discussed in Ref. [16].

2.3 Conceptual System Design

2.3.1 Definition

From the perspective of synthesis, system design is nominally comprised of conceptual, preliminary, and detail design. Conceptual design is the first and most important phase of the system design and development process. It is an early and high-level lifecycle activity with potential to establish, commit, and otherwise predetermine the function, form, cost, and development schedule of the desired system. An appropriate starting point for design at the conceptual level is the identification of a problem and associated definition of need.

The primary responsibility of conceptual design is the selection of a path forward for the design and development of a preferred system configuration, which ultimately is responsive to the identified customer requirements. A critical first step in the implementation of the SE process is to establish this early foundation, as well as to require the initial planning and development of a spectrum of manufacturing technologies. From an organizational perspective, systems engineering should take the lead in the definition of system requirements from the beginning and address them from a total integrated lifecycle perspective.

As the name implies, the outcome of the conceptual design phase is a concept or configuration which does not necessarily accompany any details. The requirements need for a specific new system first comes into focus during the conceptual design process. It is this recognition that initiates the system conceptual design process to meet these needs. Then, during the conceptual design of the system, consideration should simultaneously be given to its production and support. This gives rise to a parallel lifecycle for bringing a manufacturing capability into being.

2.3.2 Conceptual Design Flowchart

Throughout the conceptual system design phase (commencing with the need analysis), one of the major objectives is to develop and define the specific design-to requirements for the system as an entry. The results from these activities are combined, integrated, and included in a system specification. This specification constitutes the top "technical requirements" document that provides overall guidance for system design from the beginning. In general, the following steps must be performed during the conceptual design phase:

1. Identify the problem and translate it into a definition of the need for a system that will provide a solution.
2. Accomplish system planning (e.g., Gantt chart) in response to the identified need.
3. Conduct a feasibility study, making sure the system is practical and leads to the details of a technical approach for system design.
4. Develop system operational requirements describing the functions that the system must perform in accomplishing its designated mission.
5. Propose a production/maintenance plan for sustained support of the system throughout its desired lifecycle.
6. Identify and prioritize technical performance measures (TPMs) and related criteria for design.
7. Perform a system-level functional analysis and allocate requirements to the various subsystems.
8. Formulate needs and generating metrics to evaluate them.
9. Brainstorm and design a couple of concepts to address the design requirements and list their characteristics.
10. Accomplish trade-off analysis to select the best concept.
11. Develop a system specification.
12. Conduct a conceptual design review (CDR).
13. If the CDR does not confirm the concept, select a new approach and generate new concepts.

Figure 2.4 depicts these 13 steps in a flowchart. As the figure illustrates, applied research and advanced system planning begin at the conceptual design phase, but they are continued into the preliminary and detail design phases. During the applied research initiative, new technologies may be developed and born. The results of the conceptual design phase and system specifications are delivered to the next phase, the preliminary design phase.

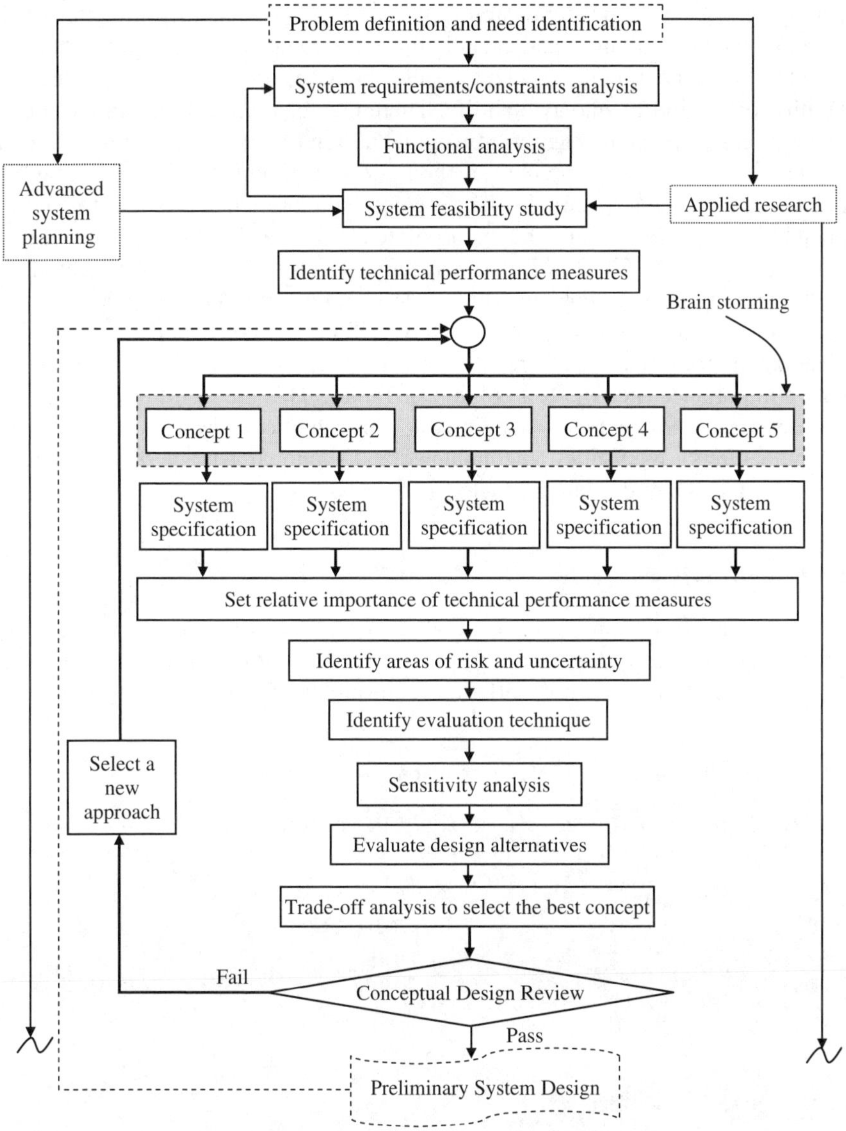

Figure 2.4 Conceptual design process

Some steps of the conceptual design phase – such as TPMs, functional analysis, system trade-off analysis, and CDR – will be reviewed briefly in the forthcoming sections.

2.3.3 Technical Performance Measures

TPMs refer to the quantitative values that describe system performance requirements. TPMs are measures of the attributes and/or characteristics which are inherent within the design (i.e., design-dependent parameters) and lead to the identification of TPMs. TPMs include such qualitative and quantitative factors as customer appeal, human factors, weight, geometry, volume, speed, process duration, operating cost, maintainability, detectability, producibility, and availability. There may be numbers for TPM values specified for a particular system during the conceptual design phase. The identification of TPMs evolves mainly from the development of the system requirements and operating cost.

In practice, some of the specified values are contradictory when it comes to determining the specific features that should be incorporated into the design. For instance, in the design of a vehicle, the TPMs of size, capacity, speed, driver comfort, manufacturing cost, and operating cost do not have the same significance. Thus, a trade-off has to be made in order to achieve a higher-level objective. Table 2.1 conveys the results of a TPM identification and prioritization effort by an auto designer team. It is interesting to note that the relative importance (i.e., the last column in Table 2.1) is the result of several meetings between designer team members with major part suppliers, dealers (competition with other manufacturers), mechanics (maintenance and support), average drivers (customers) and media (statistics), and a human factor analyzer.

2.3.4 Functional Analysis

An early essential activity in the conceptual design phase is the development of a functional description of the system to accomplish the desired mission. Establishing the need and problem statement may not sufficiently formulate the design problem for large,

Table 2.1 Prioritization of TPMs for an automobile designer

No.	Technical performance measure	Quantitative requirement	Relative importance (%)
1	Target velocity[a]	70 mile/h (112 km/h)	15
2	Driver and passenger comfort (human factor)	Relatively comfortable	30
3	Total mass	1400 kg	6
4	Manufacturing (assembly) time	5 days	9
5	Affordable/regular//luxury	Regular	12
6	Trunk size	1.5 m wide, 0.6 m high, and 1.2 m long	7
7	List price (marketability)	$17 000	12
8	Maintainability (MTBM[b])	500 km	8
Total			100

[a] This is the speed at which the car requires minimum specific fuel consumption (e.g., 35 MPG).
[b] Mean time before maintenance.

complex systems. In such a case, tasks or "functions" to be performed by the system and its components need to be described. A function refers to a specific or discrete action (or series of actions) that is necessary to achieve a given objective; that is, an operation that the system must perform. The functional analysis is used to indicate the process for identifying and describing these functions as part of formulating the design problem. The functional analysis is a process of translating system requirements into detailed design criteria and the subsequent identification of the resources required for system operation and support. The functional analysis is also performed during detailed design, since it includes breaking requirements at the system level down to the subsystem. The purpose of "functional analysis" is to present an overall integrated description of the system's functional architecture, and to provide a foundation from which all physical resource requirements are identified.

Accomplishment of the functional analysis is best facilitated through the use of a functional block diagram; that is, the application of a graphical method. The functional analysis provides the baseline from which reliability requirements, maintainability requirements, human factor requirements, supportability requirements, and manufacturability requirements are identified. The next step after functional analysis is partitioning; that is, breaking the system down into elements (or parts). Then, the design-to requirements have to be determined for each of the system elements. At this moment, the TPMs which evolved from the definition of the operational requirements must be allocated or apportioned down to the appropriate subsystems or elements that make up the system. Figure 2.5 demonstrates the function breakdown of a system into elements. To guarantee an ultimate system design configuration that meets the customer needs, there must be a top-down allocation of design criteria from the beginning; that is, during the latter stages of the conceptual design phase.

2.3.5 System Trade-Off Analysis

A design process normally contains a numerical analysis (a lot of calculations; Part 1) as well as logical selections (Part 2). The calculations part is performed by using

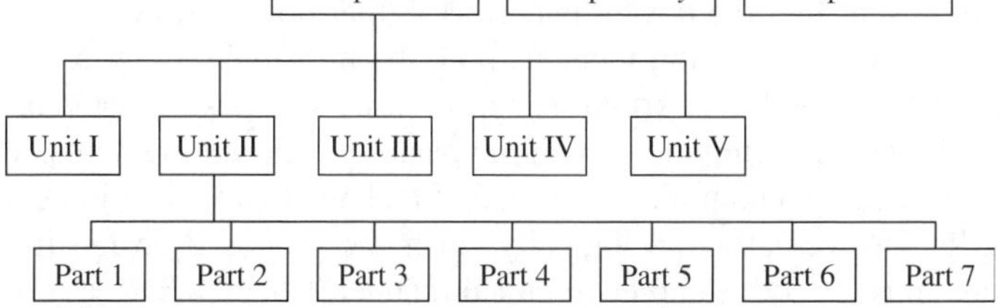

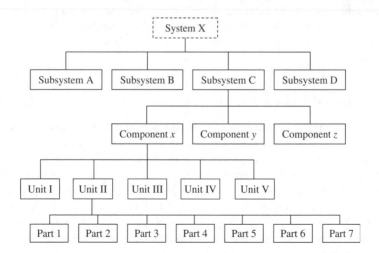

Figure 2.5 The function breakdown of a system into elements

mathematical/physical/chemical equations, relationships, formulas, and techniques. However, the selection part is accomplished during a decision-making process. A highly efficient decision-making technique based on a systems engineering approach is referred to as trade-off analysis. This technique is employed in various design stages, from the conceptual design phase to the preliminary design phase and detail design phase. Trade-off analysis is performed at all levels from system top-level to subsystem level to component level and even down to part level. As the name implies, trade-off analysis involves both gains and losses; the gains have to be maximized, and the losses must be minimized. Trade-off is a compromise made between two or more favorable alternatives.

As the system design progresses, decisions must be made regarding the evaluation and selection of appropriate configurations, technologies, materials, routines, structure, methods, and level accuracies. The sensitivity analysis is part of the trade-off study, allowing us to identify potential areas of risk and uncertainty. When solving engineering design problems, we often have to deal with multiple evaluation criteria, or multiple figures of merit (FOM). So the question is, which criterion is more important (e.g., aircraft maximum speed or operating cost). Since a system development team will make many of the decisions as a team, various group decision-making methods will replace equation solving. Hence, most design problems exhibit a trade-off, wherein one attribute improves and the other degrades. Trade-offs are caused by the interdependency of variables, typically referred to as coupling. The trade-off analysis targets the overall satisfaction of the customer, rather than one design requirement. The trade-off analysis in a complex system may be formulated as a multi-variable optimization problem [17, 18] and can be solved by employing optimization techniques.

In trade-off study, the FOM for each design alternative has to be determined and compared with other alternatives. FOM is a numerical quantity based on one or more characteristics of a system or device that represents a measure of efficiency or effectiveness. A FOM is a quantity used to characterize the performance of a device, system, or method, relative to its alternatives. The FOM is determined by summation of the values for each TPM multiplied by its relative importance. The alternative which gains the highest FOM score will be the best, and will be selected as the final design. Table 2.2 illustrates a typical trade-off analysis to compare the FOMs of five alternatives for a system. In this example, option 3 (car) has the highest score (93) and is identified as the best design (i.e., transportation vehicle). All boxes in Table 2.2 must be filled; it is left to the reader to complete this table.

2.3.6 Conceptual Design Review

At each major design phase (conceptual, preliminary, and detail), an evaluation function is accomplished to ensure that the design is correct at that point before proceeding with the next stage. The evaluation function includes both the informal day-to-day project coordination and data review, and the formal design review. The design data/characteristics is released and reviewed for compliance with the basic system requirements. The reviewing operation is performed by a committee, formed of technical and operational members. The purpose of CDR is to formally and logically cover the proposed design from the system standpoint in the most effective and economical manner. During CDR, a formalized check of the proposed system design is provided, major problems are discussed, and corrective actions are taken. In addition, it provides a common baseline for all project personnel.

Table 2.2 A sample table to compare the features of four alternatives for a transportation system

No.	Technical performance measure	Relative importance (%)	Option 1 (bicycle)	Option 2 (motorbike)	Option 3 (car)	Option 4 (train)	Option 5 (aircraft)
1	Weight	9					
2	Capacity	13					
3	Operating cost	14					
4	Market price	10					
5	Speed	26					
6	Travel time	4					
7	Producibility	10					
8	Maintainability	7					
9	Availability	5					
10	Detectability	2					
Figure of merit		100	76	64	93	68	85

The design team members are provided the opportunity to explain and justify their design approach, and reviewer committee members are provided the opportunity to ask various questions of the design team. The CDR serves as an excellent communication medium, creates a better understanding among design and support personnel, and promotes assurance and reliability. Figure 2.4 shows the importance and position of CDR at the conceptual design phase where if the committee does not approve the design, the design team has to start all over again. The CDR is usually scheduled toward the end of the conceptual design phase and prior to entering the preliminary design phase of the program.

2.4 Preliminary System Design

By the end of the conceptual design phase, design evolution continues (see Figure 2.2) by addressing some of the most fundamental system characteristics. This is accomplished during the preliminary design phase. The essential purpose of the preliminary design is to determine features of the basic components/subsystems. Some products of the preliminary design include: major technical data, design and operational trade studies, interface specifications, system mock-up and model, and plans for verification and verification tests. The preliminary design phase often includes the following steps:

- Develop design requirements for subsystems from system-level requirements.
- Prepare development, process, and materials specifications for subsystems.
- Determine performance technical measures at the subsystem level.
- Conduct functional analysis at the subsystem level.

- Establish detailed design requirements and prepare plans for their allocation.
- Identify appropriate technical design tools, software packages, and technologies.
- Accomplish a trade-off study at the subsystem level.
- Present the design output for a preliminary design review (PDR) at the end of the preliminary design phase.

The procedures for functional analysis, trade-off study, and design review at the subsystem level are very similar to what is described at the conceptual level. Sections 2.3.4–2.3.6 describe the details of such activities at the system level. Thus, the functional analysis and trade-off study must be extended from the system level down to the subsystem and below as required. Subsystem design requirements evolve from system design requirements according to operational requirements, and identification and prioritization of TPMs. This involves an iterative process of top-down/bottom-up [9] design (e.g., the "Vee" process model). The system TPM for operational requirements must be related to one or more functions of subsystems. During the preliminary design phase, the selection of hardware, software, technical staff, test facilities, data, and references is made. The subsystems, units, and modules are identified and functions are allocated to each one. The qualitative and quantitative design requirements are determined at the subsystem level.

The preliminary and detail design evaluation process can be facilitated through the application of various analytical/mathematical models. A model is defined as a mathematical representation of a real world which abstracts features of the situation relative to the problem being analyzed. The use of a mathematical model offers significant benefits. There are many interrelated elements that must be integrated into a system and not treated on an individual basis. The mathematical model allows us to deal with the problem as an entity and makes it possible to consider all major variables of the problem simultaneously. The extensiveness of the model depends on the nature of the problem, the number of variables, input parameter relationship, number of alternatives, and complexity of the operation.

There must be a top-down/bottom-up traceability of requirements throughout the overall hierarchical structure of the system. It is essential that these activities be coordinated and integrated, across the lifecycle, from the beginning. In other words, an ongoing communication process must flow throughout the development of hardware, software, and human elements. The design-related activities that occur after functional analysis at the preliminary design phase are: human factor analysis, maintenance and logistic supportability analysis, producibility analysis, disposability analysis, economic analysis, functional packaging of system elements analysis, and reliability analysis. The results of the preliminary design phase will be passed on to the detail design phase, if the PDR committee approves it as meeting the design requirements.

2.5 Detail System Design

The detail design phase represents a continuation of the iterative system development process illustrated in Figure 2.2, on bringing a system into being. The conceptual design and preliminary design phases provide a good foundation upon which to base detailed design decisions that go down to the component/part level. At this point, the system configuration as well as the specifications of subsystems, units, subassemblies, software packages, people, facilities, and elements of maintenance and support are known. The procedure for functional analysis, trade-off study, and design review at the subsystem level are very similar to what is described at the conceptual level. Sections 2.3.4–2.3.6 describe the details of such activities at the system level. Thus, the functional analysis and trade-off study must be extended from the system level down to the subsystem and below as required.

There are 10 major steps in the detail design phase as follows:

- Develop design requirements for all lower-level components of the system from subsystem requirements.
- Employ design tools and software packages.
- Plan, manage and form, and establish several design groups (based on various engineering disciplines).
- Perform extensive design operations (e.g., technical/mathematical calculations and logical selections) to fulfill all design objectives.
- Implement a trade-off analysis.
- Integrate system subsystems/components/elements/parts.
- Publish design data and documentation.
- Generate a prototype physical model.
- Plan and conduct tests and evaluations.
- Schedule and implement a detail design review (DDR).

Success in SE derives from the realization that design activity requires a "team" approach. Hence, in performing technical/mathematical calculations and logical selections, a number of design groups or teams must be established. Basically, there are two approaches: (i) the sequential approach and (ii) the concurrent approach. Both approaches are based on related engineering disciplines (e.g., mechanical, electrical, aeronautical, computer, and civil engineering). In general, the concurrent approach (i.e., simultaneous engineering) minimizes the time, but the sequential approach minimizes the cost of the design operation.

As one proceeds from the conceptual design into the preliminary design and detail design, the actual team "make-up" will vary in terms of the specific expertise required and the number of project staff assigned. Early in the conceptual and preliminary design phases, there is a need for a few highly qualified individuals with broad technical knowledge. These few people understand and believe in the systems engineering and know when to call on the appropriate disciplinary expertise. As the design progresses, the number of representatives from various individual design disciplines will often increase.

Depending on the project size, there may be relatively few individuals assigned, or there may be hundreds of people involved. Required resources may include engineering technical expertise (e.g., engineers), engineering technical support (e.g., technicians, graphics, computer programmers, and builders), and non-technical support (e.g., marketing, budgeting, and human resources). The objective is to promote the "team" culture, and to create the proper environment for the necessary communications. The sequence in the detail design phase is depicted in Figure 2.6, where a number of feedbacks carry the results of the evaluations to the design team members. The design documentation includes design drawings, materials and parts lists, and analyses and reports.

At this phase of the design, an extensive application of computer-based design aids such as computer-aided design (CAD) and computer-aided manufacturing (CAM) throughout

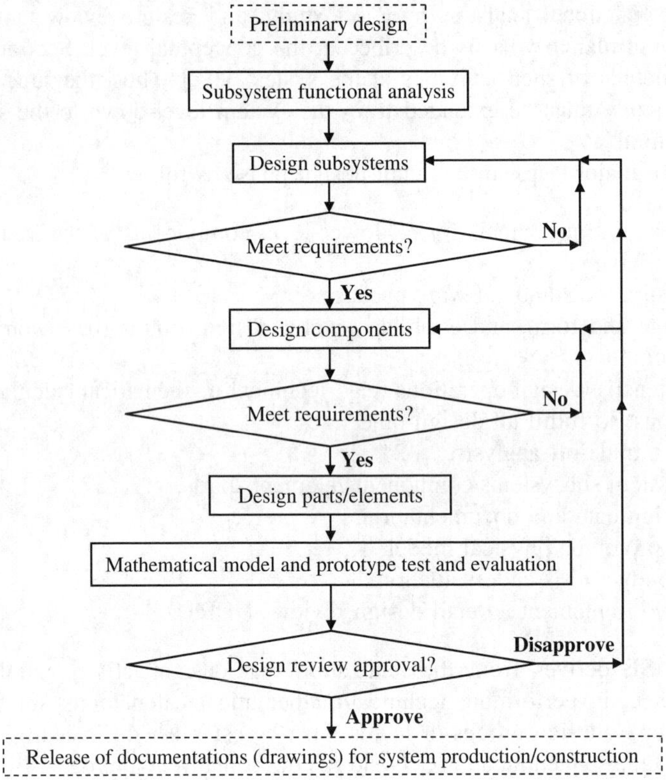

Figure 2.6 Detail design sequence

the design will facilitate the design process. They are employed to generate drawings and three-dimensional graphic displays to be submitted to the manufacturing team. The application of CAD/CAM will allow the systems engineering process to be implemented effectively, efficiently, and in a seamless manner. In order to minimize cost, it is recommended to select standard parts that are commercially available (i.e., commercial off-the-shelf items) for which there are multiple viable suppliers.

At some points in the detail design phase, a mathematical model is necessary to evaluate the design. However, further in the design, a physical model and even a prototype serves much better in the validation and/or verification of the calculation results. This is due to the fact that the incorporation of any necessary changes for corrective action will be more costly later as the design progresses toward the production/construction phase. A prototype represents the production of a system in all aspects of form, fit, and function except that it has not been fully equipped. The objective is to accomplish a specific amount of testing for the purpose of design evaluation prior to entering the production line. The basic objectives and benefits of the DDR process are similar to what is described in Section 2.3.6.

After a baseline has been established, changes are frequently initiated for any one of a number of reasons: to correct a design deficiency, improve the product, incorporate a new technology, respond to a change in operational requirements, compensate for an obsolete component, etc. So, a change may be applied from within the project, or as a result of some new externally imposed requirement. However, a change in any one item will likely have an impact on many other elements of the system. The process of the incorporation of any change must be formalized and controlled to ensure traceability from one configuration to another. A general challenge in today's environment pertains to implementing the overall system design process rapidly, in a limited amount of time, and at a minimal cost.

2.6 Design Requirements

There are specific design requirements which are required by the customer, and must be addressed by the design team. However, there are other design requirements which the customer is not necessarily aware of and may not verbally desire. In this section a list of design-related requirements is reviewed briefly, as follows:

1. performance requirements;
2. stability requirements;
3. handling requirements;
4. operational requirements;
5. affordability requirements;
6. reliability requirements;
7. maintainability requirements;
8. producibility requirements;
9. evaluability requirements;
10. usability requirements;
11. safety (airworthiness for aircraft) requirements;
12. crashworthiness requirements;
13. supportability and serviceability requirements;
14. sustainability requirements;
15. disposability requirements;
16. marketability requirements;
17. environmental requirements;
18. detectability requirements;
19. standards requirements;
20. legal requirements.

The specifications for the above listed requirements must be prepared and addressed during the design and production phases. Not all of these requirements are necessary for each system, but the designer should make sure which ones are applicable. For instance, the detectability requirement is a "must" for a stealth aircraft, but any non-home-built aircraft must follow airworthiness (i.e., safety) standards which are set by the Federal Aviation Administration. It is interesting to note that, for an airliner (civil transport aircraft), there are crashworthiness standards that must be satisfied. Accomplishing the overall design objective requires an appropriate balance between various requirements. This is very difficult to attain, since some stated requirements appear to be in opposition to others. A

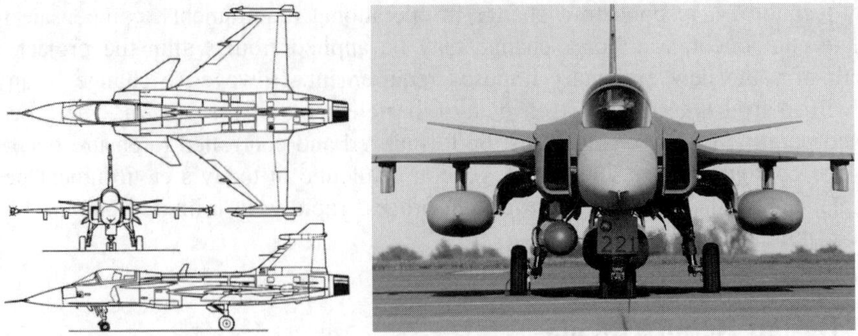

(a) (b)

Figure 2.7 Saab JAS 39 Gripen: (a) three-view; (b) aircraft during landing. Reproduced from permission of (a) Saab; (b) Antony Osborne

three-view and a front view of the lightweight single-engine fighter aircraft Saab JAS 39 Gripen are illustrated in Figure 2.7. The design requirements of modern aircraft, particularly military aircraft, are identified and finalized in a long process by a committee comprising representatives from various sectors.

2.7 Design Review, Evaluation, and Feedback

At each major design phase (conceptual, preliminary, and detail), an evaluation should be conducted to review the design and to ensure that the design is acceptable at that point before proceeding with the next stage. There is a series of formal design reviews conducted at specific times in the overall system development process. An essential technical activity within the design process is that of evaluation. Evaluation must be inherent within the systems engineering process and must be invoked regularly as the system design activity progresses. However, systems evaluation should not proceed without guidance from customer requirements and specific system design criteria. When conducted with full recognition of design criteria, evaluation is the assurance of continuous design improvement.

The evaluation process includes both the informal day-to-day project coordination and data review, and the formal design review. Therefore, there must be "checks and balances" in the form of reviews at each stage of the design progression. The purpose of the design review is to formally and logically evaluate the proposed design in the most effective and economical manner. Through subsequent review, discussion, and feedback, the proposed design is either approved or a list of recommended changes is submitted for consideration. Reference [2] shared with the test and evaluation community a few tips/tools/lessons learned by the US Air Force in terms of their systems engineering and program management success, with shrinking investment resources, and being able to achieve their strategic and tactical objectives.

The purpose of conducting any type of review is to assess if (and how well) the design configuration, as envisioned at the time, is in compliance with the initially specified quantitative and qualitative requirements. The success in conducting a formal design review is dependent on the depth of planning, organization, and preparation prior to the review itself. Each design review serves as an excellent communication medium, creates a better understanding among design and support personnel, and promotes assurance and reliability. The design data/characteristics is released and reviewed for compliance with the basic system requirements. The reviewing operation is performed by a committee formed of technical and operational members. During any design review, a formalized check of the proposed system design is provided, major problems are discussed, and corrective actions are taken. In addition, it provides a common baseline for all project personnel. The design team members are provided the opportunity to explain and justify their design approach through oral and written reports, and reviewer committee members are provided the opportunity to ask various questions from the design team.

A design review provides a formalized check of the proposed system design with respect to specification requirements. Major problems (if any) are discussed and corrective action is taken. The design review also creates a baseline for all project design members. In addition, it provides a means for solving interface problems between design groups and promotes the assurance that all system elements will be compatible. Furthermore, a group review may identify new ideas, possibly resulting in simplified processes and ultimately reduced cost. The outcome of the design project is reviewed at various stages of the design process. In principle, the specific types, titles, and scheduling of these formal reviews vary from one design project to the next. The following main four formal design reviews are recommended for a design project:

1. Conceptual Design Review (CDR);
2. Preliminary Design Review (PDR);
3. Evaluation and Test Review (ETR);
4. Critical (Final) Design Review (FDR).

Figure 2.8 shows the position of each design review in the overall design process. Design reviews are usually scheduled before each major design phase. The CDR is usually scheduled toward the end of the conceptual design phase and prior to entering the preliminary design phase of the program. The purpose of the CDR is to formally and logically cover the proposed design from the system standpoint. The PDR is usually scheduled toward the end of the preliminary design phase and prior to entering the detail design phase. The FDR is usually scheduled after the completion of the detail design phase and prior to entering the production phase. Design is essentially “fixed” at this point, and the proposed configuration is evaluated in terms of adequacy and producibility.

The ETR is usually scheduled somewhere in the middle of the detail design phase and prior to the production phase. The ETR accomplishes two major tasks: (i) finding and fixing any design problems at the subsystem/component level and then (ii) verifying and documenting the system capabilities for government certification or customer acceptance. The ETR can range from the test of a single new system for an existing system to the complete development and certification of a new system. Therefore, the duration of an ETR program can vary from a few weeks to several years. When the system is completely assembled and instrumented, it typically conducts days/weeks/months and even years of field testing.

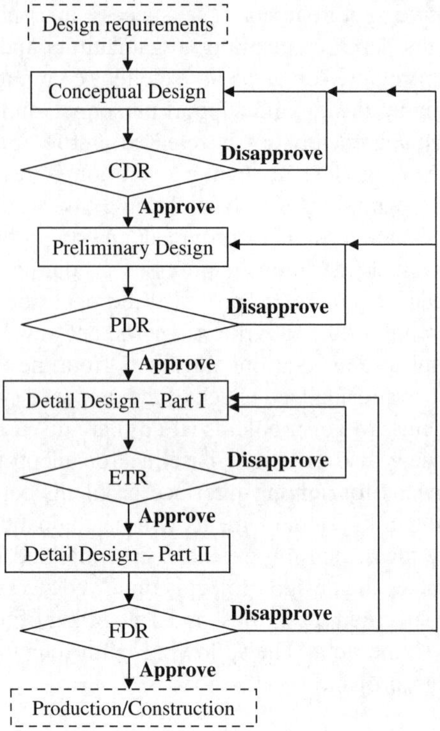

Figure 2.8 Formal design reviews

The detail design is divided into two parts, Part I and Part II regarding the ETR. For this purpose, the subsystems/components are divided into two groups: (i) primary or dominant subsystems/components and (ii) secondary or servant subsystems/components. The dominant subsystems/components are those directly responsible for the design requirements, while the servant subsystems/components are those serving the dominant subsystems/components. For instance, in an automobile, the transmission, engine, and body are assumed to be dominant subsystems, while the electric, air conditioning, and engine cooling are servant subsystems. The engine and transmission are responsible for automobile maximum speed, and the body's function is to provide space for the occupant.

As another example, in an aircraft, the wing, fuselage, tail, and engine are assumed to be dominant components, but the electric system, avionic system, air conditioning system, cabin, cockpit, aero-engine, and landing gear may be categorized as servant components. In an aircraft design project, the aircraft aerodynamic design leads the aircraft structural design, since the structure is a servant subsystem. Thus the aircraft aerodynamic design is performed in Part I of the detail design phase, but the aircraft structural design will be initiated in Part II.

After the dominant subsystems/components (e.g., wing, fuselage, tail, propulsion system) are detail designed, the evaluation and test plan are prepared. When the ETR approves the test plans, a mock-up/model/prototype is fabricated to validate the design. In the case of an aircraft design project, an aircraft model produced and employed in a wind tunnel and a prototype are utilized for flight tests. As soon as the tests are conducted and the results are satisfactory, Part II of the detail design phase begins. During this part of the detail design phase, the servant subsystems are designed. At the end of the detail design phase, the FDR is scheduled to validate/verify the final design.

2.8 Systems Engineering Approach in Aircraft Design

2.8.1 Implementation of Systems Engineering

From a historical perspective, the field of systems engineering developed well after aircraft were invented. This has led to a revolution in the aircraft design approach. In most aircraft manufacturing companies, a department is devoted to systems engineering to cooperate with design engineers. The systems engineering methods are applied to definition, design, development, production, operation, and maintenance. This, however, tends to be the case more for military than commercial aircraft. Throughout this book, systems engineering is adopted as the approach to design an aircraft. The nature of aircraft design projects (complex, multi-disciplinary, with various constraints) suggests that the systems engineering approach is the best candidate. However, the systems engineering implementation is more challenging than understanding the SE process.

The implementation of systems engineering requires a flawless interface between team members working toward a common system thinking to correctly execute the systems engineering process. Although there is a general agreement regarding the principles and objectives of systems engineering, its actual implementation will vary from one discipline to the next. The process approach and steps used will depend on the backgrounds and experience of the individuals involved. The application of systems engineering to aircraft design requires a multi-aspect study, relating aircraft design requirements and functions to systems engineering principles. A functional analysis will pave the road to determine the links between functions of aircraft components and the overall design requirements.

Table 2.3 represents the relationship between aircraft major components and design requirements. Payload has mainly two aspects: (i) weight and (ii) volume. The weight of the payload will mainly influence the aircraft maximum take-off weight, however, the payload volume and geometry affect primarily the design of the fuselage. The aircraft performance requirements may be divided into two groups: (i) range and endurance, (ii) maximum speed, rate of climb, take-off run, stall speed, ceiling, and turn performance. Range and endurance are largely fuel dependent, while other performance requirements are not primarily a function of fuel weight. Thus, endurance and range requirements will mainly influence the aircraft maximum take-off weight and required fuel weight. In contrast, other performance requirements affect engine design, landing gear design, and wing design.

Stability requirements, controllability requirements, and flying quality requirements all impact the location of the aircraft center of gravity, which in turn affects the weight distribution process. However, stability requirements will influence the design of the horizontal tail and vertical tail. In addition, control surfaces design is largely affected by the controllability and maneuverability requirements. Now we are in a position to relate the systems engineering design phases to the aircraft components, to determine what aircraft design activity must be accomplished at each design phase.

Table 2.3 Relationship between aircraft major components and design requirements

No.	Design requirements	Aircraft component/parameters affected most
1	Payload (weight) requirements	Maximum take-off weight
	Payload (volume and geometry) requirements	Fuselage
2	Performance requirements (range and endurance)	Maximum take-off weight, fuel weight
3	Performance requirements (maximum speed, rate of climb, take-off run, stall speed, ceiling, and turn performance)	Engine, landing gear, and wing
4	Stability requirements	Horizontal tail and vertical tail, weight distribution
5	Controllability requirements	Control surfaces (elevator, aileron, rudder), weight distribution
6	Flying quality requirements	Center of gravity, weight distribution
7	Airworthiness requirements	Minimum safety requirements
8	Cost requirements	Materials, engine, weight, etc.
9	Design duration requirements	Configuration optimality
10	Detectability requirements	Materials, configuration

2.8.2 Design Phases

There are a number of phases through which the system design and development process must invariably pass. Foremost among them is the identification of the customer-related need and, from that need, the determination of what the system is to do. This is followed by a feasibility analysis to discover potential technical solutions, the determination of system requirements, the design and development of system components, the construction of a prototype, and/or engineering model, and the validation of the system design through test and evaluation. According to the systems engineering approach, a total of four design phases are defined. As outlined in Sections 2.2–2.5, the system (i.e., aircraft) design process includes: (i) conceptual design, (ii) preliminary design, (iii) detail design, and (iv) test and evaluation (Figure 2.6). The details of the four phases of the integrated design of an aircraft are summarized in Table 2.4.

At the conceptual design phase, the aircraft will be designed in concert with non-precise results. In other words, almost all parameters are determined based on a decision-making process and a selection technique. Chapter 3 presents the details of the aircraft conceptual design technique. In contrast, the aircraft preliminary design phase tends to employ the outcomes of a calculation procedure. As the name implies, at the preliminary design phase, the parameters determined are not final and will be altered later. In addition,

Table 2.4 A summary of four major aircraft design phases

No.	Design phase	Design activity
1	Conceptual design	Aircraft configuration design
2	Preliminary design	Determine (i) aircraft maximum take-off weight, (ii) engine power or thrust, (iii) wing reference area
3	Detail design	Part I: Design dominant components such as wing, fuselage, tail, and propulsion system, landing gear (non-mechanical)
		Part II: Design servant components such as landing gear (mechanical), engine, structural design, cabin, cockpit, avionic system, electric system, and air conditioning system
4	Test and evaluation	Aircraft aerodynamic testing: wind tunnel test using aircraft model
		Aircraft flight dynamic testing: flight test using a prototype
		Aircraft structural testing using an aircraft structure
		Propulsion system testing using an aero-engine

at this phase, the parameters are essential and will directly influence the entire detail design phase. Therefore, ultimate care must be taken to insure the accuracy of the results of the preliminary design phase. In summary, three aircraft fundamental parameters are determined in the preliminary design: (i) aircraft maximum take-off weight (W_{TO}), (ii) wing reference area (S_{ref}), and (iii) engine power (P) if the aircraft is prop-driven or engine thrust (T) if a jet engine is selected. The details, techniques, and procedure of the aircraft preliminary design phase are developed and introduced in Chapter 4.

At the aircraft detail design phase, the technical parameters of all components (e.g., wing, fuselage, tail, landing gear, and engine) – including geometry – are calculated and finalized. The detail design phase of major components is introduced in Chapters 5. This textbook only addresses the detail design of dominant components; that is, wing, tail, fuselage, propulsion system, and non-mechanical aspects of landing gear. However, the detail design of control surfaces (e.g., elevator, aileron, and rudder) is examined in Chapter 12.

2.8.3 Design Flowchart

As emphasized in the guidelines of Section 2.2, the aircraft design process has an iterative nature due to the test and evaluation requirements. Whenever a change is applied to one component, the least consequence is that the aircraft weight and aircraft center of gravity will vary. Thus, an adjustment must be made to keep the aircraft in the correct path. Three major requirements that all customers are very sensitive to and concerned about are: (i) performance requirements, (ii) stability requirements, and (iii) controllability requirements. These design requirements necessitate three evaluations and generate three

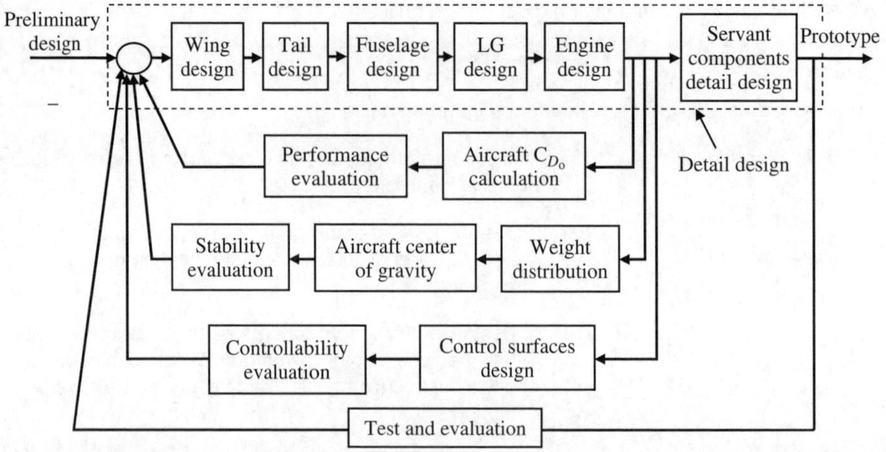

Figure 2.9 Relationship between detail design phase and design feedbacks

feedback loops, as depicted in Figure 2.9. The aircraft performance evaluation requires the calculation of the aircraft zero-lift drag coefficient (C_{D_o}). References [19, 20] are recommended for techniques to determine aircraft zero-lift drag coefficients.

Aircraft stability evaluation requires the calculation of the aircraft center of gravity, which in turn requires the aircraft weight distribution. Techniques to calculate the aircraft center of gravity are addressed in Chapter 10; and the aircraft weight distribution procedure is introduced in Chapter 11. The evaluation of aircraft controllability and maneuverability requires control surfaces design. The design of primary control surfaces – such as elevator, aileron, and rudder – is the topic of Chapter 12. The three activities of performance evaluation, stability evaluation, and controllability evaluation are part of the ETR, as discussed in Section 2.7. These evaluations are accomplished when the design of the dominant components (e.g., wing, tail, fuselage, and propulsion system) has been achieved and prior to the design of the servant components.

To see the big picture, Figure 2.10 demonstrates the interrelationships and positions of major aircraft test activities within three major design phases. Aircraft design-related sciences tend to be based on four primary areas of expertise: aerodynamics, flight dynamics, aero-structure, and propulsion. Each expertise requires an independent test and evaluation. The aerodynamic evaluation of the design is accomplished by using an aircraft model placed in a wind tunnel to measure pressure distribution and consequently lift, drag and pitching moment. The flight dynamics (i.e., performance, stability, and control) is evaluated by conducting various flight tests. However, a flight simulation employing a complete aircraft dynamic model may serve instead, but the results are not as dependable as those of flight tests using a prototype. The propulsion system is evaluated by placing an engine in a ground test station and running a variety of propulsive tests.

Ultimately, the aircraft structural tests employ a complete aircraft structure and apply various static and dynamic loads. It is interesting to note that some dynamic structural tests (e.g., fatigue) may take a very long time and possibly have to be scheduled after aircraft production. For instance, the complete structural dynamic tests of the rear fuselage

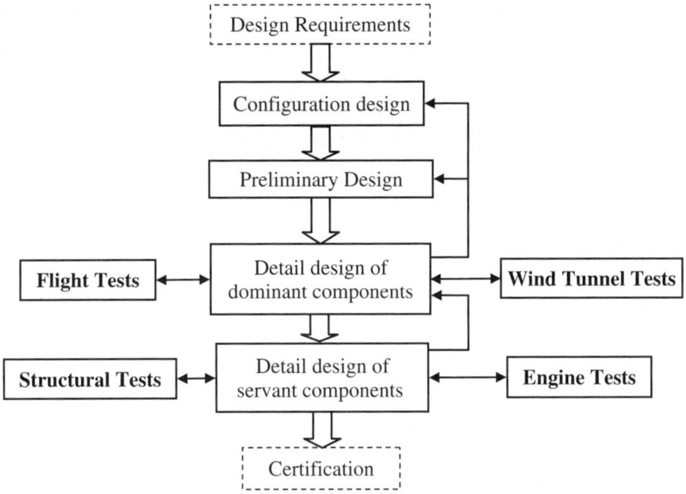

Figure 2.10 Design phases and test and evaluations

Figure 2.11 Fighter aircraft F/A-18 Hornet. Reproduced from permission of Antony Osborne

under various dynamic loads for the fighter aircraft F/A-18 Hornet (Figure 2.11) were performed several years after it had been delivered and flown by various air forces. At any rate, all four aeronautical tests have to be accomplished and the results must be satisfactory before the aircraft certification is issued and accredited. All four groups of tests are performed during and after the detail design phase.

2.8.4 Design Groups

A fundamental aircraft design activity regarding systems engineering is the planning, design management, handling process, and organization. Successful implementation of the concepts, principles, models, and methods of systems engineering and analysis requires the coordination of technical and managerial endeavors. The proper implementation of systems engineering begins with the establishment of requirements by a planning process initiated during the conceptual design phase of the aircraft. The primary objective of systems engineering management is to facilitate the timely integration of numerous design considerations into a functioning system. In planning the design organization, the format adopted may vary somewhat depending on the aircraft type, aircraft mission, and aircraft size being acquired.

An aircraft chief designer should be capable of covering and handling a broad spectrum of activities. Thus, an aircraft chief designer should have years of experience, be knowledgeable about management techniques, and preferably have full expertise and background in the area of "flight dynamics." The chief designer has a great responsibility in planning, coordinating, and conducting formal design reviews. He/she must also monitor and review the aircraft system test and evaluation activities, as well as coordinating all the formal design changes and modifications for improvement. The organization must be such as to facilitate the flow of information and technical data among various design departments. The design organization must allow the chief designer to initiate and establish the necessary ongoing liaison activities throughout the design cycle.

One of the first steps in program planning is the development of the work breakdown structure (WBS). The WBS is a product-oriented family tree that leads to the identification of the functions, activities, tasks, and work packages that must be performed for the completion of a given design program. The WBS is not an organizational chart in terms of project personnel assignment and responsibilities, but represents an organization of the work package prepared for the purpose of program planning, budgeting, contracting, and reporting. The WBS generally includes three levels of activity: level 1, level 2, and level 3.

A primary building block in organizational patterns is the functional approach, which involves the grouping of functional specialties or disciplines into separately identifiable entities. The intent is to perform similar work within one organizational group. Thus, the same organizational group will accomplish the same type of work for all ongoing projects on a concurrent basis. The ultimate objective is to establish a team approach, with the appropriate communications, enabling the application of concurrent engineering methods throughout.

There are two main approaches to handling the design activities and establishing design groups: (i) design groups based on aircraft components (Figure 2.12) and (ii) design groups based on expertise (Figure 2.13). If the approach of groups based on aircraft components is selected, the chief designer must establish the following teams: (i) wing design team, (ii) tail design team, (iii) fuselage design team, (iv) propulsion system design team, (v) landing gear design team, and (vi) equipment design team. The seventh team

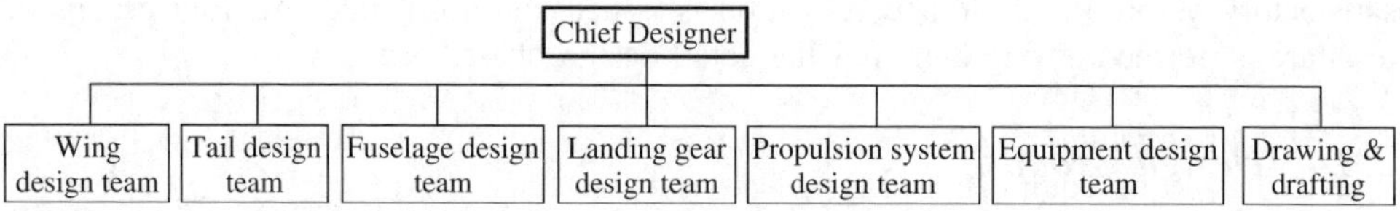

Figure 2.12 Work breakdown structure based on aircraft components during design phase

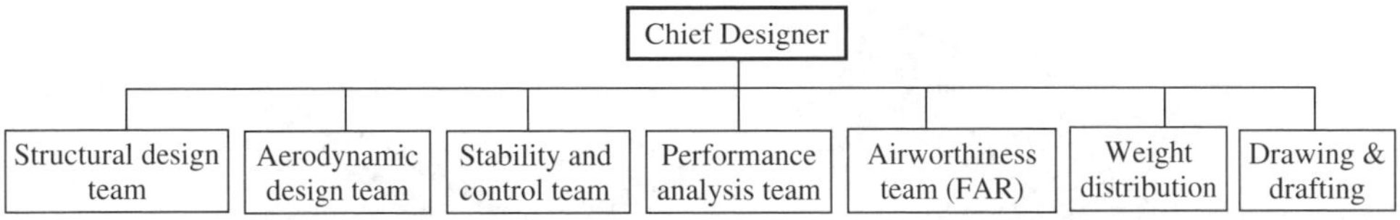

Figure 2.13 Work breakdown structure based on discipline during design phase

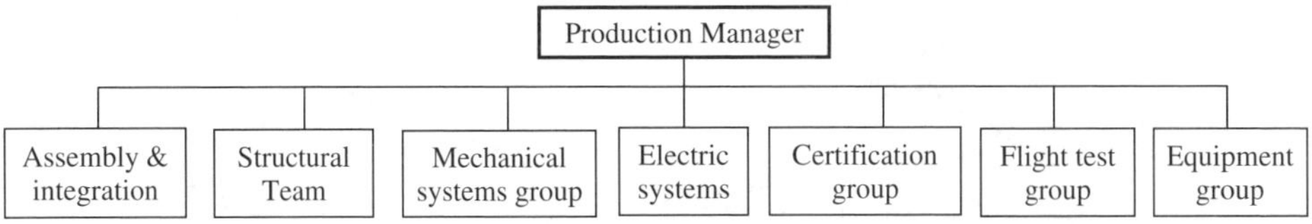

Figure 2.14 Organizational chart during fabrication phase

is established for documentation and drafting. There are various advantages and disadvantages for each of the two planning approaches in terms of ease of management, speed of communication, efficiency, and similarity of tasks. However, if the project is large – such as the design of a large transport aircraft –both groupings could be applied simultaneously. In contrast, for the design of a small model aircraft, the WBS based on aircraft components works more efficiently. Figure 2.14 represents a typical organizational chart during the fabrication phase. The difference between planning approaches during design and fabrication is evident by comparing Figures 2.11 and 2.12 with Figure 2.14.

The design of an aircraft is a team effort and calls on the extraordinary talents of engineers in each design group. The chief designer serves as the referee and will integrate everyone's efforts into the design of an air vehicle. Figure 2.15 illustrates the kind of aircraft design that might emerge if any one design group was allowed to dominate the others. The role of the chief designer is crucial to adjust the relationship among various groups and apply the appropriate limits to each group in order to achieve an optimum design.

2.8.5 Design Steps

In this section, a 47-step model of the aircraft design process is outlined briefly. Most of the steps in this model are discussed in various chapters throughout the book, as addressed earlier. However, a few general comments are in order. The 47 steps and the required data or assumptions in each step are summarized in Table 2.5. The numbers in brackets indicate the chapters where the topics are covered. When there is no bracketed number in front of a step, this means that the book does not cover that topic; so the designer must refer to the relevant textbooks or references. For instance, aircraft stability analysis and aircraft structural analysis are beyond the context of this book. The reader, however, is expected to be equipped with knowledge regarding these topics prior to learning the aircraft design techniques. In fact, such topics are a prerequisite for the aircraft design course.

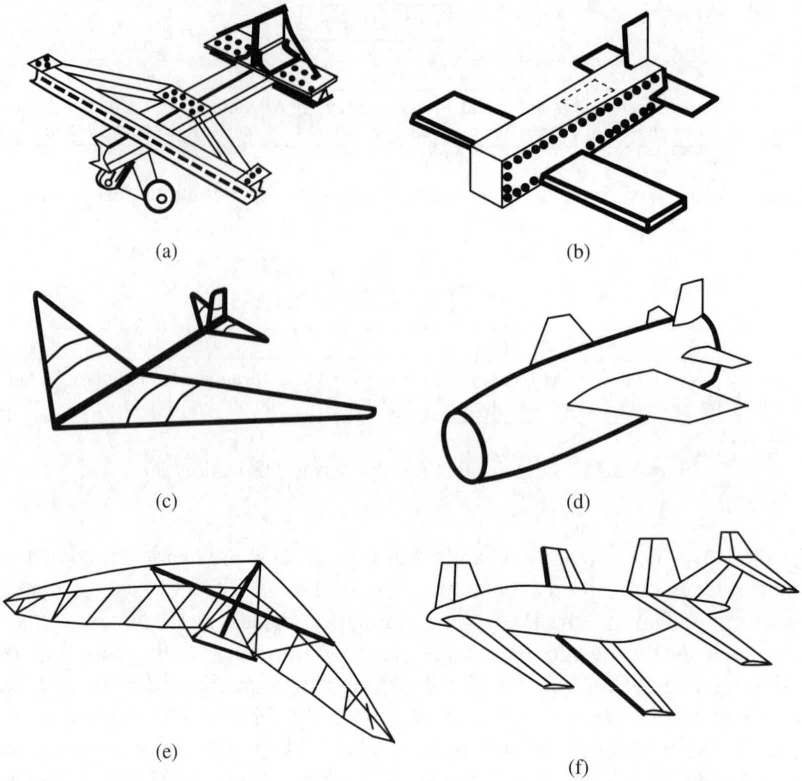

Figure 2.15 Design groups' unique visions and interests: (a) Structural group; (b) Manufacturing group; (c) Aerodynamics group; (d) Propulsion group; (e) Weight group; (f) Stability and control group

The fact that the sequence of topics presented in this book does not match up exactly with the steps in the design process model reinforces the non-sequential nature of the model. Many engineering designs are performed by teams of engineers and not every team member participates in every step of the process. Some team members may be specialists in one or more of the 46 steps. In many situations, design engineers may unconsciously blend some of these steps together. Each step may be revisited several times. However, even an experienced designer will regularly step back from his/her immersion in design details and rely on such a model to ensure that key elements are not overlooked in his/her research for a design solution. Steps 6, 11, 27, and 43 are representative of the time for reviewing the design. The design process proceeds after these steps only if the reviewer committee approves the design at that particular point.

As step 23 indicates, the design has an iterative nature. The iteration continues as long as the need continues, and ends only when the cost of continuing the design process exceeds the value of an improved design. The decision to stop the design process is difficult and requires careful thought. It may be made by the designer or the customer, or it may be the result of schedule or budget constraints. This is what is meant by saying "the design is an open-ended process"; there is frequently no readily identified closure point.

Table 2.5 Aircraft design steps

No.	Design step	Required data or assumptions
1	Identification of design requirements and constraints; definition of problem statement (Chapters 1 and 2)	Mission, customer needs
2	Prioritization of design requirements (Chapters 1 and 2)	Budget, workforce, timing, marketing, competitors info
3	Planning, breaking down the work, establishing (Chapter 2) organizational groups, preparing activity sequence chart	Management skills
4	Feasibility study, market analysis	Capability and features of current technology
5	Aircraft configuration design (Chapter 3)	Mission, criteria, preferences
6	Configuration design review (CDR) (Chapter 2)	–
7	Estimation of maximum take-off weight (MTOW) (Chapter 4)	Payload, aircraft type, statistics
8	Estimation of aircraft C_{D_o} and $C_{L_{max}}$	Statistics for similar aircraft, flap details
9	Calculation of wing reference area (S_{ref}) (Chapter 4)	Performance requirements, wing aspect ratio
10	Calculation of engine thrust/engine power (Chapter 4)	Performance requirements
11	Preliminary design review (PDR) (Chapter 2)	–
12	Wing design (Chapter 5)	S, cruise speed, stall speed
13	Fuselage design (Chapter 7)	Payload (volume and geometry), human factors
14	Horizontal tail design (Chapter 6)	Longitudinal stability requirements
15	Vertical tail design (Chapter 6)	Directional stability requirements
16	Landing gear design (Chapter 9)	Aircraft center of gravity limits
17	Propulsion system design (Chapter 8)	Engine power or engine thrust
18	Calculation of aircraft C_{D_o} and aircraft drag	–
19	Aircraft performance analysis	–
20	Redesign of propulsion system if the calculated performance does not meet the requirements	Performance requirements
21	First estimation of weight of aircraft components (Chapter 10)	–
22	Second estimation of aircraft MTOW (Chapter 10)	–

(*continued overleaf*)

Table 2.5 (*continued*)

No.	Design step	Required data or assumptions
23	Return to step 7 until the results of the first and second weight estimation are the same	–
24	Calculation of center of gravity limits (Chapter 11)	–
25	Relocation of aircraft components (i.e., weight distribution) to adjust aircraft center of gravity (Chapter 11)	Stability and controllability requirements
26	Redesign of horizontal tail and vertical tail (Chapter 6)	Trim requirements
27	Evaluation and test review (ETR) (Chapter 2)	
28	Design control surfaces (Chapter 12)	Controllability requirements
29	Calculation of interferences between wing, fuselage, engine, and tails	–
30	Modifications of wing, fuselage, engine, and tails	–
31	Stability and control analysis	–
32	Aircraft modification	–
33	Manufacturing of aircraft model	–
34	Wind tunnel test	Wind tunnel, aircraft model
35	Aircraft modification	–
36	Aircraft structural design	–
37	Calculation of weight of components and aircraft weight	–
38	Performance, stability, and control analysis	–
39	Aircraft modifications	–
40	Aircraft equipment/subsystems design (e.g., electric, pressure, power transmission)	–
41	Manufacturing of the prototype	–
42	Flight tests	Pilot, prototype
43	Modifications	–
44	Critical design review (FDR) (Chapter 2)	–
45	Optimization (Chapter 3)	–
46	Certification	Standards
47	Release documentations, drawings, and specifications for production	–

Figure 2.16 Bell-Boeing MV-22B Osprey. Reproduced from permission of Antony Osborne

Figure 2.17 Lockheed Martin F-35 Lightning II

The design histories of aircraft such as the vertical take-off and landing (VTOL) tilt-rotor military aircraft Bell-Boeing V-22 Osprey (Figure 2.16), single-seat, twin-engine, fighter aircraft Lockheed Martin/Boeing F-22 Raptor (Figure 8.21); and single-seat, single-engine stealth fighter and reconnaissance aircraft Lockheed Martin F-35 Lightning (Figure 2.17) confirms this concept. Reference [21] describes various real lessons learned in aircraft design and is a rich resource of aircraft design experiences for young designers. As an example, based on lessons learned, US congress will keep the F-18E/F in production to cover any shortfalls from Lockheed Martin F-35 Lightning II delays [22].

References

[1] MIL-STD (1969). *Systems Engineering Process*, DOD.

[2] Zadeh, S. (2010) Systems engineering: a few useful tips, tools, and lessons learned for the manager's toolbox. US Air Force T&E Days 2010, Nashville, TN, February 2–4, 2010, AIAA 2010-1758.

[3] Loren, J.R. (2004) USAF systems engineering – revitalizing fundamental processes. USAF Developmental Test and Evaluation Summit, Woodland Hills, CA, November 16–18, 2004, AIAA 2004-6855.

[4] Shishko, R. (2007) *NASA Systems Engineering Handbook*, National Aeronautics and Space Administration NASA/SP-2007-6105.

[5] Hsu, J.C., Raghunathan, S., and Curran, R. (2008) Effective learning in systems engineering. 46th AIAA Aerospace Sciences Meeting and Exhibit, Reno, NV, January 7–10, 2008, AIAA 2008-1117.

[6] Armand, J. (2010) Chaput, issues in undergraduate aerospace system engineering design, education – an outsider view from within. 10th AIAA Aviation Technology, Integration, and Operations Conference, Fort Worth, TX, September 13–15, 2010, AIAA 2010-9016.
[7] Curran, R., Tooren, M., and Dijk, L. (2009) Systems engineering as an effective educational framework for active aerospace design learning. 9th AIAA Aviation Technology, Integration, and Operations Conference, Hilton Head, SC, September 21–23, 2009, AIAA 2009-6904.
[8] Mission and Vision (2011) International Council on Systems Engineering, http://www.incose.org.
[9] Blanchard, B.S. and Fabrycky, W.J. (2006) *Systems Engineering and Analysis*, 4th edn, Prentice Hall.
[10] Buede, D.M. (2009) *The Engineering Design of Systems: Models and Methods*, 2nd edn, John Wiley & Sons, Inc.
[11] Hsu, J.C., Raghunathan, S., and Curran, R. (2009) A proposed systems engineering diagnostic method. 47th AIAA Aerospace Sciences Meeting Including The New Horizons Forum and Aerospace Exposition, Orlando, FL, January 5–8, 2009, AIAA 2009-1006.
[12] Gill, P.S., Garcia, D., and Vaughan, W.W. (2005) Engineering lessons learned and systems engineering applications. 43rd AIAA Aerospace Sciences Meeting and Exhibit, Reno, NV, January 10–13, 2005, AIAA 2005-1325.
[13] Farrell, C. (2007) Systems engineering, system architecting, and enterprise architecting – what's the difference? 45th AIAA Aerospace Sciences Meeting and Exhibit, Reno, NV, January 8–11, 2007, AIAA 2007-1192.
[14] Paul Collopy, A.D. (2010) Fundamental research into the design of large-scale complex systems. 13th AIAA/ISSMO Multidisciplinary Analysis Optimization Conference, Fort Worth, TX, September 13–15, 2010, AIAA 2010-9320.
[15] Hsu, J.C. and Raghunathan, S. (2007) Systems engineering for CDIO – conceive, design, implement and operate. 45th AIAA Aerospace Sciences Meeting and Exhibit, Reno, NV, January 8–11, 2007, AIAA 2007-591.
[16] Weaver Smith, S., Seigler, M., Smith, W.T., and Jacob, J.D. (2008) Multi-disciplinary multi-university design of a high-altitude inflatable-wing aircraft with systems engineering for aerospace workforce development. 46th AIAA Aerospace Sciences Meeting and Exhibit, Reno, NV, January 7–10, 2008, AIAA 2008-490.
[17] Onwubiko, C. (2000) *Introduction to Engineering Design Optimization*, Prentice Hall.
[18] Chong, E.K.P. and Zack, S.H. (2008) *An Introduction to Optimization*, 3rd edn, John Wiley & Sons, Inc.
[19] Sadraey, M. (2009) *Aircraft Performance Analysis*, VDM Verlag Dr. Müller.
[20] Hoak, D.E. (1978) USAF Stability and Control DATCOM, Air Force Flight Dynamics Laboratory, Wright-Patterson Air Force Base, Ohio.
[21] Roskam, J. (2007) *Lessons Learned in Aircraft Design; the Devil is in the Details*, DAR Corporation.
[22] Wilson, J.R. (2011) *F-35; A Time of Trail*, American Institute for Aeronautics and Astronautics.

3

Aircraft Conceptual Design

3.1 Introduction

As outlined in Chapter 2, in order to implement the systems engineering discipline [1], the aircraft (i.e., system) design process includes four major phases: (i) conceptual design, (ii) preliminary design, (iii) detail design, and (iv) test and evaluation. The purpose of this chapter is to present the techniques and selection processes in the aircraft conceptual design phase. Conceptual design is the first and most important phase of the aircraft system design and development process. It is an early and high-level lifecycle activity with potential to establish, commit, and otherwise predetermine the function, form, cost, and development schedule of the desired aircraft system. The identification of a problem and associated definition of need provides a valid and appropriate starting point for design at the conceptual level.

Selection of a path forward for the design and development of a preferred system configuration, which will ultimately be responsive to the identified customer requirement, is a major responsibility of conceptual design. Establishing this early foundation, as well as requiring the initial planning and evaluation of a spectrum of technologies, is a critical first step in the implementation of the systems engineering process. Systems engineering, from an organizational perspective, should take the lead in the definition of system requirements from the beginning and address them from a total integrated lifecycle perspective.

The aircraft design process generally commences with the identification of a "what" or "desire" for something and is based on a real (or perceived) deficiency. As a result, a system requirement is defined along with the priority for introduction, the date when the system capability is required for customer use, and an estimate of the resources necessary for acquiring this new system. To ensure a good start, a comprehensive statement of the problem should be presented in specific qualitative and quantitative terms, in enough detail to justify progressing to the new step. Need identification and formulation is discussed in Chapter 2.

As the name implies, the aircraft conceptual design phase is the aircraft design at the concept level. At this stage, the general design requirements are entered into a process to generate a satisfactory configuration. The primary tool at this stage of design is the

Aircraft Design: A Systems Engineering Approach, First Edition. Mohammad H. Sadraey.

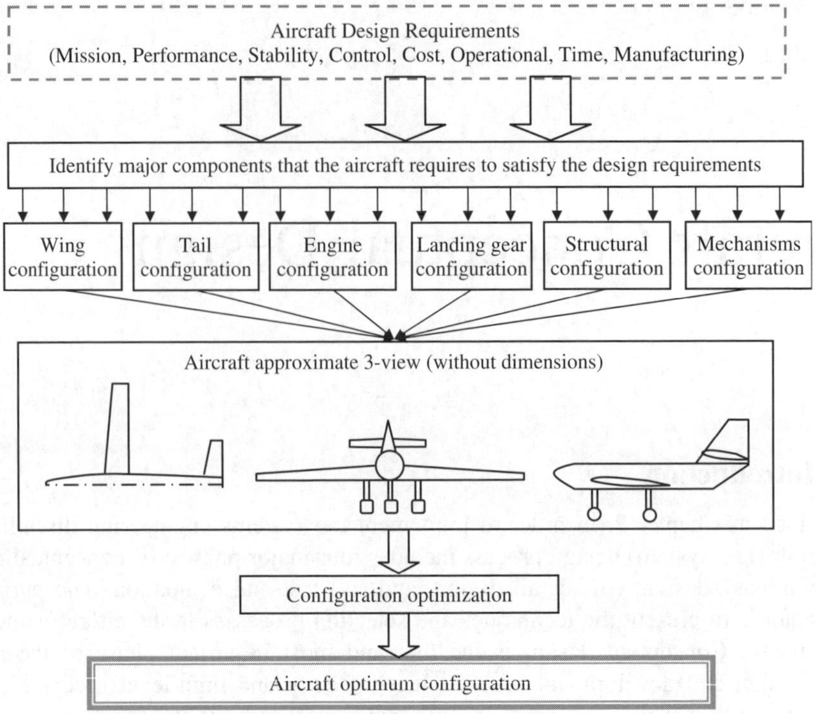

Figure 3.1 Aircraft conceptual design

selection. Although there are a variety of evaluations and analyses, there are not many calculations. The past design experience plays a crucial role in the success of this phase. Hence, the members of the conceptual design phase team must be the most experienced engineers of the corporation. The details of the advantages and disadvantages of each configuration are described in Chapters 5–11.

Figure 3.1 illustrates the major activities which are practiced in the conceptual design phase. The fundamental output of this phase is an approximate three-view of the aircraft that represents the aircraft configuration. Section 3.2 concerns the primary function and role for each aircraft component. The aircraft components (e.g., wing, fuselage, tail, landing gear, and engine) configuration alternatives are addressed in Section 3.3. Aircraft classifications from a variety of aspects are reviewed in Section 3.4. In Section 3.5, the principles of trade-off analysis to determine the most satisfactory configuration are introduced. Section 3.6 examines the conceptual design optimization with emphasis on the application of the multi-disciplinary design optimization (MDO) technique.

3.2 Primary Functions of Aircraft Components

An aircraft comprises several major components. It mainly includes the wing, horizontal tail, vertical tail (VT), fuselage, propulsion system, landing gear, and control surfaces. In order to make a decision about the configuration of each aircraft component, the designer must be fully aware of the function of each component. Each aircraft component has inter-relationships with other components and interferes with the functions of other components.

1. **Wing**. The main function of the wing is to generate the aerodynamic force of lift to keep the aircraft airborne. The wing tends to generate two other unwanted aerodynamic productions: an aerodynamic drag force plus an aerodynamic pitching moment. Furthermore, the wing is an essential component in providing the aircraft lateral stability, which is fundamentally significant for flight safety. In almost all aircraft, the aileron is arranged so as to be at the trailing edge of the outboard section. Hence, the wing is largely influential in providing the aircraft lateral control.
2. **Fuselage**. The primary function of the fuselage is to accommodate the payload which includes passengers, cargo, luggage, and other useful loads. The fuselage is often a home for the pilot and crew members, and most of the time fuel tanks and engine(s). Since the fuselage provides a moment arm to the horizontal and VT, it plays an influential role in longitudinal and directional stability and control. If the fuselage is decided to be short, a boom must be provided to allow for the tails to have sufficient arm.
3. **Horizontal tail**. The horizontal tail's primary function is to generate an aerodynamic force to longitudinally trim the aircraft. Furthermore, the VT is an essential component is providing the aircraft longitudinal stability, which is a fundamental requirement for flight safety. In the majority of aircraft, the elevator is a movable part of the horizontal tail, so longitudinal control and maneuverability are applied through the horizontal tail.
4. **Vertical tail**. The VT's primary function is to generate an aerodynamic force to directionally trim the aircraft. Furthermore, the VT is an essential component in providing the aircraft directional stability, which is a fundamental requirement for flight safety. In the majority of aircraft, the rudder is a movable part of the VT, so directional control and maneuverability are applied through the VT.
5. **Engine**. The engine is the main component in the aircraft propulsion system to generate power and/or thrust. The aircraft requires a thrust force to move forward (as in any other vehicle), so the engine's primary function is to generate the thrust. The fuel is considered to be a necessary item of the propulsion system and it sometimes constitutes a large part of the aircraft weight. An aircraft without an engine is not able to take off independently, but is capable of gliding and landing, as performed by sailplanes and gliders. Sailplanes and gliders take off with the help of other aircraft or outside devices (such as a winch), and climb with the help of wind and thermal currents.
6. **Landing gear**. The primary function of the landing gear is to facilitate take-off and landing operations. During take-off and landing operations, the fuselage, wing, tail, and aircraft components are kept away from the ground through the landing gear. The wheels of the landing gear in land-based and ship-based aircraft also play a crucial role in safe acceleration and deceleration of the aircraft. Rolling wheels as part of the landing gear allow the aircraft to accelerate without spending a considerable amount of thrust to overcome friction.

The above six components are assumed to be the fundamental components of an air vehicle. However, there are other components in an aircraft that are not assumed here as major ones. The roles of these components are described in later sections, whenever they are mentioned. Table 3.1 illustrates a summary of aircraft major components and

Table 3.1 Aircraft major components and their functions

No.	Component	Primary function	Major areas of influence
1	Fuselage	Payload accommodations	Aircraft performance, longitudinal stability, lateral stability, cost
2	Wing	Generation of lift	Aircraft performance, lateral stability
3	Horizontal tail	Longitudinal stability	Longitudinal trim and control
4	Vertical tail	Directional stability	Directional trim and control, stealth
5	Engine	Generation of thrust	Aircraft performance, stealth, cost, control
6	Landing gear	Facilitates take-off and landing	Aircraft performance, stealth, cost
7	Control surfaces	Control	Maneuverability, cost

their functions. This table also shows the secondary roles and major areas of influence of each aircraft component. This table also shows the design requirements that are affected by each component. The functions described in Table 3.1 are only the primary functions of each component, and secondary functions are not addressed. Full explanations of the function and role for each component are outlined in Chapters 5–12.

Traditional aircraft configuration design attempts to achieve improved performance and reduced operating costs by minimizing the maximum take-off weight. From the point of view of an aircraft manufacturer, however, this method does not guarantee the financial viability of an aircraft program. A better design approach would take into account not only aircraft performance and manufacturing cost, but also factors such as aircraft flying qualities and systems engineering criteria.

The historical choice of minimizing the gross take-off weight (GTOW) as the objective in aircraft design is intended to improve performance and subsequently lower operating costs, primarily through reduced fuel consumption. However, such an approach does not guarantee the optimality of a given aircraft design from the perspective of the aircraft consumer. In an increasingly competitive market for aircraft, manufacturers may wish to design for improved systems engineering of an aircraft program, as well as technical merit, before undertaking such a costly investment.

3.3 Aircraft Configuration Alternatives

When the necessary aircraft components, to satisfy design requirements, are identified and the list of major components is prepared, the step to select their configurations begins. Each major aircraft component may have several alternatives which all satisfy the design requirements. However, each alternative will carry advantages and disadvantages by which the design requirements are satisfied at different levels. Since each design requirement has a unique weight, each configuration alternative results in a different level of satisfaction. This section reviews the configuration alternatives for each major component. The description of the advantages and disadvantages for each configuration will be addressed in Chapters 5–12.

3.3.1 Wing Configuration

In general, the wing configuration alternatives from seven different aspects are as follows:

1. **Number of wings**
   a. Monoplane
   b. Biplane
   c. Triplane
2. **Wing location**
   a. High wing
   b. Mid-wing
   c. Low wing
   d. Parasol wing
3. **Wing type**
   a. Rectangular
   b. Tapered
   c. Delta
   d. Swept back
   e. Swept forward
   f. Elliptical
4. **High-lift device**
   a. Plain flap
   b. Split flap
   c. Slotted flap
   d. Kruger flap
   e. Double-slotted flap
   f. Triple-slotted flap
   g. Leading edge flap
   h. Leading edge slot
5. **Sweep configuration**
   a. Fixed wing
   b. Variable sweep
6. **Shape**
   a. Fixed shape
   b. Morphing wing
7. **Structural configuration**
   a. Cantilever
   b. Strut-braced
      i. faired
      ii. un-faired.

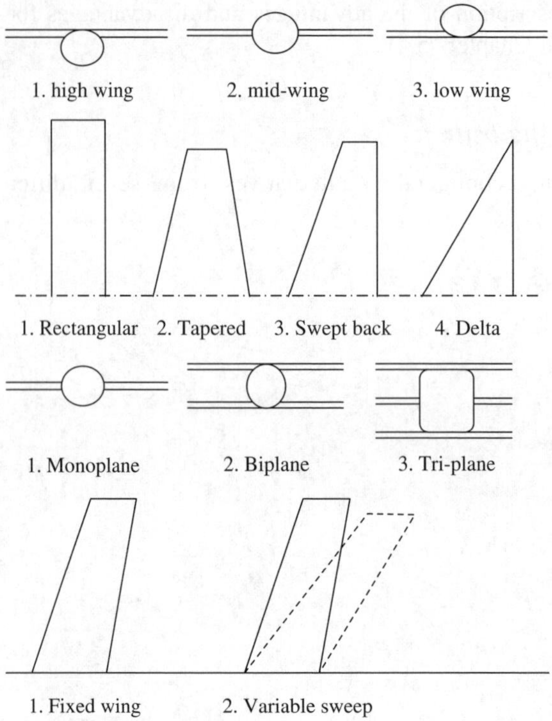

Figure 3.2 Wing configuration alternatives

The advantages and disadvantages of the wing configuration alternatives, plus the technique to select the best wing configuration alternative to meet the design requirements, are presented in Chapter 5–12. The primary impacts of the wing configuration alternatives are imposed on cost, the duration of production, ease of manufacturing, lateral stability, performance, maneuverability, and aircraft life. Figure 3.2 illustrates several wing configuration alternatives.

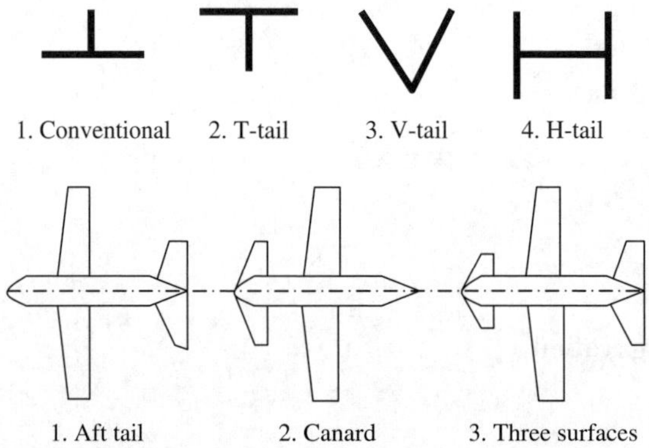

Figure 3.3 Tail configuration alternatives

3.3.2 Tail Configuration

In general, the tail configuration alternatives from three different aspects are as follows:

1. **Aft or forward**
   a. Aft conventional tail
   b. Canard (foreplane)
   c. Three surfaces
2. **Horizontal and vertical tail**
   a. Conventional
   b. V-tail
   c. T-tail
   d. H-tail
   e. Inverted U
3. **Attachment**
   a. Fixed tail
   b. Moving tail
   c. Adjustable tail.

The advantages and disadvantages of the tail configuration alternatives, plus the technique to select the best tail configuration alternative to meet the design requirements, are presented in Chapter 6. The primary impacts of the tail configuration alternatives are imposed on cost, the duration of production, ease of manufacturing, longitudinal and directional stability, longitudinal and directional maneuverability, and aircraft life. Figure 3.3 illustrates several tail configuration alternatives.

3.3.3 Propulsion System Configuration

In general, the propulsion system configuration alternatives from four different aspects are as follows:

1. **Engine type**
   a. Human-powered
   b. Solar-powered
   c. Piston prop
   d. Turboprop
   e. Turbofan
   f. Turbojet
   g. Rocket
2. **Engine and the aircraft center of gravity**
   a. Pusher
   b. Tractor
3. **Number of engines**
   a. Single-engine
   b. Twin-engine

   c. Tri-engine
   d. Four-engine
   e. Multi-engine
4. **Engine location**
   a. In front of nose (inside)
   b. Inside fuselage mid-section
   c. Inside wing
   d. Top of wing
   e. Under wing
   f. Inside vertical tail
   g. Side of fuselage at aft section
   h. Top of fuselage.

The advantages and disadvantages of the propulsion system alternatives, plus the technique to select the best engine configuration alternative to meet the design requirements, are presented in Chapter 9. The primary impacts of the engine configuration alternatives are imposed on cost of flight operation, cost of aircraft production, performance, duration of production, ease of manufacturing, maneuverability, flight time, and aircraft life. Figure 3.4 illustrates several engine configuration alternatives.

3.3.4 Landing Gear Configuration

In general, the landing gear configuration alternatives from three different aspects are as follows:

1. **Landing gear mechanism**
   a. Fixed ((i) faired and (ii) un-faired)
   b. Retractable
   c. Partially retractable
2. **Landing gear type**
   a. Tricycle (or nose gear)
   b. Tail gear (tail dragger or skid)
   c. Bicycle (tandem)
   d. Multi-wheel
   e. Bicycle (tandem)
   f. Float-equipped
   g. Removable landing gear.

   Figure 3.5 shows several landing gear configuration alternatives. Another design requirement that influences the design of the landing gear is the type of runway. There are mainly five types of runway:
3. **Runway**
   a. Land-based
   b. Sea-based

c. Amphibian
d. Ship-based
e. Shoulder-based (for small remote-controlled aircraft).

Various types of runways are introduced in Chapter 4. The runway requirements will also affect the engine design, wing design, and fuselage design. The advantages and disadvantages of the landing gear configuration alternatives, plus the technique to select the best landing gear configuration alternative to meet the design requirements, are presented in Chapter 8. The primary impacts of the landing gear configuration alternatives are imposed on cost of flight operation, cost of aircraft production, performance, duration of production, ease of manufacturing, and aircraft life.

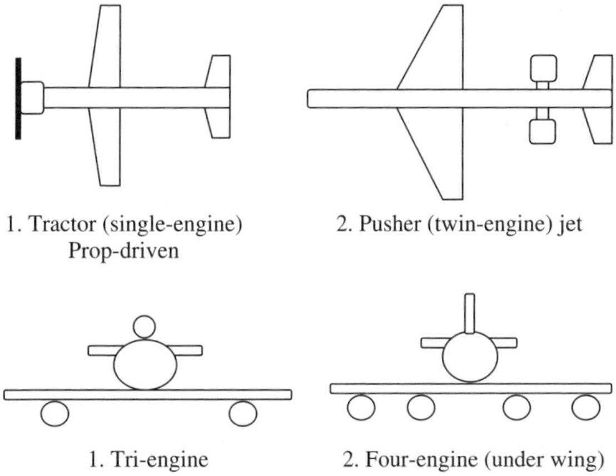

Figure 3.4 Engine configuration alternatives

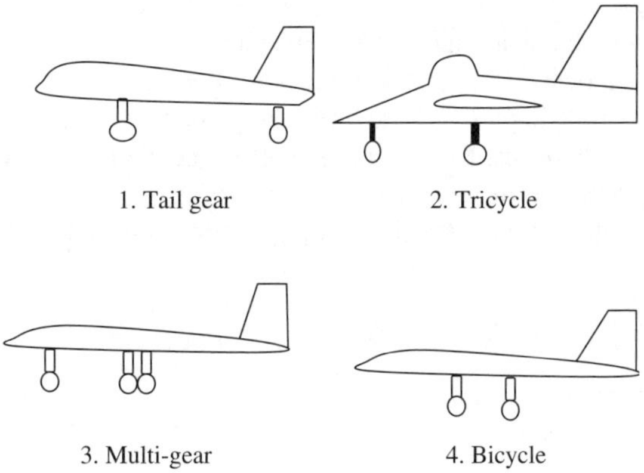

Figure 3.5 Landing gear configuration alternatives

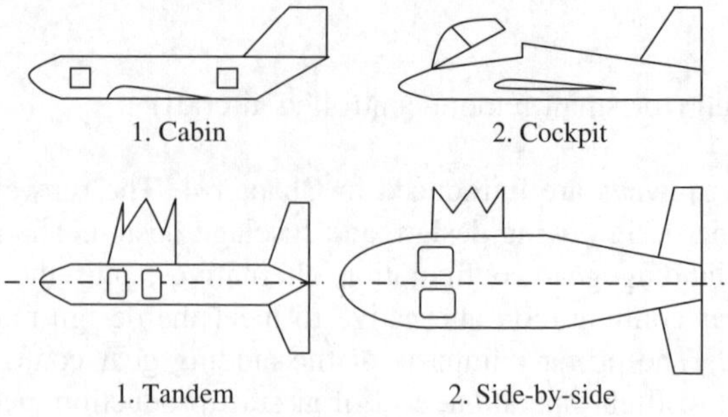

Figure 3.6 Fuselage configuration alternatives

3.3.5 *Fuselage Configuration*

In general, the fuselage configuration alternatives from three different aspects are as follows:

1. **Door**
    a. Cabin
    b. Cockpit
2. **Seat**
    a. Tandem
    b. Side-by-side
    c. *n* Seats per row
3. **Pressure system**
    a. Pressurized cabin
    b. Pressurized hose
    c. Unpressurized cabin.

The advantages and disadvantages of the fuselage configuration alternatives, plus the technique to select the best fuselage configuration alternative to meet the design requirements, are presented in Chapter 7. The primary impacts of the fuselage configuration alternatives are imposed on cost of flight operation, cost of aircraft production, performance, duration of production, ease of manufacturing, passenger comfort, and aircraft life. Figure 3.6 illustrates several fuselage configuration alternatives.

3.3.6 *Manufacturing-Related Items Configuration*

In general, the manufacturing configuration alternatives from four different aspects are as follows:

1. **Materials for structure**
    a. Metal (often aerospace aluminum)
    b. Wood and fabric

   c. Composite materials
   d. Metal and composite materials
2. **Assembly technique**
   a. Kit form (kit-plane rule: 51% amateur construction)
   b. Semi-kit form
   c. Modular
3. **Metallic components manufacturing technique**
   a. Welding
   b. Machining
   c. Casting
   d. Sheet metal work
4. **Composite materials manufacturing technique**
   a. Hand layup
   b. Machine layup
   c. Wet layup
   d. Filament winding
   e. Resin transfer molding
   f. Pultrusion
   g. Sandwich construction.

The descriptions of engineering materials and manufacturing processes are beyond the scope of this book. For details of these materials and techniques, the reader is encouraged to consult relevant references, such as [2, 3]. The primary impacts of these alternatives are imposed on cost, the duration of production, ease of manufacturing, and aircraft life.

3.3.7 Subsystems Configuration

In general, the subsystems configuration alternatives from five different aspects are as follows:

1. **Primary control surfaces**
   a. Conventional (i.e., elevator, aileron, and rudder)
   b. Elevon/rudder
   c. Aileron/ruddervator
   d. Flaperon/rudder/aileron
   e. Cross (×) or plus (+) section
2. **Secondary control surfaces**
   a. High-lift device (e.g., flap, slat, and slot)
   b. Spoiler
   c. Tab
3. **Power transmission**
   a. Mechanical
   b. Hydraulic
   c. Pneumatic

   d. Fly-by-wire
   e. Fly-by-optic
4. **Fuel tank**
   a. Inside fuselage
   b. Inside wing (both sides) ((i) between two spars and (ii) in front of main spar)
   c. Wing tip-tank
   d. External tank
5. **Store**
   a. Camera
   b. Rocket
   c. Missile
   d. Gun
   e. External tank.

The advantages and disadvantages of some of these configuration alternatives, plus the technique to select the best subsystem configuration alternative to meet the design requirements, are presented in Chapters 5–12. Table 3.2 illustrates a summary of configuration alternatives for major aircraft components.

Table 3.3 provides a list of configuration parameters and their design alternatives. These are determined and finalized by the designer. The optimization process will find and prove which configuration is best. The optimization methodology introduced will formulate a technique that enables a designer to select configuration parameters in order to meet the design requirements in an optimum fashion. These 30 groups of configurations (Table 3.3)

Table 3.2 Aircraft major components with design alternatives

No.	Component	Configuration alternatives
1	Fuselage	Geometry: lofting, cross-section Seating arrangement What to accommodate (e.g., fuel, engine, and landing gear)?
2	Wing	Type: swept, tapered, dihedral Installation: fixed, moving, adjustable Location: low wing, mid-wing, high wing, parasol
3	Horizontal tail	Type: conventional, T-tail, H-tail, V-tail, inverted V Location: aft tail, canard, three surfaces
4	Vertical tail	Single, twin, three VT, V-tail
5	Engine	Type: turbofan, turbojet, turboprop, piston-prop, rocket Location: (e.g., under fuselage, under wing, beside fuselage) Number of engines
6	Landing gear	Type: fixed, retractable, partially retractable Location: (e.g., nose, tail, multi)
7	Control surfaces	Separate vs. all moving tail, reversible vs. irreversible, conventional vs. non-conventional (e.g., elevon, ruddervator)

Table 3.3 Configuration parameters and their options (set by designer)

No.	Configuration parameter	Configuration alternatives
1	Conventionality	(i) Conventional and (ii) non-conventional
2	Power	(i) Powered and (ii) unpowered
3	Propulsion	(i) Turbojet, (ii) turbofan, (iii) turboprop, (iv) piston prop, and (v) rocket
4	Number of engine	(i) Single-engine, (ii) twin-engine, (iii) tri-engine, (iv) four-engine, and (v) multi-engine
5	Engine and aircraft cg	(i) Pusher and (ii) tractor
6	Engine installation	(i) Fixed and (ii) tilt-rotor
7	Engine location	(i) Under wing, (ii) inside wing, (iii) above wing, (iv) above fuselage, (v) beside fuselage, and (vi) inside fuselage, etc.
8	Number of wings	(i) One-wing, (ii) biplane, and (iii) tri-plane
9	Wing type	(i) Fixed-wing and (ii) rotary-wing (a. helicopter and b. gyrocopter)
10	Wing geometry	(i) Rectangular, (ii) tapered, (iii) swept, and (iv) delta
11	Wing sweep	(i) Fixed sweep angle and (ii) variable sweep
12	Wing setting angle	(i) Fixed setting angle and (ii) variable setting angle
13	Wing placement	(i) High wing, (ii) low wing, (iii) mid-wing, and (iv) parasol wing
14	Wing installation	(i) Cantilever and (ii) strut-braced
15	Tail or canard	(i) Tail, (ii) canard, and (iii) three-surfaces
16	Tail type	(i) Conventional, (ii) T shape, (iii) H shape, (iv) V shape, and (v) + shape, etc.
17	Vertical tail	(i) No vertical tail (VT), (ii) one VT at fuselage end, (iii) two VT at fuselage end, and (iv) two VT at wing tips
18	Landing gear	(i) Fixed and faired, (ii) fixed and un-faired, (iii) retractable, and (iv) partially retractable
19	Landing gear type	(i) Nose gear, (ii) tail gear, (iii) quadricycle, and (iv) multi-bogey, etc.
20	Fuselage	(i) Single short fuselage, (ii) single long fuselage, and (iii) double long fuselage, etc.
21a	Seating (in two-seat)	(i) Side-by-side and (ii) tandem
21b	Seating (with higher number of passengers)	(i) $1 \cdot n$, (ii) $2 \cdot n$, and (iii) $3 \cdot n, \ldots, 10 \cdot n$ (n = number of rows)
22	Luggage pallet	Based on types of luggage and payload, it has multiple options

(*continued overleaf*)

Table 3.3 (*continued*)

No.	Configuration parameter	Configuration alternatives
23	Cabin or cockpit	(i) Cabin and (ii) cockpit
24	Horizontal tail control surfaces	(i) Tail and elevator and (ii) all moving horizontal tail
25	Vertical tail control surfaces	(i) Vertical tail and rudder and (ii) all moving vertical tail
26	Wing control surfaces	(i) Aileron and flap and (ii) flaperon
27	Wing-tail control surfaces	(i) Conventional (elevator, aileron, and rudder), (ii) ruddervator, (iii) elevon, (iv) split rudder, and (v) thrust-vectored
28	Power system	(i) Mechanical, (ii) hydraulic, (iii) pneumatic, (iv) FBW,[a] and (v) FBOb[b]
29	Material for structure	(i) Full metal, (ii) full composite, and (iii) primary structure: metal, secondary structure: composite
30	Secondary control surfaces	(i) Trailing edge flap, (ii) leading edge slot, and (iii) leading edge slat

[a] Fly-by-wire (electrical signal).
[b] Fly-by-optic (light signal).

are an available reference for selection of alternatives by aircraft designers. It is observed that the number of design options is surprisingly large. The MDO (see Section 3.6) process is a well-established process to optimize the configuration for multi-disciplinary purposes.

3.4 Aircraft Classification and Design Constraints

One of the essential steps that a designer must take is to clarify the aircraft type with a relevant full description of specifications. This will help the design process to be straightforward and avoids confusion in the later stages. The aircraft type is primarily based on the aircraft mission, and its required specifications. This section examines the aircraft classifications and types from a variety of aspects.

One of the basic aircraft classifications is to divide aircraft groups into three large types: (i) military, (ii) civil – transport, (iii) civil – General Aviation (GA). The GA aircraft refers to all aircraft other than military, airliner, and regular cargo aircraft, both private and commercial. In terms of weight, GA aircraft have a maximum take-off weight equal to or less than 12 500 lb (for normal and acrobatic categories), or equal to or less than 19 000 lb (for utility categories). Another difference between GA aircraft and transport aircraft lies in the number of seat. The commuter category of GA aircraft is limited to propeller-driven, multi-engine airplanes that have a seating configuration, excluding pilot seats. Any non-military aircraft with a maximum take-off weight of more than 19 000 lb and more than 19 passenger seats is considered to be a transport category aircraft.

A transport aircraft is governed by Part 25 of the Federal Aviation Regulation (FAR), while GA aircraft are governed by Part 23 of FAR.

An aircraft that is ordered by a customer is accompanied with a list of requirements and constraints. In the majority of cases, there is no way to escape from these requirements, unless the designer can prove to the customer that a specific requirement is not feasible. Other than that, all requirements and constraints must be considered and met in the design process. There are other requirements as well that are imposed by airworthiness standards such as FAR, the Joint Aviation Requirements (EASA CS, formerly JAR), and Military Standards (MIL-STDs). Several of these requirements might be grouped in the aircraft classification. Aircraft configurations can be classified in many ways, based on various aspects.

One of the major steps in configuration design is to apply constraints and select the classification and type. Table 3.4 illustrates design constraints and requirements that are set by the customer. It introduces the most important classifications and can be expanded based on the situation. These constraints range from aircraft mission to payload type, to type of control, and to performance requirements. A designer initially has no influence over these requirements, unless he/she can prove that the requirements are not feasible and not practical. Otherwise, they must all be followed and met at the end of the design process. Figure 3.7 depicts a civil transport aircraft (Boeing 747), a GA aircraft (Cessna 182), and a military fighter aircraft (Eurofighter Typhoon). Figure 3.8 shows a lighter-than-air craft (Zeppelin NT) and a heavier-than-air craft (ATR-42). Figure 3.9 illustrates a manned aircraft, an unmanned aircraft, and a remote-controlled aircraft.

One of the significant design constraints originates from government regulations. In this regard, the designer has two options: (i) design an aircraft to comply with government regulations and standards and (ii) design an aircraft regardless of government regulations and standards. The designer is free to make the decision to select either of the above options, but he/she must be aware of the consequences. This decision will impact the whole design process, since this generates a totally different design environment and constraints. In general, the compliance with government regulations and standards increases the cost and makes the design harder. However, it will increase the quality of the aircraft and allows the aircraft to be sold in the US market.

An aircraft which has not been certified by the government aviation authorities is referred to as home-built or garage-built. These aircraft are usually designed by non-expert individuals and used by individual pilots. Their airworthiness has not been confirmed by the authorities, and hence the probability of a crash is much higher than for certified aircraft. Their flight permissions are limited to a few airspaces to reduce the risk of civilian casualties. Home-built aircraft are not allowed to be sold in the US market.

Several countries have established an official body to regulate the aviation issues and ratify and collect aviation standards. The US government body that regulates aviation-related issues including aircraft design and manufacture is called the Federal Aviation Administration (FAA). The civil aviation authorities of certain European countries (including the UK, France, and Germany)[1] have established common comprehensive and detailed

[1] The countries are: Austria, Belgium, Cyprus, Czech Republic, Denmark, Finland, France, Germany, Greece, Hungary, Iceland, Ireland, Italy, Latvia, Luxembourg, Malta, Monaco, Netherlands, Norway, Poland, Portugal, Romania, Slovak Republic, Slovenia, Spain, Sweden, Switzerland, Turkey, and United Kingdom.

Table 3.4 Design constraints and requirements (set by customer)

No.	Group	Design requirements and constraints
1	Standard, non-standard	(i) Standard and (ii) home-built (or garage-built)
2	General type	(i) Military (MIL-STD), (ii) civil – transport (FAR[a] 25), (iii) civil – GA (FAR 23), and (iv) very light aircraft (VLA), etc.
3	Maneuverability	(i) Normal or non-aerobatic, (ii) utility or semi-aerobatic, (iii) aerobatic or acrobatic, and (iv) highly maneuverable (e.g., fighters and anti-missile missiles)
4	GA mission	(i) General purpose, (ii) hang glider, (iii) sailplane or glider, (iv) agricultural, (v) utility, (vi) commuter, (vii) business, (viii) racer, (ix) sport, (x) touring, (xi) trainer, (xii) maneuver, and (xiii) model
5	Military mission	(i) Fighter, (ii) bomber, (iii) attack, (iv) interceptor, (v) reconnaissance, (vi) military transport, (vii) patrol, (viii) maritime surveillance, (ix) military trainer, (x) stealth, (xi) tanker, (xii) close support, (xiii) trainer, (xiv) anti-submarine, (xv) early warning, (xvi) airborne command, (xvii) communication relay, (xviii) target, (xix) missile, and (xx) rocket
6	Density	(i) Lighter-than-air craft (a. balloon, b. airship) and (ii) heavier-than-air craft
7	Pilot control	(i) Manned aircraft, (ii) unmanned aircraft, and (iii) remote control (RC)
8	Weight	(i) Model (less than 30 lb), (ii) ultralight aircraft (less than 300 kg), (iii) very light (less than 750 kg), (iv) light (less than 12 500 lb), (v) medium weight (less than 100 000 lb), and (vi) heavy or jumbo (above 100 000 lb)
9	Producibility	(i) Kit form, (ii) semi-kit form, and (iii) modular (conventional)
10	Take-off run	(i) Short take-off and landing (STOL) (runway less than 150 m), (ii) vertical take-off and landing (VTOL), and (iii) regular
11	Landing field	(i) Land-based, (ii) sea-based, (iii) ship-based, (iv) amphibian, and (v) shoulder-based
12	Stage	(i) Model, (ii) prototype, and (iii) operational
13	Term of use	(i) Long-term (regular) and (ii) experimental (X aircraft) or research
14	Payload	(i) Number of passengers, (ii) payload weight, and (iii) store, etc.
15	Aircraft subsystems	(i) Air condition, (ii) weather radar, and (iii) parachute, etc.
16	FAR and MIL requirements	(i) Number of crew, (ii) ejection seat, and (iii) reserve fuel, etc.
17	Performance	(i) Max speed, (ii) range, (iii) ceiling, (iv) rate of climb, (v) take-off run, and (vi) endurance, etc.
18	Maneuverability	(i) Turn radius, (ii) turn rate, and (iii) load factor

[a] Federal Aviation Regulations.

(a) (b) (c)

Figure 3.7 (a) Civil transport Boeing 747; (b) general aviation Cessna 182; (c) military aircraft, Eurofighter Typhoon. Reproduced from permission of (a) Anne Deus; (b) Jenny Coffey; (c) Antony Osborne.

(a) (b)

Figure 3.8 Lighter-than-air craft versus heavier-than-air craft: (a) Zeppelin NT; (b) ATR-42. Part (b) reproduced from permission of Anne Deus.

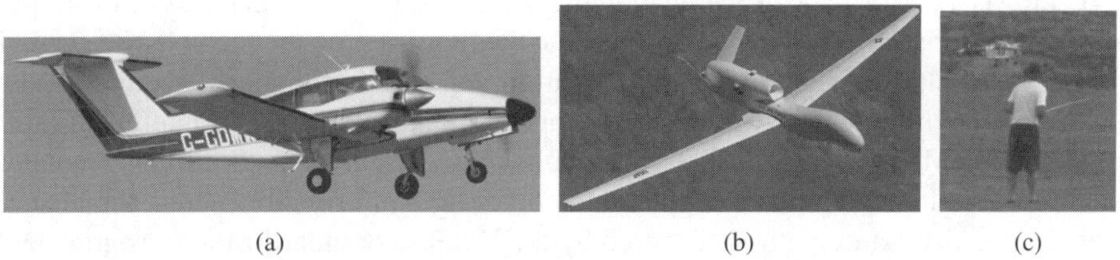

(a) (b) (c)

Figure 3.9 Manned aircraft, unmanned aircraft, and remote-controlled aircraft: (a) Beech 76 Duchess; (b) Global Hawk; (c) RC model aircraft. Part (a) reproduced from permission of Jenny Coffey.

aviation requirements (referred to as the Certification Specifications, formerly JARs) with a view to minimizing type certification problems on joint ventures, and also to facilitate the export and import of aviation products. The CSs are recognized by the civil aviation authorities of participating countries as an acceptable basis for showing compliance with their national airworthiness codes.

In the USA, the FAA [4] of the Department of Transportation regulates the aviation standards and publishes FARs. Some important parts of the FAR are:

- Part 23: Airworthiness Standards for GA aircraft.
- Part 25: Airworthiness Standards for Civil Transport aircraft.
- Part 29: Airworthiness Standards for Helicopters.
- Part 33: Airworthiness Standards for Aircraft engines.
- Part 103: Airworthiness Standards for Ultralight aircraft.

Military aircraft are required to follow and comply with MIL-STD. A United States defense standard, often referred to as a military standard, MIL-SPEC (or informally Mil Specs), is used to help achieve standardization objectives by the US Department of Defense. Although the official definitions differentiate between several types of documents, all of these documents go by the general rubric of "military standard," including defense specifications, handbooks, and standards. Strictly speaking, these documents serve different purposes. According to the Government Accountability Office, military specifications "describe the physical and/or operational characteristics of a product," while MIL-STDs "detail the processes and materials to be used to make the product." Military handbooks, on the other hand, are primarily sources of compiled information and/or guidance.

MIL-STD is a document that establishes uniform engineering and technical requirements for military-unique or substantially modified commercial processes, procedures, practices, and methods. There are five types of defense standards: interface standards, design criteria standards, manufacturing process standards, standard practices, and test method standards. There are currently more than 33 000 defense standards. Defense standards are considered reliable enough that they are often used by other government organizations and even non-government technical organizations or general industry.

MIL-PRF is a performance specification that states requirements in terms of the required results with criteria for verifying compliance, but without stating the methods for achieving the required results. A performance specification defines the functional requirements for the item, the environment in which it must operate, and interface and interchangeability characteristics. MIL-HDBK (Military Handbooks) is a guidance document containing standard procedural, technical, engineering, or design information about the material, processes, practices, and methods covered by the Defense Standardization Program. MIL-STD-962 covers the content and format for defense handbooks.

Flying models are usually what is meant by the term aero-modeling. Most flying model aircraft can be placed in one of three groups. (i) Free-flight model aircraft fly without any method of external control from the ground. This type of model pre-dates the efforts of the Wright Brothers and other pioneers. (ii) Control-line model aircraft use cables (usually two) leading from the wing to the pilot. (iii) Radio-controlled aircraft have a transmitter operated by the pilot on the ground, sending signals to a receiver in the craft. Some flying models resemble scaled-down versions of manned aircraft, while others are built with no intention of looking like piloted aircraft.

It is important to note that there are several design alternatives that, if selected, lead to some other design alternatives not being feasible any more. For instance, if a designer selects a single-engine configuration, he/she cannot select the side fuselage as the location

Table 3.5 Relationship between aircraft major components and design requirements

No.	Design requirements	Aircraft component affected most, or major design parameter
1a	Payload (weight) requirements	Maximum take-off weight
1b	Payload (volume) requirements	Fuselage
2	Performance requirements (range and endurance)	Maximum take-off weight
3	Performance requirements (maximum speed, rate of climb, take-off run, stall speed, ceiling, and turn performance)	Engine, landing gear, and wing
4	Stability requirements	Horizontal tail and vertical tail
5	Controllability requirements	Control surfaces (elevator, aileron, rudder)
6	Flying quality requirements	Center of gravity
7	Airworthiness requirements	Minimum requirements
8	Cost requirements	Materials, engine, weight, etc.
9	Timing requirements	Configuration optimality

of installation of the engine. The reason is that the aircraft becomes asymmetric, if the single engine is installed at the left or right of the fuselage. Another example is where, if a designer selects not to have any VT (for reasons of stealth), the ruddervator is not an option for the control surfaces design.

Table 3.5 shows the relationship between aircraft major components and the design requirements. The third column in Table 3.5 clarifies the aircraft component which is affected most; or the major design parameter for a design requirement. Every design requirement will normally affect more than one component, but we only consider the component that is influenced most.

For example, the payload requirement, range, and endurance will affect the maximum take-off weight (Section 4.2), engine selection, fuselage design, and flight cost. The influence of payload weight is different from that of payload volume. Thus, for optimization purposes, the designer must know exactly the payload weight and its volume. In contrast, if the payload can be divided into smaller pieces, the design constraints for the payload are easier to handle. Furthermore, the other performance parameters (e.g., maximum speed, stall speed, rate of climb, take-off run, and ceiling) will affect (Section 4.3) the wing area and engine power/thrust.

In general, design considerations are the full range of attributes and characteristics that could be exhibited by an engineered system, product, or structure. These interest both the producer and the customer. Design-dependent parameters are attributes and/or characteristics inherent in the design to be predicted or estimated (e.g., weight, design life, reliability, producibility, maintainability, and disposability). These are a subset of the design considerations for which the producer is primarily responsible. In contrast, design-independent parameters are factors external to the design that must be estimated and forecasted for use in design evaluation (e.g., fuel cost per gallon, interest rates, labor rates, and material cost per pound). These depend upon the production and operating environment of the aircraft.

3.5 Configuration Selection Process and Trade-Off Analysis

In order to select the best aircraft configuration, a trade-off analysis must be established. Many different trade-offs are possible as the aircraft design progresses. Decisions must be made regarding the evaluation and selection of appropriate components, subsystems, possible degree of automation, commercial off-the-shelf parts, various maintenance and support policies, and so on. Later in the design cycle, there may be alternative engineering materials, alternative manufacturing processes, alternative factory maintenance plans, alternative logistic support structures, and alternative methods of material phase-out, recycling, and/or disposal.

One must first define the problem, identify the design criteria or measures against which the various alternative configurations will be evaluated, the evaluation process, acquire the necessary input data, evaluate each of the candidates under consideration, perform a sensitivity analysis to identify potential areas of risk, and finally recommend a preferred approach. This process is shown in Figure 3.10, and can be tailored at any point in the lifecycle. Only the depth of the analysis and evaluation effort will vary, depending on the nature of the component.

Trade-off analysis involves synthesis, which refers to the combining and structuring of components to create an aircraft system configuration. Synthesis is design. Initially, synthesis is used in the development of preliminary concepts and to establish relationships among various components of the aircraft. Later, when sufficient functional definition and decomposition have occurred, synthesis is used to further define the *hows* at a lower level. Synthesis involves the creation of a configuration that could be representative of the form that the aircraft will ultimately take (although a final configuration should not be assumed at this early point in the design process). Given a synthesized configuration, its characteristics need to be evaluated in terms of the aircraft requirements initially specified. Changes will be incorporated as required, leading to a preferred design configuration. This iterative process of synthesis, analysis, evaluation, and design refinement leads to the establishment of the functional and product baselines.

One of the preliminary tasks in aircraft configuration design is identifying system design considerations. The definition of a need at the system level is the starting point for determining customer requirements and developing design criteria. The requirements for the system as an entity are established by describing the functions that must be performed. Design criteria constitute a set of "design-to" requirements, which can be expressed in both qualitative and quantitative terms. Design criteria are customer-specified or negotiated target values for technical performance measures. These requirements represent the bounds within which the designer must "operate" when engaged in the iterative process of synthesis, analysis, and evaluation. Both operational functions (i.e., those required to accomplish a specific mission scenario, or series of missions) and maintenance and support functions (i.e., those required to ensure that the aircraft is operational when required) must be described at the top level.

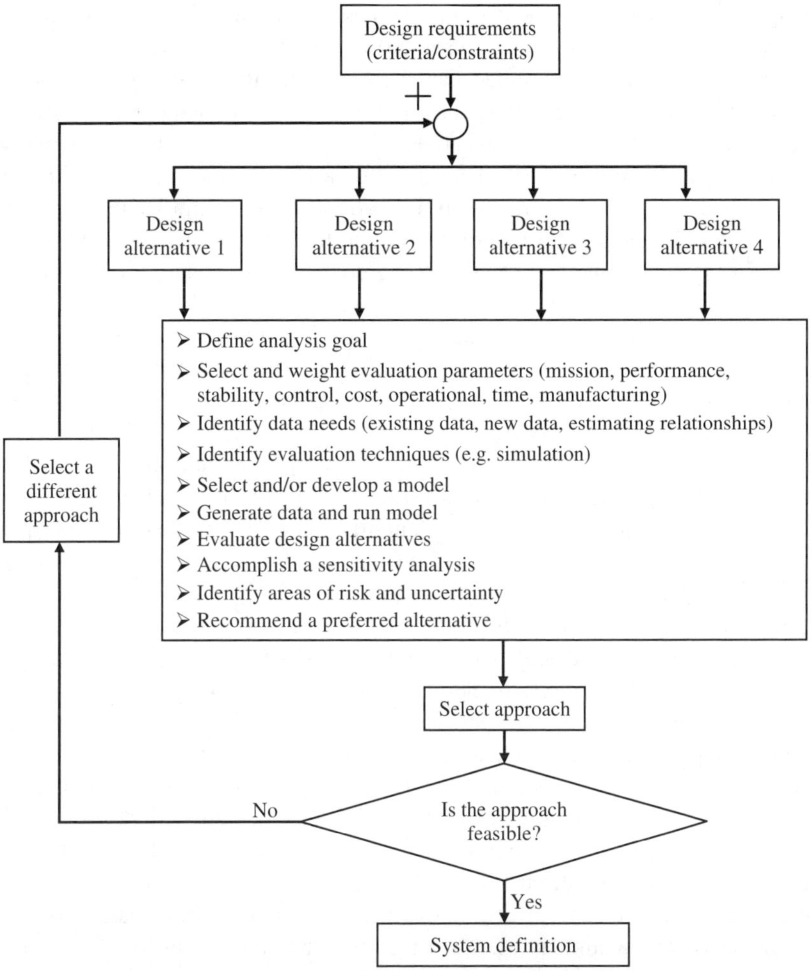

Figure 3.10 Trade-off analysis process

After a baseline configuration has been established as a result of a formal design review, changes are frequently initiated for any one of a number of reasons: to correct a design deficiency, improve a product, incorporate a new technology, respond to a change in operational requirements, compensate for an obsolete section, and so on. Changes may be initiated from within the project, or as a result of some new externally imposed requirements.

At first, it may appear that a change is relatively insignificant in nature, and that it may constitute a change in the design of a prime equipment item, a software modification, a data revision, and/or a change in some process. However, what might initially appear to be minor often turns out to have a great impact across and throughout the system hierarchical structure. For instance, a change in the design configuration of a prime component (e.g., a change in size, weight, repackaging, and added performance capability) will probably affect related components, design of test and support equipment, type and quantity of spares/repair parts, technical data, transportation and handling requirements, and so on.

A change in any one component (e.g., horizontal tail) will likely have an impact on many other components (e.g., wing, fuselage) of the aircraft. Furthermore, if there are numerous changes being incorporated at the same time, the entire system configuration may be severely compromised in terms of maintaining some degree of requirements traceability. Past experiences with a variety of systems has indicated that many of the changes incorporated are introduced late in the detail design phase, during production of construction, and early during the system utilization and sustaining support phase. While the incorporation of changes (for one reason or another) is certainly inevitable, the process for accomplishing such must be formalized and controlled to ensure traceability from one configuration baseline to another.

One of the most effective techniques in trade-off studies is MDO [5]. Researchers in academia, industry, and government continue to advance MDO and its application to practical problems of industry relevance (for instance, see [6–8]). MDO is a field of engineering that uses optimization methods to solve design problems incorporating a number of disciplines. MDO allows designers to incorporate all relevant disciplines simultaneously. The optimum solution of a simultaneous problem is superior to the design found by optimizing each discipline sequentially, since it can exploit the interactions between the disciplines. However, including all disciplines simultaneously significantly increases the complexity of the problem.

Various aircraft designers have different priorities in their design processes. These priorities are based on different objectives, requirements, and missions. There are primarily four groups of aircraft designers, namely: (i) military aircraft designers, (ii) civil transport aircraft designers, (iii) GA aircraft designers, and (iv) home-built aircraft designers. These four groups of designers have different interests, priorities, and design criteria. There are mainly 10 figures of merit for every aircraft configuration designer. They are: production cost, aircraft performance, flying qualities, design period, beauty (for civil aircraft) or scariness (for military aircraft), maintainability, producibility, aircraft weight, disposability, and stealth requirement.

Table 3.6 demonstrates the priorities of each aircraft designer against 10 figures of merit. This priority allocation is the author's idea and may be different in some cases. References [9, 10] are valuable references that describe true aircraft design stories and the lessons learned in aircraft design over 60 years. Since they introduce multiple challenges and promises of several designs, they are helpful resources to determine the priorities in the configuration design process.

Among 10 figures of merit (or criteria), grade "1" is the highest priority and grade "10" is the lowest priority. The grade "0" in this table means that this figure of merit is not a criterion for this designer. As Table 3.6 illustrates, the number one priority for a military aircraft designer is aircraft performance, while for a home-built aircraft designer cost is the number one priority. It is also interesting that stealth capability is an important priority for a military aircraft designer, while for the other three groups of designers it is not important at all. These priorities (later called weights) reflect the relative importance of the individual figures of merit in the mind of the designer.

In design evaluation, an early step that fully recognizes the design criteria is to establish a baseline against which a given alternative or design configuration may be evaluated. This baseline is determined through the iterative process of requirements analysis (i.e., identification of needs, analysis of feasibility, definition of aircraft operational requirements,

Table 3.6 Design objectives and an example of the priorities for various aircraft designers

No.	Figure of merit	Military designer	Large civil transport designer	Small GA designer	Home-built designer
1	Cost	4	2	1	1
2	Performance	1	3	2	3
3	Flying qualities	2	1	3	7
4	Period of design	5	9	8	6
5	Beauty (or scariness)	10	7	4	5
6	Maintainability	7	5	6	9
7	Producibility (ease of construction)	6	6	7	4
8	Aircraft weight	8	4	5	2
9	Disposability	9	7	9	8
10	Stealth	3	0	0	0

selection of a maintenance concept, and planning for phase-out and disposal). The mission that the aircraft must perform to satisfy a specific customer should be described, along with expectations for cycle time, frequency, speed, cost, effectiveness, and other relevant factors. Functional requirements must be met by incorporating design characteristics within the aircraft and its configuration components. As an example, Table 3.7 illustrates three scenarios of priorities (in percentage terms) for military aircraft designers.

Design criteria may be established for each level in the system hierarchical structure. Possible optimization objectives for each level are demonstrated in Table 3.8. These objectives must be formulated in order to determine the optimum design. A selected aircraft configuration would be optimum based on only one optimization function. Applicable criteria regarding the aircraft should be expressed in terms of technical performance measures and should be prioritized at the aircraft (system) level. Technical performance measures are measures for characteristics that are, or derive from, attributes inherent in the design itself. It is essential that the development of design criteria be based on an appropriate set of design considerations, considerations that lead to the identification of both design-dependent and design-independent parameters, and that support the derivation of technical performance measures.

One of the most effective techniques in trade-off studies is MDO. Most MDO techniques require large numbers of evaluations of the objectives and constraints. The disciplinary models are often very complex and can take significant amounts of time for a single evaluation. The solution can therefore be extremely time-consuming. Many of the optimization techniques are adaptable to parallel computing. Much of the current research is focused on methods of decreasing the required time. No existing solution method is guaranteed to find the global optimum of a general problem.

Table 3.7 Three scenarios of weights (%) for a military aircraft designer

No.	Figure of merit	Priority	Designer # 1 (%)	Designer # 2 (%)	Designer # 3 (%)
1	Cost	4	8	9	9
2	Performance	1	50	40	30
3	Flying qualities	2	10	15	20
4	Period of design	5	7	7	8
5	Scariness	10	1	1	2
6	Maintainability	7	4	5	5
7	Producibility	6	6	6	7
8	Weight	8	3	4	4
9	Disposability	9	2	2	3
10	Stealth	3	9	11	12
		Total →	100	100	100

In MDO an objective function subject to a set of constraints is defined and a mathematical process is used to minimize this objective function without violating the constraints. Sensitivity derivatives are usually computed as part of the optimization process. For a single mission aircraft, the formulation of the objective function might be a simpler task. But, if an aircraft is a multi-role aircraft, the formulation of a single objective function would be difficult if not impossible.

The aircraft design process has, historically, ranged from sketches on napkins to trial, error, and natural selection, to sophisticated computer-aided design programs. Because the process is so complex, involving hundreds or thousands of computer programs, and many people at many locations, it is very difficult to manage all recourses toward an optimized design. Thus most companies are continuing to improve on the strategy and develop a new approach. In the early days of airplane design, people did not do much computation. The design teams tended to be small, managed by a single "chief designer" who knew about all the design details and could make all the important decisions.

Modern design projects are often so complex that the problem has to be decomposed and each part of the problem tackled by a different team. The way in which these teams should work together is still being debated by managers and researchers. The goal of these processes, whatever form they take, is to design what is, in some sense, the best or optimum aircraft configuration.

The design process of the F/A-18E/F multi-mission fighter aircraft, including a comparison between three configurations (YF-17, F/A-18A, and F/A-18E), is described in [11, 12]. The analytical properties of three approaches to formulating and solving MDO problems

Table 3.8 Optimization criteria at group level

No.	Criteria	Objective
1	Cost	Minimum direct operating cost Minimum total manufacturing cost Minimum system cost over *X* years (lifecycle cost) Maximum profit Maximum return on investment Maximum payload per US$
2	Performance	Maximizing cruise speed Maximizing range Maximizing endurance Maximizing absolute ceiling Minimizing take-off run Maximizing rate of climb Maximizing maneuverability
3	Weight	Minimum take-off weight Minimum empty weight Maximum fuel weight
4	Flying qualities (stability and control)	Most controllable Most stable Highest flying qualities Most luxurious for passengers
5	Size	Smallest wing span Smallest fuselage length Smallest aircraft height Most specious fuselage
6	Beauty or scariness	Most attractive (civil) or most scariest (fighter)
7	Systems engineering criteria	Most maintainable Most producible Most disposable (environmental compatibility) Most flight testable Most stealthy Most flexible (growth potential) Most reliable
8	Design and operation duration	Minimum duration of design Minimum duration of manufacture Maximum aircraft operating life

are discussed, that achieve varying degrees of autonomy by distributing the problem along disciplinary lines. The external configuration design of unguided missiles is optimized in [13, 14], with MDO employed for the configuration design of a generic air-breathing aerospace vehicle considering fidelity uncertainty. An assessment of configuration design methodologies, including a detailed description of the general design configuration process – that is, preprocessing, optimization, and post-processing, is given in [15].

3.6 Conceptual Design Optimization

3.6.1 Mathematical Tools

In mathematics, the term optimization refers to the study of problems in which one seeks to minimize or maximize a real function by systematically choosing the values of real or integer variables from within an allowed set. An optimization problem is one requiring the determination of the optimal (maximum or minimum) value of a given function, called the objective function, subject to a set of stated restriction, or constraints, placed on the variables concerned. In this process, we need to first describe an optimization problem in terms of the objective function and a set of constraints. Then, algebraically manipulate and possibly graphically describe inequalities, and solve a linear programming problem in two real variables. The final action is to solve the optimization problem using a mathematical technique.

Basically, the elements of optimization are: design variable, objective function, constraints, and design space [16]. Even when there is no uncertainty, optimization can be very difficult if the number of design variables is large, the problem contains a diverse collection of design variable types, and little is known about the structure of the performance function. If we have estimates, it may not be possible to conclusively determine if one design is better than another, frustrating optimization algorithms that try to move in improving directions. The comparison of two system designs (aircraft configurations) is computationally easier than the simultaneous comparison of multiple (more than two) configurations. The dynamic optimization problem can be stated as minimizing or maximizing a cost function subject to dynamic equation constraints, control inequality constraints, interior state equality constraints, interior state inequality constraints, and specified initial and final states.

In general, a constrained single-objective optimization problem [16] is to

$$\begin{array}{l}\text{optimize } f(x)\\ \text{subject to } x \in \Omega\end{array} \tag{3.1}$$

The function $f : R^n \to R$ that we wish to optimize (maximize or minimize) is a real-valued function called the objective function or cost function. The vector x is an n-vector of independent variables: $x = [x_1, x_2, ..., x_n]^T \in R^n$. The variables $x_1, x_2, \ldots, x_n$ are often referred to as decision variables. The set Ω is a subset of R^n called the constraint set or feasible set. This optimization problem can be viewed as a decision problem that involves finding the "best" vector x of the decision variables over all possible vectors in Ω. The "best" is the one that results in the optimum (smallest or largest) value of the objective function. This vector is called the optimizer or extremizer of the f vector over Ω. Often, the constraint set Ω takes the form $\Omega = \{x : h(x) = 0, g(x) \leq 0\}$, where h and g are given functions.

Definition: suppose that $f : R^n \to R$ is a real-valued function defined on some set $\Omega \subset: R^n$. A point $x^* \in \Omega$ is a local optimizer of f over Ω if there exists $\varepsilon > 0$ such that $f(x) \geq f(x^*)$ for all $x \in \Omega \backslash \{x^*\}$ and $\|x - x^*\| < \varepsilon$. A point $x^* \in \Omega$ is a global minimize of f over Ω if $f(x) \geq f(x^*)$ for all $x \in \Omega \backslash \{x^*\}$. Strictly speaking, an optimization problem is solved only when a global minimizer (in general, extremizer) is found.

Theorem 1. First-order necessary condition: *let Ω be a subset of R^n and $f \in C^1$ a real-valued function on Ω. If x^* is a local minimizer of f over Ω, then for any feasible direction d at x^*, we have*

$$d^T \nabla f(x^*) \geq 0.^2 \tag{3.2}$$

When an optimization problem involves only one objective function, it is a single-objective optimization. Most engineering problems, including aircraft configuration design optimization, require the designer to optimize a number of conflicting objectives. The objectives are in conflict with each other if an improvement in one objective leads to deterioration in another. Multi-objective problems in which there is competition between objectives may have no single, unique optimal solution. Multi-objective optimization problems are also referred to as multi-criteria or vector optimization problems. In a multi-objective optimization problem, we are to find a decision variable that satisfies the given constraints and optimizes a vector function whose components are objective functions.

The formulation of a multi-objective optimization problem is as follows:

$$\text{minimize } f(x) = \begin{bmatrix} f_1(x_1, x_2, \ldots, x_n) \\ f_2(x_1, x_2, \ldots, x_n) \\ \cdots \\ f_1, (x_1, x_2, \ldots, x_n) \end{bmatrix} \tag{3.3}$$

$$\text{subject to } x \in \Omega$$

where $f : R^n \to R$ and $\Omega \subset: R^n$.

In general, we may have three different types of multi-objective optimization problems: (i) minimize all the objective functions, (ii) maximize all the objective functions, and (iii) minimize some and maximize other objective functions. However, any of these can be converted into an equivalent minimization problem. Analytically

$$\text{minimum} \quad f(x) = -\text{maximum} \quad [-f(x)] \tag{3.4}$$

In some cases it is possible to deal with a multi-objective optimization problem by converting the problem into a single-objective optimization problem, so that standard optimization methods can be brought to bear. One method [17] is to form a single objective function by taking a linear combination, with positive coefficients, of the components of the objective function vector ($f(x) = [f_1(x), \ldots, f_l(x)]^T$). Equivalently, we form a convex combination of the components of the objective function. In other words, we use

$$F(x) = c^T \quad f(x) \tag{3.5}$$

as the single objective function, where c is a vector of positive components. This method is called the weighted-sum method, where the coefficients of the linear combination (i.e., components of c) are called weights. These weights reflect the relative importance of the individual components in the objective vector. In general, the factors that are deemed more

[2] Proof is given in Ref. 3.

important in a given case should be weighted more heavily in the associated performance measure. This weighting process is particularly subjective if strictly objective criteria are not evident. Because of this, results obtained using optimization theory should be examined carefully from the standpoint of overall acceptability.

In configuration design, physical and economic limitations often exist which act to limit system optimization. These limitations arise for a variety of reasons and generally cannot be ignored by the decision maker. Accordingly, there may be no choice except to find the best or optimum solution subject to the constraints. The list of constraints includes: (i) time-constrained configuration design, (ii) cost-constrained configuration design, (iii) geometry-constrained configuration design, (iv) weight-constrained configuration design, (v) physically constrained configuration design, (vi) performance-constrained configuration design, and (vii) safety-constrained configuration design.

For example, consider a firefighting aircraft that is required to carry a fixed volume of water or specific liquid with fixed weight, while a particular transport aircraft may be required to carry a specific piece of equipment that has a fixed geometry beside its fixed weight. In the case of firefighting aircraft, the payload weight and total volume are fixed, but the total volume can be divided into several parts. In contrast, the transport aircraft has a fixed volume, and the payload cannot be broken into smaller parts.

Optimization is only a means for bringing mutually exclusive alternatives into comparable (or equivalent) states. When multiple criteria are present in a decision situation, neither x optimization nor y optimization are sufficient. Although necessary, these steps must be augmented with information about the degree to which each alternative meets (or exceeds) specific criteria. One means for consolidating and displaying this information is through the decision evaluation display approach [17].

The optimization problem can be classified based on: (i) existence of constraints, (ii) nature of the design variables, (iii) physical structure of the problem, (iv) nature of the equations involved, (v) permissible values of the design variables, (vi) deterministic nature of the variables, (vii) separability of the functions, and (viii) nature of the objective functions.

3.6.2 Methodology

Given a set of arbitrary objects, the configuration design corresponds to finding a suitable placement for all objects within a given space while satisfying spatial constraints and meeting or exceeding performance objectives. Most optimization practices are restricted to a solution domain defined by a selection of design variables. However, optimization theory makes a distinction between design variables and design parameters. For aircraft configuration design problems, variables specify limited differences within an aircraft configuration while parameters relate to complex variations within a configuration and inter-type differences; that is, differences in configuration. During an optimization, parameters are normally fixed and the optimization is limited to finding a combination of values for the design variables that will minimize or maximize an objective function like weight or speed. The mathematics required to optimize, at a higher level and support the choice between different concepts, emanates from the differential calculus.

The goal of this research is to derive a technique to determine a configuration that converges to an optimal solution, meets the design requirements and satisfies the constraints,

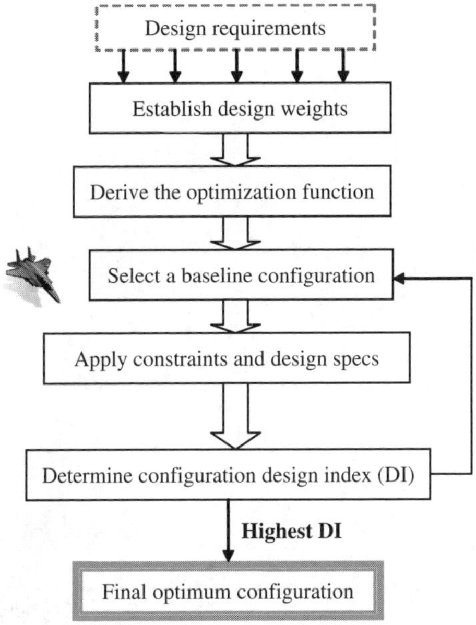

Figure 3.11 The phases in the configuration design optimization

and requires minimal time and cost. The goal here is not to find an optimum aerodynamic shape, rather to find the best configuration to yield the optimum design index. Sometimes manufacturing technologies such as casting, welding, milling, sheet metal working, riveting, or lay-up (for composite materials) will influence the design. Figure 3.11 shows the phases in the configuration design optimization. As this figure indicates, there is a feedback loop that shows the iterative nature of the configuration design process.

The methodology estimates the characteristics of systems so we can compare two designs in a quantitative way. The configuration optimization model consists of parameters and decision variables. Design parameters define the problem, but decision variables are the quantities whose numerical values will be determined in the course of obtaining the optimal configuration. These decision variables are called the design variables. The list of decision variables is illustrated in Table 3.9. The number of variables depends on the aircraft classification (Table 3.9), and as this number increases, so does the complexity of the solution.

The configuration variables may be one of three types: (i) continuous, (ii) discrete, and (iii) integer. A design variable is continuous if it is free to assume any value. When a design variable can only assume a fixed value, it is discrete. For example, landing gear can only be fixed, or retractable, or partially retractable. This would be the case when, for example, the number of engines can only be selected from a finite list (say, 1 or 2 or 3 or 4). In some situations, the number of engines can only assume integer values; these design variables are known as integer variables.

A few policies must be established and followed in order to insure that the configuration design output is feasible and reliable. Every parameter is evaluated by a number between 0 and 1. Zero means that this design parameter has no influence (or least influence) on

Table 3.9 The relationship between landing gear design options and design criteria

Design parameter	Criterion								
	Cost	**Performance**	**Flying qualities**	**Period of design**	**Beauty**	**Maintainability**	**Producibility**	**Weight**	**Disposability**
Fixed	Cheap (1)	Worst (1)	Best (10)	Short (10)	Worst (1)	Best (10)	Best (10)	Light (10)	Better (8)
Retractable	Expensive (10)	Best (10)	Worst (1)	Long (1)	Best (10)	Worst (1)	Worst (1)	Heavy (1)	Worse (3)
Partially retractable	Middle (5)	Middle (5)	Middle (5)	Middle (5)	Middle (5)	Middle (5)	Middle (5)	Middle (5)	Middle (5)

Transport aircraft Antonov An-12. Reproduced from permission of Luis David Sanchez.

a design objective. One means that this design parameter has the highest influence on a design objective. The preference percentages are divided among all preferences such that their summation is 100% (see Table 3.7). Each objective index is the summation of the contribution of each configuration parameter:

$$\mathrm{CI} = \sum_{i=1}^{27} x_{C_i} \tag{3.6}$$

$$\mathrm{PI} = \sum_{i=1}^{27} x_{P_i} \tag{3.7}$$

$$\mathrm{FI} = \sum_{i=1}^{27} x_{F_i} \tag{3.8}$$

$$\mathrm{TI} = \sum_{i=1}^{27} x_{T_i} \tag{3.9}$$

$$\mathrm{BI} = \sum_{i=1}^{27} x_{B_i} \tag{3.10}$$

$$\mathrm{MI} = \sum_{i=1}^{27} x_{M_i} \tag{3.11}$$

$$\mathrm{RI} = \sum_{i=1}^{27} x_{R_i} \tag{3.12}$$

$$\mathrm{WI} = \sum_{i=1}^{27} x_{W_i} \tag{3.13}$$

$$\mathrm{DI} = \sum_{i=1}^{27} x_{D_i} \tag{3.14}$$

$$\mathrm{SI} = \sum_{i=1}^{27} x_{S_i} \tag{3.15}$$

where CI stands for the cost index and x_{C_i} is the contribution of the ith configuration parameter on the cost index. By the same token, the other symbols are defined as PI: performance index, FI: flying qualities index, TI: period of design index, BI: beauty (or scariness) index, MI: maintainability index, RI: producibility index, WI: weight index, DI: disposability index, and SI: stealth index. Among the 10 design objectives, three objectives must be minimized, they are: cost, weight, and period of design. The other seven design objectives must be maximized, they are: performance, flying qualities, beauty (or scariness), maintainability, producibility, disposability, and stealth.

Each design option must be evaluated for features and requirements that are important to customers. It is a challenging task to compare the various design options, but the proposed methodology can simplify the task of selecting a best design. According to this methodology, a matrix (or table) is created between the criteria of selection and the design options, as shown in Table 3.9. Each design option is rated on a scale from 1 to 10 for various selection criteria. The weight assigned to each criterion depends on its significance for the application. Each rating is multiplied by a weight and totaled for the final selection. The design that yields the highest point is assumed as the best or optimum configuration.

To combine all objective indices in a comparable quantity, the design index (DI) is defined. All objectives that need to be minimized are grouped into one design index (DI_{min}), as found from the following equation:

$$DI_{min} = CI \cdot P_C + WI \cdot P_W + TI \cdot P_T \tag{3.16}$$

All objective indices that need to be maximized are grouped into another design index (DI_{max}), as found from the following equation:

$$DI_{max} = PI \cdot P_P + FI \cdot P_F + BI \cdot P_B + MI \cdot P_M + RI \cdot P_R + DI \cdot P_D + SI \cdot P_S \tag{3.17}$$

where P_x represents the priorities of objective x in the design process and can be found from Table 3.7. The summations of the priorities of all objectives that need to be minimized are:

$$P_{min} = P_C + P_W + P_T \tag{3.18}$$

The summations of the priorities of the objectives that need to be maximized are:

$$P_{max} = P_P + P_F + P_B + P_M + P_R + P_D + P_S \tag{3.19}$$

In order to determine the optimum configuration, we will consider the configuration at which the design index (DI) is at the optimum value. First, the two parameters of P_{min} and P_{max} must be considered. The design index at which the summation of the priorities of its objectives is higher is assumed as the criterion for configuration selection. There are eventually two configurations that yield the optimum design index. One configuration yields the lowest DI_{min}, and one configuration yields the highest DI_{max}.

If P_{min} is larger than P_{max}, the configuration at which its DI_{min} is the lowest will be selected as the optimum configuration. If P_{max} is larger than P_{min}, the configuration at which its DI_{max} is the highest will be selected as the optimum configuration. If the difference between P_{min} and P_{max} is not considerable (e.g., 51% and 49%), we need to follow the steps of the systems engineering process. Examples 3.1–3.3 introduce sample applications.

Example 3.1

Problem statement: A two-seat fighter aircraft is ordered to be designed to fulfill a military mission and meet the following mission requirements.

- Maximum speed: at least Mach 1.8 at 30 000 ft.
- Absolute ceiling: higher than 50 000 ft.
- Radius of action: 700 km.
- Rate of climb: more than 12 000 fpm.
- Take-off run: 600 m.
- To be able to carry a variety of military stores with a mass of 8000 kg.
- *g* limit: more than +9.
- Highly maneuverable.

Determine the optimum configuration for this aircraft.

Solution:
Initially, a baseline fighter configuration A is assumed as follows: conventional configuration, powered, turbofan engine, twin engine, tractor engine, fixed engine, engines inside fuselage, one wing, fixed wing, tapered wing, fixed sweep angle wing, fixed setting angle, low wing, cantilever, aft tail, conventional tail, twin VT at the fuselage end, retractable landing gear, nose gear, single long fuselage, tandem seating, cockpit, all moving horizontal tail, all moving VT, aileron and flap, hydraulic power system, and full metal structure. For comparison, two alternative configurations, namely B and C, with arbitrary different variables are also considered. You may assume the features of the other two configurations.

To find the design index, first the criteria index for each configuration variable is determined for all 10 figures of merit or criteria (similar to what has been done in Table 3.9). Then, the criteria index is calculated by summing up all indices for each criterion using Equations (3.6)–(3.15) (the results are shown in columns 5, 6, and 7 of Table 3.10). These indices must be compared with the other configurations. Table 3.10 demonstrates a comparison between this baseline configuration (A) and the two other configurations (B and C).

The next step is to use Equations (3.16) and (3.17) to find two design indices. The design index DI_{min} for all three configurations is determined through Equation (3.16) and the results are shown in row 5 of Table 3.10. The design index DI_{max} for all three configurations is also determined by applying Equation (3.17) and the results are shown in the last row of Table 3.10.

In contrast, the two parameters P_{min} and P_{max} are calculated (Equations (3.18) and (3.19)) as shown in column 4 (rows 5 and 13) of Table 3.10. The summation of the priorities of all objectives that need to be minimized (P_{min}) is 20%. Also, the summation of the priorities of all objectives that need to be minimized (P_{max}) is 80%. Since P_{max}

is larger than P_{min}, the configuration at which its DI_{max} is the highest (142) is selected as the optimum configuration; that is, configuration A. Thus, when the optimization methodology is carried out, the design may move from a baseline configuration to an optimized configuration. The details of the calculation are not shown here.

Table 3.10 Evaluation of three presumptive configuration alternatives for a fighter

No.	Criteria	Must be	Priority (%)	Configuration		
				A	B	C
1	Cost	Minimized	9	115	183	210
2	Weight	Minimized	4	136	163	94
3	Period of design	Minimized	7	190	176	217
DI_{min}			20	20.1	35.3	37.8
4	Performance	Maximized	40	210	195	234
5	Flying qualities	Maximized	15	183	87	137
6	Scariness	Maximized	1	87	124	95
7	Maintainability	Maximized	5	95	83	68
8	Producibility	Maximized	6	215	184	164
9	Disposability	Maximized	2	246	254	236
10	Stealth	Maximized	11	65	36	42
DI_{max}			80	142	116.5	137.7

In practice, this methodology requires large numbers of evaluations of the objectives and constraints. The disciplinary models are often very complex and can take significant amounts of time for evaluation. The solution can therefore be extremely time-consuming.

Example 3.2

Figure 3.12 shows photographs of four aircraft: Boeing 747 (transport), McDonnell Douglas F-15C Eagle (fighter), Stampe-Vertongen SV-4C (GA), and Rutan 33 VariEze (GA). By using these photographs and other reliable sources (such as [18]), identify configuration parameters of these aircraft.

Solution:

By using the photographs in Figure 3.12 and also [18], the configuration parameters of these aircraft are identified as shown in Table 3.11.

Table 3.11 Configuration features for the four aircraft of Example 3.2

No.	Attribute	Boeing 747	McDonnell Douglas F-15C Eagle	Stampe-Vertongen	Rutan 33 VariEze
1	Standard	FAR 25	MIL-STD	Home-built	Non-conventional
2	Runway	Land	Land	Land	Land
3	Materials	Mostly metal	Metal	Metal	Composite materials
4	Manufacture	Modular	Modular	Modular	Kit-form
5	Engine type	Turbofan	Turbofan	Piston-prop	Piston-prop
6	Seating (in a row)	10 seat	Single seat	Two tandem seats	Two tandem seats
7	Landing gear type	Multi-gear	Tricycle	Tail-gear	Tricycle
8	Fixed or retractable	Retractable	Retractable	Fixed	Partially retractable
9	Pusher or tractor	Pusher	Pusher	Tractor	Pusher
10	Engine location	Under wing	Inside fuselage	Fuselage nose	Rear fuselage
11	Number of engines	4	2	1	1
12	Flap	Triple slotted flap	Plain flap	Plain flap	Plain flap
13	Door	10 cabin door	Cockpit	No door	Cockpit
14	Tail or canard	Aft tail	Aft	Aft tail	Canard
15	Number of wings	Monoplane	Monoplane	Biplane	Monoplane
16	Wing location	Low wing	High wing	Low + parasol	Mid-wing
17	Wing attachment	Cantilever	Cantilever	Strut-braced	Cantilever
18	Tail configuration	Conventional	Conventional	Conventional	Canard + twin VT
19	Wing fixed/variable sweep	Fixed wing	Fixed wing	Fixed wing	Fixed wing
20	Wing configuration	Swept back	Swept back	Elliptic	Swept back

Table 3.11 (*continued*)

No.	Attribute	Boeing 747	McDonnell Douglas F-15C Eagle	Stampe-Vertongen	Rutan 33 VariEze
21	Tail attachment	Adjustable	All moving	Fixed	Fixed
22	Control surfaces	Elevator/ aileron/rudder	Elevator/ aileron/rudder	Elevator/ aileron/rudder	Elevator/ aileron/rudder
23	Power transmission	Hydraulics	Hydraulics	Mechanical	Mechanical
24	Fuel tank	Inside wing and fuselage	Inside wing and fuselage	Inside fuselage	Inside fuselage
25	Vertical tail	A VT	Twin VT	A VT	Twin VT on wing tip
26	Spoiler/tab	Spoiler and 3 tabs	No tab	No tab	No tab

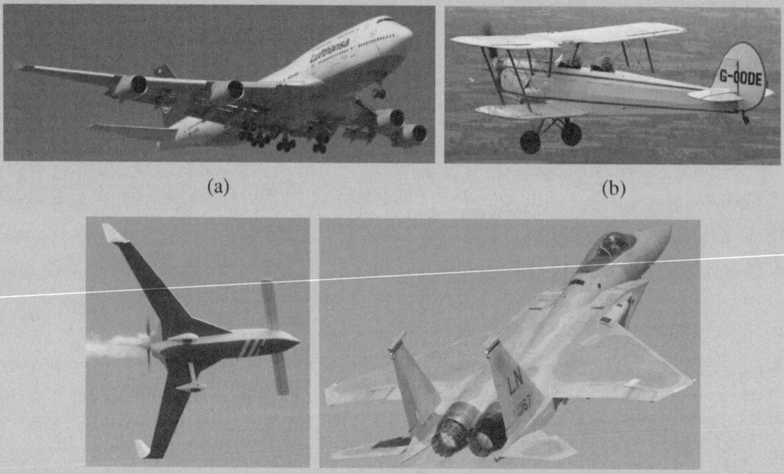

(a) (b) (c) (d)

Figure 3.12 Four aircraft to be used in Example 3.2: (a) Boeing 747; (b) Stampe-Vertongen; (c) Rutan 33 VariEze; (d) F-15C Eagle. Reproduced from permission of: (a) Anne Deus; (b, c) Jenny Coffey; (d) Antony Osborne.

Example 3.3

A university conceptual design team for a small remote-controlled aircraft is to participate in an AIAA student competition. The aircraft has to be able to carry a payload of 7 lb with different payload combinations; and also the size limitation is 4 ft by 5 ft. The performance requirements are as follows.

- Stall speed: 15 knot.
- Maximum speed: 40 knot.
- Take-off run: 80 ft.
- Endurance: 5 min.

The airplane must fly empty while carrying all payload restraint components. The objective is to complete the course profile as many times as possible within 5 minutes, while minimizing battery weight. You are a member of the wing design group, required to decide on the wing configuration, investigate monoplane, biplane, x-wing (tri-wing or higher), and a blended wing body.

Figures of merit include: weight, strength, span, take-off capability, stability, control, manufacturability, reparability, and familiarity.

If the weight of each figure of merit is:

- Weight: 20%
- Strength: 20%
- Span: 10%
- Take-off capability: 10%
- Stability and control: 10%
- Manufacturability: 10%
- Reparability: 5%
- Familiarity: 5%

determine the optimum wing configuration.

Solution:

A summary of the investigation is outlined in Table 3.12. In this table, numbers (1, 0, and −1) are employed. The number "0" indicates that this configuration does not have any influence on a particular figure of merit. The number "1" indicates that this configuration does have a positive influence on a particular figure of merit. The number "−1" indicates that this configuration does have a negative influence on a particular figure of merit.

As indicated in Table 3.12, the monoplane or biplane configuration met the design requirements at the highest level. While the monoplane would be lighter, the biplane configuration would be more structurally sound. Additionally, given the dimension restriction, more wing area could be gained (without aspect ratio penalties) by employing a biplane configuration. For a given wing area, the biplane configuration employs a smaller wing span, leaving more distance longitudinally for a tail arm to increase aircraft stability.

Table 3.12 Wing figures of merit

Figure of merit	Weight (%)	Monoplane	Biplane	X-wing	Blended wing
Weight	20	1	−1	−1	1
Strength	20	0	1	1	1
Span	10	0	0	0	0
Take-off capability	10	0	1	1	1
Stability and control	10	−1	1	1	−1
Interference	10	1	1	−1	1
Manufacturability	10	1	1	1	−1
Reparability	5	1	1	1	−1
Familiarity	5	1	0	0	−1
Total	100	0.4	0.45	0.25	0.3

Problems

1. Figure 3.13 is an image of the utility transport Canadian aircraft Vickers PBV-1A Canso A. Identify 15 different configuration parameters from this image.
2. Figure 3.14 is an image of the World War II (WWII) fighter aircraft P-51D Mustang. Identify 12 different configuration parameters from this image.

Figure 3.13 Canadian Vickers PBV-1A Canso A. Reproduced from permission of Jenny Coffey.

Figure 3.14 Commonwealth CA-18 Mustang. Reproduced from permission of Jenny Coffey.

Figure 3.15 Antonov An-140. Reproduced from permission of Antony Osborne.

3. Figure 3.15 shows a photograph of the transport aircraft Antonov An-140. Identify 12 different configuration parameters from this photograph.
4. Figure 3.16(a) is an image of the transport aircraft McDonnell Douglas MD-11. Identify 15 different configuration parameters from this three-view.
5. Figure 3.16(b) shows a photograph of the WWII fighter aircraft De Havilland Vampire T11 (DH-115). Identify 15 different configuration parameters from this image.
6. By referring to Ref. [18], identify four aircraft that have unconventional configuration.
7. By referring to Ref. [18], identify five aircraft that have canard.

(a)

(b)

Figure 3.16 (a) McDonnell Douglas MD-11 and (b) De Havilland Vampire. Reproduced from permission of (a) Anne Deus and (b) Antony Osborne.

8. By referring to Ref. [18], identify five aircraft with their engines installed above the fuselage.

9. By referring to Ref. [18], identify five transport aircraft with their engines installed beside the aft-fuselage.

10. By referring to Ref. [18], identify three aircraft that have pusher engines plus canard.

11. By referring to Ref. [18], identify two aircraft that have their landing gear partially retractable.

12. Figure 3.17 illustrates a cutaway of the GA aircraft Saab MFI-17 Supporter (T-17). Identify 15 different configuration parameters from this cutaway.

13. Figure 3.18(a) illustrates a three-view of the fighter aircraft F/A-18 Hornet. Identify 15 different configuration parameters from this three-view.

14. Figure 3.18(b) illustrates a three-view of the trainer aircraft Pilatus PC-7. Identify 15 different configuration parameters from this three-view.

15. Figure 3.18(c) illustrates a three-view of the military transport aircraft Lockheed C-130 Hercules. Identify 15 different configuration parameters from this three-view.

16. A 19-seat transport aircraft with the following design requirements is ordered to be designed:

(a) Maximum speed: at least 250 knot at 20 000 ft.
(b) Absolute ceiling: higher than 25 000 ft.
(c) Range: 700 km.
(d) Rate of climb: more than 2000 fpm.
(e) Take-off run: 1000 m.

Determine the optimum configuration for this aircraft. Then sketch its three-view by hand.

17. Figure 3.19 is an image of the solar-powered aircraft Solar Impulse with its revolutionary design. Identify 10 different configuration parameters from this three-view.

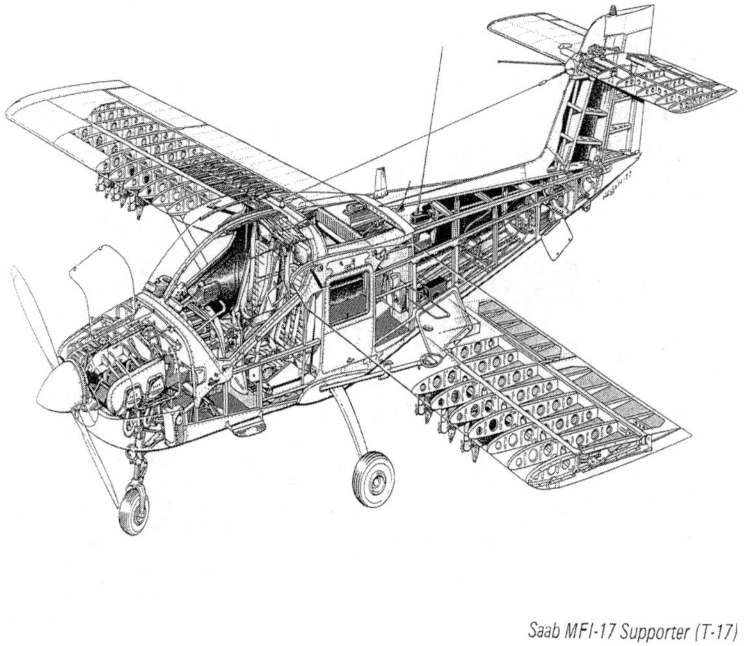

Figure 3.17 Saab MFI-17 supporter. Reproduced from permission of Saab.

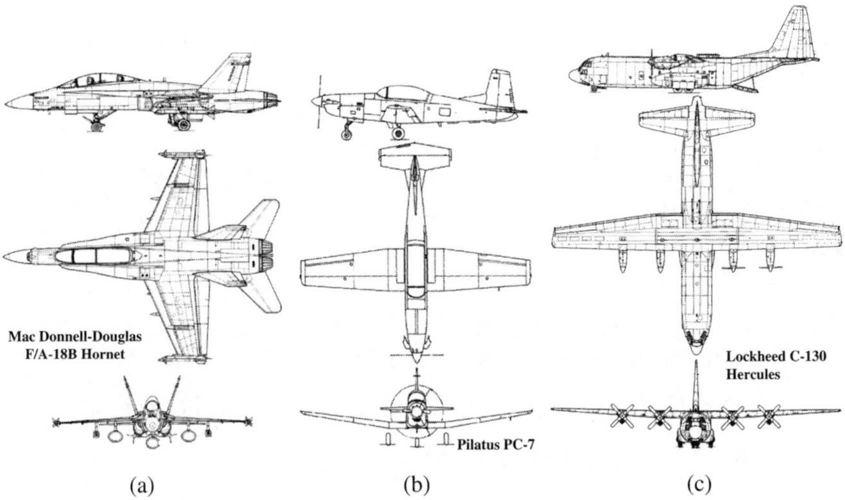

Figure 3.18 (a) F/A-18 Hornet, (b) Pilatus PC-7, and (c) Lockheed C-130 Hercules

18. The authorities of Ground Canyon National Park have ordered a touring aircraft with the following design requirements:

(a) Maximum speed: greater than 100 knot at 2000 ft.
(b) Stall speed: less than 40 knot.

(c) Absolute ceiling: higher than 12 000 ft.
(d) Range: 300 km.
(e) Rate of climb: more than 4000 fpm.
(f) Take-off run: 500 m.

The aircraft is required to carry a pilot and a tourist. Determine the optimum configuration for this aircraft. Then sketch its three-view by hand.

Figure 3.19 Solar impulse. Reproduced from permission of Vladimir Mykytarenko.

19. A civil trainer aircraft with the following design requirements is desired to be designed:

(a) Maximum speed: greater than 200 knot at 20 000 ft.
(b) Stall speed: less than 50 knot.
(c) Absolute ceiling: higher than 30 000 ft.
(d) Range: 500 km.
(e) Rate of climb: more than 3 000 fpm.
(f) Take-off run: 400 m.

The aircraft is required to carry an instructor and a student. Determine the optimum configuration for this aircraft. Then sketch its three-view by hand.

20. A cargo aircraft with the following design requirements is desired to be designed:

(a) Maximum speed: greater than 250 knot at 30 000 ft.
(b) Stall speed: less than 80 knot.
(c) Absolute ceiling: higher than 35 000 ft.
(d) Range: 10 000 km.
(e) Rate of climb: more than 2500 fpm.
(f) Take-off run: 1500 m.

The aircraft is required to carry 20 blocks of cargo, each with a volume of $3 \cdot 3 \cdot 3\ \text{m}^3$. Determine the optimum configuration for this aircraft. Then sketch its three-view by hand.

21. You are a member of a design team performing the conceptual design phase of an unmanned aircraft with the following design requirements:

(a) Maximum speed: greater than 200 knot at 30 000 ft.
(b) Stall speed: less than 70 knot.
(c) Absolute ceiling: higher than 60 000 ft.
(d) Range: 30 000 km.
(e) Rate of climb: more than 2000 fpm.
(f) Take-off run: 1000 m.

The aircraft is required to carry communication and surveillance equipments. Determine the optimum configuration for this aircraft. Then sketch its three-view by hand.

22. You are a member of a design team performing the conceptual design phase of a human-powered aircraft. The aircraft is required to carry communication and surveillance equipments. Determine the optimum configuration for this aircraft. Then sketch its three-view by hand.

23. You are a member of a design team performing the conceptual design phase of a sailplane with the following design requirements:

(a) Glide speed: 40 knot at 10 000 ft.
(b) Stall speed: less than 30 knot.
(c) Take-off run (when towed by another aircraft): 300 m.
(d) Endurance (when flight begins from 10 000 ft): 2 hours.

The aircraft is required to have two seats. Determine the optimum configuration for this aircraft. Then sketch its three-view by hand.

24. Sketch by hand a four-seat aircraft with the following configuration features:
Monoplane, high wing, canard, pusher piston-prop engine, fixed tail gear, tapered wing, and tip-tank.

25. Sketch by hand a two-seat aircraft with the following configuration features:
Monoplane, low wing, T-tail, twin turboprop engines on the wing, retractable nose gear, and rectangular wing.

26. Sketch by hand a cargo aircraft with the following configuration features:
High rectangular wing, conventional tail, four turboprop engines on the wing, and retractable multi-gear landing gear.

27. Sketch by hand a transport aircraft with the following configuration features:
Low swept back wing, T-tail, two turbofan engines beside rear fuselage, and retractable tricycle landing gear.

28. Sketch by hand a single-seat fighter aircraft with the following configuration features:
Monoplane, low wing, canard, twin vertical tail, single turbofan engine inside fuselage, retractable tricycle landing gear, and variable sweep.

References

[1] Blanchard, B.S. and Fabrycky, W.J. (2006) *Systems Engineering and Analysis*, 4th edn, Prentice Hall.
[2] Niu Michael, C.Y. (2005) *Composite Airframe Structures*, 5th edn, Conmilit Press.
[3] Groover, M.P. (2010) *Fundamentals of Modern Manufacturing: Materials, Processes, and Systems*, 4th edn, John Wiley & Sons, Inc.
[4] Federal Aviation Regulations (2011), Federal Aviation Administration, Department of Transportation, www.faa.gov.
[5] Eschenauer, H., Koski, J., and Osyczka, A. (1990) *Multicriteria Design Optimization: Procedures and Applications*, Springer.
[6] Padula, S.L., Alexandrov, N.M., and Green, L.L. (1996) MDO test suite at NASA Langley Research Center. 6th AIAA/NASA/ISSMO Symposium on Multidisciplinary Analysis and Optimization, Bellevue, WA.
[7] Kroo, I., Altus, S., Braun, R. *et al.* (1994) Multidisciplinary Optimization Methods for Aircraft Preliminary Design, AIAA 94–4325.
[8] Rao, C., Tsai, H., and Ray, T. (2004) Aircraft configuration design using a multidisciplinary optimization approach. 42nd AIAA Aerospace Sciences Meeting and Exhibit, Reno, NV, January 5–8, 2004, AIAA-2004-536.
[9] Roskam, J. (2007) *Lessons Learned in Aircraft Design*, DAR Corporation.
[10] Roskam, J. (2006) *Roskam's Airplane War Stories*, DAR Corporation.
[11] Young, J.A., Anderson, R.D., and Yurkovich, R.N. (1998) A Description of the F/A-18E/F Design and Design Process, AIAA-98-4701.
[12] Alexandrov, N.M. and Lewis, R.M. (2000) Analytical and computational properties of distributed approaches to MDO. 8th AIAA/USAF/NASA/ISSMO Symposium on Multidisciplinary Analysis & Optimization, Long Beach, CA, September 6–8, 2000, AIAA 2000-4718.
[13] Umakant, J., Sudhakar, K., Mujumdar, P.M., and Panneerselvam, S. (2004) Configuration design of a generic air-breathing aerospace vehicle considering fidelity uncertainty. 10th AIAA/ISSMO Multidisciplinary Analysis and Optimization Conference, Albany, NY, August 30–September 1, 2004, AIAA 2004-4543.
[14] Tanrikulu, O. and Ercan, V. (1997) Optimal external configuration design of unguided missiles. AIAA Atmospheric Flight Mechanics Conference, New Orleans, LA, August 11–13, 1997, AIAA-1997-3725.
[15] YorkBlouin, V.Y., Miao, Y., Zhou, X., and Fadel, G.M. (2004) An assessment of configuration design methodologies. 10th AIAA/ISSMO Multidisciplinary Analysis and Optimization Conference, Albany, NY, August 30–September 1, 2004, AIAA 2004-4430.
[16] Onwubiko, C. (2000) *Introduction to Engineering Design Optimization*, Prentice Hall.
[17] Chong, E.K.P. and Zack, S.H. (2008) *An Introduction to Optimization*, 3rd edn, John Wiley & Sons, Inc.
[18] Jackson, P. *Jane's All the World's Aircraft*, Jane's Information Group, various years (1996–2011).

4

Preliminary Design

4.1 Introduction

The purpose of this chapter is to describe the preliminary design phase of an aircraft. Based on the systems engineering approach, an aircraft will be designed during three phases: (i) conceptual design phase, (ii) preliminary design phase, and (iii) detail design phase. At the conceptual design phase, the aircraft will be designed in concept without precise calculations. In other words, almost all the parameters are determined based on a decision-making process and selection technique. In contrast, the preliminary design phase tends to employ the outcomes of a calculation procedure. As the name implies, at the preliminary design phase, the parameters determined are not final and will be altered later. In addition, at this phase, parameters are essential and will influence the entire detail design phase directly. Therefore, ultimate care must be taken to insure the accuracy of the results of the preliminary design phase.

Three fundamental aircraft parameters determined during the preliminary design phase are: (i) aircraft maximum take-off weight (MTOW or W_{TO}), (ii) wing reference area (S_W or S_{ref} or S), and (iii) engine thrust (T_E or T) or engine power (P_E or P). Hence, three primary aircraft parameters of W_{TO}, S, and T (or P) form the output of the preliminary design phase. These three parameters will govern the aircraft size, the manufacturing cost, and the complexity of calculations. If, during the conceptual design phase, a jet engine is selected, the engine thrust is calculated during this phase. But if, during the conceptual design phase, a prop-driven engine is selected, the engine power is calculated during this phase. A few other non-important aircraft parameters such as aircraft zero-lift drag coefficient and aircraft maximum lift coefficient are estimated during this phase too.

The preliminary design phase is performed in two steps:

- **Step 1**. Estimate aircraft MTOW.
- **Step 2**. Determine wing area and engine thrust (or power) simultaneously.

In this chapter, two design techniques are developed. First, a technique based on the statistics is developed to determine the wing reference area and engine thrust (or power).

Aircraft Design: A Systems Engineering Approach, First Edition. Mohammad H. Sadraey.

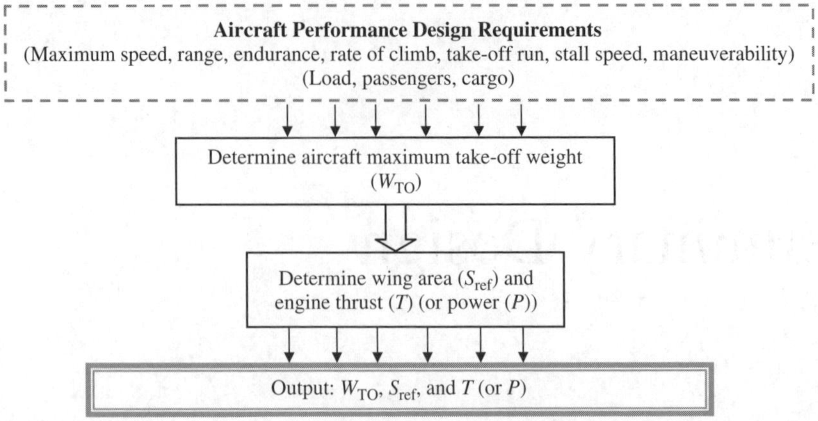

Figure 4.1 Preliminary design procedure

Second, another technique is developed based on the air craft performance requirements (such as maximum speed, range, and take-off run) to determine the wing area and the engine thrust (or power). This technique is sometimes referred to as the matching plot or matching chart, due to its graphical nature. In some references, this process and this design phase are referred to as *initial sizing*. This is due to the nature of the process, which literally determines the size of three fundamental features of the aircraft.

Figure 4.1 illustrates a summary of the preliminary design process. In general, the first technique is not accurate (in fact, it is an estimation) and the approach may carry some inaccuracies, while the second technique is very accurate and the results are reliable.

4.2 Maximum Take-Off Weight Estimation

4.2.1 The General Technique

The purpose of this section is to introduce a technique to obtain the first estimate of the MTOW (or all-up weight) for an aircraft before it is designed and built. The word "estimation" is selected intentionally to indicate the degree of accuracy and reliability of the output. Hence, the value for the MTOW is not final and must be revised in the later design phases. The result of this step may have up to about 20% inaccuracies, since it is not based on its own aircraft data. But the calculation relies on other aircraft data with similar configuration and mission. Thus, we adopt past history as the major source of information for the calculations in this step. At the end of the preliminary design phase, the take-off weight estimation is repeated by using another more accurate technique which will be introduced in Chapter 10. As described in Chapter 1, the aircraft design nature is iterative, thus new data for the MTOW requires a new round of calculations and new designs for all aircraft components such as wing, tail, and fuselage.

Since the accuracy of the result of this design step depends largely on the past history, one must be careful to utilize only aircraft data that are current, with aircraft that are similar in configuration and mission. The currency of data and similarity play a vital role, as there are many aspects to compare. As the years pass, the science of materials and also manufacturing technologies are changing and improving. For instance, every year, new engineering materials are introduced to the market which are lighter and stronger. New materials such as composite materials have caused a revolution in the aircraft industry. In addition, new power transmission technologies such as fly-by-wire allow aircraft to be much lighter than expected. The trend is continuing, therefore, more current data results in a more reliable estimation.

Due to the fact that various aircraft manufacturing industries employ different approaches in their products, data from more than one aircraft must be obtained. The suggestion is to use data from at least five different aircraft to estimate the take-off weight of your aircraft. Aircraft manufacturing companies such as Boeing, Airbus, Lockheed, Grumman, Cessna, Raytheon, Bombardier, Dassult, Emberaer, Learjet, and Jetstream each have different management systems, design techniques, and market approaches. Thus, their aircraft productions have several differences, including MTOW. When you are selecting several aircraft for data applications, select aircraft from different companies and even from different regions of the world. Another recommendation is to choose aircraft data from recent years. For example, a comparison among fighters in the World War I era (e.g., Avro 504), the World War II era (e.g., Mustang and Spitfire), and the current modern advanced fighters (e.g., F-16 Fighting Falcon) demonstrates how much lighter the current aircraft are compared with older ones.

4.2.2 Weight Build-up

An aircraft has a range of weights from minimum to maximum depending upon the number of pilots and crew, fuel, and payloads (passengers, loads, luggage, and cargo). As the aircraft flies, the fuel is burning and the aircraft weight is decreasing. The most important weight in the design of an aircraft is the maximum allowable weight of the aircraft during take-off operation. This is also referred to as the all-up weight. The design MTOW or W_{TO} is the total weight of an aircraft when it begins the mission for which it was designed. The maximum design take-off weight is not necessarily the same as the maximum nominal take-off weight, since some aircraft can be overloaded beyond design weight in an emergency situation, but will suffer a reduced performance and reduced stability. Unless specifically stated, MTOW is the design weight. It means every aircraft component (e.g., wing, tail) is designed to support this weight.

The general technique to estimate the MTOW is as follows: the aircraft weight is broken into several parts. Some parts are determined based on statistics, but some are calculated from performance equations.

The MTOW is broken into four elements:

1. Payload weight (W_{PL}).
2. Crew weight (W_C).
3. Fuel weight (W_f).
4. Empty weight (W_E).

$$W_{TO} = W_{PL} + W_C + W_F + W_E \tag{4.1}$$

The payload weight and crew weight are mostly known and determined from the given data (by customer and standards), and not dependent on the aircraft take-off weight. In contrast, the empty weight and fuel weight are both functions of the MTOW. Hence, to simplify the calculations, both the fuel weight and the empty weight are expressed as fractions of the MTOW. Hence:

$$W_{TO} = W_{PL} + W_C + \left(\frac{W_f}{W_{TO}}\right) W_{TO} + \left(\frac{W_E}{W_{TO}}\right) W_{TO} \tag{4.2}$$

This can be solved for W_{TO} as follows:

$$W_{TO} - \left(\frac{W_f}{W_{TO}}\right) W_{TO} - \left(\frac{W_E}{W_{TO}}\right) W_{TO} = W_{PL} + W_C \tag{4.3}$$

The take-off weight can be factored out:

$$W_{TO}\left[1 - \left(\frac{W_f}{W_{TO}}\right) - \left(\frac{W_E}{W_{TO}}\right)\right] = W_{PL} + W_C \tag{4.4}$$

Thus:

$$W_{TO} = \frac{W_{PL} + W_C}{1 - \left(\frac{W_f}{W_{TO}}\right) - \left(\frac{W_E}{W_{TO}}\right)} \tag{4.5}$$

In order to find W_{TO}, one needs to determine the four variables of W_{PL}, W_C, W_f/W_{TO}, and W_E/W_{TO}. The first three parameters, namely payload, crew, and fuel fraction, are determined fairly accurately, but the last parameter (i.e., empty weight fraction) is estimated from statistics.

4.2.3 Payload Weight

The payload is the net carrying capacity of an aircraft. An aircraft is originally required and designed to carry the payload or useful load. The payload includes luggage, cargo, passenger, baggage, store, military equipments, and other intended loads. Thus, the name payload has a broad meaning. For instance, sometimes the Space Shuttle cannot successfully land at Kennedy Space Center in Florida due to poor weather conditions. So, the Shuttle will first land at another runway such as one at Edward Air Force Base in California, and then it will be carried out by a Boeing 747 (Figures 3.7, 3.12, and 9.4) to Florida. Thus, the Space Shuttle is called the payload for the Boeing 747 in this mission.

In case of a passenger aircraft, the passengers' weight is to be determined. Actual passenger weights must be used in computing the weight of an aircraft with a limited seating capacity. Allowance must be made for heavy winter clothing when such is worn. There is no standard human, since every kind of passenger (such as infants, young, and senior) may get into the plane. To make the calculation easy, one might assume a number as the tentative weight for a typical passenger and then multiply this value by the number of passengers. There are several references in human factors and ergonomic engineering areas that have these numbers. The Federal Aviation Administration (FAA) [1] has regulated this topic and the reader is encouraged to consult its publications; see the Federal Aviation Regulations (FAR). For example, FAR Part 25 (which regulates airworthiness standards for transport aircraft) asks the aircraft designers to consider the

Table 4.1 Standard average passenger weights [2]

No	Passenger	Weight per passenger (lb)	
		Summer	Winter
1	Average adult	190	195
2	Average adult male	200	205
3	Average adult female	179	184
4	Child weight (2 years to less than 13 years of age)	82	87

reasonable numbers for an average passenger. The following is a suggested value for passenger weight based on published data:

$$W_{pass} = 180\,\text{lb} \tag{4.6}$$

Note that this number is updated every year (due to obesity and other issues), so it is recommended to consult the FAA publications for accurate data. For instance, the FAA in 2005 issued an Advisory Circular [2] and had several recommendations for airlines. One example is illustrated in Table 4.1. In this table, the standard average passenger weight includes 5 pounds for summer clothing, 10 pounds for winter clothing, and a 16-pound allowance for personal items and carry-on bags. Where no gender is given, the standard average passenger weights are based on the assumption that 50% of passengers are male and 50% of passengers are female. The weight of children under the age of 2 has been factored into the standard average and segmented adult passenger weights.

In determining the total weight of passengers, it is wise to consider the worst-case scenario, which is the heaviest possible case. This means that all passengers are considered to be adult and male. Although this is a rare case, it guarantees flight safety. In a passenger aircraft, the water and food supply must be carried on long trips. However, these are included in the empty weight.

The weight of luggage and carry-on bags is another item that must be decided. The FAA has some recommendations about the weight of bags and luggage on a passenger aircraft. Due to high rising fuel costs, airlines have regulated the weight themselves. For instance, the majority of airlines currently accept two bags of 70 lb for international flights and one bag of 50 lb for domestic flights. There is some suggestion that these numbers are going to drop in the near future.

4.2.4 Crew Weight

Another part of the aircraft weight is the weight of the people who are responsible for conducting the flight operations and serving passengers and payload. A human-piloted aircraft needs at least one human to conduct the flight. In case of a large passenger aircraft, more staff (e.g., copilot, flight engineer, navigation pilot) may be needed. Moreover, one or more crew is necessary to serve the passengers. In case of a large cargo aircraft, several officers are needed to locate the loads and secure them in the right place.

In a large transport aircraft, this weight counts for almost nothing compared with the aircraft all-up weight. In a hang glider, however, the weight of the pilot counts for more than 70% of the aircraft weight. Therefore, in the smaller aircraft, more attention must be paid in determining the weight of the pilot. Two parameters must be determined in this part: (i) number of pilots and crew members and (ii) weight of each crew member.

In a small GA (General Aviation) or a fighter aircraft, the number of pilots is given to the designer, but in large passenger and cargo aircraft, more pilots and crew are needed to conduct the flight operations safely. In the 1960s, a large transport aircraft was required to have two pilots plus one flight engineer and one navigation engineer. Due to advances in avionic systems, the last two jobs have been cancelled, leaving the pilot and copilot to take care of them – this is due to the fact that more and more measurement devices are becoming electronic and integrated, and illustrated in one large display. In the 1950s, a large transport aircraft such as Boeing 727 had about 200 gauges, instruments, knobs, switches, lights, display, and handles to be monitored and controlled throughout the flight operation. However, thanks to digital electronics and modern computers, at the moment, one pilot can not only conduct the flight safely, but is also able to monitor tens of flight variables and aircraft motions through a display and a control platform simultaneously.

If the aircraft is under commercial flight operations, it would be operating under Parts 119 and 125. The flight attendant's weight is designated as 119.3. In Subpart I of Part 125, there are pilot-in-command and second-in-command qualifications. There may be space on the aircraft for more crew members, but based on the language of the document, two flight crew members is the minimum allowed.

The FAA [1] has regulated the number of crew for transport aircraft. Based on FAR Part 125, Section 125.269, for airplanes having more than 100 passengers, two flight attendants plus one additional flight attendant for each unit of 50 passengers above 100 passengers are required:

> Each certificate holder shall provide at least the following flight attendants on each passenger-carrying airplane used:
>
> (1) For airplanes having more than 19 but less than 51 passengers – one flight attendant.
>
> (2) For airplanes having more than 50 but less than 101 passengers – two flight attendants.
>
> (3) For airplanes having more than 100 passengers – two flight attendants plus one additional flight attendant for each unit (or part of a unit) of 50 passengers above 100 passengers.

Therefore, for instance, a large passenger aircraft is required to have two pilots plus eight flight attendants. The followings regulations are reproduced [1] from FAR Part 119, Section 119.3:

> Crew – for each crew member required by the Federal Aviation Regulations –
>
> (A) For male flight crew members – 180 pounds.
>
> (B) For female flight crew members – 140 pounds.
>
> (C) For male flight attendants – 180 pounds.
>
> (D) For female flight attendants – 130 pounds.
>
> (E) For flight attendants not identified by gender – 140 pounds.

The following sentence is also reproduced [1] from FAR Part 125, Section 125.9:

Crew – 200 pounds for each crew member required under this chapter.

The reader is encouraged to observe the particular FAA standards which apply to the case.

For military aircraft, particularly fighters, pilots are usually equipped with helmet, goggles, g-suit, and other special equipment (such as pressure system). Not only is the fighter pilot often heavier than a civil pilot, but each equipment weight must also be added to the pilot's weight. For more information the reader is encouraged to consult the military standards. Reference [3] has some useful information and standards. The general rule to determine the weight of each pilot, flight attendant, or crew is similar to what is introduced in Section 4.2.3 (i.e., Equation (4.6)). In order to obtain the certificate, the designer must follow FAA regulations [1].

In case of a home-built or special mission aircraft (such as the non-stop globe-circling aircraft Voyager, or the aircraft to carry another aircraft to space for the first time Space Ship One), the weight of each pilot is obtained exactly by weighting the specified pilot on scale. Table 4.2 demonstrates typical values of the crew weight fraction for several aircraft.

Table 4.2 Typical values for the crew weight fraction [4]

No.	Aircraft	W_C/W_{TO} (%)
1	Hang glider/kite/paraglider	70–80
2	Single-seat glider/sailplane	10–20
3	Two-seat motor glider	10–30
4	Ultralight	30–50
5	Microlight	20–40
6	Very light aircraft (VLA)	15–25
7	GA single-seat piston engine	10–20
8	GA multi-seat	10–30
9	Agriculture	2–3
10	Business jet	1.5–3
11	Jet trainer	4–8
12	Large transport aircraft	0.04–0.8
13	Fighter	0.2–0.4
14	Bomber	0.1–0.5

4.2.5 Fuel Weight

Another part of the aircraft MTOW is the fuel weight. The required amount of the total fuel weight necessary for a complete flight operation depends upon the mission to be followed, the aerodynamic characteristics of the aircraft, and the engine specific fuel consumption (SFC). The mission specification is normally given to the designer and must be known. The aircraft aerodynamic model and the SFC may be estimated from the aircraft configuration that is designed in the conceptual design phase. Recall from Equation (4.5) that we are looking for the fuel fraction (W_f/W_{TO}).

The first step to determine the total fuel weight is to define the flight mission segments. Three typical mission profiles are demonstrated in Figure 4.2 for three typical aircraft; that is, transport, fighter, and reconnaissance. A typical flight mission for a GA aircraft is often very similar to a flight mission of a transport aircraft, but the duration is shorter. For other types of aircraft such as trainer, agriculture, bomber, the designer can build the mission profile based on the given information from the customer.

Each flight mission consists of several segments, but usually one of them takes the longest time. The main feature of the flight of a transport aircraft is "cruise," that makes up the longest segment of the flight. The main feature of the flight of a reconnaissance/

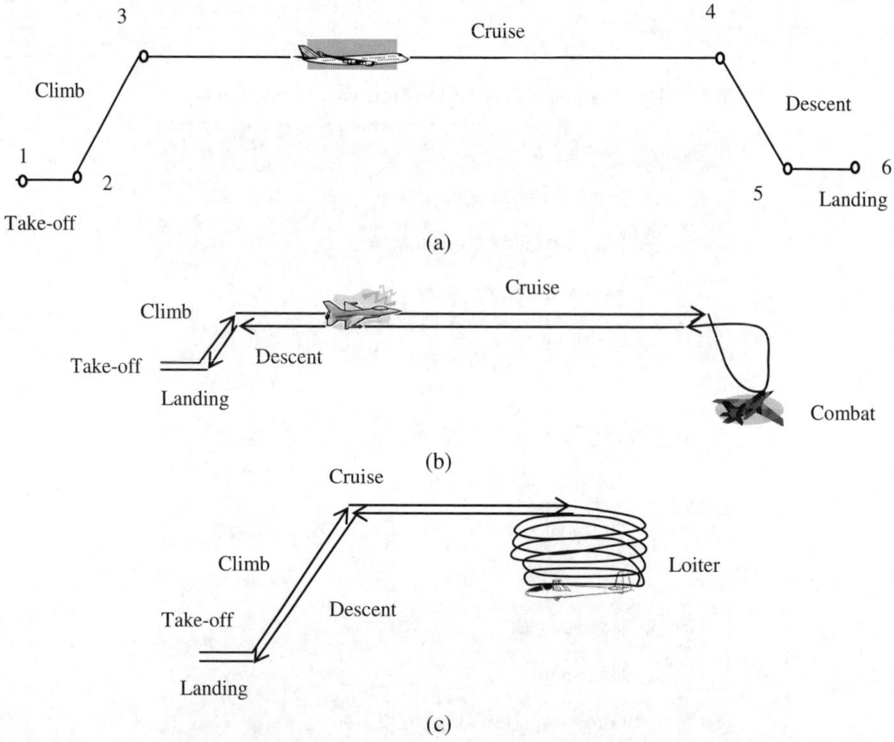

Figure 4.2 Typical mission profiles for three typical aircraft: (a) transport aircraft, (b) fighter, and (c) reconnaissance

patrol/monitor/relay aircraft is loitering, that makes up the longest segment of the flight. The main feature of the flight of a fighter aircraft is "dash," that makes up the longest segment of the flight. In terms of flight mechanics, the cruising flight is measured by *range*, a loitering flight is measured by *endurance*, and a dash is measured by *radius of action*.

For analysis, each mission segment is numbered; 1 denotes the beginning of take-off and 2 is the end of take-off. For example, in the case of a regular flight of a transport aircraft, segments could be numbered as follows: 1. taxi/take-off, 2. climb, 3. cruise, 4. descent, 5. landing. In a similar fashion, the aircraft weight at each phase of the flight mission can be numbered. Hence, W_1 is the aircraft weight at the beginning of take-off (i.e., MTOW). W_2 is the aircraft weight at the end of take-off, which is the beginning of the climb phase. W_3 is the aircraft weight at the end of the climb phase, which is the beginning of the cruising phase. W_4 is the aircraft weight at the end of the cruising phase, which is the beginning of the descending phase. W_5 is the aircraft weight at the end of the descending phase, which is the beginning of the landing phase. Finally, W_6 is the aircraft weight at the end of the landing phase. Thus, for any mission segment i, the mission segment weight fraction is expressed as (W_{i+1}/W_i). If these weight fractions can be estimated for all the segments, they can be multiplied together to find the ratio of the aircraft weight at the end of flight operations, divided by the initial weight; that is, MTOW. This ratio would then be employed to determine the total fuel fraction.

During each segment, the fuel is burnt and the aircraft loses weight. If an aircraft has a mission to drop load or parachute, the technique must be applied with a slight correction. The aircraft weight at the end of a segment divided by its weight at the beginning of that segment is called the segment weight fraction. For instance, W_4/W_3 in the flight mission of Figure 4.2(a) is the fuel fraction during the cruise segment. This will make a basis for estimating the required fuel weight and fuel fraction during a flight operation. The difference between the aircraft weight at the end of the flight (i.e., landing) and the aircraft weight at the beginning of the flight (i.e., take-off) is exactly equal to the fuel weight:

$$W_{\mathrm{TO}} - W_{\mathrm{landing}} = W_{\mathrm{f}} \tag{4.7}$$

Thus, in a regular flight mission, the ratio between the aircraft weight at the end of the flight to the aircraft weight at the beginning of the flight is:

$$\frac{W_{\mathrm{landing}}}{W_{\mathrm{TO}}} = \frac{W_{\mathrm{TO}} - W_{\mathrm{f}}}{W_{\mathrm{TO}}} \tag{4.8}$$

Therefore, for the case of a mission with five segments as shown in Figure 4.2(a), the fuel weight fraction is obtained as follows:

$$\frac{W_{\mathrm{f}}}{W_{\mathrm{TO}}} = 1 - \frac{W_6}{W_1} \tag{4.9}$$

where $\frac{W_6}{W_1}$ can be written as:

$$\frac{W_6}{W_1} = \frac{W_2}{W_1}\frac{W_3}{W_2}\frac{W_4}{W_3}\frac{W_5}{W_4}\frac{W_6}{W_5} \tag{4.10}$$

Table 4.3 Typical average segment weight fractions

No.	Mission segment	W_{i+1}/W_i
1	Taxi and take-off	0.98
2	Climb	0.97
3	Descent	0.99
4	Approach and landing	0.997

For other flight missions, the reader is required to identify the segments and to build a similar numbering system to derive a similar equation. For the sake of flight safety, it is recommended to carry reserve fuel in case the intended airport is closed, so the aircraft has to land at another nearby airport. FAA regulation requires a transport aircraft to carry 20% more fuel than needed on a flight of 45 minutes to observe airworthiness standards. The extra fuel required for safety purposes is almost 5% of the aircraft total weight, so it is applied as follows:

$$\frac{W_f}{W_{TO}} = 1.05\left(1 - \frac{W_6}{W_1}\right) \tag{4.11}$$

Therefore, in order to find the fuel weight fraction, one must first determine these weight fractions for all the mission segments (e.g., $\frac{W_2}{W_1}, \frac{W_3}{W_2}, \frac{W_4}{W_3}, \frac{W_5}{W_4}, \frac{W_6}{W_5}$). There are primarily six flight segments: take-off, climb, cruise, loiter, descent, and landing. These flight phases or segments can be divided into two groups:

1. The segments during which the fuel weight that is burnt is almost nothing and negligible compared with the MTOW. These include taxi, take-off, climb, descent, approach, and landing. The fuel weight fractions for these mission segments are estimated based on the statistics. Table 4.3 illustrates typical average values for fuel fractions of take-off, climb, descent, and landing.
2. The segments during which the fuel weight that is burnt is considerable. These include cruise and loiter and are determined through mathematical calculations.

Table 4.4 shows the fuel weight fractions for several aircraft.

4.2.5.1 Cruise Weight Fraction for Jet Aircraft

The fuel weight fraction for the cruise segment is determined by employing the Breguet range equation. By definition, the range is the total distance that an aircraft can fly with a full fuel tank and without refueling. This consists of take-off, climb, cruise, descent, and landing and does not include the wind effect (either positive or negative). Since this definition is not applicable for our case, we resort to the gross still air range, which does

Table 4.4 Fuel weight fraction for several aircraft

No.	Aircraft	Type	Range (km)	S (m^2)	m_{TO} (kg)	m_f (kg)	$\frac{m_f}{m_{TO}}$
1	MIT Daedalus 88	Man-powered	N/C[a]	29.98	104	0	0
2	Volmer VJ-25 Sunfun	Hang glider/kite	N/C	15.14	140.5	50	0
3	Manta Fledge III	Sailplane/glider	N/C	14.95	133	0	0
4	Merlin E-Z	Ultralight	–	15.33	476	163	0.342
5	Pilatus PC-12	Turboprop transport	3 378	25.81	4 100	1 200	0.293
6	C-130J Hercules	Military transport	5 250	162.12	70 305	17 075	0.243
7	Beech super king air B200	Light transport	2 204	28.18	5 670	1 653	0.292
8	Hawkeye E-2C	Early warning	2 854	65.03	24 687	5 624	0.228
9	MD-95 ER	Jet transport	3 705	92.97	54 885	10 433	0.19
10	Airbus 380-841	Wide-bodied airliner	15 200	845	590 000	247 502	0.419
11	Boeing 777	Airliner	10 556	427.8	229 520	94 210	0.41
12	Beechcraft 390	Light business jet	1 457	22.95	5 670	1 758	0.31
13	F-16C	Fighter	2 742	27.87	19 187	3 104	0.16
14	Voyager	Circumnavigation	39 000	30.1	4 398	3 168	0.72
15	Global Hawk	Unmanned reconnaissance	24 985	50.2	10 387	6 536	0.629

[a] Not constant.

not include any segment other than cruising flight. In order to cruise, basically there are three flight programs that satisfy trim requirements. They are:

- **Flight program 1**. *Constant-altitude, constant-lift* coefficient flight.
- **Flight program 2**. *Constant-airspeed, constant-lift* coefficient flight.
- **Flight program 3**. *Constant-altitude, constant-airspeed* flight.

Each flight program has a unique range equation, but for simplicity, we use the second flight program, since its equation is easiest to apply in our preliminary design phase. The range equation for a jet aircraft is slightly different from that of a prop-driven aircraft. The origin of the difference is that the jet engine is generating thrust (T), while a prop-driven engine produces power (P). Thus, they are covered separately.

For an aircraft with a jet engine (i.e., turbojet and turbofan), the optimum range equation [5] with the specified speed of $V_{(L/D)_{\max}}$ is:

$$R_{\max} = \frac{V_{(L/D)_{\max}}}{C}\left(\frac{L}{D}\right)_{\max} \ln\left(\frac{W_i}{W_{i+1}}\right) \tag{4.12}$$

where W_i denotes the aircraft weight at the beginning of cruise, and W_{i+1} is the aircraft weight at the end of the cruising flight. Thus, the term $\frac{W_i}{W_{i+1}}$ indicates the fuel weight fraction for the cruise segment. Also, the parameter C is the engine SFC and L/D is the lift-to-drag ratio. The cruising speed is usually a performance requirement and is given. But the two parameters of C and $(L/D)_{\max}$ are unknown at this moment, since we are in the preliminary design phase and the aerodynamic aspect of the aircraft and also the propulsion system are not determined. Again, we resort to historical values and employ data for similar aircraft. Table 4.5 shows typical values for maximum lift-to-drag ratios of several aircraft. The supersonic transport aircraft Concorde (Figures 7.24 and 11.15) tends to have a lift-to-drag ratio of 7.1 at a speed of Mach 2.

Table 4.5 The typical maximum lift-to-drag ratio for several aircraft

No.	Aircraft type	$(L/D)_{\max}$
1	Sailplane (glider)	20–35
2	Jet transport	12–20
3	GA	10–15
4	Subsonic military	8–11
5	Supersonic fighter	5–8
6	Helicopter	2–4
7	Home-built	6–14
8	Ultralight	8–15

From flight mechanics the reader may recall that there are three differences between an economic cruising flight and a flight to maximize range.

1. Almost no aircraft is cruising to maximize the range, since it ends up having a longer trip and some operational difficulties. Most transport aircraft are recommended to fly with a Carson's speed 32% higher than the speed for maximizing range:

$$V_{\mathrm{C}} = 1.32V_{(L/D)_{\max}} \tag{4.13}$$

2. In contrast, in a cruising flight with Carson's speed, the lift-to-drag ratio is slightly less than the maximum lift-to-drag ratio. That is:

$$\left(\frac{L}{D}\right)_{\mathrm{cruise}} = \frac{\sqrt{3}}{2}\left(\frac{L}{D}\right)_{\max} = 0.866\left(\frac{L}{D}\right)_{\max} \tag{4.14}$$

3. In a cruising flight, the maximum engine thrust is not normally employed. This is to reduce the cost and the engine SFC (C).

   For more details the reader is referred to Ref. [5]. By taking into account these above-mentioned economic and operational considerations, Equation (4.12) is modified as follows:

$$R = 0.866\frac{V_{\mathrm{C}}}{C}\left(\frac{L}{D}\right)_{\max}\ln\left(\frac{W_i}{W_{i+1}}\right) \tag{4.15}$$

   Therefore, the cruise fuel weight ratio is determined as:

$$\frac{W_{i+1}}{W_i} = \mathrm{e}^{\frac{-RC}{0.866V(L/D)_{\max}}} \tag{4.16}$$

   The definition and typical values for the variable C are presented in Section 4.2.5.5.

4.2.5.2 Cruise Weight Fraction for Prop-Driven Aircraft

The definition and flight approaches to satisfy a trimmed operation for a specified range are discussed in Section 4.5.2.1. Since the type of propulsion system is prop-driven, the engine is generating power, and the propeller efficiency influences the overall thrust. As in the case of a jet aircraft, there are three flight approaches to hold the aircraft trim despite the loss of weight due to fuel burn. For the sake of simplicity and due to the expected accuracy in the preliminary design phase, we select only one of them. If the design requirements specify that the aircraft must have a different approach, one needs to employ the relevant equation.

For an aircraft with a prop-driven engine (i.e., piston-prop or turboprop), the optimum range will be achieved when the aircraft is flying with the minimum drag speed. Thus the range equation [5] is:

$$R_{\max} = \frac{\eta_{\mathrm{P}}\,(L/D)_{\max}}{C}\ln\left(\frac{W_i}{W_{i+1}}\right) \tag{4.17}$$

This is for the case where the lift coefficient (C_L) or angle of attack (α) is held constant. In other words, either the flight speed is decreasing or the flight altitude is increasing (air density is decreased) to compensate for the loss of aircraft weight. This is referred to as the *Breguet* range equation for prop-driven aircraft. Therefore, the cruise fuel weight ratio is determined as:

$$\frac{W_{i+1}}{W_i} = e^{\frac{-RC}{\eta_P (L/D)_{max}}} \tag{4.18}$$

The definition and typical values for the variable C are presented in Section 4.2.5.5. In this equation, all parameters except C are without unit. Since the unit of range is in terms of length (such as m, km, ft, nm), the unit of C must be converted into the reciprocal of length (such as 1/m, 1/km, 1/ft, 1/nm). Recall that the unit of C is initially lb/(h lb) or N/(h N).

4.2.5.3 Loiter Weight Fraction for Jet Aircraft

The aircraft performance criterion loiter is measured with a parameter called the endurance. In order to determine the fuel fraction for loitering flight, the equation for endurance is used. Endurance (E) is the length of time that an aircraft can remain airborne for a given expenditure of fuel and for a specified set of flight conditions. For some aircraft (such as reconnaissance, surveillance, and border monitoring), the most important performance parameter of their mission is to be airborne as long as possible. Several technical aspects of endurance and range are similar. The only difference is to consider how long (time) the aircraft can fly rather than how far (distance) it can travel. The objective for this flight is to minimize the fuel consumption, because the aircraft has limited fuel. A *loiter* is a flight condition where the endurance is its primary objective. For more information and its derivation, the reader is encouraged to consult Ref. [5]. The endurance equation [5] for a jet aircraft is:

$$E_{max} = \frac{(L/D)_{max}}{C} \ln\left(\frac{W_i}{W_{i+1}}\right) \tag{4.19}$$

Therefore, the fuel weight ratio for a loitering flight is determined as:

$$\frac{W_{i+1}}{W_i} = e^{\frac{-EC}{(L/D)_{max}}} \tag{4.20}$$

The definition and typical values for the variable C are presented in Section 4.2.5.5. Since the unit of E is in terms of time (such as s, h), the unit of C must be converted into the reciprocal of time (such as 1/s, 1/h). Recall that the unit of C is initially lb/(h lb) or N/(h N).

4.2.5.4 Loiter Weight Fraction for Prop-Driven Aircraft

The definition and flight approaches to satisfy a trimmed operation for a specified loiter are discussed in Section 4.5.2.3. Since the type of propulsion system is prop-driven, the engine is generating power, and the propeller efficiency influences the overall thrust. As in the case of a jet aircraft, there are three flight approaches to hold the aircraft trim despite the loss of weight due to fuel burn. For the sake of simplicity and due to the expected accuracy in the preliminary design phase, we select only the case where the flight speed is decreasing (i.e., constant altitude/constant lift coefficient flight). If the design requirements specify that the aircraft must have a different approach, one needs to employ the relevant endurance equation.

For an aircraft with a prop-driven engine (i.e., piston-prop or turboprop), the optimum endurance will be achieved when the aircraft is flying with the minimum drag speed. Thus the range equation [5] is:

$$E_{\max} = \frac{(L/D)_{E_{\max}} \eta_P}{CV_{E_{\max}}} \ln\left(\frac{W_i}{W_{i+1}}\right) \quad (4.21)$$

For a prop-driven aircraft, the endurance will be maximized when the ratio $\left(C_L^{3/2} / C_D\right)$ is at its maximum. In other words:

$$(L/D)_{E_{\max}} = 0.866\,(L/D)_{\max} \quad (4.22)$$

Then:

$$E_{\max} = \frac{0.866\,(L/D)_{\max}\,\eta_P}{CV_{E_{\max}}} \ln\left(\frac{W_i}{W_{i+1}}\right) \quad (4.23)$$

Therefore, for a prop-driven aircraft, the fuel weight fraction for a loitering flight is determined as:

$$\frac{W_{i+1}}{W_i} = e^{\frac{-ECV_{E_{\max}}}{0.866\eta_P\,(L/D)_{\max}}} \quad (4.24)$$

The speed for maximum endurance ($V_{E_{\max}}$) for a prop-driven aircraft [5] happens when the aircraft is flying with the minimum power speed (i.e.,$V_{P_{\min}}$). Since the aircraft has not yet been fully designed at the preliminary design phase, the calculation of minimum power speed cannot be implemented. Hence, the recommendation is to use a reasonable approximation. The minimum power speed for most prop-driven aircraft is about 20–40% higher than the stall speed. Then:

$$V_{E_{\max}} = V_{P_{\min}} \approx 1.2V_s - 1.4V_s \quad (4.25)$$

The definition and typical values for the variable C are presented in Section 4.2.5.5. Since the unit of E is in terms of time (such as s, h), and the unit of speed in distance per time, the unit of C must be converted into the reciprocal of distance (such as 1/m, 1/ft). Recall that the unit of C is initially lb/(h lb) or N/(h N).

4.2.5.5 Specific Fuel Consumption

The remaining unknown in the range and endurance relationships (Equations (4.16), (4.18), (4.20), and (4.24)) is C or the SFC. The SFC is a technical figure of merit for an engine that indicates how efficiently the engine is burning fuel and converting it to thrust. SFC depends on the type and design technology of the engine and also the type of fuel. SFC is used to describe the fuel efficiency of an engine with respect to its mechanical output.

Various grades of fuel have evolved during the development of jet engines in an effort to ensure both satisfactory performance and adequate supply. JP-8 is the most commonly used fuel for US Air Force jet aircraft. The US Navy uses JP-5, a denser, less volatile fuel than JP-8, which allows it to be safely stored in the skin tanks of ships. The most common commercial aircraft fuel is Jet A and Jet A-1. In general, piston engine fuels are about 10% lighter than jet fuels.

The SFC for jet engines (turbojet and turbofan) is defined as the weight (sometimes mass) of fuel needed to provide a given thrust for a given period (e.g., lb/h/lb or g/s/N in SI units). In propeller-driven engines (piston, turboprop, and turboshaft), SFC measures the mass of fuel needed to provide a given thrust or power for a given period. The common unit of measure in British units is lb/hp/h (i.e., lb/(hp h)); that is, pounds of fuel consumed for every unit of horsepower generated during 1 hour of operation (or kg/kW/h in SI units). Therefore, a lower number indicates better efficiency.

The unit of C can be converted readily between SI and British units. For instance, a typical piston engine has a SFC of about 0.5 lb/hp/h (or 0.3 kg/kW/h or 83 g/MJ), regardless of the design of any particular engine. As an example, if a piston engine consumes 400 lb of fuel to produce 200 hp for 4 hours, its SFC will be as follows:

$$\mathrm{SFC} = \frac{400\,\mathrm{lb}}{4\,\mathrm{h}\cdot 200\,\mathrm{hp}} = 0.5\,\frac{\mathrm{lb}}{\mathrm{h\,hp}} = 2.98\,\frac{\mathrm{N}}{\mathrm{h\,kW}}$$

Table 4.6 shows typical values of SFC for various engines. It is very important to use consistent units in the range and endurance equations. In general, the unit of C in the range equation must be 1 over the time unit (e.g., 1/s). If SI units are used (e.g., km/h for cruising speed), the unit of C must be 1/h. If British units are utilized (e.g., ft/s for the cruising speed), the unit of C must be 1/s; moreover, the unit of C in the endurance equation must be 1 over the unit of distance (e.g., 1/m or 1/ft). The following two examples demonstrate how to convert the unit of lb/hp h to 1/ft, and convert the unit of lb/h lb to 1/s. Recall that 1 hp is equivalent to 550 lb ft/s, and 1 hour contains 3600 seconds.

$$\mathrm{SFC} = 0.5\,\frac{\mathrm{lb}}{\mathrm{h\,hp}} = 0.5\,\frac{\mathrm{lb}}{(3600\,\mathrm{s})\cdot\left(550\,\frac{\mathrm{lb\,ft}}{\mathrm{s}}\right)}$$

$$= \frac{0.5}{3600\cdot 550}\,\frac{1}{\mathrm{ft}} = \frac{0.5}{1\,980\,000}\,\frac{1}{\mathrm{ft}} = 2.52\cdot 10^{-7}\,\frac{1}{\mathrm{ft}}$$

$$\mathrm{SFC} = 0.7\,\frac{\mathrm{lb}}{\mathrm{h\,lb}} = 0.7\,\frac{1}{3600\,\mathrm{s}} = 0.000194\frac{1}{\mathrm{s}}$$

4.2.6 Empty Weight

The last term in determining the MTOW in Equation (4.5) is the empty weight fraction $\left(\frac{W_E}{W_{TO}}\right)$. At this moment (preliminary design phase), the aircraft has been designed only conceptually, hence there is no geometry or sizing. Therefore, the empty weight fraction cannot be calculated analytically. The only way is to use past history and statistics.

Table 4.6 Typical values of SFC for various engines

No.	Engine type	SFC in cruise	SFC in loiter	Unit (British units)
1	Turbojet	0.9	0.8	lb/h/lb
2	Low-bypass ratio turbofan	0.7	0.8	lb/h/lb
3	High-bypass ratio turbofan	0.4	0.5	lb/h/lb
4	Turboprop	0.5–0.8	0.6–0.8	lb/h/hp
5	Piston (fixed pitch)	0.4–0.8	0.5–0.7	lb/h/hp
6	Piston (variable pitch)	0.4–0.8	0.4–0.7	lb/h/hp

Table 4.7 shows the empty weight fraction for several aircraft. The only known information about the aircraft is the configuration and aircraft type based on the mission. According to this data, the author has developed a series of empirical equations to determine the empty weight fraction. The equations are based on the published data taken from Ref. [4] and other sources. In general, the empty weight fraction varies from about 0.2 to about 0.75. Figure 4.3 shows the human-powered aircraft Daedalus with an empty weight-to-take-off weight ratio of 0.3.

$$\frac{W_E}{W_{TO}} = aW_{TO} + b \tag{4.26}$$

where a and b are found in Table 4.8. Note that Equation (4.26) is curve-fitted in the British units system. Thus, the unit for MTOW and empty weight is lb. Table 4.8 illustrates statistical curve-fit values for the trends demonstrated in aircraft data as shown in Table 4.7. Note that the unit of W_{TO} in Table 4.8 is lb. This is included due to the fact that all data in FAR publications are in British units.

In Table 4.8, the assumption is that either the entire aircraft structure or the majority of aircraft components are made up of aluminum. The preceding take-off weight calculations have thus implicitly assumed that the new aircraft would also be constructed of aluminum. In case the aircraft is expected to be made up of composite material, the value of $\frac{W_E}{W_{TO}}$ must be multiplied by 0.9. The values for GA aircraft in Table 4.8 are for normal aircraft. If a GA aircraft is of utility type, the value of $\frac{W_E}{W_{TO}}$ must be multiplied by 1.03. If a GA aircraft is of acrobatic type, the value of $\frac{W_E}{W_{TO}}$ must be multiplied by 1.06.

Figure 4.4(a) illustrates the British fighter aircraft Aerospace Harrier GR9 with a thrust-to-weight ratio of 1.13, and Figure 4.4(b) the transport aircraft Antonov An-124 with a thrust-to-weight ratio of 0.231. Figure 4.5 shows the non-conventional composite aircraft Voyager with an empty weight-to-take-off weight ratio of 0.23, while Figure 4.6 demonstrates the fighter aircraft General Dynamics F-16C Fighting Falcon with an empty weight-to-take-off weight ratio of 0.69.

Table 4.7 Empty weight fraction for several aircraft [4]

No.	Aircraft	Type	Engine	S (m^2)	m_{TO} (kg)	m_E (kg)	$\frac{W_E}{W_{TO}}$
1	Voyager	Circumnavigation	piston	30.1	4 398	1 020	0.23
2	Questair Spirit	Sport home-built	Piston	6.74	771	465	0.6
3	SkystarKitfox V	Kit-built	Piston	12.16	544	216	0.397
4	Beech Bonanza A36	Utility	Piston	16.8	1 655	1 047	0.63
5	Air & Space 20A	Autogyro	Piston	11.33[a]	907	615	0.68
6	Stemme S10	Motor glider	Piston	18.7	850	640	0.75
7	BN2B Islander	Multi-role transport	Turboprop	30.19	2 993	1 866	0.62
8	C-130H Hercules	Tactical transport	Turboprop	162.12	70 305	34 686	0.493
9	Saab 2000	Regional transport	Turboprop	55.74	22 800	13 800	0.605
10	ATR 42	Regional transport	Turboprop	54.5	16 700	10 285	0.616
11	Air Tractor AT-602	Agricultural	Turboprop	31.22	5 443	2 471	0.454
12	Cessna 750	Business jet	Turbofan	48.96	16 011	8 341	0.52
13	Gulfstream V	Business jet	Turbofan	105.63	40 370	21 228	0.523
14	Falcon 2000	Business transport	Turbofan	49.02	16 238	9 405	0.58
15	Airbus A340	Wide-bodied airliner	Turbofan	363.1	257 000	123 085	0.48
16	MD-90	Airliner	Turbofan	112.3	70 760	39 916	0.564
17	Beechjet	Military trainer	Turbofan	22.43	7 303	4 819	0.66
18	Boeing 777-300	Wide-bodied airliner	Turbofan	427.8	299 370	157 215	0.525
19	Airbus 380-841	Wide-bodied airliner	Turbofan	845	590 000	270 015	0.485
20	BAe Sea Harrier	Fighter and attack	Turbofan	18.68	11 880	6 374	0.536
21	F-16C Falcon	Fighter	Turbofan	27.87	12 331	8 273	0.67
22	Eurofighter 2000	Fighter	Turbofan	50	21 000	9 750	0.46
23	Volmer VJ-25 Sunfun	Hang glider/kite	No engine	15.14	140.5	50	0.35
24	Manta Fledge III	Sailplane/glider	No engine	14.95	133	33	0.25
25	MIT Daedalus 88	Man-powered	Prop-human	29.98	104	32	0.307
26	Global Hawk	Unmanned	Turbofan	50.2	10 387	3 851	0.371

[a] The value is for the area of rotor disk.

Figure 4.3 Human-powered aircraft Daedalus. Reproduced from permission of NASA

Table 4.8 The coefficients *a* and *b* for the empirical Equation (4.26)

No.	Aircraft	*a*	*b*
1	Hang glider	$6.53 \cdot 10^{-3}$	−1.663
2	Man-powered	$-1.05 \cdot 10^{-5}$	0.31
3	Glider/sailplane	$-2.3 \cdot 10^{-4}$	0.59
4	Motor-glider	$-1.95 \cdot 10^{-4}$	1.12
5	Microlight	$-7.22 \cdot 10^{-5}$	0.481
6	Home-built	$-4.6 \cdot 10^{-5}$	0.68
7	Agricultural	$3.36 \cdot 10^{-4}$	−3.57
8	GA-single engine	$1.543 \cdot 10^{-5}$	0.57
9	GA-twin engine	$2.73 \cdot 10^{-4}$	−9.08
10	Twin turboprop	$-8.2 \cdot 10^{-7}$	0.65
11	Jet trainer	$1.39 \cdot 10^{-6}$	0.64
12	Jet transport	$-7.754 \cdot 10^{-8}$	0.576
13	Business jet	$1.13 \cdot 10^{-6}$	0.48
14	Fighter	$-1.1 \cdot 10^{-5}$	0.97
15	Long range, long endurance	$-1.21 \cdot 10^{-5}$	0.95

(a) (b)

Figure 4.4 (a) British Aerospace Harrier GR9 with a thrust-to-weight ratio of 1.13 and (b) Antonov An-124 with a thrust-to-weight ratio of 0.231. Reproduced from permission of Antony Osborne

Figure 4.5 Voyager aircraft with an empty weight-to-take-off weight ratio of 0.23. Reproduced from permission of NASA

4.2.7 Practical Steps of the Technique

The technique to determine the aircraft MTOW has 11 steps, as follows:

- **Step 1**. Establish the flight mission profile and identify the mission segments (similar to Figure 4.2).
- **Step 2**. Determine the number of flight crew members.
- **Step 3**. Determine the number of flight attendants.
- **Step 4**. Determine the overall weight of flight crew and flight attendants and also flight crew and attendants weight ratio.

Figure 4.6 F-16 Falcon with an empty weight-to-take-off weight ratio of 0.69. Reproduced from permission of Antony Osborne

- **Step 5**. Determine the overall weight of payloads (i.e., passengers, luggage, bag, cargo, store, loads, etc.).
- **Step 6**. Determine fuel weight ratios for the segments of taxi, take-off, climb, descent, approach, and landing (use Table 4.3).
- **Step 7**. Determine fuel weight ratios for the segments of range and loiter using equations introduced in Section 4.2.5.
- **Step 8**. Find the overall fuel weight ratio using equations similar to Equations (4.10) and (4.11).
- **Step 9**. Substitute the value of overall fuel weight ratio into Equation (4.5).
- **Step 10**. Establish the empty weight ratio by using Equation (4.26) and Table 4.8.
- **Step 11**. Finally, Equation (4.5) (derived in step 9) and Equation (4.26) (derived in step 10) must be solved simultaneously to find the two unknowns of W_{TO} and $\frac{W_E}{W_{TO}}$.
  The primary unknown that we are looking for is W_{TO}, which is the aircraft MTOW. These two equations form a set of non-linear algebraic equations and may be solved by employing an engineering software package such as MathCad[1] or MATLAB.[2] If you do not have access to such software packages, a trial-and-error technique can be employed to solve the equations.

A fully solved example in Section 4.4 demonstrates the application of the technique to estimate the aircraft MTOW.

4.3 Wing Area and Engine Sizing

4.3.1 Summary of the Technique

In the first step of the aircraft preliminary design phase, the aircraft's most fundamental parameter (i.e., aircraft MTOW, W_{TO}) is determined. The technique was introduced in

[1] Mathcad is a registered trademark of Mathsoft, Inc.
[2] MATLAB is a registered trademark of Mathworks, Inc.

Section 4.2. The second crucial step in the aircraft preliminary design phase is to determine the wing reference area (S_{ref}) plus engine thrust (T). However, if the aircraft propulsion system has been chosen to be prop-driven, the engine power will be determined. Hence, two major outputs of this design step are:

1. wing reference area (S or S_{ref});
2. engine thrust (T) or engine power (P).

Unlike the first step in the preliminary design phase during which the main reference was statistics, this phase is solely dependent upon the aircraft performance requirements and employs flight mechanics theories. Hence, the technique is an analytical approach and the results are highly reliable without inaccuracy. The aircraft performance requirements utilized to size the aircraft in this step are:

- stall speed (V_s);
- maximum speed (V_{max});
- maximum rate of climb (ROC_{max});
- take-off run (S_{TO});
- ceiling (h_c);
- turn requirements (turn radius and turn rate).

Recall that the following aircraft performance requirements have been used to determine aircraft MTOW (Section 4.2):

- range (R);
- endurance (E).

Hence, they will not be utilized again in this technique. There are a few aircraft parameters (such as aircraft maximum lift coefficient) which may be needed throughout the technique, but they have not been calculated analytically prior to this preliminary design phase. These parameters will currently be estimated based on the statistics that will be provided in this section. However, in the later design phase, where their exact values are determined, these calculations will be repeated to correct the inaccuracies. References [5–7] introduce techniques of aircraft performance analysis.

In this section, three new parameters appear in almost all equations. So, we need to define them first:

1. **Wing loading**. The ratio between aircraft weight and wing area is referred to as wing loading and represented by W/S. This parameter indicates how much load (i.e., weight) is held by each unit area of the wing.
2. **Thrust-to-weight ratio**. The ratio between aircraft engine thrust and aircraft weight is referred to as thrust loading and is represented by T/W. This parameter indicates how heavy the aircraft is with respect to engine thrust. The term thrust-to-weight ratio is associated with jet aircraft (e.g., turbofan or turbojet engines). Reference [8] refers to this term as the thrust loading. Although this designated name is convenient to use, it does not seem to fit very well with the concept related to thrust and weight. However, W/T seems to be a more convenient symbol to refer to as the thrust loading; which means how much weight is carried by each unit of thrust.

3. **Power loading**. The ratio between aircraft weight and engine power is referred to as power loading and is represented by W/P. This parameter indicates how heavy the aircraft is with respect to engine power. A better name for this parameter is weight-to-power ratio. This term is associated with propeller-driven aircraft (turboprop or piston engines).

Table 4.9 illustrates wing loading, power loading, and thrust loading for several aircraft. In general, two desired parameters (S and T (or P)) are determined in ***six*** steps. The following are the steps to determine the wing area and engine power for a prop-driven aircraft. If the aircraft is jet-driven, substitute the words thrust loading instead of power loading. The principles and steps of the technique are similar for both types of aircraft.

1. Derive one equation for each aircraft performance requirement (e.g., V_s, V_{max}, ROC, S_{TO}, h_c, R_{turn}, ω_{turn}). If the aircraft is prop-driven, the equations are in the form of W/P as functions of W/S, as follows:

$$\left(\frac{W}{P}\right)_{V_s} = f_1\left(\frac{W}{S}, V_s\right) \tag{4.27a}$$

$$\left(\frac{W}{P}\right)_{V_{max}} = f_2\left(\frac{W}{S}, V_{max}\right) \tag{4.27b}$$

$$\left(\frac{W}{P}\right)_{S_{TO}} = f_3\left(\frac{W}{S}, S_{TO}\right) \tag{4.27c}$$

$$\left(\frac{W}{P}\right)_{ROC} = f_4\left(\frac{W}{S}, \mathrm{ROC}\right) \tag{4.27d}$$

$$\left(\frac{W}{P}\right)_{ceiling} = f_5\left(\frac{W}{S}, h_c\right) \tag{4.27e}$$

$$\left(\frac{W}{P}\right)_{turn} = f_6\left(\frac{W}{S}, R_{turn}, \omega_{turn}\right) \tag{4.27f}$$

However, if the aircraft is jet-driven, the equations are in the form of T/W as functions of W/S. The details of the derivation are presented in the next sections.
2. Sketch all derived equations in one plot. The vertical axis is power loading (W/P) and the horizontal axis is wing loading (W/S). Thus, the plot illustrates the variations of power loading with respect to wing loading. These graphs will intersect each other at several points and may produce several regions.
3. Identify the acceptable region inside the regions that are produced by the axes and the graphs. The acceptable region is the region that meets all aircraft performance requirements. A typical diagram is shown in Figure 4.7. The acceptable region is recognized by the fact that as a performance variable (say V_{max}) is varied inside the permissible limits, the power loading must behave to confirm that trend. For instance, consider the graph of power loading versus wing loading for maximum speed. Assume that the power loading is inversely proportional to the maximum speed. Now, if the maximum speed is increased – which is inside the permissible limits – the power loading

Table 4.9 Wing loading, power loading, and thrust loading (in British units) of several aircraft

1. Prop-driven aircraft							
No.	Aircraft	Type	W_{TO} (lb)	S (ft^2)	P (hp)	W/S (lb/ft^2)	W/P (lb/hp)
1	C-130 Hercules	Large transport	155 000	1 754	4 · 4508	88.37	8.59
2	Beech Bonanza	Utility-piston prop	2 725	178	285	15.3	9.5
3	Gomolzig RF-9	Motor glider	1 642	193.7	80	8.5	20.5
4	Piaggio P180 Avanti	Transport	10 510	172.2	2 · 800	61	6.5
5	Canadair CL-215T	Amphibian	43 500	1 080	2 · 2100	40.3	10.3
6	Socata TB30 Epsilon	Military trainer	2 756	97	300	28.4	9.2
7	DHC-8 Dash 8-100	Short-range transport	34 500	585	2 · 2000	59	8.6
8	Beechcraft King Air 350	Utility twin turboprop	15 000	310	2 · 1050	48.4	7.14

2. Jet aircraft							
No.	Aircraft	Type	W_T (lb)	S (ft^2)	T (lp)	W/S (lb/ft^2)	T/W (lb/lb)
1	Paragon spirit	Business jet	5 500	140	1 900	39.3	0.345
2	Cessna 650 Citation VII	Business jet	22 450	312	2 · 4080	71.9	0.36
3	F-15 Eagle	Fighter	81 000	608	2 · 23450	133.2	0.58
4	Lockheed C-5 Galaxy	Transport	840 000	6 200	4 · 43000	135.5	0.205
5	Boeing 747-400	Airliner	800 000	5 825	4 · 56750	137.3	0.28
6	F-5A Freedom Fighter	Fighter	24 700	186	2 · 3500	132.3	0.283
7	AV-8B Harrier II	VTOL fighter	20 750	243.4	23 500	85.2	1.133
8	F-16C Falcon	Fighter	27 185	300	29 588	90.6	1.09
9	B-2 Spirit	Bomber	336 500	5 000	4 · 17300	67.3	0.206
10	Eurofighter	Fighter	46 297	538	2 · 16000	86	0.691
11	Embraer EMB 190	Regional jet	105 359	996	2 · 14200	195.8	0.27

is decreased. So the reduction in power loading is acceptable. Hence, the lower region of this particular graph is acceptable.

4. Determine the design point (i.e., the optimum selection). The design point on the plot is only one point that yields the smallest engine in terms of power (i.e., the lowest cost). For the case of the jet aircraft, the design point yields an engine with the smallest thrust.
5. From the design point, obtain two numbers: corresponding wing loading $(W/S)_d$ and corresponding power loading $(W/P)_d$. A typical graphical technique is illustrated in Figure 4.7 (for prop-driven aircraft) and Figure 4.8 (for jet aircraft). For the case of the jet aircraft, read corresponding thrust loading $(T/W)_d$.
6. Calculate the wing area and engine power from these two values, since the aircraft MTOW (W_{TO}) has been determined previously. The wing area is calculated by dividing the aircraft take-off weight by the wing loading. The engine power is also calculated by dividing the aircraft take-off weight by the power loading:

$$S = W_{TO} \Big/ \left(\frac{W}{S}\right)_d \tag{4.28}$$

$$P = W_{TO} \Big/ \left(\frac{W}{P}\right)_d \tag{4.29}$$

While, in the case of a jet aircraft, the engine thrust is calculated by multiplying the aircraft take-off weight by the thrust loading:

$$T = W_{TO} \cdot \left(\frac{T}{W}\right)_d \tag{4.28a}$$

The principles of the technique were originally introduced in a NASA technical report [9] and later developed by Roskam [8]. The technique is further developed by the author

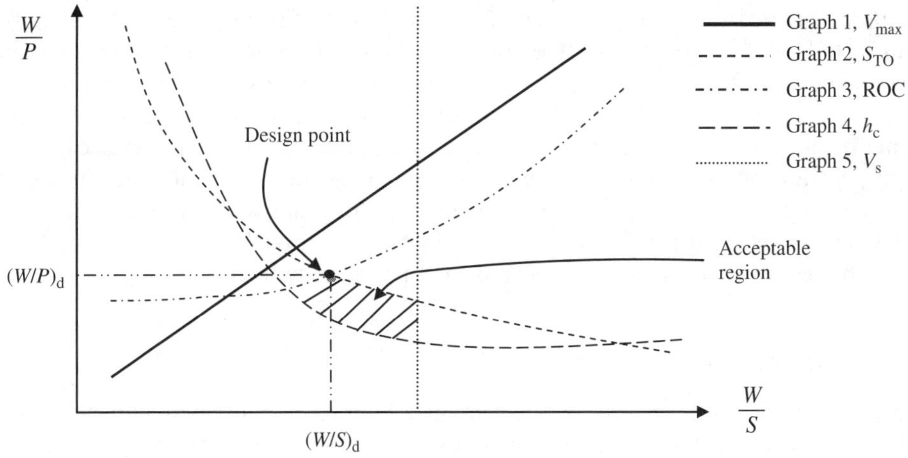

Figure 4.7 Matching plot for a prop-driven aircraft

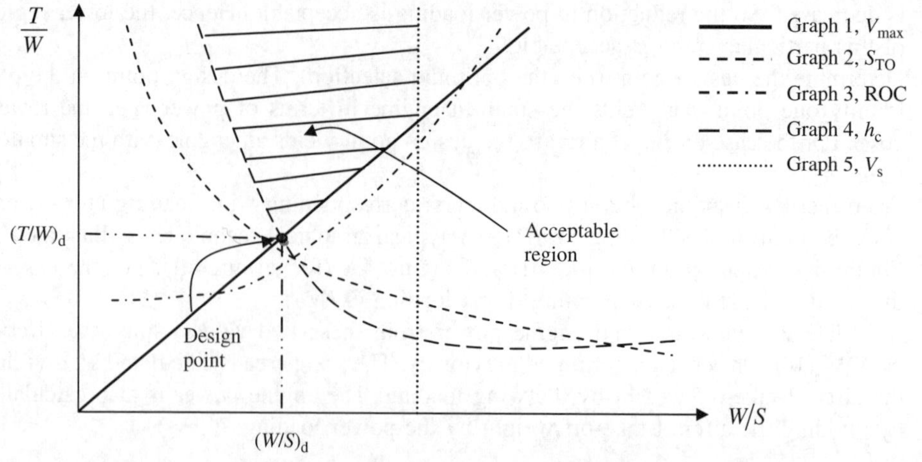

Figure 4.8 Matching plot for a jet aircraft

in this section. This graph that contains several performance charts is sometimes referred to as the matching plot, matching chart, or matching diagram.

It must be mentioned that there is an analytical solution to this set of equations. One can write a computer program and apply all limits and inequalities. The results will be the values for two required unknowns (S and T (or P)). Extreme caution must be taken to use consistent units in the application process. If British units are used, convert the unit of W/P to lb/hp to make the comparison more convenient. Since in some of the equations W/S is on the denominator, do not begin the horizontal axis of the plot from zero. This is to avoid the value of W/P going toward infinity. So, it is suggested to have values of W/S from, say, 5 lb/ft^2 to 100 lb/ft^2 (in British units).

If the performance requirements are completely consistent, the acceptable region would be only one point; that is, the design point. However, as the performance requirements are more scattered, the acceptable region becomes wider. For instance, if the aircraft is required to have a rate of climb (ROC) of 10 000 fpm, but the absolute ceiling is required to be only 15 000 ft, this is assumed to be a group of non-consistent design requirements. The reason is that a 10 000 fpm ROC requires a powerful engine, but an 15 000 ft absolute ceiling requires a low-thrust engine. It is clear that a powerful engine is easily capable of satisfying a low-altitude absolute ceiling. This type of performance requirement makes the acceptable region in the matching chart a wide one. An example application is presented in Section 4.4. Now, the derivations of equations for each performance requirement are carried out using the mathematical methods and practical methods.

4.3.2 Stall Speed

One of the aircraft performance requirements is a limit to the minimum allowable speed. Only helicopters and VTOL aircraft (or rotary wing aircraft) are able to fly (i.e., hover) with a zero forward airspeed. The other conventional (i.e., fixed-wing) aircraft need to have a minimum airspeed in order to be airborne. For most aircraft the mission demands a stall speed not higher than some minimum value. In such a case, the mission specification includes a requirement for a minimum speed. From the lift equation, as the aircraft speed is decreased, the aircraft lift coefficient must be increased, until the aircraft stalls. Hence, the minimum speed that an aircraft can fly with is referred to as the stall speed (V_s).

An aircraft must be longitudinally trimmed at any cruising flight condition including at any flight speed. The range of acceptable speeds is between the stall speed and the maximum speed. In a cruising flight with the stall speed, the aircraft weight must be balanced with the lift (L):

$$L = W = \frac{1}{2}\rho V_s^2 S C_{L_{max}} \tag{4.30}$$

where ρ denotes the air density at the specified altitude, and $C_{L_{max}}$ is the aircraft maximum lift coefficient. From Equation (4.30) we can derive the following when dividing both sides by S:

$$\left(\frac{W}{S}\right)_{V_s} = \frac{1}{2}\rho V_s^2 C_{L_{max}} \tag{4.31}$$

This provides the first graph in the matching plot. The wing sizing based on stall speed requirements is represented by Equation (4.31) as the variations of wing loading versus stall speed. As can be seen from Equation (4.31), neither power loading (W/P) nor thrust loading (T/W) makes a contribution to wing loading in this case. In other words, the wing loading to satisfy the stall speed requirements is not a function of power loading or thrust loading. Therefore, the graph of power loading or thrust loading versus wing loading is always a vertical line in a matching plot as sketched in Figure 4.9.

In general, a low stall speed is desirable, since the lower stall speed results in a safer aircraft. When an unfortunate aircraft crash happens, a lower stall speed normally causes less damage and fewer casualties. In contrast, a lower stall speed results in a safer take-off and a safer landing, since an aircraft at a lower take-off and landing speed is more controllable. This is due to the fact that the take-off speed and the landing speed are often slightly higher than the stall speed (normally 1.1–1.3 times stall speed). Hence, in theory, any stall speed less than the stall speed specified by the mission requirements is acceptable. Therefore, the left side of the graph in Figure 4.8 or Figure 4.9 is an

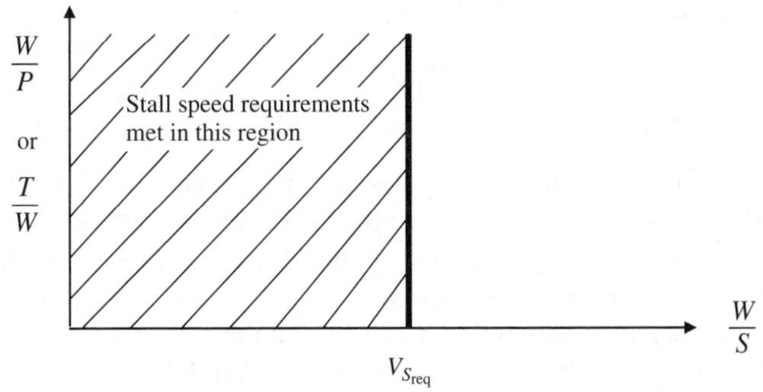

Figure 4.9 Stall speed contribution in constructing a matching plot

acceptable region and on the right side, the stall requirement is not met. So, by specifying a maximum allowable stall speed, Equation (4.31) provides a maximum allowable wing loading for a given value of $C_{L_{max}}$.

Based on FAR Part 23, a single-engine aircraft and also a multi-engine aircraft with a MTOW of less than 6000 lb may not have a stall speed greater than 61 knot. A very light aircraft (VLA) that is certified with EASA[3] may not have a stall speed greater than 45 knot:

$$V_s \leq 61 \text{ knot} \quad \text{(FAR 23)} \tag{4.32}$$

$$V_s \leq 45 \text{ knot} \quad \text{(EASA CS-VLA)} \tag{4.33}$$

There are no maximum stall speed requirements for transport aircraft that are certified by FAR Part 25. It is clear that the stall speed requirements can be met with flap-up configuration, since flap deflection allows for a higher lift coefficient and thus lower stall speed. An example application is presented in Section 4.4.

Equation (4.31) has two unknowns (ρ and $C_{L_{max}}$) which often are not provided by the customer, so must be determined by the aircraft designer. The air density must be chosen to be at sea level ($\rho = 1.225\,\text{kg/m}^3$, or $0.002378\ \text{slug/ft}^3$), since it provides the highest air density, which results in the lowest stall speed. This selection helps to satisfy more closely the stall speed requirement.

The aircraft maximum lift coefficient is mainly a function of wing and airfoil design, and also high-lift device. The wing and airfoil design, and also the high-lift device selection, are discussed in Chapter 5. At this moment of the design phase (preliminary design) – where the wing has not been designed yet and the high-lift device has not yet been finalized – it is recommended to select a reasonable value for the maximum lift coefficient. Table 4.10 presents the maximum lift coefficient for several aircraft. This table also provides the aircraft stall speed for your information. If the stall speed is not given by the aircraft customer, use this table as a useful reference. This selection must be honored in the wing design (Chapter 5), hence, select a reasonable maximum lift coefficient. Table 4.11 presents typical values of maximum lift coefficient and stall speed for different types of aircraft.

Employ extreme caution in using the units for variables. In the SI system, the unit of W is N, the unit of S is m^2, the unit of V_s is m/s, and the unit of ρ is kg/m^3. However, in the British system, the unit of W is lb, the unit of S is ft^2, the unit of V_s is ft/s, and the unit of ρ is slug/ft^3.

4.3.3 Maximum Speed

Another very important performance requirement, particularly for fighter aircraft, is the maximum speed. Two major contributors, other than aircraft weight, to the satisfaction of this requirement are the wing area and engine thrust (or power). In this section, the relevant equations are derived for the application in the matching plot. The derivations are presented in two separate subsections; one subsection for jet aircraft (Section 4.3.3.1) and another subsection for prop-driven aircraft (Section 4.3.3.2).

[3] European Aviation Safety Agency.

Table 4.10 Maximum lift coefficient for several aircraft [4]

No.	Aircraft	Type	m_{TO} (kg)	S (m^2)	V_s (knot)	$C_{L_{max}}$
1	Volmer VJ-25 Sunfun	Hang glider/kite	140.5	15.14	13	3.3
2	Manta Fledge III	Sailplane/glider	133	14.95	15	2.4
3	Euro Wing Zephyr II	Microlight	340	15.33	25	2.15
4	Campana AN4	Very light	540	14.31	34	1.97
5	Jurca MJ5 Sirocco	GA two seat	760	10	59	1.32
6	Piper Cherokee	GA single engine	975	15.14	47.3	1.74
7	Cessna 208-L	GA single turboprop	3 629	25.96	61	2.27
8	Short Skyvan 3	Twin turboprop	5 670	35.12	60	2.71
9	Gulfstream II	Business twin jet	29 700	75.2	115	1.8
10	Learjet 25	Business twin jet	6 800	21.5	104	1.77
11	Hawkeye E-2C	Early warning	24 687	65.03	92	2.7
12	DC-9-50	Jet airliner	54 900	86.8	126	2.4
13	Boeing 727-200	Jet airliner	95 000	153.3	117	2.75
14	Airbus 300	Jet airliner	165 000	260	113	3
15	F-14 Tomcat	Fighter	33 720	54.5	110	3.1

Table 4.11 Typical values of maximum lift coefficient and stall speed for different types of aircraft

No.	Aircraft type	$C_{L_{max}}$	V_s (knot)
1	Hang glider/kite	2.5–3.5	10–15
2	Sailplane/glider	1.8–2.5	12–25
3	Microlight	1.8–2.4	20–30
4	Very light	1.6–2.2	30–45
5	GA light	1.6–2.2	40–61
6	Agricultural	1.5–2	45–61
7	Home-built	1.2–1.8	40–70
8	Business jet	1.6–2.6	70–120
9	Jet transport	2.2–3.2	95–130
10	Supersonic fighter	1.8–3.2	100–120

4.3.3.1 Jet Aircraft

Consider a jet aircraft which is flying with the maximum constant speed at a specified altitude (ρ_{alt}). The aircraft is in longitudinal trim; hence, the maximum engine thrust (T_{max}) must be equal to the maximum aircraft drag (D_{max}) and the aircraft weight (W) must be equal to the lift (L):

$$T_{\text{max}} = D_{\text{max}} \tag{4.34}$$

$$W = L \tag{4.35}$$

where lift and drag are two aerodynamic forces and are defined as:

$$D = \frac{1}{2}\rho V_{\text{max}}^2 S C_D \tag{4.36}$$

$$L = \frac{1}{2}\rho V_{\text{max}}^2 S C_L \tag{4.37}$$

On the other hand, the engine thrust is decreasing with increasing aircraft altitude. This requires knowledge of how the engine thrust of an aircraft varies with airspeed and altitude. A general relationship between engine thrust and altitude, which is represented by air density (ρ), is:

$$T_{\text{alt}} = T_{\text{SL}}\left(\frac{\rho}{\rho_{\text{o}}}\right) = T_{\text{SL}}\sigma \tag{4.38}$$

where ρ_{o} is the sea-level air density, T_{alt} is the engine thrust at altitude, and T_{SL} is the engine thrust at sea level. By substituting Equations (4.36) and (4.38) into Equation (4.34), we have:

$$T_{\text{SL}}\sigma = \frac{1}{2}\rho V_{\text{max}}^2 S C_D \tag{4.39}$$

The aircraft drag coefficient has two contributors, the zero-lift drag coefficient ($C_{D_{\text{o}}}$) and the induced drag coefficient ($C_{D_{\text{i}}}$):

$$C_D = C_{D_{\text{o}}} + C_{D_{\text{i}}} = C_{D_{\text{o}}} + K \cdot C_L^2 \tag{4.40}$$

where K is referred to as the induced drag factor and is determined by:

$$K = \frac{1}{\pi \cdot e \cdot \text{AR}} \tag{4.41}$$

Typical values for e (Oswald span efficiency factor) are between 0.7 and 0.95. The typical values for wing aspect ratio (AR) are given in Table 5.8. Substitution of Equation (4.40) into Equation (4.39) yields:

$$T_{\text{SL}}\sigma = \frac{1}{2}\rho V_{\text{max}}^2 S\left(C_{D_{\text{o}}} + K \cdot C_L^2\right) \tag{4.42}$$

From Equations (4.35) and (4.37), the aircraft lift coefficient can be derived as:

$$C_L = \frac{2W}{\rho V_{\text{max}}^2 S} \tag{4.43}$$

Substituting this lift coefficient into Equation (4.42) provides:

$$T_{SL}\sigma = \frac{1}{2}\rho V_{max}^2 S\left(C_{D_o} + K \cdot \left[\frac{2W}{\rho V_{max}^2 S}\right]^2\right) \tag{4.44}$$

Now, we can simplify this equation as:

$$T_{SL}\sigma = \frac{1}{2}\rho V_{max}^2 S C_{D_o} + \frac{1}{2}\rho V_{max}^2 S \frac{K\,(2W)^2}{\left(\rho V_{max}^2 S\right)^2} = \frac{1}{2}\rho V_{max}^2 S C_{D_o} + \frac{2KW^2}{\rho V_{max}^2 S} \tag{4.45}$$

Both sides of this equation can be divided by aircraft weight (W) as:

$$\frac{T_{SL}}{W}\sigma = \frac{1}{2}\rho V_{max}^2 \frac{S}{W} C_{D_o} + \frac{2KW^2}{\rho V_{max}^2 S W} \tag{4.46}$$

This can also be written as:

$$\left(\frac{T_{SL}}{W}\right)_{V_{max}} = \rho_o V_{max}^2 C_{D_o} \frac{1}{2\left(\frac{W}{S}\right)} + \frac{2K}{\rho\sigma V_{max}^2}\left(\frac{W}{S}\right) \tag{4.47}$$

Thus, thrust loading (T/W) is a non-linear function of wing loading (W/S) in terms of maximum speed, and may be simplified as:

$$\left(\frac{T}{W}\right) = \frac{aV_{max}^2}{\left(\frac{W}{S}\right)} + \frac{b}{V_{max}^2}\left(\frac{W}{S}\right) \tag{4.48}$$

The wing and engine sizing based on maximum speed requirements are represented by Equation (4.47) as the variations of thrust loading versus wing loading. This variation of T/W as a function of W/S based on V_{max} can be sketched by using Equation (4.47) in constructing the matching plot as shown in Figure 4.10. In order to determine the acceptable region, we need to find what side of this graph is satisfying the maximum speed requirement. As the value of V_{max} in Equation (4.47) is increased, the value of thrust-to-weight ratio (T/W) is increased too. Since any value of V_{max} greater than the

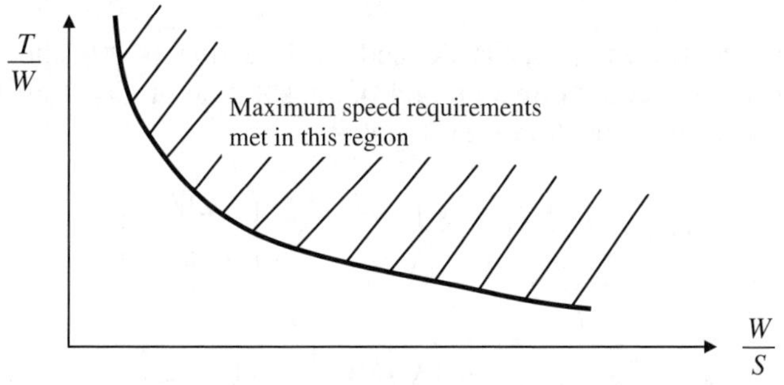

Figure 4.10 Maximum speed contribution in constructing a matching plot for a jet aircraft

specified maximum speed is satisfying the maximum speed requirement, so the region above the graph is acceptable.

Employ extreme care to use a consistent unit when applying Equation (4.47) (either the SI system or the British system). In the SI system, the unit of V_{max} is m/s, the unit of W is N, the unit of T is N, the unit of S is m^2, and the unit of ρ is kg/m^3. However, in the British system, the unit of V_{max} is ft/s, the unit of W is lb, the unit of T is lb, the unit of S is ft^2, and the unit of ρ is slug/ft^3. An example application is presented in Section 4.4.

If, instead of the maximum speed, the cruising speed is given as a design requirement, assume that the maximum speed is about 20–30% greater than the cruise speed. This is due to the fact that cruise speeds for jet aircraft are usually calculated at 75–80% thrust:

$$V_{max} = 1.2V_C \text{ to } 1.3V_C \tag{4.49}$$

Section 4.3.3.3 provides a technique to estimate the aircraft zero-lift drag coefficient (C_{D_o}).

4.3.3.2 Prop-Driven Aircraft

Consider a prop-driven aircraft which is flying with the maximum constant speed at a specified altitude (ρ_{alt} or simply ρ). The aircraft is in longitudinal trim; hence, the maximum available engine power (P_{max}) must be equal to the maximum required power (P_{req}), which is thrust multiplied by maximum speed:

$$P_{avl} = P_{req} \Rightarrow \eta_P P_{max} = TV_{max} \tag{4.50}$$

where engine thrust (T) must be equal to aircraft drag (D); Equation (4.36). In contrast, the engine power is decreasing with increasing aircraft altitude. This requires knowledge of how the engine power of an aircraft varies with airspeed and altitude. A general relationship between engine power and altitude, which is represented by air density (ρ), is:

$$P_{alt} = P_{SL}\left(\frac{\rho}{\rho_o}\right) = P_{SL}\sigma \tag{4.51}$$

where P_{alt} is the engine power at altitude, and P_{SL} is the engine power at sea level. By substituting Equations (4.36) and (4.51) into Equation (4.50), we obtain:

$$\eta_P P_{SL}\sigma = \frac{1}{2}\rho V_{max}^2 SC_D \cdot V_{max} = \frac{1}{2}\rho V_{max}^3 SC_D \tag{4.52}$$

The aircraft drag coefficient (C_D) is defined by Equation (4.40) and the aircraft lift coefficient (C_L) is provided by Equation (4.43). Substitution of C_D (Equation (4.40)) and C_L (Equation (4.43)) into Equation (4.52) yields:

$$\eta_P P_{SL}\sigma = \frac{1}{2}\rho V_{max}^3 S\left(C_{D_o} + K \cdot \left[\frac{2W}{\rho V_{max}^2 S}\right]^2\right) \tag{4.53}$$

or:

$$\eta_P P_{SL}\sigma = \frac{1}{2}\rho V_{max}^3 SC_{D_o} + \frac{1}{2}\rho V_{max}^3 S\frac{K\,(2W)^2}{\left(\rho V_{max}^2 S\right)^2} = \frac{1}{2}\rho V_{max}^3 SC_{D_o} + \frac{2KW^2}{\rho SV_{max}} \tag{4.54}$$

Both sides of this equation can be divided by the aircraft weight (W) as:

$$\frac{P_{SL}}{W}\eta_P\sigma = \frac{1}{2}\rho V_{max}^3\frac{S}{W}C_{D_o} + \frac{2KW^2}{\rho V_{max}SW} = \frac{1}{2}\rho C_{D_o}V_{max}^3\frac{1}{\left(\frac{W}{S}\right)} + \frac{2K}{\rho V_{max}}\left(\frac{W}{S}\right) \quad (4.55)$$

This equation can be inverted as follows:

$$\left(\frac{W}{P_{SL}}\right) = \frac{\sigma\eta_P}{\frac{1}{2}\rho V_{max}^3 C_{D_o}\frac{1}{\left(\frac{W}{S}\right)} + \frac{2K}{\rho V_{max}}\left(\frac{W}{S}\right)} \Rightarrow$$

$$\left(\frac{W}{P_{SL}}\right)_{V_{max}} = \frac{\eta_P}{\frac{1}{2}\rho_o V_{max}^3 C_{D_o}\frac{1}{\left(\frac{W}{S}\right)} + \frac{2K}{\rho\sigma V_{max}}\left(\frac{W}{S}\right)} \quad (4.56)$$

Thus, the power loading (W/P) is a non-linear function of the wing loading (W/S) in terms of maximum speed, and may be simplified as:

$$\left(\frac{W}{P}\right) = \frac{\eta_P}{\frac{aV_{max}^3}{\left(\frac{W}{S}\right)} + \frac{b}{V_{max}}\left(\frac{W}{S}\right)} \quad (4.57)$$

The wing and engine sizing based on maximum speed requirements are represented by Equation (4.57) as the variations of power loading versus wing loading. The variation of W/P as a function of W/S based on V_{max} for a prop-driven aircraft can be sketched by using Equation (4.56) in constructing the matching plot as shown in Figure 4.11. In order to determine the acceptable region, we need to find what side of this graph is satisfying the maximum speed requirement. As the value of V_{max} in Equation (4.56) is increased, the value of power loading (P/W) is decreased. This is due to the fact that the first term

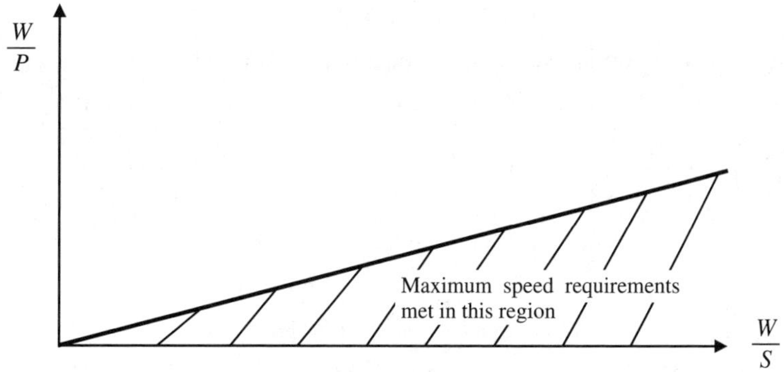

Figure 4.11 Maximum speed contribution in constructing a matching plot for a prop-driven aircraft

in the denominator of Equation (4.56) is V_{max}^3. Since any value of V_{max} greater than the specified maximum speed is satisfying the maximum speed requirement, so the region below the graph is acceptable. Extreme caution must be taken to use consistent units in the application process. If British units are used, convert the unit of W/P to lb/hp to make the comparison more convenient.

Employ extreme caution to use a consistent unit when applying Equation (4.56) (either the SI system or the British system). In the SI system, the unit of V_{max} is m/s, the unit of W is N, the unit of P is W, the unit of S is m^2, and the unit of ρ is kg/m^3. However, in the British system, the unit of V_{max} is ft/s, the unit of W is lb, the unit of P is lb ft/s, the unit of S is ft^2, and the unit of ρ is slug/ft^3. If the British units are used, convert the unit of W/P to lb/hp to make the comparison more convenient. Recall that each unit of horsepower (hp) is equivalent to 550 lb ft/s. An example application is presented in Section 4.4.

If, instead of the maximum speed, the cruising speed is given as a design requirement, assume that the maximum speed is about 20–30% greater than the cruise speed:

$$V_{max} = 1.2V_C \text{ to } 1.3V_C \tag{4.58}$$

This is due to the fact that cruise speeds for prop-driven aircraft are usually calculated at 75–80% power.

4.3.3.3 Aircraft C_{D_o} Estimation

An important aircraft parameter that must be known and is necessary in constructing the matching plot is the aircraft zero-lift drag coefficient (C_{D_o}). Although the aircraft is not aerodynamically designed yet at this phase of design, there is a reliable way to estimate this parameter. The technique is primarily based on statistics. However, in most references (such as [4]), the aircraft C_{D_o} is not given, but it can readily be determined based on aircraft performance which is often given.

Consider a jet aircraft that is flying with its maximum speed at a specified altitude. The governing trim equations are introduced in Section 4.3.3.1 and the relationships are expanded until we obtain the following equation:

$$\left(\frac{T_{SL}}{W}\right) = \rho_o V_{max}^2 C_{D_o} \frac{1}{2\left(\frac{W}{S}\right)} + \frac{2K}{\rho\sigma V_{max}^2}\left(\frac{W}{S}\right) \tag{4.47}$$

The aircraft C_{D_o} can be obtained from this equation as follows:

$$\left(\frac{T_{SL}}{W}\right) - \frac{2KW}{\rho\sigma V_{max}^2 S} = \rho_o V_{max}^2 C_{D_o} \frac{S}{2W} \Rightarrow C_{D_o} = \frac{\frac{T_{SL}}{W} - \frac{2KW}{\rho\sigma V_{max}^2 S}}{\rho_o V_{max}^2 \frac{S}{2W}} \tag{4.59}$$

or:

$$C_{D_o} = \frac{2T_{SL_{max}} - \frac{4KW^2}{\rho\sigma V_{max}^2 S}}{\rho_o V_{max}^2 S} \tag{4.60}$$

If the aircraft is prop-driven, the engine thrust is a function of engine power, airspeed, and propeller efficiency (η_p), so:

$$T_{max} = \frac{P_{max} \cdot \eta_P}{V_{max}} \tag{4.61}$$

where prop efficiency is about 0.7–0.85 when an aircraft is cruising with its maximum speed.

Equation (4.61) can be substituted into Equation (4.60):

$$C_{D_o} = \frac{2\frac{P_{SL_{max}} \cdot \eta_P}{V_{max}} - \frac{4KW^2}{\rho\sigma V_{max}^2 S}}{\rho_o V_{max}^2 S} \tag{4.62}$$

Equations (4.60) and (4.62) are employed to determine the aircraft C_{D_o} for jet and prop-driven aircraft respectively. In these equations, $T_{SL_{max}}$ is the maximum engine thrust at sea level, and $P_{SL_{max}}$ is the maximum engine power at sea level, ρ is the air density at flight altitude, and σ is the relative air density at flight altitude. Make sure to use a consistent unit for all parameters (either in metric units or British units).

In order to estimate the C_{D_o} for the aircraft which is under preliminary design, calculate the C_{D_o} of several aircraft which have similar performance characteristics and similar configuration. Then find the average C_{D_o} of those aircraft. If you have selected five similar aircraft, then the C_{D_o} of the aircraft under preliminary design is determined as follows:

$$C_{D_o} = \frac{C_{D_{o1}} + C_{D_{o2}} + C_{D_{o3}} + C_{D_{o4}} + C_{D_{o5}}}{5} \tag{4.63}$$

where $C_{D_{oi}}$ is the C_{D_o} of the ith aircraft. Table 4.12 presents typical values of C_{D_o} for different types of aircraft, and Examples 4.1 and 4.2 give an example C_{D_o} calculation and estimation respectively. References [5, 10] present details of the technique to calculate the complete C_{D_o} of an aircraft.

Table 4.12 Typical values of C_{D_o} for different types of aircraft

No.	Aircraft type	C_{D_o}
1	Jet transport	0.015–0.02
2	Turboprop transport	0.018–0.024
3	Twin-engine piston prop	0.022–0.028
4	Small GA with retractable landing gear	0.02–0.03
5	Small GA with fixed landing gear	0.025–0.04
6	Agricultural	0.04–0.07
7	Sailplane/glider	0.012–0.015
8	Supersonic fighter	0.018–0.035
9	Home-built	0.025–0.04
10	Microlight	0.02–0.035

Example 4.1: C_{D_o} Calculation

Determine the zero-lift drag coefficient (C_{D_o}) of the fighter aircraft F/A-18 Hornet (Figures 2.11, 6.12) which is flying with a maximum speed of Mach 1.8 at 30 000 ft. This fighter has the following characteristics:

$$T_{SL_{max}} = 2 \cdot 71170 \text{ N}, m_{TO} = 16\ 651 \text{ kg}, S = 37.16 \text{ m}^2, \text{AR} = 3.5, e = 0.7$$

Solution:

We need to first find out the maximum speed in terms of m/s. The air density at 30 000 ft is 0.000892 slug/ft^3 or 0.46 kg/m^3, and the air temperature is 229 K. From physics, we know that the speed of sound is a function of air temperature. Thus:

$$a = \sqrt{\gamma RT} = \sqrt{1.4 \cdot 287 \cdot 229} = 303.3 \text{ m/s} \tag{4.64}$$

From aerodynamics, we know that the Mach number is the ratio between airspeed and the speed of sound. Hence, the aircraft maximum speed is:

$$M = \frac{V}{a} \Rightarrow V_{max} = M_{max} \cdot a = 1.8 \cdot 303.3 = 546 \frac{\text{m}}{\text{s}} \tag{4.65}$$

The induced drag factor is:

$$K = \frac{1}{\pi \cdot e \cdot \text{AR}} = \frac{1}{3.14 \cdot 0.7 \cdot 3.5} \Rightarrow K = 0.13 \tag{4.41}$$

Then:

$$C_{D_o} = \frac{2T_{SL_{max}} - \dfrac{4KW^2}{\rho\sigma V_{max}^2 S}}{\rho_o V_{max}^2 S}$$

$$= \frac{2 \cdot 2 \cdot 71170 - \dfrac{4 \cdot 0.13 \cdot (16651 \cdot 9.81)}{0.46 \cdot \left(\dfrac{0.46}{1.225}\right) \cdot (546)^2 \cdot (37.16)}}{1.225 \cdot (546)^2 \cdot (37.16)} = 0.02 \tag{4.60}$$

Thus, the zero-lift drag coefficient (C_{D_o}) of the fighter aircraft F/A-18 Hornet at 30 000 ft is 0.02.

Example 4.2: Aircraft C_{D_o} Estimation

You are a member of a team that is designing a transport aircraft which is required to carry 45 passengers with the following performance features:

1. Max speed: at least 300 knots at sea level.
2. Max range: at least 1500 km.

3. Max ROC: at least 2500 fpm.
4. Absolute ceiling: at least 28 000 ft.
5. Take-off run: less than 4000 ft.

In the preliminary design phase, you are required to estimate the zero-lift drag coefficient (C_{D_o}) of such an aircraft. Identify five current similar aircraft and, based on their statistics, estimate the C_{D_o} of the aircraft being designed.

Solution:
Reference [4] is a reliable source to look for similar aircraft in terms of performance characteristics. Table 4.13 illustrates five aircraft with similar performance requirements

Table 4.13 Characteristics of five aircraft with similar performance

No.	Name	P_{max}	V_{max} (knot)	Range (km)	ROC (fpm)	S_{TO} (ft)	Ceiling (ft)
1	DHC-8 Dash 8-300B	50	287	1 711	1 800	3 600	25 000
2	Antonov 140 (Figure 3.15)	46	310 @ 23 620 ft	1 721	1 345	2 890	25 000
3	Embraer 145MP	50	410 @ 37 000 ft	3 000	1 750	6 465	37 000
4	Bombardier Challenger 604	19	471 @ 17 000 ft	4 274	3 395	2 910	41 000
5	Saab 340 (Figure 8.21)	35	280 @ 20 000 ft	1 750	2 000	4 325	25 000

as the aircraft that is being designed. There are three turboprops and two jets, so either engine configuration may be satisfactory. All of them are twin engines, and have retractable gear. There are no mid-wing aircraft listed here. The wing areas are very similar, ranging from 450 to 605 ft^2. Except for the Bombardier Challenger 604 which can carry 19 passengers (the minimum requirement for the aircraft being designed), the other four listed aircraft can accommodate ~50 passengers. The weights of the aircraft vary, with the Challenger 800 weighing the most. The powers of the prop-driven aircraft are all around 2000 hp/engine, and the thrust for the jet aircraft is around 8000 lb/engine.

In order to calculate the C_{D_o} of each aircraft, Equation (4.60) is employed for the jet aircraft and Equation (4.62) is used for the prop-driven aircraft.

$$C_{D_o} = \frac{2T_{SL_{max}} - \dfrac{4KW^2}{\rho\sigma V_{max}^2 S}}{\rho_o V_{max}^2 S} \tag{4.60}$$

$$C_{D_o} = \frac{2\frac{P_{SL_{max}} \cdot \eta_P}{V_{max}} - \frac{4KW^2}{\rho\sigma V_{max}^2 S}}{\rho_o V_{max}^2 S} \tag{4.62}$$

The Oswald span efficiency factor was assumed to be 0.85, and the prop efficiencies for the propeller aircraft were assumed to be 0.82. Example 4.1 shows the application of Equation (4.60) for a jet aircraft, the following is the application of Equation (4.62) for Saab 340, a turbo prop-driven airliner. The cruise altitude for Saab 340 is 20 000 ft, so the air density at 20 000 ft is 0.001267 slug/ft^3 and the relative air density at 20 000 ft is 0.533.

$$K = \frac{1}{\pi \cdot e \cdot \text{AR}} = \frac{1}{3.14 \cdot 0.85 \cdot 11} = 0.034 \tag{4.41}$$

$$C_{D_o} = \frac{2\frac{2 \cdot 1750 \cdot 550 \cdot 0.82}{280 \cdot 1.688} - \frac{4 \cdot 0.034 \cdot (29\,000)^2}{0.001267 \cdot 0.533 \cdot (280 \cdot 1.688)^2 \cdot 450}}{0.002378 \cdot (280 \cdot 1.688)^2 \cdot 450}$$

$$\Rightarrow C_{D_o} = 0.021 \tag{4.62}$$

where 1 knot is equivalent to 1.688 ft/s and 1 hp is equivalent to 550 lb ft/s.

Table 4.14 C_{D_o} of five similar aircraft

No.	Aircraft	Type	W_o (lb)	P (hp)/T (lb)	S (ft^2)	AR	C_{D_o}
1	DHC-8 Dash 8-300B	Twin-turboprop	41 100	2 · 2500 hp	605	13.4	0.02
2	Antonov An-140	Twin-turboprop	42 220	2 · 2466 hp	549	11.5	0.016
3	Embraer EMB-145	Regional jet	42 328	2 · 7040 lb	551	7.9	0.034
4	Bombardier Challenger 604	Business jet	47 600	2 · 9220 lb	520	8	0.042
5	Saab 340	Twin-turboprop	29 000	2 · 1750 hp	450	11	0.021

The aircraft geometries, engine powers, and C_{D_o} and also the results of the calculation are shown in Table 4.14. The zero-lift drag coefficient for two turboprop aircraft is very similar, 0.02 or 0.021 and one is only 0.016. This coefficient for jet aircraft is higher, 0.034 and 0.042. It seems that these three numbers (0.016, 0.034, and 0.042) are unrealistic; therefore, some of the published data are not reliable. The estimation of C_{D_o} of the aircraft being designed is determined by taking the average of five C_{D_o}'s:

$$C_{D_o} = \frac{C_{D_{o1}} + C_{D_{o2}} + C_{D_{o3}} + C_{D_{o4}} + C_{D_{o5}}}{5} = \frac{0.02 + 0.016 + 0.034 + 0.042 + 0.021}{5}$$

$$\Rightarrow C_{D_o} = 0.027 \tag{4.63}$$

Therefore, the C_{D_o} for the aircraft under preliminary design will be assumed to be 0.027.

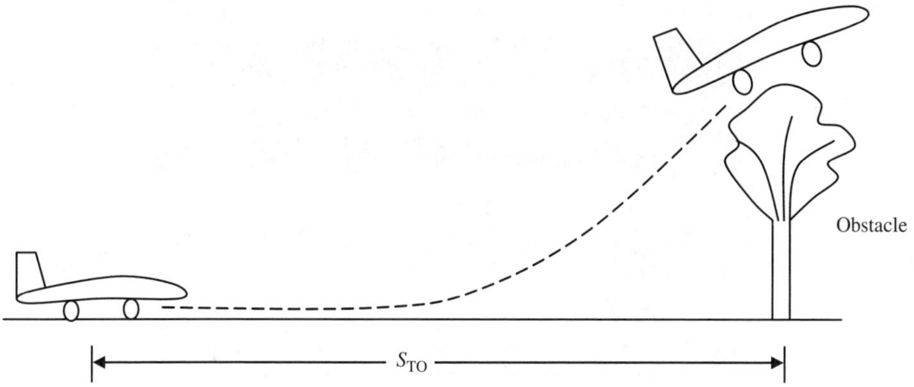

Figure 4.12 The definition of take-off run

4.3.4 Take-Off Run

The take-off run (S_{TO}) is another significant factor in aircraft performance and will be employed in constructing a matching chart to determine wing area and engine thrust (or power). The take-off requirements are frequently spelled out in terms of minimum ground run requirements, since every airport has a limited runway. The take-off run is defined as the distance between the take-off starting point to the location of a standard obstacle that the aircraft must clear (Figure 4.12). The aircraft is required to clear an imaginary obstacle at the end of the airborne section, so the take-off run includes a ground section plus an airborne section. The obstacle height is determined by airworthiness standards. Based on FAR Part 25, the obstacle height (h_o) is 35 ft for passenger aircraft, and based on FAR Part 23 Section 23.53, the obstacle height is 50 ft for GA aircraft. There is no FAR requirement for take-off run; instead, FAR has a number of regulations on the balanced field length.

4.3.4.1 Jet Aircraft

Based on Ref. [5], the take-off run for a jet aircraft is determined by the following equation:

$$S_{TO} = \frac{1.65W}{\rho g S C_{D_G}} \ln\left[\frac{\frac{T}{W} - \mu}{\frac{T}{W} - \mu - \frac{C_{D_G}}{C_{L_R}}}\right] \tag{4.66}$$

where μ is the friction coefficient of the runway (see Table 4.15) and C_{D_G} is defined as:

$$C_{D_G} = \left(C_{D_{TO}} - \mu C_{L_{TO}}\right) \tag{4.67a}$$

The parameter C_{L_R} is the aircraft lift coefficient at take-off rotation and is obtained from:

$$C_{L_R} = \frac{2mg}{\rho S V_R^2} \tag{4.67b}$$

Table 4.15 Friction coefficients for various runway surfaces

No.	Surface	Friction coefficient (μ)
1	Dry concrete/asphalt	0.03–0.05
2	Wet concrete/asphalt	0.05
3	Icy concrete/asphalt	0.02
4	Turf	0.04–0.07
5	Grass	0.05–0.1
6	Soft ground	0.1–0.3

where V_R is the aircraft speed at rotation, which is about $1.1V_s$ to $1.2V_s$. The aircraft drag coefficient at take-off configuration ($C_{D_{TO}}$) is:

$$C_{D_{TO}} = C_{D_{oTO}} + KC_{L_{TO}}^2 \tag{4.68}$$

where the aircraft zero-lift drag coefficient at take-off configuration ($C_{D_{oTO}}$) is:

$$C_{D_{oTO}} = C_{D_o} + C_{D_{oLG}} + C_{D_{oHLD\_TO}} \tag{4.69a}$$

where C_{D_o} is the clean aircraft zero-lift drag coefficient (see Table 4.12), $C_{D_{oLG}}$ is the landing gear drag coefficient, and $C_{D_{oHLD\_TO}}$ is the high-lift device (e.g., flap) drag coefficient at take-off configuration. The typical values for $C_{D_{oLG}}$ and $C_{D_{oHLD\_TO}}$ are as follows:

$$\begin{aligned} C_{D_{oLG}} &= 0.006 \text{ to } 0.012 \\ C_{D_{oHLD\_TO}} &= 0.003 \text{ to } 0.008 \end{aligned} \tag{4.69b}$$

where the take-off lift coefficient is determined as:

$$C_{L_{TO}} = C_{L_C} + \Delta C_{L_{flapTO}} \tag{4.69c}$$

where C_{L_C} is the aircraft cruise lift coefficient and $\Delta C_{L_{flapTO}}$ is the additional lift coefficient that is generated by flap at take-off configuration. The typical value for aircraft cruise lift coefficient is about 0.3 for a subsonic aircraft and 0.05 for a supersonic aircraft. The typical value for take-off flap lift coefficient ($\Delta C_{L_{flapTO}}$) is about 0.3–0.8. Equation (4.66) can be manipulated to be formatted as the thrust loading (T/W) in terms of wing loading (W/S) and take-off run. The derivation is as follows:

$$\frac{\rho g S C_{D_G} S_{TO}}{1.65W} = \ln\left[\frac{\frac{T}{W} - \mu}{\frac{T}{W} - \mu - \frac{C_{D_G}}{C_{L_R}}}\right] \Rightarrow \frac{\frac{T}{W} - \mu}{\frac{T}{W} - \mu - \frac{C_{D_G}}{C_{L_R}}} = \exp\left(0.6\rho g C_{D_G} S_{TO} \frac{S}{W}\right)$$

$$\Rightarrow \frac{T}{W} - \mu = \left(\frac{T}{W} - \mu - \frac{C_{D_G}}{C_{L_{TR}}}\right)\left[\exp\left(0.6\rho g C_{D_G} S_{TO} \frac{S}{W}\right)\right]$$

$$\Rightarrow \frac{T}{W} - \mu = \left(\frac{T}{W}\right)\left[\exp\left(0.6\rho g C_{D_G} S_{TO} \frac{S}{W}\right)\right] - \left(\mu + \frac{C_{D_G}}{C_{L_R}}\right)\left[\exp\left(0.6\rho g C_{D_G} S_{TO} \frac{S}{W}\right)\right]$$

$$\Rightarrow \frac{T}{W} - \left(\frac{T}{W}\right)\left[\exp\left(0.6\rho g C_{D_G} S_{TO} \frac{S}{W}\right)\right] = \mu - \left(\mu + \frac{C_{D_G}}{C_{L_R}}\right)\left[\exp\left(0.6\rho g C_{D_G} S_{TO} \frac{S}{W}\right)\right]$$

$$\Rightarrow \frac{T}{W}\left[1 - \left[\exp\left(0.6\rho g C_{D_G} S_{TO} \frac{S}{W}\right)\right]\right] = \mu - \left(\mu + \frac{C_{D_G}}{C_{L_R}}\right)\left[\exp\left(0.6\rho g C_{D_G} S_{TO} \frac{S}{W}\right)\right] \tag{4.70}$$

Finally:

$$\left(\frac{T}{W}\right)_{S_{TO}} = \frac{\mu - \left(\mu + \frac{C_{D_G}}{C_{L_R}}\right)\left[\exp\left(0.6\rho g C_{D_G} S_{TO} \frac{1}{W/S}\right)\right]}{1 - \exp\left(0.6\rho g C_{D_G} S_{TO} \frac{1}{W/S}\right)} \tag{4.71}$$

The wing and engine sizing based on take-off run requirements are represented by Equation (4.71) as the variations of thrust loading versus wing loading. The variation of T/W as a function of W/S based on S_{TO} for a jet aircraft can be sketched using Equation (4.71) in constructing the matching plot, as shown in Figure 4.13. In order to determine the acceptable region, we need to find what side of this graph is satisfying the take-off run requirement. Both the numerator and the denominator of Equation (4.71) contain an exponential term with a positive power that includes the parameter S_{TO}.

As the value of the take-off run (S_{TO}) in Equation (4.71) is *increased*, the value of the thrust-to-weight ratio (T/W) would *drop*. Since any value of S_{TO} greater than the specified take-off run is not satisfying the take-off run requirement, so the region *below* the graph (Figure 4.13) is not acceptable. Employ extreme care to use a consistent unit when applying Equation (4.71) (either the SI system or the British system). In the SI

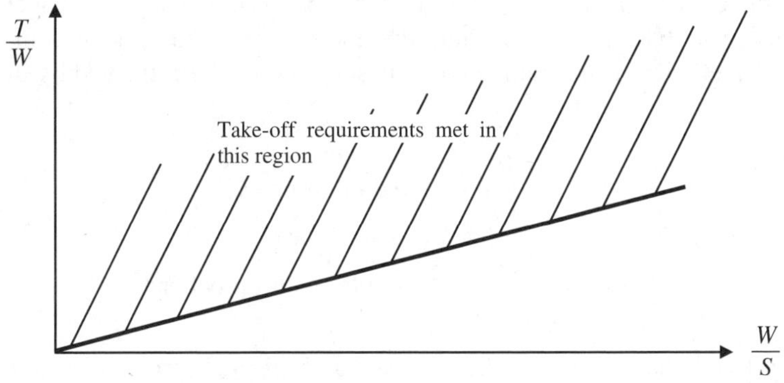

Figure 4.13 Take-off run contribution in constructing a matching plot for a jet aircraft

system, the unit of S_{TO} is m, the unit of W is N, the unit of T is N, g is 9.81 m/s^2, the unit of S is m^2, and the unit of ρ is kg/m^3. However, in the British system, the unit of S_{TO} is ft, the unit of W is lb, the unit of T is lb, g is 32.17 ft/s^2, the unit of S is ft^2, and the unit of ρ is slug/ft^3. An example application is presented in Section 4.4.

4.3.4.2 Prop-Driven Aircraft

In a prop-driven aircraft, the engine thrust is a function of the propeller efficiency and the aircraft speed. However, take-off operation is considered as an accelerating motion, so the aircraft speed is not constant. The aircraft speed varies quickly from zero to rotation speed and then to take-off speed. The take-off speed (V_{TO}) is normally slightly greater than the stall speed (V_s):

$$V_{TO} = 1.1V_s \text{ to } 1.3V_s \tag{4.72}$$

The following is reproduced directly from FAR 23.51:

> For normal, utility, and acrobatic category airplanes, the speed at 50 feet above the takeoff surface level must not be less than:
>
> (1) or multiengine airplanes, the highest of
>
> (i) A speed that is shown to be safe for continued flight (or emergency landing, if applicable) under all reasonably expected conditions, including turbulence and complete failure of the critical engine;
>
> (ii) 1.10 V_{MC}; or
>
> (iii) 1.20 V_{S1}.
>
> (2) For single-engine airplanes, the higher of
>
> (i) A speed that is shown to be safe under all reasonably expected conditions, including turbulence and complete engine failure; or
>
> (ii) 1.20 V_{S1}.

Furthermore, the prop efficiency is not constant and is much lower than its maximum attainable efficiency. If the prop is of fixed-pitch type, its efficiency is considerably higher than that of a variable pitch. To include the above-mentioned variations in the aircraft speed and prop efficiency, the engine thrust is suggested to be estimated by the following equations:

$$T_{TO} = \frac{0.5P_{max}}{V_{TO}} \text{ (fixed-pitch propeller)} \tag{4.73a}$$

$$T_{TO} = \frac{0.6P_{max}}{V_{TO}} \text{ (variable-pitch propeller)} \tag{4.73b}$$

This demonstrates that the prop efficiency for a fixed-pitch propeller is 0.5, and for a variable-pitch propeller it is 0.6. The above thrust estimation works for the majority of aero-engines. A better thrust model might be found through engine manufacturers.

By substituting Equation (4.73) into Equation (4.71), we obtain:

$$\left(\frac{\frac{\eta_P P_{max}}{V_{TO}}}{W}\right)_{S_{TO}} = \frac{\mu - \left(\mu + \frac{C_{D_G}}{C_{L_R}}\right)\left[\exp\left(0.6\rho g C_{D_G} S_{TO} \frac{1}{W/S}\right)\right]}{1 - \exp\left(0.6\rho g C_{D_G} S_{TO} \frac{1}{W/S}\right)} \tag{4.74}$$

or:

$$\left(\frac{P}{W}\right)_{S_{TO}} = \frac{V_{TO}}{\eta_P} \frac{\mu - \left(\mu + \frac{C_{D_G}}{C_{L_R}}\right)\left[\exp\left(0.6\rho g C_{D_G} S_{TO} \frac{1}{W/S}\right)\right]}{1 - \exp\left(0.6\rho g C_{D_G} S_{TO} \frac{1}{W/S}\right)} \tag{4.75}$$

This equation can be inverted and written as follows:

$$\left(\frac{W}{P}\right)_{S_{TO}} = \frac{1 - \exp\left(0.6\rho g C_{D_G} S_{TO} \frac{1}{W/S}\right)}{\mu - \left(\mu + \frac{C_{D_G}}{C_{L_R}}\right)\left[\exp\left(0.6\rho g C_{D_G} S_{TO} \frac{1}{W/S}\right)\right]} \frac{\eta_P}{V_{TO}} \tag{4.76}$$

The wing and engine sizing based on take-off run requirement is represented by Equation (4.76) as the variations of power loading versus wing loading. Remember that the prop efficiency is 0.5 for a fixed-pitch propeller and 0.6 for a variable-pitch propeller. The variation of W/P as a function of W/S based on S_{TO} for a prop-driven aircraft can be sketched using Equation (4.76) in constructing the matching plot, as shown in Figure 4.14. In order to determine the acceptable region, we need to find what side of this graph is satisfying the take-off run requirement. Both the numerator and the denominator of Equation (4.76) contain an exponential term with a positive power that includes the parameter S_{TO}. As the take-off run is increased, the magnitude of the exponential term will increase.

As the value of S_{TO} in Equation (4.76) is *increased*, the value of power loading (W/P) is going *up*. Since any value of S_{TO} greater than the specified take-off run is not satisfying

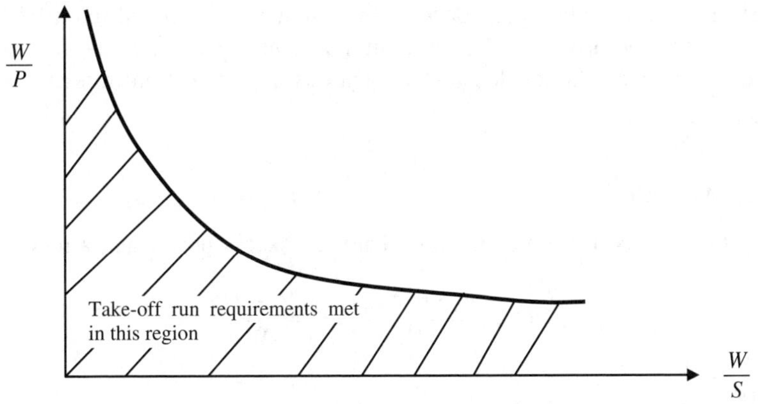

Figure 4.14 Take-off run contribution in constructing a matching plot for a prop-driven aircraft

the take-off run requirement, so the region *above* the graph is not acceptable. Employ extreme caution to use a consistent unit when applying Equation (4.76) (either the SI system or the British system).

In the SI system, the unit of S_{TO} is m, the unit of W is N, the unit of P is W, the unit of S is m^2, the unit of V_{TO} is m/s, the variable g is 9.81 m/s^2, and the unit of ρ is kg/m^3. However, in the British system, the unit of S_{TO} is ft, the unit of W is lb, the unit of P is lb ft/s, the unit of S is ft^2, the unit of V_{TO} is ft/s, the variable g is 32.17 ft/s^2, and the unit of ρ is slug/ft^3. If British units are used, convert the unit of W/P to lb/hp to make the comparison more convenient. Recall that each unit of horsepower (hp) is equivalent to 550 lb ft/s. An example application is presented in Section 4.4.

4.3.5 *Rate of Climb*

Every type of aircraft must meet certain ROC requirements. For civil aircraft, the climb requirements of FAR[4] Part 23 (for GA aircraft), or FAR Part 25 (for transport aircraft) must be met. For military aircraft, the requirements specified by military standards, handbooks, and specifications[5] must be met. In some instances, climb requirements are spelled out in terms of time-to-climb, but this can readily be translated into ROC requirements. The rate of climb is defined as the aircraft speed in the vertical axis or the vertical component of the aircraft airspeed. Hence, ROC is about how fast an aircraft gains height.

Based on FAR Part 23 Section 23.65, there are requirements for gradient of climb as follows:

1. Each normal, utility, and acrobatic category reciprocating engine-powered airplane of 6000 lb or less maximum weight must have a steady climb gradient at sea level of at least 8.3% for landplanes or 6.7% for seaplanes and amphibians.
2. Each normal, utility, and acrobatic category reciprocating engine-powered airplane of more than 6000 lb maximum weight and turbine engine-powered airplanes in the normal, utility, and acrobatic category must have a steady gradient of climb after take-off of at least 4%.

The derivation of an expression for wing and engine sizing based upon ROC requirements for jet and prop-driven aircraft is examined separately. Since the maximum ROC is obtained at sea level, the air density in equations in this section implies the sea-level air density.

4.3.5.1 Jet Aircraft

In general, the ROC is defined as the ratio between excess power and aircraft weight:

$$\text{ROC} = \frac{P_{\text{avl}} - P_{\text{req}}}{W} = \frac{(TV - DV)}{W} \tag{4.77}$$

[4] Reference [1].

[5] For instance, see MIL-C-005011B (USAF), Military specification charts: Standard aircraft characteristics and performance, piloted aircraft, 1977.

This can be written as:

$$\mathrm{ROC} = V\left[\frac{T}{W} - \frac{D}{W}\right] = V\left[\frac{T}{W} - \frac{D}{L}\right] = V\left[\frac{T}{W} - \frac{1}{L/D}\right] \tag{4.78}$$

In order to maximize ROC, both the engine thrust and the lift-to-drag ratio must be maximized. This is to maximize the magnitude of the term inside the brackets in Equation (4.78):

$$ROC_{\max} = V_{\mathrm{ROC}_{\max}}\left[\frac{T_{\max}}{W} - \frac{1}{(L/D)_{\max}}\right] \tag{4.79}$$

In order to maximize the lift-to-drag ratio, the climb speed must be such that the aircraft drag is minimized, as outlined by Sadraey [5] as follows:

$$V_{\mathrm{ROC}_{\max}} = V_{D_{\min}} = \sqrt{\frac{2W}{\rho S\sqrt{\frac{C_{D_o}}{K}}}} \tag{4.80}$$

Substituting Equation (4.80) into Equation (4.79) yields:

$$\mathrm{ROC}_{\max} = \sqrt{\frac{2W}{\rho S\sqrt{\frac{C_{D_o}}{K}}}}\left[\frac{T_{\max}}{W} - \frac{1}{(L/D)_{\max}}\right] \tag{4.81}$$

This equation can be manipulated further to be in the form of thrust loading as a function of wing loading. Hence:

$$\left[\frac{T_{\max}}{W} - \frac{1}{(L/D)_{\max}}\right] = \frac{\mathrm{ROC}_{\max}}{\sqrt{\frac{2W}{\rho S\sqrt{\frac{C_{D_o}}{K}}}}} \Rightarrow \frac{T_{\max}}{W} = \frac{\mathrm{ROC}_{\max}}{\sqrt{\frac{2W}{\rho S\sqrt{\frac{C_{D_o}}{K}}}}} + \frac{1}{(L/D)_{\max}} \tag{4.82}$$

Thus:

$$\left(\frac{T}{W}\right)_{\mathrm{ROC}} = \frac{\mathrm{ROC}}{\sqrt{\frac{2}{\rho\sqrt{\frac{C_{D_o}}{K}}}\left(\frac{W}{S}\right)}} + \frac{1}{(L/D)_{\max}} \tag{4.83}$$

The wing and engine sizing based on ROC requirements is represented in Equation (4.83) as the variations of thrust loading versus wing loading. Since the fastest climb is obtained at sea level, where the engine thrust is at its maximum value, the air density must be considered at sea level.

The variation of T/W as a function of W/S based on ROC for a jet aircraft can be sketched using Equation (4.83) in constructing the matching plot, as shown in Figure 4.15. In order to determine the acceptable region, we need to find what side of this graph is satisfying the take-off run requirement. Since the ROC is in the denominator, as the rate of

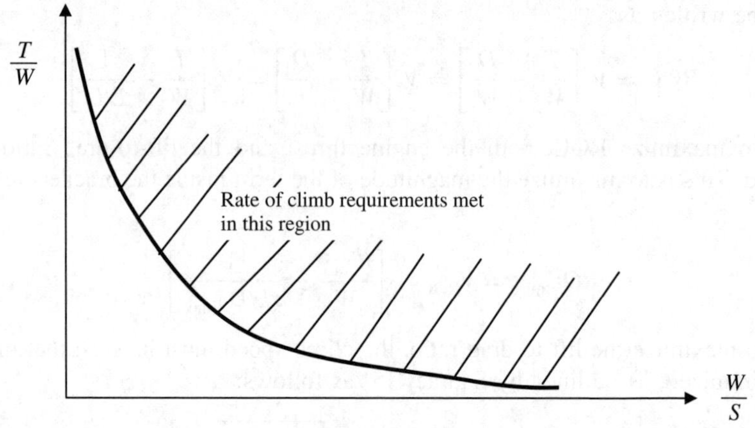

Figure 4.15 Rate of climb contribution in constructing a matching plot for a jet aircraft

climb in Equation (4.83) is *increased*, the value of the thrust loading (T/W) is going *up*. Since any value of ROC greater than the specified ROC is satisfying the ROC requirement, so the region *above* the graph is acceptable. Employ extreme care to use a consistent unit when applying Equation (4.83) (either the SI system or the British system). Typical values of maximum lift-to-drag ratio for several types of aircraft are given in Table 4.5.

Employ extreme caution to use a consistent unit when applying Equation (4.83) (either the SI system or the British system). In the SI system, the unit of ROC is m/s, the unit of W is N, the unit of T is N, the unit of S is m^2, and the unit of ρ is kg/m^3. However, in the British system, the unit of ROC is ft/s, the unit of W is lb, the unit of T is lb, the unit of S is ft^2, and the unit of ρ is slug/ft^3. An example application is presented in Section 4.4.

4.3.5.2 Prop-Driven Aircraft

Returning to the definition of ROC in Section 4.3.5.1, and noting that the available power is the engine power times the prop efficiency, we have:

$$\text{ROC} = \frac{P_{\text{avl}} - P_{\text{req}}}{W} = \frac{\eta_{\text{P}} P - DV}{W} \tag{4.84}$$

where the speed to obtain the maximum ROC for a prop-driven aircraft [5] is:

$$V_{\text{ROC}_{\max}} = \sqrt{\frac{2W}{\rho S \sqrt{\frac{3C_{D_o}}{K}}}} \tag{4.85}$$

By substituting Equation (4.85) into Equation (4.84), we obtain:

$$\text{ROC}_{\max} = \frac{\eta_{\text{P}} P_{\max}}{W} - \frac{D}{W}\sqrt{\frac{2W}{\rho S \sqrt{\frac{3C_{D_o}}{K}}}} \tag{4.86}$$

However, aircraft drag is a function of aircraft speed and wing area, as follows:

$$D = \frac{1}{2}\rho V^2 S C_D \tag{4.36}$$

An expression for wing loading is obtained by inserting Equation (4.36) into Equation (4.86), as follows:

$$\mathrm{ROC}_{\max} = \frac{\eta_P P_{\max}}{W} - \frac{\frac{1}{2}\rho V^2 S C_D}{W}\sqrt{\frac{2W}{\rho S\sqrt{\frac{3C_{D_o}}{K}}}} \tag{4.87}$$

This equation can be simplified further,[6] as follows:

$$\mathrm{ROC}_{\max} = \frac{\eta_P P_{\max}}{W} - \sqrt{\frac{2}{\rho\sqrt{\frac{3C_{D_o}}{K}}}\left(\frac{W}{S}\right)}\left(\frac{1.155}{(L/D)_{\max}}\right) \tag{4.88}$$

This equation may be manipulated and inverted to obtain the power loading as follows:

$$\frac{P_{\max}}{W} = \frac{\mathrm{ROC}_{\max}}{\eta_P} + \sqrt{\frac{2}{\rho\sqrt{\frac{3C_{D_o}}{K}}}\left(\frac{W}{S}\right)}\left(\frac{1.155}{(L/D)_{\max}\,\eta_P}\right) \Rightarrow$$

$$\left(\frac{W}{P}\right)_{\mathrm{ROC}} = \frac{1}{\dfrac{\mathrm{ROC}}{\eta_P} + \sqrt{\dfrac{2}{\rho\sqrt{\frac{3C_{D_o}}{K}}}\left(\dfrac{W}{S}\right)}\left(\dfrac{1.155}{(L/D)_{\max}\,\eta_P}\right)} \tag{4.89}$$

where the prop efficiency (η_P) in climbing flight is about 0.7. Typical values of maximum lift-to-drag ratio for several types of aircraft are given in Table 4.5.

The wing and engine sizing based on ROC requirements is represented in Equation (4.89) as the variations of power loading versus wing loading. The prop efficiency in climbing flight is about 0.5–0.6. The variation of W/P as a function of W/S based on ROC for a prop-driven aircraft can be sketched using Equation (4.89) in constructing the matching plot, as shown in Figure 4.16. In order to determine the acceptable region, we need to find what side of this graph is satisfying the climb requirements.

We note that the ROC is in the denominator; hence, as the value of ROC in Equation (4.89) is *increased*, the value of thrust loading (W/P) will *drop*. Since any value of ROC greater than the specified ROC is satisfying the climb requirement, so the region *below* the graph is acceptable.

Employ extreme caution to use a consistent unit when applying Equation (4.89) (either the SI system or the British system). In the SI system, the unit of ROC is m/s, the unit of W is N, the unit of P is W, the unit of S is m^2, and the unit of ρ is kg/m^3. However,

[6] The simplification is given in Ref. [5].

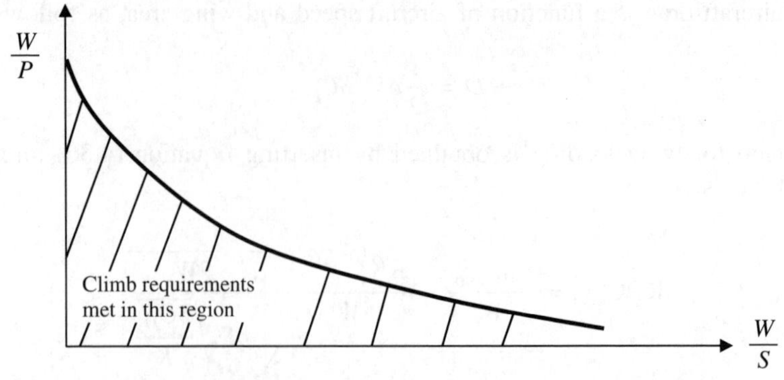

Figure 4.16 Rate of climb contribution in constructing a matching plot for a prop-driven aircraft

in the British system, the unit of ROC is ft/s, the unit of W is lb, the unit of P is lb ft/s, the unit of S is ft^2, and the unit of ρ is slug/ft^3. If British units are used, convert the unit of W/P to lb/hp to make the comparison more convenient. Recall that each unit of horsepower (hp) is equivalent to 550 lb ft/s. An example application is presented in Section 4.4.

4.3.6 Ceiling

Another performance requirement that influences the wing and engine sizing is the ceiling. The ceiling is defined as the highest altitude that an aircraft can safely have a straight level flight. Another definition is the highest altitude that an aircraft can reach by its own engine and have sustained flight. For many aircraft, the ceiling is not a crucial requirement, but for others such as the reconnaissance aircraft SR-71 Black Bird, a ceiling of about 65 000 ft was the most difficult performance requirement to meet. This design requirement made the designers design and invent a special engine for this mission. In general, there are four types of ceiling:

1. **Absolute ceiling (h_{ac}).** As the name implies, the absolute ceiling is the absolute maximum altitude that an aircraft can ever maintain level flight. In other terms, the ceiling is the altitude at which the ROC is zero.
2. **Service ceiling (h_{sc}).** The service ceiling is defined as the highest altitude at which the aircraft can climb with a rate of 100 ft/min (i.e., 0.5 m/s). The service ceiling is lower than the absolute ceiling.
3. **Cruise ceiling (h_{cc}).** The cruise ceiling is defined as the altitude at which the aircraft can climb with a rate of 300 ft/min (i.e., 1.5 m/s). The cruise ceiling is lower than the service ceiling.
4. **Combat ceiling (h_{cc}).** The combat ceiling is defined as the altitude at which a fighter can climb with a rate of 500 ft/min (i.e., 5 m/s). The combat ceiling is lower than the cruise ceiling. This ceiling is defined only for fighter aircraft.

These four definitions are summarized as follows:

$$\begin{aligned} \text{ROC}_{\text{AC}} &= 0 \\ \text{ROC}_{\text{SC}} &= 100\,\text{fpm} \\ \text{ROC}_{\text{CrC}} &= 300\,\text{fpm} \\ \text{ROC}_{\text{CoC}} &= 500\,\text{fpm} \end{aligned} \tag{4.90}$$

In this section, an expression for wing and engine sizing based on ceiling requirements is derived in two sections: (i) jet aircraft and (ii) prop-driven aircraft. Since the ceiling requirements are defined based on the ROC requirements, the equations developed in Section 4.3.5 are employed.

4.3.6.1 Jet Aircraft

An expression for the thrust loading (T/W), as a function of wing loading (W/S) and ROC, was derived in Equation (4.83). It can also be applied to the ceiling altitude, as follows:

$$\left(\frac{T_{\text{C}}}{W}\right) = \frac{\text{ROC}_{\text{C}}}{\sqrt{\dfrac{2}{\rho_{\text{C}}\sqrt{\dfrac{C_{D_o}}{K}}}\left(\dfrac{W}{S}\right)}} + \frac{1}{(L/D)_{\max}} \tag{4.91}$$

where ROC_{C} is the ROC at ceiling, and T_{C} is the engine maximum thrust at ceiling. In contrast, the engine thrust is a function of altitude, or air density. The exact relationship depends upon the engine type, engine technology, engine installation, and airspeed. At this moment of the design phase, where the aircraft is not completely designed, the following approximate relationship (as introduced in Section 4.3) is utilized:

$$T_{\text{C}} = T_{\text{SL}}\left(\frac{\rho_{\text{C}}}{\rho_o}\right) = T_{\text{SL}}\sigma_{\text{C}} \tag{4.92}$$

Inserting this equation into Equation (4.91) yields the following:

$$\left(\frac{T_{\text{SL}}\sigma_{\text{C}}}{W}\right) = \frac{\text{ROC}_{\text{C}}}{\sqrt{\dfrac{2}{\rho_{\text{C}}\sqrt{\dfrac{C_{D_o}}{K}}}\left(\dfrac{W}{S}\right)}} + \frac{1}{(L/D)_{\max}} \tag{4.93}$$

By modeling the atmosphere, one can derive an expression for the relative air density (σ) as a function of altitude (h). The following are reproduced from Ref. [11]:

$$\sigma = \left(1 - 6.873 \cdot 10^{-6}\,\text{h}\right)^{4.26} \quad \text{(from 0 to 36 000 ft)} \tag{4.94a}$$

$$\sigma = 0.2967 \exp\left(1.7355 - 4.8075 \cdot 10^{-5}\,\text{h}\right) \quad \text{(from 36 000 to 65 000 ft)} \tag{4.94b}$$

This pair of equations is in British units, that is, the unit of h is ft. For the atmospheric model in SI units, refer to references such as [5, 7]. Appendices A and B illustrate the pressure, temperature, and air density at various altitudes in SI and British units, respectively. By moving σ in Equation (4.93) to the right-hand side, the following is obtained:

$$\left(\frac{T}{W}\right)_{h_C} = \frac{\text{ROC}_C}{\sigma_C\sqrt{\dfrac{2}{\rho_C\sqrt{\dfrac{C_{D_o}}{K}}}\left(\dfrac{W}{S}\right)}} + \frac{1}{\sigma_C\,(L/D)_{\max}} \tag{4.95}$$

Since at the absolute ceiling (h_{AC}) the ROC is zero ($\text{ROC}_{AC} = 0$), the corresponding expression for the thrust loading will be obtained by eliminating the first term of Equation (4.95):

$$\left(\frac{T}{W}\right)_{h_{AC}} = \frac{1}{\sigma_{AC}\,(L/D)_{\max}} \tag{4.96}$$

where σ_C is the relative air density at the required ceiling, σ_{AC} is the relative air density at the required absolute ceiling, and ROC_C is the ROC at the required ceiling (h_C). The wing and engine sizing based on ceiling requirements (h_C or h_{AC}) is represented in Equations (4.95) and (4.96) as relative air density (σ and σ_{AC}) and can be obtained by Equation (4.94). The ROC at different ceilings is defined at the beginning of this section (Equation (4.90)). Typical values of maximum lift-to-drag ratio for several types of aircraft are given in Table 4.5.

Equation (4.95) represents the contribution of cruise, service, or combat ceiling (h_C) to size engine and wing. However, Equation (4.96) represents the contribution of absolute ceiling (h_{AC}) to size engine and wing. Equations (4.95) and (4.96) represent the non-linear variations of thrust loading versus wing loading as a function of ceiling. The variations of T/W as a function of W/S based on h_C or h_{AC} for a jet aircraft can be sketched using Equations (4.95) and (4.96) in constructing the matching plot, as shown in Figure 4.17.

In order to determine the acceptable region, we need to find out what side of this graph is satisfying the ceiling run requirement. Equation (4.95) has two positive terms; one includes ROC_C and σ_C, while the other includes only σ_C. The ceiling rate of climb (ROC_C) is in the numerator of the first term, and in the denominator of both terms; so, as the rate of climb in Equation (4.95) is *decreased*, the value of the thrust loading (T/W) *drops*. Since any value of ROC greater than the specified ROC_C, or any altitude higher than the required ceiling, is satisfying the ceiling requirement, so the region *above* the graph is acceptable.

Employ extreme care to use a consistent unit when applying Equations (4.95) and (4.96) (either the SI system or the British system). In the SI system, the unit of ROC is m/s, the unit of W is N, the unit of T is N, the unit of S is m^2, and the unit of ρ is kg/m^3. However, in the British system, the unit of ROC is ft/s, the unit of W is lb, the unit of T is lb, the unit of S is ft^2, and the unit of ρ is slug/ft^3. An example application is presented in Section 4.4.

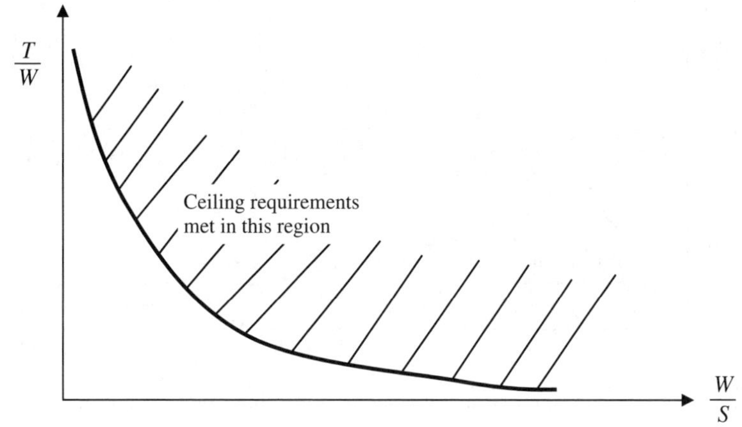

Figure 4.17 Ceiling contribution in constructing a matching plot for a jet aircraft

4.3.6.2 Prop-Driven Aircraft

An expression for the power loading (W/P), as a function of wing loading (W/S) and ROC, was derived in Equation (4.89). It can also be applied to the ceiling altitude, as follows:

$$\left(\frac{W}{P_C}\right) = \frac{1}{\dfrac{ROC_C}{\eta_P} + \sqrt{\dfrac{2}{\rho_C\sqrt{\dfrac{3C_{D_o}}{K}}}\left(\dfrac{W}{S}\right)}\left(\dfrac{1.155}{(L/D)_{max}\,\eta_P}\right)} \tag{4.97}$$

where ROC_C is the ROC at ceiling, ρ_C is the air density at ceiling, and P_C is the engine maximum thrust at ceiling. In contrast, the engine power is a function of altitude, or air density. The exact relationship depends upon the engine type, engine technology, engine installation, and airspeed. At this moment of the design phase, where the aircraft is not completely designed, the following approximate relationship (as introduced in Section 4.3) is utilized:

$$P_C = P_{SL}\left(\frac{\rho_C}{\rho_o}\right) = P_{SL}\sigma_C \tag{4.98}$$

Inserting this equation into Equation (4.97) yields the following:

$$\left(\frac{W}{P_{SL}\sigma_C}\right) = \frac{1}{\dfrac{ROC_C}{\eta_P} + \sqrt{\dfrac{2}{\rho_C\sqrt{\dfrac{3C_{D_o}}{K}}}\left(\dfrac{W}{S}\right)}\left(\dfrac{1.155}{(L/D)_{max}\,\eta_P}\right)} \tag{4.99}$$

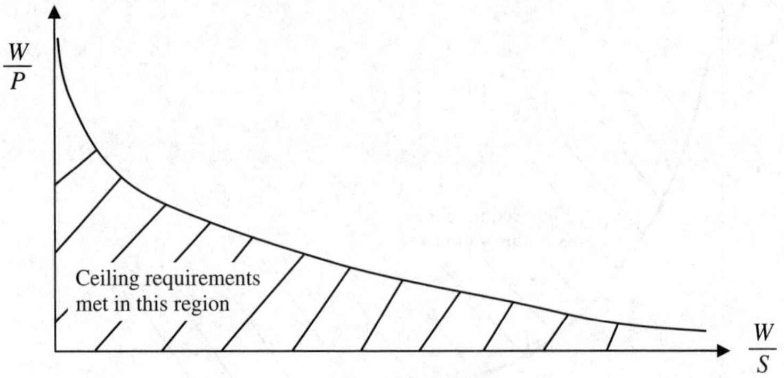

Figure 4.18 Ceiling contribution in constructing a matching plot for a prop-driven aircraft

By moving σ_C in Equation (4.99) to the right-hand side, the following is obtained:

$$\left(\frac{W}{P_{SL}}\right)_C = \frac{\sigma_C}{\frac{ROC_C}{\eta_P} + \sqrt{\frac{2}{\rho_C\sqrt{\frac{3C_{D_o}}{K}}}\left(\frac{W}{S}\right)}\left(\frac{1.155}{(L/D)_{max}\,\eta_P}\right)} \tag{4.100}$$

Since, at the absolute ceiling (h_{AC}), the ROC is zero ($ROC_{AC} = 0$), the corresponding expression for the power loading will be obtained by eliminating the first term of the denominator of Equation (4.100):

$$\left(\frac{W}{P_{SL}}\right)_{AC} = \frac{\sigma_{AC}}{\sqrt{\frac{2}{\rho_{AC}\sqrt{\frac{3C_{D_o}}{K}}}\left(\frac{W}{S}\right)}\left(\frac{1.155}{(L/D)_{max}\,\eta_P}\right)} \tag{4.101}$$

where σ_C is the relative air density at the required ceiling, σ_{AC} is the relative air density at the required absolute ceiling, and ROC_C is the ROC at the required ceiling (h_C). The wing and engine sizing based on ceiling requirements (h_C or h_{AC}) is represented in Equations (4.100) and (4.101). The relative air density (σ and σ_{AC}), as a function of ceiling, can be obtained through Equation (4.94). The ROC at different types of ceiling is defined at the beginning of this section (Equation (4.90)). Typical values of the maximum lift-to-drag ratio for several types of aircraft are given in Table 4.5.

The variation of W/P as a function of W/S based on h_C or h_{AC} for a prop-driven aircraft can be sketched using Equation (4.100) or (4.101) in constructing the matching plot, as shown in Figure 4.18. In order to determine the acceptable region, we need to find what side of this graph is satisfying the climb requirements.

Equation (4.101) has σ_C in the numerator, while it has ρ_C in the numerator of the numerator. As the altitude is increased, the air density (ρ) and the relative air density (σ) are decreased. Hence, by increasing the altitude, the magnitude of the right-hand side in Equation (4.101) is *decreased* and the value of the thrust loading (T/W) *drops*. Since any value of h greater than the specified h_C, or any altitude higher than the required ceiling, is satisfying the ceiling requirement, so the region *below* the graph is acceptable.

Employ extreme caution to use a consistent unit when applying Equations (4.100) and (4.101) (either the SI system or the British system). In the SI system, the unit of ROC is m/s, the unit of W is N, the unit of P is W, the unit of S is m^2, and the unit of ρ is kg/m^3. However, in the British system, the unit of ROC is ft/s, the unit of W is lb, the unit of P is lb ft/s or hp, the unit of S is ft^2, and the unit of ρ is slug/ft^3. If British units are used, convert the unit of W/P to lb/hp to make the comparison more convenient. Recall that each unit of horsepower (hp) is equivalent to 550 lb ft/s. An example application is presented in Section 4.4.

4.4 Design Examples

In this section, two fully solved design examples are provided: Example 4.3 estimates MTOW (W_{TO}) and Example 4.4 determines the wing reference area (S) and engine power (P).

Example 4.3: Maximum Take-Off Weight

Problem statement: You are to design a conventional civil transport aircraft that can carry 700 passengers plus their luggage. The aircraft must be able to fly with a cruise speed of Mach 0.8, and have a range of 9500 km. At this point, you are only required to estimate the aircraft MTOW. You need to follow FAA regulations and standards. Assume that the aircraft is equipped with two high-bypass ratio turbofan engines and is cruising at 35 000 ft altitude.

Solution:

Hint: Since FAR values are in British units, we convert all units to British units.

- **Step 1**. The aircraft is stated to be civil transport and to carry 700 passengers. Hence, the aircraft must follow FAR Part 25. Therefore, all selections must be based on FAR. The regular mission profile for this aircraft consists of taxi and take-off, climb, cruise, descent, loiter, and landing (see Figure 4.19).

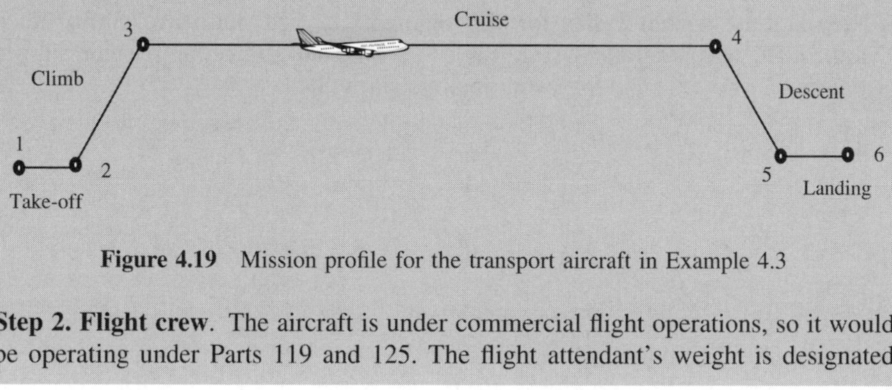

Figure 4.19 Mission profile for the transport aircraft in Example 4.3

- **Step 2. Flight crew**. The aircraft is under commercial flight operations, so it would be operating under Parts 119 and 125. The flight attendant's weight is designated to be 119.3. In Subpart I of Part 125, there are pilot-in-command and second-in-command qualifications. There may be space on the aircraft for more crew members, but based on the language of the document, two flight crew members is the minimum allowed. Also, the criteria for determining minimum flight crew could be found from Appendix D of FAR Part 25. In order to have flight crew to perform the basic workload functions (listed in Appendix D of FAR Part 25 and in Section 119.3) safely and comfortably, we designate two crew members as one pilot and one copilot.

- **Step 3. Flight attendants**. The number of flight attendants is regulated by FAR Part 125 Section 125.269:

  For airplanes having more than 100 passengers – two flight attendants plus one additional flight attendant for each unit (or part of a unit) of 50 passengers above 100 passengers.

  Since there are 700 passengers, the number of flight attendants must be 14:

$$700 = 100 + (12 \cdot 50) \Rightarrow 2 + (12 \cdot 1) = 14$$

- **Step 4. Weight of flight crew and attendants**. As defined in Section 125.9 Definitions, flight crew members are assumed to have a weight of 200 lb. In contrast, the flight attendant's weight is designated to be 119.3 and requires that 140 lb be allocated for a flight attendant whose sex is unknown. Thus, the total weight of flight crew members and flight attendants is:

$$200 + 200 + (14 \cdot 140) \Rightarrow W_C = 2360\,\text{lb}$$

- **Step 5. The weight of payloads**. The payload for a passenger aircraft primarily includes passengers and their luggage and baggage. In reality, passengers could be a combination of adult males, adult females, children, and infants. Table 4.1 shows the nominal weight for each category. To observe the reality and to be on the safe side, an average weight of 180 lb is selected. This weight includes the allowance for personal items and carry-on bags. In contrast, 100 lb of luggage is considered for each passenger. So the total payload would be:

$$(700 \cdot 180) + (700 \cdot 100) \Rightarrow W_{PL} = 196\,000\,\text{lb}$$

- **Step 6. Fuel weight ratios for the segments of taxi, take-off, climb, descent, approach, and landing**. Using Table 4.3 and the numbering system shown in Figure 4.2, we have the following fuel weight ratios:

  Taxi, take-off: $\frac{W_2}{W_1} = 0.98$

  Climb: $\frac{W_3}{W_2} = 0.97$

  Descent: $\frac{W_5}{W_4} = 0.99$

  Approach and landing: $\frac{W_6}{W_5} = 0.997$

- **Step 7. Fuel weight ratio for the segment of range**. The aircraft has jet (turbofan) engine, so Equation (4.16) must be employed. In this flight mission, cruise is the third phase of flight.

$$\frac{W_4}{W_3} = e^{\frac{-RC}{0.866V(L/D)_{max}}} \quad (4.16)$$

where range (R) is 9500 km, C is 0.4 lb/h/lb (from Table 4.6) or 4/3600 1/s, and $(L/D)_{max}$ is 17 (chosen from Table 4.5). The aircraft speed (V) would be the Mach number times the speed of sound [5]:

$$V = M \cdot a = 0.8 \cdot 296.6 = 237.3\,\text{m/s} = 778.5\,\text{m/s} \quad (4.65)$$

where the speed of sound at 35 000 ft altitude is 296.6 m/s or 973.1 ft/s. Thus,

$$\frac{W_4}{W_3} = e^{\frac{-RC}{0.866V(L/D)_{max}}} = e^{\frac{-9500\,000 \cdot 3.28 \cdot \frac{0.4}{3600}}{0.866 \cdot 973.1 \cdot 17}} = e^{-0.3053} \Rightarrow \frac{W_4}{W_3} = 0.737 \quad (4.16)$$

- **Step 8. Overall fuel weight ratio**. By using equations similar to Equations (4.10) and (4.11), we obtain:

$$\frac{W_6}{W_1} = \frac{W_2}{W_1}\frac{W_3}{W_2}\frac{W_4}{W_3}\frac{W_5}{W_4}\frac{W_6}{W_5} = 0.98 \cdot 0.97 \cdot 0.737 \cdot 0.99 \cdot 0.997 \Rightarrow \frac{W_6}{W_1} = 0.692 \quad (4.10)$$

$$\frac{W_f}{W_{TO}} = 1.05\left(1 - \frac{W_6}{W_1}\right) = 1.05\,(1 - 0.692) \Rightarrow \frac{W_f}{W_{TO}} = 0.323 \quad (4.11)$$

- **Step 9. Substitution**. The known values are substituted into Equation (4.5):

$$W_{TO} = \frac{W_{PL} + W_C}{1 - \left(\frac{W_f}{W_{TO}}\right) - \left(\frac{W_E}{W_{TO}}\right)} = \frac{196\,000 + 2360}{1 - 0.323 - \left(\frac{W_E}{W_{TO}}\right)} = \frac{198360}{0.677 - \left(\frac{W_E}{W_{TO}}\right)} \quad (4.5)$$

- **Step 10. Empty weight ratio**. The empty weight ratio is established by using Equation (4.26), where the coefficients a and b are taken from Table 4.8:

$$a = -7.754 \cdot 10^{-8},\ b = 0.576\ \text{(Table 4.8)}$$

Thus:

$$\frac{W_E}{W_{TO}} = aW_{TO} + b \Rightarrow \frac{W_E}{W_{TO}} = -7.754 \cdot 10^{-8} W_{TO} + 0.576 \quad (4.26)$$

- **Step 11. Final step**. The following two equations (one from step 9 and one from step 10) must be solved simultaneously:

$$W_{TO} = \frac{198360}{0.677 - \left(\frac{W_E}{W_{TO}}\right)} \quad \text{(1) (step 9)}$$

$$\frac{W_E}{W_{TO}} = -7.754 \cdot 10^{-8} W_{TO} + 0.576 \qquad (2)\ (\text{step } 10)$$

MathCad software may be used to solve this set of two non-linear algebraic equations, as follows:

assumption: $x := 0.6 \ \ y := 1\,000\,000$

Given

$$y = \frac{198\,360}{0.677 - x} \qquad x = -7.754 \cdot 10^{-8} \cdot y + 0.576$$

$$\text{Find }(x,\ y) = \begin{pmatrix} 0.493 \\ 1075664.161 \end{pmatrix}$$

Thus, the empty weight ratio is 0.493 and the MTOW is:

$$W_{TO} = 1\ 075\ 664\,\text{lb} = 4\ 784\ 792\ \text{N}$$

So, the maximum take-off mass is:

$$m_{TO} = 487\,913\,\text{kg}$$

An alternative way to find W_{TO} is the trial-and-error technique, as shown in Table 4.16. It is observed that after seven trials, the error reduces to only 0.4%, which is acceptable. This technique produces a similar result ($W_{TO} = 1\ 074\ 201$).

The third alternative is to solve the equations analytically. We first manipulate Equation (4.1) as follows:

$$W_{TO} = \frac{198\,360}{0.677 - \left(\frac{W_E}{W_{TO}}\right)} \Rightarrow 0.677 - \left(\frac{W_E}{W_{TO}}\right) = \frac{198360}{W_{TO}} \Rightarrow \left(\frac{W_E}{W_{TO}}\right)$$
$$= 0.677 - \frac{198360}{W_{TO}}$$

Then, we need to substitute the right-hand side into Equation (4.2) and simplify:

$$0.677 - \frac{198360}{W_{TO}} = -7.754 \cdot 10^{-8} W_{TO} + 0.576$$
$$\Rightarrow 7.754 \cdot 10^{-8} W_{TO} + 0.576 - 0.677 + \frac{198360}{W_{TO}} = 0$$
$$\Rightarrow -7.754 \cdot 10^{-8} W_{TO} + \frac{198360}{W_{TO}} - 0.101 = 0$$

This non-linear algebraic equation has one unknown (W_{TO}) and only one acceptable (reasonable) solution. This alternative technique also produces the same result. For comparison, it is interesting to note that the MTOW of the giant transport aircraft

Table 4.16 Trial-and-error technique to determine maximum take-off weight of the aircraft in Example 4.3

Iteration	Step 1	Step 2	Step 3	Error (%)
	Guess W_{TO} (lb)	Substitute W_{TO} of Step 1 into the first equation: $\frac{W_E}{W_{TO}} = -7.754 \cdot 10^{-8} W_{TO} + 0.576$	Substitute W_E/W_{TO} of Step 2 into the second equation: $W_{TO} = \frac{198.360}{0.677 - \left(\frac{W_E}{W_{TO}}\right)}$	
1	1,500,000	0.456	912,797 lb	−64
2	912,797	0.505	1,154,744 lb	20.9
3	1,154,744	0.486	1,041,047 lb	−10.9
4	1,041,047	0.495	1,091,552 lb	4.6
5	1,091,552	0.491	1,068,525 lb	−2.1
6	1,068,525	0.493	1,078,902 lb	0.96
7	1,078,902	0.4923	1,074,201 lb	−0.4

Airbus 380 with 853 passengers is 1 300 700 lb. Thus, the aircraft maximum aircraft weight would be:

$$W_{TO} = 1\,074\,201\ \text{lb} \Rightarrow m_{TO} = 487\,249\ \text{kg}$$

Example 4.4: Wing and Engine Sizing

Problem statement: In the preliminary design phase of a turboprop transport aircraft, the MTOW is determined to be 20 000 lb and the aircraft C_{D_o} is determined to be 0.025. The hob airport is located at a city with an elevation of 3000 ft. Using the matching plot technique, determine the wing area (S) and engine power (P) of the aircraft that is required to have the following performance capabilities:

1. Maximum speed: 350 KTAS at 30 000 ft.
2. Stall speed: less than 70 KEAS.
3. ROC: more than 2700 fpm at sea level.
4. Take-off run: less than 1200 ft (on a dry concrete runway).
5. Service ceiling: more than 35 000 ft.
6. Range: 4000 nm.
7. Endurance: 2 hours.

Assume any other parameters that you may need for this aircraft.

Solution:

First, it must be noted that the range and endurance requirements do not have any effect on the engine power or wing area, so we ignore them at this design phase. The air density at 3000 ft is 0.002175 slug/ft^3 and at 30 000 ft is 0.00089 slug/ft^3.

The matching plot is constructed by deriving five equations:

- **Stall speed**. The stall speed is required to be greater than 70 KEAS. The wing sizing based on stall speed requirements is represented by Equation (4.31). From Table 4.11, the aircraft maximum lift coefficient is selected to be 2.7.

$$\left(\frac{W}{S}\right)_{V_s} = \frac{1}{2}\rho V_s^2 C_{L_{max}} = \frac{1}{2}\cdot 0.002378 \cdot (70\cdot 1.688)^2 \cdot 2.7 = 44.8\,\frac{\text{lb}}{\text{ft}^2} \quad \text{(4.31) or (E-1)}$$

where 1 knot is equivalent to 1.688 ft/s.

- **Maximum speed**. The maximum speed is required to be greater than 350 KTAS at 30 000 ft. The wing and engine sizing based on maximum speed requirements for a prop-driven aircraft is represented by Equation (4.41).

$$\left(\frac{W}{P_{SL}}\right)_{V_{max}} = \frac{\eta_P}{\frac{1}{2}\rho_o V_{max}^3 C_{D_o}\frac{1}{\left(\frac{W}{S}\right)} + \frac{2K}{\rho\sigma V_{max}}\left(\frac{W}{S}\right)} \tag{4.56}$$

From Table 5.8, the wing AR is selected to be 12. From Section 4.3.3, the Oswald span efficiency factor is considered to be 0.85. Thus:

$$K = \frac{1}{\pi\cdot e\cdot \text{AR}} = \frac{1}{3.14\cdot 0.85\cdot 12} = 0.031 \tag{4.41}$$

The air relative density (σ) at 30 000 ft is 0.00089/0.002378 or 0.374. The substitution yields:

$$\left(\frac{W}{P_{SL}}\right)_{V_{max}} = \frac{0.7\cdot 550}{0.5\cdot 0.002378\cdot(350\cdot 1.688)^3\cdot 0.025\frac{1}{\left(\frac{W}{S}\right)} + \frac{2\cdot 0.031}{0.00089\cdot 0.374\cdot(350\cdot 1.688)}\left(\frac{W}{S}\right)}$$

or:

$$\left(\frac{W}{P_{SL}}\right)_{V_{max}} = \frac{385}{6129.7\frac{1}{\left(\frac{W}{S}\right)} + 0.317\left(\frac{W}{S}\right)}\left(\frac{\text{lb}}{\text{hp}}\right) \tag{E-2}$$

where the whole term is multiplied by 550 to convert lb/(lb ft/s) to lb/hp, and the prop efficiency is assumed to be 0.7.

- **Take-off run**. The take-off run is required to be greater than 1200 ft at an elevation of 3000 ft. The wing and engine sizing based on take-off run requirements for a prop-driven aircraft is represented by Equation (4.76). Recall that the air density at 3000 ft is 0.002175 slug/ft^3.

$$\left(\frac{W}{P}\right)_{S_{\mathrm{TO}}} = \frac{1-\exp\left(0.6\rho g C_{D_\mathrm{G}} S_{\mathrm{TO}}\frac{1}{W/S}\right)}{\mu-\left(\mu+\frac{C_{D_\mathrm{G}}}{C_{L_\mathrm{R}}}\right)\left[\exp\left(0.6\rho g C_{D_\mathrm{G}} S_{\mathrm{TO}}\frac{1}{W/S}\right)\right]}\frac{\eta_\mathrm{P}}{V_{\mathrm{TO}}} \tag{4.76}$$

where, based on Table 4.15, μ is 0.04. The take-off speed is assumed to be:

$$V_{\mathrm{TO}} = 1.1V_\mathrm{s} = 1.1\cdot 70 = 77\mathrm{KEAS} \tag{4.72}$$

The take-off lift and drag coefficients are:

$$C_{L_{\mathrm{TO}}} = C_{L_C} + \Delta C_{L_{\mathrm{flap\_TO}}} \tag{4.69c}$$

where the aircraft lift coefficient C_{L_C} is assumed to be 0.3 and $\Delta C_{L_{\mathrm{flap\_TO}}}$ to be 0.6. Thus:

$$C_{L_{\mathrm{TO}}} = 0.3 + 0.6 = 0.9 \tag{4.69c}$$

$$C_{D_{\mathrm{oLG}}} = 0.009$$

$$C_{D_{\mathrm{oHLD\_TO}}} = 0.005 \tag{4.69a}$$

$$C_{D_{\mathrm{oTO}}} = C_{D_\mathrm{o}} + C_{D_{\mathrm{oLG}}} + C_{D_{\mathrm{oHLD\_TO}}} = 0.025 + 0.009 + 0.005 = 0.039 \tag{4.69}$$

$$C_{D_{\mathrm{TO}}} = C_{D_{\mathrm{oTO}}} + KC_{L_{\mathrm{TO}}}^2 = 0.039 + 0.031\,(0.9)^2 = 0.064 \tag{4.68}$$

The take-off rotation lift coefficients is:

$$C_{L_\mathrm{R}} = \frac{C_{L_{\max}}}{(1.1)^2} = \frac{C_{L_{\max}}}{1.21} = \frac{2.7}{1.21} = 2.231 \tag{4.69b}$$

The variable C_{D_G} is:

$$C_{D_\mathrm{G}} = \left(C_{D_{\mathrm{TO}}} - \mu C_{L_{\mathrm{TO}}}\right) = 0.064 - 0.04\cdot 0.9 = 0.028 \tag{4.67}$$

It is assumed that the propeller is of variable-pitch type, so based on Equation (4.73b) the prop efficiency is 0.6. The substitution yields:

$$\left(\frac{W}{P}\right)_{S_{\mathrm{TO}}} = \frac{1-\exp\left(0.6\rho g C_{D_\mathrm{G}} S_{\mathrm{TO}}\frac{1}{W/S}\right)}{\mu-\left(\mu+\frac{C_{D_\mathrm{G}}}{C_{L_\mathrm{R}}}\right)\left[\exp\left(0.6\rho g C_{D_\mathrm{G}} S_{\mathrm{TO}}\frac{1}{W/S}\right)\right]}\frac{\eta_\mathrm{P}}{V_{\mathrm{TO}}} \tag{4.76}$$

$$\left(\frac{W}{P}\right)_{S_{\mathrm{TO}}}=\frac{\left[1-\exp\left(0.6\cdot 0.002175\cdot 32.2\cdot 0.028\cdot 1,200\frac{1}{(W/S)}\right)\right]}{0.04-\left(0.04+\frac{0.028}{2.231}\right)\left[\exp\left(0.6\cdot 0.002175\cdot 32.2\cdot 0.028\cdot 1200\frac{1}{(W/S)}\right)\right]}$$

$$\cdot\left(\frac{0.6}{77\cdot 1.688}\right)\cdot 550$$

or:

$$\left(\frac{W}{P}\right)_{S_{\mathrm{TO}}}=\frac{\left[1-\exp\left(\frac{1.426}{(W/S)}\right)\right]}{0.04-(0.053)\left[\exp\left(\frac{1.426}{(W/S)}\right)\right]}(0.0046\cdot 550)\ \frac{\mathrm{lb}}{\mathrm{hp}} \tag{E-3}$$

Again, the whole term is multiplied by 550 to convert lb/(lb ft/s) to lb/hp.

- **Rate of Climb**. The ROC run is required to be greater than 2700 fpm (or 45 ft/s) at sea level. The wing and engine sizing based on ROC requirements for a prop-driven aircraft is represented by Equation (4.89). Based on Table 4.5, the maximum lift-to-drag ratio is selected to be 18:

$$\left(\frac{W}{P}\right)_{\mathrm{ROC}}=\frac{1}{\frac{\mathrm{ROC}}{\eta_{\mathrm{P}}}+\sqrt{\frac{2}{\rho\sqrt{\frac{3C_{D_o}}{K}}}\left(\frac{W}{S}\right)}\left(\frac{1.155}{(L/D)_{\max}\,\eta_{\mathrm{P}}}\right)} \tag{4.89}$$

The substitution yields:

$$\left(\frac{W}{P}\right)_{\mathrm{ROC}}=\frac{1\cdot 550}{\frac{2700}{60\cdot 0.7}+\sqrt{\frac{2}{0.002378\sqrt{\frac{3\cdot 0.025}{0.031}}}\left(\frac{W}{S}\right)}\left(\frac{1.155}{18\cdot 0.7}\right)} \tag{4.89}$$

$$\left(\frac{W}{P}\right)_{\mathrm{ROC}}=\frac{1\cdot 550}{64.3+\left(\sqrt{540.7\left(\frac{W}{S}\right)}\right)(0.092)} \tag{E-4}$$

And again, the whole term is multiplied by 550 to convert lb/(lb ft/s) to lb/hp.

- **Service ceiling**. The service ceiling is required to be greater than 35 000 ft. The wing and engine sizing based on service ceiling requirements for a prop-driven aircraft is represented by Equation (4.100). At service ceiling, the ROC is required to be 100 ft/min (or 1.667 ft/s). At 35 000 ft altitude, the air density is 0.000738 slug/ft^3

(Appendix B); so the relative air density is 0.31. The substitution yields:

$$\left(\frac{W}{P_{SL}}\right)_C = \frac{\sigma_C}{\frac{ROC_C}{\eta_P} + \sqrt{\frac{2}{\rho_C\sqrt{\frac{3C_{D_o}}{K}}}\left(\frac{W}{S}\right)}\left(\frac{1.155}{(L/D)_{max}\eta_P}\right)} \tag{4.100}$$

$$\left(\frac{W}{P}\right)_C = \frac{0.31 \cdot 550}{\frac{100}{60 \cdot 0.7} + \sqrt{\frac{2}{0.000738\sqrt{\frac{3 \cdot 0.025}{0.031}}}\left(\frac{W}{S}\right)}\left(\frac{1.155}{18 \cdot 0.7}\right)} \tag{4.100}$$

or:

$$\left(\frac{W}{P}\right)_C = \frac{170.5}{2.38 + \left(\sqrt{1742.3\left(\frac{W}{S}\right)}\right)(0.092)} \tag{E-5}$$

And again, the whole term is multiplied by 550 to convert lb/(lb ft/s) to lb/hp.

- **Construction of matching plot**. Now, we have five equations (E-1), (E-2), (E-3), (E-4), and (E-5). In all of them, power loading is defined as functions of wing loading. When we plot all of them in one graph, Figure 4.20 will be produced. Recall in this example that the unit of W/S is lb/ft^2 and the unit of W/P is lb/hp.

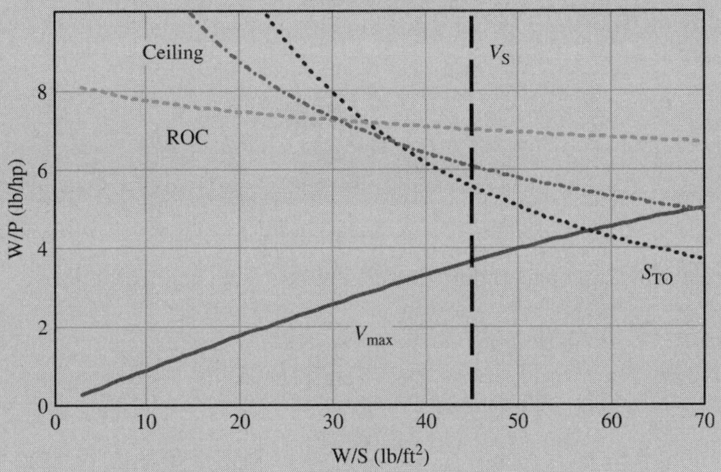

Figure 4.20 Matching plot for Example 4.4

Now, we need to recognize the acceptable regions. As discussed in Section 4.3, the region below each graph is satisfying the performance requirements. In other words, the region above each graph is not satisfying the performance requirements. For the case of stall speed, the region on the left side of the graph is satisfying the stall speed requirements (see Figure 4.21). Hence, the region between the graphs of maximum speed, take-off run, and stall speed is the target area.

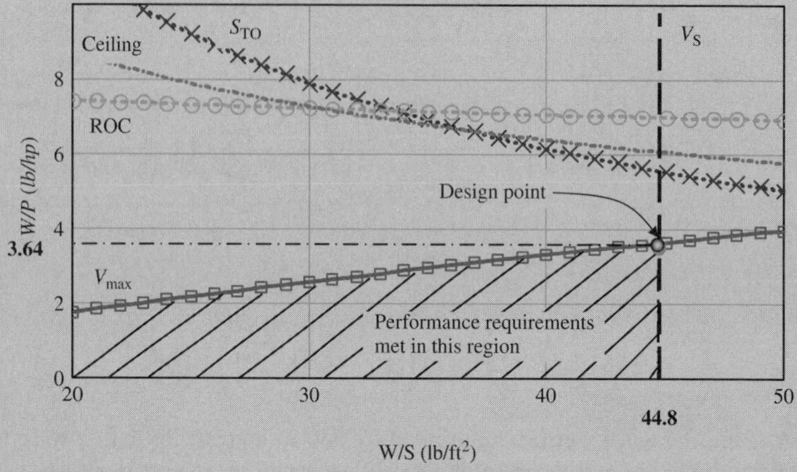

Figure 4.21 Acceptable regions in the matching plot for Example 4.4

In this region, we are looking for the smallest engine (lowest power) that has the lowest operating cost. Thus the highest point (Figure 4.21) of this region is the design point. Therefore the wing loading and power loading will be extracted from Figure 4.21 as:

$$\left(\frac{W}{P}\right)_d = 3.64$$

$$\left(\frac{W}{S}\right)_d = 44.8$$

Then, the wing area and engine power will be calculated as follows:

$$S = W_{TO} \Big/ \left(\frac{W}{S}\right)_d = \frac{20\,000\,\text{lb}}{44.8\frac{\text{lb}}{\text{ft}^2}} = 446.4\text{ft}^2 = 41.47\text{m}^2 \tag{4.27}$$

$$P = W_{TO} \Big/ \left(\frac{W}{P}\right)_d = \frac{20\,000\,\text{lb}}{3.64\frac{\text{lb}}{\text{hp}}} = 5495.5\,\text{hp} = 4097.2\,\text{kW} \tag{4.28}$$

Therefore, the wing area and engine power will be:

$$S = 446.4\,\text{ft}^2, \quad P = 5495.5\,\text{hp}$$

Problems

1. Determine the zero-lift drag coefficient (C_{D_o}) of the two-seat ultralight aircraft Scheibe SF 40 which is flying with a maximum cruising speed of 81 knot at sea level. This aircraft has one piston engine and the following characteristics:

$$P_{SL_{max}} = 44.7\,\text{kW},\ m_{TO} = 400\,\text{lb},\ S = 13.4\,\text{m}^2,\ \text{AR} = 8.7,\ e = 0.88,\ \eta_P = 0.75$$

2. Determine the zero-lift drag coefficient (C_{D_o}) of the fighter aircraft F-16C Falcon which is flying with a maximum speed of Mach 2.2 at 40 000 ft. This fighter has a turbofan engine and the following characteristics:

$$T_{SL_{max}} = 29\ 588\,\text{lb},\ W_{TO} = 27\ 185\,\text{lb},\ S = 300\,\text{ft}^2,\ \text{AR} = 3.2,\ e = 0.76$$

3. Determine the zero-lift drag coefficient (C_{D_o}) of the jet fighter aircraft F-15 Eagle which is flying with a maximum speed of Mach 2.5 at 35 000 ft. This fighter has two turbofan engines and the following characteristics:

$$T_{SL_{max}} = 2 \cdot 23\,450\,\text{lb},\ W_{TO} = 81\ 000\,\text{lb},\ S = 608\,\text{ft}^2,\ \text{AR} = 3,\ e = 0.78$$

4. Determine the zero-lift drag coefficient (C_{D_o}) of the transport aircraft Boeing 747-400 which is flying with a maximum speed of Mach 0.92 at 35 000 ft. This aircraft has four turbofan engines and the following characteristics:

$$T_{SL_{max}} = 4 \cdot 56\,750,\ W_{TO} = 800\ \ 000\,\text{lb},\ S = 5825\,\text{ft}^2,\ \text{AR} = 10.2,\ e = 0.85$$

5. Determine the zero-lift drag coefficient (C_{D_o}) of the fighter aircraft Eurofighter which is flying with a maximum speed of Mach 2 at 35 000 ft. This fighter has two turbofan engines and the following characteristics:

$$T_{SL_{max}} = 2 \cdot 16\,000,\ W_{TO} = 46\ 297\,\text{lb},\ S = 538\,\text{ft}^2,\ \text{AR} = 2.2,\ e = 0.75$$

6. Determine the zero-lift drag coefficient (C_{D_o}) of the bomber aircraft B-2 Spirit which is flying with a maximum speed of Mach 0.95 at 20 000 ft. This aircraft has four turbofan engines and the following characteristics:

$$T_{SL_{max}} = 4 \cdot 17\,300\,\text{lb},\ W_{TO} = 336\ \ 500\,\text{lb},\ S = 5\ 000\,\text{ft}^2,\ \text{AR} = 6.7,\ e = 0.73$$

7. Determine the zero-lift drag coefficient (C_{D_o}) of the military transport aircraft C-130 Hercules which is flying with a maximum speed of 315 knot at 23 000 ft. This aircraft has four turboprop engines and the following characteristics:

$$P_{SL_{max}} = 4 \cdot 4508\,hp,\ W_{TO} = 155\,000\,\text{lb},\ S = 1\ 754\,\text{ft}^2,\ \text{AR} = 10.1,\ e = 0.92,$$

$$\eta_P = 0.81$$

8. Determine the zero-lift drag coefficient (C_{D_o}) of the transport aircraft Piaggio P180 Avanti which is flying with a maximum speed of 395 knot at 20 000 ft. This aircraft has two turboprop engines and the following characteristics:

$$P_{SL_{max}} = 2 \cdot 800\,hp, W_{TO} = 10\ 510\,\text{lb},\ S = 172.2\,\text{ft}^2,\ \text{AR} = 12.1,\ e = 0.88,$$
$$\eta_P = 0.84$$

9. Determine the zero-lift drag coefficient (C_{D_o}) of the small utility aircraft Beech Bonanza which is flying with a maximum speed of 166 knot at sea level. This aircraft has one piston engine and the following characteristics:

$$P_{SL_{max}} = 285\,\text{hp},\ W_{TO} = 2725\,\text{lb},\ S = 178\,\text{ft}^2,\ \text{AR} = 6,\ e = 0.87,\ \eta_P = 0.76$$

10. Determine the zero-lift drag coefficient (C_{D_o}) of the multi-mission aircraft Cessna 208 Caravan which is flying with a maximum cruising speed of 184 knot at 10 000 ft. This aircraft has one turboprop engine and the following characteristics:

$$P_{SL_{max}} = 505\,\text{kW},\ m_{TO} = 3970\,\text{kg},\ S = 26\,\text{m}^2,\ \text{AR} = 9.7,\ e = 0.91,\ \eta_P = 0.75$$

11. You are a member of a team that is designing a GA aircraft which is required to have four seats and the following performance features:

(a) Maximum speed: at least 150 knots at sea level.
(b) Maximum range: at least 700 km.
(c) Maximum rate of climb: at least 1800 fpm.
(d) Absolute ceiling: at least 25 000 ft.
(e) Take-off run: less than 1200 ft.

At the preliminary design phase, you are required to estimate the zero-lift drag coefficient (C_{D_o}) of such an aircraft. Identify five current similar aircraft and based on their statistics, estimate the C_{D_o} of the aircraft being designed.

12. You are a member of a team that is designing a business jet aircraft which is required to carry 12 passengers and have the following performance features:

(a) Maximum speed: at least 280 knots at sea level.
(b) Maximum range: at least 1000 km.
(c) Maximum rate of climb: at least 3000 fpm.
(d) Absolute ceiling: at least 35 000 ft.
(e) Take-off run: less than 2000 ft.

At the preliminary design phase, you are required to estimate the zero-lift drag coefficient (C_{D_o}) of such an aircraft. Identify five current similar aircraft and based on their statistics, estimate the C_{D_o} of the aircraft being designed.

13. You are a member of a team that is designing a fighter aircraft which is required to carry two pilots and have the following performance features:

(a) **Maximum speed**: at least Mach 1.8 at 30 000 ft.
(b) **Maximum range**: at least 1500 km.
(c) **Maximum rate of climb**: at least 10 000 fpm.
(d) **Absolute ceiling**: at least 45 000 ft.
(e) **Take-off run**: less than 2800 ft.

At the preliminary design phase, you are required to estimate the zero-lift drag coefficient (C_{D_0}) of such an aircraft. Identify five current similar aircraft and based on their statistics, estimate the C_{D_0} of the aircraft being designed.

14. You are involved in the design of a civil transport aircraft which can carry 200 passengers plus their luggage. The aircraft must be able to fly with a cruise speed of Mach 0.8, and have a range of 10 000 km. At this point, you are only required to estimate the aircraft maximum take-off weight. You need to follow FAA regulations and standards. Assume that the aircraft is equipped with two high-bypass ratio turbofan engines and is required to cruise at 37 000 ft altitude.

15. You are to design a surveillance/observation aircraft which can carry four crew members. The aircraft must be able to fly with a cruise speed of Mach 0.3, and have a range of 2000 km and an endurance of 15 h. At this point, you are only required to estimate the aircraft maximum take-off weight. Assume that the aircraft is equipped with two turboprop engines and is required to cruise at 8000 m altitude.

16. You are involved in the design of a jet trainer aircraft that can carry one pilot and one student. The aircraft must be able to fly with a cruise speed of Mach 0.4, and have a range of 1500 km. At this point, you are only required to estimate the aircraft maximum take-off weight. Assume that the aircraft is equipped with one turboprop engine and is required to cruise at 20 000 ft altitude.

17. At the preliminary design phase of a GA (normal) aircraft, the maximum take-off weight is determined to be 2000 lb and the aircraft C_{D_0} is determined to be 0.027; the engine is selected to be one piston-prop. By using the matching plot technique, determine the wing area (S) and engine power (P) of the aircraft that is required to have the following performance capabilities:

(a) **Maximum speed**: 180 KTAS at 20 000 ft.
(b) **Stall speed**: less than 50 KEAS.
(c) **Rate of climb**: more than 1200 fpm at sea level.
(d) **Take-off run**: less than 800 ft (on a dry concrete runway).
(e) **Service ceiling**: more than 25 000 ft.
(f) **Range**: 1000 nm.
(g) **Endurance**: 1 h.

Assume any other parameters that you may need for this aircraft.

18. At the preliminary design phase of a jet transport aircraft, the maximum take-off weight is determined to be 120 000 lb and the aircraft C_{D_0} is determined to be 0.022. The hob airport is located at a city with an elevation of 5000 ft. By using the matching plot technique, determine the wing area (S) and engine thrust (T) of the aircraft that is required to have the following performance capabilities:

(a) **Maximum speed**: 370 KTAS at 27 000 ft.
(b) **Stall speed**: less than 90 KEAS.
(c) **Rate of climb**: more than 3200 fpm at sea level.
(d) **Take-off run**: less than 3000 ft (on a dry concrete runway).
(e) **Service ceiling**: more than 40 000 ft.
(f) **Range**: 8000 nm.
(g) **Endurance**: 5 h.

Assume any other parameters that you may need for this aircraft.

19. At the preliminary design phase of a fighter aircraft, the maximum take-off mass is determined to be 12 000 kg and the aircraft C_{D_o} is determined to be 0.028. By using the matching plot technique, determine the wing area (S) and engine thrust (T) of the aircraft that is required to have the following performance capabilities:

(a) **Maximum speed**: Mach 1.9 at 10 000 m.
(b) **Stall speed**: less than 50 m/s.
(c) **Rate of climb**: more than 50 m/s at sea level.
(d) **Take-off run**: less than 1000 m (on a dry concrete runway).
(e) **Service ceiling**: more than 15 000 m.
(f) **Radius of action**: 4000 km.

Assume any other parameters that you may need for this aircraft.

20. At the preliminary design phase of a twin-turboprop regional transport aircraft, the maximum take-off mass is determined to be 16 000 kg and the aircraft C_{D_o} is determined to be 0.019. By using the matching plot technique, determine the wing area (S) and engine power (P) of the aircraft that is required to have the following performance capabilities:

(a) **Maximum speed**: Mach 0.6 at 2500 m.
(b) **Stall speed**: less than 190 km/h.
(c) **Rate of climb**: more than 640 m/min at sea level.
(d) **Take-off run**: less than 1100 m (on a dry concrete runway).
(e) **Service ceiling**: more than 9000 m.
(f) **Range**: 7000 km.

Assume any other parameters that you may need for this aircraft.

References

[1] Federal Aviation Regulations, Federal Aviation Administration, Department of Transportation, 2011, www.faa.gov.
[2] Advisory Circular (2005) Aircraft Weight and Balance Control, FAA, AC 120-27E.
[3] MIL-STD 1797A (2004) *Flying Qualities of Piloted Aircraft*, Department of Defense Interface Standard.
[4] Jackson, P. *Jane's All the World's Aircraft*, Jane's Information Group, various years 1996 to 2011.
[5] Sadraey, M. (2009) *Aircraft Performance Analysis*, VDM Verlag Dr. Müller.
[6] Lan, E.C.T. and Roskam, J. (2003) *Airplane Aerodynamics and Performance*, DAR Corporation.

[7] Anderson, J.D. (1999) *Aircraft Performance and Design*, McGraw-Hill.
[8] Roskam, J. (2005) *Airplane Design*, vol. **I**, DAR Corporation.
[9] Loftin, L.K. (1980) *Subsonic Aircraft: Evolution and the Matching of Size to Performance*, NASA, Reference Publication 1060.
[10] Hoak, D.E., Ellison, D.E., and Fink, R.D. (1978) USAF stability and control DATCOM. Flight Control Division, Air Force Flight Dynamics Laboratory, Wright-Patterson AFB, Ohio.
[11] Bertin, L.J. and Cummings, R.M. (2009) *Aerodynamics for Engineers*, 5th edn, Pearson/Prentice Hall.

5

Wing Design

5.1 Introduction

In Chapter 4, aircraft preliminary design – the second step in the design process – was introduced. Three parameters were determined during preliminary design, namely: aircraft maximum take-off weight (W_{TO}), engine power (P) or engine thrust (T), and wing reference area (S_{ref}). The third step in the design process is the detail design. During detail design, major aircraft components such as the wing, fuselage, horizontal tail, vertical tail, propulsion system, landing gear, and control surfaces are designed one by one. Each aircraft component is designed as an individual entity at this step, but in later design steps, they are integrated as one system – aircraft and their interactions are considered.

This chapter focuses on the detail design of the wing. The wing may be considered as the most important component of an aircraft, since a fixed-wing aircraft is not able to fly without it. Since the wing geometry and its features influence all other aircraft components, we begin the detail design process with wing design. The primary function of the wing is to generate sufficient lift force or simply lift (L). However, the wing has two other productions, namely the drag force or drag (D) and nose-down pitching moment (M). While a wing designer is looking to maximize the lift, the other two (drag and pitching moment) must be minimized. In fact, a wing is considered as a lifting surface where lift is produced due to the pressure difference between the lower and upper surfaces. Aerodynamics textbooks are a good source to consult for information about mathematical techniques to calculate the pressure distribution over the wing and for determining the flow variables.

Basically, the principles and methodologies of systems engineering are followed in the wing design process. Limiting factors in the wing design approach originate from design requirements such as performance requirements, stability and control requirements, producibility requirements, operational requirements, cost, and flight safety. Major performance requirements include stall speed, maximum speed, take-off run, range, and endurance. Primary stability and control requirements include lateral-directional static stability, lateral-directional dynamic stability, and aircraft controllability during probable wing stall.

Aircraft Design: A Systems Engineering Approach, First Edition. Mohammad H. Sadraey.

During the wing design process, 18 parameters must be determined. They are as follows:

1. wing reference (or planform) area (S_W or S_{ref} or S);
2. number of wings;
3. vertical position relative to the fuselage (high, mid-, or low wing);
4. horizontal position relative to the fuselage;
5. cross-section (or airfoil);
6. aspect ratio (AR);
7. taper ratio (λ);
8. tip chord (C_t);
9. root chord (C_r);
10. mean aerodynamic chord (MAC or C);
11. span (b);
12. twist angle (or washout) (α_t);
13. sweep angle (Λ);
14. dihedral angle (Γ);
15. incidence (i_w) (or setting angle, α_{set});
16. high-lifting devices such as flap;
17. aileron;
18. other wing accessories.

Of the above long list, only the first one (i.e., planform area) has been calculated so far (during the preliminary design step). In this chapter, the approach to calculate or select the other 17 wing parameters is examined. The aileron design (item 17) is a rich topic in the wing design process and has a variety of design requirements, so it will not be discussed in this chapter. Chapter 12 is devoted to the control surfaces design and the aileron design technique (as one control surface) will be presented in that chapter. Horizontal wing position relative to the fuselage will be discussed later in Chapter 7, when the fuselage and tail have been designed.

Thus, the wing design begins with one known variable (S), and considering all design requirements, the other 17 wing parameters are obtained. The wing must produce sufficient lift while generating minimum drag, and minimum pitching moment. These design goals must be satisfied collectively throughout all flight operations and missions. There are other wing parameters that could be added to this list, such as wing tip, winglet, engine installation, fairing, vortex generator, and wing structural considerations. Wing tip, winglet, fairing, and vortex generator will be discussed in Section 5.15; and engine installation will be addressed in Chapter 8. The topic of wing structural considerations is beyond the scope of this text. Figure 5.1 illustrates the flowchart of wing design. It starts with the known variable (S) and ends with optimization. The details of the design steps for each box will be explained later in this chapter.

As Figure 5.1 implies, the wing design is an iterative process and the selections/calculations are usually repeated several times. For instance, 76 various wings were designed for the Boeing 767 (Figure 5.4) in 1986 until the best wing was finalized. However, only 11 wings were designed for the Boeing 787 Dreamliner (Figure 1.10) in 2008. A reduction in the number of iterations is evident, which is partially due to advances in software/hardware in recent years, and partly due to the years of experience of Boeing designers.

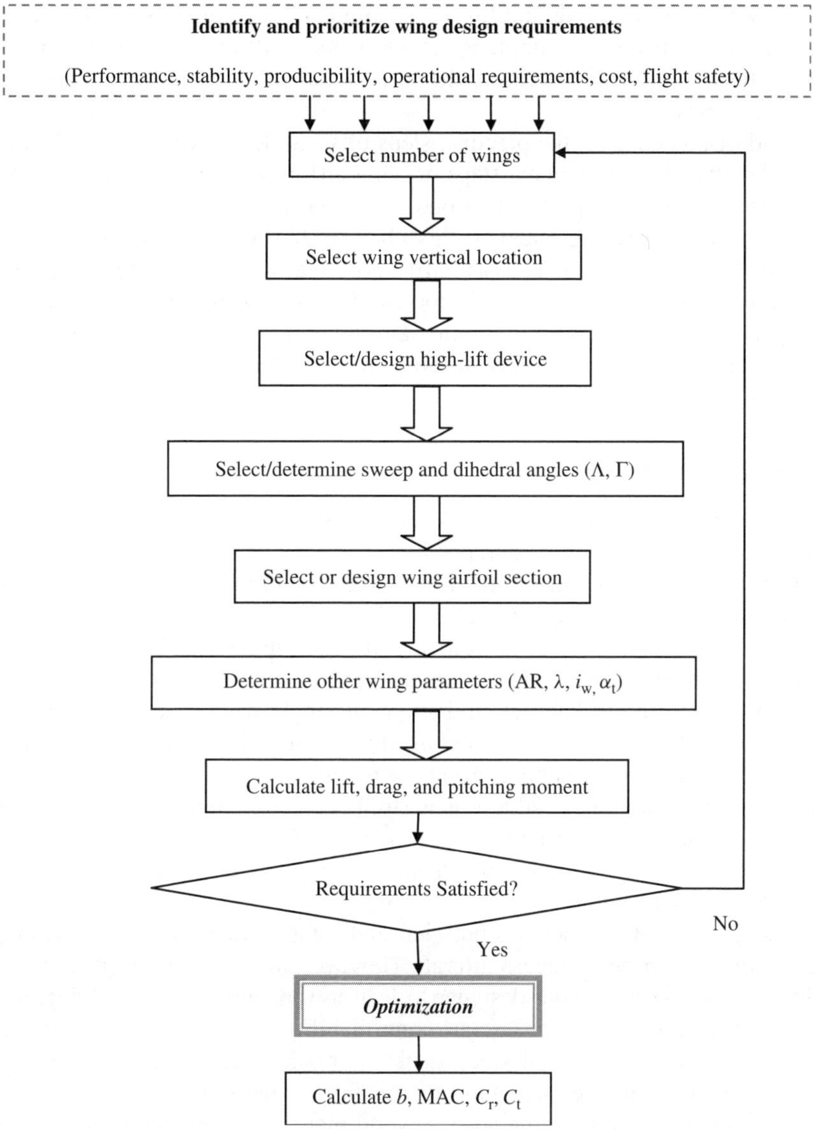

Figure 5.1 Wing design procedure

One of the necessary tools in the wing design process is an aerodynamic technique to calculate wing lift, wing drag, and wing pitching moment. With the progress in the science of aerodynamics, there are a variety of techniques and tools to accomplish this time-consuming job. A variety of tools and software based on aerodynamics and numerical methods has been developed in past decades. Computational fluid dynamics (CFD) software based on the solution of Navier–Stokes equations, the vortex lattice method, thin airfoil theory, and circulation are available in the market. The application of such software packages – which is expensive and time-consuming – at this early stage of wing design seems unnecessary. Instead, a simple approach, namely lifting-line theory, is introduced. Using this theory, one can determine the three wing productions (L, D, and M) with acceptable accuracy.

At the end of this chapter, the practical steps of wing design will be introduced. In the middle of the chapter, the practical steps of wing airfoil selection will also be presented. Two fully solved example problems – one about wing airfoil selection and one about whole wing design – are presented in this chapter. It should be emphasized again, as discussed in Chapter 3, that it is essential to note that the wing design is a box in the iterative process of the aircraft design process. The procedure described in this chapter will be repeated several times until all other aircraft components are at an optimum point. Thus, the wing parameters will vary several times until the combinations of all design requirements are met.

5.2 Number of Wings

One of the decisions a designer must make is to select the number of wings. The options are described in the following paragraphs.

A number of wings higher than three is not practical. Figure 5.2 illustrates a front view of three aircraft with various configurations. Nowadays, modern aircraft almost all have a monoplane. Currently, there are a few aircraft that employ a biplane, but no modern aircraft is found to have three wings. In the past, the major reason to select more than one wing was manufacturing technology limitations. A single wing usually has a longer wing span compared with two wings (with the same total area). Old manufacturing technologies were not able to structurally support a long wing, to stay level and rigid. With advances in manufacturing technologies and also new strong aerospace materials (such as advanced light aluminum and composite materials), this reason is no longer valid. Another reason was the limitations on the aircraft wing span. Hence a way to reduce the wing span is to increase the number of wings.

Thus, a single wing (that includes both left and right sections) is almost the only practical option in conventional modern aircraft. However, a few other design considerations may still force the modern wing designer to lean toward more than one wing. The most significant is the requirement for aircraft controllability. An aircraft with a shorter wing span delivers higher roll control, since it has a smaller mass moment of inertia about the x-axis. Therefore, if one is looking to roll faster, one option is to have more than one wing leading to a shorter wing span. Several maneuverable aircraft in the 1940s and 1950s had a biplane and even three wings. In contrast, the disadvantages of an option

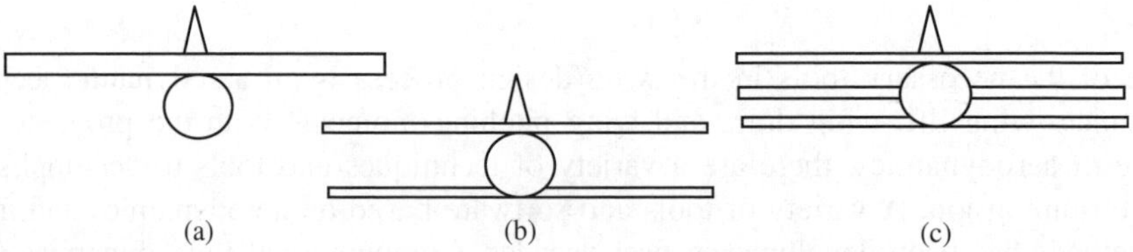

Figure 5.2 Three options in number of wings (front view): (a) Monoplane; (b) Biplane; and (c) Tri-wing

other than a monoplane include higher weight, lower lift, and pilot visibility limits. The recommendation is to begin with a monoplane, and if the design requirements are not satisfied, resort to a higher number of wings.

5.3 Wing Vertical Location

One of the wing parameters that could be determined at the early stages of the wing design process is the wing vertical location relative to the fuselage center line. This wing parameter will influence the design of other aircraft components directly, including aircraft tail design, landing gear design, and center of gravity. In principle, there are four options for the vertical location of the wing.

Figure 5.3 shows the schematics of these four options. In this figure, only the front views of the aircraft fuselage and wing are shown. In general, cargo aircraft and some GA (General Aviation) aircraft have a high wing, while most passenger aircraft have a low wing. In contrast, most fighter airplanes and some GA aircraft have a mid-wing, while hang gliders and most amphibian aircraft have a parasol wing. The primary criterion to select the wing location originates from operational requirements, while other requirements such as stability and producibility are the influencing factors in some design cases.

Figure 5.4 illustrates four aircraft in which various wing locations are shown: (i) cargo aircraft Lockheed Martin C-130J Hercules (high wing); (ii) passenger aircraft Boeing 767 (low wing); (iii) home-built aircraft Pietenpol Air Camper-2 (parasol wing); and (iv) military aircraft Hawker Sea Hawk FGA6 (mid-wing). In this section, the advantages and disadvantages of each option are examined. The final selection will be made based on the summations of all advantages and disadvantages when incorporated into the design requirements. Since each option has a weight relative to the design requirements, the summation of all weights derives the final choice.

5.3.1 High Wing

The high-wing configuration (Figures 5.3(a) and 5.4(a)) has several advantages and disadvantages that make it suitable for some flight operations, but unsuitable for other flight missions. In the following subsections, these advantages and disadvantages are presented.

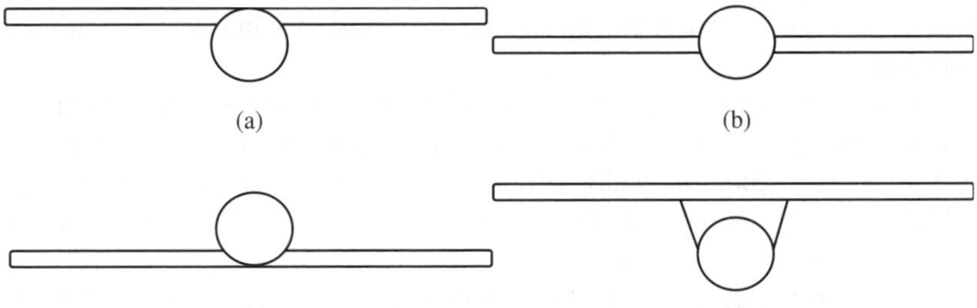

Figure 5.3 Options in vertical wing positions: (a) High wing; (b) Mid-wing; (c) Low wing; and (d) Parasol wing

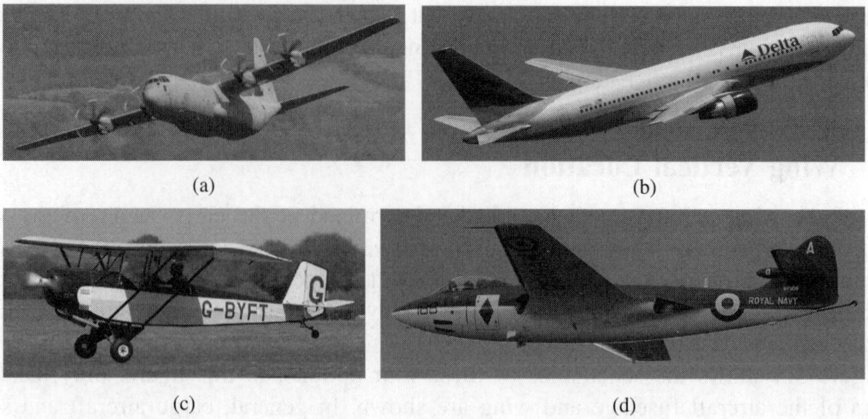

Figure 5.4 Four aircraft with different wing vertical positions: (a) cargo aircraft Lockheed Martin C-130J Hercules (high wing); (b) passenger aircraft Boeing 767 (low wing); (c) home-built aircraft Pietenpol Air Camper-2 (parasol wing); (d) military aircraft Hawker Sea Hawk (mid-wing) (Reproduced from permission of (a, d) Antony Osborne; (b) Anne Deus; (c) Jenny Coffey.)

5.3.1.1 Advantages

1. Eases and facilitates the loading and unloading of loads and cargo into and out of cargo aircraft. For instance, truck and other load lifter vehicles can easily move around aircraft and under the wing without the anxiety of hitting and breaking the wing.
2. Facilitates the installation of an engine on the wing, since the engine (and propeller) clearance is higher (and safer) compared with a low-wing configuration.
3. Saves the wing from high-temperature exit gases in a vertical take-off and landing (VTOL) aircraft (e.g., Harrier GR9 (Figure 4.4) and BAe Sea Harrier (Figure 5.51)). The reason is that the hot gases bounce back when they hit the ground, so they wash the wing afterward. Even with a high wing, this will severely reduce the lift of the wing structure. Thus, the higher the wing is the farther it will be from the hot gases.
4. Facilitates the installation of a strut. This is based on the fact that a strut (rod or tube) can handle higher tensile stress compared with compression stress. In a high wing, the strut has to withstand tensile stress, while the strut in a low wing must bear compression stress. Figure 5.3(d) shows a sketch of a parasol wing with a strut. Figure 5.56(c) illustrates the strut-braced wing of a GA aircraft Piper Super Cub.
5. The aircraft structure is lighter when struts are employed (as item 4 implies).
6. Facilitates taking off and landing from sea. In a sea-based or amphibian aircraft, during a take-off operation, water will splash around the aircraft. An engine installed on a high wing will receive less water compared with a low wing. Thus, the possibility of engine shut-off is much lower.
7. Facilitates aircraft control for a hang glider pilot, since the aircraft center of gravity is lower than the wing.
8. Increases the dihedral effect (C_{l_β}). It makes the aircraft laterally more stable. The reason lies in the higher contribution of the fuselage to the wing dihedral effect ($C_{l_{\beta W}}$).

9. The wing will produce more lift compared with a mid- and low wing, since two parts of the wing are attached at least on the top part.
10. For the same reason as in item 9, the aircraft will have a lower stall speed, since $C_{L_{max}}$ will be higher.
11. The pilot has a better view in lower-than-horizon. A fighter pilot has a full view under the aircraft.
12. For an engine that is installed under the wing, there is less possibility of sand and debris entering the engine and damaging the blades and propellers.
13. There is a lower possibility of human accident in hitting the propeller and being pulled into the engine inlet. In a few rare accidents, several careless people have died (by hitting the rotating propeller or being pulled into the jet engine inlet).
14. The aerodynamic shape of the fuselage lower section can be smoother.
15. There is more space inside the fuselage for cargo, luggage, or passengers.
16. The wing drag produces a nose-up pitching moment, so it is longitudinally destabilizing. This is due to the higher location of the wing drag line relative to the aircraft center of gravity ($M_{D_{cg}} > 0$).

5.3.1.2 Disadvantages

1. The aircraft tends to have more frontal area (compared with mid-wing). This will increase aircraft drag.
2. The ground effect is lower, compared with low wing. During take-off and landing operations, the ground will influence the wing pressure distribution. The wing lift will be slightly lower than for the low-wing configuration. This will increase the take-off run slightly. Thus, a high-wing configuration is not the right option for short take-off and landing (STOL) aircraft.
3. The landing gear is longer if connected to the wing. This makes the landing gear heavier and requires more space inside the wing for the retraction system. This will further make the wing structure heavier.
4. The pilot has less higher-than-horizon view. The wing above the pilot will obscure part of the sky for a fighter pilot.
5. If the landing gear is connected to the fuselage and there is not sufficient space for the retraction system, extra space must be provided to house the landing gear after retraction. This will increase the fuselage frontal area and thus will increase the aircraft drag.
6. The wing produces more induced drag (D_i) due to the higher lift coefficient.
7. The horizontal tail area of an aircraft with a high wing is about 20% larger than the horizontal tail area with a low wing. This is due to more downwash of a high wing on the tail.
8. A high wing is structurally about 20% heavier than a low wing.
9. The retraction of the landing gear inside the wing is not usually an option, due to the long required length of the landing gear.
10. The aircraft lateral control is weaker compared with mid-wing and low wing, since the aircraft has more laterally dynamic stability.

Although the high wing has more advantages than disadvantages, not all the items have the same weighing factor. It depends on which design objectives are more significant than the others in the eyes of the customer. Systems engineering delivers an approach to determine the best option for a specific aircraft, using a comparison table.

5.3.2 Low Wing

In this section, the advantages and disadvantages of a low-wing configuration (Figures 5.3(c) and 5.4(b)) will be presented. Since the reasons for several items are similar to the reasons for a high-wing configuration, they are not repeated here. In the majority of cases, the specifications of a low wing are compared with a high-wing configuration.

5.3.2.1 Advantages

1. The aircraft take-off performance is better, compared with a high-wing configuration, due to the ground effect.
2. The pilot has a better higher-than-horizon view, since he/she is above the wing.
3. The retraction system inside the wing is an option, along with inside the fuselage.
4. The landing gear is shorter if connected to the wing. This makes the landing gear lighter and requires less space inside the wing for the retraction system. This will further make the wing structure lighter.
5. In a light GA aircraft, the pilot can walk on the wing in order to get into the cockpit.
6. The aircraft is lighter compared with a high-wing structure.
7. The aircraft frontal area is less.
8. The application of a wing strut is usually no longer an option for the wing structure.
9. Item 8 implies that the aircraft structure is lighter since no strut is utilized.
10. Due to item 8, the aircraft drag is lower.
11. The wing has less induced drag.
12. It is more attractive to the eyes of a regular viewer.
13. The aircraft has higher lateral control compared with a high-wing configuration, since the aircraft has less lateral static stability, due to the fuselage contribution to the wing dihedral effect ($C_{l_{\beta_W}}$).
14. The wing has less downwash on the tail, so the tail is more effective.
15. The tail is lighter, compared with a high-wing configuration.
16. The wing drag produces a nose-down pitching moment, so a low wing is longitudinally stabilizing. This is due to the lower position of the wing drag line relative to the aircraft center of gravity ($M_{D_{cg}} < 0$).

5.3.2.2 Disadvantages

1. The wing generates less lift, compared with a high-wing configuration, since the wing has two separate sections.
2. With the same token as item 1, the aircraft will have a higher stall speed compared with a high-wing configuration, due to a lower $C_{L_{max}}$.
3. Due to item 2, the take-off run is longer.
4. The aircraft has lower airworthiness due to a higher stall speed.
5. Due to item 1, the wing produces less induced drag.

6. The wing makes a lower contribution to the aircraft dihedral effect, thus the aircraft is laterally dynamically less stable.
7. Due to item 4, the aircraft is laterally more controllable, and thus more maneuverable.
8. The aircraft has a lower landing performance, since it needs more landing run.
9. The pilot has a lower-than-horizon view. The wing below the pilot will obscure part of the sky for a fighter pilot.

Although the low wing has more advantages than disadvantages, not all the items have the same weighing factors. It depends on which design objectives are more significant than the others in the eyes of the customer. Systems engineering delivers an approach to determine the best option for a specific aircraft.

5.3.3 Mid-Wing

In general, the features of the mid-wing configuration (Figures 5.3(b) and 5.4(d)) stand somewhere between the features of a high-wing configuration and the features of a low-wing configuration. The major difference lies in the necessity to cut the wing spar in half in order to save space inside the fuselage. However, another alternative is not to cut the wing spar and to let it pass through the fuselage, which leads to an occupied space of the fuselage. Both alternatives carry a few disadvantages. Other than those features that can easily be derived from the two previous sections, some new features of a mid-wing configuration are as follows:

1. The aircraft structure is heavier, due to the necessity of reinforcing the wing root at the intersection with the fuselage.
2. The mid-wing is more expensive compared with high- and low-wing configurations.
3. The mid-wing is more attractive compared with the two other configurations.
4. The mid-wing is aerodynamically streamlined compared with the two other configurations.
5. A strut is usually not used to reinforce the wing structure.
6. The pilot can get into the cockpit using the wing as a step in a small GA aircraft.
7. The mid-wing has less interference drag than the low wing or high wing.

5.3.4 Parasol Wing

This wing configuration is usually employed in hang gliders plus amphibian aircraft. In several areas, the features are similar to a high-wing configuration. The reader is referred to the above items for more details, and is expected to be able to derive conclusions by comparing various configurations. Since the wing utilizes longer struts, it is heavier and has more drag compared with a high-wing configuration.

5.3.5 The Selection Process

The best approach to select the wing vertical location is to produce a table (such as Table 5.1) which consists of the weight of each option for various design objectives. The

Table 5.1 A sample table to compare the features of four wing vertical locations

Design objectives	Weight (%)	High wing	Low wing	Mid-wing	Parasol wing
Stability requirements	20				
Control requirements	15				
Cost	10				
Producibility requirements	10				
Operational requirements	40				
Other requirements	5				
Summation	100	93	76	64	68

weight of each design objective must usually be designated such that the summation adds up to 100%. A comparison between the summations of points among four options leads the designer to the best configuration. Table 5.1 illustrates a sample table to compare four wing configurations in the wing design process for a cargo aircraft. All elements of this table must carefully be filled with numbers. The last row is the summation of all numbers in each column. In the case of this table, the high wing has gained the highest number of points (93), so the high wing seems to be the best candidate for the sample problem. As observed, even the high-wing configuration does not fully satisfy all the design requirements, but it is an optimum option among the four available options. Reference [1] is a rich resource of procedures for selection techniques.

5.4 Airfoil Section

This section is devoted to the process of determining the airfoil section for a wing. It is appropriate to claim that the airfoil section is the second most important wing parameter, after the wing planform area. The airfoil section is responsible for the generation of the optimum pressure distribution on the top and bottom surfaces of the wing such that the required lift is created with the lowest aerodynamic cost (i.e., drag and pitching moment). Although every aircraft designer has some basic knowledge of aerodynamics and the basics of airfoils to have a uniform starting point, the concept of an airfoil and its governing equations will be reviewed. The section begins with a discussion of airfoil selection or airfoil design. Then the basics of airfoil, airfoil parameters, and the most important factor of airfoil section will be presented. A review of the National Advisory Committee for Aeronautics (NACA) – the predecessor of the present National Administration for Aeronautics and Astronautics (NASA) – airfoils will be presented later, since the focus in this section is on airfoil selection. The criteria for airfoil selection will be introduced and finally, the procedure to select the best airfoil will be introduced. The section ends with a fully solved example to select an airfoil for a candidate wing.

5.4.1 Airfoil Design or Airfoil Selection

The primary function of the wing is to generate lift force. This will be generated by a special wing cross-section called the airfoil. A wing is a three-dimensional component, while an airfoil is a two-dimensional section. Because of the airfoil section, two other outputs of the airfoil, and consequently the wing, are drag and pitching moment. The wing may have a constant or a non-constant cross-section across the wing. This topic will be discussed in Section 5.9.

There are two ways to determine the wing airfoil section:

1. airfoil design;
2. airfoil selection.

The design of the airfoil is a complex and time-consuming process and needs expertise in the fundamentals of aerodynamics at graduate level. Since the airfoil needs to be verified by testing it in a wind tunnel, it is expensive too. Large aircraft production companies such as Boeing and Airbus have sufficient human experts (aerodynamicists) and budget to design their own airfoil for every aircraft, but small aircraft companies, experimental aircraft producers, and home-built manufacturers cannot afford to design their own airfoils. Instead, they select the best airfoils among the currently available airfoils found in several books or websites.

With the advent of high-speed and powerful computers, the design of an airfoil is not as hard as it was 30 years ago. There are currently a couple of aerodynamic software packages (CFD) in the market that can be used to design airfoils for a variety of needs. Not only aircraft designers need to design airfoils – there are many other areas where airfoils need to be designed for products. This list includes jet engine axial compressor blades, jet engine axial turbine blades, steam power plant axial turbine blades, wind turbine propellers, centrifugal and axial pump impeller blades, turboprop engine propellers, centrifugal and axial compressor impeller blades, and large and small fans. The efficiencies of all these industrial mechanical or aerospace devices rely heavily on the section of their blades; that is, the *airfoil*.

If you have enough time, budget, and manpower – and decide to design an airfoil for your aircraft – you are referred to the references listed at the end of this textbook. But remember, the airfoil design is a design project in itself and needs to be integrated into the aircraft design process properly. If you are a junior aircraft designer with limited resources, you are recommended to select an airfoil from the available airfoil database.

Any aerodynamics textbook introduces several theories to analyze flow around an airfoil. The application of potential-flow theory together with boundary-layer theory to airfoil design and analysis was accomplished many years ago. Since then, potential-flow and boundary-layer theories have been improved steadily. With the advent of computers, these theories have been used increasingly to complement wind-tunnel tests. Today, computing costs are so low that a complete potential-flow and boundary-layer analysis of an airfoil costs considerably less than 1% of the equivalent wind-tunnel test. Accordingly, the tendency today is toward more and more commonly applicable computer codes. These codes reduce the amount of required wind-tunnel testing and allow airfoils to be tailored to each specific application.

One of the oldest and most reliable airfoil designers is Eppler [2] in Germany. Eppler has developed an airfoil design code that is based on conformal mapping. The Epplercode has been developed over the past 45 years. It combines a conformal-mapping method for the design of airfoils with prescribed velocity-distribution characteristics, a panel method for the analysis of the potential flow about given airfoils, and an integral boundary-layer method. The code contains an option that allows aircraft-oriented boundary-layer developments to be computed, where the Reynolds number and the Mach number vary with the aircraft lift coefficient and the local wing chord. In addition, a local twist angle can be input. Aircraft drag polar, including the induced drag and aircraft parasite drag, can also be computed.

The code will execute on almost any personal computer, workstation, or server, with run times varying accordingly. The most computationally intensive part of the code, the analysis method, takes only a few seconds to run on a personal computer. The code is written in standard FORTRAN 77 and, therefore, a FORTRAN compiler is required to translate the supplied source code into executable code. A sample input and output case is included. All the graphics routines are contained in a separate, plot-post-processing code that is also supplied. The post-processing code generates an output file that can be sent directly to a printer. The user can adapt the post-processing code to other plotting devices, including the screen. It is very efficient and has been applied successfully at Reynolds numbers from $3 \cdot 10^4$ to $5 \cdot 10^7$. A compressibility correction to the velocity distributions, which is valid as long as the local flow is not supersonic, has been incorporated into the code. The code is available, for a fee, in North America exclusively from Mark D. Maughmer.[1]

If you are not ready to design your own airfoil, you are recommended to select a proper airfoil from the previously designed and published airfoil sections. Two reliable airfoil resources are NACA and Eppler. The details of Eppler airfoils have been published in Ref. [2]. NACA airfoils have been published in a book published by Abbott and Von Donehoff [3]. The book was first published in the 1950s, but has been reprinted and is still available in almost every aerospace-related library. Both references present the airfoil coordinates plus pressure distribution and a few other graphs such as C_{l}, C_{d}, and C_{m} for a range of angles of attack. Eppler airfoil names begin with the letter "E" followed by three numbers. More details on NACA airfoils will be presented in Section 5.3.4. In general, the Eppler airfoils are for very low Reynolds number, Wortman airfoils for low (sailplane-ish) Reynolds number, and the NASA Low-Speed airfoils (e.g., LS(1)-0413) and Mid-Speed airfoils (e.g., MS(1)-0313) are for "moderate" Reynolds numbers.

A regular flight operation consists of take-off, climb, cruise, turn, maneuver, descent, approach, and landing. Basically, the airfoil's optimum function is in cruise, as an aircraft spends much of its flight time in this flight phase. At a cruising flight, lift (L) is equal to aircraft weight (W), and drag (D) is equal to engine thrust (T). Thus the wing must produce sufficient lift coefficient, while the drag coefficient must be minimum. Both of these coefficients come mainly from the airfoil section. Thus, two governing equations for a cruising flight are:

$$L = W \Rightarrow \frac{1}{2}\rho V^2 S C_L = mg \tag{5.1}$$

[1] RR 1, Box 965 Petersburg, PA 16669, USA.

$$D = T \Rightarrow \frac{1}{2}\rho V^2 S C_D = nT_{\max} \text{ (jet engine)} \tag{5.2}$$

$$D = T \Rightarrow \frac{1}{2}\rho V^2 S C_D = \frac{n\eta_P P_{\max}}{V_C} \text{ (prop-driven engine)} \tag{5.3}$$

Equation (5.2) is for an aircraft with a jet engine, but Equation (5.3) is for an aircraft with a prop-driven engine. The variable n ranges between 0.6 and 0.9. It means that only a partial engine throttle is used in a cruising flight and maximum engine power or engine thrust is not employed. The exact value for n will be determined in later design steps. For the airfoil initial design, it is suggested to use 0.75. The maximum engine power or engine thrust is only used during take-off or when cruising with maximum speed. Since a major criterion for airfoil design is to satisfy cruising flight requirements, Equations (5.1)–(5.3) are used in airfoil design, as explained later in this section. In the following section, the wing airfoil selection procedure is described.

5.4.2 General Features of an Airfoil

Any section of the wing cut by a plane parallel to the aircraft's xz plane is called an airfoil. It usually looks like a positive-cambered section with the thicker part in front of the airfoil. An airfoil-shaped body moved through the air will vary the static pressure on the top surface and on the bottom surface of the airfoil. A typical airfoil section is shown in Figure 5.5, where several geometric parameters are illustrated. If the mean camber line is a straight line, the airfoil is referred to as a symmetric airfoil, otherwise it is called a cambered airfoil. The camber of an airfoil is usually positive. In a positive-cambered airfoil, the upper surface static pressure in less than the ambient pressure, while the lower surface static pressure is higher than the ambient pressure. This is due to higher airspeed at the upper surface and lower speed at the lower surface of the airfoil (see Figures 5.6 and 5.7). As the airfoil angle of attack increases, the pressure difference between the upper and lower surfaces will be higher (see Ref. [4]).

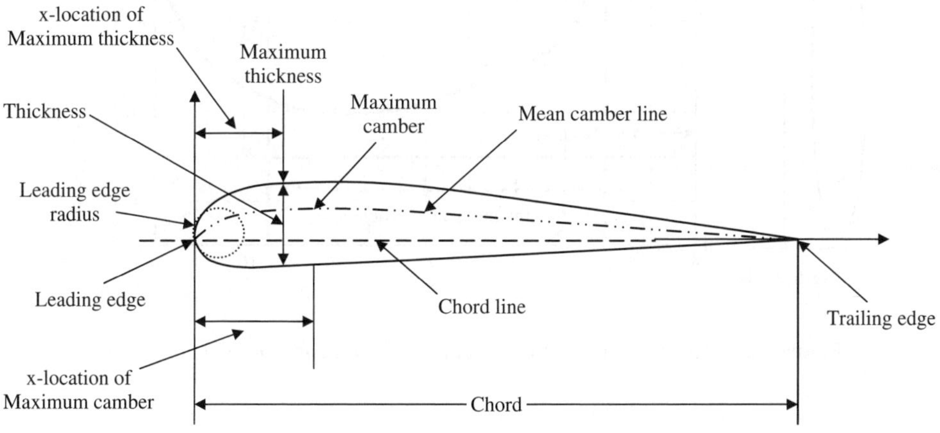

Figure 5.5 Airfoil geometric parameters

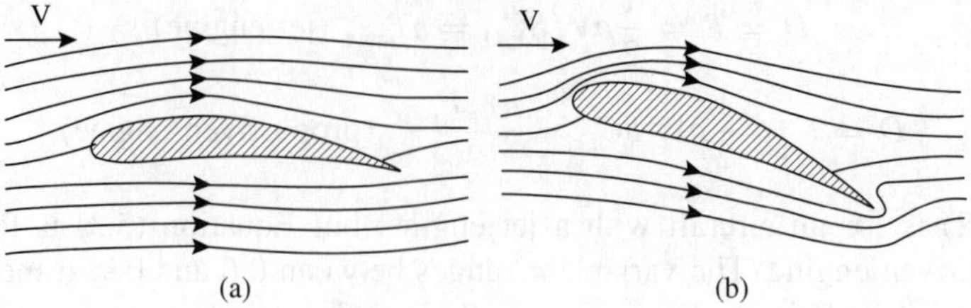

Figure 5.6 Flow around an airfoil: (a) Small angle of attack; (b) Large angle of attack

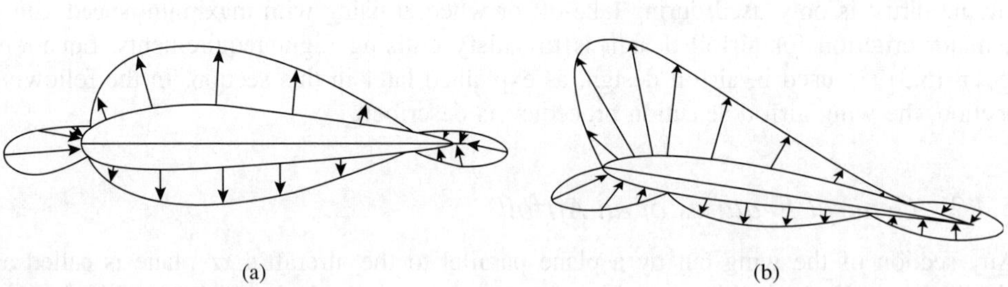

Figure 5.7 Pressure distribution around an airfoil: (a) Small angle of attack; (b) Large angle of attack

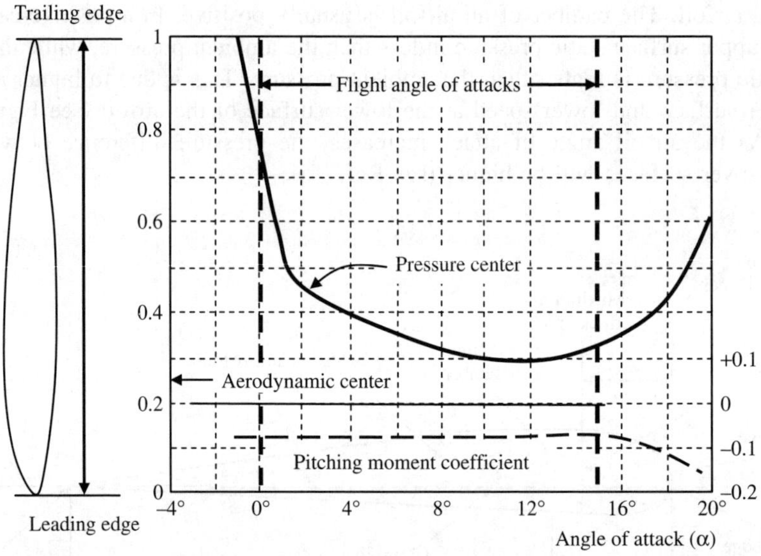

Figure 5.8 The pressure center movement as a function of angle of attack

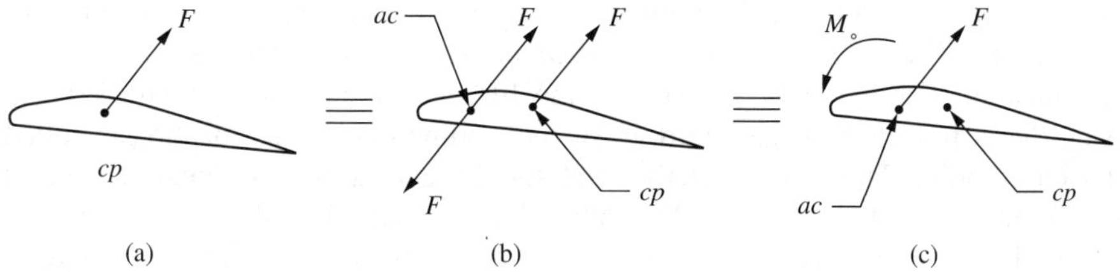

Figure 5.9 The movement of resultant force to aerodynamic center: (a) Force on pressure center; (b) Addition of two equal forces; and (c) Force on aerodynamic center

The force divided by the area is called the pressure, so the aerodynamic force generated by an airfoil in a flow field may be calculated by multiplication of total pressure by area. The total pressure is simply determined by integration of pressure over the entire surface. The magnitude, location, and direction of this aerodynamic force are functions of airfoil geometry, angle of attack, flow property, and airspeed relative to the airfoil.

The location of this resultant force out of the integration is called the center of pressure (cp). The location of this center depends on the aircraft speed plus the airfoil's angle of attack. As the aircraft speed increases, the center of pressure moves rearward (see Figure 5.8). At lower speeds, the cp location is close to the leading edge and at higher speeds, it moves toward the trailing edge. There is a location on the airfoil that has significant features in aircraft stability and control. The aerodynamic center is a useful concept for the study of stability and control. In fact, the force and moment system on a wing can be specified completely by the lift and drag acting through the aerodynamic center, plus the moment about the aerodynamic center, as sketched in Figure 5.9.

It is convenient to move the location of the resultant force to a new location – the aerodynamic center – that is almost stable. By the operation of adding two equal forces – one at the center of pressure and another at the aerodynamic center – we can move the location of the resultant force. By doing so, we have to account for introducing an aerodynamic pitching moment (see Figure 5.10). This will add a moment to our aerodynamic force. Therefore we can conclude that the pressure and shear stress distributions over a wing produce a pitching moment. This moment can be taken about any arbitrary point (the leading edge, the trailing edge, the quarter chord, etc.). The moment can be visualized as being produced by the resultant lift acting at a particular distance back from the leading edge. As a fraction of the chord, the distance to this point is known as the center of pressure.

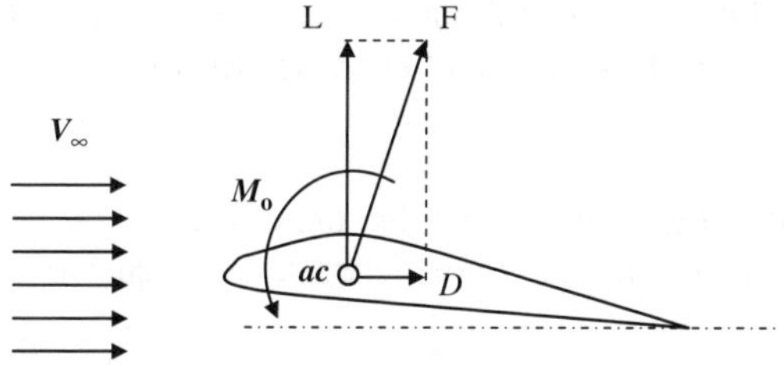

Figure 5.10 The aerodynamic lift, drag, and pitching moment

However, there exists a particular point about which the moments are independent of the angle of attack. This point is defined as the *aerodynamic center* for the wing.

The subsonic airfoil theory shows that the lift due to the angle of attack acts at a point on the airfoil 25% of the chord aft of the leading edge. This location is called the quarter-chord point. The point through which this lift acts is the aerodynamic center (ac). In wind-tunnel tests, the ac is usually within 1 or 2% chord of the quarter-chord point until the Mach number increases to within a small percentage of the drag divergence Mach number. The aerodynamic center then slowly moves aft as the Mach number is increased further.

Thus, the pressure and shear stress distributions over the airfoil generate an aerodynamic force. However, this resultant force is replaced with two aerodynamic forces and one aerodynamic moment, as shown by the vector in Figure 5.10. In other words, the aerodynamic force can be resolved into two forces, perpendicular (lift) and parallel (drag) to the relative wind. The lift is always defined as the component of the aerodynamic force perpendicular to the relative wind. The drag is always defined as the component of the aerodynamic force parallel to the relative wind.

5.4.3 Characteristic Graphs of an Airfoil

In the process of wing airfoil selection, we do not look at airfoil geometry only, or its pressure distribution. Instead, we examine the airfoil operational outputs that are more informative to satisfy design requirements. There are several graphs that illustrate the characteristics of each airfoil when compared to other airfoils in the wing airfoil selection process. These are mainly the variations of non-dimensionalized lift, drag, and pitching moment relative with angle of attack. Two aerodynamic forces and one aerodynamic pitching moment are usually non-dimensionalized[2] by dividing them to appropriate parameters as follows.

$$C_\mathrm{l} = \frac{l}{\frac{1}{2}\rho V^2 (C \cdot 1)} \tag{5.4}$$

$$C_\mathrm{d} = \frac{d}{\frac{1}{2}\rho V^2 (C \cdot 1)} \tag{5.5}$$

$$C_\mathrm{m} = \frac{m}{\frac{1}{2}\rho V^2 (C \cdot 1) \cdot C} \tag{5.6}$$

where l, d, and m are lift, drag, and pitching moment of a two-dimensional airfoil. The area $(C \cdot 1)$ is assumed to be the airfoil chord times the unit span ($b = 1$).

Thus, we evaluate the performance and characteristics of an airfoil by looking at the following graphs:

1. The variations of lift coefficient versus the angle of attack.
2. The variations of pitching moment coefficient about a quarter-chord versus the angle of attack.

[2] The technique was first introduced by Edger Buckingham (1867–1940) as Buckingham Π Theorem. The details may be found in most fluid mechanics textbooks.

3. The variations of pitching moment coefficient about the aerodynamic center versus the lift coefficient.
4. The variations of the drag coefficient versus the lift coefficient.
5. The variations of the lift-to-drag ratio versus the angle of attack.

These graphs have several critical features that are essential to the airfoil selection process. Let's first review these graphs.

5.4.3.1 The Graph of Lift Coefficient (C_l) versus Angle of Attack (α)

Figure 5.11 shows the typical variations of lift coefficient versus angle of attack for a positive-cambered airfoil. Seven significant features of this graph are: stall angle (α_s), maximum lift coefficient ($C_{l_{max}}$), zero lift angle of attack (α_o), ideal lift coefficient (C_{l_i}) and angle of attack corresponding to ideal lift coefficient (α_{Cl_i}), lift coefficient at zero angle of attack (C_{l_o}), and lift curve slope (C_{l_α}). These are critical to identify the performance of an airfoil.

1. The stall angle (α_s) is the angle of attack at which the airfoil stalls; that is, the lift coefficient will no longer increase with increasing angle of attack. The maximum lift coefficient that corresponds to the stall angle is the maximum angle of attack. The stall angle is related directly to flight safety, since the aircraft will lose the balance of forces in a cruising flight. If the stall is not controlled properly, the aircraft may enter a spin and eventually crash. In general, the higher the stall angle, the safer the aircraft – thus a high stall angle is sought in airfoil selection. The typical stall angles for the majority of airfoils are between 12 and 16 deg. This means that the pilot is not allowed to increase the angle of attack to more than about 16 deg. Therefore an airfoil with a higher stall angle is more desirable.
2. The maximum lift coefficient ($C_{l_{max}}$) is the maximum capacity of an airfoil to produce non-dimensional lift; that is, the capacity of an aircraft to lift a load (i.e., aircraft

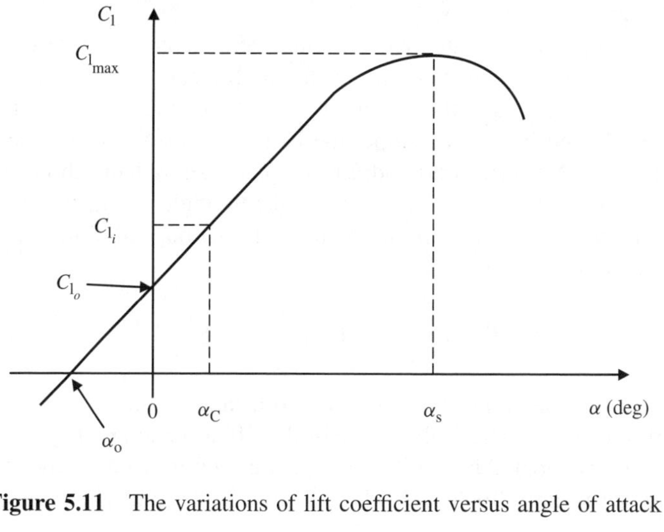

Figure 5.11 The variations of lift coefficient versus angle of attack

weight). The maximum lift coefficient usually occurs at the stall angle. The stall speed (V_s) is inversely a function of the maximum lift coefficient, thus the higher $C_{l_{max}}$ leads to the lower V_s. Thus the higher $C_{l_{max}}$ results in a safer flight. Therefore, the higher maximum lift coefficient is desired in an airfoil selection process.

3. The zero lift angle of attack (α_o) is the airfoil angle of attack at which the lift coefficient is zero. A typical number for α_o is around -2 deg when no high-lift device (HLD) is employed. However, when a HLD is employed (such as -40 deg of flap down), the α_o increases to about -12 deg. The design objective is to have a higher α_o (more negative), since this leaves the capacity to have more lift at zero angle of attack. This is essential for a cruising flight, since the fuselage center line is aimed to be level (i.e., zero fuselage angle of attack) for a variety of flight reasons – such as the comfort of passengers.
4. The ideal lift coefficient (C_{l_i}) is the lift coefficient at which the drag coefficient does not vary significantly with the slight variations of the angle of attack. The ideal lift coefficient usually corresponds to the minimum drag coefficient. This is critical in airfoil selection, since a lower drag coefficient means a lower flight cost. Thus, the design objective is to cruise at flight situations such that the cruise lift coefficient is as close as possible to the ideal lift coefficient. The value of this C_{l_i} will be clear when the graph of variation of drag coefficient versus lift coefficient is discussed. The typical value of ideal lift coefficient for a GA aircraft is about 0.1–0.4, and for a supersonic aircraft about 0.01–0.05.
5. The angle of attack corresponding to the ideal lift coefficient ($\alpha_{C_{l_i}}$) is self-explanatory. The wing setting angle is often selected to be the same as this angle, since it will result in a minimum drag. In contrast, the minimum drag corresponds to minimum engine thrust, which means minimum flight cost. This will be discussed in more detail when the wing setting angle is discussed. The typical value of $\alpha_{C_{l_i}}$ is around 2–5 deg. Thus, such an angle will be an optimum candidate for the cruising angle of attack.
6. The lift coefficient at zero angle of attack (C_{l_o}) is the lift coefficient when the angle of attack is zero. From a design point of view, the higher C_{l_o} is the better, since it implies we can produce a positive lift even at zero angle of attack. Thus, the higher C_{l_o} is the better.
7. The lift curve slope (C_{l_α}) is another important performance feature of an airfoil. The lift curve slope is the slope of variation of lift coefficient with respect to the change in the angle of attack, and its unit is 1/deg or 1/rad. Since the main function of an airfoil is to produce lift, the higher the slope, the better the airfoil. The typical value of lift curve slope of a 2D airfoil is around 2π (or 6.28) per radian (about 0.1 per deg). It implies that for each 1 deg of change in the airfoil angle of attack, the lift coefficient will be increased by 0.1. The lift curve slope (1/rad) may be found by the following empirical equation:

$$C_{l_\alpha} = \frac{dC_l}{d\alpha} = 1.8\pi\left(1 + 0.8\frac{t_{max}}{c}\right) \tag{5.7}$$

where t_{max}/c is the maximum thickness-to-chord ratio of the airfoil.

8. Another airfoil characteristic is the shape of the lift curve at and beyond the stall angle of attack (stall behavior). An airfoil with a gentle drop in lift after the stall, rather than an abrupt or sharp rapid lift loss, leads to a safer stall from which the pilot can more

easily recover (see Figure 5.12). Although the sudden airfoil stall behavior does not necessarily imply sudden wing stall behavior, a careful wing design can significantly modify the airfoil tendency to rapid stall. In general, airfoils with thickness or camber, in which the separation is associated with the adverse gradient on the aft portion rather than the nose pressure peak, have a more gradual loss of lift. Unfortunately, the best airfoils in this regard tend to have lower maximum lift coefficient.

As observed, there are several parameters to judge the acceptability of an airfoil. In the next section, the technique to select the best airfoil based on these performance characteristics will be introduced.

5.4.3.2 The Variations of Pitching Moment Coefficient versus Angle of Attack

Figure 5.13 shows the typical variations of pitching moment coefficient about a quarter chord versus the angle of attack for a positive-cambered airfoil. The slope of this graph is usually negative and it is in the region of negative C_m for a typical range angle of attacks. The negative slope is desirable, since it stabilizes the flight if the angle of attack is disturbed by a gust. The negative C_m is sometimes referred to as nose-down pitching moment. This is due to its negative direction about the y-axis, which means the aircraft's nose will be pitched down by such a moment.

Figure 5.14 also illustrates the typical variations of pitching moment coefficient about the aerodynamic center versus lift coefficient for a positive-cambered airfoil. The magnitude of C_m is constant (recall the definition of aerodynamic center) for a typical range of lift coefficient. The typical magnitude is usually about −0.02 to −0.05. However, when

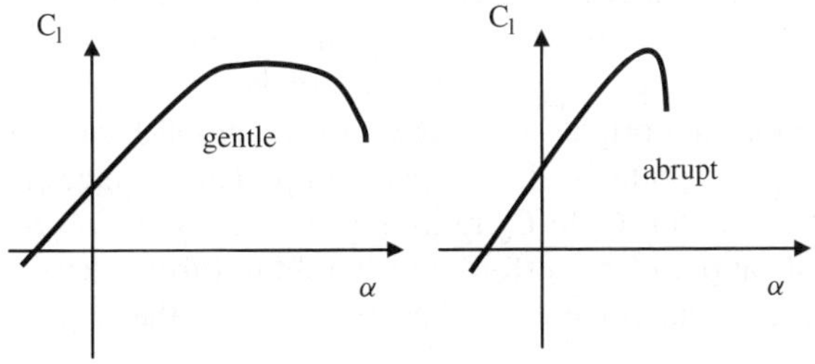

Figure 5.12 Stall characteristics

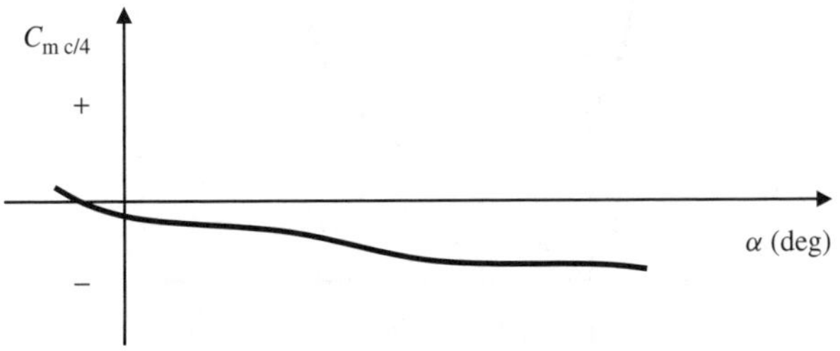

Figure 5.13 The variations of pitching moment coefficient versus angle of attack

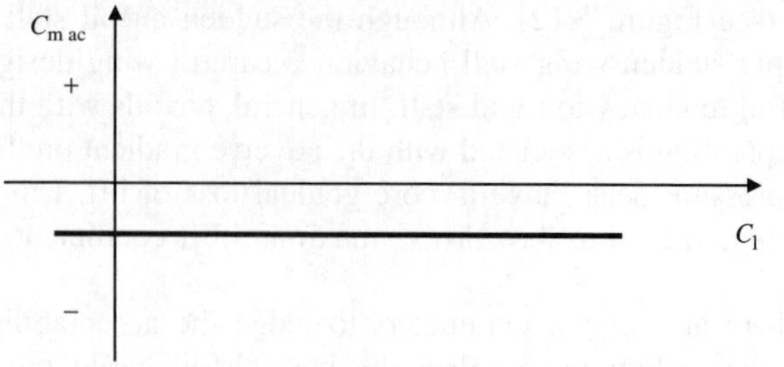

Figure 5.14 The variations of pitching moment coefficient versus lift coefficient

C_m is transferred from ac to another point (such as c/4), it will not be constant any more. The design objective is to have C_m close to zero as far as possible. The reason is that the aircraft must be in equilibrium in cruising flight. This pitching moment must be nullified by another component of the aircraft, such as the tail. Thus, a higher C_m (more negative) results in a larger tail, which means a heavier aircraft. Therefore an airfoil with lower C_m is more desirable. It is interesting to note that the pitching moment coefficient for a symmetrical airfoil section is zero.

5.4.3.3 The Variations of Drag Coefficient as a Function of Lift Coefficient

Figure 5.15 shows typical variations of the drag coefficient as a function of lift coefficient for a positive-cambered airfoil. The lowest point of this graph is called the minimum drag coefficient ($C_{d_{min}}$). The corresponding lift coefficient to the minimum drag coefficient is called $C_{l_{min}}$. As the drag is related directly to the cost of flight, the $C_{d_{min}}$ is of great importance in airfoil design or airfoil selection. A typical value for $C_{d_{min}}$ is about 0.003–0.006. Therefore an airfoil with lower $C_{d_{min}}$ is more desirable.

A line drawn through the origin and tangent to the graph locates a point that denotes the minimum slope. This point is also of great importance, since it indicates the flight situation where the maximum C_l-to-C_d ratio is generated, since $(C_d/C_l)_{min} = (C_l/C_d)_{max}$. This is an important output of an airfoil, and is referred to as the maximum lift-to-drag ratio. In addition to the requirement for the lowest $C_{d_{min}}$, the highest $(C_l/C_d)_{max}$ is also

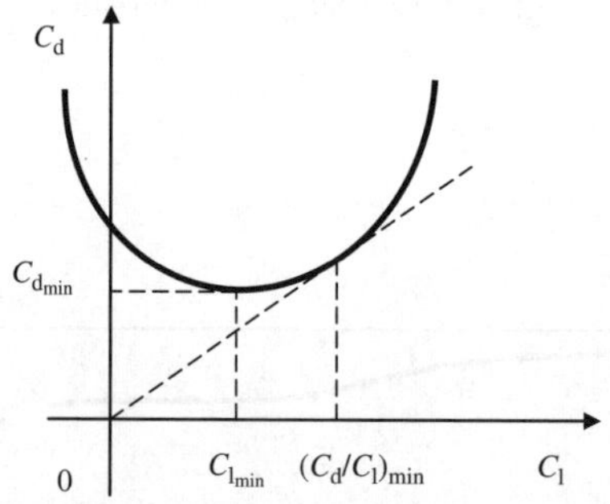

Figure 5.15 The typical variations of drag coefficient versus lift coefficient

desired. These two objectives may not happen at the same time in one airfoil, but based on aircraft mission and weight of each design requirement, one of them gets more attention.

The variation of the drag coefficient as a function of lift coefficient (Figure 5.15) may be modeled mathematically by the following second-order equation:

$$C_d = C_{d_{min}} + K\left(C_l - C_{l_{min}}\right)^2 \tag{5.8}$$

where K is called the section drag factor. The parameter K can be determined by selecting a point on the graph (C_{l_1} and C_{d_1}) and plugging in Equation (5.8).

Figure 5.16 shows the typical variations of drag coefficient as a function of lift coefficient for a laminar airfoil, such as 6-series NACA airfoils. This graph has a unique feature which is the bucket, due to the bucket shape of the lower portion of the graph. The unique aspect of the bucket is that the $C_{d_{min}}$ will not vary for a limited range of C_l. This is very significant, since it implies that the pilot can stay at the lowest drag point while changing the angle of attack. This situation matches with the cruising flight, since the aircraft weight is reduced as the fuel is burned. Hence, the pilot can bring the aircraft nose down (decrease the angle of attack) without being worried about an increase in the aircraft drag. Therefore it is possible to keep the engine throttle low during cruising flight.

The middle point of the bucket is called the ideal lift coefficient (C_{l_i}), while the highest C_l in the bucket region is referred to as the design lift coefficient (C_{l_d}). These two points are among the list of significant criteria to select/design an airfoil. Remember that the design lift coefficient occurs at the point whose C_d/C_l is minimum or C_l/C_d is maximum. For some flight operations (such as cruising flight), flying at the point where the lift coefficient is equivalent to C_{l_i} is the goal, while for some other flight operations (such as loiter), the objective is to fly at the point where the lift coefficient is equivalent to C_{l_d}. This airfoil lift coefficient is a function of the aircraft cruise lift coefficient (C_{L_i}), as will be discussed later in this chapter.

5.4.3.4 The Variations of Lift-to-Drag Ratio (C_l/C_d) as a Function of Angle of Attack

The last interesting graph that is utilized in the process of airfoil selection is the variations of lift-to-drag ratio (C_l/C_d) as a function of angle of attack. Figure 5.17 illustrates the

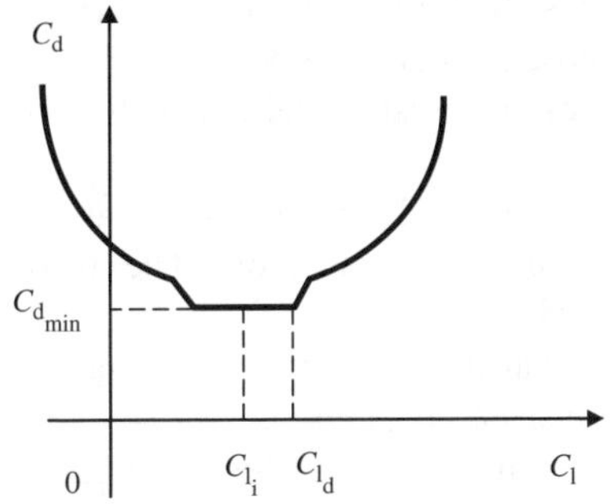

Figure 5.16 The variations of C_l versus C_d for a laminar airfoil.

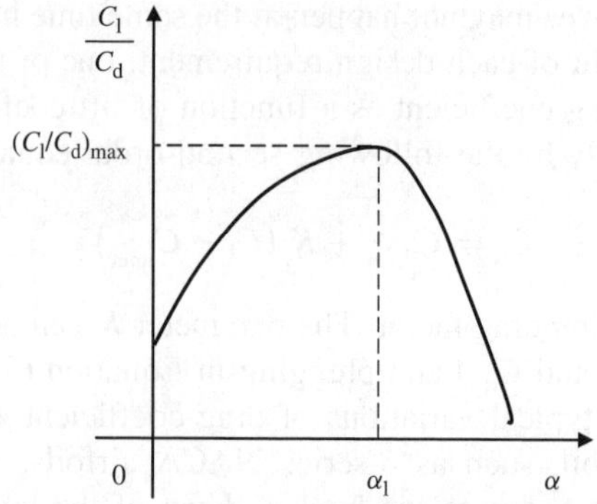

Figure 5.17 The typical variations of lift-to-drag ratio versus angle of attack

typical variations of lift-to-drag ratio versus angle of attack. As noted, this graph has one maximum point where the value of the lift-to-drag ratio is the highest at this point. The angle of attack corresponding to this point is an optimum candidate for a loitering flight (α_l).

The application of these four graphs and 12 parameters in the airfoil selection process will be introduced in later sections.

5.4.4 Airfoil Selection Criteria

Selecting an airfoil is part of the overall wing design. Selection of an airfoil for a wing begins with the clear statement of the flight requirements. For instance, subsonic flight design requirements are very much different from supersonic flight design objectives. In contrast, flight in the transonic region requires a special airfoil that meets Mach divergence requirements. The designer must also consider other requirements such as airworthiness, structural, manufacturability, and cost requirements. In general, the following are the criteria to select an airfoil for a wing with a collection of design requirements:

1. The airfoil with the highest maximum lift coefficient ($C_{l_{max}}$).
2. The airfoil with the proper ideal or design lift coefficient (C_{l_d} or C_{l_i}).
3. The airfoil with the lowest minimum drag coefficient ($C_{d_{min}}$).
4. The airfoil with the highest lift-to-drag ratio ($(C_l/C_d)_{max}$).
5. The airfoil with the highest lift curve slope ($C_{l_{\alpha max}}$).
6. The airfoil with the lowest (closest to zero; negative or positive) pitching moment coefficient (C_m).
7. The proper stall quality in the stall region (the variation must be gentle, not sharp).
8. The airfoil must be structurally reinforceable. The airfoil should not be so thin that spars cannot be placed inside.
9. The airfoil must be such that the cross-section is manufacturable.
10. The cost requirements must be considered.
11. Other design requirements must be considered. For instance, if the fuel tank has been designated to be placed inside the wing inboard section, the airfoil must allow sufficient space for this purpose.

12. If more than one airfoil is considered for a wing, the integration of two airfoils in one wing must be observed. This item will be discussed in more detail in Section 5.8.

Usually, there is no unique airfoil that has the optimum values for all the above-mentioned requirements. For example, you may find an airfoil that has the highest $C_{l_{max}}$, but not the highest $\left(\frac{C_l}{C_d}\right)_{max}$. In such cases, there must be compromise through a weighting process, since not all design requirements have the same importance. The weighting process will be discussed later in this chapter.

As a guide, typical values for the airfoil maximum thickness-to-chord ratio of the majority of aircraft are about 6–18%.

1. For a low-speed aircraft with a high lift requirement (such as cargo aircraft), the typical wing $(t/c)_{max}$ is about 15–18%.
2. For a high-speed aircraft with a low lift requirement (such as high subsonic passenger aircraft), the typical wing $(t/c)_{max}$ is about 9–12%.
3. For supersonic aircraft, the typical wing $(t/c)_{max}$ is about 3–9%.

The details of the airfoil selection procedure will be presented in Section 5.3.7. Figure 5.18 illustrates a few sample airfoils.

5.4.5 NACA Airfoils

The main focus of this section is how to select a wing airfoil from the available list of NACA airfoils, so this section is dedicated to the introduction of NACA airfoils. One of the most reliable resources and widely used databases is the list of airfoils that have

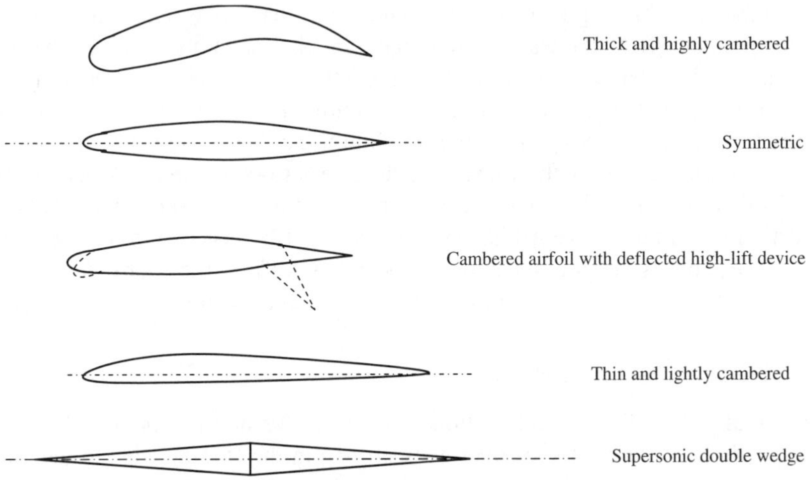

Figure 5.18 Five sample airfoil sections

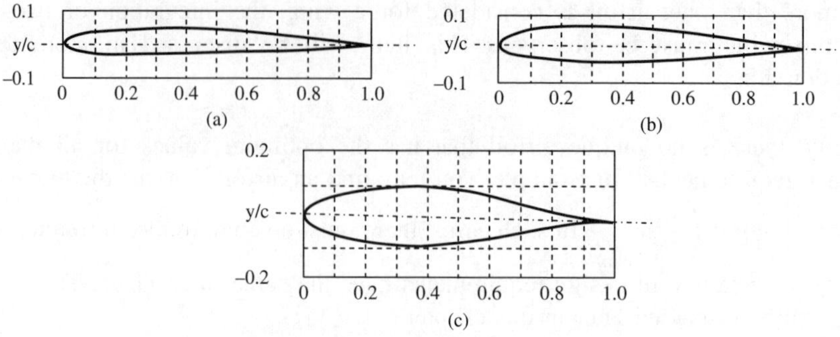

Figure 5.19 A four-digit, a five-digit, and a 6-series airfoil section [3]: (a) NACA 1408 airfoil section; (b) NACA 23012 airfoil section; and (c) NACA 63_3-218 airfoil section. Reproduced from permission of Dover Publications, Inc.

been developed by the NACA (predecessor of NASA) in the 1930s and 1940s. The three following groups of NACA airfoils are most interesting:

- four-digit NACA airfoils;
- five-digit NACA airfoils;
- 6-series NACA airfoils.

As the names imply, four-digit airfoils are named using four digits (such as 2415), five-digit airfoils are named using five digits (such as 23018), but 6-series airfoil names begin with the number 6 (in fact, they have five main digits). Figure 5.19 illustrates a four-digit, a five-digit, and a 6-series airfoil.

5.4.5.1 Four-Digit NACA Airfoils

The four-digit NACA airfoil sections are the oldest and simplest NACA airfoils to generate. The camber of a four-digit airfoil is made up of two parabolas. One parabola generates the camber geometry from the leading edge to the maximum camber, and another parabola produces the camber shape from the maximum camber to the trailing edge. In a four-digit NACA airfoil, the first digit indicates the maximum camber in a percentage chord. The second digit indicates the position of the maximum camber in tenths of a chord length. The last two digits represent the maximum thickness-to-chord ratio. A zero in the first digit means that this airfoil is a symmetrical airfoil section. For example, the NACA 1408 airfoil section (see Figure 5.19(a)) has an 8% $(t/c)_{\max}$ (the last two digits), its maximum camber is 10%, and its maximum camber is located at 40% of the chord length. Although these airfoils are easy to produce, they generate high drag compared with new airfoils.

5.4.5.2 Five-Digit NACA Airfoils

The camber of a five-digit airfoil section is made up of one parabola and one straight line. The parabola generates the camber geometry from the leading edge to the maximum camber, and then a straight line connects the end point of the parabola to the trailing edge. In a five-digit NACA airfoil section the first digit represents 2/3 of the ideal lift coefficient (see Figure 5.16) in tenths. It is an approximate representation of the maximum camber in a percentage chord. The second digit indicates the position of the maximum camber in two hundredths of a chord length. The last two digits represent the maximum thickness-to-chord ratio. A zero in the first digit means that this airfoil is a symmetrical airfoil section. For example, the NACA 23012 airfoil section (see Figure 5.19(b)) has a 12% maximum thickness-to-chord ratio $(t/c)_{max}$. The ideal lift coefficient of this airfoil is 0.3 (the second digit), since $2/3 \cdot C_{l_i} = 2/10$, thus, $C_{l_i} = 0.2/(2/3) = 0.3$. Finally its maximum camber is located at 12% of the chord length.

5.4.5.3 The 6-Series NACA Airfoils

The four- and five-digit airfoil sections were designed simply using parabola and line. They were not supposed to satisfy major aerodynamic design requirements, such as laminar flow and no flow separation. When it became clear that the four- and five-digit airfoils had not been carefully designed, NACA researchers began an investigation to develop a new series of airfoils driven by design requirements. In contrast, newly designed faster aircraft require more efficient airfoil sections. Several series of airfoils were designed at that time, but the 6-series were found to be the best. The 6-series airfoils were designed to maintain laminar flow over a large part of the chord, thus they maintain a lower $C_{d_{min}}$ compared with four- and five-digit airfoils. The 6-series NACA airfoils are designated by five main digits and begin with the number 6. Some 6-series airfoils have a subscript number after the second digit. There is also a "-" between the second digit and the third digit.

The meaning of each digit is as follows. The first digit is always 6; that is, the series designation. The second digit represents the chord-wise position of the minimum pressure in tenths of a chord for the basic symmetrical section at zero lift. The third digit indicates the ideal lift coefficient in tenths. The last two digits represent the maximum thickness-to-chord ratio. In case the airfoil name has a subscript after the second digit, this indicates the lift coefficient range in tenths above and below the value of the ideal lift coefficient in which a favorable pressure gradient and low drag exist. A zero in the third digit means that this airfoil is a symmetrical airfoil section.

For example, the NACA 63_3-218 airfoil section (see Figure 5.19(c)) has 18% thickness-to-chord ratio. The position of the minimum pressure in this airfoil is located at 30% of the chord (the second digit). The ideal lift coefficient of the airfoil is 0.2 (the third digit). Finally, the lift coefficient range above and below the value of the ideal lift coefficient is 0.3 (the subscript). It demonstrates that the bucket in the C_d-C_l diagram (see Figure 5.20) begins from the lift coefficient of 0 (since $0.3 - 0.3 = 0$) and ends at 0.6 (since $0.3 + 0.3 = 0.6$).

Figure 5.20 shows a general comparison between four-digit, five-digit, and 6-series airfoil sections. Figure 5.21 demonstrates C_l-α, C_m-α, and C_d-C_l graphs for the NACA 63_2-615 airfoil section. There are two groups of graphs, one for flap up and another for flap down (60 deg split flap). As noted, the flap deflection has doubled the airfoil drag (in fact $C_{d_{min}}$), increased the pitching moment tremendously, but at the same time has increased the lift coefficient by 1.2. Example 5.1 illustrates how to extract various airfoil characteristics (e.g. α_s, $C_{l_{max}}$, and α_o) from C_l-α, C_l-C_d, and C_m-α graphs.

Besides NACA airfoil sections, there are a variety of other airfoil sections that have been designed in the past several decades for different purposes. A few examples are peaky,

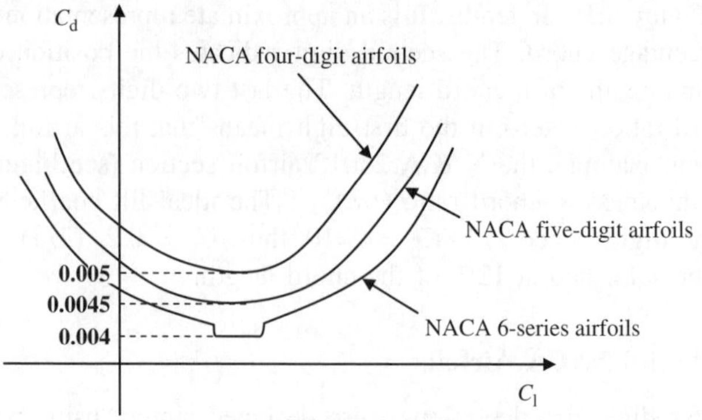

Figure 5.20 A general comparison between four-digit, five-digit, and 6-series airfoil sections

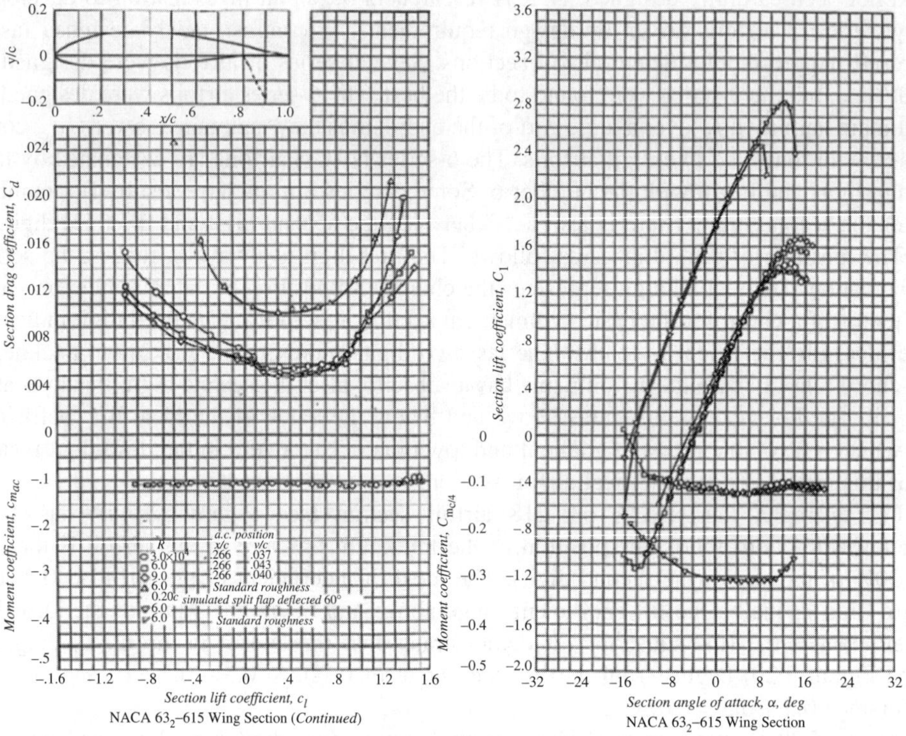

Figure 5.21 C_l-α, C_m-α, and C_d-C_l graphs of NACA 63_2-615 airfoil section [3]. Reproduced from permission of Dover Publications, Inc.

supercritical, modern, and supersonic airfoils. Table 5.2 illustrates the characteristics of several NACA airfoil sections. Table 5.3 illustrates the wing airfoil sections for several prop-driven aircraft. Table 5.4 illustrates the wing airfoil sections for several jet aircraft. As noted, all employ NACA airfoils, from GA aircraft Cessna 182 to fighter aircraft F-16 Falcon (Figure 4.6).

Example 5.1

Identify C_{l_i}, $C_{d_{min}}$, C_m, $(C_l/C_d)_{max}$, α_o (deg), α_s (deg), $C_{l_{max}}$, C_{l_α} (1/rad), and $(t/c)_{max}$ of the NACA 63-209 airfoil section (flap up). You need to indicate the locations of all parameters on the airfoil graphs.

Solution:

By referring to Figure 5.22, the required values for all parameters are determined as follows:

C_{l_i}	$C_{d_{min}}$	C_m	$(C_l/C_d)_{max}$	α_o (deg)	α_s (deg)	$C_{l_{max}}$	C_{l_α} (1/rad)	$(t/c)_{max}$
0.2	0.0045	−0.03	118	−1.5	12	1.45	5.73	9%

The locations of all points of interest are illustrated in Figure 5.22.

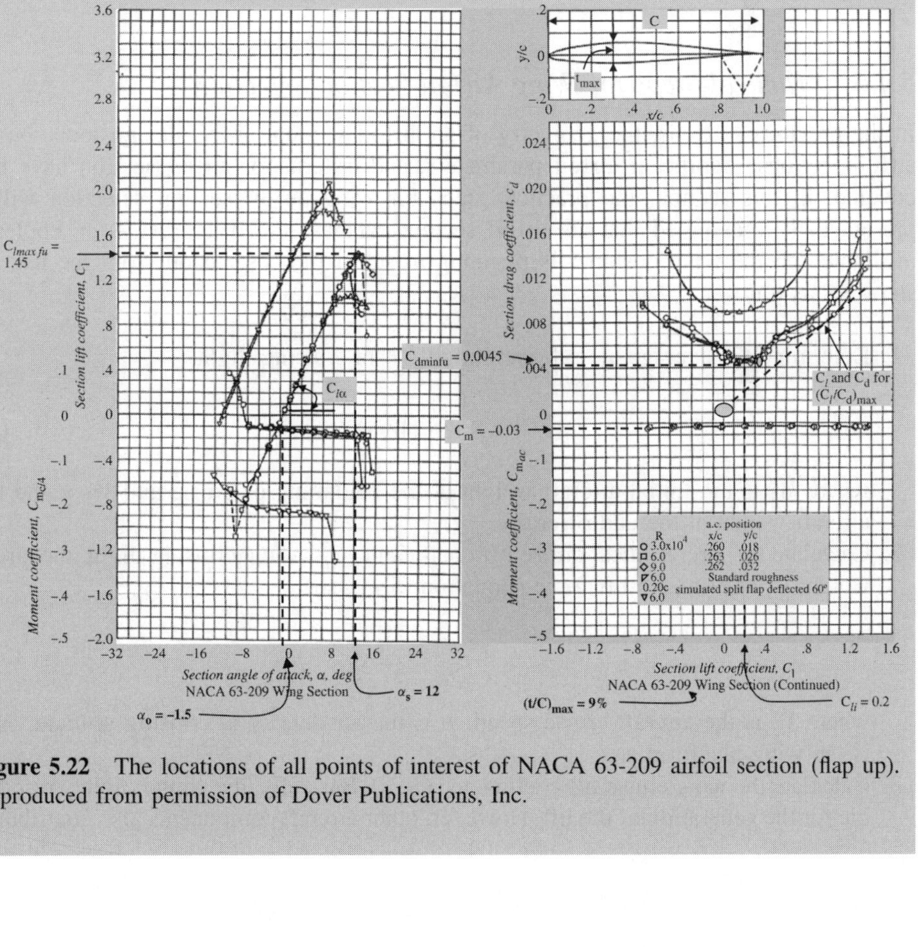

Figure 5.22 The locations of all points of interest of NACA 63-209 airfoil section (flap up). Reproduced from permission of Dover Publications, Inc.

Table 5.2 Characteristics of several NACA airfoil sections

No.	Airfoil section	$C_{l_{max}}$ at Re $=3\cdot10^6$	α_s (deg)	C_{m_o}	$(C_l/C_d)_{max}$	C_{l_i}	$C_{d_{min}}$	$(t/c)_{max}$ (%)
1	0009	1.25	13	0	112	0	0.0052	9
2	4412	1.5	13	−0.09	125	0.4	0.006	12
3	2415	1.4	14	−0.05	122	0.3	0.0065	15
4	23012	1.6	16	−0.013	120	0.3	0.006	12
5	23015	1.5	15	−0.008	118	0.1	0.0063	15
6	63_1-212	1.55	14	−0.004	100	0.2	0.0045	12
7	63_2-015	1.4	14	0	101	0	0.005	15
8	63_3-218	1.3	14	−0.03	103	0.2	0.005	18
9	64-210	1.4	12	−0.042	97	0.2	0.004	10
10	65_4-221	1.1	16	−0.025	120	0.2	0.0048	21

5.4.6 Practical Steps for Wing Airfoil Section Selection

In the previous sections, the geometry of an airfoil section, airfoil design tools, NACA airfoil sections, significant airfoil parameters, and criteria for airfoil section have been covered. In this section, the practical steps for wing airfoil section selection will be presented. It is assumed that an airfoil section database (such as NACA or Eppler) is available and the wing designer is planning to select the best airfoil from the list. The steps are as follows:

1. Determine the average aircraft weight (W_{avg}) in cruising flight:

$$W_{avg} = \frac{1}{2}\left(W_i + W_f\right) \tag{5.9}$$

   where W_i is the initial aircraft weight at the beginning of cruise and W_f is the final aircraft weight at the end of cruise.
2. Calculate the aircraft ideal cruise lift coefficient (C_{L_C}). In a cruising flight, the aircraft weight is equal to the lift force (Equation (5.1)), so:

$$C_{L_C} = \frac{2W_{ave}}{\rho V_c^2 S} \tag{5.10}$$

   where V_c is the aircraft cruise speed, ρ is the air density at cruising altitude, and S is the wing planform area.
3. Calculate the wing cruise lift coefficient ($C_{L_{C_w}}$). Basically, the wing is solely responsible for the generation of the lift. However, other aircraft components also contribute to

Table 5.3 The wing airfoil section of several prop-driven aircraft [5]

No.	Aircraft name	First flight	Max speed (knot)	Root airfoil	Tip airfoil	Average $(t/c)_{max}$ (%)
1	Cessna 550	1994	275	23014	23012	13
2	Beech Bonanza	1945	127	23016.5	23015	15.75
3	Cessna 150	1957	106	2412	2412	12
4	Piper Cherokee	1960	132	65_2-415	65_2-415	15
5	Dornier Do-27	1955	145	23018	23018	18
6	Fokker F-27	1955	227	64_4-421	64_4-421	21
7	Lockheed L100	1954	297	64A-318	64A-412	15
8	Pilatus PC-7	1978	270	64_2-415	64_1-415	15
9	Hawker Siddely	1960	225	23018	4412	15
10	Beagle 206	1967	140	23015	4412	13.5
11	Beech Super king	1970	294	23018-23016.5	23012	14.5
12	Lockheed Orion	1958	411	0014	0012	13
13	Mooney M20J	1976	175	63_2-215	641-412	13.5
14	Lockheed Hercules	1951	315	64A318	64A412	15
15	Thurston TA16	1980	152	64_2-A215	64_2-A215	15
16	ATR 42	1981	269	43 series (18%)	43 series (13%)	15.5
17	AIRTECH CN-235	1983	228	65_3-218	65_3-218	18
18	Fokker 50	1987	282	64_4-421	64_4-415	18

the total lift (negatively or positively); sometimes as much as 20%. Thus the relation between aircraft cruise lift coefficient and wing cruise lift coefficient is a function of aircraft configuration. The contribution of fuselage, tail, and other components will determine the wing contribution to the aircraft lift coefficient. If you are at the preliminary design phase and the geometry of the other components has not yet been determined, the following approximate relationship is recommended:

$$C_{L_{C_w}} = \frac{C_{L_C}}{0.95} \tag{5.11}$$

Table 5.4 The wing airfoil sections of several jet aircraft [5]

No.	Aircraft name	First flight	Max speed (knot)	Root airfoil	Tip airfoil	Average $(t/c)_{max}$ (%)
1	F-15E Strike Eagle	1982	Mach 2.5	64A (6.6%)	64A (3%)	4.8
2	Beech Starship	1988	468	13.2%	11.3%	12.25
3	Lockheed L-300	1963	493	0013	0010	11.5
4	Cessna 500 Citation Bravo	1994	275	23014	23012	13
5	Cessna 318	1954	441	2418	2412	15
6	Gates Learjet 25	1969	333	64A-109	64_4-109	9
7	Aero Commander	1963	360	64_1-212	64_1-212	12
8	Lockheed Jetstar	1957	383	63A-112	63A-309	10.5
9	Airbus 310	1982	595	15.2%	10.8%	13
10	Rockwell/DASA X-31A	1990	1485	Transonic airfoil	Transonic airfoil	5.5
11	Kawasaki T-4	1988	560	Supercritical airfoil (10.3%)	Supercritical airfoil (7.3%)	8.8
12	Gulfstream IV-SP	1985	340	Sonic roof top (10%)	Sonic roof top (8.6%)	9.3
13	Lockheed F-16	1975	Mach 2.1	64A-204	64A-204	4
14	Fokker 50	1985	282	64_4-421	64_4-415	18

In the later design phases – when the other components are designed – this relationship must be clarified. A CFD software package is a reliable tool to determine this relationship.

4. Calculate the wing airfoil ideal lift coefficient (C_{l_i}). The wing is a three-dimensional body, while an airfoil is a two-dimensional section. If the wing chord is constant, with no sweep angle, no dihedral, and the wing span is assumed to be infinite, theoretically the wing lift coefficient should be the same as the wing airfoil lift coefficient. However, at this moment, the wing has not been designed yet, and we have to resort to an approximate relationship. In reality, the span is limited, and in most cases, the wing has a sweep angle and a non-constant chord, so the wing lift coefficient will be slightly less than the airfoil lift coefficient. For this purpose, the

following approximate equation[3] is recommended at this moment:

$$C_{l_i} = \frac{C_{L_{C_w}}}{0.9} \tag{5.12}$$

In the later design phases, using aerodynamic theories and tools, this approximate relation must be modified to include the wing geometry to the required airfoil ideal coefficient.

5. Calculate the aircraft maximum lift coefficient ($C_{L_{max}}$):

$$C_{L_{max}} = \frac{2W_{TO}}{\rho_o V_s^2 S} \tag{5.13}$$

where V_s is the aircraft stall speed, ρ_o is the air density at sea level, and W_{TO} is the aircraft maximum take-off weight.

6. Calculate the wing maximum lift coefficient ($C_{L_{max\_w}}$). With the same logic that was described in step 3, the following relationship is recommended:

$$C_{L_{max\_w}} = \frac{C_{L_{max}}}{0.95} \tag{5.14}$$

7. Calculate the wing airfoil gross maximum lift coefficient ($C_{l_{max\_gross}}$):

$$C_{l_{max\_gross}} = \frac{C_{L_{max\_w}}}{0.9} \tag{5.15}$$

where the wing airfoil *gross* maximum lift coefficient is the airfoil maximum lift coefficient in which the effect of HLD (e.g., flap) is included.

8. Select/design the HLD (type, geometry, and maximum deflection). This step will be discussed in detail in Section 5.12.
9. Determine the HLD contribution to the wing maximum lift coefficient ($\Delta C_{l_{HLD}}$). This step will also be discussed in detail in Section 5.12.
10. Calculate the wing airfoil *net* maximum lift coefficient ($C_{l_{max}}$):

$$C_{l_{max}} = C_{l_{max\_gross}} - \Delta C_{l_{HLD}} \tag{5.16}$$

11. Identify airfoil section alternatives that deliver the desired C_{l_i} (step 4) and $C_{l_{max}}$ (step 10). This is an essential step. Figure 5.23 shows a collection of C_{l_i} and $C_{l_{max}}$ for several NACA airfoil sections in just one graph. The horizontal axis represents the airfoil ideal lift coefficient while the vertical axis represents the airfoil maximum lift coefficient. Every black circle represents one NACA airfoil section. For C_{l_i} and $C_{l_{max}}$ of other airfoil sections, refer to Refs [3, 4]. If there is no airfoil section that delivers the desired C_{l_i} and $C_{l_{max}}$, select the airfoil section that is nearest to the design point (desired C_{l_i} and $C_{l_{max}}$).
12. If the wing is designed for a high subsonic passenger aircraft, select the thinnest airfoil (the lowest $(t/c)_{max}$). The reason is to reduce the critical Mach number (M_{cr})

[3] Please note that the subscript L is used for the 3D application (wing), but the subscript l is employed for the 2D application (airfoil).

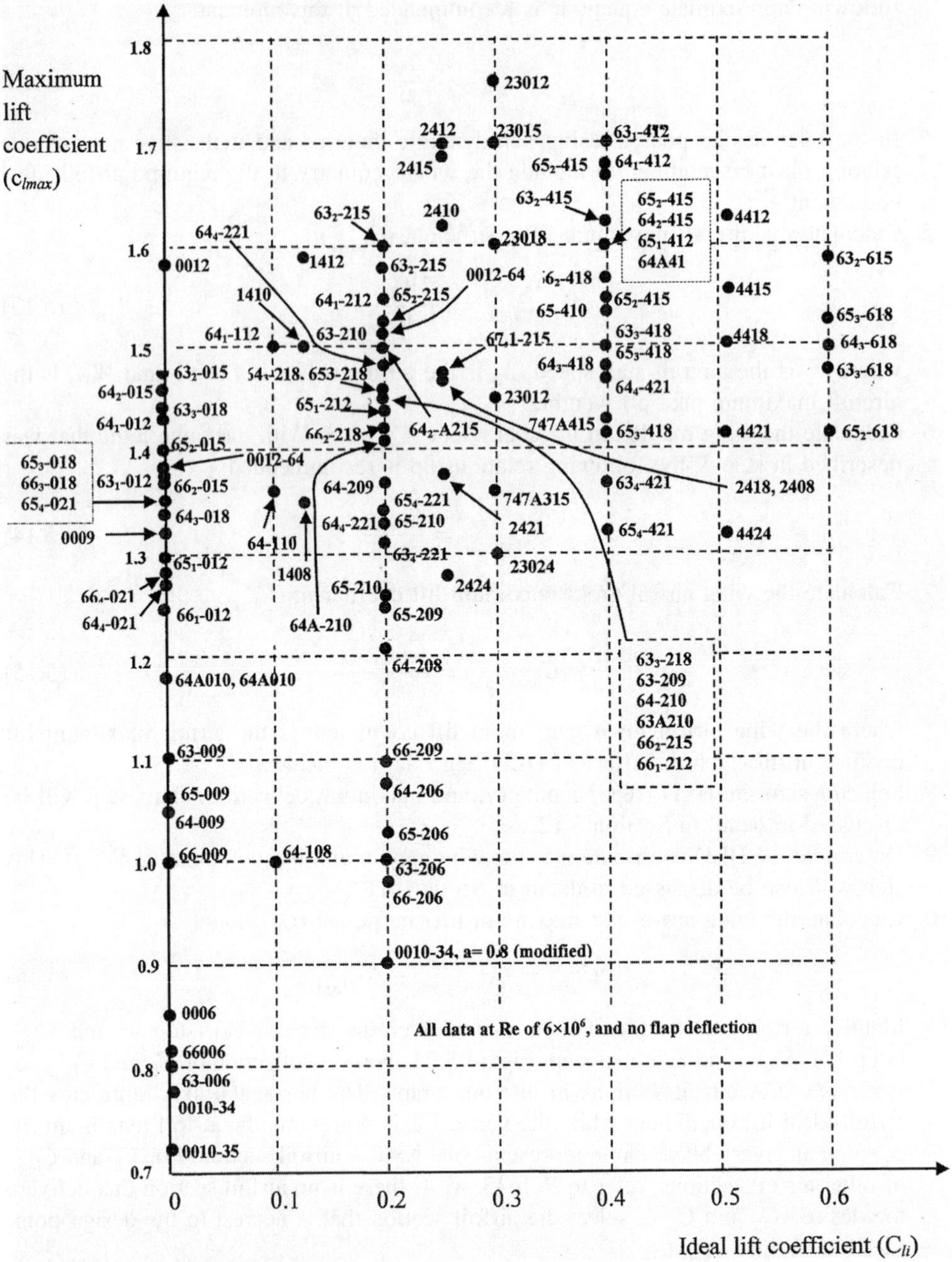

Figure 5.23 Maximum lift coefficient versus ideal lift coefficient for several NACA airfoil sections. Reproduced from permission of Dover Publications, Inc.

and the drag-divergent[4] Mach number (M_{dd}). This allows the aircraft to fly closer to Mach one before the drag rise is encountered. In general, a thinner airfoil will have a higher M_{cr} than a thicker airfoil [6]. Figure 5.24 shows the typical variation of the wing zero-lift and wave-drag coefficient versus Mach number for four wings with airfoil thickness ratio as a parameter. As noted, the M_{dd} of the wing with 9% thickness-to-chord ratio occurs at a value of about 0.88. By reducing the wing $(t/c)_{max}$ to 6% and 4%, the magnitude of the drag rise is progressively reduced, and the value of M_{dd} is increased, moving closer to Mach one.

13. Among several acceptable alternatives, select the optimum airfoil section by using a comparison table. A typical comparison table which includes a typical weight for each design requirement is shown in Table 5.5. Reference [1] is a rich resource for the systematic procedure of the selection technique and table construction.

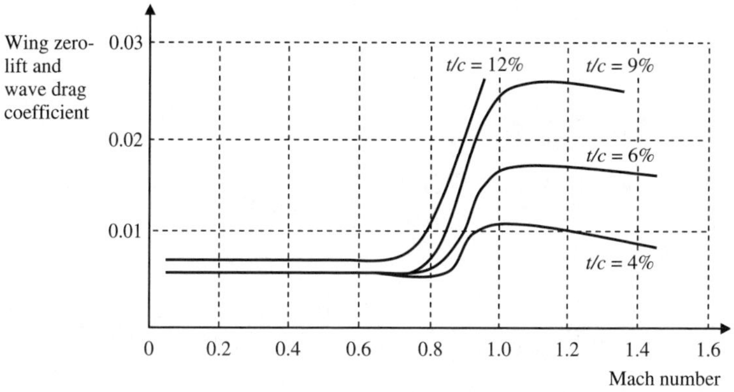

Figure 5.24 Variation of wing zero-lift and wave drag coefficient versus Mach number for various airfoil thickness ratios

Table 5.5 A sample table to compare the features of five airfoil sections

Design objectives	Weight	Airfoil 1	Airfoil 2	Airfoil 3	Airfoil 4	Airfoil 5
$C_{d_{min}}$	25%					
C_{m_o}	15%					
α_s	15					
α_o	10					
$(C_l/C_d)_{max}$	10%					
C_{l_α}	5%					
Stall quality	20%					
Summation	100%	64	76	93	68	68

[4] M_{dd} is defined as the Mach number at which the slope of the curve of C_D versus M is 0.05 (Ref. [6]).

Example 5.2 gives a sample calculation.

Example 5.2

Select a NACA airfoil section for the wing for a non-maneuverable jet GA aircraft with the following characteristics:

$m_{TO} = 4000$ kg, $S = 30$ m^2, $V_c = 250$ knot (at 3000 m), $V_s = 65$ knot (at sea level)

The HLD (split flap) will provide $\Delta C_L = 0.8$ when deflected.

Solution:
Ideal lift coefficient:

$$C_{L_C} = \frac{2W_{ave}}{\rho V_c^2 S} = \frac{2 \cdot 4000 \cdot 9.81}{0.9 \cdot (250 \cdot 0.514)^2 \cdot 30} = 0.176 \quad (5.10)$$

$$C_{L_{C_w}} = \frac{C_{L_C}}{0.95} = \frac{0.176}{0.95} = 0.185 \quad (5.11)$$

$$C_{l_i} = \frac{C_{L_{C_w}}}{0.9} = \frac{0.185}{0.9} = 0.205 \cong 0.2 \quad (5.12)$$

Maximum lift coefficient:

$$C_{L_{max}} = \frac{2W_{TO}}{\rho_o V_s^2 S} = \frac{2 \cdot 4000 \cdot 9.81}{1.225 \cdot (65 \cdot 0.514)^2 \cdot 30} = 1.909 \quad (5.13)$$

$$C_{L_{max\_w}} = \frac{C_{L_{max}}}{0.95} = \frac{1.909}{0.95} = 2.01 \quad (5.14)$$

$$C_{l_{max\_gross}} = \frac{C_{L_{max\_w}}}{0.9} = \frac{2.01}{0.9} = 2.233 \quad (5.15)$$

$$C_{l_{max}} = C_{l_{max\_gross}} - \Delta C_{l_{max\_HLD}} = 2.233 - 0.8 = 1.433 \quad (5.16)$$

Thus, we need to look for NACA airfoil sections that yield an ideal lift coefficient of 0.2 and a net maximum lift coefficient of 1.433. Referring to Figure 5.23, we find the following airfoils whose characteristics match our design requirements (all have $C_{li} = 0.2$, $C_{lmax} = 1.43$): 63_3-218, 64-210, 66_1-212, 66_2-215, 65_3-218.

Now we need to compare these airfoils to see which one is the best, as demonstrated in Table 5.6. The best airfoil is the airfoil whose C_{m_o} is the lowest, $C_{d_{min}}$ is the lowest, α_s is the highest, $(C_l/C_d)_{max}$ is the highest, and stall quality is docile.

By comparing the numbers in the table, we can conclude the following:

1. The NACA airfoil section 66_1-212 yields the highest maximum speed, since it has the lowest $C_{d_{min}}$ (0.0032).
2. The NACA airfoil section 65_3-218 yields the lowest stall speed, since it has the highest stall angle (16 deg).

3. The NACA airfoil section 65_3-218 yields the highest endurance, since it has the highest $(C_l/C_d)_{max}$ (111).
4. The NACA 63_3-218 yields the safest flight, due to its docile stall quality.
5. The NACA airfoil sections 63_3-218, 66_2-215, and 65_3-218 deliver the lowest control problems in flight, due to the lowest C_{m_o} (−0.028).

Table 5.6 A comparison among five airfoil candidates for use in the wing of Example 5.2

No.	NACA	$C_{d_{min}}$	C_{m_o}	α_s (deg) flap up	α_o (deg) $\delta_f = 60$ deg	$(C_l/C_d)_{max}$	Stall quality
1	63_3-218	0.005	−0.028	12	−12	100	Docile
2	64-210	0.004	−0.040	12	−13	75	Moderate
3	66_1-212	0.0032	−0.030	12	−13	86	Sharp
4	66_2-215	0.0035	−0.028	14	−13.5	86	Sharp
5	65_3-218	0.0045	−0.028	16	−13	111	Moderate

Since the aircraft is a non-maneuverable GA aircraft, the stall quality cannot be sharp; hence NACA airfoil sections 66_1-212 and 66_2-215 are not acceptable. If safety is the highest requirement, the best airfoil is NACA 63_2-218. However, if low cost is the most important requirement, NACA 64-210 with the lowest $C_{d_{min}}$ is the best. If aircraft performance (stall speed, endurance, or maximum speed) is of greatest importance, the NACA airfoil sections 65_3-218, 65_3-218, or 66_1-212 are the best, respectively. This may be determined using a comparison table incorporating the weighted design requirements.

5.5 Wing Incidence

The wing incidence (i_w) is the angle between the fuselage center line and the wing chord line at its root (see Figure 5.25). It is sometimes referred to as the wing setting angle (α_{set}). The fuselage center line lies in the plane of symmetry and is usually defined parallel to the cabin floor. This angle could be selected to be variable during a flight operation, or be constant throughout all flight operations. If it is selected to vary during flight, there is no need to determine the wing setting angle for the purpose of aircraft manufacture. However, in this case, the mechanism to vary the wing incidence during flight phases must be designed. Thus the required wing incidence for every flight phase must be calculated. A variable wing incidence is not recommended, since there are huge safety and operational concerns. To allow for the wing having a variable setting angle, there must be a single shaft around which the wing is rotated by pilot control. Such a mechanism is not 100% reliable for aviation purposes, due to fatigue, weight, and stress concentration concerns. In the history of aviation, there is only one aircraft (Vought f 8 u Crusader) whose wing had variable incidence. A flying wing, such as that of the Northrop Grumman B-2 Spirit

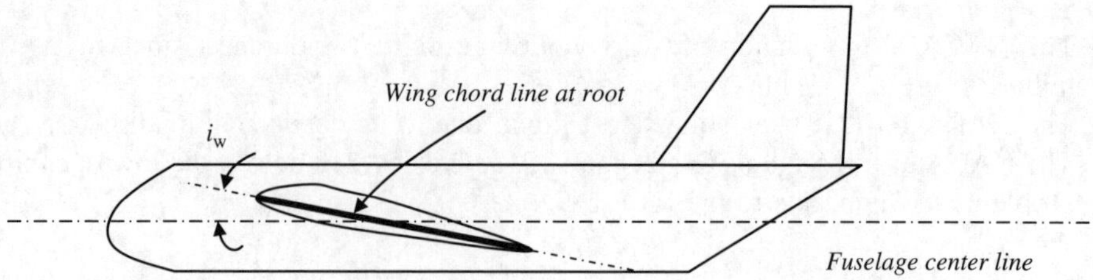

Figure 5.25 Wing setting (incidence) angle

(Figure 6.8) has no wing incidence, since there is no fuselage, however the wing angle of attack must be determined for operational purposes.

A second, very convenient option is to have a constant wing setting angle. The wing can be attached to the fuselage via welding, screw, or other manufacturing technique at the specified setting angle. This is much safer compared with a variable setting angle. For this option, the designer must determine the angle at which the wing is attached to the fuselage. The wing incidence must satisfy the following design requirements:

1. The wing must be able to generate the desired lift coefficient during cruising flight.
2. The wing must produce minimum drag during cruising flight.
3. The wing setting angle must be such that the wing angle of attack could be varied safely (in fact increased) during take-off operation.
4. The wing setting angle must be such that the fuselage generates minimum drag during cruising flight (i.e., the fuselage angle of attack must be zero in cruise).

These design requirements naturally match with the wing airfoil angle of attack corresponding to the airfoil ideal lift coefficient (see Figure 5.26). Therefore, as soon as the wing ideal lift coefficient is determined, a reference to the $C_l - \alpha$ graph demonstrates the wing setting angle. Table 5.7 illustrates the wing incidence for several aircraft.

The typical wing incidence number for the majority of aircraft is between 0 and 4 deg. As a general guide, the wing setting angle in supersonic fighters is between 0 and 1 deg; in GA aircraft between 2 and 4 deg; and in jet transport aircraft between 3 and 5 deg. It is very hard to have the exact same incidence on both left and right wing sections. Due

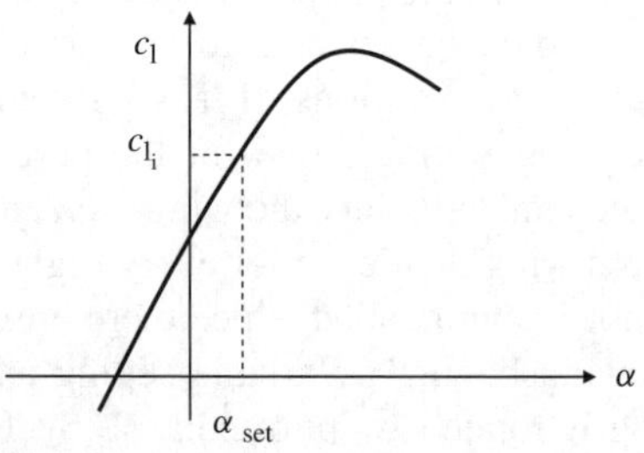

Figure 5.26 Wing setting angle corresponds with ideal lift coefficient

Table 5.7 Wing setting angle for several aircraft [5]

No.	Aircraft	Type	Wing incidence	Cruising speed (knot)
1	Airbus 310	Jet transport	5° 30′	Mach 0.8
2	Fokker 50	Prop-driven transport	3° 30′	282
3	Sukhoi Su-27	Jet fighter	0°	Mach 2.35
4	Embraer FMB-120 Brasilia	Prop-driven transport	2°	272
5	Embraer Tucano	Turboprop trainer	1° 25′	222
6	Antonov An-26	Turboprop transport	3°	235
7	BAe Jetstream 31	Turboprop business	3°	282
8	BAe Harrier	V/STOL close support	1° 45′	570
9	Lockheed P-3C Orion	Prop-driven transport	3°	328
10	Rockwell/DASA X-31A	Jet combat research	0°	1485
11	Kawasaki	Prop-driven transport	0°	560
12	ATR 42	Prop-driven transport	2°	265
13	Beech Super King Air B200	Turboprop transport	3° 48′	289
14	SAAB 340B	Turboprop transport	2°	250
15	AVRO RJ	Jet transport	3° 6′	412
16	McDonnell MD-11	Jet transport	5° 51′	Mach 0.87
17	F-15J Eagle	Fighter	0	>Mach 2.2

to this fact, when there is an inboard stall, the aircraft will roll. The wing outboard stall is unacceptable; if a transport aircraft is at approach, and an outboard stall occurs, it is a disaster. The reason is that the ailerons are not effective to apply roll control.

The wing setting angle may be modified as the design process progresses. For instance, a fuselage with large unsweep over the rear portion to accept aft cargo doors may have its minimum drag at a small positive angle of attack. In such cases, the wing incidence will be reduced accordingly. Another, less fundamental, consideration is the stopping performance during landing operation to get as much weight on the braked wheels as possible. Thus, there is a benefit to reducing the wing incidence slightly to the extent that the change is not felt significantly in the cabin. Reducing the nose gear length will do the same thing. This technique is limited in passenger aircraft because a level cabin floor is desirable on the ground. But, for fighter aircraft, a level floor is not a design consideration.

5.6 Aspect Ratio

The aspect ratio (AR)[5] is defined as the ratio between the wing span b (see Figure 5.33) and the wing MAC or $\overline{C}$:

$$AR = \frac{b}{\overline{C}} \tag{5.17}$$

The wing planform area with a rectangular or straight tapered shape is defined as the span times the MAC:

$$S = b \cdot \overline{C} \tag{5.18}$$

Thus, the aspect ratio shall be redefined as:

$$AR = \frac{bb}{\overline{C}b} = \frac{b^2}{S} \tag{5.19}$$

This equation is not to be used for a wing with geometry other than rectangular – such as a triangle, trapezoid, or ellipse – except when the span is redefined. Example 5.4 clarifies this point. At this point, only the wing planform area is known. The designer has infinite options to select the wing geometry. For instance, consider an aircraft whose wing reference area has been determined to be 30 m^2. A few design options are as follows:

1. A rectangular wing with a 30 m span and a 1 m chord (AR = 30).
2. A rectangular wing with a 20 m span and a 1.5 m chord (AR = 13.333).
3. A rectangular wing with a 15 m span and a 2 m chord (AR = 7.5).
4. A rectangular wing with a 10 m span and a 3 m chord (AR = 3.333).
5. A rectangular wing with a 7.5 m span and a 4 m chord (AR = 1.875).
6. A rectangular wing with a 6 m span and a 5 m chord (AR = 1.2).
7. A rectangular wing with a 3 m span and a 10 m chord (AR = 0.3).
8. A triangular (Delta) wing with a 20 m span and a 3 m root chord (AR = 13.33; please note that the wing has two sections (left and right)).
9. A triangular (Delta) wing with a 10 m span and a 6 m root chord (AR = 3.33).

There are other options too but since we have not discussed the parameter of taper ratio, we will not address them at this moment. Figure 5.27 depicts several rectangular wings with different aspect ratio. These wings have the same planform area, but their spans and chords are different. In terms of lift equation (Equation (5.1)), all are expected to generate the same lift, provided they have the same lift coefficient. However, the wing lift coefficient is not a function of wing area; rather, it is a function of non-dimensional aerodynamic characteristics of the wing such as airfoil and aspect ratio. It is interesting to note that the aspect ratio of the 1903 Wright Flyer was 6.

The question for a wing designer is how to select the aspect ratio, or which wing geometry is the best. To address this question, we need to discuss the effects of aspect ratio on various flight features such as aircraft performance, stability, control, cost, and manufacturability.

[5] Some textbooks use the symbol A instead of AR.

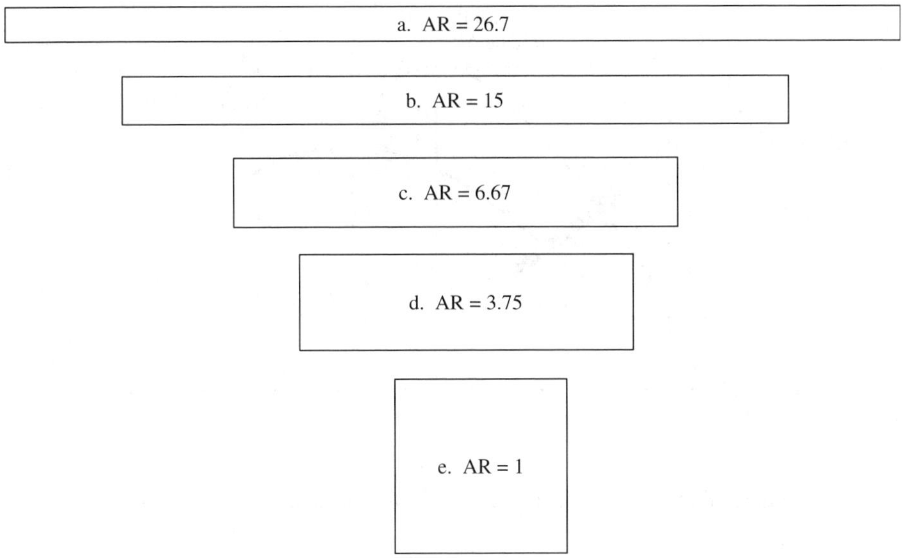

Figure 5.27 Several rectangular wings with the same planform area but different aspect ratio

1. From aerodynamic points of view, as the AR is increased, the aerodynamic features of a three-dimensional wing (such as $C_{L\alpha}$, α_o, α_s, $C_{L_{max}}$, and $C_{D_{min}}$) are getting closer to its two-dimensional airfoil section (such as C_{l_α}, α_o, α_s, $C_{l_{max}}$, and $C_{d_{min}}$). This is due to reduction of the influence of the wing tip vortex. The flow near the wing tips tends to curl around the tip, being forced from the high-pressure region just underneath the tips to the low-pressure region on top [4]. As a result, on the top surface of the wing, there is generally a spanwise component of flow from the tip toward the wing root, causing the streamlines over the top surface to bend toward the root. Similarly, on the bottom surface of the wing, there is generally a spanwise component of flow from the root toward the wing tip, causing the streamlines over the bottom surface to bend toward the tip.
2. Due to the first item, as the AR is increased, the wing lift curve slope (C_{L_α}) is increased toward the maximum theoretical limit of 2π 1/rad (see Figure 5.28). The relationship [4] between 3D wing lift curve slope (C_{L_α}) and 2D airfoil lift curve slope (C_{l_α}) is as follows:

$$C_{L_\alpha} = \frac{dC_L}{d\alpha} = \frac{C_{l_\alpha}}{1 + \frac{C_{l_\alpha}}{\pi \cdot AR}} \quad (5.20)$$

   For this reason, a high-AR (longer) wing is desired.
3. As the AR is increased, the wing stall angle (α_s) is decreased toward the airfoil stall angle since the wing effective angle of attack is increased (see Figure 5.28). For this reason, the horizontal tail is required to have an aspect ratio lower than the wing aspect ratio to allow for a higher tail stall angle. This will result in the tail stalling after the wing has stalled, and allow for a safe recovery. For the same reason, a

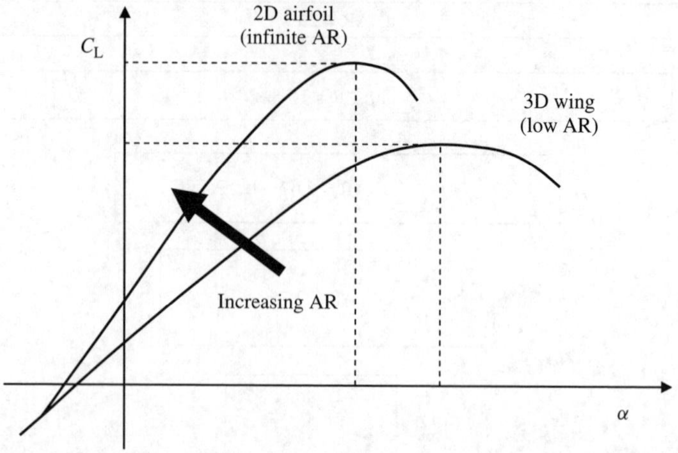

Figure 5.28 The effect of AR on C_L versus angle of attack graph

canard is desired to have an aspect ratio more than the wing aspect ratio. For this reason, a high-AR (longer) wing is desired.

4. Due to the third item, as the AR is increased, the wing maximum lift coefficient ($C_{L_{max}}$) is increased toward the airfoil maximum lift coefficient ($C_{l_{max}}$). This is due to the fact that the wing effective angle of attack is increased (see Figure 5.28). For this reason, a high-AR (longer) wing is desired.
5. As the AR is increased, the wing will be heavier. The reason lies in the requirement for structural stiffness. As the wing gets longer, the wing weight (W_w) bending moment (M) gets larger (since $M = \frac{W_w}{2}\frac{b}{2}$), and the wing root will have a higher bending stress. Thus, the wing root must be stronger to hold the long wing. This requires a heavier wing. The greater weight of the wing translates into more cost. For this reason, a low-AR (shorter) wing is desired.
6. As the $\sqrt{AR}$ is increased, the aircraft maximum lift-to-drag ratio is increased. Since

$$\left(\frac{L}{D}\right)_{max} = \frac{1}{2\sqrt{KC_{D_o}}} \tag{5.21}$$

where:

$$K = \frac{1}{\pi \cdot e \cdot AR} \tag{5.22}$$

where K is the wing induced drag factor, e is the Oswald span efficiency factor, and C_{D_o} is the aircraft zero-lift drag coefficient [7, 8]. For the derivation of these two equations, you are referred to Ref. [7]. For this reason, a high-AR (longer) wing is desired. This is one of the reasons that the gliders have large aspect ratio and long wing. For this reason, a high-AR (longer) wing is desired.

7. As the AR is increased, the wing induced drag is decreased, since the induced drag (C_{D_i}) is inversely proportional to the aspect ratio. For this reason, a low-AR (shorter) wing is desired:

$$C_{D_i} = \frac{C_L^2}{\pi \cdot e \cdot \mathrm{AR}} \tag{5.23}$$

8. As the AR is increased, the effect of wing tip vortex on the horizontal tail is decreased. As explained in item 1, the tendency for the flow to leak around the wing tips establishes a circulation which trails downstream of the wing; that is, a trailing vortex is created at each wing tip. This downward component is called downwash. If the tail is in the region of downwash, the tail effective angle of attack is reduced by downwash. This will influence the longitudinal stability and longitudinal control [9] of the aircraft.
9. As the AR increases, the aileron arm will be increased, since the aileron are installed outboard of the wing. This means that the aircraft has more lateral control.
10. As the AR increases, the aircraft mass moment of inertia around the x-axis [10] will be increased. This means that it takes longer to roll. In other words, this will reduce the maneuverability of the aircraft in roll [9]. For instance, the Bomber aircraft Boeing B-52 (Figures 8.6 and 9.4), which has a very long span, takes several seconds to roll at low speed, while the fighter aircraft F-16 Falcon (Figure 4.6) takes a fraction of a second to roll. For this reason, a low-AR (shorter) wing is desired for a maneuverable aircraft. The tactical supersonic missiles have a low AR of around 1 to enable them to roll and maneuver as fast as possible.
11. If the fuel tank is supposed to be inside the wing, it is desirable to have a low aspect ratio wing. This helps to have a more concentrated fuel system. For this reason, a low-AR (shorter) wing is desired.
12. As the aspect ratio is increased, the wing stiffness around the y-axis is decreased. This means that the tendency of the wing tips to drop during a take-off is increased, while the tendency to rise during high-speed flight is increased. In practice, the manufacture of a very high aspect ratio wing with sufficient structural strength is difficult.

    This wing behavior was observed during the flight of Voyager aircraft (Figure 4.5) with AR of 38 and wingspan of 33.8 m in 1986 during its record-breaking flight to circle around the globe without refueling. The Voyager wing tip drop was more than 5 ft during take-off (low-speed flight), while the wing tips raised more than 4 ft during cruising (high-speed flight). During Voyager's take-off, as the plane accelerated, the tips of the wings (which were heavily loaded with fuel) were damaged as they scraped against the runway, ultimately causing pieces of winglets to break off at both ends. The aircraft accelerated very slowly and needed approximately 14 200 ft of the runway to gain enough speed to lift from the ground, the wings arching up dramatically just before take-off. The plane also continuously reminded the pilots of its pitch instability and fragility. They had to maneuver around bad weather numerous times.

    Another example is the transport aircraft Boeing 747 (Figures 3.7, 3.12, and 9.4) with an AR of 7.7 and wingspan of 59.6 m, whose wing tips drop about 1 ft while the aircraft in on the ground prior to take-off. The wing tip drop is not desirable, especially for a take-off maneuver, since the wing tip clearance is of great importance for safety. For this reason, a low-AR (shorter) wing is desired. A shorter wing is easier to build compared with a long wing. For the manufacturability reason, a low-AR (shorter) wing is desired.

13. A shorter wing needs lower cost to build compared with a long wing. For the cost reason, a low-AR (shorter) wing is desired.
14. As the AR is increased, the occurrence of the aileron reversal [9] is expected more, since the wing will be more flexible. The aileron reversal is not a desirable phenomenon for a maneuverable aircraft. For this reason, a low-AR (shorter) wing is desired.
15. In general, a wing with rectangular shape and high AR is gust sensitive.

As noted, the aspect ratio has several influences over the aircraft features. For some design requirements, a low-aspect ratio wing is favorable, while for other design requirements, a high-aspect ratio wing is desirable. The exact value of the AR will be determined through a thorough investigation and lots of calculation over aircraft performance, stability, control, manufacturability, and cost.

A systems engineering technique [1] using a weighted parametric table must be employed to determine the exact value of the aspect ratio. Table 5.8 illustrates the typical values of aspect ratio for different aircraft types. Table 5.9 illustrates the aspect ratio for several aircraft. As noted, the aspect ratio ranges from 2.2 for fighter aircraft Eurofighter 2000 (Figure 3.7) to 32.9 for high-altitude, long-endurance (HALE) aircraft Socata. Figure 5.56(b,d) illustrates the fighter aircraft MiG-29 with a low-AR wing, and Sailplane Schleicher ASK-18 with a high-AR wing respectively.

Table 5.8 Typical values of wing aspect ratio

No.	Aircraft type	Aspect ratio
1	Hang glider	4–8
2	Glider (sailplane)	20–40
3	Home-built	4–7
4	General aviation	5–9
5	Jet trainer	4–8
6	Low-subsonic transport	6–9
7	High-subsonic transport	8–12
8	Supersonic fighter	2–4
9	Tactical missile	0.3–1
10	Hypersonic aircraft	1–3

Table 5.9 Aspect ratio and taper ratio for several aircraft

No.	Aircraft	Type	Engine	V_{max} (knot)	S (m²)	AR	λ
1	Cessna 172	GA	Piston	121	16.2	7.52	0.67
2	Air Tractor AT-402B	Agricultural	Turboprop	174	27.3	8.9	1
3	Piper Comanche	GA	Piston	170	16.5	7.3	0.46
4	McDonnell DC-9	Transport	Turbofan	Mach 0.84	86.8	8.56	0.25
5	Lockheed L-1011	Transport	Turbofan	Mach 0.86	321	7.16	0.29
6	Boeing 747-400	Transport	Turbofan	Mach 0.92	525	6.96	0.3
7	Tucano	Trainer	Turboprop	Mach 0.4	19.2	6.4	0.465
8	Airbus 310	Transport	Turbofan	Mach 0.9	219	8.8	0.26
9	Jet stream 41	Regional Airliner	Turboprop	295	32.59	10.3	0.365
10	Lockheed F-16 Falcon	Fighter	Turbofan	>Mach 2	27.87	3.2	0.3
11	SAAB 39 Gripen	Fighter	Turbofan	>Mach 2	27	2.6	0.25
12	Grumman B-2 Spirit	Bomber	Turbofan	550	465.5	5.92	0.24
13	Schweizer SA 2-38A	Surveillance	Piston	157	21	18.2	0.4
14	Grob G 850 Strato 2C	Surveillance	Piston	280	145	22	0.25
15	Stemme S10	Motor glider	Piston	97	18.7	28.2	0.26
16	Socata HALE	Surveillance	Turboprop	162	70	32.9	0.6
17	Voyager	Circle the globe	Piston	106	30.1	38	0.25
18	Eurofighter 2000	Fighter	Turbofan	Mach 2	50	2.2	0.19
19	Dassault Mirage 2000	Fighter	Turbofan	Mach 2.2	41	2	0.08

5.7 Taper Ratio

The taper ratio (λ) is defined as the ratio between the tip chord (C_t) and the root chord (C_r).[6] This definition is applied to the wing, as well as the horizontal tail and the vertical tail. Root chord and tip chord are illustrated in Figure 5.31:

$$\lambda = \frac{C_t}{C_r} \tag{5.24}$$

[6] In some older textbooks, the taper ratio was defined as the ratio between the root chord and the tip chord.

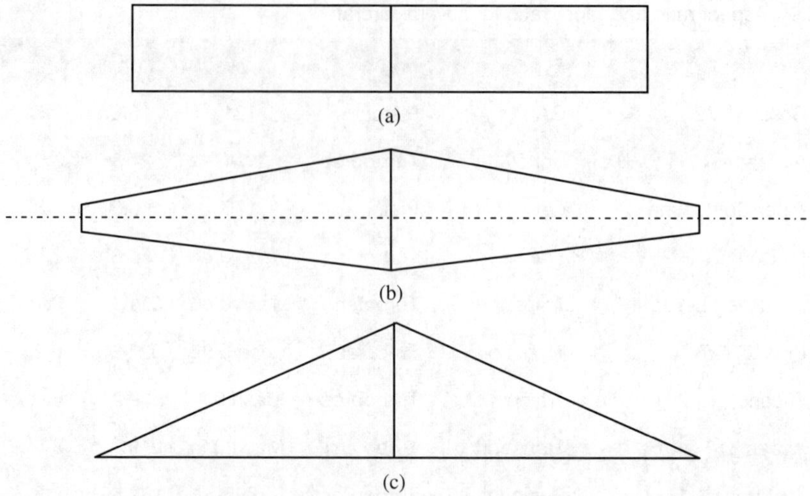

Figure 5.29 Wings with various taper ratios: (a) Rectangle ($\lambda = 1$); (b) Trapezoid $0 < \lambda < 1$ (straight tapered); and (c) Triangle (delta) $\lambda = 0$

The geometric result of taper is a smaller tip chord. In general, the taper ratio varies between zero and one:

$$0 \leq \lambda \leq 1$$

where three major planform geometries relating to taper ratio are rectangular, trapezoidal, and delta shape (see Figure 5.29).

In general, a rectangular wing planform is aerodynamically inefficient, while it has a few advantages, such as performance, cost, and ease of manufacture. A wing with a rectangular planform has a larger downwash angle at the tip than at the root. Therefore, the effective angle of attack at the tip is reduced compared with that at the root. Thus, the wing tip will tend to stall later than the root. The spanwise lift distribution is far from elliptical, where it is highly desirable to minimize the induced drag. Hence, one of the reasons to taper the planform is to reduce the induced drag.

In addition, since the tip chord is smaller than the root chord, the tip Reynolds number will be lower, as well as a lower tip induced downwash angle. Both effects will lower the angle of attack at which stall occurs. This may result in the tip stalling before the root. This is undesirable from the viewpoint of lateral stability and lateral control. In contrast, a rectangular wing planform is structurally inefficient, since there is a lot of area outboard, which supports very little lift. Wing taper will help resolve this problem as well. The effect of wing taper can be summarized as follows:

1. The wing taper will change the wing **lift distribution**. This is assumed to be an advantage of the taper, since it is a technical tool to improve the lift distribution. One of the wing design objectives is to generate lift such that the spanwise lift distribution is elliptical. The significance of elliptical lift distribution will be examined in the next section. Based on this item, the exact value for the taper ratio will be determined by the lift distribution requirement.

2. The wing taper will increase the **cost** of wing manufacture, since the wing ribs will have different shapes. Unlike a rectangular planform where all ribs are similar, each rib will have a different size. If the cost is of major issue (such as for home-built aircraft), do not taper the wing.
3. The taper will reduce the wing **weight**, since the center of gravity of each wing section (left and right) will move toward the fuselage center line. This results in a lower bending moment at the wing root. This is an advantage of the taper. Thus, to reduce the weight of the wing, more taper (toward 0) is desired.
4. Due to item 3, the wing mass moment of inertia about the x-axis (longitudinal axis) will be decreased. Consequently, this will improve the aircraft **lateral control**. In this regard, the best taper is to have a delta wing ($\lambda = 0$).
5. The taper will influence the aircraft static **lateral stability** (C_{l_β}), since the taper usually generates a sweep angle (either on the leading edge or on a quarter chord line). The effect of the sweep angle on the aircraft stability will be discussed in Section 5.8.

As noted, the taper ratio has mixed influences over the aircraft features. The aspect ratio of a conventional aircraft is a compromise between conflicting aerodynamic, structural, performance, stability, cost, and manufacturability requirements. For some design requirements (e.g., cost, manufacturability), a no-taper ratio wing is favorable; while for other design requirements (such as stability, performance, and safety), a tapered wing is desirable. The first estimate of the taper ratio will be determined by lift distribution calculations, as introduced in the next section. The exact value of the taper ratio will be finalized through a thorough investigation and lots of calculation over aircraft performance, stability, control, manufacturability, and cost. A systems engineering technique [1] using a weighted parametric table must be employed to determine the exact value of the taper ratio. Table 5.9 illustrates the taper ratio for several aircraft. The typical effect of taper ratio on the lift distribution is sketched in Figure 5.30.

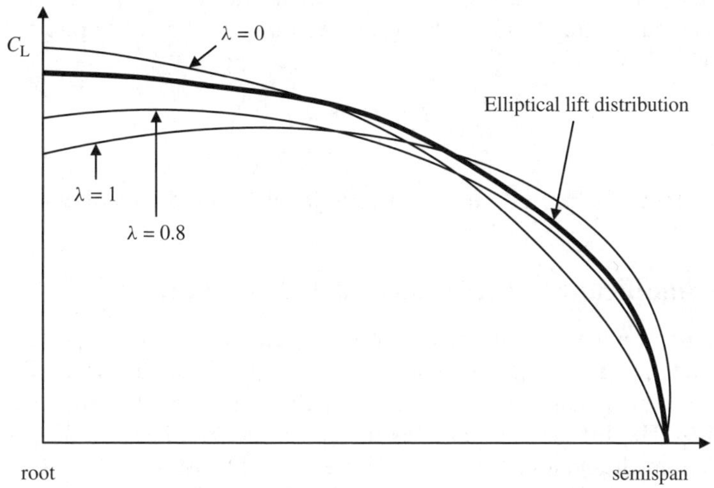

Figure 5.30 The typical effect of taper ratio on the lift distribution

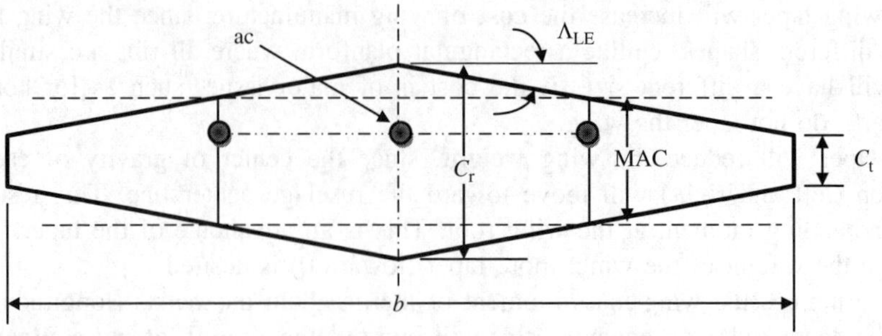

Figure 5.31 Mean aerodynamic chord and aerodynamic center in a straight wing

In the normal flight range, the resultant aerodynamic forces acting on any lifting surface (e.g., lift, tail) can be represented as a lift and drag acting at the ac, together with a pitching moment which is independent of angle of attack. Methods for determining planform aerodynamic center locations may be found in most aerodynamic textbooks. Until compressibility effects begin to play a role, it is experienced that the planform aerodynamic center ranges from 25% to about 30% of MAC or $\overline{C}$. In the transonic and supersonic speed range, the ac tends to move aft, such that at transonic speeds, the ac moves close to the 50% chord point on the MAC. The aerodynamic center lies in the plane of symmetry of the wing. However, in determining MAC, it is convenient to work with the half wing. For a general planform, the location of length of the MAC can be determined using the following integral:

$$\overline{C} = \frac{2}{S}\int_{0}^{b/2} c^2(y)\mathrm{d}y \tag{5.25}$$

where c is the local chord and y is the aircraft lateral axis. For a constant-taper and constant-sweep angle (trapezoidal) planform (see the geometry of Figure 5.31), MAC is determined [11] as follows:

$$\overline{C} = \frac{2}{3}C_r\left(\frac{1+\lambda+\lambda^2}{1+\lambda}\right) \tag{5.26}$$

Table 5.9 illustrates the aspect ratio for several jet and prop-driven aircraft.

5.8 The Significance of Lift and Load Distributions

The distribution of wing non-dimensional lift (i.e., lift coefficient C_L) per unit span along the wing is referred to as *lift distribution*. Each unit area of the wing along the span produces a specific amount of lift. The total lift is equal to the summation of these individual lifts. The lift distribution goes to zero at the tips, because there is a pressure equalization from the bottom to the top of the wing precisely at $y = -b/2$ and $+b/2$. Hence no lift is generated at these two points. In addition, the variation of "lift coefficient times sectional chord ($C \cdot C_L$)" along the span is referred to as the *load distribution*. Both

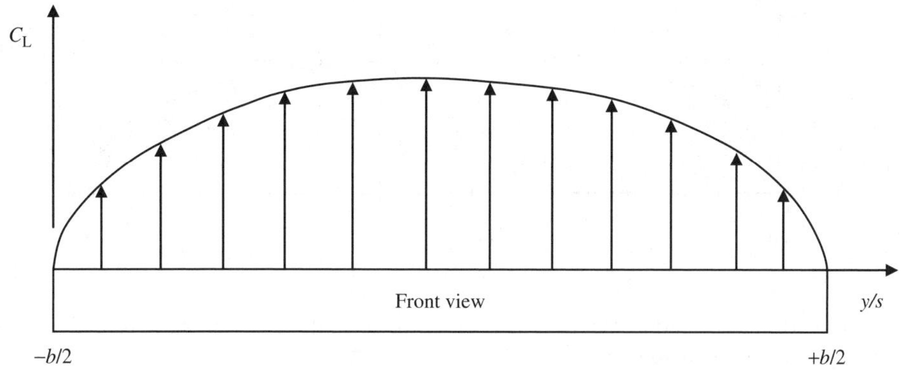

Figure 5.32 Elliptical lift distribution over the wing

lift distribution and load distribution are of great importance in the wing design process. The major application of lift distribution is in aerodynamic calculation, while the primary application of the load distribution is in wing structural design as well as controllability analysis.

In the past (1930s), it was thought that for an elliptic lift distribution, the chord must vary elliptically along the span. The direct result of such logic was that the wing planform must be elliptical. For this reason, several aircraft wing planforms such as that of the Supermarine Spitfire (Figure 8.3), a famous British World War II fighter, were made elliptic. But today, we know that there are various parameters that make the lift distribution elliptic, thus, there is no need for the wing planform to be elliptic.

The type of both lift distribution and load distribution is very important in wing design; and will influence the aircraft performance, airworthiness, stability, control, and cost. Ideally, both lift distribution and load distribution are preferred to be elliptical. For the above-mentioned reasons, the elliptical lift distribution and the elliptical load distribution are ideal and are the design objectives in the wing design process. An elliptical lift distribution is sketched in Figure 5.32, where a front view of the wing is illustrated. The horizontal axis in Figure 5.32 is y/s, where y is the location in the y-axis, and s denotes the semispan ($s = b/2$). In this figure, no HLD (e.g., flap) is deflected and the effect of the fuselage is ignored. The elliptical lift distribution and elliptical load distribution have the following desirable properties:

1. If the wing tends to stall ($C_{L_{\max}}$), the wing root is stalled before the wing tip ($C_{L_{\text{root}}} = C_{L_{\max}}$ while $C_{L_{\text{tip}}} < C_{L_{\max}}$). In a conventional aircraft, the flaps are located inboard, while the ailerons are installed outboard of the wing. In such a situation, ailerons are active, since the flow over the wing outboard section is healthy. This is of greater importance for spin recovery (which often happens after stall); since the aileron (in addition to the rudder) application is critical to stop the autorotation. Thus, the elliptical lift distribution provision guarantees flight safety in the event of stall (see Figure 5.33).
2. The bending moment at the wing root is a function of load distribution. If the load distribution is concentrated near the root, the bending moment is considerably less than when it is concentrated near the tip. The center of an elliptical load distribution

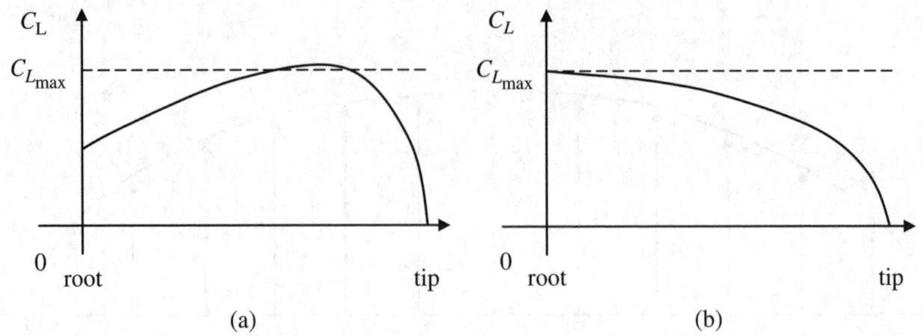

Figure 5.33 Lift distribution over the half wing: (a) non-elliptical (tip stalls before the root); (b) elliptical (root stalls before the tip)

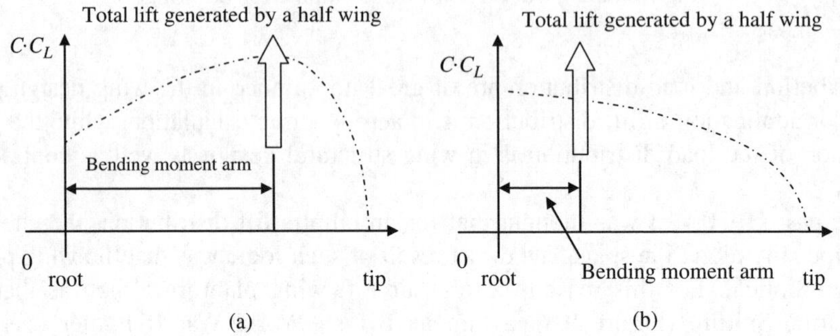

Figure 5.34 Load distribution over a half wing: (a) non-elliptical (load is farther from the root); (b) elliptical (load is closer to the root)

is closer to the wing root, thus it leads to a lower bending moment, which results in less bending stress and less stress concentration at the wing root (see Figure 5.34). This means a lighter wing spar and a lighter wing structure that is always one of the design requirements. The load distribution is a function of the lift distribution.

3. The center of gravity of each wing section (left or right) for an elliptical load distribution is closer to the fuselage center line. This means a lower wing mass moment of inertia about the x-axis, which is an advantage in lateral control. Basically, an aircraft rolls faster when the aircraft mass moment of inertia is smaller.
4. The downwash is constant over the span for an elliptical lift distribution [4]. This will influence the horizontal tail effective angle of attack.
5. For an elliptical lift distribution, the induced angle of attack is also constant along the span. An elliptical lift distribution also yields the minimum induced drag.
6. The variation of lift over the span for an elliptical lift distribution is steady (gradually increasing from tip (zero) to root (maximum)). This will simplify the wing spar(s) design.

The reader may have noticed that if the contribution of the fuselage is added to the wing lift distribution, the distribution may not be elliptical due to negligible fuselage lift contribution. This is true, and more realistic, since in a conventional aircraft, the

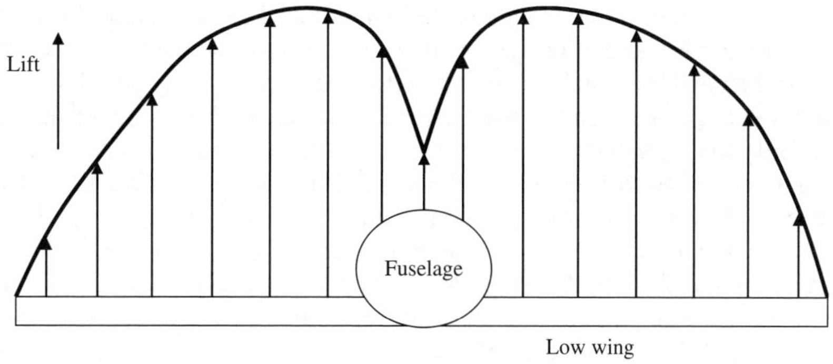

Figure 5.35 The fuselage contribution to the lift distribution of a low-wing configuration

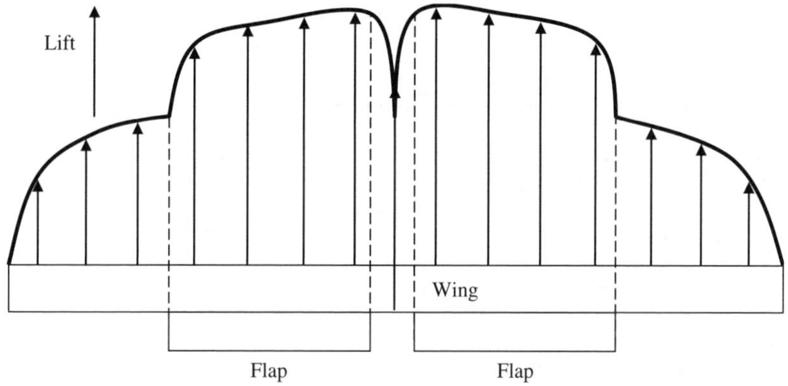

Figure 5.36 The flap contribution to the lift distribution

wing is attached to the fuselage. What we examined here in this section is an ideal case, and the reader may modify the lift distribution by considering the fuselage contribution. Figure 5.35 depicts the fuselage contribution to a low-wing configuration. A similar case may be made for the effect of flap on the lift distribution when deflected. Figure 5.36 illustrates the flap contribution to the wing lift distribution. In principle, the goal in the wing design is to obtain an elliptical wing distribution without considering the contributions of fuselage, flap, or other components.

In Section 5.15, a mathematical technique will be introduced to determine the lift and load distribution along the wing.

5.9 Sweep Angle

Consider the top view of an aircraft. The angle between a constant percentage chord line along the semispan of the wing and the lateral axis perpendicular to the aircraft center line (y-axis) is called the leading edge sweep (Λ_{LE}). The angle between the wing leading edge and the y-axis of the aircraft is called the leading edge sweep (Λ_{LE}). Similarly, the angle between the wing trailing edge and the longitudinal axis (y-axis) of the aircraft is called the trailing edge sweep (Λ_{TE}). In the same fashion, the angle between the wing quarter chord line and the y-axis of the aircraft is called the quarter chord sweep ($\Lambda_{C/4}$). And finally, the angle between the wing 50% chord line and the y-axis of the aircraft is the 50% chord sweep ($\Lambda_{C/2}$).

If the angle is greater than zero (i.e., the wing is inclined toward the tail), it is called aft sweep or simply sweep; otherwise it is referred to as forward sweep. Figure 5.37 shows five wings with various sweep angles. Figure 5.37(a) illustrates a wing without sweep, while Figure 5.37(b–d) shows four swept wings. The leading edge sweep is depicted in the wing of Figure 5.37(b), while the trailing edge sweep is shown in the wing of Figure 5.37(e). In addition, the quarter chord sweep is illustrated in the wing

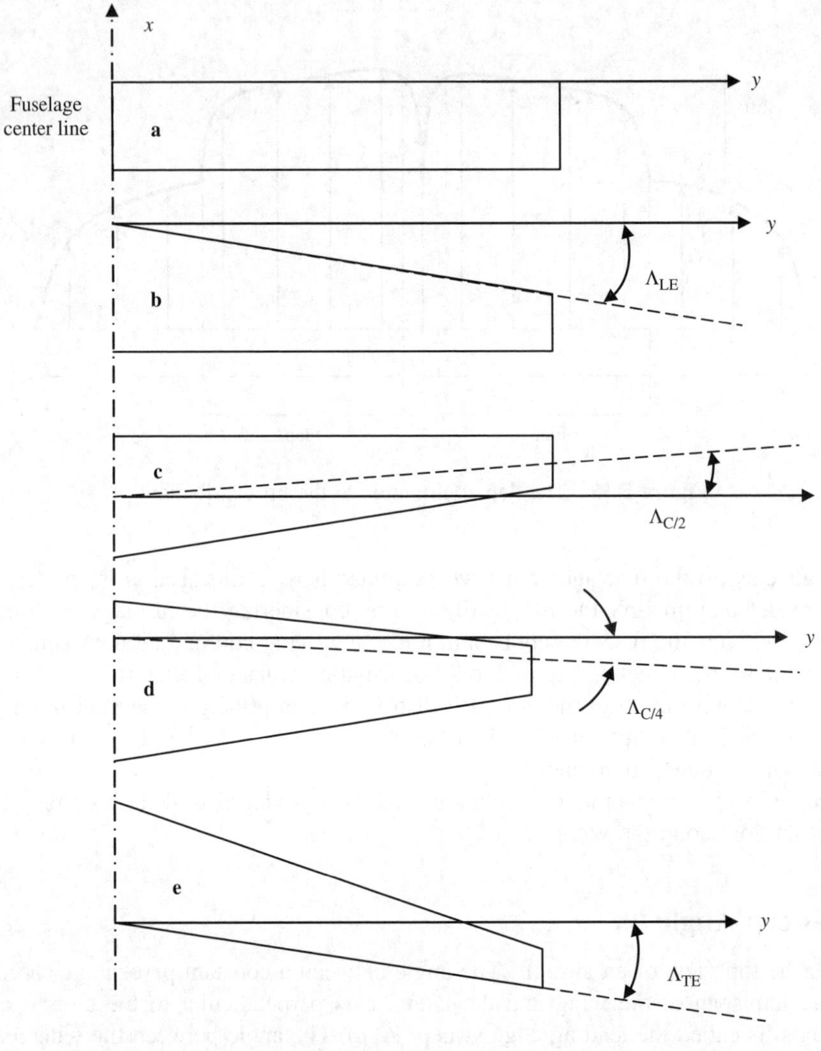

Figure 5.37 Five wings with different sweep angles

of Figure 5.37(d), and the 50% chord sweep is illustrated in the wing of Figure 5.37(c). Most high-speed airplanes designed since the mid-1940s – such as North American F-86 Sabre – have swept wings. On a sweptback tapered wing, typical of almost all high-speed aircraft, the leading edge has more sweep than the trailing edge.

With reference to the definition of sweep angle, a particular wing may have aft leading edge sweep, while it has forward trailing edge sweep. Among four types of sweep angles, the quarter chord sweep and leading edge sweep are the most important ones. The subsonic lift due to angle of attack normally acts at the quarter chord. In addition, the crest is usually close to the quarter chord. The discussion in this section regarding the characteristics (advantages and disadvantages) of the sweep angle is mostly about the leading edge sweep angle, unless otherwise stated.

Basically, a wing is being swept for the following five design goals:

1. Improving the wing aerodynamic features (lift, drag, and pitching moment) at transonic, supersonic, and hypersonic speeds by delaying the compressibility effects.
2. Adjusting the aircraft center of gravity.
3. Improving static lateral stability.
4. Impacting longitudinal and directional stability.
5. Increasing pilot view (especially for fighter pilots).

These items will be described in more detail in this section. For more information, the reader needs to refer to technical textbooks that are listed at the end of this chapter. The practical influence of the sweep angle on various flight features is as follows:

1. The sweep angle, in practice, tends to increase the distance between the leading edge and the trailing edge. Accordingly, the pressure distribution will vary.
2. The effective chord length of a swept wing is longer (see Figure 5.38) by a factor of $1/\cos(\Lambda)$. This makes the effective thickness-to-chord ratio thinner, since the thickness remains constant.
3. Item 2 can also be translated into the reduction of Mach number (M_n) normal to the wing leading edge to $M \cos(\Lambda)$. Hence, by sweeping the wing, the flow behaves as if the airfoil section is thinner, with a consequent increase in the critical Mach number of the wing. For this reason, a classic design feature used to increase M_{cr} is to sweep the wing [6].
4. The effect of the swept wing is to curve the streamline flow over the wing as shown in Figure 5.38. The curvature is due to the deceleration and acceleration of flow in the plane perpendicular to the quarter chord line. Near the wing tip the flow around the tip from the lower surface to the upper surface obviously alters the effect of sweep. The effect is to unsweep the spanwise constant-pressure lines; isobar. To compensate, the wing tip may be given additional structural sweep.
5. The wing aerodynamic center (ac) is moved aft by the wing aft sweep at about a few percent. The aft movement of the ac with increase in sweptback angle occurs because the effect of the downwash pattern associated with a swept wing is to raise the lift coefficient on the outer wing panel relative to the inboard lift coefficient. Since sweep movers the outer panel aft relative to the inner portion of the wing, the effect on the center of lift is an aftward movement. The effect of wing sweep on ac

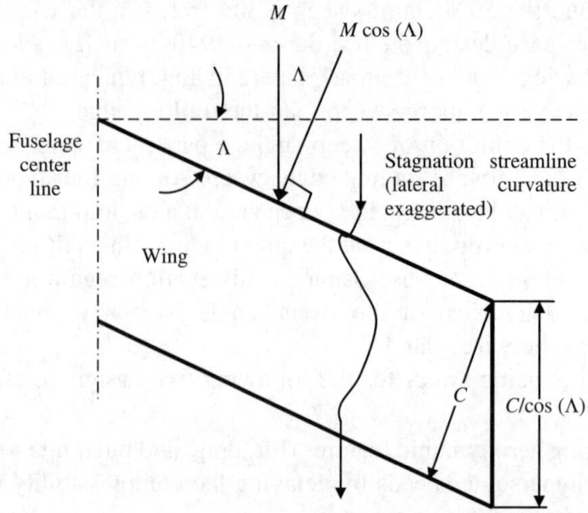

Figure 5.38 The effect of the sweep angle on the normal Mach number

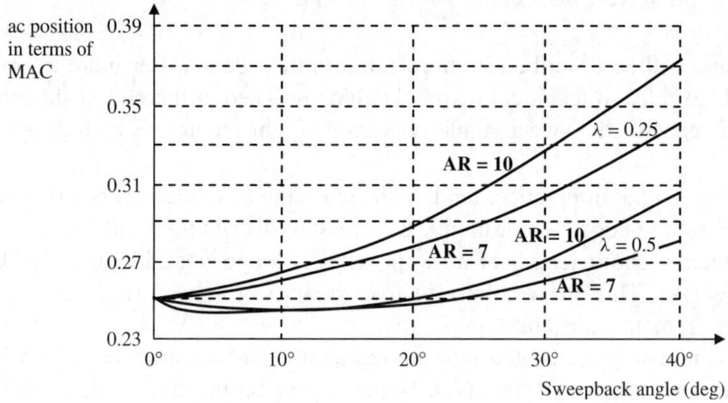

Figure 5.39 Effect of wing sweepback on ac position for several combinations of AR and λ

position is shown in Figure 5.39 for aspect ratios of 7 and 10 and for taper ratios of 0.25 and 0.5.

6. The effective dynamic pressure is reduced, although not by as much as in cruise.
7. The sweep angle tends to change the lift distribution as sketched in Figure 5.40. The reason becomes clear by looking at the explanation in item 5. As the sweep angle is increased, the Oswald efficiency factor (e) will decrease (Equation (5.25)).

The Oswald span efficiency for a straight wing and a swept wing is given respectively by Equation (5.27a,b) [12]:

$$e = 1.78\left(1 - 0.045\mathrm{AR}^{0.68}\right) - 0.64 \tag{5.27a}$$

$$e = 4.61\left(1 - 0.045\mathrm{AR}^{0.68}\right)\left[\cos(\Lambda_{\mathrm{LE}})\right]^{0.15} - 3.1 \tag{5.27b}$$

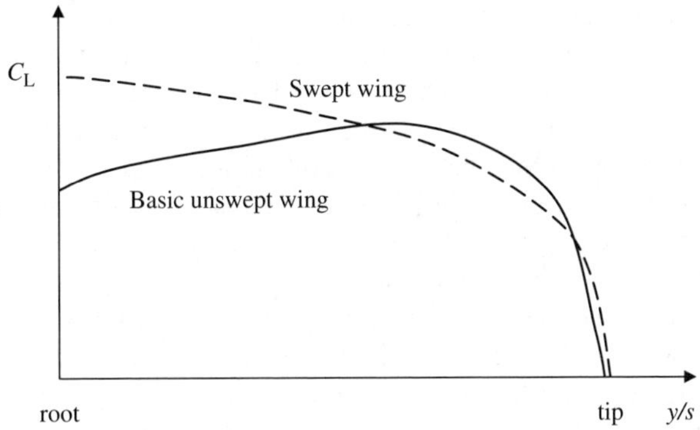

Figure 5.40 Typical effect of sweep angle on lift distribution

Equation (5.27a) is for a straight wing and Equation (5.27b) is for a swept wing where the sweep angle is more than 30 deg. When the Oswald span efficiency is equal to 1, it indicates that the lift distribution is elliptic, otherwise it is non-elliptic. Equation 5.27 is not valid for low aspect ratio wings (AR less than 6).

8. The wing maximum lift coefficient can actually increase with increasing sweep angle. However, the maximum useful lift coefficient actually decreases with increasing sweep angle, due to the loss of control in a pitch-up situation. Whether or not pitch-up occurs depends not only on the combination of sweep angle and aspect ratio, but also on airfoil type, twist angle, and taper ratio. Thus, the sweep angle tends to increase the stall speed (V_s).

   The maximum lift coefficient of the basic wing without HLD is governed by the following semi-empirical relationship [13]:

$$C_{L_{\max(\Lambda \neq 0)}} = C_{l_{\max}} [0.86 - 0.002(\Lambda)] \tag{5.28}$$

   where the sweep angle (Λ) is in degrees and $C_{l_{\max}}$ denotes the maximum lift coefficient for the outer panel airfoil section.

9. Wing sweep tends to reduce the wing lift curve slope (C_{L_α}). A modified equation based on the Prandtl–Glauert approximation is introduced by Shevell [13] as follows:

$$C_{L_\alpha} = \frac{2\pi \mathrm{AR}}{2 + \sqrt{\mathrm{AR}^2\left(1 + \tan^2\Lambda - M^2\right) + 4}} \tag{5.29}$$

10. The aircraft pitching moment will be increased, provided the aircraft cg is forward of the aircraft ac. The reason is that the wing aerodynamic center is moving aft with an increase in sweep angle.
11. An aft swept wing tends to have tip stall because of the tendency toward outboard, spanwise flow. This causes the boundary layer to thicken as it approaches the tips. For a similar reason, a swept forward wing would tend toward root stall. This tends to have an influence opposite to that of wing twist.
12. On most aft swept wing aircraft, the wing tips are located behind the aircraft center of gravity. Therefore, any loss of lift at the wing tips causes the wing center of pressure to move forward. This in turn will cause the aircraft nose to pitch up. This pitch-up

tendency can cause the aircraft angle of attack to increase even further. This may result in a loss of aircraft longitudinal control. For a similar reason, a forward swept wing aircraft would exhibit a pitch-down tendency in a similar situation.

13. Tip stall on a swept wing is very serious. If the outboard section of a swept wing stalls, the lift loss is behind the wing aerodynamic center. The inboard portion of the wing ahead of the aerodynamic center maintains its lift and produces a strong pitch-up moment, tending to throw the aircraft deeper into the stall. Combined with the effect of tip stall on the pitching moment produced by the tail, this effect is very dangerous and must be avoided by options such as wing twist.
14. A swept wing produces a negative rolling moment because of a difference in velocity components normal to the leading edge between the left and right wing sections [14]. The rolling moment due to aft sweep is proportional to the sine of twice the leading edge sweep angle:

$$C_{l_\beta} \propto \sin(2\Lambda_{LE}) \tag{5.30}$$

This makes the dihedral effect (C_{l_β}) more negative and it means that a swept wing has an inherent dihedral effect. Hence, a swept wing may not need a dihedral or anhedral to satisfy lateral-directional stability requirements. Thus, the sweep angle tends to reinforce the dihedral effect. It is interesting to note that making the dihedral effect (C_{l_β}) more negative will make an aircraft more spirally stable. At the same time, the dutch-roll damping ratio tends to decrease. This presents a design conflict [14] which must be resolved through some compromise.

15. In supersonic flight, the sweep angle tends to reduce the shock wave drag. The drag generated by the oblique shock wave is referred to as the wave drag, which is inherently related to the loss of total pressure and increase of entropy across the oblique shock waves created by the wing. For this purpose, the sweep angle must be greater (see Figure 5.41) than the Mach angle, μ [6]:

$$\mu = \sin^{-1}\left(\frac{1}{M}\right) \tag{5.31}$$

$$\Lambda = 1.2 \cdot (90 - \mu) \tag{5.32}$$

where M is the aircraft's cruising Mach number. A 20% higher sweep angle will guarantee the low-wave drag at supersonic speeds.

16. A wing with high wing loading (W/S) and a high quarter-chord sweep ($\Lambda_{c/4}$) exhibits a good ride in turbulence.

At hypersonic speed (e.g., the Space Shuttle), if the oblique shock wave is very close to the wing leading edge due to a low sweep angle, it generates very high temperature due to aerodynamic heating (about 3000 °F) such that the wing's leading edge surface may be melted. Thus, the sweep angle must be such that the wing leading edge surface survives very high temperature. This ensures that the wing is located inside the Mach cone.

17. With the application of the sweep angle, the wing effective span (b_{eff}) will be shorter than the original theoretical span. This results in a lower wing mass moment of inertia about the x-axis, which increases the lateral controllability of the aircraft. Hence, a higher sweep angle allows for better maneuverability.

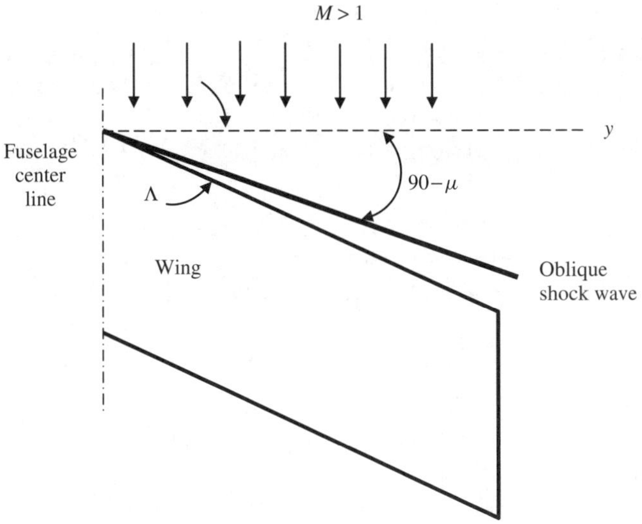

Figure 5.41 The sweep angle and Mach angle in supersonic flight

Sweep angle selection guideline: As noted, the sweep angle has several advantages and disadvantages that can only be balanced via compromise. The following guidelines help the reader to select the initial value and update the value throughout the design iterative process.

1. **Low subsonic aircraft**. If the aircraft maximum speed is less than Mach 0.3 (the borderline to include the compressibility effect), no sweep angle is recommended for the wing, since its disadvantages will negate all the improvement produced. For instance, by using 5 deg of sweep angle, you may have reduced the aircraft drag by say 2% but you will have increased the cost by say 15% as well as adding complexity to the wing manufacture. Thus a straight wing is recommended.
2. **High subsonic and supersonic aircraft**. The initial value can be determined through Equation (5.32) as a function of aircraft cruising speed. However, the final value will be finalized after a series of calculations and analysis on aerodynamics, performance, stability, control, structure, as well as cost and manufacturability. Remember, if the wing is tapered, it must have a sweep angle anyway.

Table 5.10 shows sweep angles of several aircraft along with their maximum speeds. As noted, as the maximum speed is increased, so is the sweep angle.

The following practical comments (including a few drawbacks) will help the designer to make the right decision on the wing sweep angle:

1. **Variable sweep**. If the aircraft needs to have different sweep angles at various flight conditions, an ideal option is to select a variable-sweep wing. This is an ideal objective from a few design aspects, however, it generates design problems. The example is a

Table 5.10 Sweep angles for several low- and high-speed aircraft

No.	Aircraft	Type	First flight	Max speed (Mach, knot)	Λ_{LE} (deg)
1	Cessna 172	Single-piston engine GA	1955	121 knot	0
2	Tucano	Turboprop trainer	1983	247 knot	4
3	AIRTECH	Turboprop transport	1981	228 knot	3° 51′ 36″
4	ATR 42	Turboprop transport	1984	265 knot	3° 6′
5	Jetstream 31	Turboprop business	1967	Mach 0.4	5° 34′
6	Beech Starship	Turboprop business	1991	Mach 0.78	20
7	DC-9 series 10	Jet passenger	1965	Mach 0.84	24
8	Falcon 900B	Business jet	1986	Mach 0.87	24° 30′
9	Gulfstream V	Business jet	1996	Mach 0.9	27
10	Boeing 777	Jet transport	1994	Mach 0.87	31.6
11	B-2A Spirit	Strategic bomber	1989	Mach 0.95	33
12	MD-11	Jet transport	2001	Mach 0.945	35
13	Boeing 747	Jet transport	1969	Mach 0.92	37° 30′
14	Airbus 340	Jet transport	1991	Mach 0.9	30
15	F-16	Fighter	1974	>Mach 2	40
16	F/A-18	Fighter	1992	>Mach 1.8	28
17	Mig-31	Fighter	1991	Mach 2.83	40
18	Su-34	Fighter	1996	Mach 2.35	42
19	Eurofighter Typhoon	Fighter	1986	Mach 2	53
20	Mirage 2000	Fighter	1975	Mach 2.2	58
21	Concorde	Supersonic jet transport	1969	Mach 2.2	75 inboard 32 outboard
22	Space Shuttle	Spacecraft (flies in air during return mission)	1981	Mach 21	81 inboard 44 outboard

fighter aircraft that spends the vast majority of its flight time at subsonic speeds, using its supersonic capability for short "supersonic dashes," depending on its mission.

A variable-sweep wing is a wing that may be swept back and then returned to its original position during flight. It allows the wing's geometry to be modified in flight. Typically, a swept wing is more suitable for high speeds (e.g., cruise), while an unswept wing is more suitable for lower speeds (e.g., take-off and landing), allowing the aircraft to carry more fuel and payload, as well as improving field performance. A variable-sweep wing allows a pilot to select the exact wing configuration for the intended speed. The variable-sweep wing is most useful for those aircraft that are expected to function at both low and high speed, thus it has been used primarily in fighters.

A number of successful designs (with variable sweep) – such as the Bell X-5, Grumman F-14 Tomcat (Figure 5.44(a)), General Dynamics F-111, Rockwell supersonic Bomber B-1B, Mikoyan Mig-23, Panavia Tornado (Figure 6.18), and Sukhoi Su-27

(Figure 6.7) – were introduced from the 1940s through the 1970s. However, the recent advances in flight control technology and structural materials have allowed designers to closely tailor the aerodynamics and structure of aircraft, removing the need for variable geometry to achieve the required performance. Aerodynamically, the exact sweep angle will generate the lowest possible drag while producing the highest possible lift and control. The drawback is the loose structural integrity as well as the sweep angle control mechanism problems (manual or automatic). The last variable-sweep wing military aircraft to date was the Soviet Tu-160 "Blackjack," which first flew in 1980.

2. **Wing/fuselage interference**. It is at the wing root that the straight fuselage sides more seriously degrade the sweep effect by interfering with the curved flow of Figure 3.36. Wing airfoils are often modified near the root to change the basic pressure distribution to compensate for the distortion to the swept wing flow. Since the fuselage effect is to increase the effective airfoil camber, the modification is to reduce the root airfoil camber and in some cases to use negative camber. The influence of the fuselage then changes the altered root airfoil pressure back to the desired positive-camber pressure distribution existing farther out along the wing span [13]. This same swept wing root compensation can be achieved by adjusting the fuselage shape to match the natural swept wing streamlines. This imposes serious manufacturing burdens and passenger cabin arrangement problems. Thus the airfoil approach is preferred for transport aircraft. Instead, the employment of large fillet or even fuselage shape variation is appropriate for fighter aircraft.
3. **Non-constant sweep**. In some cases, one sweep angle cannot satisfy all design requirements. For instance, a very high sweep angle wing satisfies high-speed cruise requirements, however, at low subsonic speed, the aircraft is not satisfactorily controllable or laterally stable. One solution is to divide the wing sections into inboard plane and outboard plane, each having different sweep angles (see Figure 5.42). The supersonic transport aircraft Concorde and Space Shuttle have such a feature.
4. **Control surfaces**. The sweep angle will influence the performance of a HLD (such as a flap) as well as control surfaces (such as ailerons). In practice, since both a HLD and control surfaces need to have sweep angles (with slightly different values), their lifting forces will be spoiled. Consequently, the HLD's contribution to generate lift at low speed will be reduced. With the same logic, it can be shown that the aileron will also produce less lateral control. To compensate for these shortcomings, both control surfaces and HLD must have slightly larger areas.

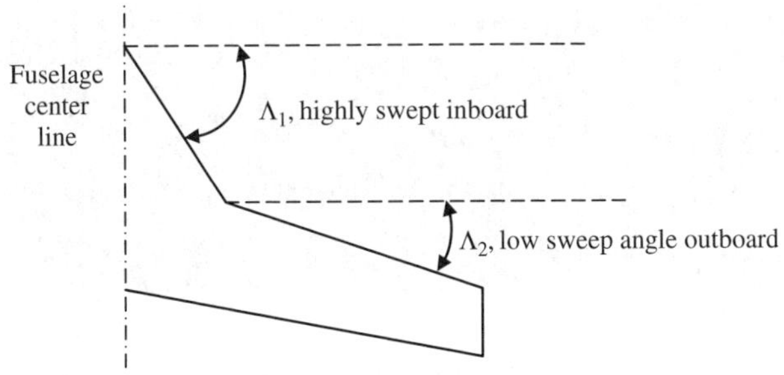

Figure 5.42 Top view of a wing with two sweep angles

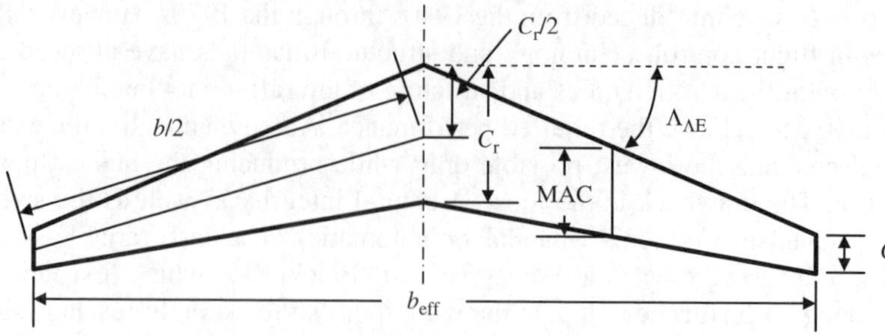

Figure 5.43 Effective wing span in a swept wing

5. **Spar**. When the wing has a sweep angle, the wing spar can no longer be in one piece, since two wing sections (left and right) have opposite sweep angles. This is assumed to be a disadvantage of the sweep angle, since the wing's structural integrity will be negatively influenced. This adds to the complexity of the wing manufacture as well.
6. **Effective span (b_{eff}) and effective aspect ratio (AR_{eff})**. With the presence of a sweep angle, the wing span (b) will have slightly different meaning, so the new parameter of effective span (b_{eff}) is introduced. When the 50% chord line sweep angle is not zero, the wing span will be greater than the wing effective span. The wing span in a straight wing is basically defined as the distance between two wing tips parallel to the aircraft lateral axis (y-axis). However, in a swept wing, the wing span is defined as twice the distance between one wing tip and the fuselage center line parallel to the 50% sweep chord line. Thus, the effective wing span in a swept wing is defined as the distance between wing tips parallel to the aircraft lateral axis (y-axis). Figure 5.43 depicts the difference between span and effective span. This indicates that the wing sweep angle alters the wing span to an effective span which is smaller:

$$AR_{eff} = \frac{{b_{eff}}^2}{S} \tag{5.33}$$

The technique to determine the effective span is based on the laws of a triangle. The application of the technique is illustrated in Examples 5.3 and 5.4. Figure 5.44 illustrates the sweep angles of fighter aircraft Grumman F-14D, GA aircraft Pilatus PC-21, and transport aircraft Fokker 70.

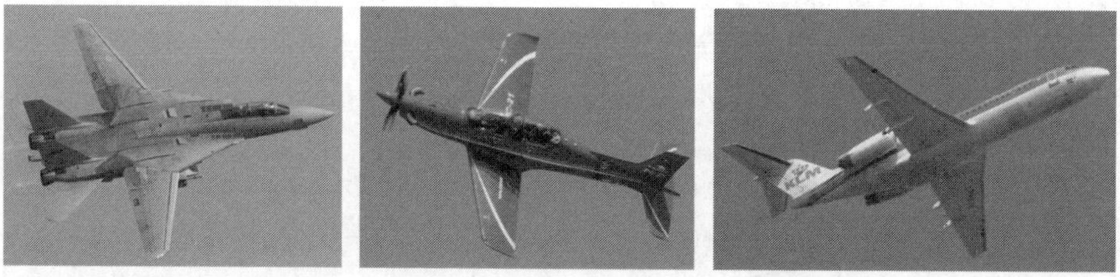

(a) (b) (c)

Figure 5.44 Sweep angles for three aircraft: (a) Grumman F-14D; (b) Pilatus PC-21; (c) Fokker 70. (Reproduced from permission of (a, b) Antony Osborne, (c) Anne Deus.)

Example 5.3

An aircraft has a wing area of $S = 20\ \text{m}^2$, aspect ratio $\text{AR} = 8$, and taper ratio $\lambda = 0.6$. It is required that the 50% chord line sweep angle be zero. Determine the tip chord, root chord, MAC, and span, as well as the leading edge sweep, trailing edge sweep, and quarter chord sweep angles.

Solution:
To determine the unknown variables, we first employ the following equations:

$$\text{AR} = \frac{b^2}{S} \Rightarrow b = \sqrt{S \cdot \text{AR}} = \sqrt{20 \cdot 8} \Rightarrow b = 12.65\ \text{m} \tag{5.19}$$

$$\text{AR} = \frac{b}{\overline{C}} \Rightarrow \overline{C} = \frac{b}{\text{AR}} = \frac{12.65}{8} \Rightarrow \overline{C} = 1.58\ \text{m} \tag{5.17}$$

$$\overline{C} = \frac{2}{3} C_{\text{r}} \left(\frac{1 + \lambda + \lambda^2}{1 + \lambda} \right) \Rightarrow 1.58 = \frac{2}{3} C_{\text{r}} \left(\frac{1 + 0.6 + 0.6^2}{1 + 0.6} \right) \Rightarrow C_{\text{r}} = 1.936\ \text{m} \tag{5.26}$$

$$\lambda = \frac{C_{\text{t}}}{C_{\text{r}}} \Rightarrow 0.6 = \frac{C_{\text{t}}}{1.935} \Rightarrow C_{\text{t}} = 1.161\ \text{m} \tag{5.24}$$

Since the 50% chord line sweep angle is zero ($\Lambda_{C/2} = 0$), the leading edge, trailing edge, and quarter chord sweep angles are determined using the triangle law in triangle ABC (see Figure 5.45) as follows:

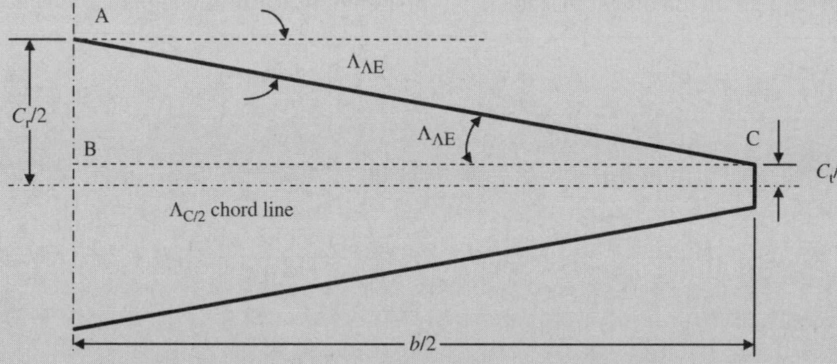

Figure 5.45 The wing of Example 5.3 (λ and angles are exaggerated)

$$\tan\left(\Lambda_{\text{LE}}\right) = \frac{\text{AB}}{\text{BC}} \Rightarrow \Lambda_{\text{LE}} = \tan^{-1}\left(\frac{\frac{C_{\text{r}}}{2} - \frac{C_{\text{t}}}{2}}{b/2} \right) = \tan^{-1}\left(\frac{\frac{1.936 - 1.161}{2}}{12.65/2} \right)$$

$$\Rightarrow \Lambda_{\text{LE}} = 3.5\ \text{deg (sweep back)}$$

The wing is straight, thus the trailing edge sweep angle would be:

$$\Lambda_{TE} = -3.5 \text{ deg (sweep forward)}$$

The quarter chord sweep angle is determined using the tangent law in a similar triangle as follows:

$$\Lambda_{C/4} = \tan^{-1}\left(\frac{\frac{C_r - C_t}{4}}{b/2}\right) = \tan^{-1}\left(\frac{\frac{1.936 - 1.161}{4}}{12.65/2}\right)$$

$$\Rightarrow \Lambda_{C/4} = 1.753 \text{ deg (sweep back)}$$

It is interesting to note that, although the wing is straight ($\Lambda_{C/2} = 0$), the leading edge, trailing edge, and quarter chord line all are swept.

Example 5.4

An aircraft has a wing area of $S = 20 \text{ m}^2$, aspect ratio AR = 8, and taper ratio $\lambda = 0.6$. It is required that the 50% chord line sweep angle be 30 deg. Determine the tip chord, root chord, MAC, span, and effective span, as well as the leading edge sweep, trailing edge sweep, and quarter chord sweep angles.

Solution:

To determine the unknown variables, we first employ the following equations:

$$\text{AR} = \frac{b^2}{S} \Rightarrow b = \sqrt{S \cdot \text{AR}} = \sqrt{20 \cdot 8} \Rightarrow b = 12.65 \text{ m} \tag{5.19}$$

$$\text{AR} = \frac{b}{\overline{C}} \Rightarrow \overline{C} = \frac{b}{\text{AR}} = \frac{12.65}{8} \Rightarrow \overline{C} = 1.58 \text{ m} \tag{5.17}$$

$$\overline{C} = \frac{2}{3}C_r\left(\frac{1+\lambda+\lambda^2}{1+\lambda}\right) \Rightarrow 1.58 = \frac{2}{3}C_r\left(\frac{1+0.6+0.6^2}{1+0.6}\right) \Rightarrow C_r = 1.936 \text{ m} \tag{5.26}$$

$$\lambda = \frac{C_t}{C_r} \Rightarrow 0.6 = \frac{C_t}{1.935} \Rightarrow C_t = 1.161 \text{ m} \tag{5.24}$$

Since the 50% chord line sweep angle is 30 deg ($\Lambda_{C/2} = 30$ deg), the leading edge, trailing edge, and quarter chord sweep angles are determined using the triangle law (see Figure 5.46). But we first need to calculate a few parameters.

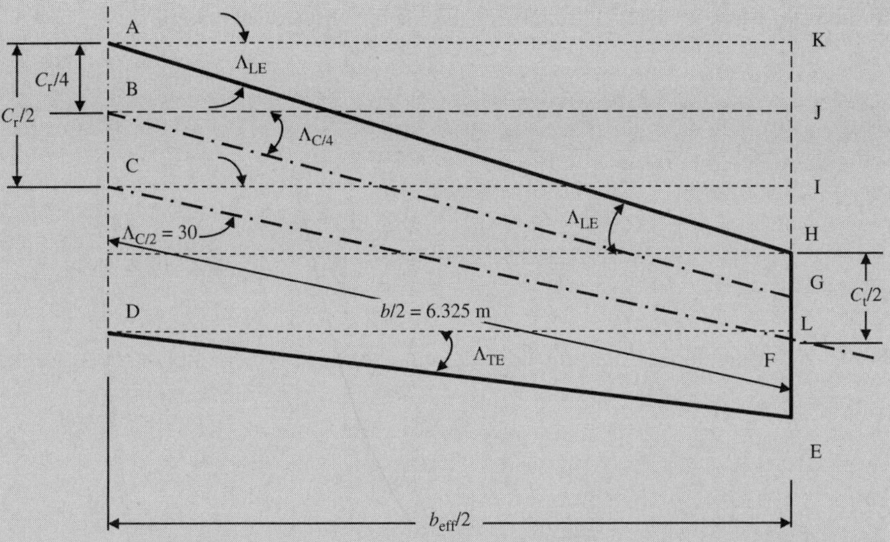

Figure 5.46 The top view of the right wing of Example 5.4

In the right triangle CIF that includes the 50% chord sweep angle ($\Lambda_{C/2}$), we can write:

$$\sin\left(\Lambda_{C/2}\right) = \frac{\text{FI}}{b/2} \Rightarrow \text{FI} = \frac{12.65}{2}\sin(30) = 3.1625 \text{ m}$$

$$(\text{CI})^2 + (\text{FI})^2 = (\text{CF})^2 \Rightarrow \text{CI} = \sqrt{(\text{CF})^2 - (\text{FI})^2} \Rightarrow \frac{b_{\text{eff}}}{2}$$

$$= \sqrt{\left(\frac{12.65}{2}\right)^2 - 3.1625^2} \Rightarrow b_{\text{eff}} = 10.955 \text{ m}$$

Hence, the effective span is less than the regular span. Consequently, the effective aspect ratio is reduced to:

$$\text{AR}_{\text{eff}} = \frac{{b_{\text{eff}}}^2}{S} = \frac{10.955^2}{20} \Rightarrow \text{AR}_{\text{eff}} = 6 \tag{5.33}$$

It is noted that the AR has been reduced from 8 to 6. The length of IH is:

$$\text{IH} = \text{FI} - \frac{C_t}{2} = 3.1625 - \frac{1.161}{2} = 2.582\,\text{m}$$

In the right triangle AKH that includes the leading edge sweep angle (Λ_{LE}), we have:

$$\tan\left(\Lambda_{LE}\right) = \frac{KH}{AK} = \frac{KI + IH}{\frac{b_{eff}}{2}} = \frac{\frac{C_r}{2} + 2.582}{\frac{10.955}{2}}$$

$$= \frac{\frac{1.936}{2} + 2.582}{\frac{10.955}{2}} = 0.648 \Rightarrow \Lambda_{LE} = 33 \text{ deg (aft sweep)}$$

In the right triangle GJB that includes the quarter chord sweep angle ($\Lambda_{C/4}$), we have:

$$\tan\left(\Lambda_{C/4}\right) = \frac{GJ}{BJ} = \frac{GH + JH}{\frac{b_{eff}}{2}} = \frac{\frac{C_t}{4} + KH - KJ}{\frac{b_{eff}}{2}} = \frac{\frac{C_t}{4} + (KI + IH) - KJ}{\frac{b_{eff}}{2}}$$

$$= \frac{\frac{C_t}{4} + \left(\frac{C_r}{2} + 2.582\right) - \frac{C_r}{4}}{\frac{b_{eff}}{2}} = \frac{\frac{1.161}{4} + \left(\frac{1.936}{2} + 2.582\right) - \frac{1.936}{4}}{\frac{10.955}{2}}$$

$$= 0.613 \Rightarrow \Lambda_{C/4} = 31.5 \text{ deg (aft sweep)}$$

This reveals that both the leading edge sweep and the quarter chord sweep angles are greater than the 50% chord line sweep angle.

Finally, in the right triangle DLE that includes the trailing edge sweep angle (Λ_{TE}), we have:

$$\tan\left(\Lambda_{TE}\right) = \frac{EL}{LD} = \frac{EK - KL}{\frac{b_{eff}}{2}} = \frac{EK - C_r}{\frac{b_{eff}}{2}} = \frac{\frac{C_t}{2} + KH - C_r}{\frac{b_{eff}}{2}}$$

$$= \frac{\frac{C_t}{2} + (KI + IH) - C_r}{\frac{b_{eff}}{2}} = \frac{\frac{C_t}{2} + \left(\frac{C_r}{2} + 2.582\right) - C_r}{\frac{b_{eff}}{2}}$$

$$= \frac{\frac{1.161}{2} + (2.582) - \frac{1.936}{2}}{\frac{10.955}{2}} = 0.401 \Rightarrow \Lambda_{TE} = 21.85 \text{ deg (aft sweep)}$$

The trailing edge sweep angle is considerably less than the 50% chord line sweep angle.

5.10 Twist Angle

If the wing tip is at a lower incidence than the wing root, the wing is said to have negative twist or simply twist (α_t) or washout. In contrast, if the wing tip is at a higher incidence than the wing root, the wing is said to have positive twist or washin. The twist is usually negative, which means the wing tip angle of attack is lower than the root angle of attack, as sketched in Figure 5.47(a). This indicates that the wing angle of attack is reduced along the span. The wings on a number of modern aircraft have different airfoil sections along the span, with different values of zero lift angle of attack; this is called the aerodynamic twist. The wing tip airfoil section is often thinner than the root airfoil section, as sketched in Figure 5.47(b). Sometimes, the tip and root airfoil sections have the same thickness-to-chord ratio, but the root airfoil section has a higher zero-lift angle of attack (i.e., more negative) than the tip airfoil section.

When the tip incidence and root incidence are not the same, the twist is referred to as *geometric twist*. However, if the tip airfoil section and root airfoil section are not the same, the twist is referred to as *aerodynamic twist*. Both types of twist have advantages and disadvantages, by which the designer must establish a selection that satisfies the design requirements. The application of twist is a selection at decision making, but the amount of twist is determined via calculations. In this section, both items will be discussed.

In practice, the application of aerodynamic twist is more convenient than geometric twist. The reason is that in aerodynamic twist, one part of the wing has different ribs from another part, while all parts of the wing have the same incidence. The difficulty

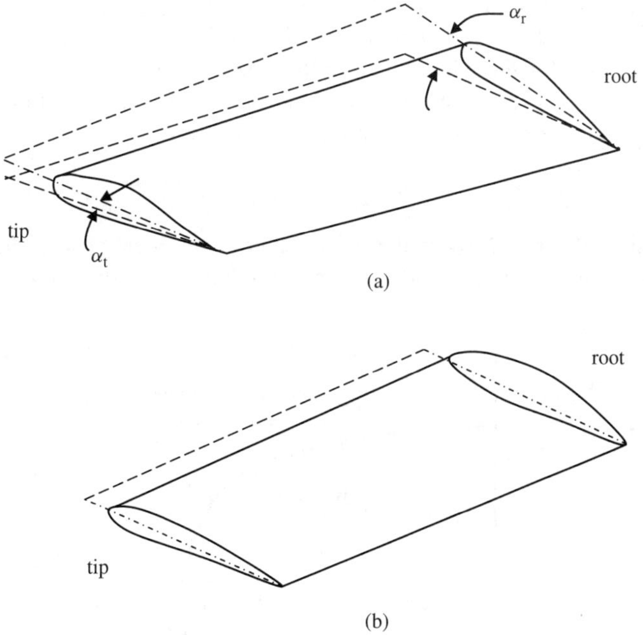

Figure 5.47 Wing twist: (a) Geometric twist; (b) Aerodynamic twist

in the application of geometric twist arises from a manufacturing point of view. Every portion of the wing has a unique incidence, since the angle of attack must be decreased (usually linearly) from the wing setting angle i_w (at the root) to a new value at the tip. This technique is applied by twisting the main wing spar, through which the wing (rib) twist is automatically applied. The alternative solution is to divide each section of the wing (left and right) into two portions, an inboard portion and an outboard portion. Then, the inboard portion has an incidence equal to the wing setting angle, while the outboard portion has a value such that the twist is produced. If the situation allows, both geometric and aerodynamic twist may be employed.

There are two major goals for employing the twist in a wing design process:

1. Avoiding tip stall before root stall.
2. Modification of the lift distribution to an elliptical one.

In addition to the two above-mentioned desired goals, there is another one unwanted output in twist:

3. Reduction in lift.

When the wing root enters the stall before the wing tip, the pilot is able to utilize the aileron to control the aircraft, since the fair low at the outboard section has not yet been stalled. This provision improves the safety of the aircraft on the advent of wing stall. The significance of the elliptical lift distribution has been described in Section 5.7. The major drawback in twist is the loss of lift, since the twist is usually negative. As the angle of attack of a wing section is decreased, the lift coefficient will be decreased too. The criterion and the limit for the wing twist are that the twist angle must not be so high that it results in a negative lift in the outer wing portions. Since any section has a zero-lift angle of attack (α_o), the criterion is formulated as follows:

$$|\alpha_t| + i_w \geqslant |\alpha_o| \tag{5.34}$$

When a portion of the outboard of the wing generates a negative lift, the overall lift is decreased. This is not desirable and must be avoided in the twist angle determination

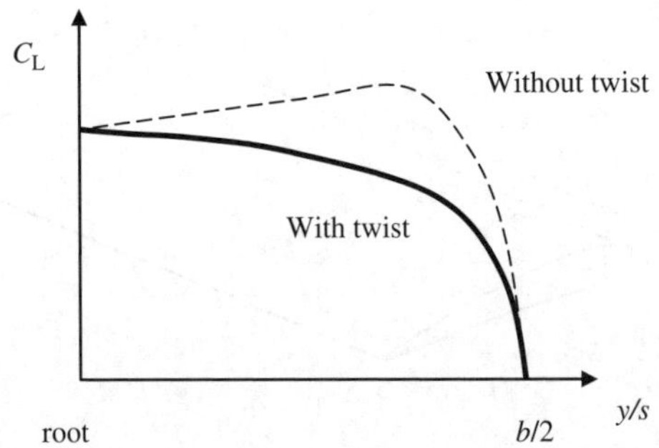

Figure 5.48 The typical effect of a (negative) twist angle on the lift distribution

process. A typical value for the geometric twist is between −1 and −4 deg (i.e., negative twist). The exact value of the twist angle must be determined such that the tip stalls after the root, as well as the lift distribution being elliptic. Figure 5.48 illustrates the typical effect of a (negative) twist angle on the lift distribution. Table 5.11 shows twist angles for several aircraft. As noted, several aircraft such as the Cessna 208, Beech 1900D, Beechjet 400A, AVRO RJ100, and Lockheed C-130 Hercules (Figure 5.4) have both geometric and aerodynamic twists.

Table 5.11 Twist angles for several aircraft
(a) Geometric twist [5, 15]

No.	Aircraft	MTOW (lb)	Wing incidence at root (i_w) (deg)	Wing angle at tip (deg)	Twist (deg)
1	Fokker 50	20 800	+3.5	+1.5	-2
2	Cessna 310	4 600	+2.5	-0.5	-3
3	Cessna Citation I	11 850	+2.5	-0.5	-3
4	Beech King Air	11 800	+4.8	0	-4.8
5	Beech T-1A JayHawk	16 100	+3	-3.3	-6.3
6	Beech T-34C	4 300	+4	+1	-3
7	Cessna StationAir 6	3 600	+1.5	-1.5	-3
8	Gulfstream IV	73 000	+3.5	-2	-5.5
9	Northrop-Grumman E-2C Hawkeye	55 000	+4	+1	-3
10	Piper Cheyenne	11 200	+1.5	-1	-2.5
11	Beech SuperKing	12 500	+3° 48′	-1° 7′	4.55′
12	Beech starship	14 900	+3	-5	-3.5
13	Cessna 208	8 000	+2° 37′	-3° 6′	-5° 31′
14	Beech 1900D	16 950	+3° 29′	-1° 4′	-4° 25′
15	Beech jet 400A	16 100	+3	-3° 30′	-6° 30′
16	AVRO RJ100	101 500	+3° 6′	0	-3° 6′
17	Lockheed C-130 Hercules	155 000	+3	0	-3
18	Pilatus PC-9	4 960	+1	-1	-2
19	Piper PA-28-161 Warrior	2 440	+2	-1	-3

(b) Aerodynamic twist [5]

No.	Aircraft	MTOW (lb)	Root airfoil section	Tip airfoil section	$\Delta t/C$ (%)
1	Cessna 208	8 000	NACA 23017.424	NACA 23012	5
2	Beech 1900D	16 950	NACA 23018	NACA 23012	6
3	Beechjet 400A	16 100	$t/C = 13.2\%$	$t/C = 11.3\%$	1.9
4	AVRO RJ100	101 500	$t/C = 15.3\%$	$t/C = 12.2\%$	3.1
5	Lockheed C-130 Hercules	155 000	NACA 64A318	NACA 64A412	6
6	Gulfstream IV-SP	74 600	$t/C = 10\%$	$t/C = 8.6\%$	1.4
7	Boeing 767	412 000	$t/C = 15.1\%$	$t/C = 10.3\%$	4.8
8	Harrier II	31 000	$t/C = 11.5\%$	$t/C = 7.5\%$	4
9	BAE Sea Harrier	26 200	$t/C = 10\%$	$t/C = 5\%$	5
10	Kawasaki T-4	12 544	$t/C = 10.3\%$	$t/C = 7.3\%$	3
11	F/A-18 Hornet	52000	NACA 65A-005	NACA 65A-003	2

5.11 Dihedral Angle

When you look at the front view of an aircraft, the angle between the chord line plane of a wing with the *xy* plane is referred to as the wing dihedral (Γ). The chord line plane of the wing is an imaginary plane that is generated by connecting all chord lines across the span. If the wing tip is higher than the *xy* plane, the angle is called positive dihedral or simply dihedral, but when the wing tip is lower than the *xy* plane, the angle is called negative dihedral or anhedral (see Figure 5.49). For the purpose of aircraft symmetry, both right and left sections of a wing must have the same dihedral angle. There are several advantages and disadvantages for the dihedral angle. In this section, these characteristics are introduced, followed by the design recommendations to determine the dihedral angle.

The primary reason for applying a wing dihedral is to improve the lateral stability of the aircraft. The lateral stability is mainly the tendency of an aircraft to return to its original trim level-wing flight condition if disturbed by a gust and rolls around the x-axis. In some references, it is called *dihedral stability*, since a wing dihedral angle provides

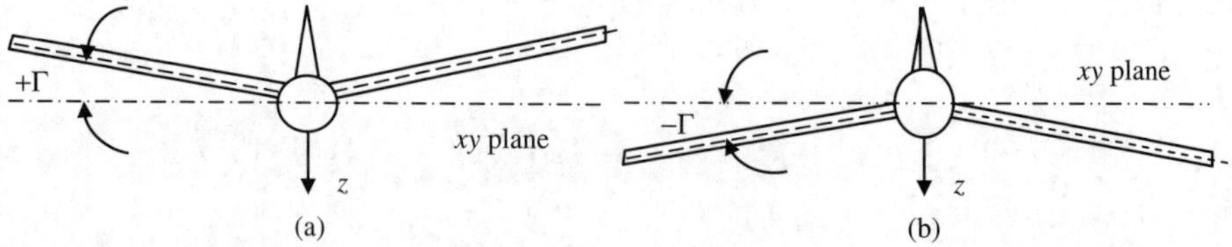

Figure 5.49 (a) Dihedral and (b) anhedral (aircraft front view)

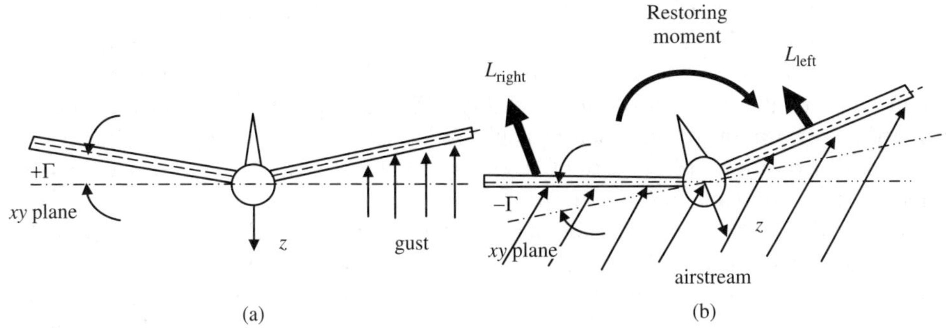

Figure 5.50 The effect of dihedral angle on a disturbance in roll (aircraft front view): (a) before gust; (b) after gust

the necessary restoring rolling moment. The lateral static stability is primarily represented by a stability derivative called the aircraft dihedral effect ($C_{l_\beta} = \frac{dC_l}{d\beta}$) that is the change in aircraft rolling moment coefficient due to a change in aircraft sideslip angle (β).

Observe a level-wing aircraft that has experienced a disturbance (see Figure 5.50) which has produced an undesired rolling moment (e.g., a gust under one side of the wing). When the aircraft rolls, one side of the wing (say the left) goes up, while the other side (say the right) goes down. This is called a positive roll. The right wing section that has dropped has temporarily lost a small percentage of its lift. Consequently, the aircraft will accelerate and slip down toward the right wing, which produces a sideslip angle (β). This is equivalent to a wing approaching from the right of the aircraft; the sideslip angle is positive. In response, a laterally statically stable aircraft must produce a negative rolling moment to return to the original wing-level situation. This is technically translated into a negative dihedral effect ($C_{l_\beta} < 0$). The role of the wing dihedral angle is to induce a positive increase in angle of attack ($\Delta\alpha$). This function of the wing dihedral angle is carried out by producing a normal velocity ($V_n = V\Gamma$):

$$\Delta\alpha \approx \frac{V\Gamma}{U} \approx \frac{U\beta\Gamma}{U} \approx \beta\Gamma \tag{5.35}$$

where U is the airspeed component along the x-axis and V is the airspeed component along the y-axis. It is this increment in the angle of attack that produces a corresponding increment in the lift. This in turn results in a negative rolling moment contribution. It is interesting that the left wing section experiences exactly the opposite effect, which also results in a negative rolling moment. Therefore, the rolling moment due to sideslip from a geometric wing dihedral is proportional to the dihedral angle. Basically, a positive wing geometric dihedral causes the rolling moment due to the sideslip derivative C_{l_β} to be negative. Aircraft must have a certain minimum amount of negative rolling moment due to sideslip; the dihedral effect. This is needed to prevent excessive spiral instability. Too much dihedral effect tends to lower the dutch-roll damping. A more negative C_{l_β} means more spiral stability, but at the same time less dutch-roll stability.

The anhedral has exactly the opposite function. In other words, the anhedral is laterally destabilizing. The reason for using an anhedral in some configurations is to balance between the roles of wing parameters (such as sweep angle and wing vertical position) in lateral stability. The reason is that a more laterally stable aircraft means a less rolling controllable aircraft. In the wing design, one must be careful to determine the wing parameters such that they satisfy both stability and controllability requirements. Since the primary reason for the wing dihedral angle is lateral stability, wing sweep angles and wing vertical position are driven not only by lateral stability, but also by performance requirements and operational requirements.

For instance, a cargo aircraft usually has a high wing to satisfy the loading and unloading operational requirements. The high-wing contribution to the lateral stability is highly positive, which means the aircraft is laterally more stable than necessary. In order to make the aircraft less laterally stable, one of the designer's options is to add an anhedral to the wing. This decision does not alter the operational characteristics of the aircraft, but improves the rolling controllability of the aircraft. In general, high-wing aircraft have an inherent dihedral effect while low-wing aircraft tend to be deficient in their inherent dihedral effect C_{l_β}. For this reason, low-wing aircraft tend to have considerably greater dihedral angle than high-wing aircraft. In contrast, swept wing aircraft tend to have too much dihedral effect C_{l_β} due to the sweep angle. This can be offset in high-wing aircraft by giving the wing a negative dihedral (i.e., anhedral). The balance between lateral stability and roll control is a major criterion for the determination of dihedral angle.

Another effect of the wing dihedral is to alter the ground and water clearance, since aircraft wings, nacelles, and propellers must have a minimum amount of ground and water clearance. It is clear that a dihedral would increase ground and water clearance, while an anhedral would decrease ground and water clearance. In aircraft with a high aspect ratio and highly elastic wings (such as the record-breaking Voyager), the elastic deformation of the wing in flight generates extra dihedral angle. This must be considered in the wing design of such aircraft.

When the dihedral angle is applied on a wing, the wing effective planform area (S_{eff}) is reduced. This in turn will reduce the lift generated by the wing without dihedral, which is undesirable. If you need to apply the dihedral angle to a wing, consider the lowest value for the dihedral to minimize the lift reduction. The effective wing planform area as a function of dihedral angle is determined as follows:

$$S_{\text{eff}} = S_{\text{ref}} \cos(\Gamma) \tag{5.36}$$

Table 5.12 illustrates dihedral (and anhedral) angles for several aircraft along with their wing vertical position. As noted, the typical dihedral angle is a value between −15 and +10 deg. Figure 5.51 illustrates two aircraft with different dihedral angles. Table 5.13 shows typical values of dihedral angle for swept or unswept wings of various wing vertical positions. This table is a recommended reference for the starting point. You can select an initial value for the dihedral angle from this table. However, the exact value of the dihedral angle is determined during the stability and control analysis of whole aircraft. When other aircraft components (e.g., fuselage, tail) are designed, evaluate the lateral stability of the whole aircraft.

The suggested value for the aircraft dihedral effect (C_{l_β}) to have an acceptable lateral controllability and lateral stability is a value between −0.1 and +0.4 1/rad. Then you can adjust the dihedral angle to satisfy all the design requirements. If one dihedral angle for the whole wing does not satisfy all the design requirements, you may divide

Table 5.12 Dihedral (or anhedral) angles for several aircraft

No.	Aircraft	Type	Wing position	Dihedral (deg)
1	Pilatus PC-9	Turboprop trainer	Low wing	7 (outboard)
2	MD-11	Jet transport	Low wing	6
3	Cessna 750 Citation X	Business jet	Low wing	3
4	Kawasaki T-4	Jet trainer	High wing	−7
5	Boeing 767	Jet transport	Low wing	$4^\circ\ 15'$
6	Falcon 900 B	Business jet transport	Low wing	$0^\circ\ 30'$
7	C-130 Hercules	Turboprop cargo	High wing	$2^\circ\ 30'$
8	Antonov An-74	Jet STOL transport	Parasol wing	−10
9	Cessna 208	Piston engine GA	High wing	3
10	Boeing 747	Jet transport	Low wing	7
11	Airbus 310	Jet transport	Low wing	$11^\circ\ 8'$
12	F-16 Fighting Falcon	Fighter	Mid-wing	0
13	BAE Sea Harrier	V/STOL fighter	High wing	−12
14	MD/BAe Harrier II	V/STOL close support	High wing	−14.6
15	F-15J Eagle	Fighter	High wing	−2.4
16	Fairchild SA227	Turboprop commuter	Low wing	4.7
17	Fokker 50	Turboprop transport	High wing	3.5
18	AVRO RJ	Jet transport	High wing	−3
19	MIG-29	Fighter	Mid-wing	−2

(a)

(b)

Figure 5.51 Two aircraft with different dihedral angles: (a) Airbus A330, dihedral; (b) British Aerospace Sea Harrier, anhedral (Reproduced from permission of (a) A J Best, (b) Jenny Coffey.)

Table 5.13 Typical values of dihedral angle for various wing configurations

No.	Wing	Low wing	Mid-wing	High wing	Parasol wing
1	Unswept	5 to 10	3–6	−4 to −10	−5 to −12
2	Low-subsonic swept	2 to 5	−3 to +3	−3 to −6	−4 to −8
3	High-subsonic swept	3 to 8	−4 to +2	−5 to −10	−6 to −12
4	Supersonic swept	0 to −3	1 to −4	0 to −5	NA
5	Hypersonic swept	1 to 0	0 to −1	−1 to −2	NA

the wing into inboard and outboard sections, each with a different dihedral angle. For instance, you may apply a dihedral angle to the outboard plane, in order to keep the wing level in the inboard plane.

5.12 High-Lift Device

5.12.1 The Functions of a High-Lift Device

One of the design goals in wing design is to maximize the capability of the wing in the generation of the lift. This design objective is technically shown as the maximum lift coefficient ($C_{L_{max}}$). In a trimmed cruising flight, the lift is equal to the weight. When the aircraft generates its maximum lift coefficient, the airspeed is referred to as stall speed:

$$L = W \Rightarrow \frac{1}{2}\rho V_s^2 S C_{L_{max}} = mg \tag{5.37}$$

Two design objectives among the list of objectives are: (i) maximizing the payload weight and (ii) minimizing the stall speed (V_s). As Equation (5.36) indicates, increasing $C_{L_{max}}$ tends to increase the payload weight (W) and decrease the stall speed. The lower stall speed is desirable since a safe take-off and landing requires a lower stall speed. In contrast, the higher payload weight will increase the efficiency of the aircraft and reduce the cost of flight. A higher $C_{L_{max}}$ allows the aircraft to have a smaller wing area that results in a lighter wing. Hence, in a wing design, the designer must find a way to maximize the $C_{L_{max}}$. In order to increase the lift coefficient, the only in-flight method is to temporarily vary (increase) the wing camber. This will happen only when the HLD is deflected downward. In the 1970s the maximum lift coefficient at take-off was 2.8, while the record currently belongs to the Airbus A-320 with a magnitude of 3.2.

The primary applications of HLDs are during take-off and landing operations. Since the airspeed is very low compared with the cruising speed, the wing must produce a bigger lift coefficient. The aircraft speed during take-off and landing is slightly greater than the stall speed. Airworthiness standards specify the relationship between take-off speed and landing speed with stall speed. As a general rule, we have:

$$V_{TO} = k \cdot V_s \tag{5.38}$$

where k is about 1.1 for fighter aircraft, and about 1.2 for jet transports and GA aircraft.

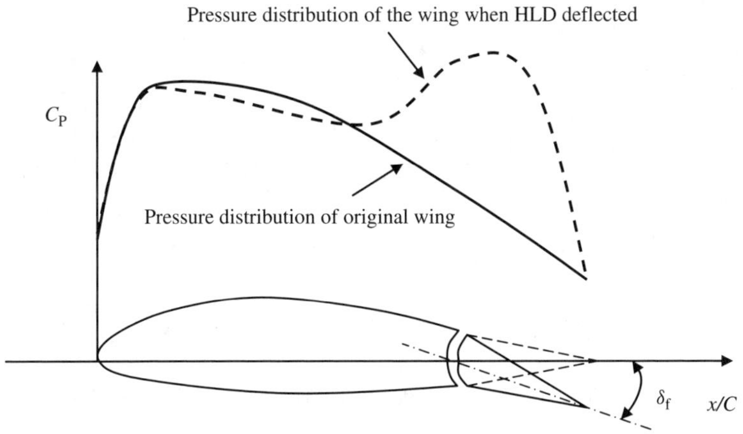

Figure 5.52 Example of pressure distribution with the application of a high-lift device

The application of the HLD tends to change the airfoil section's and wing's camber (in fact the camber will be positively increased). This in turn will change the pressure distribution along the wing chord as sketched in Figure 5.52. In this figure, C_P denotes the pressure coefficient.

In contrast, the leading edge high-lift device (LEHLD) tends to improve the boundary layer energy of the wing. Some type of HLD has been used on almost every aircraft designed since the early 1930s. HLDs are the means to obtain the sufficient increase in $C_{L_{max}}$.

At the airfoil level, a HLD deflection tends to cause the following six changes in the airfoil features:

1. Lift coefficient (C_l) is increased.
2. Maximum lift coefficient ($C_{l_{max}}$) is increased.
3. Zero-lift angle of attack (α_o) is changed.
4. Stall angle (α_s) is changed.
5. Pitching moment coefficient is changed.
6. Drag coefficient is increased.
7. Lift curve slope is increased.

These effects are illustrated in Figure 5.53. Along with three desirable advantages (first two items) to the application of HLDs, there are a few negative side-effects (the last five items) as well. A plain flap tends to decrease the stall angle, while a slotted flap and leading edge slat tend to increase the stall angle. In addition, among all types of flaps, the Fowler flap and leading edge slat tend to increase the lift curve slope (C_{L_α}). In contrast, the leading edge flap tends to increase (shift to the right) the zero-lift angle of attack (α_o).

A reduction in stall angle is undesirable, since the wing may stall at a lower angle of attack. During the take-off and landing operation, a high angle of attack is required to successfully take off and land. The high angle of attack will also tend to reduce the take-off run and landing run that are desirable in an airport with limited runway length.

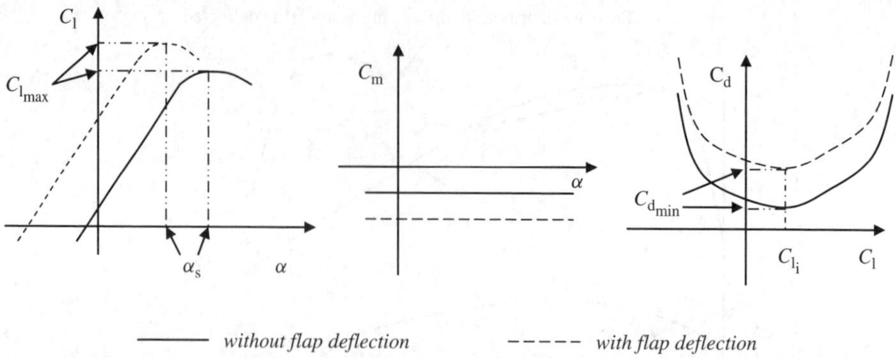

Figure 5.53 Typical effects of a high-lift device on wing airfoil section features

Table 5.14 Maximum lift coefficient for several aircraft

$C_{L\max}$	Cessna 172	Piper Cherokee	Short Skyvan 3	Gulfstream II	DC-9	Boeing 727	Airbus 300	Learjet 25
Take-off	1.5	1.3	2.07	1.4	1.9	2.35	2.7	1.37
Landing	2.1	1.74	2.71	1.8	2.4	2.75	3	1.37

An increase in pitching moment coefficient requires a higher horizontal tail area to balance the aircraft. An increase in drag coefficient decreases the acceleration during take-off and landing. Although the application of HLD generates three undesirable side-effects, the advantages outweigh the disadvantages.

If the natural value of $C_{L_{max}}$ for an aircraft is not high enough for safe take-off and landing, it can be increased temporarily by mechanical HLDs. Thus, employing the same airfoil section, one is able to increase $C_{L_{max}}$ temporarily as needed without actually pitching the aircraft. Two flight operations at which the $C_{L_{max}}$ needs to be increased are take-off and landing. Table 5.14 shows the maximum lift coefficient for several aircraft at take-off and landing configurations.

In a cruising flight, there is no need to utilize the maximum lift coefficient since the speed is high. These mechanical devices are referred to as high-lift devices. HLDs are parts of wings to increase the lift when deflected down. They are located at an inboard section of the wing and usually employed during take-off and landing.

5.12.2 High-Lift Device Classification

Two main groups of HLDs are:

1. leading edge high-lift device (LEHLD) and
2. trailing edge high-lift device (TEHLD or flap).

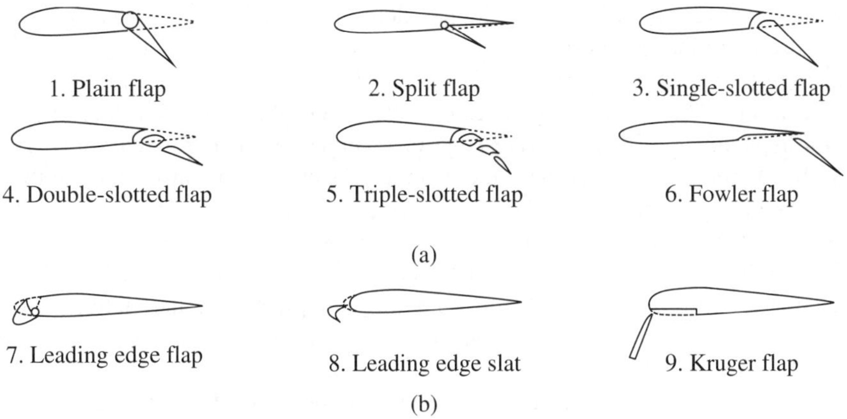

Figure 5.54 Various types of high-lift device: (a) Trailing edge high-lift device; (b) Leading edge high-lift device

There are many types of wing trailing edge flaps but the most common are split flap, plain flap, single-slotted flap, double-slotted flap, triple-slotted flap, and Fowler flap as illustrated in Figure 5.54(a). They are all deflected downward to increase the camber of the wing, so $C_{L_{max}}$ will be increased. The most common leading edge devices are leading edge flap, leading edge slat, and Kruger flap as shown in Figure 5.54(b).

A common problem with the application of HLDs is how to deal with the gap between the HLD and the main wing. This gap can be either sealed or left untouched. In both cases, there are undesirable side-effects. If the gap is left open, the airflow from the downside escapes to the upper surface which in turn degrades the pressure distribution. In contrast, if the gap is sealed by means such as a diaphragm, it may be blocked by ice during flights into colder humid air. In both cases, it needs special attention as an operational problem. In the following, the technical features of various HLDs are discussed.

1. The *plain flap* (Figure 5.54-1) is the simplest and earliest type of HLD. It is an airfoil shape that is hinged at the wing trailing edge such that it can be rotated downward and upward. However, the downward deflection is considered only. A plain flap increases the lift simply by mechanically increasing the effective camber of the wing section. In terms of cost, a plain flap is the cheapest HLD. In terms of manufacturing, the plain flap is the easiest one to build. Most home-built aircraft and many GA aircraft employ the plain flap. The increment in lift coefficient for a plain flap at 60 deg of deflection (full extension) is about 0.9. If it is deflected at a lower rate, the C_L increment will be lower. Some old GA aircraft – such as Piper 23 Aztec D – have a plain flap. It is interesting to know that modern fighters such as the F-15E Eagle (Figure 9.14) and MIG-29 (Figure 5.56) aircraft also employ plain flaps.
2. In the *split flap* (Figure 5.54-2), only the bottom surface of the flap is hinged so that it can be rotated downward. The split flap performs almost the same function as a plain flap. However, the split flap produces more drag and less change in the pitching moment compared to a plain flap. The split flap was invented by Orville Wright in 1920, and it was employed because of its simplicity on many of the 1930s and 1950s

aircraft. However, because of the higher drag associated with a split flap, they are rarely used on modern aircraft.

3. The *single-slotted flap* (Figure 5.54-3) is very similar to a plain flap, except it has two modifications. First, the leading edges of these two trailing edge flaps are different, as shown in Figure 5.51. The leading edge of a single-slotted flap is carefully designed such that it modifies and stabilizes the boundary layer over the top surface of the wing. A low pressure is created on the leading edge that allows a new boundary layer to form over the flap, which in turn causes the flow to remain attached to a very high flap deflection. The second modification is to allow the flap to move rearward during the deflection (i.e., the slot). The aft movement of a single-slotted flap actually increases the effective chord of the wing, which in turn increases the effective wing planform area. The larger wing planform area naturally generates more lift.

   Thus, a single-slotted flap generates considerably higher lift than a plain and split flap. The main disadvantage is the higher cost and the higher degree of complexity in the manufacturing process associated with the single-slotted flap. Single-slotted flaps are in common use on modern GA light aircraft. In general, the stall angle is increased by the application of the slotted flap. Several modern GA light aircraft such as the Beech Bonanza F33A and several turboprop transport aircraft such as the Beech 1900D and Saab 2000 have deployed single-slotted flaps.
4. The *double-slotted flap* is similar to a single-slotted flap, except it has two slots; that is, the flap is divided into two segments, each with a slot as sketched in Figure 5.54-4. A flap with two slots almost doubles the advantages of a single-slotted flap. This benefit is achieved at the cost of increased mechanical complexity and higher cost. Most modern turboprop transport aircraft such as the ATR-42 (Figure 3.8) and several jet aircraft such as the jet trainer Kawasaki T-4 employ a double-slotted flap. The jet transport aircraft Boeing 767 (Figure 5.4) has a single-slotted outboard flap and a double-slotted inboard flap. It is common practice to deflect the first segment (slot) of the flap during a take-off operation, but employ full deflection (both segments) during landing. The reason is that more lift coefficient is needed during landing than at take-off.
5. A *triple-slotted flap* (Figure 5.54-5) is an extension of a double-slotted flap; that is, it has three slots. This flap is mechanically the most complex, and the most expensive flap in design and operation. However, a triple-slotted flap produces the highest increment in lift coefficient. It is used mainly in heavyweight transport aircraft which have high wing loading. The jet transport aircraft Boeing 747 (Figures 3.7, 3.12, and 9.4) has employed a triple-slotted flap.
6. A *Fowler flap* (Figure 5.54-6) has a special mechanism such that when deployed, it not only deflects downward but also translates or tracks to the trailing edge of the wing. The second feature increases the exposed wing area, which means a further increase in lift. Because of this benefit, the concept of the Fowler flap may be combined with the double-slotted and triple-slotted flaps. For instance, the jet transport aircraft Boeing B-747 (Figures 3.7, 3.12, and 9.4) has utilized a triple-slotted Fowler flap. In general, the wing lift curve slope is increased slightly by application of the Fowler flap. The maritime patrol aircraft Lockheed Orion P-3 with four turboprop engines has a Fowler flap.
7. A *leading edge flap* (or *droop*) is illustrated in Figure 5.54-7. This flap is similar to a trailing edge plain flap, except it is installed at the leading edge of the wing. Hence, the leading edge pivots downward, increasing the effective camber. A feature

of the leading edge flap is that the gap between the flap and the main wing body is sealed with no slot. In general, the wing zero-lift angle of attack is shifted to the right by the application of a leading edge flap. Since the leading edge flap has a lower chord compared with the trailing edge flaps, it generates a lower increment in the lift coefficient (ΔC_L is about 0.3).

8. The *leading edge slat* (see Figure 5.54-8) is a small, highly cambered section, located slightly forward of the leading edge of the wing body. When deflected, a slat is basically a flap at the leading edge, but with an unsealed gap between the flap and the leading edge. In addition to the primary airflow over the wing, there is a secondary flow that takes place through the gap between the slat and the wing leading edge. The function of a leading edge slat is primarily to modify the pressure distribution over the top surface of the wing. The slat itself, being highly cambered, experiences a much lower pressure over its top surface but the flow interaction results in a higher pressure over the top surface of the main wing body. Thus it delays flow separation over the wing and mitigates to some extent the otherwise strong adverse pressure gradient that would exist over the main wing section.

   By such a process, the lift coefficient is increased with no significant increase in drag. Since the leading edge slat has a lower chord compared with the trailing edge flaps, it generates a lower increment in the lift coefficient (ΔC_L is about 0.2). Several modern jet aircraft, such as the two-seat fighter aircraft Dassault Rafale (Figure 6.8), Eurofighter 2000 (Figure 3.7), Bombardier BD 701 Global Express, McDonnell Douglas MD-88 (Figure 9.4), and Airbus A-330 (Figures 5.51 and 9.14), have a leading edge slat. In general, the wing lift curve slope is increased slightly by the application of a leading edge slat.
9. A *Kruger flap* is demonstrated in Figure 5.54-9. This LEHLD is essentially a leading edge slat which is thinner, and lies flush with the bottom surface of the wing when not deflected. Therefore, it is suitable for use with thinner wing sections. The most effective method used on all large transport aircraft is the leading edge slat. A variant on the leading edge slat is a variable camber slotted Kruger flap used on the Boeing 747 (Figures 3.7, 3.12, and 9.4). Aerodynamically this is a slat, but mechanically it is a Kruger flap.

As a general comparison, Table 5.15 shows the typical values of maximum wing lift coefficient for various types of HLDs. In this table, C_f/C denotes the ratio between the chord of a HLD to the chord of the main wing body as shown in Figure 5.55. Table 5.16 demonstrates various features for HLDs of several aircraft.

5.12.3 Design Technique

In designing the HLD for a wing, the following items must be determined:

1. HLD location along the span.
2. The type of HLD (among the list in Figure 5.54).
3. HLD chord (C_f).
4. HLD span (b_f).
5. HLD maximum deflection (down) ($\delta_{f_{max}}$).

Table 5.15 Lift coefficient increment by various types of high-lift device (when deflected 60 deg)

No.	High-lift device	ΔC_L
1	Plain flap	0.7–0.9
2	Split flap	0.7–0.9
3	Fowler flap	1–1.3
4	Slotted flap	1.3 C_f/C
5	Double-slotted flap	1.6 C_f/C
6	Triple-slotted flap	1.9 C_f/C
7	Leading edge flap	0.2–0.3
8	Leading edge slat	0.3–0.4
9	Kruger flap	0.3–0.4

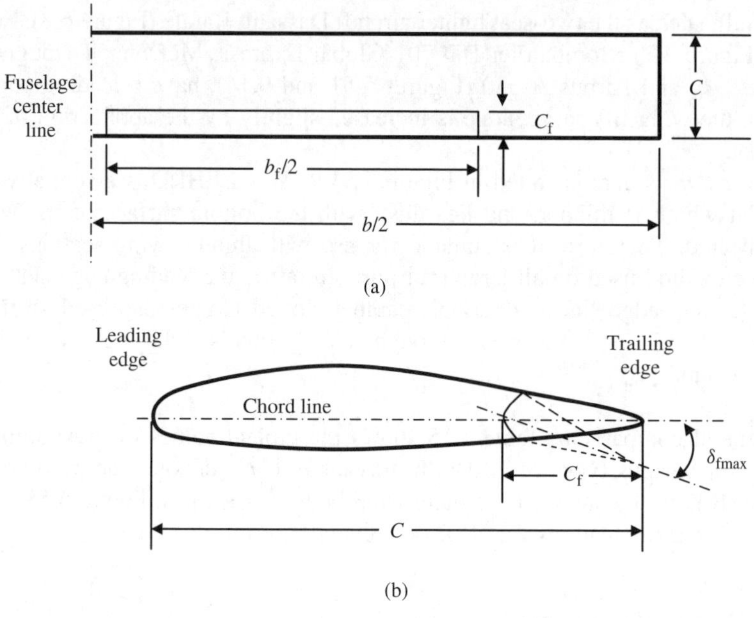

Figure 5.55 High-lift device parameters: (a) Top view of the right wing; (b) Side view of the inboard wing (flap deflected)

The last three parameters are sketched in Figure 5.55. The first and second items must be selected through an evaluation and analysis technique considering all advantages and disadvantages of each option regarding design requirements. However, the last three parameters must be determined through a series of calculations. In the following, the design technique for HLD to determine the above five items will be presented.

Table 5.16 Characteristics of high-lift devices for several aircraft

No.	Aircraft	Engine	HLD	C_f/C	b_f/b	$\delta_{f_{max}}$	
						TO	Landing
1	Cessna 172	Piston	Single-slotted	0.33	0.46	20	40
2	Piper Cherokee	Piston	Single-slotted	0.17	0.57	25	50
3	Lake LA-250	Piston	Single-slotted	0.22	0.57	20	40
4	Short Skyvan 3	Turboprop	Double-slotted	0.3	0.69	18	45
5	Fokker 27	Turboprop	Single-slotted	0.313	0.69	16	40
6	Lockheed L-100	Turboprop	Fowler	0.3	0.7	18	36
7	Jetstream 41	Turboprop	Double-slotted	0.35	0.55	24	45
8	Boeing 727	Turbofan	Triple-slotted + LE flap	0.3	0.74	25	40
9	Airbus A-300	Turbofan	Double-slotted + LE flap	0.32	0.82	15	35
10	Learjet 25	Turbofan	Single-slotted	0.28	0.61	20	40
11	Gulfstream II	Turbofan	Fowler	0.3	0.73	20	40
12	McDonnell DC-9	Turbofan	Double-slotted	0.36	0.67	15	50
13	Antonov 74	Turbofan	Double-slotted + triple-slotted + LE flap	0.24	0.7	25	40
14	McDonnell F-15E Eagle	Turbofan	Plain flap	0.25	0.3	–	–
15	Mikoyan MIG-29	Turbofan	Plain flap + LE flap	0.35	0.3 + 1	–	–
16	X-38	Rocket	Split flap	Lifting body		NA	30

5.12.3.1 HLD Location

The best location for a HLD is the inboard portion of both the left and right wing sections. When a HLD is applied symmetrically on the left and right wing sections, it will prevent any rolling moment; hence the aircraft will remain laterally trimmed. The deflection of a HLD will increase the lift on both inboard sections, but since they are generated symmetrically, both lift increments will cancel each other's rolling moments.

There are two reasons for the selection of an inboard section. First of all, it produces a lower bending moment on the wing root. This makes the wing structure lighter and causes less fatigue on the wing in the long run. The second reason is that it allows the aileron to have a large arm, which is employed on the outboard wing trailing edge. The larger arm for the aileron, when installed on the outboard panels, means higher lateral control and a faster roll. The design of the aileron will be discussed in Chapter 12.

5.12.3.2 Type of High-Lift Device

The options for the HLD are introduced in Section 5.11.2. Several design requirements will affect the decision on the type of HLD. They include, but are not limited to: (i) performance requirements (i.e., the required lift coefficient (ΔC_L) increment during take-off and landing); (ii) cost considerations; (iii) manufacturing limitations; (iv) operational requirements; (v) safety considerations; and (vi) control requirements. The following guidelines will help the designer to make the right decision.

The final decision is the outcome of a compromise among all options using a table including the weighted design requirements. For a home-built aircraft designer, low cost is the number one priority while for a fighter aircraft designer, performance is the first priority. A large transport passenger aircraft designer may believe that airworthiness should be at the top of their list of priorities.

The following are several guidelines that relate the HLD options to the design requirements:

1. A more powerful HLD (higher ΔC_L) is usually more expensive. For instance, a double-slotted flap is more expensive than a split flap.
2. A more powerful HLD (higher ΔC_L) is usually more complex to build. For example, a triple-slotted flap is more complex in manufacture than a single-slotted flap.
3. A more powerful HLD (higher ΔC_L) is usually heavier. For instance, a double-slotted flap is heavier than a single-slotted flap.
4. The more powerful HLD (higher ΔC_L) results in a smaller wing area.
5. The more powerful HLD (higher ΔC_L) results in a slower stall speed, which consequently means a safer flight.
6. A heavier aircraft requires a more powerful HLD (higher ΔC_L).
7. A more powerful HLD results in a shorter runway length during take-off and landing.
8. A more powerful HLD (higher ΔC_L) allows a more powerful aileron.
9. A simple HLD requires a simpler mechanism to operate (deflect or retract) compared with a more complex HLD such as a triple-slotted flap.

When low cost is the number one priority, select the least expensive HLD (that is, the plain flap). If performance is the number one priority, select the HLD that satisfies the performance requirements. If only one HLD such as a single-slotted flap does not satisfy the performance requirements, add another HLD such as a leading edge flap to meet the design requirements. The other option is to combine two HLDs into one new HLD. For instance, the business jet Gulf Stream IV (Figure 11.15) and Dassault Falcon 900 (Figure 6.12) employ a single-slotted Fowler flap that is a combination of the single-slotted flap and the Fowler flap.

All large aircraft use some form of slotted flap. The drag and lift of slotted flaps depend on the shape and dimensions of the vanes and flaps, their relative position, and the slot geometry. Mounting hinges and structure may seriously degrade flap performance if not carefully designed to minimize flow separation. Typical examples are the McDonnell Douglas DC-8 original flap hinges and the McDonnell Douglas DC-9 original slat design, both of which were redesigned during the flight test stage to obtain the required $C_{L_{max}}$ and low drag.

The triple-slotted flap is almost the ultimate in mechanical complexity. For this reason, in the interests of lower design and production costs, some recent aircraft designs have returned to simpler mechanisms. For example, the Boeing 767 (Figure 5.4) has a single-slotted outboard flap and a double-slotted inboard flap.

LEHLDs such as slats function very differently compared with trailing edge HLDs. The lift coefficient at a given angle of attack is increased very little, but the stall angle is increased greatly. One disadvantage of slats is that the aircraft must be designed to fly at a high angle of attack for take-off and landing to utilize the high available lift increment. This clearly affects the design of the windshield, because of the pilot's visibility requirements. Despite the disadvantages of slats, they are so powerful in high lift that all high-speed transport aircraft designed since about 1964 use some form of slat in addition to trailing edge flaps. If leading edge devices serve simply to shorten take-off and/or landing runway lengths below the required values and the wing area cannot be reduced (say because of fuel tank requirements), the weight and complexity due to the application of a leading edge device are not justified.

Leading edge devices intended to raise the $C_{L_{max}}$ substantially must extend along the entire leading edge except for a small cutout near the fuselage to trigger the inboard stall. Some designs utilize a less powerful device, such as a Kruger flap, on the inboard part of the wing to insure inboard initial stall. Table 5.16 illustrates the type of HLD for several aircraft.

5.12.3.3 HLD Span

The spanwise extent of HLDs depends on the amount of span required for ailerons. In general, the outer limit of the flap is at the spanwise station where the aileron begins. The exact span needed for ailerons depends on the aircraft's lateral controllability requirements. Low-speed GA aircraft utilize about 30% of the total semispan for an aileron. This means that flaps can start at the side of the fuselage and extend to the 70% semispan station. In large transport aircraft, a small inboard aileron is often provided for gentle maneuver at high speeds, and this serves to reduce the effective span of the flaps. However, in fighter aircraft which are highly maneuverable, ailerons require all wing span stations, so there is theoretically no space for flaps. This leads to the idea of a flaperon that serves as an aileron as well as a flap. The HLD span is usually introduced as the ratio to the wing span (i.e., b_f/b). In some references, b_f/b refers to the ratio between flap span and net wing span (i.e., from root to tip, not from center line to tip).

Table 5.16 illustrates the ratio of the HLD span to the wing span for several aircraft. As an initial value, it is recommended to allocate 70% of the wing span to the HLD. The exact value must be determined through the calculation of lift increment due to this span (b_f) for the HLD. There are several aerodynamic tools to accomplish this analysis. An aerodynamic technique called the *lifting-line theory* will be introduced in Section 5.13. Such a technique can be employed to calculate the lift increment for each HLD span. You

Figure 5.56 Four aircraft with various wing characteristics: (a) Panavia Tornado GR4 with its long span flap; (b) Mikoyan-Gurevich MiG-29 with a low AR and high sweep angle; (c) Piper Super Cub with strut-braced wing; (d) Sailpane Schleicher ASK-18 with high AR (Reproduced from permission of (a, b) Antony Osborne, (c) Jenny Coffey, (d) Akira Uekawa.)

can then adjust the HLD span (b_f) to achieve the required lift increment. An example at the end of this chapter will illustrate the application. Figure 5.56(a) illustrates the fighter aircraft Panavia Tornado GR4 with its long span flap that leaves no span for ailerons on the wing.

5.12.3.4 HLD Chord

Since the HLD is employed temporarily in a regular flight mission during take-off and landing, the least amount of wing chord must be intended for a HLD. The wing structural integrity must be considered when allocating part of the wing chord to a HLD. The chord of the HLD is often introduced as the ratio to the wing chord (i.e., C_f/C). It is important to note that the deflection of a HLD will increase the wing drag. Hence, the HLD chord must not be so high that the drag increment, due to its deflection, nullifies its advantages. In contrast, as the HLD chord is increased, the power required to deflect the device is increased. If the pilot uses manual power to move the HLD, a longer HLD chord requires more pilot power. Therefore, a shorter HLD is better in many respects.

Another consideration for the HLD chord is that the designer can extend the chord up to the rear spar of the wing. Since, in most aircraft, the rear spar is important for the wing's structural integrity, do not try to cut the rear spar in order to extend the HLD chord. The HLD chord and span can be interchanged to some extent. If you have to reduce the span of the HLD, due to aileron requirements, you can increase the HLD chord instead. The opposite is also true. If you have to reduce the chord of the HLD, due to structural considerations, you can increase the HLD span instead. If the wing is tapered, you may taper the flap as well. So the HLD chord does not have a constant chord.

Table 5.16 illustrates the ratio of the HLD chord to the wing chord (C_f/C) for several aircraft. As an initial value, it is recommended to allocate 20% of the wing chord to the HLD. The exact value must be determined through the calculation of the lift increment due to this chord for a HLD. There are several aerodynamic tools to accomplish this analysis. Such aerodynamic techniques as the *lifting-line theory* can be employed to calculate the lift increment for each HLD chord. You can then adjust the HLD chord (C_f) to achieve the required lift increment. An example at the end of this chapter will illustrate the application.

5.12.3.5 HLD Maximum Deflection

Another parameter that must be determined in the design of the HLD is the amount of its deflection ($\delta_{f_{max}}$). The exact value of the deflection must be determined through the calculation of the lift increment due to the HLD deflection. Table 5.16 illustrates the HLD deflection ($\delta_{f_{max}}$) for several aircraft. As an initial value, it is recommended to consider a deflection of 20 deg during take-off and 50 deg for landing. There are several aerodynamic tools to accomplish this analysis. Such aerodynamic techniques as the *lifting-line theory* can be employed to calculate the lift increment for each HLD deflection. You can then adjust the HLD chord (C_f) to achieve the required lift increment. An example at the end of this chapter will illustrate the application.

In using aerodynamic techniques to calculate the incremental lift due to the extension of the trailing edge flap, it is necessary to determine the increment in the wing zero-lift angle of attack ($\Delta\alpha_o$). The following is an empirical equation that allows for such approximation:

$$\Delta\alpha_o \approx -1.15 \cdot \frac{C_f}{C}\delta_f \tag{5.39}$$

This equation provides the section's incremental zero-lift angle of attack ($\Delta\alpha_o$) as a function of flap-to-wing chord ratio and flap deflection.

5.13 Aileron

An aileron is very similar to a trailing edge plain flap except it is deflected both up and down. An aileron is located at the outboard portion of the left and right sections of a wing. Unlike a flap, ailerons are deflected differentially, left up and right down or left down and right up. Lateral control is applied on an aircraft through the differential motions of ailerons. Aileron design is part of wing design, but because of the importance and great amount of material that needs to be covered for aileron design, it will be discussed in a separate chapter (Chapter 12).

In this section, it is mainly emphasized not to consume all the wing's trailing edge for a flap and to leave about 30% of the wing outboard for ailerons. Figure 5.57 illustrates the typical location of the aileron on the wing. Three major parameters that need to be determined in the aileron design process are: aileron chord, aileron span, and aileron deflection (up and down). The primary design requirements in aileron design originate

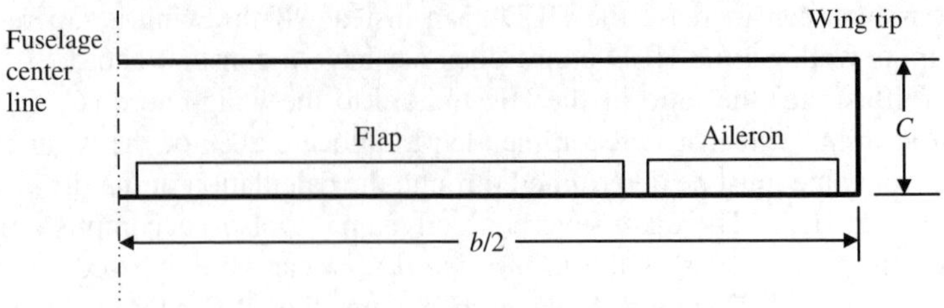

Figure 5.57 Typical location of the aileron on the wing

from the roll controllability of the aircraft. A full discussion of aileron design and aileron design techniques will be covered in Chapter 12.

5.14 Lifting-Line Theory

In Section 5.7 it is explained that in the wing design process, the designer must calculate the lift force that a wing is generating. Then, by changing the wing parameters, one can finalize the wing parameters to achieve the design goals while satisfying all the design requirements. The technique comes essentially from the area of aerodynamics, however, in order to complete the discussion of the wing design, a rather straightforward but at the same time relatively accurate technique is introduced. A wing designer must have a solid background in aerodynamics, so this section plays the role of a review for you. For this reason, materials in this section are covered without proof. For more and detailed information, you are referred to Ref. [16].

The technique introduced in this section allows the reader to determine the amount of lift that is generated by a wing without using sophisticated CFD software. You need to have all the wing data to hand, such as wing area, airfoil section and its features, aspect ratio, taper ratio, wing incidence, and HLD type and data. By solving several aerodynamic equations simultaneously, one can determine the amount of lift that a wing is producing. Furthermore, the technique will generate the lift distribution along the span, hence one can make sure if the lift distribution is elliptical or not.

The technique was initially introduced by Ludwig Prandtl and called the *lifting-line theory* in 1918. Almost every aerodynamics textbook has details of this simple and remarkably accurate technique. The major weakness of this classical technique is that it is a linear theory; thus, it does not predict stall. Therefore, if you know the airfoil section's stall angle, do not employ this approach beyond the airfoil's stall angle. The technique can be applied for a wing with both flap up and flap down (i.e., deflected). In the following, the steps to calculate the lift distribution along the span plus the total wing lift coefficient will be presented. Since a wing has a symmetric geometry, we only need to consider one half of the wing. The technique can later be extended to both the left and right wing halves. The application of the technique will be demonstrated at the end of this chapter.

- **Step 1**. Divide one half of the wing (semispan) into several (say N) segments. The segments along the semispan could have equal span, but it is recommended to have smaller segments in the regions closer to the wing tip. A higher number of segments (N) is desired, since it yields a higher accuracy. As an example, in Figure 5.58, a wing is shown that is divided into seven equal segments. As noted, each segment has a unique chord and may have a unique span. You have the option to consider a unique airfoil section for each segment as well (recall the aerodynamic twist). Then, identify the geometry (e.g., chord and span) and aerodynamic properties (e.g., α, α_o, and C_{l_α}) of each segment for future application.

- **Step 2**. Calculate the corresponding angle (θ) for each section. These angles are functions of lift distribution along the semispan, as depicted in Figure 5.59. Each angle (θ) is defined as the angle between the horizontal axis and the intersection between the lift distribution curve and the segment line. In fact, we originally assume that the lift distribution along the semispan is elliptical. This assumption will be corrected later.

  The angle θ varies between 0 for the last segment and a number close to 90 deg for the first segment. The value of the angle θ for other segments may be determined from the corresponding triangle as shown in Figure 5.59. For instance in Figure 5.59, the angle θ_6 is the angle corresponding to the segment 6.

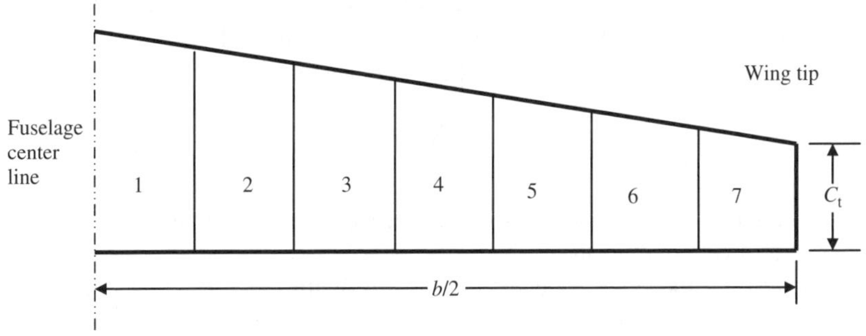

Figure 5.58 Dividing a wing into several sections

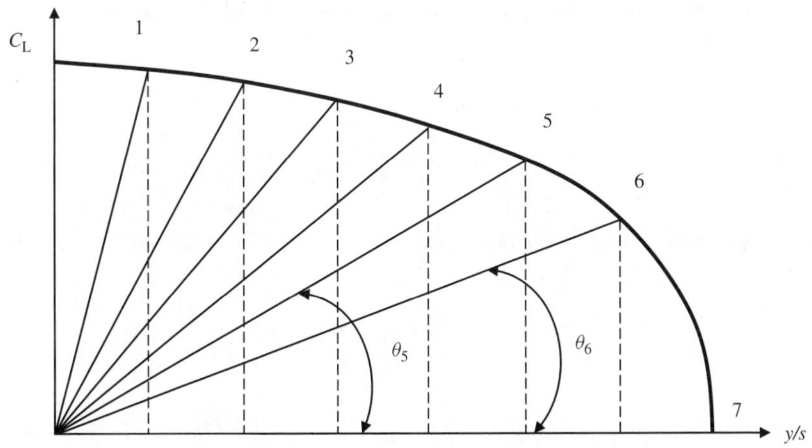

Figure 5.59 Angles corresponding to each segment in lifting-line theory

- **Step 3**. Solve the following group of equations to find A_1 to A_n:

$$\mu\left(\alpha_{\rm o}-\alpha\right)=\sum_{n=1}^{N}A_n\sin\left(n\theta\right)\left(1+\frac{\mu n}{\sin\left(\theta\right)}\right) \tag{5.40}$$

  This equation lies at the heart of the theory and is referred to as the *lifting-line* equation or *monoplane* equation. The equation was initially developed by Prandtl. In this equation, N denotes the number of segments, α the segment's angle of attack, $\alpha_{\rm o}$ the segment's zero-lift angle of attack, and coefficients A_n are the intermediate unknowns. The parameter μ is defined as follows:

$$\mu=\frac{\overline{C}_{\rm i}\cdot C_{{\rm l}_\alpha}}{4b} \tag{5.41}$$

  where $\overline{C}_{\rm i}$ denotes the segment's mean geometric chord, $C_{{\rm l}_\alpha}$ the segment's lift curve slope (1/rad), and b the wing span. If the wing has a twist ($\alpha_{\rm t}$), the twist angle must be applied to all segments linearly. Thus, the angle of attack for each segment is reduced by deducting the corresponding twist angle from the wing setting angle. If the theory is applied to a wing in take-off operation, where the flap is deflected, the inboard segments have a larger zero-lift angle of attack ($\alpha_{\rm o}$) than the outboard segments.
- **Step 4**. Determine each segment's lift coefficient using the following equation:

$$C_{L_{\rm i}}=\frac{4b}{\overline{C}_{\rm i}}\sum A_n\sin\left(n\theta\right) \tag{5.42}$$

  Now you can plot the variation of each segment's lift coefficient ($C_{\rm L}$) versus the semispan (i.e., lift distribution).
- **Step 5**. Determine the wing total lift coefficient using the following equation:

$$C_{L_{\rm w}}=\pi\cdot{\rm AR}\cdot A_1 \tag{5.43}$$

  where AR is the wing aspect ratio.

  Please note that the lifting-line theory has other useful features, but they are not covered or used here.

Example 5.5 gives a sample calculation.

Example 5.5

Determine and plot the lift distribution for a wing with the following characteristics. Divide the half wing into 10 sections.

$$S=25\,{\rm m}^2,\ {\rm AR}=8,\ \lambda=0.6,\ i_{\rm w}=2\,{\rm deg},\ \alpha_{\rm t}=-1\,{\rm deg}$$

Airfoil section: NACA 63-209

If the aircraft is flying at an altitude of 5000 m ($\rho=0.736$ kg/m^3) with a speed of 180 knot, how much lift is produced?

Solution:

By using Ref. [3], we can find the airfoil section's features. A copy of the airfoil graphs is shown in Figure 5.22. Based on the C_l/α graph, we have the following data:

$$\alpha_o = -1.5 \text{ deg}, \ C_{l_\alpha} = 6.3 \text{ 1/rad}$$

The application of the lifting-line theory is formulated through the following MATLAB m-file:

```
clc
clear
N = 9; % (number of segments - 1)
S = 25; % m^2
AR = 8; % Aspect ratio
lambda = 0.6; % Taper ratio
alpha_twist = -1; % Twist angle (deg)
i_w = 2; % wing setting angle (deg)
a_2d = 6.3; % lift curve slope (1/rad)
alpha_0 = -1.5; % zero-lift angle of attack (deg)
b = sqrt(AR*S); % wing span (m)
MAC = S/b; % Mean Aerodynamic Chord (m)
Croot = (1.5*(1+lambda)*MAC)/(1+lambda+lambda^2); % root chord (m)
theta = pi/(2*N):pi/(2*N):pi/2;
alpha = i_w+alpha_twist:-alpha_twist/(N-1):i_w;
  % segment's angle of attack
z = (b/2)*cos(theta);
c = Croot * (1 - (1-lambda)*cos(theta)); % Mean Aerodynamics
  Chord at each segment (m)
mu = c * a_2d / (4 * b);
LHS = mu .* (alpha-alpha_0)/57.3; % Left Hand Side
% Solving N equations to find coefficients A(i):
for i=1:N
    for j=1:N
    B(i,j) =  sin((2*j-1) * theta(i)) * (1 + (mu(i) * (2*j-1)) /
      sin(theta(i)));
    end
end
A=B\transpose(LHS);
for i = 1:N
    sum1(i) = 0;
    sum2(i) = 0;
    for j = 1 : N
        sum1(i) = sum1(i) + (2*j-1) * A(j)*sin((2*j-1)*theta(i));
        sum2(i) = sum2(i) + A(j)*sin((2*j-1)*theta(i));
    end
  end
CL = 4*b*sum2 ./ c;
CL1=[0 CL(1) CL(2) CL(3) CL(4) CL(5) CL(6) CL(7) CL(8) CL(9)];
y_s=[b/2 z(1) z(2) z(3) z(4) z(5) z(6) z(7) z(8) z(9)];
```

```
plot(y_s,CL1,'-o')
 grid
title('Lift distribution')
xlabel('Semi-span location (m)')
ylabel ('Lift coefficient')
CL_wing = pi * AR * A(1)
```

Figure 5.60 shows the lift distribution of the example wing as an output of the m-file.

As noted, the distribution in this wing is not elliptical, so it is not ideal. The wing needs some modification (such as increasing the wing twist) to produce an acceptable output. The total lift coefficient of the wing is $C_L = 0.268$. The lift generated by this wing is as follows:

$$L = \frac{1}{2}\rho V^2 S C_L = \frac{1}{2} \cdot 0.736 \cdot (180 \cdot 0.5144)^2 \cdot 25 \cdot 0.268 = 21\,169.2 \text{ N} \qquad (5.1)$$

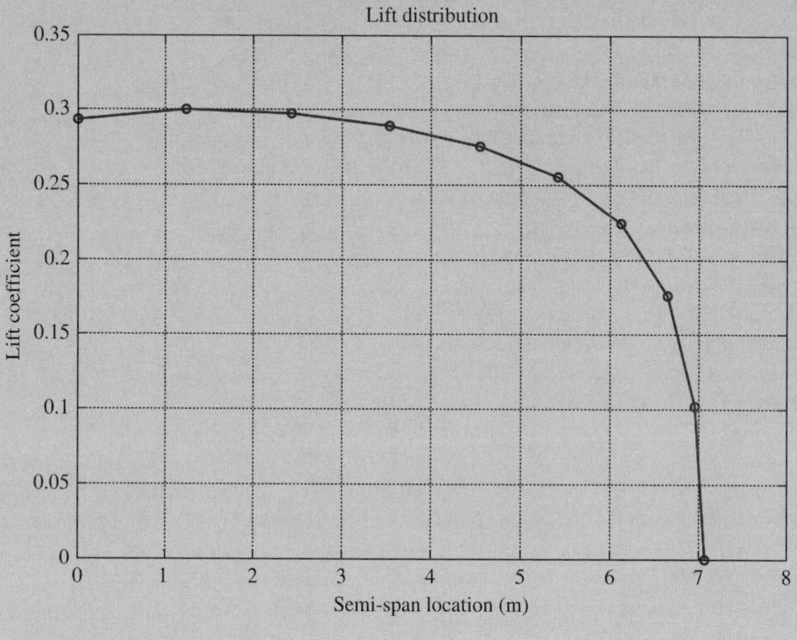

Figure 5.60 The lift distribution of the wing in Example 5.5

5.15 Accessories

Depending upon the aircraft type and flight conditions, the wing may have a few accessories to improve the flow over the wing. Accessories such as wing tip, fence, vortex generator, stall stripes, and strake are employed to increase the wing efficiency. In this section, a few practical considerations will be introduced.

5.15.1 Strake

A strake (also known as a leading edge extension) is an aerodynamic surface generally mounted on the fuselage of an aircraft to finetune the airflow and control the vortex over the wing. In order to increase lift and improve directional stability and maneuverability at high angles of attack, highly swept strakes along the fuselage forebody may be employed to join the wing sections. Aircraft designers choose the location, angle, and shape of the strake to produce the desired interaction. Fighter aircraft F-16 and F-18 have employed strakes to improve the wing efficiency at high angles of attack. In addition, the provision of strakes on the fuselage, in front of the tail, will increase the fuselage damping which consequently improves the spin recovery characteristics of the aircraft. The design of the strake needs a high-fidelity CFD software package and is beyond the scope of this book.

5.15.2 Fence

Stall fences are used in swept wings to prevent the boundary layer drifting outboard toward the wing tips. Boundary layers on swept wings tend to drift because of the spanwise pressure gradient of a swept wing. A swept wing often has a leading edge fence of some sort, usually at about 35% of the span from the fuselage center line as shown in Figure 5.61. The cross-flow creates a side lift on the fence that produces a strong trailing

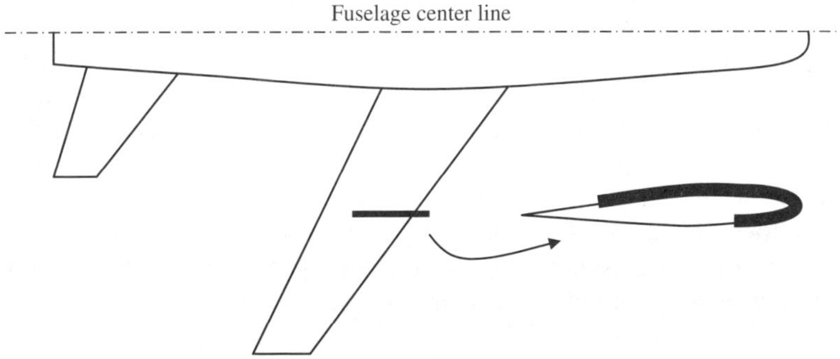

(a)

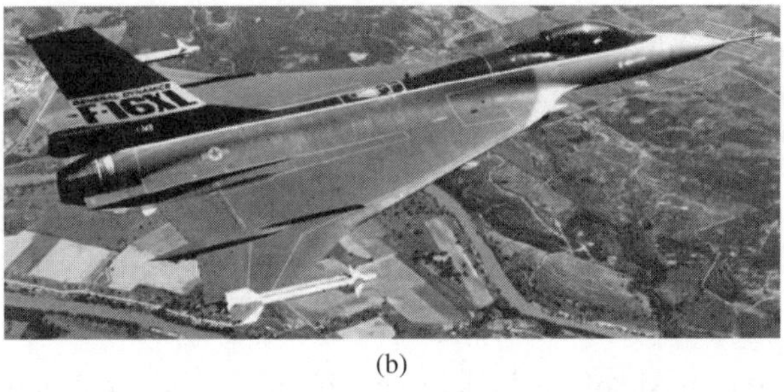

(b)

Figure 5.61 Example of a stall fence: (a) Fence over the wing; (b) Fence over the wing of General Dynamics F-16XL vortex. This vortex is carried over the top surface of the wing, mixing fresh air into the boundary layer and sweeping the boundary layer off the wing and into the outside flow. The result is a reduction in the amount of boundary layer air flowing outboard at the rear of the wing. This improves the outer panel maximum lift coefficient.

Similar results can be achieved with a leading edge snag. Such snags tend to create a vortex which acts like a boundary layer fence. The ideal device is the under-wing fence, referred to as a *vertilon*. Pylons supporting the engines under the wing, in practice, serve the purpose of leading edge fences. Several high-subsonic transport aircraft such as the McDonnell Douglas DC-9 and Beech Starship have utilized a fence on their swept lifting surfaces. The design of the fence needs a high-fidelity CFD software and is beyond the scope of this book.

5.15.3 Vortex Generator

Vortex generators are very small, low-aspect ratio wings placed vertically at some local angle of attack on the wing, fuselage, or tail surfaces of aircraft. The span of the vortex generator is typically selected such that they are just outside the local edge of the boundary layer. Since they are some type of lifting surface, they will produce lift and therefore tip vortices near the edge of the boundary layer. These vortices will then mix with the high-energy air to raise the kinetic energy level of the flow inside the boundary layer. Hence, this process allows the boundary layer to advance further into an adverse pressure gradient before separating. Vortex generators are employed in many different sizes and shapes.

Most of today's high-subsonic jet transport aircraft have a large number of vortex generators on wings, tails, and even nacelles. Even though vortex generators are beneficial in delaying local wing stall, they can generate a considerable increase in aircraft drag. The precise number and orientation of vortex generators are often determined in a series of sequential flight tests. For this reason, they are sometimes referred to as "aerodynamic after thoughts." Vortex generators are usually added to an aircraft after tests have indicated certain flow separations. The Northrop Grumman B-2A Spirit (Figure 6.8) strategic penetration bomber utilizes small, drop-down spoiler panels ahead of weapon bay doors to generate vortices to ensure clean weapon release.

5.15.4 Winglet

Since there is a considerable pressure difference between the lower and upper surfaces of a wing, tip vortices are produced at the wing tips. These tip vortices will then roll up and get around the local edges of a wing. This phenomenon will reduce the lift at the wing tip station, so they can be represented as a reduction in effective wing span. Experiments have shown that wings with square or sharp edges have the widest effective span. To compensate this loss, three solutions are tip-tank, extra wing span, and winglet. Winglets are small, nearly vertical lifting surfaces, mounted rearward, and/or downward relative to the wing tips.

The aerodynamic analysis of a winglet (e.g., lift, drag, and local flow circulation) may be performed by classical aerodynamic techniques. The necessity of wing tips depends on the mission and the configuration of an aircraft, since they will add to the aircraft's weight. Several small and large transport aircraft such as Pilatus PC-12, Boeing 747-400

(Figures 3.7, 3.12, and 9.4), McDonnell Douglas C-17A Globemaster III (Figure 9.9), and Airbus 340-300 (Figure 8.7) have winglets.

5.16 Wing Design Steps

At this stage, we are in a position to summarize the chapter. In this section, the practical steps in a wing design process are introduced (see Figure 5.1) as follows.

Primary function: Generation of the lift

1. Select number of wings (e.g., monoplane, biplane) (see Section 5.2).
2. Select wing vertical location (e.g., high, mid-, low) (see Section 5.3).
3. Select wing configuration (e.g., straight, swept, tapered, delta).
4. Calculate average aircraft weight at cruise:

$$W_{\text{ave}} = \frac{1}{2}\left(W_{\text{i}} + W_{\text{f}}\right) \tag{5.44}$$

   where W_{i} is the aircraft at the beginning of cruise and W_{f} is the aircraft at the end of cruising flight.
5. Calculate required aircraft cruise lift coefficient (with average weight):

$$C_{L_{\text{c}}} = \frac{2W_{\text{ave}}}{\rho V_{\text{c}}^2 S} \tag{5.45}$$

6. Calculate the required aircraft take-off lift coefficient:

$$C_{L_{\text{TO}}} = 0.85\frac{2W_{\text{TO}}}{\rho V_{\text{TO}}^2 S} \tag{5.46}$$

   The coefficient 0.85 originates from the fact that during take-off, the aircraft has a take-off angle (say about 10 deg). Thus about 15% of the lift is maintained by the vertical component (sin (10)) of the engine thrust.
7. Select the HLD type and its location on the wing. See Section 5.12.
8. Determine the HLD geometry (span, chord, and maximum deflection). See Section 5.12.
9. Select/design the airfoil (you can select a different airfoil for the tip and root). This procedure was introduced in Section 5.4.
10. Determine the wing incidence or setting angle (i_{w}). This corresponds to the airfoil ideal lift coefficient $C_{l_{\text{i}}}$ (where the airfoil drag coefficient is minimum). See Section 5.5.
11. Select the sweep angle ($\Lambda_{0.5\text{C}}$) and the dihedral angle (Γ). See Sections 5.9 and 5.11.
12. Select other wing parameters such as the aspect ratio (AR), taper ratio (λ), and wing twist angle (α_{twist}). See Sections 5.6, 5.7, and 5.10.
13. Calculate the lift distribution at cruise (without flap, or flap up). Use tools such as the lifting-line theory (see Section 5.14) and CFD.
14. Check the lift distribution at cruise is elliptic. Otherwise, return to step 13 and change a few parameters.
15. Calculate the wing lift at cruise ($C_{L_{\text{w}}}$). Recall that HLDs are not employed at cruise.

16. The wing lift coefficient at cruise (C_{L_w}) must be equal to the required cruise lift coefficient (step 5). If not, return to step 10 and change the wing setting angle.
17. Calculate the wing lift coefficient at take-off ($C_{L_{w\_TO}}$). Employ a flap at take-off with deflection δ_f and wing angle of attack $\alpha_w = \alpha_{s_{TO}} - 1$. Note that α_s at take-off is usually smaller than α_s at cruise. Please note that the minus one (-1) is for safety.
18. The wing lift coefficient at take-off ($C_{L_{w\_TO}}$) must be equal to the take-off lift coefficient (step 6). If not, first play with the flap deflection (δ_f) and the geometry (C_f, b_f), otherwise return to step 7 and select another HLD. You can have more than one for greater safety.
19. Calculate the wing drag (D_w).
20. Play with the wing parameters to minimize the wing drag.
21. Calculate the wing pitching moment (M_{o_w}). This moment will be used in the tail design process.
22. Optimize the wing to minimize wing drag and wing pitching moment.

A fully solved example will demonstrate the application of these steps in the next section.

5.17 Wing Design Example

In this section, a major wing design example with full solution is presented (Example 5.6). To avoid lengthening the section, a few details are not described and are left to the reader to discover. These details are very similar to the solutions explained in other examples of this chapter.

Example 5.6

Design a wing for a normal category GA aircraft with the following features:

$$S = 18.1\,\text{m}^2,\ m = 1800\,\text{kg},\ V_c = 130\ \text{knot (at sea level)},\ V_s = 60\ \text{knot}$$

Assume the aircraft has a monoplane high wing and employs a split flap.

Solution:
The number of wings and wing vertical position are stated in the problem statement, so we do not need to investigate these two parameters.

1. **Dihedral angle**. Since the aircraft is a high-wing, low-subsonic, and mono-wing aircraft, based on Table 5.13, a -5 deg of anhedral is selected. This value will be revised and optimized when other aircraft components are designed during lateral stability analysis.
2. **Sweep angle**. The aircraft is a low-subsonic prop-driven normal category aircraft. To keep the cost low in the manufacturing process, we select no sweep angle at 50% of wing chord. However, we may need to taper the wing; hence the leading edge and trailing edge may have sweep angles.

3. **Airfoil**. To be fast in the wing design, we select an airfoil from NACA selections. The design of an airfoil is beyond the scope of this book. The selection process of an airfoil for the wing requires some calculation as follows.

Section's ideal lift coefficient:

$$C_{L_C} = \frac{2W_{\text{ave}}}{\rho V_c^2 S} = \frac{2 \cdot 1800 \cdot 9.81}{1.225 \cdot (130 \cdot 0.514)^2 \cdot 18.1} = 0.356 \quad (5.10)$$

$$C_{L_{C\_w}} = \frac{C_{L_C}}{0.95} = \frac{0.356}{0.95} = 0.375 \quad (5.11)$$

$$C_{l_i} = \frac{C_{L_{C\_w}}}{0.9} = \frac{0.375}{0.9} = 0.416 \quad (5.12)$$

Section's maximum lift coefficient:

$$C_{L_{\max}} = \frac{2W_{\text{TO}}}{\rho_o V_s^2 S} = \frac{2 \cdot 1800 \cdot 9.81}{1.225 \cdot (60 \cdot 0.514)^2 \cdot 18.1} = 1.672 \quad (5.13)$$

$$C_{L_{\max\_w}} = \frac{C_{L_{\max}}}{0.95} = \frac{1.672}{0.95} = 1.76 \quad (5.14)$$

$$C_{l_{\max\_gross}} = \frac{C_{L_{\max\_w}}}{0.9} = \frac{1.76}{0.9} = 1.95 \quad (5.15)$$

The aircraft has a split flap, and the split flap generates a ΔC_L of 0.45 when deflected 30 deg. Thus:

$$C_{l_{\max}} = C_{l_{\max\_gross}} - \Delta C_{l_{\max\_HLD}} = 1.95 - 0.45 = 1.5 \quad (5.16)$$

Thus, we need to look for NACA airfoil sections that yield an ideal lift coefficient of 0.4 and a net maximum lift coefficient of 1.5:

$$C_{l_i} = 0.416 \approx 0.4$$

$$C_{l_{\max}} = 1.95 \quad \text{(flap down)}$$

$$C_{l_{\max}} = 1.5 \quad \text{(flap up)}$$

By referring to Ref. [3] and Figure 5.23, we find the following seven airfoil sections whose characteristics match or are close to our design requirements (all have $C_{l_i} = 0.4, C_{l_{\max}} \approx 1.5$):

$$63_1\text{-}412,\ 63_2\text{-}415,\ 64_1\text{-}412,\ 64_2\text{-}415,\ 66_2\text{-}415, 4412, 4418$$

Now we need to compare these airfoil sections to see which one is best. Table 5.17 compares the characteristics of the seven candidates. The best airfoil is the airfoil whose C_{m_o} is the lowest, $C_{d_{\min}}$ is the lowest, α_s is the highest, $(C_l/C_d)_{\max}$ is the

Table 5.17 A comparison between seven airfoil candidates for the wing in Example 5.6

No.	NACA	$C_{d_{min}}$	C_{m_o}	α_s (deg) flap up	α_o (deg) δ_f = 60 deg	$(C_l/C_d)_{max}$	C_{l_i}	$C_{l_{max}}$ δ_f = 30 deg	Stall quality
1	63_1-412	0.0049	−0.075	11	−13.8	120	0.4	2	Moderate
2	63_2-415	0.0049	−0.063	12	−13.8	120	0.4	1.8	Docile
3	64_1-412	0.005	−0.074	12	−14	111	0.4	1.8	Sharp
4	64_2-415	0.005	−0.056	12	−13.9	120	0.4	2.1	Docile
5	66_2-415	0.0044	−0.068	17.6	−9	150	0.4	1.9	Moderate
6	4412	0.006	−0.1	14	−15	133	0.4	2	Moderate
7	4418	0.007	−0.085	14	−16	100	0.4	2	Moderate

highest, and stall quality is docile. By comparing the numbers in the table, we can conclude the following:

(a) The NACA airfoil section 66_2-415 yields the highest maximum speed, since it has the lowest $C_{d_{min}}$ (i.e., 0.0044).
(b) The NACA airfoil section 64_2-415 yields the lowest stall speed, since it has the highest maximum lift coefficient (i.e., 2.1).
(c) The NACA airfoil section 66_2-415 yields the highest endurance, since it has the highest $(C_l/C_d)_{max}$ (i.e., 150).
(d) The NACA airfoil sections 63_2-415 and 64_2-415 yield the safest flight, due to their docile stall quality.
(e) The NACA airfoil section 64_2-415 delivers the lowest longitudinal control effort in flight, due to the lowest C_{m_o} (i.e., −0.056).

Since the aircraft is a non-maneuverable GA aircraft, the stall quality cannot be sharp; hence NACA 64_1-412 is not acceptable. If safety is the highest requirement, the best airfoil is NACA 64_2-415 due to its high $C_{l_{max}}$. When maximum endurance is the highest priority, NACA airfoil section 66_2-415 is the best due to its high $(C_l/C_d)_{max}$. In contrast, if low cost is the most important requirement, NACA 66_2-415 with the lowest $C_{d_{min}}$ is best. However, if aircraft stall speed, stall quality, and lowest longitudinal control power are of greatest importance, NACA airfoil section 64_2-415 is best. This may be determined by using a comparison table incorporating the weighted design requirements.

Due to the fact that the NACA airfoil section 64_2-415 is the best in terms of three criteria, we select it as the most suitable airfoil section for this wing. Figure 5.62 illustrates the characteristic graphs of this airfoil.

4. **Wing setting angle**. The wing setting angle is initially determined to be the angle corresponding to the airfoil ideal lift coefficient. Since the airfoil ideal lift coefficient

is 0.416, Figure 5.62 (left) reads the corresponding angle to be 2 deg. The value ($i_w = 2$ deg) may need to be revised based on calculation to satisfy the design requirements later.

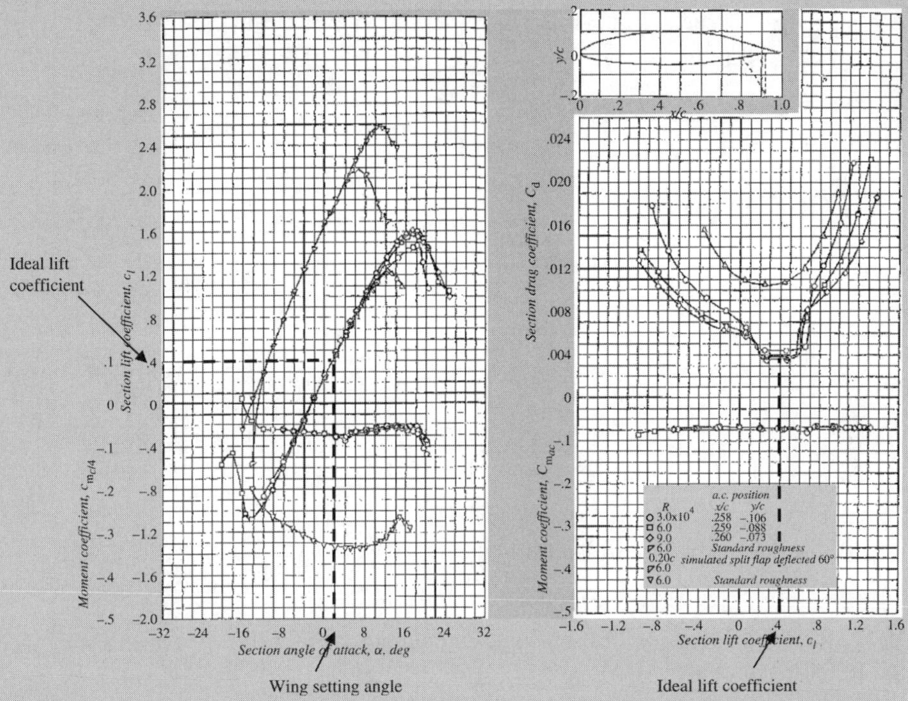

Figure 5.62 Airfoil section NACA 66_2-415. Reproduced from permission of Dover Publications, Inc.

5. **Aspect ratio, taper ratio, and twist angle**. The three parameters of aspect ratio, taper ratio, and twist angle are determined concurrently, since they are all influential for the lift distribution. Several combinations of these three parameters might yield a desirable lift distribution which is elliptical. Based on Table 5.6, the aspect ratio is selected to be 7 (AR = 7). No twist is assumed ($\alpha_t = 0$) at this time to keep the manufacture easy and low cost. The taper ratio is tentatively considered to be 0.3 ($\lambda = 3$). Now we need to find out (i) if the lift distribution is elliptical and (ii) if the lift created by this wing at cruise is equal to the aircraft weight. The lifting-line theory is employed to determine the lift distribution and wing lift coefficient.

A MATLAB m-file is developed similar to that shown in Example 5.5. The application of the lifting-line theory is formulated through this m-file. Figure 5.63

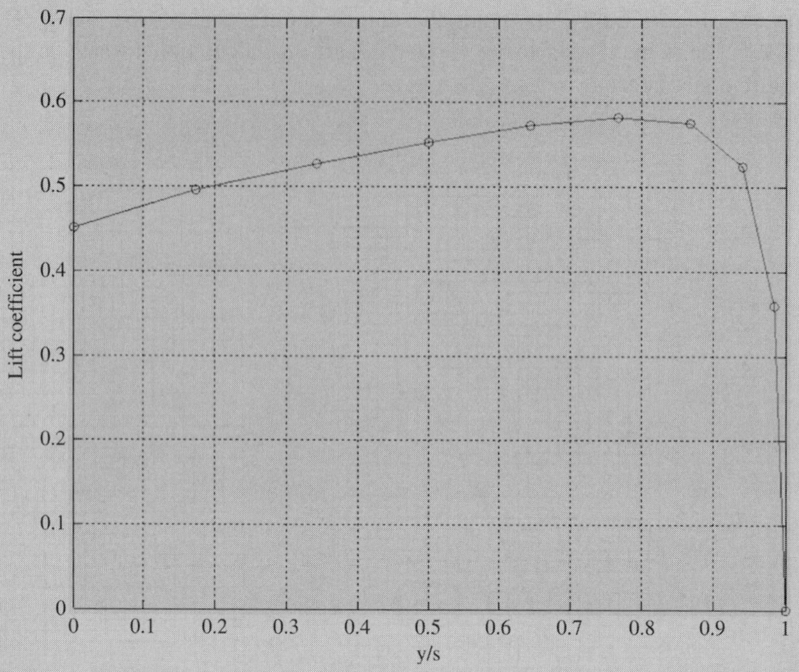

Figure 5.63 The lift distribution of the wing (AR = 7, $\lambda = 0.3$, $\alpha_t = 0$, $i_w = 2$ deg)

shows the lift distribution of the wing as an output of the m-file. The m-file also yields the lift coefficient as:

$$C_L = 0.4557$$

Two observations can be made from the results: (i) the lift coefficient is slightly higher than what is needed ($0.4557 > 0.356$) and (ii) the lift distribution is not elliptical. Therefore, some wing features must be changed to correct both outcomes.

After several runs of trial and error, the following wing specifications are found to satisfy the design requirements:

$$AR = 7,\ \lambda = 0.8,\ \alpha_t = -1.5\ \text{deg},\ i_w = 1.86\ \text{deg}$$

By using the same m-file and these new parameters, the following results are obtained:

$a - C_L = 0.359$

b – Elliptical lift distribution as shown in Figure 5.64.

Hence, this wing with the above parameters satisfies the aircraft cruise requirements. Now, we need to proceed to design the flap and determine the flap parameters to satisfy the take-off requirements.

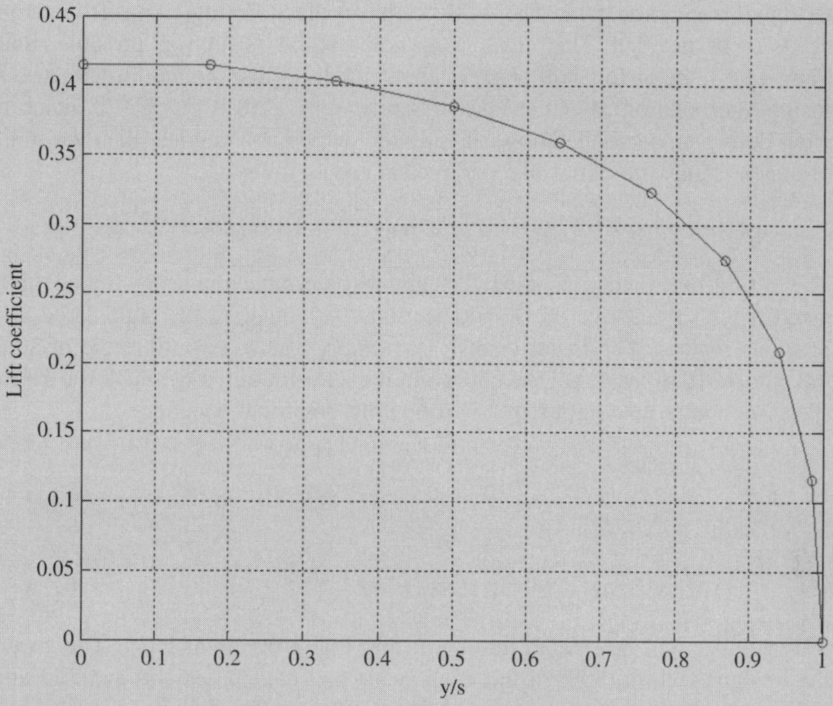

Figure 5.64 The lift distribution of the wing (AR = 7, $\lambda = 0.8$, $\alpha_t = -1.5$, $i_w = 1.86$ deg)

6. **Flap parameters**. A flap is usually employed during take-off and landing operations. We design the flap based on the take-off requirements and adjust it for the landing requirements. The take-off speed for a GA aircraft is about 20% faster than the stall speed:

$$V_{TO} = 1.2 \cdot V_S = 1.2 \cdot 60 = 72 \text{ knot} = 37 \text{ m/s} \tag{5.38}$$

Hence the wing, while the flap is deflected, must generate the following lift coefficient during take-off:

$$C_{L_{TO}} = \frac{2W_{TO}}{\rho_o V_{TO}^2 S} = \frac{2 \cdot 1800 \cdot 9.81}{1.225 \cdot (37)^2 \cdot 18.1} = 1.161 \tag{5.46}$$

As the problem statement indicates, the wing employs a split flap. We need to determine the flap chord, flap span, and flap deflection during take-off and landing. The flap chord is tentatively set to be 20% of the wing chord. The flap span is tentatively set to be 60% of the wing span. This leaves about 40% of the wing span for an aileron in future design applications. The flap deflection for take-off operation is tentatively set to be 13 deg. The reasons for these three selections can be found in Section 5.12. The wing angle of attack during take-off operation also needs to be decided. This angle is assumed to be as high as possible. Based on Figure 5.58, the airfoil stall angle is about 12 deg when the flap is deflected 13 deg (using interpolation). For the sake of safety, only a 10 deg angle of attack for the wing during take-off operation is employed, which is 2 deg less than the stall angle of attack. Thus, the initial flap parameters are as follows:

$$b_f/b = 0.6;\ C_f/C = 0.2,\ \alpha_{TO_{wing}} = 10\ \text{deg},\ \delta_f = 13\ \text{deg}$$

The lifting-line theory is utilized again to determine the wing lift coefficient at take-off with the above HLD specifications. A similar m-file is prepared as in the previous section. The major change is to apply a new zero-lift angle of attack for the inboard (flap) section. The change in the zero-lift angle of attack for the inboard (flap) section is determined by the following empirical equation:

$$\Delta\alpha_{o_{flap}} \approx -1.15 \cdot \frac{C_f}{C}\delta_f \tag{5.39}$$

or:

$$\Delta\alpha_{o_{flap}} \approx -1.15 \cdot 0.2 \cdot 13 = -2.99 \approx -3\ \text{deg} \tag{5.39}$$

This number will be entered into the lifting line program as input. This means that the inboard section (60% of the wing span) will have a zero-lift angle of attack of −6 (i.e., (−3) + (−3) = −6) due to flap deflection. The following is the MATLAB m-file to calculate the wing lift coefficient while the flap is deflected down during the take-off operation:

```
clc
clear
N = 9; % (number of segments-1)
S = 18.1; % m^2
AR = 7; % Aspect ratio
lambda = 0.8; % Taper ratio
alpha_twist = -1.5; % Twist angle (deg)
i_w = 10; % wing setting angle (deg)
a_2d = 6.3; % lift curve slope (1/rad)
a_0 = -3; % flap up zero-lift angle of attack (deg)
a_0_fd = -6; % flap down zero-lift angle of attack (deg)
b = sqrt(AR*S); % wing span
bf_b=0.6; flap-to-wing span ratio
MAC = S/b; % Mean Aerodynamic Chord
Croot = (1.5*(1+lambda)*MAC)/(1+lambda+lambda^2); % root chord
theta = pi/(2*N):pi/(2*N):pi/2;
alpha=i_w+alpha_twist:-alpha_twist/(N-1):i_w;
    % segment's angle of attack
for i=1:N
if (i/N)>(1-bf_b)
        alpha_0(i)=a_0_fd; %flap down zero lift AOA
```

```
else
      alpha_0(i)=a_0; %flap up zero lift AOA
 end
end
z = (b/2)*cos(theta);
c = Croot * (1 - (1-lambda)*cos(theta)); % MAC at each segment
mu = c * a_2d / (4 * b);
LHS = mu .* (alpha-alpha_0)/57.3; % Left Hand Side
% Solving N equations to find coefficients A(i):
for i=1:N
    for j=1:N
    B(i,j) =  sin((2*j-1) * theta(i)) * (1 + (mu(i) *
      (2*j-1)) / sin(theta(i)));
    end
end
A=B\transpose(LHS);
for i = 1:N
    sum1(i) = 0;
    sum2(i) = 0;
    for j = 1 : N
        sum1(i) = sum1(i) + (2*j-1) * A(j)
           *sin((2*j-1)*theta(i));
        sum2(i) = sum2(i) + A(j)*sin((2*j-1)*theta(i));
    end
   end
CL_TO = pi * AR * A(1)
```

In take-off, the lift distribution is not a concern, since the flap increases the wing inboard lift coefficient. The m-file yields the following results:

$$C_{L_{\mathrm{TO}}} = 1.254$$

Since the wing-generated take-off lift coefficient is slightly higher than the required take-off lift coefficient, one or more of the wing or flap parameters must be changed. The easiest change is to reduce the wing angle of attack during take-off. Other options are to reduce the size of the flap and reduce the flap deflection. By trial and error, it is determined that by reducing the wing angle of attack to 8.88 deg, the wing will generate the required lift coefficient of 1.16:

$$C_{L_{\mathrm{TO}}} = 1.16$$

Since the wing has a setting angle of 1.86 deg, the fuselage will be pitched up 7 deg during take-off, since 8.88 – 1.86 = 7.02. Thus:

$$i_{\mathrm{w}} = 1.86\ \mathrm{deg},\ \alpha_{\mathrm{TO_{wing}}} = 8.88\ \mathrm{deg},\ \alpha_{\mathrm{TO_{fus}}} = 7.02\ \mathrm{deg},\ \delta_{\mathrm{f_{TO}}} = 13\,\mathrm{deg}$$

At this moment, it is noted that the wing satisfies the design requirements both at cruise and at take-off.

7. **Other wing parameters**. To determine the other wing parameters (i.e., wing span (b), root chord (C_r), tip chord (C_t), and MAC), we have to solve the following four equations simultaneously:

$$S = b \cdot \overline{C} \tag{5.18}$$

$$\text{AR} = \frac{b^2}{S} \tag{5.17}$$

$$\lambda = \frac{C_t}{C_r} \tag{5.24}$$

$$\overline{C} = \frac{2}{3} C_r \left(\frac{1 + \lambda + \lambda^2}{1 + \lambda} \right) \tag{5.26}$$

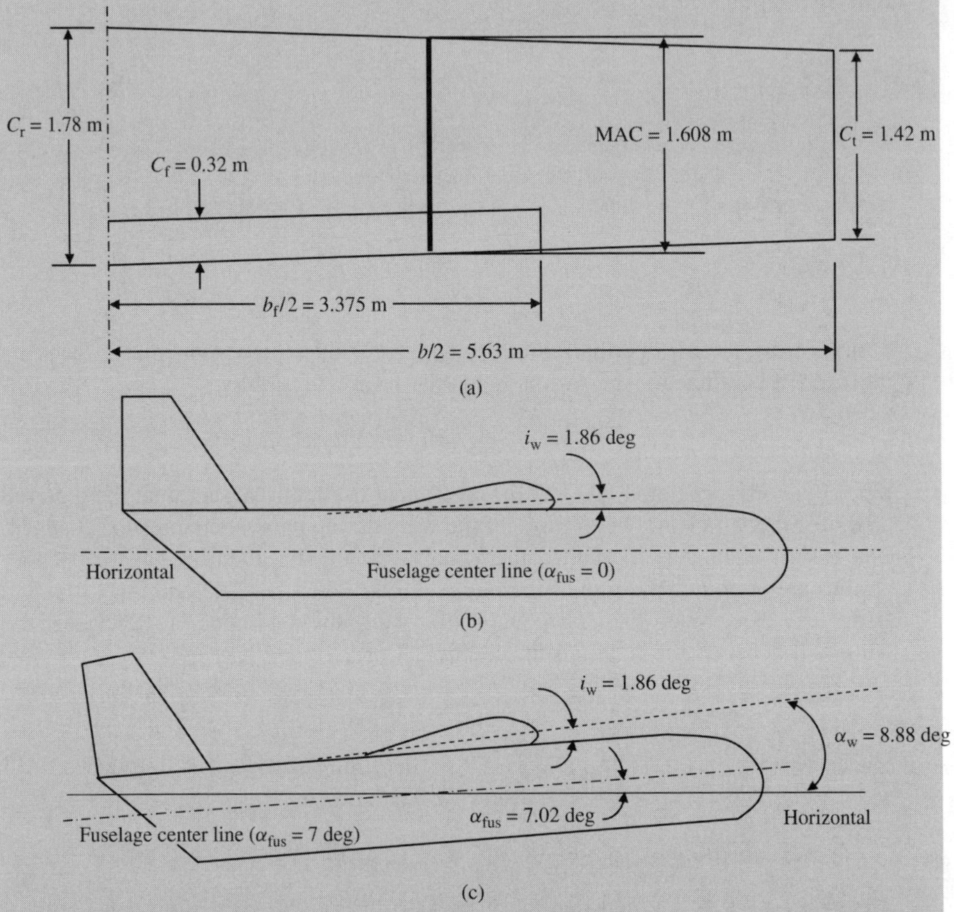

Figure 5.65 Wing parameters of Example 5.6: (a) Top view of the right half wing; (b) Side view of the aircraft in cruising flight; and (c) Side view of the aircraft in take-off

Solution of these equations simultaneously yields the following results:

$$b = 11.256\,\text{m},\quad \text{MAC} = 1.608\,\text{m},\quad C_r = 1.78\,\text{m},\quad C_t = 1.42\,\text{m}$$

Consequently, the other flap parameters are determined as follows:

$$\frac{b_f}{b} = 0.6 \Rightarrow b_f = 0.6 \cdot 11.256 = 6.75\ \text{m}$$

$$\frac{C_f}{C} = 0.2 \Rightarrow C_f = 0.2 \cdot 1.608 = 0.32\ \text{m}$$

Figure 5.65 illustrates the right half wing with the wing and flap parameters of Example 5.6.

The next step in the wing design process is to optimize the wing parameters such that the wing drag and pitching moment are minimized. This step is not shown in this example to reduce the length of the chapter.

Problems

1. Identify C_{l_i}, $C_{d_{min}}$, C_m, $(C_l/C_d)_{max}$, $\alpha_o\,(\text{deg})$, $\alpha_s\,(\text{deg})$, $C_{l_{max}}$, $a_o\,(1/\text{rad})$, and $(t/c)_{max}$ of the NACA 2415 airfoil section (flap up). You need to indicate the locations of all parameters on the airfoil graphs as shown in Figure 5.66.

2. Identify C_{l_i}, $C_{d_{min}}$, C_m, $(C_l/C_d)_{max}$, $\alpha_o\,(\text{deg})$, $\alpha_s\,(\text{deg})$, $C_{l_{max}}$, $a_o\,(1/\text{rad})$, and $(t/c)_{max}$ of the NACA 63_2-615 airfoil section (flap up). You need to indicate the locations of all parameters on the airfoil graphs as shown in Figure 5.21.

3. A NACA airfoil has thickness-to-chord ratio of 18%. Estimate the lift curve slope for this airfoil in 1/rad.

4. Select a NACA airfoil section for the wing for a prop-driven normal category GA aircraft with the following characteristics:

$$m_{TO} = 3500\,\text{kg},\ S = 26\,\text{m}^2,\ V_c = 220\,\text{knot (at 4000 m)},$$
$$V_s = 68\,\text{knot (at sea level)}$$

The high-lift device (plain flap) will provide $\Delta C_L = 0.4$ when deflected.

5. Select a NACA airfoil section for the wing for a prop-driven transport aircraft with the following characteristics:

$$m_{TO} = 23\,000\,\text{kg},\ S = 56\,\text{m}^2,\ V_c = 370\,\text{knot (at 25 000 ft)},$$
$$V_s = 85\,\text{knot (at sea level)}$$

The high-lift device (single-slotted flap) will provide $\Delta C_L = 0.9$ when deflected.

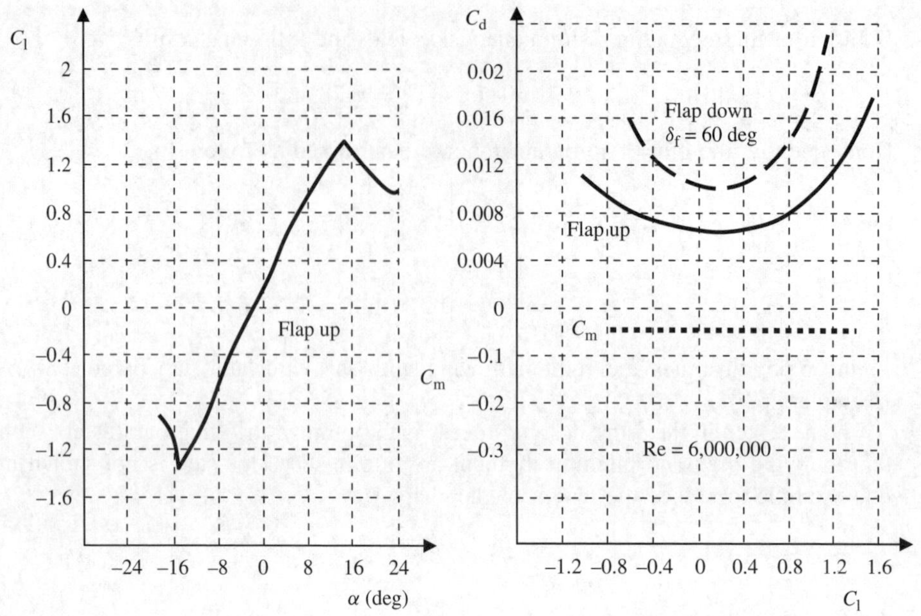

Figure 5.66 Airfoil section NACA 2415. Reproduced from permission of Dover Publications, Inc.

6. Select a NACA airfoil section for the wing for a business jet aircraft with the following characteristics:

$$m_{TO} = 4800\,\text{kg},\ S = 22.3\ \text{m}^2,\ V_c = 380\,\text{knot (at 33 000 ft)},$$
$$V_s = 81\,\text{knot (at sea level)}$$

The high-lift device (double-slotted flap) will provide $\Delta C_L = 1.1$ when deflected.

7. Select a NACA airfoil section for the wing for a jet transport aircraft with the following characteristics:

$$m_{TO} = 136\ 000\,\text{kg},\ S = 428\,\text{m}^2,\ V_c = 295\,\text{m/s (at 42 000 ft)},$$
$$V_s = 88\,\text{knot (at sea level)}$$

The high-lift device (triple-slotted flap) will provide $\Delta C_L = 1.3$ when deflected.

8. Select a NACA airfoil section for the wing for a fighter jet aircraft with the following characteristics:

$$m_{TO} = 30\ 000\,\text{kg},\ S = 47\,\text{m}^2,\ V_c = 1200\,\text{knot (at 40 000 ft)},$$
$$V_s = 95\,\text{knot (at sea level)}$$

The high-lift device (plain flap) will provide $\Delta C_L = 0.8$ when deflected.

9. A designer has selected a NACA 2415 (Figure 5.66) for an aircraft wing during a design process. Determine the wing setting angle.

10. The airfoil section of a wing with aspect ratio of 9 is NACA 2415 (Figure 5.66). Determine the wing lift curve slope in terms of 1/rad.

11. Determine the Oswald span efficiency for a wing with aspect ratio of 12 and sweep angle of 15 deg.

12. Determine the Oswald span efficiency for a wing with aspect ratio of 4.6 and sweep angle of 40 deg.

13. A straight rectangular wing has a span of 25 m and MAC of 2.5 m. If the wing is swept back by 30 deg, determine the effective span of the wing.

14. A trainer aircraft has a wing area of $S = 32\ \text{m}^2$, aspect ratio AR = 9.3, and taper ratio $\lambda = 0.48$. It is required that the 50% chord line sweep angle be zero. Determine the tip chord, root chord, mean aerodynamic chord, and span, as well as the leading edge sweep, trailing edge sweep, and quarter chord sweep angles.

15. A cargo aircraft has a wing area of $S = 256\ \text{m}^2$, aspect ratio AR = 12.4, and taper ratio $\lambda = 0.63$. It is required that the 50% chord line sweep angle be zero. Determine the tip chord, root chord, mean aerodynamic chord, and span, as well as the leading edge sweep, trailing edge sweep, and quarter chord sweep angles.

16. A jet fighter aircraft has a wing area of $S = 47\ \text{m}^2$, aspect ratio AR = 7, and taper ratio $\lambda = 0.8$. It is required that the 50% chord line sweep angle be 42 deg. Determine the tip chord, root chord, mean aerodynamic chord, span, and effective span, as well as the leading edge sweep, trailing edge sweep, and quarter chord sweep angles.

17. A business jet aircraft has a wing area of $S = 120\ \text{m}^2$, aspect ratio AR = 11.5, and taper ratio $\lambda = 0.55$. It is required that the 50% chord line sweep angle be 37 deg. Determine the tip chord, root chord, mean aerodynamic chord, span, and effective span, as well as the leading edge sweep, trailing edge sweep, and quarter chord sweep angles.

18. Sketch the wing for Problem 16.

19. Sketch the wing for Problem 17.

20. A fighter aircraft has a straight wing with a planform area of $50\ \text{m}^2$, aspect ratio of 4.2, and taper ratio of 0.6. Determine the wing span, root chord, tip chord, and mean aerodynamic chord. Then sketch the wing.

21. A hang glider has a swept wing with a planform area of $12\ \text{m}^2$, aspect ratio of 7, and taper ratio of 0.3. Determine the wing span, root chord, tip chord, and mean aerodynamic chord. Then sketch the wing, if the sweep angle is 35 deg.

22. The planform area for a cargo aircraft is $182\ \text{m}^2$. The wing has an anhedral of −8 deg; determine the effective wing planform area of the aircraft.

23. A jet transport aircraft has the following characteristics:

$$m_{TO} = 140\,000\,\text{kg},\ S = 410\,\text{m}^2,\ V_s = 118\,\text{knot (at sea level)},\ \text{AR} = 12, \lambda = 0.7$$
$$b_{A_{in}}/b = 0.7,\ i_w = 3.4\,\text{deg},\ \alpha_t = -2\,\text{deg}$$

Airfoil section: NACA 63_2-615 (Figure 5.21)

Design the high-lift device (determine type, b_f, C_f, and δ_f) for this aircraft to be able to take off with a speed of 102 knot while the fuselage is pitched up 10 deg.

24. A twin-engine GA aircraft has the following characteristics:

$$m_{TO} = 4500\,\text{kg},\ S = 24\,\text{m}^2,\ \text{AR} = 8.3,\ \lambda = 0.5,\ b_{A_{in}}/b = 0.6,$$
$$i_w = 2.8\,\text{deg},\ \alpha_t = -1\,\text{deg}$$

Airfoil section: NACA 63_2-615 (Figure 5.21)

Design the high-lift device (determine type, b_f, C_f, and δ_f) for this aircraft to be able to take off with a speed of 85 knot while the fuselage is pitched up 10 deg.

25. Determine and plot the lift distribution for a business aircraft with a wing with the following characteristics. Divide the half wing into 12 sections.

$$S = 28\,\text{m}^2,\ \text{AR} = 9.2,\ \lambda = 0.4,\ i_w = 3.5\,\text{deg},\ \alpha_t = -2\,\text{deg}$$

Airfoil section: NACA 63-209

If the aircraft is flying at an altitude of 10 000 ft with a speed of 180 knot, how much lift is produced?

26. Determine and plot the lift distribution for a cargo aircraft with a wing with the following characteristics. Divide the half wing into 12 sections.

$$S = 104\,\text{m}^2,\ \text{AR} = 11.6,\ \lambda = 0.72,\ i_w = 4.7\,\text{deg},\ \alpha_t = -1.4\,\text{deg},$$

Airfoil section: NACA 4412

If the aircraft is flying at an altitude of 25 000 ft with a speed of 250 knot, how much lift is produced?

27. Consider the aircraft in Problem 25. Determine the lift coefficient at take-off when the following high-lift device is employed:

$$\text{single-slotted flap},\ b_f/b = 0.65,\ C_f/C = 0.22,\ \delta_f = 15\,\text{deg},\ \alpha_{TO_{wing}} = 9\,\text{deg}$$

28. Consider the aircraft in Problem 26. Determine the lift coefficient at take-off when the following high-lift device is employed:

$$\text{triple-slotted flap},\ b_f/b = 0.72,\ C_f/C = 0.24,\ \delta_f = 25\,\text{deg},\ \alpha_{TO_{wing}} = 12\,\text{deg}$$

29. Consider the aircraft in Problem 28. How much flap needs to be deflected in landing, if the fuselage is allowed to pitch up only 7 deg with a speed of 95 knot?

30. Design a wing for a utility category GA aircraft with the following features:

$$S = 22\,\text{m}^2,\ m = 2100\,\text{kg},\ V_c = 152\,\text{knot (at 20 000 ft)},\ V_s = 67\,\text{knot (at sea level)}$$

The aircraft has a monoplane low wing and employs a plain flap. Determine the airfoil section, aspect ratio, taper ratio, tip chord, root, chord, MAC, span, twist angle, sweep angle, dihedral angle, incidence, high-lifting device type, flap span, flap chord, flap deflection, and wing angle of attack at take-off. Plot the lift distribution at cruise and sketch the wing including dimensions.

31. Design a wing for a jet cargo aircraft with the following features:

$$S = 415\,\text{m}^2,\ m = 150\ 000\,\text{kg},\ V_c = 520\,\text{knot (at 30 000 ft)},$$
$$V_s = 125\,\text{knot (at sea level)}$$

The aircraft has a monoplane high wing and employs a triple-slotted flap. Determine the airfoil section, aspect ratio, taper ratio, tip chord, root, chord, MAC, span, twist angle, sweep angle, dihedral angle, incidence, high-lifting device type, flap span, flap chord, flap deflection, and wing angle of attack at take-off. Plot the lift distribution at cruise and sketch the wing including dimensions.

32. Design a wing for a supersonic fighter aircraft with the following features:

$$S = 62\,\text{m}^2,\ m = 33\ 000\,\text{kg},\ V_c = 1350\,\text{knot (at 45 000 ft)},$$
$$V_s = 105\,\text{knot (at sea level)}$$

Controllability and high performance are two high priorities in this aircraft. Determine the wing vertical position, airfoil section, aspect ratio, taper ratio, tip chord, root chord, MAC, span, twist angle, sweep angle, dihedral angle, incidence, high-lifting device type, HLD span, HLD chord, HLD deflection, and wing angle of attack at take-off. Plot the lift distribution at cruise and sketch the wing including dimensions.

33. Determine and plot the lift distribution for the aircraft Cessna 304A at cruising flight. The characteristics of this aircraft are given below. Then determine the lift coefficient at cruise. $S = 17.1\,\text{m}^2$, $m_{TO} = 2717\,\text{kg}$, $V_C = 233$ knot (at 24500 ft), $\lambda = 0.7$, $AR = 7.2$, $\alpha_t = -5.9\,\text{deg}$, $i_w = 2°3'$, airfoil section: NACA 23018 (root), NACA 23015 (tip).

34. Determine and plot the lift distribution for the aircraft Scottish Aviation SA-3-120 at cruising flight. The characteristics of this aircraft are given below. Then determine the lift coefficient at cruise. $S = 12.52\,\text{m}^2$, $m_{TO} = 1066\,\text{kg}$, $V_C = 120$ knot (at 4000 ft), $\lambda = 0.6$, $AR = 8.4$, $\alpha_t = 0\,\text{deg}$, $i_w = 1.15\,\text{deg}$, airfoil section: NACA 63_2-615

35. Determine and plot the lift distribution for the aircraft Bellanca 19-25 at cruising flight. The characteristics of this aircraft are given below. Then determine the lift coefficient at cruise. $S = 16.9\,\text{m}^2$, $m_{TO} = 1860\,\text{kg}$, $V_C = 262$ knot (at 24000 ft), $\lambda = 0.7$, $AR = 6.7$, $\alpha_t = 0\,\text{deg}$, $i_w = 2\,\text{deg}$, airfoil section: NACA 63_2-215

References

[1] Blanchard, B.S. and Fabrycky, W.J. (2006) *Systems Engineering and Analysis*, 3rd edn, Prentice Hall.
[2] Eppler, R. (1990) *Airfoil Design and Data*, Springer-Verlag.
[3] Abbott, I.H. and Von Donehoff, A.F. (1959) *Theory of Wing Sections*, Dover.
[4] Anderson, J.D. (2010) *Fundamentals of Aerodynamics*, 5th edn, McGraw-Hill.
[5] Jackson, P. (1995) *Jane's All the World's Aircraft*, Jane's Information Group, various years.
[6] Anderson, J.D. (2003) *Modern Compressible Flow*, 3rd edn, McGraw-Hill.
[7] Sadraey, M. (2009) *Aircraft Performance Analysis*, VDM Verlag Dr. Müller.
[8] Anderson, J.D. (1999) *Aircraft Performance and Design*, McGraw-Hill.
[9] Stevens, B.L. and Lewis, F.L. (2003) *Aircraft Control and Simulation*, 2nd edn, Wiley-VCH Verlag GmbH.
[10] Hibbeler, R.C. (2001) *Engineering Mechanics, Dynamics*, 9th edn, Prentice Hall.
[11] Etkin, B. and Reid, L.D. (1996) *Dynamics of Flight, Stability and Control*, 3rd edn, Wiley-VCH Verlag GmbH.
[12] Cavallok, B. (1966) Subsonic Drag Estimation Methods, US Naval Air Development Center, NADC-AW-6604.
[13] Shevell, R.S. (1989) *Fundamentals of Flight*, 2nd edn, Prentice Hall.
[14] Roskam, J. (2007) *Airplane Flight Dynamics and Automatic Flight Control, Part I*, DAR Corporation.
[15] Lan, E.C.T. and Roskam, J. (2003) *Airplane Aerodynamics and Performance*, DAR Corporation.
[16] Houghton, E.L. and Carpenter, P.W. (2003) *Aerodynamics for Engineering Students*, 5th edn, Elsevier.

6

Tail Design

6.1 Introduction

As introduced in Chapter 2, the next appropriate step after wing design would be tail design. In this chapter, after describing the tail's primary functions and introducing the fundamentals that govern tail performance, the techniques and procedure to design a horizontal tail and a vertical tail will be provided. At the end of this chapter a fully solved example illustrates the implementation of the design technique.

The horizontal tail and vertical tail (i.e., tails) along with the wing are referred to as lifting surfaces. This name differentiates tails and wing from control surfaces – namely aileron, elevator, and rudder. Several design parameters associated with tails and wing – such as airfoil, planform area, and angle of attack – are similar. Thus, several tail parameters are discussed in brief. The major difference between wing design and tail design originates from the primary function of the tail that is different from the wing. The primary function of the wing is to generate the maximum amount of lift, while the tail is supposed to use a fraction of its ability to generate lift. If, at any instance of a flight mission, the tail nears its maximum angle of attack (i.e., tail stall angle) this indicates that there was a mistake in the tail design process. In some texts and references, the tail is referred to as an empennage.

The tail in a conventional aircraft often has two components, horizontal tail and vertical tail, and carries two primary functions:

1. Trim (longitudinal and directional).
2. Stability (longitudinal and directional).
   Since two conventional control surfaces (i.e., elevator and rudder) are indeed parts of the tail to implement control, it is proper to add the following item as the third function of a tail:
3. Control (longitudinal and directional).

These three functions are described in brief here; however, more details are presented in later sections. The first and primary function of a horizontal tail is longitudinal trim; also referred to as equilibrium or balance. But the first and primary function of a vertical

Aircraft Design: A Systems Engineering Approach, First Edition. Mohammad H. Sadraey.

tail is directional stability. The reason is that an aircraft is usually symmetric about the xz plane, while the pitching moment of the wing about the aircraft center of gravity must be balanced via a component.

Longitudinal trim in a conventional aircraft is applied through the horizontal tail. Several pitching moments, namely longitudinal moment of the wing's lift about the aircraft center of gravity, wing aerodynamic pitching moment, and sometimes engine thrust's longitudinal moment, need to be trimmed about the y-axis. The summation of these three moments about the aircraft center of gravity is often negative; hence the horizontal tail often generates a negative lift to counteract the moment. For this reason, the horizontal tail setting angle is often negative. Since the aircraft center of gravity is moving along the x-axis (due to fuel burn during flight duration), the horizontal tail is responsible for longitudinal trim throughout the flight time. To support the longitudinal trimability of the aircraft, conventional aircraft employ an elevator as part of their horizontal tail.

Since conventional aircraft are almost always manufactured symmetrically about the xz plane, the trim is not a major function for a vertical tail. However, in a few instances, a vertical tail has the primary function of directional trim or lateral trim. In a multi-engine aircraft, the vertical tail has great responsibility during one-engine inoperative (OEI) situations in order to maintain directional trim. The vertical tail must generate a yawing moment to balance the aircraft against the yawing moment generated by active engines. Even in single-engine prop-driven aircraft, the vertical tail has to counteract the rolling moment generated by propeller rotation. This is to maintain aircraft lateral trim and prevent an unwanted roll. In this case, the vertical tail is often installed at a few degrees relative to the xz plane. The aircraft trim requirement provides the main design requirement in the tail design process. The derivation of design requirements based on the trim will be discussed in detail in Section 6.2.

The second function of the tails is to provide stability. The horizontal tail is responsible for maintaining longitudinal stability, while the vertical tail is responsible for maintaining directional stability. Aircraft stability is defined as the tendency of an aircraft to return to the original trim conditions if diverted by a disturbance. The major disturbance source is atmospheric phenomena, such as gusts. The stability requirement must also be included in the list of tail design requirements. This topic will be discussed in detail in Section 6.3.

The third major function of the tails is *control*. The elevator as part of the horizontal tail is designed to provide longitudinal control, while the rudder as part of the vertical tail is responsible for providing directional control. Tails must be powerful enough to control the aircraft, such that the aircraft is able to change flight conditions from one trim condition (say cruise) to another new trim condition (say take-off and landing). For instance, during take-off, the tail must be able to lift up the fuselage nose in a specified pitch rate.

In general, the tail is designed based on trim requirements, but later revised based on stability and control requirements. The following are the tail parameters which need to be determined during the design process:

1. Tail configuration (horizontal tail – horizontal location with respect to the fuselage, aft tail or canard)

   (a) Horizontal tail
2. Planform area (S_h)

3. Tail arm (l_h)
4. Airfoil section
5. Aspect ratio (AR_h)
6. Taper ratio (λ_h)
7. Tip chord ($C_{h_{tip}}$)
8. Root chord ($C_{h_{root}}$)
9. Mean aerodynamic chord (MAC_h or C_h)
10. Span (b_h)
11. Sweep angle (Λ_h)
12. Dihedral angle (Γ_h)
13. Tail installation
14. Incidence (i_h)

(b) Vertical tail

15. Planform area (S_v)
16. Tail arm (l_v)
17. Airfoil section
18. Aspect ratio (AR_v)
19. Taper ratio (λ_v)
20. Tip chord (C_{t_v})
21. Root chord (C_{r_v})
22. Mean aerodynamic chord (MAC_v or C_v)
23. Span (b_v)
24. Sweep angle (Λ_v)
25. Dihedral angle (Γ_v)
26. Incidence (i_v).

All 26 tail parameters listed above must be determined in the tail design process. The majority of the parameters are finalized through technical calculations, while a few parameters are decided via an engineering selection approach. There are a few other intermediate parameters such as downwash angle, sidewash angle, and effective angle of attack that will be used to calculate some tail parameters. These are determined in the design process, but not employed in the manufacturing period.

As discussed in Chapter 2, the systems engineering approach has been adopted as the basic technique to design the tail. The tail design technique has been developed with this approach to satisfy all the design requirements while maintaining low cost in an optimum fashion. Figure 6.1 illustrates a block diagram of the tail design process. As explained in Chapter 2, the aircraft design is an iterative process; therefore, this procedure (tail design) will be repeated several times until the optimum aircraft configuration has been achieved. The design of the vertical and horizontal tails might be performed almost in parallel. However, there is one step in the vertical tail design (i.e., spin recovery) where the effect of a horizontal tail into a vertical tail is investigated. The details of each step will be introduced in later sections. The purpose of this chapter is to provide design considerations, design technique, and design examples for the preliminary design of the aircraft tail.

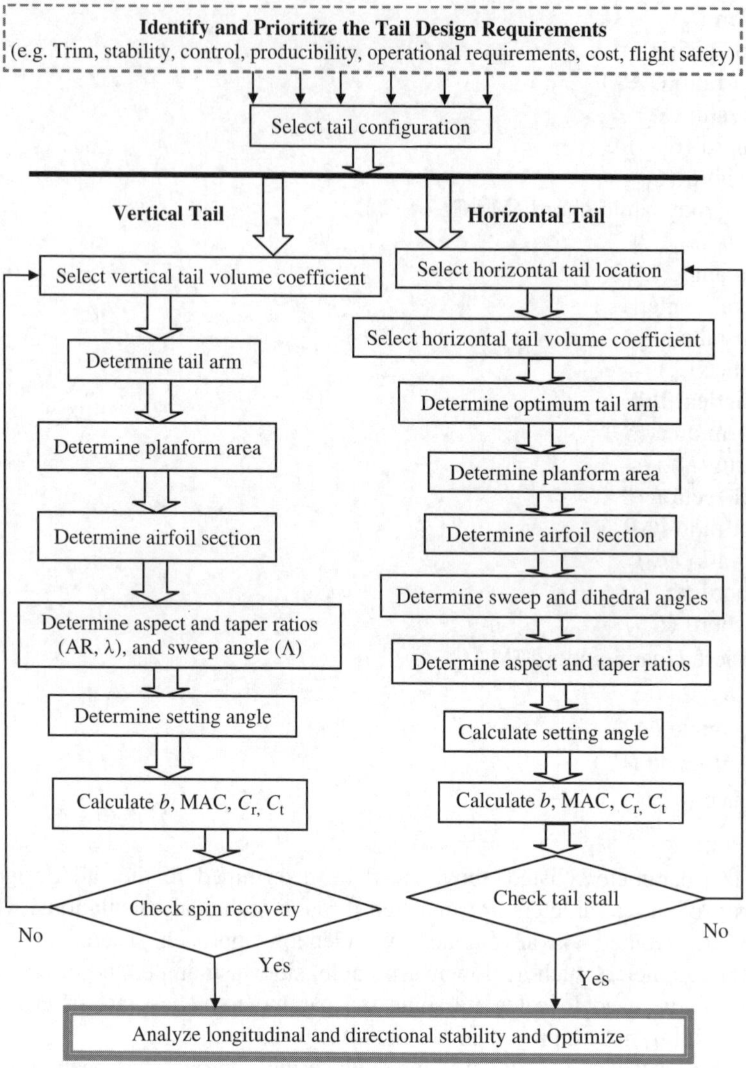

Figure 6.1 The tail design procedure

6.2 Aircraft Trim Requirements

Trim is one of the inevitable requirements of a safe flight. When an aircraft is at trim, the aircraft will not rotate about its center of gravity (cg), and the aircraft will either keep moving in a desired direction or will move in a desired circular motion. In other words, when the summations of all forces and moments are zero, the aircraft is said to be in trim:

$$\sum F = 0 \tag{6.1}$$

$$\sum M = 0 \tag{6.2}$$

The aircraft trim must be maintained about three axes (x, y, and z): (i) the lateral axis (x), (ii) the longitudinal axis (y), and (iii) the directional axis (z). When the summation of all forces in the x direction (such as drag and thrust) is zero, and the summation of all moments including aerodynamic pitching moment about the y-axis is zero, the aircraft is said to have longitudinal trim:

$$\sum F_x = 0 \tag{6.3}$$

$$\sum M_{cg} = 0 \tag{6.4}$$

The horizontal tail is responsible for maintaining longitudinal trim and making the summations zero, by generating a necessary horizontal tail lift and contributing in the summation of moments about the y-axis. A horizontal tail can be installed behind the fuselage or close to the fuselage nose. The first is called a conventional tail or aft tail, while the second is referred to as a first tail, fore plane, or canard. Equation (6.4) will be used in the horizontal tail design. When the summation of all forces in the y direction (such as side forces) is zero, and the summation of all moments including aerodynamic yawing moment about the z-axis is zero, the aircraft is said to have directional trim:

$$\sum F_y = 0 \tag{6.5}$$

$$\sum N_{cg} = 0 \tag{6.6}$$

The vertical tail is responsible for maintaining directional trim and making the summations zero, by generating a necessary vertical tail lift and contributing in the summation of moments about the y-axis. Equation (6.6) will be used in the vertical tail design. When the summation of all forces in the z direction (such as lift and weight) is zero, and the summation of all moments including aerodynamic rolling moment about the x-axis is zero, the aircraft is said to have directional trim:

$$\sum F_z = 0 \tag{6.7}$$

$$\sum L_{cg} = 0 \tag{6.8}$$

The vertical tail is responsible for maintaining directional trim and making the summation of moments zero, by generating a necessary vertical tail lift and contributing to the summation of moments about the z-axis. Equation (6.8) will also be used in the vertical tail design. More details can be found in most flight dynamics textbooks. As an example, the interested reader is referred to Refs [1–3].

A major design requirements' reference is the Federal Aviation Administration [4]. The following is reproduced from Section 161 of PAR 23 of the Federal Aviation Regulations (FAR) concerning lateral-directional and longitudinal trim of a General Aviation (GA) aircraft:

> (a) General. Each airplane must meet the trim requirements of this section after being trimmed and without further pressure upon, or movement of, the primary controls or their corresponding trim controls by the pilot or the automatic pilot. In addition, it must be possible, in other

conditions of loading, configuration, speed and power to ensure that the pilot will not be unduly fatigued or distracted by the need to apply residual control forces exceeding those for prolonged application of §23.143(c). This applies in normal operation of the airplane and, if applicable, to those conditions associated with the failure of one engine for which performance characteristics are established.

(b) Lateral and directional trim. The airplane must maintain lateral and directional trim in level flight with the landing gear and wing flaps retracted as follows:

(1) For normal, utility, and acrobatic category airplanes, at a speed of 0.9 V_H, V_C, or V_{MO}/M_O, whichever is lowest; and

(2) For commuter category airplanes, at all speeds from 1.4 V_{S1} to the lesser of V_H or V_{MO}/M_{MO}.

(c) Longitudinal trim. The airplane must maintain longitudinal trim under each of the following conditions: (1) A climb, (2) Level flight at all speeds, (3) A descent, (4) Approach

(d) In addition, each multiple airplane must maintain longitudinal and directional trim, and the lateral control force must not exceed 5 pounds at the speed used in complying with §23.67(a), (b)(2), or (c)(3).

For other types of aircraft, the reader is encouraged to refer to other parts of FAR; for instance, for transport aircraft, the reference is Part 25.

6.2.1 Longitudinal Trim

For the horizontal tail design process, we need to develop a few equations; hence, the longitudinal trim will be described in more detail. Consider the side view of a conventional aircraft (i.e., with aft tail) in Figure 6.2 that is in longitudinal trim. Figure 6.2(a) depicts the aircraft when its cg is behind the wing/fuselage aerodynamic center (ac_{wf}).[1] In Figure 6.2(b), the aircraft is depicted when its cg is forward of the wing/fuselage aerodynamic center. There are several moments about the y-axis (cg) that must be balanced by the horizontal tail's lift. Two of these are: (i) wing/fuselage aerodynamic pitching moment and (ii) the moment of lift about the aircraft center of gravity. Other sources of moments about the cg could be engine thrust, wing drag, landing gear drag, and store drag. For the sake of simplicity, these moments are not included in this figure. The reader is expected to be able to follow the discussion when other moments are present and/or the aircraft has a canard instead of an aft tail.

The wing/fuselage lift (L_{wf}) is the wing lift (L_w) when the contribution of the fuselage lift (L_f) is included. The fuselage lift is usually assumed to be about 10% of the wing lift. Reference [1] can be consulted for the exact calculation. When the cg is aft of the ac_{wf} (as in Figure 6.2(a)), this moment of the wing/fuselage lift (L_{wf}) is positive, while when the cg is forward of the ac_{wf} (as in Figure 6.2(b)), this moment of the wing/fuselage lift

[1] The wing/fuselage aerodynamic center is simply the wing aerodynamic center when the contribution of the fuselage is added. The fuselage contribution for most conventional aircraft is usually about ±5% $\overline{C}$. Since the wing aerodynamic center is often located at about quarter MAC (i.e. 25% $\overline{C}$), the wing/fuselage aerodynamic center is often located between 20% MAC and 30% MAC or $\overline{C}$. The reader is referred to Ref. [1] for more information.

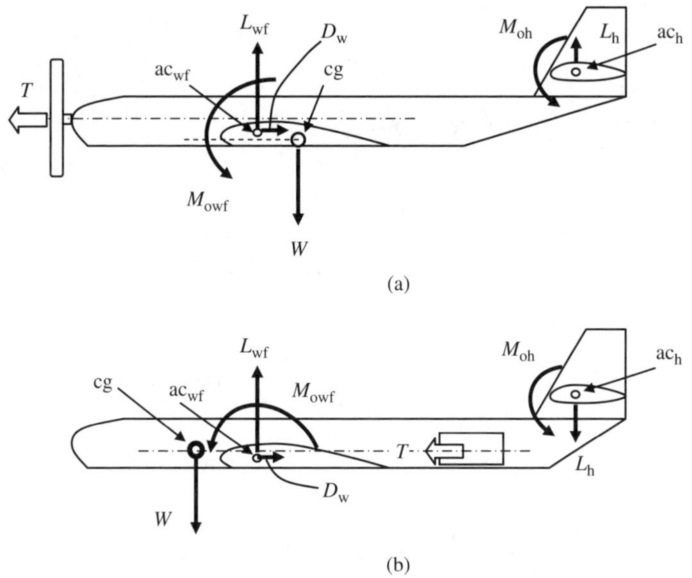

Figure 6.2 A conventional aircraft in longitudinal trim. (a) cg aft of ac_{wf}; (b) cg forward of ac_{wf}

is negative. Recall from flight dynamics that the clockwise direction is assumed to be positive, and the y-axis is located at the cg and is directed into the page.

Another moment is referred to as the wing/fuselage aerodynamic pitching moment (i.e., $M_{o_{wf}}$). The wing/fuselage aerodynamic pitching moment ($M_{o_{wf}}$) is the wing aerodynamic pitching moment (M_{o_w}) when the contribution of the fuselage (M_f) is included. The subscript "o" denotes that the aerodynamic moment is measured relative to the wing aerodynamic center. This aerodynamic moment is often negative (as sketched in Figure 6.2) so it is often called a nose-down pitching moment, due to its desire to pitch down the fuselage nose. Often, the summation of these two moments (i.e., the wing/fuselage aerodynamic pitching moment and the wing/fuselage lift generated moment) is not zero. Hence, the horizontal tail is employed to generate a lift in order to balance these moments and make the summation zero. This function maintains the aircraft longitudinal trim.

In a similar fashion, a discussion about the directional trim can be addressed. In this case, despite the symmetry of a conventional aircraft about the xz plane, there are forces such as asymmetric engine thrust (when one engine is inoperative in a multi-engine aircraft) that disturb the directional trim of an aircraft. In such a situation, the vertical tail is required to generate a lift force in the y direction (i.e., side force) to maintain the directional trim about the z-axis. The details of this case are left to the reader.

Now, consider the aircraft in Figure 6.3 where the tail aerodynamic pitching moment is neglected. Please note that in this case, the thrust line passes through the aircraft cg, so the engine thrust tends to impose no influence on the aircraft longitudinal trim. Although the wing/fuselage lift is positive in a normal flight situation, the moment of the lift about the cg might be positive or negative due to the relationship between cg and ac_{wf}. Thus,

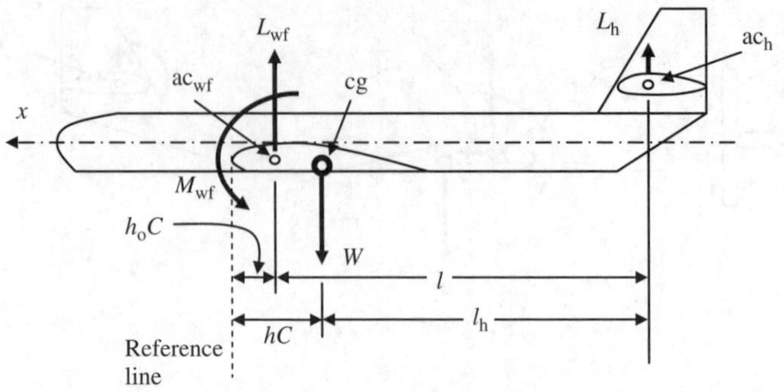

Figure 6.3 The distance between cg, ac_t, ac_{wf} and the reference line

the horizontal tail could be negative or positive. The application of the trim equation leads to the following:[2]

$$\sum M_{cg} = 0 \Rightarrow M_{o_{wf}} + M_{L_{wf}} + M_{L_h} = 0 \tag{6.9}$$

Recall that the aircraft weight generates no moment about the aircraft cg. If the engine thrust line does not pass through the aircraft cg, Equation (6.9) must be modified. To make this equation more convenient to apply, we need to non-dimensionalize it. In order to non-dimensionalize the parameters, it is often customary to measure the distances in the x direction as a factor of MAC ($\overline{C}$ or simply C). Moreover, a reference line (or point) must be selected to measure all distances with respect to. Here, we select the fuselage nose as the reference line. Hence, the distance between ac_{wf} and the reference line is h_o times $\overline{C}$ (i.e., $h_o\overline{C}$), while the distance between cg and the reference line is h times $\overline{C}$ (i.e., $h\overline{C}$). Both parameters are shown in Figure 6.3. The distance between the horizontal tail aerodynamic center and the wing/fuselage aerodynamic center is denoted l, while the distance between the horizontal tail aerodynamic center and the aircraft center of gravity is denoted l_t. Now, we can substitute the values of two moments into Equation (6.9):

$$M_{o_{wf}} + L_{wf}\left(h\overline{C} - h_o\overline{C}\right) - L_h \cdot l_h = 0 \tag{6.10}$$

To expand the equation, we need to define the variables of wing/fuselage lift (L_{wf}), horizontal tail lift (L_h), and wing/fuselage aerodynamic pitching moment (M_{owf}):

$$L_{wf} = \frac{1}{2}\rho V^2 S C_{L_{wf}} \tag{6.11}$$

$$L_h = \frac{1}{2}\rho V^2 S_h C_{L_h} \tag{6.12}$$

$$M_{o_{wf}} = \frac{1}{2}\rho V^2 S C_{m_{o\_wf}} \overline{C} \tag{6.13}$$

where $C_{L_{wf}}$ denotes the wing/fuselage lift coefficient, C_{L_h} denotes the horizontal tail lift coefficient, $C_{m_{o\_wf}}$ denotes the wing/fuselage aerodynamic pitching moment coefficient, S

[2] The horizontal tail aerodynamic pitching moment is ignored, due to its small value.

denotes the wing planform area, S_h denotes the horizontal tail planform area, V denotes the aircraft airspeed, and ρ denotes the air density.

By substituting Equations (6.11)–(6.13) into Equation (6.10), we have the following:

$$\frac{1}{2}\rho V^2 S C_{m_{o\_wf}} \overline{C} + \frac{1}{2}\rho V^2 S C_{L_{wf}} \left(h\overline{C} - h_o\overline{C}\right) - \frac{1}{2}\rho V^2 S_h C_{L_h} \cdot l_h = 0 \quad (6.14)$$

This equation is then non-dimensionalized by dividing into $\frac{1}{2}\rho V^2 S\overline{C}$. Thus, the following is obtained:

$$C_{m_{o\_wf}} + C_{L_{wf}}\left(h - h_o\right) - \frac{l_h}{\overline{C}} C_{L_h} \frac{S_h}{S} = 0 \quad (6.15)$$

Now return to Figure 6.3. The distance between the horizontal tail aerodynamic center and the reference line can be written in two ways:

$$l + h_o\overline{C} = l_h + h\overline{C} \quad (6.16)$$

or

$$\frac{l_h}{\overline{C}} = \frac{l}{\overline{C}} - \left(h_o - h\right) \quad (6.17)$$

Substituting Equation (6.17) into Equation (6.15) yields:

$$C_{m_{o\_wf}} + C_{L_{wf}}\left(h - h_o\right) - \left[\frac{l}{\overline{C}} - \left(h_o - h\right)\right] C_{L_h} \frac{S_h}{S} = 0 \quad (6.18)$$

This can be further simplified as:

$$C_{m_{o\_wf}} + \left(C_{L_{wf}} + C_{L_h}\frac{S_h}{S}\right)\left(h - h_o\right) - \frac{l}{\overline{C}}\frac{S_h}{S} C_{L_h} = 0 \quad (6.19)$$

In contrast, the aircraft total lift is the summation of the wing/fuselage lift and the horizontal tail lift:

$$L = L_{wf} + L_h \quad (6.20)$$

which leads to:

$$\frac{1}{2}\rho V^2 S C_L = \frac{1}{2}\rho V^2 S C_{L_{wf}} + \frac{1}{2}\rho V^2 S_h C_{L_h} \quad (6.21)$$

This equation is non-dimensionalized as follows:

$$C_L = C_{L_{wf}} + C_{L_h}\frac{S_h}{S} \quad (6.22)$$

Now, Equation (6.22) can be substituted into Equation (6.19):

$$C_{m_{o\_wf}} + C_L\left(h - h_o\right) - \frac{l}{\overline{C}}\frac{S_h}{S} C_{L_h} = 0 \quad (6.23)$$

The combination $\frac{l}{\overline{C}}\frac{S_h}{S}$ in Equation (6.23) is an important non-dimensional parameter in the horizontal tail design, and is referred to as the *horizontal tail volume coefficient*. The name originates from the fact that both the numerator and denominator have the unit of volume (e.g., m^3). The numerator is a function of the horizontal tail parameters, while the denominator is a function of the wing parameters. Thus, the parameter is the ratio of the horizontal tail geometries to the wing geometries. It is denoted by the symbol $\overline{V}_{\mathrm{H}}$:

$$\overline{V}_{\mathrm{H}} = \frac{l}{\overline{C}}\frac{S_{\mathrm{h}}}{S} \tag{6.24}$$

Thus, Equation (6.23) is further simplified as follows:

$$C_{\mathrm{m_{o\_wf}}} + C_L\left(h - h_o\right) - \overline{V}_{\mathrm{H}} C_{L_{\mathrm{h}}} = 0 \tag{6.25}$$

This non-dimensional longitudinal trim equation provides a critical tool in the design of the horizontal tail. The importance of this equation will be explained later, and its application will be described in later sections of the chapter. This non-dimensional parameter $\overline{V}_{\mathrm{H}}$ has a limited range of values and also is not a function of the aircraft size or weight. From a small aircraft such as Cessna 172 (Figure 11.15) to a large jumbo jet aircraft such as Boeing 747 (Figures 3.7, 3.12, and 9.4), all have similar tail volume coefficient. Table 6.1 illustrates the tail volume coefficients for several aircraft.

Table 6.4 shows typical values of tail volume coefficient for several aircraft types. The tail volume coefficient is an indication of the handling quality in longitudinal stability and longitudinal control. As $\overline{V}_{\mathrm{H}}$ increases, the aircraft tends to be more longitudinally stable and less longitudinally controllable. Fighter aircraft that are highly maneuverable tend to have a very low tail volume coefficient, namely about 0.2. In contrast, jet transport aircraft which must be highly safe and stable tend to have a high tail volume coefficient, namely about 1.1. This parameter is a crucial variable in horizontal tail design and must

Table 6.1 Tail volume coefficients of several aircraft [5]

No.	Aircraft	Type	Mass (kg)	Wing area (m^2)	Overall length (m)	$\overline{V}_{\mathrm{H}}$
1	Cessna 172	Light GA (piston)	1 100	16.2	7.9	0.76
2	Piper PA-46-350P	Light transport (piston)	1 950	16.26	8.72	0.66
3	Alenia G222	Turboprop transport	28 000	82	22.7	0.85
4	Fokker 100	Jet transport	44 000	93.5	35.5	1.07
5	Lake LA-250	Amphibian	1 424	15.24	9.04	0.8
6	Boeing 747-400	Jet transport	362 000	541	73.6	0.81
7	Airbus 340-200	Jet transport	257 000	363.1	59.39	1.11
8	Pilatus PC-12	Turboprop transport	4 100	25.81	14.4	1.08
9	Eurofighter 2000	Fighter	21 000	50	15.96	0.063
10	F/A-18 Hornet	Fighter	29 937	46.45	18.31	0.49

be selected at the early stages of tail plane design. Although the primary function of the horizontal tail is longitudinal stability, the tail volume coefficient serves as a significant parameter both in longitudinal stability and longitudinal trim issues.

The wing/fuselage pitching moment coefficient ($C_{m_{o\_wf}}$) in Equation (6.25) can be estimated via the following equation [6]:

$$C_{m_{o\_wf}} = C_{m_{af}} \frac{AR \cos^2(\Lambda)}{AR + 2\cos(\Lambda)} + 0.01\alpha_t \quad (6.26)$$

where $C_{m_{af}}$ is the wing airfoil section pitching moment coefficient, AR is the wing aspect ratio, Λ is the wing sweep angle, and α_t is the wing twist angle (in degrees). Please note that α_t is often a negative number. The value of $C_{m_{o\_wf}}$ can be determined using airfoil graphs, an example of which is shown in Figure 5.21 for the NACA 63_2-615 airfoil section. For instance, the value of $C_{m_{af}}$ for this airfoil is -0.11.

The parameter C_L in Equation (6.25) is the aircraft cruise lift coefficient that is determined by the following equation:

$$C_L = \frac{2W_{avg}}{\rho V_c^2 S} \quad (6.27)$$

where V_c is the cruising speed and W_{avg} is the average aircraft weight during the cruising flight. If the wing has been designed prior to the design of the horizontal tail, and the aircraft center of gravity (h) was decided, Equation (5.25) has only two unknowns, namely C_{L_h} and $\overline{V}_H$. However, in practice, the design of the wing and the location of the cg are not independent of the tail design. Hence, this is an ideal case, and the tail design is indeed an iterative process. The longitudinal trim equation (i.e., Equation (5.26)) must be valid in every possible flight condition. This includes all aircraft allowable load weights, all feasible flight speeds, all aircraft designated configurations (e.g., flap and landing gear, up and down), all allowable cg locations, and all possible flight altitudes. These various flight possibilities can be summarized to be between the following two extreme critical conditions:

1. The first unknown flight condition at which the horizontal tail is required to generate the greatest positive pitching moment about the aircraft cg.
2. The second unknown flight condition at which the horizontal tail is required to generate the greatest negative pitching moment about the aircraft cg.

These two critical flight conditions for the horizontal tail are unknown at this moment, but will be clear later on in the design process. The change in the sign of the tail pitching moment about the aircraft cg indicates the necessity of a change in the tail lift coefficient from positive to negative. Two possible solutions are:

1. The application of a moving horizontal tail.
2. The application of a fixed horizontal tail, plus a control surface (i.e., elevator).

In the early stage of horizontal tail design, the design is performed without considering the elevator. The criterion is to design a horizontal tail to satisfy the cruising flight longitudinal trim requirements. The reason is that the aircraft spends the majority of its flight mission time in cruising flight.

Due to the effect of wing and fuselage on the horizontal tail (i.e., downwash and sidewash), a new parameter is added to Equation (6.25). This new parameter is the ratio between the dynamic pressure at the tail and the aircraft dynamic pressure, and is called the tail efficiency (η_h) and defined as follows:

$$\eta_h = \frac{q_t}{q} = \frac{0.5\rho V_h^2}{0.5\rho V^2} = \left(\frac{V_h}{V}\right)^2 \tag{6.28}$$

where V is the aircraft airspeed and V_h is the effective airspeed at the horizontal tail region. The typical value of the tail efficiency for an aircraft with a conventional tail varies from 0.85 to 0.95. For an aircraft with a T-tail, the tail efficiency can be considered to be 1, which means the wing and fuselage have no effect on the tail dynamic pressure. The horizontal tail of a T-tail is usually out of the region of wing wake and downwash during cruising flight. Applying the tail efficiency in Equation (6.25) yields a revised version:

$$C_{m_{o\_wf}} + C_L\left(h - h_o\right) - \eta_h \overline{V}_H C_{L_h} = 0 \tag{6.29}$$

This is the most important equation in the design of a horizontal tail and implies the requirements for the longitudinal trim. It will be used in both conventional aft tail and canard configurations. The equation is derived in this section, but its application technique will be presented in Sections 6.6 and 6.8. One of the four parameters in the tail volume coefficient is the distance from the wing aerodynamic center to the horizontal tail aerodynamic center (l). This distance has a statistical relationship with the aircraft overall length (L). The ratio between the distance from the wing aerodynamic center to the horizontal tail aerodynamic center and the aircraft overall length is illustrated in Table 6.2 for several aircraft configurations. It may be employed in the early stage of horizontal tail design as a starting point. The value will be revised and optimized in the later design steps when more data are available.

6.2.2 Directional and Lateral Trim

One of the primary functions for the vertical tail is directional trim. Moreover, the vertical tail tends to have a considerable contribution in the aircraft lateral trim. In this section, the role of the vertical tail in the aircraft directional and lateral trim is examined. Two aircraft are illustrated in Figure 6.4, one in directional trim and another in lateral trim.

Table 6.2 Typical values of l/L for various aircraft configurations

No.	Aircraft configuration/type	l/L
1	An aircraft whose engine is installed at the nose and has an aft tail	0.6
2	An aircraft whose engine(s) are installed above the wing and has an aft tail	0.55
3	An aircraft whose engine is installed at the aft fuselage and has an aft tail	0.45
4	An aircraft whose engine is installed under the wing and has an aft tail	0.5
5	Glider (with an aft tail)	0.65
6	Canard aircraft	0.4
7	An aircraft whose engine is inside the fuselage (e.g., fighter) and has an aft tail	0.3

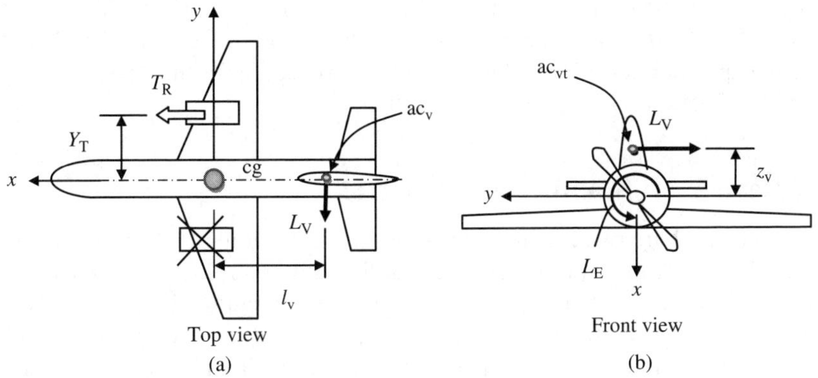

Figure 6.4 Vertical tail role in the aircraft lateral and directional trim. (a) One engine inoperative (directional trim); (b) Single propeller engine (lateral trim)

In Figure 6.4(a) the top view of an aircraft is shown where the vertical tail is generating a yawing moment to nullify the yawing moment created by asymmetric thrust of the right engine. In addition, in Figure 6.4(b) the front view of an aircraft is shown where the vertical tail is generating a rolling moment to nullify the rolling moment created by the rotation of the propeller of the engine. In both cases, the primary production of the vertical tail is an aerodynamic lift in the direction of the y-axis.

When an aircraft is in directional trim, the summation of all moments about the z-axis must be zero.

$$\sum N_{cg} = 0 \tag{6.6}$$

When an aircraft is in lateral trim, the summation of all moments about the x-axis must be zero:

$$\sum L_{cg} = 0 \tag{6.8}$$

In maintaining the directional and lateral trim, an aerodynamic force along the y-axis (lift, L_v) needs to be created by the vertical tail. Thus, the directional and lateral trim equations are:

$$\sum N_{cg} = 0 \Rightarrow T_R Y_T + L_v l_{vt} = 0 \tag{6.30}$$

$$\sum L_{cg} = 0 \Rightarrow L_E + L_v z_v = 0 \tag{6.31}$$

where T_R denotes the right engine thrust, Y_T is the distance between the thrust line and the aircraft cg in the xy plane, l_v is the distance between the vertical tail aerodynamic center and the aircraft cg, L_E is the yawing moment generated by the prop rotation, and z_v denotes the distance between the vertical tail aerodynamic center and the aircraft cg in the yz plane. The vertical tail lift is obtained from:

$$L_v = \frac{1}{2}\rho V^2 S_v C_{L_v} \tag{6.32}$$

where S_v is the vertical tail area and C_{L_v} is the vertical tail lift coefficient. The four unknowns of S_v, C_{L_v}, l_v, and z_v form the basis of the design of the vertical tail. Section 6.8 examines the application of this technique and the procedure for the design of the vertical tail to satisfy directional and lateral trim requirements.

6.3 A Review on Stability and Control

Stability and control are two requirements of a safe flight. Both the horizontal tail and the vertical tail have a strong role in aircraft stability and control. Although the horizontal tail and vertical tail are initially designed to satisfy the longitudinal and directional trim requirements, in the later stages of design the longitudinal and directional stability and control requirements must also be implemented. Thus, the initial design of the horizontal tail and vertical tail will be revised to make sure that longitudinal and directional stability and control requirements have been satisfied. In this section, a brief introduction to aircraft stability and control will be provided. This will pave and clarify the way to the design of the horizontal tail and vertical tail. Due to the stability requirements for the tail, the horizontal tail is sometimes referred to as a horizontal stabilizer and the vertical tail as a vertical stabilizer.

6.3.1 Stability

The second function of the tail is stability, and the third function of the tail is control. Due to this role, the tail is sometimes referred to as a stabilizer or stabilator. Stability is defined as the tendency of an aircraft to oppose a disturbance (e.g., gust) and return to its initial steady-state trim condition if disturbed. Stability is often divided into two branches:

1. static stability;
2. dynamic stability.

Static stability is defined as the initial tendency of an aircraft, without pilot assistance, to develop forces and/or moments which oppose an instantaneous perturbation of a motion variable from a steady-state flight condition. Dynamic stability is defined as the tendency of an aircraft, without pilot assistance, to return to the initial steady-state trim condition after a disturbance disturbs the trim values. Dynamic stability concerns the entire history of the motion, in particular the rate at which the motion damps out. As a general rule, an aircraft must have some form of dynamic stability even though certain mild disabilities can be tolerated under certain conditions. When an aircraft is dynamically stable, it definitely has static stability. However, if an aircraft is statically stable, there is no guarantee that it has dynamic stability.

An aircraft motion (flight) has six degrees of freedom (6 DOF), due to two types of freedom (one linear and one rotational) about each three axes of x, y, and z. Therefore, stability is measured about these three axes:

1. lateral stability;
2. longitudinal stability;
3. directional stability.

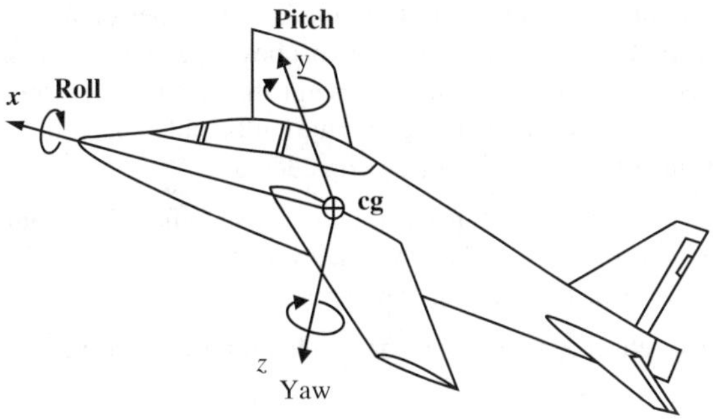

Figure 6.5 Body coordinate system and three rotational motions of roll, pitch, and yaw

Lateral stability is defined as the stability of any rotational motion about the x-axis (i.e., roll) and any corresponding linear motion along the yz plane (i.e., side motion). Longitudinal stability is defined as the stability of any rotational motion about the y-axis (i.e., pitch) and any linear motion along the xz plane (i.e., forward and aft, up and down). Directional stability is defined as the stability of any rotational motion about the z-axis (e.g., yaw) and any corresponding linear motion along the xy plane (e.g., sideslip). Figure 6.5 provides an aircraft body coordinate system, plus three rotational motions of roll, pitch, and yaw. The convention is that the clockwise rotation about any axis, when you look from the pilot's seat, is assumed as positive rotation.

The requirements for aircraft static and dynamic stability (longitudinal, lateral, and directional) are different. When the aircraft derivative C_{m_α} is negative, the aircraft is said to be statically longitudinally stable. An aircraft is said to be statically laterally stable when the aircraft derivative C_{l_β} (known as the dihedral effect) is negative. When the aircraft derivative C_{n_β} is positive, the aircraft is said to be statically directionally stable. For an aircraft to be dynamically longitudinally stable, both short-period and long-period (phugoid) modes must be damped (damping ratio greater than zero). When all modes and oscillations (including dutch-roll, spiral, and roll) are damped, an aircraft is said to be dynamically laterally directionally stable. Some dynamic longitudinal, lateral, and directional stabilities are tabulated in Section 12.3.

Among major aircraft components, the horizontal tail has the largest contribution to the aircraft longitudinal stability. The reason is that the horizontal tail is able to generate the counter-pitching moment in order to restore the longitudinal trim position. In contrast, the vertical tail has the largest contribution to the aircraft directional stability. The vertical tail is able to generate the counter-yawing moment in order to restore the directional trim position. Both the horizontal tail and the vertical tail make a significant contribution to the aircraft lateral stability, since both are capable of generating counter-rolling moments in order to restore the lateral trim position. Since the chapter is concerned with tail design, only longitudinal and directional stability requirements are emphasized.

The following is reproduced from Section 173 of PAR 23 of FAR [4], which concerns static longitudinal stability of a GA aircraft:

> Under the conditions specified in §23.175 and with the airplane trimmed as indicated, the characteristics of the elevator control forces and the friction within the control system must be as follows:

(a) A pull must be required to obtain and maintain speeds below the specified trim speed and a push required to obtain and maintain speeds above the specified trim speed. This must be shown at any speed that can be obtained, except that speeds requiring a control force in excess of 40 pounds or speeds above the maximum allowable speed or below the minimum speed for steady unstalled flight need not be considered.

(b) The airspeed must return to within the tolerances specified for applicable categories of airplanes when the control force is slowly released at any speed within the speed range specified in paragraph (a) of this section. The applicable tolerances are

(1) The airspeed must return to within plus or minus 10 percent of the original trim airspeed; and

(2) For commuter category airplanes, the airspeed must return to within plus or minus 7.5 percent of the original trim airspeed for the cruising condition specified in §23.175(b).

(c) The stick force must vary with speed so that any substantial speed change results in a stick force clearly perceptible to the pilot.

The following is reproduced from Section 177 of PAR 23 of FAR [4], which concerns static directional stability of a GA aircraft:

(a) The static directional stability, as shown by the tendency to recover from a wings level sideslip with the rudder free, must be positive for any landing gear and flap position appropriate to the takeoff, climb, cruise, approach, and landing configurations. This must be shown with symmetrical power up to maximum continuous power, and at speeds from 1.2 V_{S1} up to the maximum allowable speed for the condition being investigated. The angel of sideslip for these tests must be appropriate to the type of airplane. At larger angles of sideslip, up to that at which full rudder is used or a control force limit in §23.143 is reached, whichever occurs first, and at speeds from 1.2 V_{S1} to V_O, the rudder pedal force must not reverse.

(b) The static lateral stability, as shown by the tendency to raise the low wing in a sideslip, must be positive for all landing gear and flap positions. This must be shown with symmetrical power up to 75 percent of maximum continuous power at speeds above 1.2 V_{S1} in the take-off configuration(s) and at speeds above 1.3 V_{S1} in other configurations, up to the maximum allowable speed for the configuration being investigated, in the takeoff, climb, cruise, and approach configurations. For the landing configuration, the power must be that necessary to maintain a 3 degree angle of descent in coordinated flight. The static lateral stability must not be negative at 1.2 V_{S1} in the takeoff configuration, or at 1.3 V_{S1} in other configurations. The angle of sideslip for these tests must be appropriate to the type of airplane, but in no case may the constant heading sideslip angle be less than that obtainable with a 10 degree bank, or if less, the maximum bank angle obtainable with full rudder deflection or 150 pound rudder force.

The following is reproduced from Section 181 of PAR 23 of FAR [4], which concerns dynamic lateral/directional/longitudinal stability of a GA aircraft:

(a) Any short period oscillation not including combined lateral-directional oscillations occurring between the stalling speed and the maximum allowable speed appropriate to the configuration of the airplane must be heavily damped with the primary controls

(1) Free; and

(2) In a fixed position.

(b) Any combined lateral-directional oscillations ("Dutch roll") occurring between the stalling speed and the maximum allowable speed appropriate to the configuration of the airplane must be damped to 1/10 amplitude in 7 cycles with the primary controls

(1) Free; and

(2) In a fixed position.

(c) If it is determined that the function of a stability augmentation system, reference §23.672, is needed to meet the flight characteristic requirements of this part, the primary control requirements of paragraphs (a)(2) and (b)(2) of this section are not applicable to the tests needed to verify the acceptability of that system.

(d) During the conditions as specified in §23.175, when the longitudinal control force required to maintain speeds differing from the trim speed by at least plus and minus 15 percent is suddenly released, the response of the airplane must not exhibit any dangerous characteristics nor be excessive in relation to the magnitude of the control force released. Any long-period oscillation of flight path, phugoid oscillation, that results must not be so unstable as to increase the pilot's workload or otherwise endanger the airplane.

Since the longitudinal stability is concerned with a motion in pitch, the pertinent dynamic characteristic is the variation of the pitching moment with respect to the angle of attack (α). Thus, the primary stability derivative that determines the static longitudinal stability is C_{m_α}. Moreover, the primary stability derivative that influences the dynamic longitudinal stability is C_{m_q}. The derivative C_{m_α} is the rate of change of the pitching moment coefficient (C_m) with respect to change in the angle of attack (α). The derivative C_{m_q} is the rate of change of pitching moment coefficient (C_m) with respect to the change in pitch rate (q):

$$C_{\mathrm{m}_\alpha} = \frac{\partial C_\mathrm{m}}{\partial \alpha} \tag{6.33}$$

$$C_{\mathrm{m}_q} = \frac{\partial C_\mathrm{m}}{\partial q} \tag{6.34}$$

These two stability derivatives are most influential in the design of a horizontal tail. A statically longitudinally stable aircraft requires C_{m_α} to be negative. The typical value for most aircraft is about -0.3 to -1.5 1/rad. A dynamically longitudinally stable aircraft requires that the real parts of the roots of the longitudinal characteristic equation be negative. One of the major contributors to this requirement is C_{m_q}, such that a negative value has a strong stabilizing impact. The typical value of C_{m_q} for most aircraft is about -5 to -30 1/rad.

It is interesting to note that the horizontal tail volume coefficient ($\overline{V}_\mathrm{H}$) is the most important parameter affecting both C_{m_α} and C_{m_q}. Figure 6.6(a) provides a graphical representation of the stability derivative C_{m_α}. Details of the technique to determine the derivatives C_{m_α} and C_{m_q} are available in Ref. [6]. Another very important parameter that can be employed to determine the aircraft longitudinal static stability is the aircraft neutral point. Some textbooks refer to this point as the aircraft aerodynamic center (ac_A). If the aircraft neutral point is behind the aircraft center of gravity, the aircraft is said to have longitudinal static stability. In this situation, the static margin (i.e., the non-dimensional distance between the aircraft neutral point and the aircraft cg) is said to be positive. Details of the technique to determine the aircraft neutral point and static margin may be found in Refs [1] and [6].

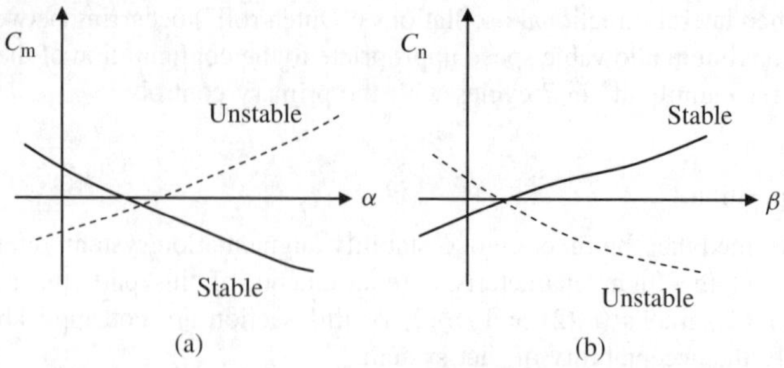

Figure 6.6 Graphical representations of derivatives C_{m_α} and C_{n_β}. (a) C_m versus α curve; (b) C_n versus β curve

The directional stability is mainly concerned with motion in yaw, so the pertinent dynamic characteristic is the variation of the yawing moment with respect to the sideslip angle (β). Thus, the primary stability derivative that determines the static directional stability is C_{n_β}. Moreover, the primary stability derivative that influences the dynamic directional stability is C_{n_r}. The derivative C_{n_β} is the rate of change of yawing moment coefficient (C_n) with respect to change in the sideslip angle (β). The derivative C_{n_r} is the rate of change of yawing moment coefficient (C_n) with respect to change in the yaw rate (r):

$$C_{n_\beta} = \frac{\partial C_n}{\partial \beta} \tag{6.35}$$

$$C_{n_r} = \frac{\partial C_n}{\partial r} \tag{6.36}$$

These two stability derivatives are most influential in the design of a vertical tail. A statically directionally stable aircraft requires C_{n_β} to be positive. The typical value for most aircraft is about +0.1 to +0.4 1/rad. A dynamically directionally stable aircraft requires that the real parts of the roots of the lateral-directional characteristic equation be negative. One of the major contributors to this requirement is C_{n_r}, such that a negative value has a strong stabilizing impact. The typical value for most aircraft is about −0.1 to −1 1/rad. These two derivatives are among the influential parameters in the design of a vertical tail. Table 6.3 summarizes the requirements for static and dynamic longitudinal and directional stability. Figure 6.6(b) provides a graphical representation of the stability derivative C_{n_β}. The technique to determine the derivatives C_{n_β} and C_{n_r} is available in Ref. [6].

Almost all GA and transport aircraft are longitudinally and directionally stable. For military aircraft only the advanced fighters are an exception, which means fighters are the only military aircraft that may not be longitudinally and/or directionally stable. The reason lies behind their tough fighting mission. In order to provide a highly maneuverable fighter aircraft, the stability requirements are relaxed, and the safety of flight is left to the pilot plus the fighter advanced automatic control system. Thus, we primarily design the horizontal and vertical tail to satisfy longitudinal and directional requirements.

6.3.2 Control

Control is defined as the ability of an aircraft to vary the aircraft condition from trim condition 1 (say cruise) to trim condition 2 (say climb). Due to three axes in the aircraft coordinate system, there are three branches in aircraft control:

Table 6.3 The static and dynamic longitudinal and directional stability requirements

No.	Requirements	Stability derivatives	Symbol	Typical value (1/rad)
1a	Static longitudinal stability	Rate of change of pitching moment coefficient with respect to angle of attack	$C_{m\alpha}$	−0.3 to −1.5
1b	Static longitudinal stability	Static margin	$h_{np}-h_{cg}$	0.1–0.3
2	Dynamic longitudinal stability	Rate of change of pitching moment coefficient with respect to pitch rate	C_{m_q}	−5 to −40
3	Static directional stability	Rate of change of yawing moment coefficient with respect to sideslip angle	C_{n_β}	+0.05 to +0.4
4	Dynamic directional stability	Rate of change of yawing moment coefficient with respect to yaw rate	C_{n_r}	−0.1 to −1

1. lateral control;
2. longitudinal control;
3. directional control.

Lateral control is the control of an aircraft about the x-axis, longitudinal control is the control of an aircraft about the y-axis, and directional control is the control of an aircraft about the z-axis. In a conventional aircraft lateral control is applied though an aileron, longitudinal control is applied though an elevator, and directional control is applied though a rudder. Since the elevator is part of the horizontal tail, and the rudder is part of the vertical tail, the tail designer must make sure that the horizontal tail and vertical tail are large enough to satisfy longitudinal and directional controllability requirements.

Based on Section 145 of PAR 23 of FAR [4], which concerns longitudinal control of GA aircraft:

> With the airplane as nearly as possible in trim at 1.3 V_{S1}, it must be possible, at speeds below the trim speed, to pitch the nose downward so that the rate of increase in airspeed allows prompt acceleration to the trim speed.

The following is reproduced from Section 147 of PAR 23 of FAR [4], which concerns directional and lateral control of GA aircraft:

> (a) For each multiengine airplane, it must be possible, while holding the wings level within five degrees, to make sudden changes in heading safely in both directions. This ability must be shown at 1.4 V_{S1} with heading changes up to 15 degrees, except that the heading change at which the rudder force corresponds to the limits specified in §23.143 need not be exceeded,
>
> (b) For each multiengine airplane, it must be possible to regain full control of the airplane without exceeding a bank angle of 45 degrees, reaching a dangerous attitude or encountering

dangerous characteristics, in the event of a sudden and complete failure of the critical engine, making allowance for a delay of two seconds in the initiation of recovery action appropriate to the situation, with the airplane initially in trim.

(c) For all airplanes, it must be shown that the airplane is safely controllable without the use of the primary lateral control system in any all-engine configuration(s) and at any speed or altitude within the approved operating envelope. It must also be shown that the airplane's flight characteristics are not impaired below a level needed to permit continued safe flight and the ability to maintain attitudes suitable for a controlled landing without exceeding the operational and structural limitations of the airplane. If a single failure of any one connecting or transmitting link in the lateral control system would also cause the loss of additional control system(s), compliance with the above requirement must be shown with those additional systems also assumed to be inoperative.

Since the design of control surfaces is covered in detail in Chapter 12, more information about controllability requirements can be found there. In case a horizontal tail design satisfies the longitudinal trim and stability requirements but is unable to satisfy the longitudinal control requirements, the horizontal tail parameters must be revised. In a similar fashion, if a vertical tail design satisfies the directional trim and stability requirements but is unable to satisfy the directional control requirements, the vertical tail parameters must be revised.

6.3.3 Handling Qualities

Stability and control are at odds with each other. The reinforcement of stability in an aircraft design weakens the aircraft controllability, while the improvement of controllability of an aircraft has a negative effect on the aircraft stability. As the stability features of an aircraft are improved, its controllability features are degraded. A highly stable aircraft (such as a passenger aircraft) tends to be less controllable, while a highly maneuverable aircraft (such as a fighter or a missile) tends to be less stable or even not stable. The decision about the extent of stability and controllability is very hard, and crucial to make for an aircraft designer. The provision of longitudinal and directional stability is almost straightforward compared with lateral stability, which tends to negatively influence other desired aspects of an aircraft. In the majority of cases the provision of lateral stability is very hard to achieve, so the majority of aircraft (even transport aircraft) suffer from a lack of sufficient lateral stability.

The determination of the borderline between stability and control of an aircraft is executed through a topic referred to as *handling qualities*. The degree of stability and the degree of controllability have been investigated and established by standards such as the FAR standards, or Military Standards (MIL-STDs). The handling qualities (sometimes called flying qualities) are determined to guarantee the comfort of the pilot and passengers as well as airworthiness standards. The handling quality requirements largely influence several aspects of the horizontal and vertical tail. The initial selection of tail parameters (such as tail volume coefficient) must include a satisfactory achievement of handling quality requirements. If your customer has not requested specific and unique handling qualities, you can trust and follow the published standards such as FAR and MIL-STD. More details of handling qualities are presented in Section 12.3. The technique outlined in this chapter considers the public aviation standards that are available to aircraft designers in libraries and on official government websites.

6.4 Tail Configuration

6.4.1 Basic Tail Configuration

The purpose of this section is to present design requirements and design information related to the selection of the tail configuration. The term *tail* in this section means the combination of horizontal and vertical tail. The first step in the tail design is the selection of the tail configuration. The choice of the tail configuration is the output of a selection process, not the result of a mathematical calculation. The decision for the selection of the tail configuration must be made based on the reasoning, logic, and evaluation of various configurations against the design requirements.

The list of design requirements that must be considered and satisfied in the selection of tail configurations is as follows:

1. longitudinal trim;
2. directional trim;
3. lateral trim;
4. longitudinal stability;
5. directional stability;
6. lateral stability;
7. manufacturability and controllability;
8. handling qualities (e.g., passenger comfort);
9. stealth (only in some specific military aircraft);
10. operational requirements (e.g., pilot view);
11. airworthiness (e.g., safety, tail stall, and deep stall);
12. survivability (e.g., spin recovery);
13. cost;
14. competitiveness (in the market);
15. size limits (for example, an aircraft may be required to have a limited height because of hangar space limits. This will influence the vertical tail configuration).

The technical details of these requirements must be established prior to the selection of the tail configuration. Often, no single tail configuration can satisfy all the design requirements; hence, a compromise must be made. After a few acceptable candidates have been prepared, a table based on the systems engineering approach must be provided to determine the final selection; that is, the best choice. Sometimes a design requirement (such as lateral stability) is completely ignored (i.e., sacrificed) in order to satisfy other more important design requirements (such as maneuverability or stealth requirements).

In general, the following tail configurations are available that are capable of satisfying the design requirements in one way or another:

1. aft tail and one aft vertical tail;
2. aft tail and twin aft vertical tails;
3. canard and aft vertical tail;
4. canard and twin-wing vertical tail;
5. triplane (i.e., aft tail as aft plane, and canard as fore plane plus wing as third plane);
6. tailless (delta wing with one vertical tail);
7. no formal tail (also known as *flying wing*, such as the B-2 Spirit (see Figure 6.8)).

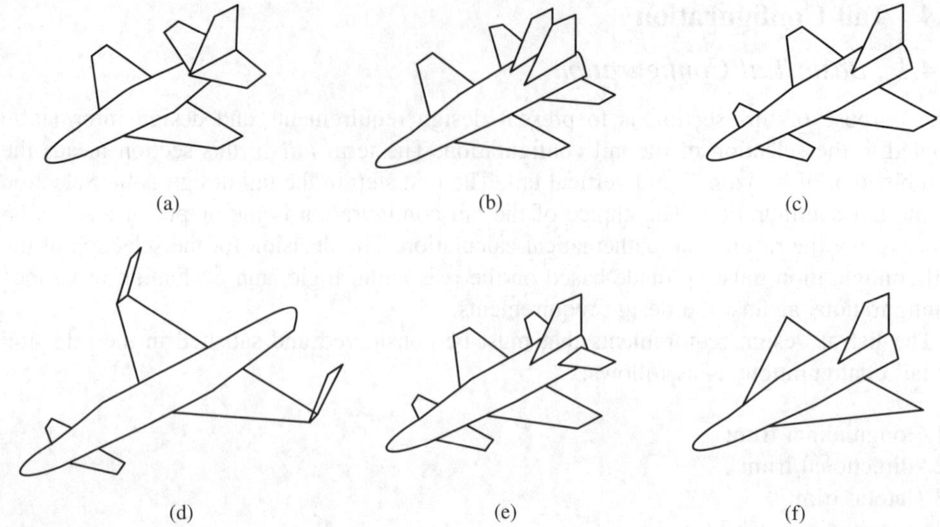

Figure 6.7 Basic tail configurations. (a) Aft tail and one aft vertical tail; (b) Aft tail and two aft vertical tails; (c) Canard and aft vertical tail; (d) Canard and two wing vertical tail; (e) Triplane; (f) Delta wing with one vertical tail

Figure 6.7 depicts these configurations. Based on the statistics, the majority of aircraft designers (about 85%) select the aft tail configuration. About 10% of current aircraft have a canard. About 5% of today's aircraft have other configurations that could be called unconventional tail configurations. The general characteristics of the canard will be described in Section 6.5.

The first configuration (aft tail and one aft vertical tail) has several sub-configurations that will be examined in Section 6.4.2. In the first three configurations (see Figure 6.7(a–c)), a vertical tail is installed at the aft of the fuselage, while in the fourth configuration (see Figure 6.7(d)), two vertical tails are installed at the wing tips. The features of the canard configuration will be examined in Section 6.5. The selection of a twin vertical tail (VT) largely originates from the fact that it provides high directional control, while it does not degrade the roll control. Two short-span vertical tails (see Figure 6.7(d)) tend to have a lower mass moment of inertia about the x-axis compared with one long-span vertical tail. Figure 6.8(f) illustrates the aircraft Piaggio P-180 with a triplane configuration.

The primary functions of the tail in an aircraft with no tail configuration are performed via other components or automatic control systems. For instance, in hang gliders, the longitudinal trim of the aircraft is employed by the pilot by moving his/her body in order to vary the cg of the aircraft. Furthermore, the longitudinal stability requirements are satisfied through a particular wing airfoil section that has a negative camber at the trailing edge (i.e., reflexed trailing edge), as sketched in Figure 6.9. Moreover, the pilot is able to continuously control and make considerable changes to the wing airfoil section via a manual mechanism provided for him/her. This technique is typically employed in hang gliders.

The majority of GA aircraft have a conventional aft horizontal tail, and an aft vertical tail configuration. The majority of fighter aircraft have one aft tail and twin vertical

Figure 6.8 Several aircraft with various tail configurations: (a) Aero Designs Pulsar (aft tail); (b) Dassault Rafale (canard); (c) B-2 Spirit (flying wing); (d) Lockheed F-117 Nighthawk (V-tail); (e) Velocity 173 Elite (canard and twin VT); (f) Piaggio P-180 (triplane); (g) De Havilland DH-110 Sea Vixen (unconventional twin VT); (h) PZL-Mielec M-28 Bryza (H-tail). Reproduced from permission of (a, e, g, h) Jenny Coffey; (b, d) Antony Osborne; (f) Hanseuli Krapf.

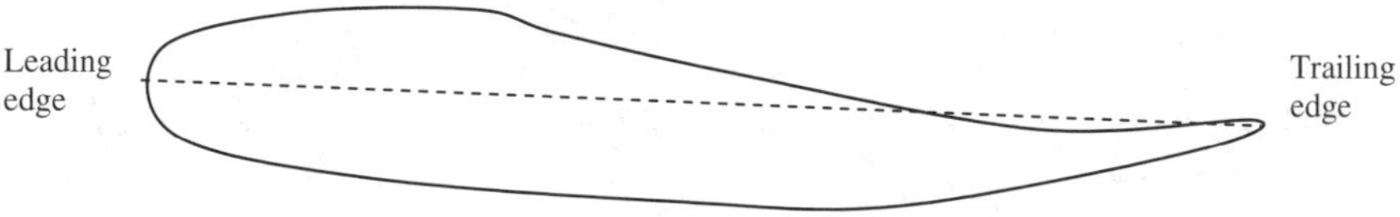

Figure 6.9 A wing airfoil section with reflexed trailing edge

tails, due to their maneuverability requirements. Some European fighters (mainly French fighters such as the Dassault Rafale) have a canard configuration (see Figure 6.8(b)). The primary reason for the Bomber aircraft B-2 Spirit's flying wing (Figure 6.8(c)) is stealth requirements. Most hang gliders do not employ a horizontal tail, however, they satisfy the longitudinal stability requirements through a wing reflex trailing edge.

In some cases, some aircraft configurations impose limits on the tail configuration. For instance, when a prop-driven engine is considered to be installed inside an aft fuselage (i.e., a pusher aircraft as seen in MQ-9 Reaper UAV (Figure 6.12)), the aft horizontal tail is not a proper option. The reason is that the horizontal tail will be under continuous wake effect of the engine, and its efficiency will be degraded. By the same reasoning a canard is not a good option if a prop-driven engine is inside a fuselage nose (e.g., the Aero Designs Pulsar as shown in Figure 6.8(a)). The main disadvantage of a higher number of tails, such as a triplane (Figure 6.8(f)) or two vertical tails (Figure 6.8(g,h)), is the higher cost of manufacturing and the complexity of the design. Figure 6.8(h) illustrates the PZL-Mielec M-28 Bryza with an H-tail.

The basic rule for the selection of the tail configuration is as follows. In general, the conventional aft tail configuration (Figure 6.7(a)) is often able to satisfy all the design requirements, unless one or more requirements imply another configuration. Thus, it is recommended to begin with a conventional aft tail configuration and then to evaluate its features against the design requirements. If one or more requirements are not satisfied, change to a new configuration nearest the current configuration until all the requirements can be satisfied. If the aircraft is in the manufacturing phase and a change is needed to improve the longitudinal and directional stability, one can utilize a smaller auxiliary horizontal tail (sometimes referred to as an stabilon) and a ventral strake. These tricks are employed in the twin-turboprop regional transport aircraft Beech 1900D.

6.4.2 Aft Tail Configuration

An aft tail has several configurations that are all able to satisfy the design configurations. Each has unique advantages and disadvantages. The purpose of this section is to provide a comparison between these configurations to enable an aircraft designer to make a decision and select the best one. The aft tail configurations are as follows: (i) conventional, (ii) T-shape, (iii) cruciform (+), (iv) H-shape, (v) triple-tail, (vi) V-tail, (vii) inverted V-tail, (viii) improved V-tail, (ix) Y-tail, (x) twin vertical tail, (xi) boom-mounted, (xii) inverted boom-mounted, (xiii) ring-shape, (xiv) twin T, (xv) half T, and (xvi) U-tail. Figure 6.10 provides several aft tail configurations.

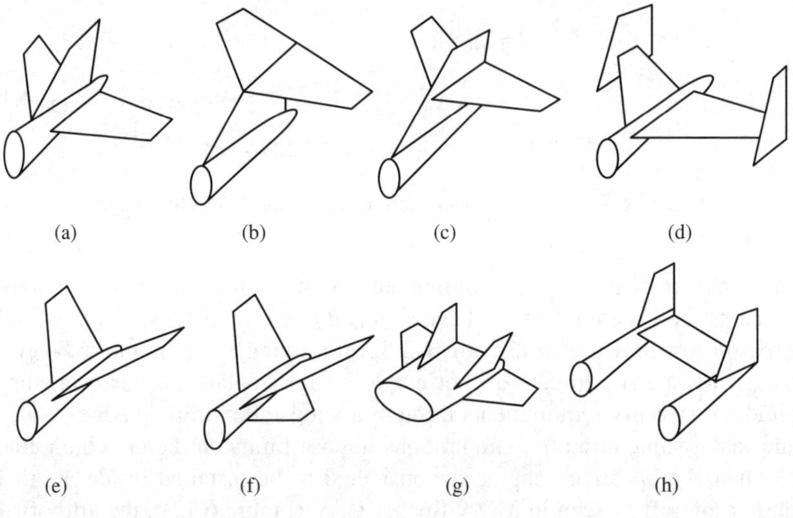

Figure 6.10 Several aft tail configurations. (a) Conventional; (b) T-tail; (c) Cruciform; (d) H-tail; (e)V-tail; (f) Y-tail; (g) Twin vertical tail; (h) Boom mounted

6.4.2.1 Conventional

The conventional tail or inverted T-shape configuration (see Figure 6.10(a)) is the simplest configuration and the most convenient to perform all tail functions (i.e., trim, stability, and control). The analysis and evaluation of the performance of a conventional tail is straightforward. This configuration includes one horizontal tail (two left and right sections) located on the aft fuselage, and one vertical tail (one section) located on top of the aft fuselage. Both horizontal and vertical tails are located and mounted to the aft of the fuselage. The horizontal tail is mainly employed to satisfy the longitudinal trim and stability requirements, while the vertical tail is mainly used to satisfy the directional trim and stability requirements. If the designer has little experience, it is recommended to initially select the conventional tail configuration.

Almost all flight dynamics textbooks examine the features of a conventional tail, but not every flight dynamics textbook discusses the characteristics of other tail configurations. The designer must be professional and skillful in the area of trim analysis, stability analysis, and control analysis if other configurations are selected. This is one of the reasons that about 60% of current aircraft in service have a conventional tail. Furthermore, such a tail is lightweight, efficient, and performs under regular flight conditions. GA aircraft such as the Cessna 172 (Figure 11.15), Cessna 560 Citation, Beech King Air C90B, Learjet 60, Embraer EMB-314 Super Tucano (Figure 10.6), Socata TBM 700, and Pilatus PC-9; large transport aircraft such as the Fokker 60, Boeing 747 (Figures 3.7, 3.12, and 9.4), Boeing 777 (Figure 6.12(a)), and Airbus 340 (Figure 8.6); and fighter aircraft such as the F-15 Eagle (Figure 3.12), Harrier GR. Mk 7 (Figure 4.19), and Panavia Tornado F. Mk3 (Figure 5.56) all have a conventional tail. Figure 6.8(b) illustrates the aircraft Aero Designs Pulsar with a conventional tail configuration.

6.4.2.2 T-tail

A T-tail is an aft tail configuration (see Figure 6.10(b)) that looks like the letter "T," which implies that the vertical tail is located on top of the horizontal tail. The T-tail configuration is another aft tail configuration that provides a few advantages, while it has few disadvantages. The major advantage of a T-tail configuration is that it is out of the regions of wing wake, wing downwash, wing vortices, and engine exit flow (i.e., hot and turbulent high-speed gas). This allows the horizontal tail to provide a higher efficiency, and a safer structure. The lower influence from the wing results in a smaller horizontal tail area, and the lower effect from the engine leads to less tail vibration and buffet. The lesser tail vibration increases the life of the tail, with lower fatigue problems. Furthermore, another advantage of the T-tail is the positive influence of a horizontal tail over a vertical tail. This is referred to as the end-plate effect and results in a smaller vertical tail area.

In contrast, the disadvantages associated with a T-tail are: (i) heavier vertical tail structure and (ii) deep stall. The bending moment created by the horizontal tail must be transferred to the fuselage through the vertical tail. This structural behavior requires the vertical tail main spar to be stronger, which causes the vertical tail to be heavier.

Aircraft with a T-tail are subject to a dangerous condition known as deep stall [7], which is a stalled condition at an angle of attack far above the original stall angle. T-tail aircraft often suffer a sever pitching moment instability at angles well above the initial stall angle of about 13 deg without a wing leading edge high-lift device, or about 18

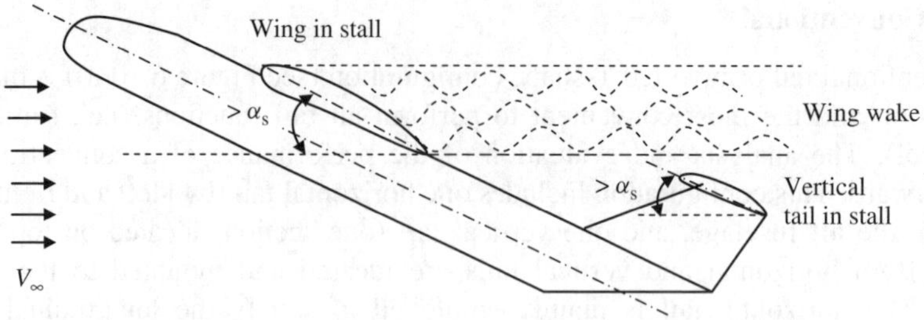

Figure 6.11 Deep stall in a T-tail configuration aircraft

deg with a wing leading edge high-lift device. If the pilot allows the aircraft to enter this unstable region, it might rapidly pitch up to a higher angle of about 40 deg. The causes of the instability are fuselage vortices, shed from the forward portion of the fuselage at high angles of attack, and the wing and engine wakes. Thus the horizontal tail contribution to the longitudinal stability is largely reduced. Eventually, at a higher angle of attack, the horizontal tail exits the wing and nacelle wakes and the aircraft becomes longitudinally stable (see Figure 6.11).

This condition may be assumed as a stable condition, but it accompanies an enormous drag along with a resulting high rate of descent. At this moment, the elevator and aileron effectiveness have been severely reduced because both wing and horizontal tail are stalled at a very high angle of attack. This is known as a locked-in deep stall, a potentially fatal state. The design solutions to avoid a deep stall in a T-tail configuration are to: (i) ensure a stable pitch down at the initial stall, (ii) extend the horizontal tail span substantially beyond the nacelles, and (iii) employ a mechanism to enable full down elevator angles if a deep stall occurs. In addition, the aircraft must be well protected from the initial stall by devices such as a stick shaker, lights, and stall horn.

Despite the above-mentioned disadvantages of a T-tail, it is becoming more and more popular among aircraft designers. About 25% of today's aircraft employ a T-tail configuration. It is interesting to note that the GA aircraft Piper Cherokee has two versions: the Cherokee III (Figure 7.4) with conventional tail and the Cherokee IV with T-tail. The aircraft has a single-piston engine at the nose and a low-wing configuration. Several GA and transport aircraft, such as the Grob Starto 2C, Cessna 525 CitationJet, Beech Super King Air B200, Beechjet T-1A Jayhawk, Learjet 60, Gulfstream IV (Figure 11.15), MD-90, Boeing 727, Fokker 100 (Figure 10.6), AVRO RJ115, Bombardier BD 701 Global Express, Dassault Falcon 900 (Figure 6.12), Sky Arrow 1450L (Figure 6.12(c)), Embraer EMB-120, Airbus A400M (Figure 8.3), and Boeing (formerly McDonnell Douglas) C-17 Globemaster III (Figure 9.9), employ a T-tail configuration.

6.4.2.3 Cruciform

Some tail designers have combined the advantages of a conventional tail and a T-tail and come up with a new configuration known as the cruciform (see Figure 6.10(c)). Thus, the disadvantages of both configurations are considerably released. The cruciform, as the name implies, is a combination of horizontal tail and vertical tail such that it looks like a cross or "+" sign. This means that the horizontal tail is installed at almost the middle of

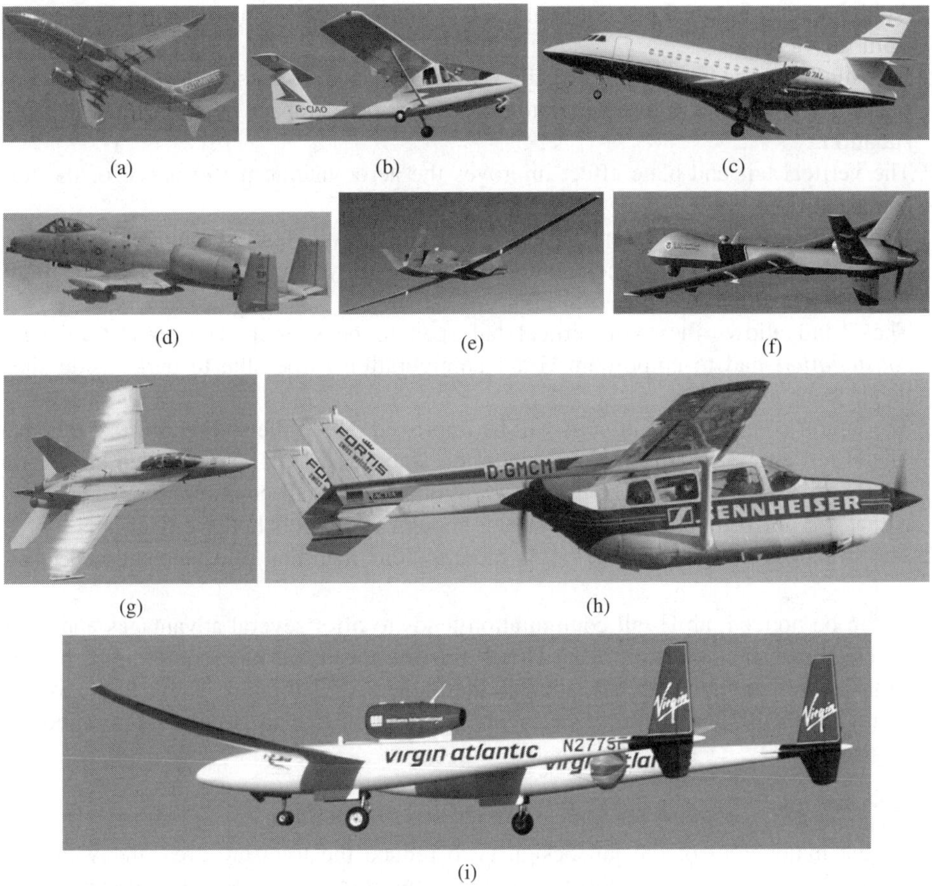

Figure 6.12 Several aircraft with various aft tail configurations: (a) Boeing 737 (conventional); (b) Sky Arrow 1450L (T-tail); (c) Dassault Falcon 900 (cruciform); (d) Fairchild A-10 Thunderbolt (H-tail); (e) Global Hawk UAV (V-tail); (f) MQ-9 Reaper UAV (Y-tail); (g) F-18 Hornet (twin VT); (h) Reims F337F Super Skymaster (boom mounted); (i) Global Flyer (unconventional tail). Reproduced from permission of: (a) Anne Deus; (b, c, h) Jenny Coffey; (d–g) Antony Osborne; (i) NASA.

the vertical tail. The location of the horizontal tail (i.e., its height relative to the fuselage) must be carefully determined such that deep stall does not occur and at the same time, the vertical tail does not get too heavy. Several aircraft, such as the Thurston TA16, Dassault Falcon 2000, ATR 42-400 (Figure 3.8), Dassault Falcon 900B (Figure 6.12(c)), Jetstream 41, Hawker 100, and Mirage 2000D (Figure 9.12), employ the cruciform tail configuration.

6.4.2.4 H-tail

The H-tail (see Figure 6.10(d)), as the name implies, looks like the letter "H." An H-tail is comprised of one horizontal tail in between two vertical tails. The features associated with an H-tail are as follows:

1. At high angles of attack, the vertical tail is not influenced by the turbulent flow coming from the fuselage.
2. In a multi-engine turboprop aircraft, vertical tails are located behind the prop wash region. This causes the vertical tail to have higher performance in an inoperative engine situation.
3. The vertical tail end-plate effect improves the aerodynamic performance of the horizontal tail.
4. In military aircraft, the engine's very hot exhaust gasses could be hidden from radars or infrared missiles. This technique has been employed in the close support aircraft Fairchild A-10 Thunderbolt (see Figure 6.12(d)).
5. The H-tail allows the twin vertical tail span to be shorter. The aircraft *Lockheed constellation* had to employ an H-tail configuration to be able to park inside short-height hangars.
6. The lateral control of the aircraft will be improved due to the shorter vertical tail span.
7. The H-tail allows the fuselage to be shorter, since the tail can be installed on a boom.
8. The H-tail is slightly heavier than conventional and T-tail configurations. The reason is that the horizontal tail must be strong enough to support both vertical tails.
9. The structural design of the H-tail is more tedious than that of a conventional tail.

As can be noticed, an H-tail configuration tends to offer several advantages and disadvantages; hence, the selection of an H-tail must be the result of a compromise process. Several GA and military aircraft, such as the Sadler A-22 Piranha, T-46, Short Skyvan, and Fairchild A-10 Thunderbolt (see Figure 6.12(d)), utilize an H-tail configuration.

6.4.2.5 V-tail

When the major goal of the tail design is to reduce the total tail area, the V-tail (see Figure 6.10(e)) is a proper candidate. As the name implies, the V-tail configuration has two sections which form a shape that looks like the letter "V." In other words, a V-tail is similar to a horizontal tail with high anhedral angle and without any vertical tail. Two sections of a V-tail act as both horizontal and vertical tails. Due to the angle of each section the lift perpendicular to each section has two components, one in the y direction and one in the z direction. If no controller is deflected, two components in the y direction cancel each other, while two lift components in the z direction are added together. The V-tail may perform the longitudinal and directional trim role satisfactorily, but it has deficiencies in maintaining the aircraft longitudinal and directional stability. In addition, the V-tail design is more susceptible to Dutch-roll tendencies than a conventional tail, and total reduction in drag is minimal.

The V-tail design utilizes two slanted tail surfaces to perform the same functions as the surfaces of a conventional elevator and rudder configuration. The movable surfaces, which are usually called ruddervators, are connected through a special linkage that allows the control wheel to move both surfaces simultaneously. In contrast, displacement of the rudder pedals moves the surfaces differentially, thereby providing directional control. When both rudder and elevator controls are moved by the pilot, a control mixing mechanism moves each surface the appropriate amount. The control system for the V-tail is more complex than that required for a conventional tail. A ruddervator induces the undesirable phenomenon of adverse roll/yaw coupling. The solution could be an inverted V-tail configuration, which has other disadvantages. A few aircraft, such as the Beechcraft Bonanza V35, Robin ATL Club, Aviation Farm J5 Marco, high-altitude, long-endurance unmanned aerial reconnaissance vehicle Global Hawk (Figure 6.12(e)), and Lockheed F-117 Nighthawk (Figure 6.8(d)), employ a V-tail. The unmanned aircraft General Atomic MQ-1 Predator has an inverted V-tail plus a vertical tail under the aft fuselage.

6.4.2.6 Y-tail

The Y-tail (see Figure 6.10(f)) is an extension of the V-tail, since it has an extra surface located under the aft fuselage. This extra surface reduces the tail contribution in the aircraft dihedral effect. The lower section plays the role of a vertical tail, while the two upper sections play the role of a horizontal tail. Therefore, the lower surface has a rudder and the control surface of the upper section plays the role of the elevator. Thus, the complexity of the Y-tail is much lower than that of the V-tail. One of the reasons this tail configuration is used is to keep the tail out of the wing wake at high angles of attack. The lower section may limit the performance of the aircraft during take-off and landing, since the tail hitting the ground must be avoided. This configuration is not popular, and only a few old aircraft had this configuration. The unmanned aircraft General Atomic MQ-9 Reaper (see Figure 6.12(f)) employs a Y-tail configuration.

6.4.2.7 Twin Vertical Tail

A twin vertical tail configuration (see Figure 6.10(g)) has a regular horizontal tail, but two separate and often parallel vertical tails. The twin vertical tail largely improves the directional controllability of an aircraft. Two short-span vertical tails have smaller mass moment of inertia about the x-axis compared with a long-span vertical tail. Thus, a twin tail has the same directional control power, while it has a less negative effect on the roll control. In addition, both rudders are almost out of the fuselage wake region, since they are not located along the fuselage center line. A disadvantage of this configuration is that they have a slightly heavier weight compared with the conventional tail. Several modern fighter aircraft, such as the F-14 Phantom (Figure 5.46), McDonnell Douglas F-15 Eagle (Figure 4.21), and F/A-18 Hornet (Figure 6.12), employ a twin-tail configuration.

6.4.2.8 Boom-Mounted

Sometimes some specific design requirements do not allow the aircraft designer to select the conventional tail configuration. For instance, if a prop-driven engine must be installed at the rear of the fuselage, a conventional tail will tend to have a low efficiency. The reason is the interference between the propeller flow and the tail. One of the options is to use two booms and install the tail at the end of the booms (see Figure 6.10(h)). This option, in turn, allows the use of a shorter fuselage, but the overall aircraft weight would be slightly heavier. Two options are: (i) U-tail and (ii) inverted U-tail. The reconnaissance aircraft Reims F337F Super Skymaster (Figure 6.12(h)) and RutanVoyager (Figure 4.5) employ a boom-mounted U-tail. The twin-turboprop light utility aircraft Partenavia PD.90 Tapete Air Truck employs a boom-mounted inverted U-tail configuration which allows for an integrated loading ramp/air-stair.

6.4.2.9 Other Configurations

There is a variety of other unconventional tail configurations which are usually the forced options of a designer. For instance, sometimes some specific mission requirements – such as loading, operational, structural, and engine requirements – remove the conventional or T-tail configuration from the list of possible options. Thus, the designer must come up with a new configuration to make an aircraft trimmed and stable throughout flight. A few invented unconventional configurations are as follows: (i) boom-mounted twin vertical tails plus canard (e.g., Rutan Voyager), (ii) boom-mounted twin vertical tails plus two separated horizontal tails (e.g., Space Ship One (Figure 6.12(i))), (iii) twin T-tail (e.g., Global Flyer (Figure 6.12(i))), (iv) T-tail plus two fins and an auxiliary fixed horizontal tail (e.g., Beech 1900 D of Continental Express), (v) ring tail (e.g., Cagny 2000), and (vi) triple vertical tail.

6.5 Canard or Aft Tail

One of the critical issues in the design of a horizontal tail is the selection of the location of the horizontal tail. The options are: (i) aft tail (sometimes referred to as tail aft) and (ii) fore plane or canard[3] (sometimes referred to as tail first). As discussed before, the primary function of the horizontal tail is longitudinal trim, and then longitudinal stability. Both the aft tail and canard are capable of satisfactorily fulfilling both mission requirements. However, there are several aspects of flight features that are influenced differently by these two options. It is interesting to note that the first aircraft in history (i.e., Wright Flyer) had a canard configuration. The canard configuration is not as popular as the aft tail, but several GA and military (and a few transport) aircraft employ a canard. Examples are RutanVariEze (Figure 3.12), Rutan Voyager (Figure 4.5), Mirage 2000, Dassault Rafale (Figure 6.8), Eurofighter Typhoon (Figure 3.7), B-1B Lancer, Saab Viggen, Grumman X-29, Piaggio P-180 Avanti (Figure 6.8(f)), XB-70 Valkyrie, and Beechcraft Starship (Figure 6.13).

(a)

(b)

Figure 6.13 Two aircraft with canard configuration: (a) Beech Starship; (b) Saab JAS-39B Gripen. Reproduced from permission of (a) Ken Mist; (b) Antony Osborne.

[3] Canard is originally a French word, meaning "duck." Some early aircraft such as the French Canard Vision had a tail-first configuration that observers thought resembled a flying duck.

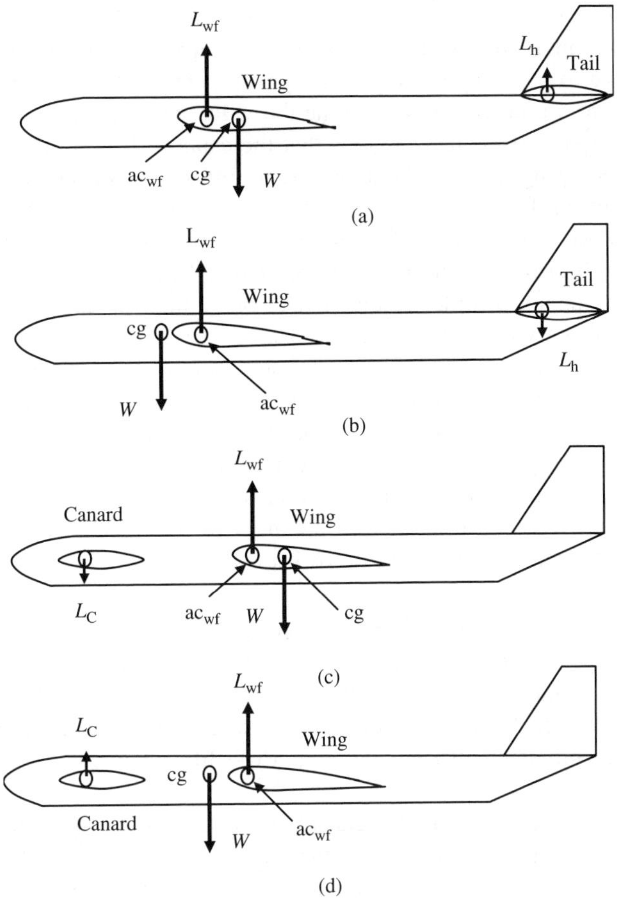

Figure 6.14 The lift of the tail (or canard) in four configurations. (a) Positive tail lift; (b) Negative tail lift. (c) Negative canard lift; (d) Positive canard lift

To comprehend the fundamental differences between an aft tail and a canard, consider four aircraft configurations as shown in Figure 6.14, where two aircraft have an aft tail while the other two have a canard. In this figure, the wing nose-down pitching moment is not shown for simplicity. The difference between each two figures is the location of the cg compared with the wing/fuselage aerodynamic center. This simple difference causes a variety of advantages and disadvantages for the canard over the conventional aft tail. In all four configurations, the longitudinal trim must hold:

$$\sum M_{cg} = 0 \Rightarrow M_{o_{wf}} + L_h \cdot l_h + L_{wf}\left(h - h_o\right)\overline{C} = 0 \quad \text{(aft tail configuration)} \quad (6.37a)$$

$$\sum M_{cg} = 0 \Rightarrow M_{o_{wf}} + L_C \cdot l_C + L_{wf}\left(h - h_o\right)\overline{C} = 0 \quad \text{(canard configuration)} \quad (6.37b)$$

$$\sum F_z = 0 \Rightarrow W = L_{wf} + L_h \quad \text{(aft tail configuration)} \quad (6.38a)$$

$$\sum F_z = 0 \Rightarrow W = L_{wf} + L_C \quad \text{(canard configuration)} \quad (6.38b)$$

where L_C denotes the canard lift. Equations (6.37) and (6.38) indicate that the aft tail lift or canard lift might be positive, or negative, depending upon the location of the aircraft cg relative to the wing/fuselage aerodynamic center (see Figure 6.14). Equations (6.37b) and (6.38b) are utilized to determine the value and direction of the canard lift to satisfy the trim requirements. It is obvious that the canard lift is sometimes negative (see Figure 6.14(c)). Keeping in mind the above basic difference between aft tail and canard, a comparison between features of the canard as compared with the aft tail is presented.

The canard avoids deep stall 100%. This is interesting when we note that about 23% of all world aircraft crashes relate to deep stall. Consider a pilot who intends to increase the wing angle of attack in order to either take off, climb, or land. Since the canard is located forward of the wing, the canard will stall first (i.e., before the wing stalls). This causes the canard to drop and exit out of the stall before the wing enters the stall. The canard drop is due to the fact that when it stalls, its lift is reduced and this causes the aircraft nose to drop. This is regarded as one of the major advantages of a canard, and makes the canard configuration much safer that the aft tail configuration.

Since the canard stalls before the main wing, the wing can never reach its maximum lift capability. Hence, the main wing must be larger than in a conventional configuration, which increases its weight and also the zero-lift drag.

1. A canard has a higher efficiency when compared with an aft tail. The reason is that it is located in front of the wing, so the wing wake does not influence the canard aerodynamic characteristics. The wing, however, is located aft of the canard; hence, it is negatively affected by the canard wake. Thus a wing in a canard configuration has a lower aerodynamic efficiency (i.e., a lower lift) when compared with an aircraft with aft tail configuration.
2. It is not appropriate to employ a canard when the engine is a pusher and located at the fuselage nose. The reason is that the aircraft nose will be heavy and the cg adjustment is difficult. Moreover, the structural design of the fuselage nose is somewhat complicated, since it must hold both the engine and the canard.
3. An aircraft with a canard configuration tends to have a smaller static margin compared with an aircraft with a conventional aft tail configuration. In other words, the distance between the aircraft neutral point and the aircraft center of gravity is shorter. This makes the canard aircraft longitudinally statically less stable. This feature is regarded as a disadvantage of the canard configuration.
4. The center of gravity range in an aircraft with a canard configuration tends to be wider; hence, it is more flexible in the load transportation area.
5. Due to the forward location of a canard, the aircraft cg moves slightly forward compared with an aircraft with a conventional aft tail configuration. This feature requires a slightly larger vertical tail for directional trim and stability.
6. A canard tends to generate a lower "trim drag" compared with an aft tail. In other words, a canard aircraft produces less lift-dependent drag to longitudinally trim the aircraft. However, this feature may lead to a larger wetted area (S_{wet}).
7. One of the potential design challenges in a canard aircraft is to optimally locate the fuel tank. The general rule is to place the fuel tank near the aircraft center of gravity, as close as possible in order to avoid a large movement of the cg during flight operation. The aircraft cg in a canard configuration, if the fuel tank is inside

the wing, is often forward of the fuel tank. To improve the cg location, designers would rather place the fuel tank in the fuselage, which in turn increases the possibility of an aircraft fire. Another solution is to considerably increase the wing root chord (i.e., employing a strake) and place the fuel tank in the wing root. But this technique increases the wing wetted area and reduces the cruise efficiency. The canard aircraft Beechcraft Starship (Figure 6.13) has a wing strake and utilizes this technique.

8. A canard obscures the view of the pilot. This is another disadvantage of the canard configuration.
9. Often the canard generates a positive lift (see Figure 6.14(d)) while a conventional tail often produces a negative lift (see Figure 6.14(b)). The reason is that the aircraft cg in a canard configuration is often forward of the wing/fuselage ac. The aircraft cg in a conventional tail configuration is typically aft of the wing/fuselage ac. Recall that the cg moves during flight as the fuel burns. The cg range, in a modern aircraft with a conventional tail or a canard, is usually determined such that the cg is mostly forward of the wing aerodynamic center. However, in a fewer instances of cruising flight, the cg is aft of the wing aerodynamic center. Thus, in an aircraft with a conventional tail, during cruising flight, the cg usually moves from the most forward location toward the most aft location. However, in an aircraft with a canard, during cruising flight, the cg often moves from the most aft location toward the most forward location.

   Thus, a canard often generates part of the aircraft lift, while a tail mostly cancels part of the lift generated by the wing. This feature tends to reduce the aircraft weight and increase the aircraft cruising speed. In addition, during a take-off in which the wing nose-down pitching moment is large, the canard lift is higher. Using the same logic, it can be shown that the canard lift is higher during supersonic speeds. Recall that at a supersonic speed, the wing aerodynamic center moves aft toward about 50% of the MAC. This is one of the reasons that some European supersonic fighters, such as the Mirage 2000 (Figure 9.12), have employed the canard configuration.
10. Item 9 results in the following conclusion: an aircraft with a canard is slightly lighter than an aircraft with a conventional tail.
11. In general, the canard aerodynamic and stability analysis techniques are considerably more complicated than the technique to evaluate the aerodynamic feature and stability analysis of a conventional tail configuration aircraft. Literature surveys include a variety of published materials regarding conventional tails, while many fewer papers and technical reports are available for canard analysis. Thus, the design of a canard is more time-intensive and complicated than the conventional tail design.
12. A canard configuration seems to be more stylish and more attractive than a conventional tail.
13. A canard is more efficient for fulfilling the longitudinal trim requirements, while a conventional tail tends to be more efficient for fulfilling the longitudinal control requirements.

In general, canard designs fall into two main categories: the lifting canard and the control canard. As the name implies, in a lifting canard the weight of the aircraft is shared between the main wing and the canard wing. The upward canard lift tends to increase the overall lift capability of the configuration. With a lifting canard, the main wing must be located further aft of the cg range than with a conventional aft tail, and this increases the pitching moment caused by trailing edge flaps. The first airplane to fly, the Wright Flyer, and the X-29 had a lifting canard. Figure 6.13 depicts two aircraft ((Beech Starship and Saab Gripen) with canard configuration. It is interesting to know that about 98% of American aircraft are conventional, not canard.

In the control canard, most of the weight of the aircraft is carried by the main wing and the canard wing serves primarily as the longitudinal control device. A control canard could be all-moving or could have a large elevator. The control canard often has a higher aspect ratio and employs a thicker airfoil section than a lifting canard. A control canard mostly operates at zero angle of attack. Fighter aircraft with a canard configuration, such as the Eurofighter Typhoon (Figure 3.7), typically have a control canard. One benefit obtainable from a control canard is avoidance of pitch-up. An all-moving canard capable of a significant nose-down deflection will protect against pitch-up. Control canards have poor stealth characteristics, because they present large moving surfaces forward of the wing.

The pros and cons of the canard versus a conventional tail configuration are numerous and complex, and it is hard to say which is superior without considering a specific design requirement. One must use systems engineering techniques to compromise, and to decide on the tail configuration. In the preliminary design phase, the suggestion is to begin with a conventional tail, unless the designer has a solid reason to employ a canard configuration.

6.6 Optimum Tail Arm

One of the tail parameters that must be determined during the tail design process is the tail arm (l_t), which is the distance between the tail aerodynamic center and the aircraft center of gravity. The tail arm serves as the arm for the tail pitching moment (i.e., tail lift multiplied by tail arm) about the aircraft cg to maintain the longitudinal trim. To determine the tail arm one must establish the criteria based on the design requirements. Two basic tail parameters which interact most are the tail arm and tail area; the latter is responsible for generation of the tail lift. As the tail arm is increased, the tail area must be decreased, while as the tail arm is reduced, the tail area must be increased. Both short arms (as in fighters) and long arms (as in most transport aircraft) are capable of satisfying longitudinal trim requirements, given the appropriate necessary tail area. But the question is, what tail arm is optimum? To answer this question, one must look at the other design requirements.

Two very significant aircraft general design requirements are low aircraft weight and low drag. Both of these may be combined and translated into the requirement for a low aircraft wetted area. As the horizontal tail arm is increased, the fuselage wetted area is increased, but the horizontal tail wetted area is decreased. Also, as the horizontal tail arm is decreased, the fuselage wetted area is decreased, but the horizontal tail wetted area is increased. Hence, we are looking to determine the optimum tail arm to minimize drag, which in turn means to minimize the total wetted area of the aft portion of the aircraft. The following is a general educational approach to determine the optimum tail arm; each designer must develop his/her own technique and derive a more accurate equation based on the suggested approach. The approach is based on the fact that the aircraft zero-lift drag is essentially a function of the aircraft wetted area. Therefore, if the total wetted area is minimized, the aircraft zero-lift drag will be minimized. Moreover, the technique will influence the fuselage length, since the aft portion of the fuselage must structurally support the tail.

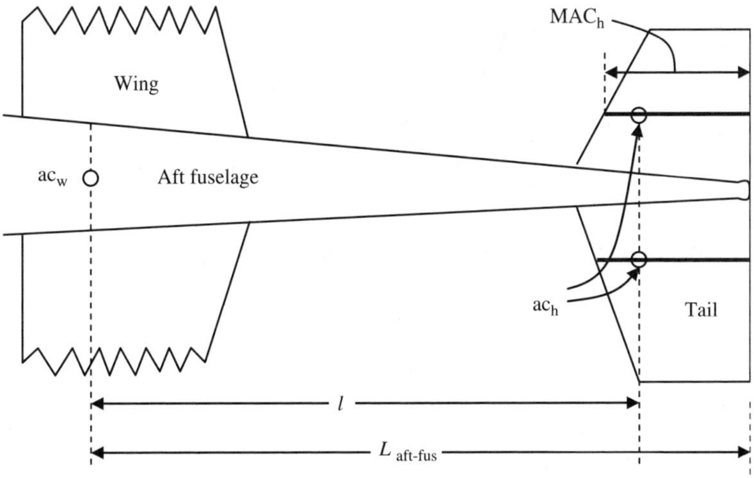

Figure 6.15 Top view of aft portion of the aircraft

Consider the top view of an aft aircraft (see Figure 6.15) that includes the aft portion of the fuselage plus the horizontal tail. The wetted area of the aft portion of the aircraft is the summation of the wetted area of the aft portion of the fuselage ($S_{\text{wet}_{\text{aft\_fus}}}$) plus the wetted area of the horizontal tail ($S_{\text{wet}_{\text{ht}}}$):

$$S_{\text{wet}_{\text{aft}}} = S_{\text{wet}_{\text{aft\_fus}}} + S_{\text{wet}_{\text{h}}} \tag{6.39}$$

Here we assume that the aft portion of the fuselage is conical. Hence, the wetted area of the aft portion of the fuselage is:

$$S_{\text{wet}_{\text{aft\_fus}}} = \frac{1}{2}\pi \cdot D_{\text{f}} L_{\text{fus}_{\text{aft}}} \tag{6.40}$$

where D_{f} is the maximum fuselage diameter and $L_{\text{fus}_{\text{aft}}}$ is the length of the aft portion of the fuselage. At the moment, it is assumed that $L_{\text{fus}_{\text{aft}}}$ is equal to half the fuselage length (L_{f}). In contrast, the wetted area of the horizontal tail is about twice the tail planform area:

$$S_{\text{wet}_{\text{t}}} \approx 2S_{\text{h}} \tag{6.41}$$

But, the tail volume coefficient is defined as in Equation (6.24), so:

$$\overline{V}_{\text{H}} = \frac{l}{\overline{C}}\frac{S_{\text{h}}}{S} \Rightarrow S_{\text{h}} = \frac{\overline{C} \cdot S \cdot \overline{V}_{\text{H}}}{l} \tag{6.42}$$

So:

$$S_{\text{wet}_{\text{h}}} \approx 2\frac{\overline{C}S\overline{V}_{\text{H}}}{l} \tag{6.43}$$

Substituting Equations (6.41) and (6.43) into Equation (6.39) yields:

$$S_{\text{wet}_{\text{aft}}} = \frac{1}{2}\pi D_{\text{f}} L_{\text{f}_{\text{aft}}} + 2\frac{\overline{C}S\overline{V}_{\text{H}}}{l} \tag{6.44}$$

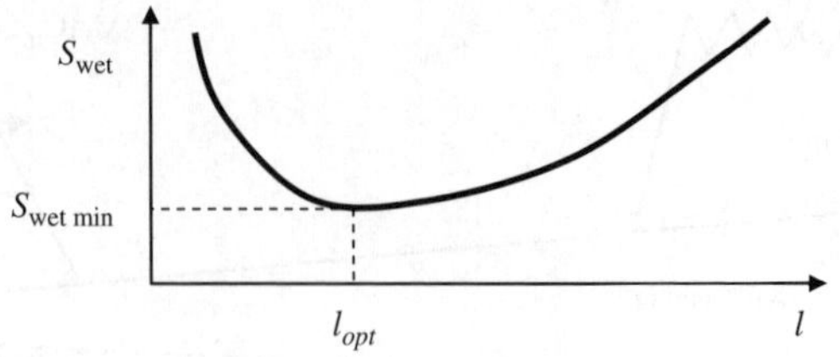

Figure 6.16 The variation of wetted area with respect to tail arm

The relationship between $L_{fus_{aft}}$ and l depends upon the location of the horizontal tail (see Figure 6.15). We simply assume they are equal ($L_{fus_{aft}} = l$). This assumption is not accurate for every aircraft configuration, but is reasonable based on the data of Table 6.2. This assumption will be modified later. In order to minimize the zero-lift drag of the aft part of the aircraft, we have to differentiate the wetted area of the aft part of the aircraft with respect to the tail arm (see Figure 6.16) and then set it equal to zero. The differentiation yields:

$$\frac{\partial S_{wet_{aft}}}{\partial l} = \frac{1}{2}\pi D_f + 2\frac{\overline{C}S\overline{V}_H}{l^2} = 0 \tag{6.45}$$

The optimum tail arm is obtained by solving this equation as follows:

$$l_{opt} = \sqrt{\frac{4\overline{C}S\overline{V}_H}{\pi D_f}} \tag{6.46}$$

To compensate for our inaccurate assumption, we add a fudge factor as follows:

$$l_{opt} = K_c\sqrt{\frac{4\overline{C}S\overline{V}_H}{\pi D_f}} \tag{6.47}$$

where K_c is a correction factor and varies between 1 and 1.4 depending on the aircraft configuration. $K_c = 1$ is used when the aft portion of the fuselage has a conical shape. As the shape of the aft portion of the fuselage goes further away from a conical shape, the K_c factor is increased up to 1.4. As a general rule, for a single-seat single-engine prop-driven GA aircraft, the factor K_c is assumed to be 1.1, but for a transport aircraft, K_c will be 1.4. Note that in a large transport aircraft, most of the fuselage shape is cylindrical, and only its very aft portion has a conical shape. Therefore, if the horizontal tail is located at l_{opt}, the wetted area of the aft part of the aircraft will be minimized, so the drag of the aft part of the aircraft will be minimized. When the horizontal tail arm is less than three times the wing MAC ($3\overline{C}$), the aircraft is said to be short-coupled. An aircraft with such a tail configuration possesses the longitudinal trim penalty (e.g., fighters). Example 6.1 provides a sample calculation.

Example 6.1

Consider a twin-seat GA aircraft whose wing reference area is 10 m^2 and wing MAC is 1 m. The longitudinal stability requirements dictate the tail volume coefficient to be 0.6. If the maximum fuselage diameter is 117 cm, determine the optimum tail arm and then calculate the horizontal tail area. Assume that the aft portion of the fuselage is conical.

Solution:
The aircraft is a GA and has two seats, so the factor K_c is assumed to be 1.4. Using Equation (6.47), we have:

$$l_{opt} = K_c\sqrt{\frac{4\overline{C}S\overline{V}_H}{\pi D_f}} = 1.4 \cdot \sqrt{\frac{4 \cdot 1 \cdot 10 \cdot 0.6}{\pi \cdot 1.17}} \Rightarrow l_{opt} = 3.577\,\text{m} \quad (6.47)$$

The horizontal tail area is calculated by employing the tail volume coefficient equation as follows:

$$\overline{V}_H = \frac{l}{\overline{C}}\frac{S_h}{S} \Rightarrow S_h = \frac{\overline{V}_H\overline{C}S}{l} = \frac{0.6 \cdot 1 \cdot 10}{3.577} = 1.677\text{ m}^2 \quad (6.24)$$

6.7 Horizontal Tail Parameters

After the tail configuration is determined, the horizontal tail and vertical tail can be designed almost independently. This section presents the technique to design the horizontal tail and the method to determine the horizontal tail parameters. Since the horizontal tail is a lifting surface and also several characteristics of the wing and tail are similar (as discussed in Chapter 5), some aspects of the horizontal tail (such as the taper ratio, sweep angle, dihedral angle, and airfoil section) are discussed in brief. The horizontal tail design is also an iterative process, and a strong function of several wing parameters and a few fuselage parameters. Hence, as soon as the major wing and fuselage parameters are changed, the tail must be redesigned and its parameters need to be updated.

6.7.1 Horizontal Tail Design Fundamental Governing Equation

The horizontal tail design fundamental governing equation must be driven based on the primary function of the horizontal tail (i.e., longitudinal trim). Figure 6.2 depicts a general case of an aircraft along with the sources of forces along the x- and z-axes, and moments about the y-axis which influence the aircraft longitudinal trim. The longitudinal trim requires that the summation of all moments about the y-axis must be zero:

$$\sum M_{cg} = 0 \Rightarrow M_{o_{wf}} + M_{L_{wf}} + M_{L_h} + M_{o_h} + M_{T_{eng}} + M_{D_w} = 0 \quad (6.48)$$

where $M_{o_{wf}}$ denotes the nose-down wing/fuselage aerodynamic pitching moment, $M_{L_{wf}}$ denotes the pitching moment generated by the wing/fuselage lift, M_{L_h} denotes the pitching moment generated by the horizontal tail lift, M_{o_h} denotes the nose-down horizontal tail aerodynamic pitching moment, $M_{T_{eng}}$ denotes the pitching moment generated by the engine thrust, and M_{D_w} denotes the pitching moment generated by the wing drag. The sign of each pitching moment depends upon the location of the source force relative to the aircraft center of gravity. This equation must hold at all flight conditions, but the horizontal tail is designed for cruising flight, since the aircraft spends much of its flight time in cruise. For other flight conditions, a control surface such as the elevator will contribute.

Based on the aerodynamics fundamentals, two aerodynamic pitching moments of the wing and horizontal tail are always nose down (i.e., negative). The sign of the wing drag moment depends on the wing configuration. For instance, a high wing generates a nose-up pitching moment, while a low wing generates a nose-down pitching moment. The sign of the engine thrust moment depends on the thrust line and engine incidence. If the engine has a setting angle other than zero, both horizontal and vertical components will contribute to the longitudinal trim. The major unknown in this equation is the horizontal tail lift. Another requirement for the longitudinal trim is that the summations of all forces along the x- and z-axes must be zero. Only the summation of forces along the z-axis contributes to the tail design:

$$\sum F_z = 0 \Rightarrow L_{wf} + T\sin(i_T) + L_h = 0 \tag{6.49}$$

where T is the engine thrust and i_T is the engine thrust setting angle (i.e., the angle between the thrust line and the x-axis). This angle is almost always non-zero. The reason is the engine thrust contribution to the aircraft longitudinal stability. The typical engine setting angle is about 2–4 deg. The horizontal tail designer should expand both Equations (6.48) and (6.49) and solve simultaneously for the two unknowns of wing lift and horizontal tail lift. The latter is employed in the horizontal tail design. The derivation is left to the reader.

It is presumed that the horizontal tail designer is familiar with the flight dynamics principles and is capable of deriving the complete set of longitudinal trim equations based on the aircraft configuration. Since the goal of this textbook is educational, so a simple version of the longitudinal trim equation is employed. If the pitching moments of engine thrust, wing drag, and horizontal tail pitching moment are ignored (as shown in Figure 6.3), the non-dimensional horizontal tail design principle equation is as derived earlier:

$$C_{m_{o\_wf}} + C_L\left(h - h_o\right) - \eta_h \overline{V}_H C_{L_h} = 0 \tag{6.29}$$

The full derivation has been introduced is Section 6.2. This equation has three terms, the last of which is the horizontal tail contribution to the aircraft longitudinal trim. The cruising flight is considered for horizontal tail design application. The equation has only two unknowns (i.e., $\overline{V}_H$ and C_{L_h}). The first unknown (horizontal tail volume coefficient, $\overline{V}_H$) is determined primarily based on the longitudinal stability requirements. The longitudinal flying qualities requirements govern this parameter. The reader is encouraged to consult Refs [1] and [6] for full guidance. However, Chapter 12 presents a summary of the longitudinal flying qualities requirements. A higher value for $\overline{V}_H$ results in a longer fuselage, and/or a smaller wing, and/or a larger horizontal tail.

As the value of $\overline{V}_H$ is increased, the aircraft becomes longitudinally more stable. In contrast, a more stable aircraft means a less controllable flight vehicle. Hence, a lower

Table 6.4 Typical values for horizontal and vertical tail volume coefficients

No.	Aircraft	Horizontal tail volume coefficient ($\overline{V}_H$)	Vertical tail volume coefficient ($\overline{V}_v$)
1	Glider and motor glider	0.6	0.03
2	Home-built	0.5	0.04
3	GA single prop-driven engine	0.7	0.04
4	GA twin prop-driven engine	0.8	0.07
5	GA with canard	0.6	0.05
6	Agricultural	0.5	0.04
7	Twin turboprop	0.9	0.08
8	Jet trainer	0.7	0.06
9	Fighter aircraft	0.4	0.07
10	Fighter (with canard)	0.1	0.06
11	Bomber/military transport	1	0.08
12	Jet transport	1.1	0.09

value for $\overline{V}_H$ causes the aircraft to become longitudinally more controllable and less stable. If the horizontal tail design is at the preliminary design phase, and the other aircraft components have not yet been designed, a typical value for $\overline{V}_H$ must be selected. Table 6.4 illustrates the typical values for horizontal and vertical tail volume coefficients. The values are driven from the current successful aircraft statistics. A number from this table based on the aircraft mission and configuration is recommended at the early design phase. When the other aircraft components are designed and their data are available, a more accurate value for $\overline{V}_H$ may be determined.

The variable h_o denotes the non-dimensional wing/fuselage aerodynamic center $\left(\frac{X_{ac_{wf}}}{\overline{C}}\right)$ position. A typical value for h_o is about 0.2–0.25 for the majority of aircraft configurations. References [1] and [6] introduce a precise technique to evaluate the value of h_o. Another significant parameter in Equation (6.29) is h. The parameter h denotes the non-dimensional aircraft cg position $\left(\frac{X_{cg}}{\overline{C}}\right)$. The value for h must be known prior to the horizontal tail design.

Chapter 11 is dedicated to the techniques and methods to determine the aircraft cg position, provided the details of geometries of all aircraft components. However, if at the early stages of the horizontal tail design the other aircraft components such as fuselage, engine, and landing gear have not yet been designed, the only option is to pick a value for h. The best value is a mid-value between the most forward and the most aft position of the aircraft cg. This minimizes the aircraft trim drag while in cruise. This is based on a logical assumption that the aircraft cg is at one end of the extreme position (say most forward) at the beginning of the cruise, and moves to another end of the extreme position (say most aft) at the end of the cruise.

In contrast, in order to reduce the longitudinal control effort during a cruising flight, the aircraft cg is recommended to be close to the wing/fuselage aerodynamic center. The aircraft non-dimensional center of gravity limit (Δh) is the difference between the most forward and the most aft position of the aircraft cg. The typical values for the aircraft non-dimensional center of gravity limit are:

$$\Delta h = 0.1 \text{ to } 0.3 \tag{6.50}$$

This means that a typical value for the most forward position of the aircraft cg is about 10% of the wing MAC. In addition, a typical value for the most aft position of the aircraft cg is about 30% of the wing MAC. Therefore, a proper assumption for the value of h at the early stage of the horizontal tail design would be about 0.2. As soon as a more realistic value for the aircraft cg position (h) is available, the horizontal tail design must be updated. The value for the aircraft lift coefficient (C_L) in Equation (6.29) is determined based on the cruising velocity, cruise altitude, and the aircraft average weight (Equation (5.10)). Finally, by solving Equation (6.29), the only unknown (C_{L_h}), is determined.

At this moment, three horizontal tail parameters are decided (i.e., $\overline{V}_H$, C_{L_h}, and l). In contrast, since the tail volume coefficient is a function of the horizontal tail area (S_h), the horizontal tail area is readily determined using Equation (6.24). By the technique that has just been introduced, the three horizontal tail parameters that have been determined are as follows:

1. horizontal tail planform area (S_h);
2. horizontal tail moment arm (l);
3. horizontal tail cruise lift coefficient (C_{L_h}).

It is important to remember that the design is an iterated process, so as soon as any assumption (such as aircraft cg) is changed, the horizontal tail design must be revised.

6.7.2 Fixed, All-Moving, or Adjustable

Due to the fact that the aircraft has numerous flight conditions – such as various speeds, cg locations, weights, and altitudes, the longitudinal trim requirements are satisfied only through a change in the horizontal tail lift. Since the horizontal tail has a fixed planform area and fixed airfoil section, the only way to change the tail lift is to vary its angle of attack (α_h). There are three tail setting configurations (as sketched in Figure 6.17) to fulfill a change in the angle of attack:

1. fixed horizontal tail;
2. adjustable tail;
3. all-moving tail.

A fixed tail is permanently attached to the fuselage by some joining technique such as a screw and nut or welding. A fixed tail angle of attack cannot be varied unless by pitching

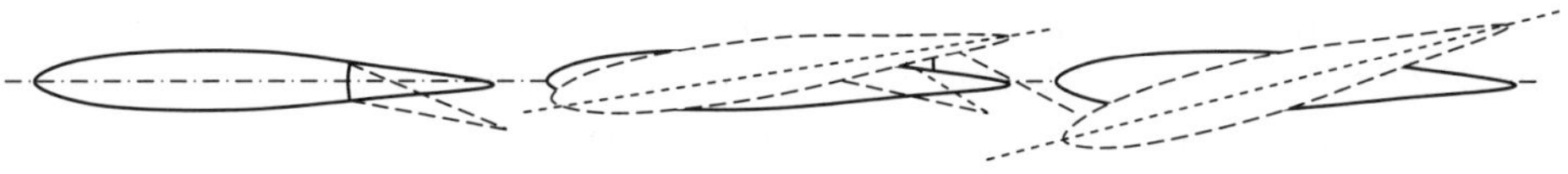

(a) (b) (c)

Figure 6.17 Three horizontal tail setting configurations. (a) Fixed; (b) Adjustable; (c) All moving

up or down the fuselage nose. In contrast, the angle of attack of an all-moving tail is easily changed by the pilot using the forward or aft motion of the stick inside the cockpit.

There are several basic differences between these options. First of all, a fixed tail is much lighter, cheaper, and structurally easier to design compared with an all-moving tail. Moreover, a fixed tail is safer than an all-moving tail due to the possibility of failure of a moving mechanism. In contrast, an aircraft with an all-moving tail (such as the fighter aircraft Dassault Rafale shown in Figure 6.8) is more controllable and maneuverable than an aircraft with a fixed tail. One difference between these two tails is that a fixed tail is equipped with a longitudinal control surface (i.e., an elevator) while an all-moving tail does not have any separate deflectable section. In general, the trim drag of a fixed tail is higher than that of an all-moving tail. An all-moving tail is sometimes referred to as a variable incidence tailplane.

A tail option which has some advantages of a fixed tail and some advantages of a moving tail is referred to as an adjustable tail (such as in the Fairchild C-26A Metro III shown in Figure 6.18(a)). As the name implies, an adjustable tail allows the pilot to adjust its setting angle for a long time. The adjustment process usually happens before the flight; however, a pilot is allowed to adjust the tail setting angle during the flight operation. An adjustable tail employs an elevator, but a major difference between an adjustable tail and an all-moving tail lies in the tail rotation mechanism. An all-moving tail is readily and rapidly (in a fraction of a second) rotated about its hinge by the pilot. However, the angle

(a) (b)

Figure 6.18 An adjustable tail and an all-moving tail: (a) adjustable horizontal tail in Fairchild C-26A Metro III; (b) all-moving tail in Panavia Tornado. Reproduced from permission of (a) Luis David Sanchez; (b) Antony Osborne.

of attack adjustment process for an adjustable tail takes time (maybe a few seconds). The range of deflections of an adjustable tail (about +5 to −12 deg) is considerably less than that of an all-moving tail (about +15 to −15 deg). For instance, the tailplane deflection for transport aircraft Boeing 777 is 4° up and 11° down.

If longitudinal maneuverability is not a desired design requirement, it is recommended to employ a fixed tail configuration. But if the aircraft is required to be able to perform fast maneuver, the appropriate option is an all-moving tail. In contrast, if flight cost is a significant issue in the design requirements list, it is better to employ an adjustable tail. In general, most GA and small transport aircraft (e.g., Cessna 172 (Figure 11.15), Jetstream 41) have a fixed tail, most large transport aircraft (e.g., Boeing 767 (Figure 5.4), Airbus 340 (Figure 8.6)) utilize an adjustable tail, and most fighter aircraft (e.g., F/A-18 Hornet (Figures 2.11 and 6.12), F-15 Eagle (Figure 3.12), and Harrier GR. Mk 7 (Figure 4.4)) employ an all-moving tail. Table 6.5 shows the setting configuration of a horizontal tail for several aircraft. Figure 6.18 demonstrates the adjustable horizontal tail of a Fairchild C-26A Metro III, and the all-moving horizontal tail of a Panavia Tornado.

6.7.3 Airfoil Section

A horizontal tailplane is a lifting surface (similar to the wing) and requires a special airfoil section. The basic fundamentals of airfoil sections (definition, parameters, selection criteria, and related calculations) have been presented in Section 5.4, hence they are not repeated here. In summary, a tailplane requires an airfoil section that is able to generate the required lift with minimum drag and minimum pitching moment. The specific horizontal tail airfoil requirements are described in this section.

Basically, the tailplane airfoil lift curve slope ($C_{L_{\alpha\_t}}$) must be as large as possible along with a considerably wide usable angle of attack. Since the aircraft center of gravity moves during cruising flight, the airfoil section must be able to create sometimes a positive ($+L_h$) and sometimes a negative lift ($-L_h$). This requirement necessitates the tailplane behaving similarly in both positive and negative angles of attack. For this reason, a symmetric airfoil section is a suitable candidate for a horizontal tail.

Recall from Chapter 5 that the indication of a symmetric airfoil is that the second digit in a four-digit and the third digit in a five-digit and 6-series NACA airfoil section is zero. This denotes that the airfoil design lift coefficient and zero-lift angle of attack are both zero. NACA airfoil sections such as 0009, 0010, 0012, 63-006, 63-009, 63-012, 63-015, 63-018, 64-006, 64-012, 64A010, 65-009, 65-015, 66-012, 66-018, and 66-021 are all symmetric airfoils. Reference [8] is a rich collection for NACA airfoil sections.

In several GA aircraft, NACA airfoil sections 0009 or 0012 (with 9% or 12% maximum thickness-to-chord ratio) are employed for the horizontal tail. Both these NACA airfoil sections are symmetric. Moreover, it is desired that the horizontal tail never stalls, and the wing must stall before the tail. Hence, the stall feature of the tail airfoil section (sharp or docile) is not significant.

In addition, another tail requirement is that the horizontal tail must be clean of compressibility effect. In order for the tail to be beyond the compressibility effect, the tail lift coefficient is determined to be less than the wing lift coefficient. To insure this requirement, the flow Mach number at the tail must be less than the flow Mach number at the wing. This objective will be realized by selecting a horizontal tail airfoil section to be

Table 6.5 Horizontal tail characteristics for several aircraft

No.	Aircraft	m_{TO} (kg)	Tail type	Airfoil	$(t/C)_{max}$ (%)	$\overline{V}_H$	S_h/S	AR_h	λ_h	Λ_h (deg)	Γ_h (deg)	i_h (deg) +	i_h (deg) –
1	Wright Flyer	420	Moving	Cambered plate	Low	−0.36	0.16	5.7	1	0	0	–	–
2	Cessna 177	1 100	Fixed	NACA 0012/0009	10.5	0.6	0.2	4	1	0	0	–	–
3	Cessna Citation I	5 375	Fixed	NACA 0010/0008	9	0.75	0.26	5.2	0.5	–	9	–	–
4	Beech Starship	6 759	Fixed	–	–	−0.96	0.22	10.2	0.5	33	3	–	–
5	Fokker F-27	19 773	Fixed	NACA 63A-014	14	0.96	0.23	6	0.4	0	6	–	–
6	Boeing 737-100	50 300	Adjustable	12 to 9%	10.5	1.14	0.32	4.16	0.38	30	7	–	–
7	Boeing 707-320	151 320	Adjustable	BAC 317	11.6	0.63	0.216	3.37	0.42	35	7	0.5	14
8	Boeing 747-100	333 390	Adjustable	–	9	1	0.267	3.6	0.26	37	8.5	1	12
9	DC-8-10	141 000	Adjustable	DSMA-89-90	8.75	0.59	0.203	4.04	0.33	35	10	2	10
10	Airbus 300B	165 000	Adjustable	–	–	1.07	0.27	4.13	0.5	32.5	6	3	12
11	Lockheed C-130 Hercules	70 305	Fixed	Inverted NACA	12	1	0.313	5.2	0.36	7.5	0	–	–
12	Lockheed L-1011	211 000	Adjustable	–	8	0.928	0.37	4	0.33	35	3	0	14
13	Lockheed C-5A	381 000	Adjustable	–	10	0.7	0.156	4.9	0.36	24.5	−4.5	4	12
14	Eurofighter 2000	21 000	Movable	–	–	−0.1	0.048	3.4	0.34	45	17	–	–
15	F-15 Eagle	36 741	Movable	–	–	0.24	0.183	2.3	0.36	48	0	–	–

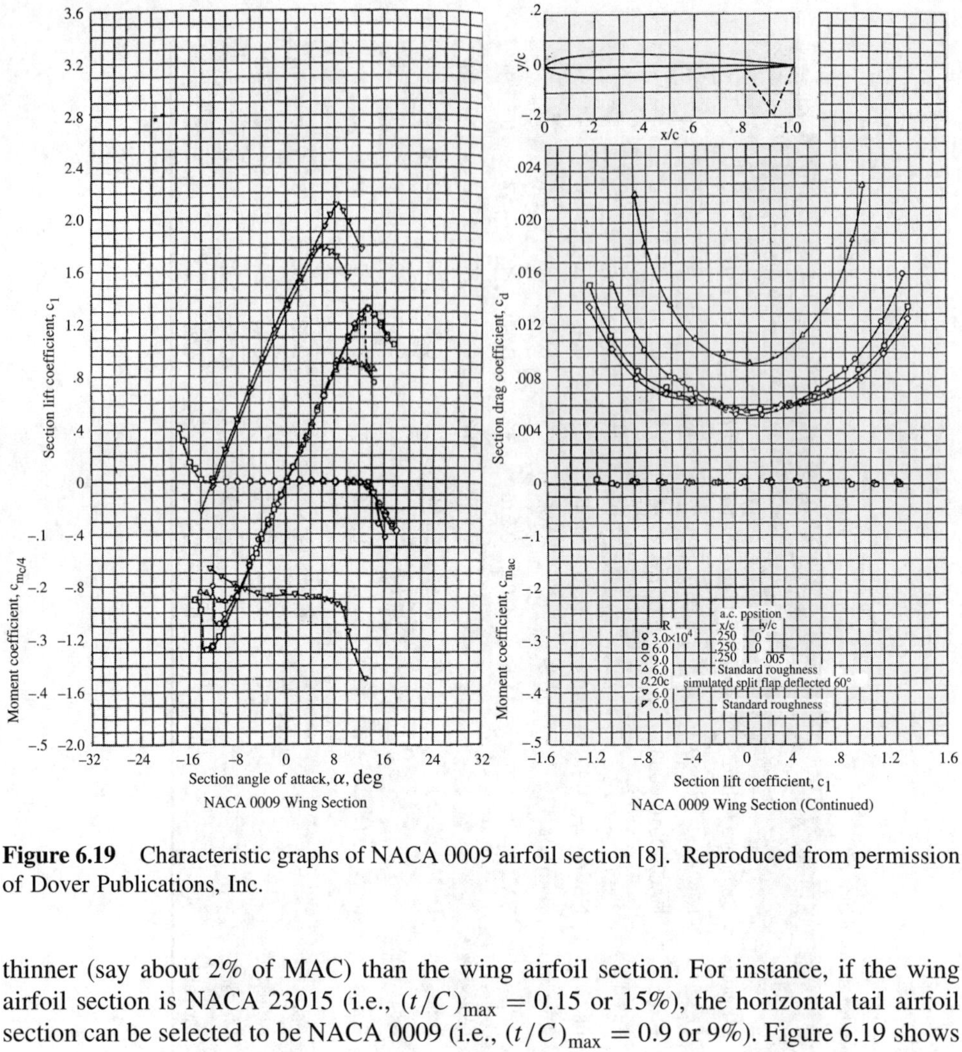

Figure 6.19 Characteristic graphs of NACA 0009 airfoil section [8]. Reproduced from permission of Dover Publications, Inc.

thinner (say about 2% of MAC) than the wing airfoil section. For instance, if the wing airfoil section is NACA 23015 (i.e., $(t/C)_{\max} = 0.15$ or 15%), the horizontal tail airfoil section can be selected to be NACA 0009 (i.e., $(t/C)_{\max} = 0.9$ or 9%). Figure 6.19 shows the characteristic graphs of the NACA 0009 airfoil section.

In an aircraft with an aft tail configuration, when the center of gravity (most of the time) is behind the wing/fuselage aerodynamic center, the horizontal tail must produce a negative lift to longitudinally trim the aircraft. If the aircraft center of gravity range is such that the tail must produce a negative lift coefficient most of the time, an inverted non-symmetric airfoil section may be utilized. This is the case for the cargo aircraft Lockheed C-130B tail airfoil section.

6.7.4 *Tail Incidence*

When a fixed tail configuration is adopted, the horizontal tail setting angle (i.e., tail incidence) i_t must be determined. The tail setting angle's (i_t) primary requirement is to nullify the pitching moment about cg at cruising flight. This is the longitudinal trim requirement through which the tail is generating a lift to counteract all other aircraft pitching moments. The tail incidence is determined to satisfy the trim design requirement when no control surface (i.e., elevator) is deflected. Although this fixed setting angle satisfies only one flight condition, it must be such that a mild change (through the application of an elevator) is necessary to trim the aircraft in other flight situations.

Looking at the C_L-α graph of the tail airfoil section (such as in Figure 6.19), it is noticed that the tail angle of attack is simply a function of the tail lift coefficient. Therefore, as soon as the tail lift coefficient is known, the tail incidence is readily determined by using this graph as the corresponding angle. As already discussed in Section 6.2, the tail lift coefficient is obtained from the non-dimensional longitudinal trim equation such as Equation (6.29):

$$C_{m_{o\_wf}} + C_L\left(h - h_o\right) - \eta_h \overline{V}_H C_{L_h} = 0 \tag{6.29}$$

In summary, the desired tail lift coefficient is calculated through Equation (6.29), and then the tail incidence will be determined by using the C_L-α graph of the tail airfoil section:

$$C_{L_{\alpha\_h}} = \frac{C_{L_h}}{\alpha_h} \Rightarrow \alpha_h = \frac{C_{L_h}}{C_{L_{\alpha\_h}}} \tag{6.51}$$

This is an initial value for the setting angle and will be revised in the later design phases. The typical value would be about -1 deg. In case the tail configuration is adjustable, the highest incidence (usually a positive angle) and the lowest incidence (usually a negative angle) must be determined. For instance, the large transport aircraft Boeing 727 has an adjustable tail with $+4$ deg for most positive incidence and -12.5 deg for most negative incidence. Table 6.5 introduces the horizontal tail setting angles for several aircraft. So the horizontal tail angle of attack in this aircraft is negative most of the time.

Another factor influencing the value of the tail setting angle is the requirement for longitudinal static stability. Several parameters will affect the aircraft longitudinal static stability, but it can be shown that the "longitudinal dihedral" will have a positive impact on the longitudinal static stability. The term *longitudinal dihedral* was invented by tail designers to transfer the technical meaning of the wing dihedral angle (Γ) from the yz plane to a similar angle in the aircraft xz plane. As the aircraft lateral stability benefits from the wing and tail dihedral angles, the aircraft longitudinal stability will be improved by a geometry referred to as the aircraft longitudinal dihedral angle. When the horizontal tail chord line and wing chord line can form a V-shape, it is said that the aircraft has longitudinal dihedral.

There are a few other technical interpretations for the longitudinal dihedral as follows:

1. When the wing (or fore plane, such as a canard) setting angle is positive and the horizontal tail (or aft plane, such as the wing in a canard configuration) angle is negative, the aircraft is said to have longitudinal dihedral:

$$i_w > i_h$$

2. When the wing (or fore plane) lift coefficient is higher than that of the horizontal tail (or fore plane), the aircraft is said to have longitudinal dihedral:

$$C_{L_w} > C_{L_h}$$

3. When the wing (or fore plane) zero-lift angle of attack is higher than that of the horizontal tail (or aft plane), the aircraft is said to have longitudinal dihedral:

$$\alpha_{o_w} = \alpha_{o_h}$$

4. When the wing (or fore plane) effective angle of attack is higher than that of the horizontal tail (or aft plane), the aircraft is said to have longitudinal dihedral.

These four above-mentioned definitions are very similar, but it seems that the last one (see Figure 6.20) is technically more accurate. Hence, in determining the horizontal tail setting angle, make sure that the aircraft has longitudinal dihedral. So this requirement is as follows:

$$\begin{aligned} &\alpha_{\text{eff}_w} > \alpha_{\text{eff}_t} \quad \text{(conventional configuration)} \\ &\alpha_{\text{eff}_c} > \alpha_{\text{eff}_w} \quad \text{(canard configuration)} \end{aligned} \tag{6.52}$$

The difference between the tail setting angle and the effective tail angle of attack needs to be clarified. Due to the presence of the downwash at the horizontal tail location, the tail effective angle of attack is defined as follows:

$$\alpha_h = \alpha_f + i_h - \varepsilon \tag{6.53}$$

where α_f is the fuselage angle of attack and ε is the downwash at the tail (see Figure 6.21).

The fuselage angle of attack is defined as the angle between the fuselage center line and the aircraft flight path (V_∞). The downwash is the effect of the wing trailing vortices on the flow field after passing through the wing airfoil section. Each trailing vortex causes a downflow at and behind the wing and an upflow outboard of the wing. The downwash is constant along the span of a wing with elliptical lift distribution. The downwash is a function of wing angle of attack (α_w) and is determined [2] as follows:

$$\varepsilon = \varepsilon_o + \frac{\partial \varepsilon}{\partial \alpha}\alpha_w \tag{6.54}$$

where ε_o (downwash angle at zero angle of attack) and $\frac{\partial \varepsilon}{\partial \alpha}$ (downwash slope) are found as:

$$\varepsilon_o = \frac{2C_{L_w}}{\pi \cdot \text{AR}} \tag{6.55}$$

$$\frac{\partial \varepsilon}{\partial \alpha} = \frac{2C_{L_{\alpha\_w}}}{\pi \cdot \text{AR}} \tag{6.56}$$

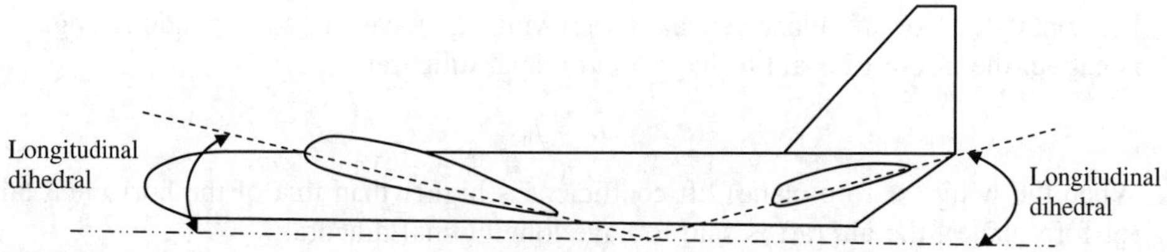

Figure 6.20 Longitudinal dihedral (angle is exaggerated)

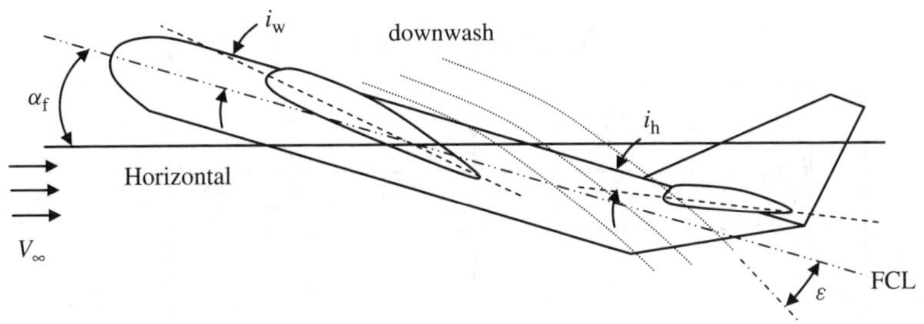

Figure 6.21 Horizontal tail effective angle of attack (downwash is exaggerated)

The wing lift curve slope ($C_{L_{\alpha\_w}}$) is in 1/rad and ε is in rad. The parameter C_{L_w} is the wing lift coefficient. The typical value for ε_o is about 1 deg and $\frac{\partial \varepsilon}{\partial \alpha}$ is about 0.3 rad/rad. The ideal value for the horizontal tail setting angle (i_h) is zero; however, it is usually a few degrees close to zero (+ or −). The exact value for i_h is obtained in the calculation process as described in this section.

An intermediate horizontal tail parameter that must be determined is its lift curve slope ($C_{L_{\alpha_h}}$). Since the horizontal tail is a lifting surface, similar to the wing, the horizontal tail lift curve slope (3D) is determined [9], 10] as follows:

$$C_{L_{\alpha\_h}} = \frac{dC_{L_h}}{d\alpha_h} = \frac{C_{l_{\alpha\_h}}}{1 + \dfrac{C_{l_{\alpha\_h}}}{\pi \cdot AR_h}} \tag{6.57}$$

where $C_{l_{\alpha\_h}}$ denotes the horizontal tail airfoil section lift curve slope (2D).

6.7.5 Aspect Ratio

The definition, the benefits, and the parameters affecting the aspect ratio were explained in Section 5.6, so they are not repeated here. The tail aspect ratio has influences on the aircraft lateral stability and control, aircraft performance, tail aerodynamic efficiency, and aircraft center of gravity. Most of the tail aspect ratio benefits are very similar to those of the wing benefits, but on a smaller scale. The tail designer is encouraged to consult with Section 5.6 for more information. Similar to the wing, the tail aspect ratio is defined as the ratio between the tail span and the tail MAC:

$$AR_h = \frac{b_h}{\overline{C}_h} \tag{6.58}$$

The tail aspect ratio (AR_h) tends to have a direct effect on the tail lift curve slope. As the tail aspect ratio is increased, the tail lift curve slope is increased. There are several similarities between the wing and the horizontal tail in terms of the aspect ratio, but on a smaller scale. The differences are as follows:

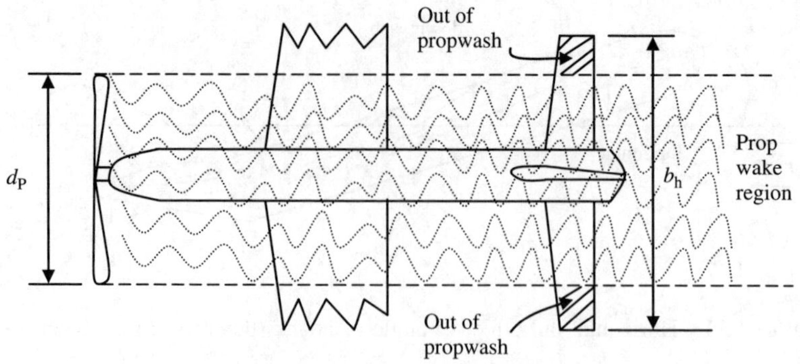

Figure 6.22 The tail span and propwash

1. The elliptical lift distribution is not required for the tail.
2. A lower aspect ratio is desirable for the tail, compared with that of the wing. The reason is that the deflection of the elevator creates a large bending moment at the tail root. Hence, a lower aspect ratio results in a smaller bending moment.
3. In a single-engine prop-driven aircraft, it is recommended to have an aspect ratio such that the tail span (b_h) is longer than the propeller diameter (d_P) (see Figure 6.22). This provision insures that the tail flow field is fresh and clean of wake and out of the propellor wash area. Therefore, the efficiency of the tail (η_h) will be increased.

Based on the above reasoning, an initial value for the tail aspect ratio may be determined as follows:

$$AR_h = \frac{2}{3} AR_w \tag{6.59}$$

A typical value for the horizontal tail aspect ratio is about 3–5. Table 6.5 illustrates the horizontal tail aspect ratio for several aircraft. The final value for the tail aspect ratio will be determined based on the aircraft stability and control, cost, and performance analysis evaluations after the other aircraft components have been designed.

6.7.6 Taper Ratio

The definition, the benefits, and the parameters affecting the taper ratio were explained in Section 5.7, so they are not repeated here. The tail taper ratio has influences on the aircraft lateral stability and control, aircraft performance, tail aerodynamic efficiency, and aircraft weight and center of gravity. Most of the tail taper ratio benefits are very similar to those of the wing benefits, but on a smaller scale. The tail designer is encouraged to consult with Section 5.7 for more information. Similar to the wing, the tail taper ratio (λ_h) is defined as the ratio between the tail tip chord and the tail root chord:

$$\lambda_h = \frac{C_{h_{tip}}}{C_{h_{root}}} \tag{6.60}$$

Thus, the value is between zero and one. The major difference from the wing taper ratio is that the elliptical lift distribution is not a requirement for the tail. Thus the main motivation behind the value for the tail taper ratio is to lower the tail weight.

For this reason, the tail taper ratio is typically smaller than the wing taper ratio. The tail taper ratio is typically between 0.7 and 1 for GA aircraft and between 0.4 and 0.7 for transport aircraft. For instance, the transport aircraft Boeing B-727 and Boeing B-737 (Figure 6.12) have a tail taper ratio of 0.4 and the Airbus A-300 has a tail taper ratio of 0.5. Table 6.5 shows the horizontal tail taper ratio for several aircraft. The final value for the tail taper ratio will be determined based on the aircraft stability and control, cost, and performance analysis evaluations after the other aircraft components have been designed.

6.7.7 Sweep Angle

The definition, benefits, and parameters affecting the sweep angle were explained in Section 5.9, so they are not repeated here. The sweep angle is normally measured either relative to the leading edge or relative to the quarter chord line. Similar to the wing, the tail leading edge sweep angle ($\Lambda_{h_{LE}}$) is defined as the angle between the tail leading edge and the y-axis in the xy plane. The horizontal tail sweep angle has influences on the aircraft longitudinal and lateral stability and control, aircraft performance, tail aerodynamic efficiency, and aircraft center of gravity. Most of the tail sweep angle effects are very similar to those of the wing effects, but on a smaller scale. The tail designer is encouraged to consult Section 5.9 for more information. The value of the horizontal tail sweep angle is often the same as the wing sweep angle.

Table 6.5 shows the horizontal tail sweep angle for several aircraft. As an initial selection in the preliminary design phase, select the value of the tail sweep angle to be the same as the wing sweep angle. The final value for the tail sweep angle will be determined based on the aircraft stability and control, cost, and performance analysis evaluations after the other aircraft components have been designed.

6.7.8 Dihedral Angle

The definition, benefits, and parameters affecting the dihedral angle were explained in Section 5.11, so they are not repeated here. Similar to the wing, the tail dihedral angle (Γ_h) is defined as the angle between each tail half section and the y-axis in the yz plane. The horizontal tail dihedral angle makes a contribution to the aircraft lateral stability and control, aircraft performance, and tail aerodynamic efficiency. Most of the tail dihedral angle contributions are very similar to those of the wing effects, but on a smaller scale. The tail designer is encouraged to consult Section 5.11 for more information.

The value of the horizontal tail dihedral angle is often the same as the wing sweep angle. In some cases, the tail dihedral angle is totally different from the wing dihedral angle. There are several reasons for this difference, including a need for aircraft lateral stability adjustment (e.g., a few transport aircraft, such as the tail dihedral of −3 deg for the Boeing 727), a need for lateral control adjustment (e.g., fighters such as the McDonnell Douglas F-4 Phantom), and a need for a reduction in aircraft height and operational requirements (e.g., unmanned aircraft Predator). Table 6.5 shows the tail dihedral angle for several aircraft. In some aircraft instances, the manufacturing limits and considerations force the designer not to employ any dihedral for the wing. So the need for lateral stability requires a large dihedral for the tail. As an initial selection in the preliminary design phase, select the value of the tail dihedral angle to be the same as the wing dihedral angle. The final value for the tail dihedral angle will be determined based on the aircraft stability and control, and performance analysis evaluations after the other aircraft components have been designed.

6.7.9 Tail Vertical Location

In an aircraft with aft tail configuration, the height of the horizontal tail relative to the wing chord line must be decided. In a conventional aircraft, the horizontal tail has two options for installation: (i) at the fuselage aft section and (ii) at the vertical tail. Beside the structural considerations and complexities, the horizontal tail efficiency and its contribution to aircraft longitudinal and lateral stability must be analyzed. Unlike the wing vertical location, there are no locations for the tail such as low tail, mid-tail or high tail. However, the low tail implies a conventional tail, the high tail implies a T-tail, and the mid-tail implies a cruciform tail.

A complete aircraft computational fluid dynamic model allows the designer to find the best location in order to increase the effectiveness of the tail. There are a few components that are sources of interference with the tail effectiveness. These include the wing, fuselage, and engine.

The wing influences the horizontal tail via downwash, wake, and tailing vortices. In general, the wing downwash decreases the tail effective angle of attack. Moreover, the wing wake degrades the tail efficiency, reduces the tail efficiency (h_t), and decreases the tail dynamic pressure. The most important consideration in the location of the horizontal tail relative to the wing is the prevention of deep stall. The horizontal tail location must not be in the wing wake region when wing stall happens. As Figure 6.23 illustrates, there are three major regions for tail installation behind the wing: (i) out of the wake region and downwash, (ii) inside the wake region but out of the wing downwash, and (iii) out of the wake region but affected by the downwash. In terms of deep stall avoidance criteria,

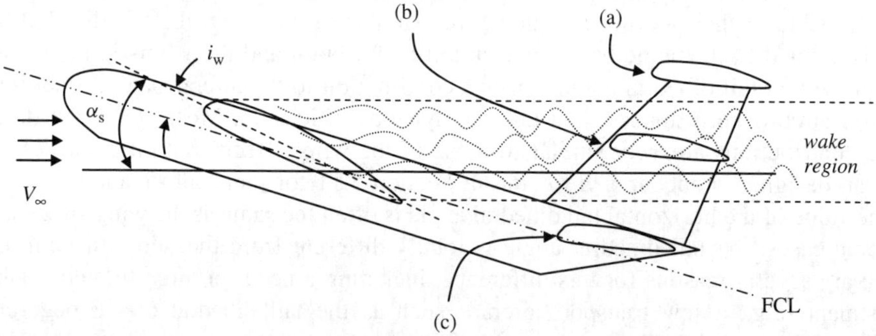

Figure 6.23 An aircraft with three tail installation locations when the wing stalls. (a) Out of wake region and downwash; (b) Inside wake region but out of wing downwash; (c) Out of wake region but affected by downwash

region (i) is the best and safest. Region (iii) is safe from deep stall and pitch-up, but the tail is not efficient. Region (ii) is not safe and not recommended for horizontal tail installation.

The decision about the vertical height of the horizontal tail must be made after a thorough analysis, since a variety of parameters – including wing airfoil, tail airfoil, wing/fuselage aerodynamic pitching moment, and tail arm – plus manufacturing considerations contribute. The following experimental equations are recommended for the initial approximation of the horizontal tail vertical height:

$$h_t > l \cdot \tan\left(\alpha_s - i_w + 3\right) \tag{6.61}$$

$$h_t < l \cdot \tan\left(\alpha_s - i_w - 3\right) \tag{6.62}$$

where h_t is the vertical height of the horizontal tail relative to the wing aerodynamic center, l is the horizontal tail moment arm, α_s is the wing stall angle (in degrees), and i_w denotes the wing incidence (in degrees).

The fuselage interferes with the tail through fuselage wake and sidewash. The reader is referred to aerodynamic textbooks for the details. In a multi-engine jet aircraft, the engine's hot and high-speed gasses have both positive and negative effects. High-speed gas increases the tail dynamic pressure, while hot gas creates a fatigue problem for the tail structure. If the tail is made of composite materials, make sure that the tail is out of the engine exhaust area. Hence, the horizontal tail location is the output of a compromise process to satisfy all design requirements.

6.7.10 Other Tail Geometries

Other horizontal tail geometries include the tail span (b_h), tail tip chord ($C_{h_{tip}}$), tail root chord ($C_{h_{root}}$), and tail MAC ($\overline{C}_h$ or MAC_h). These four tail parameters are sketched in Figure 6.24, which shows the top view of an aircraft aft section. These unknowns are determined by solving the following four equations simultaneously:

$$AR_h = \frac{b_h}{\overline{C}_h} \tag{6.63}$$

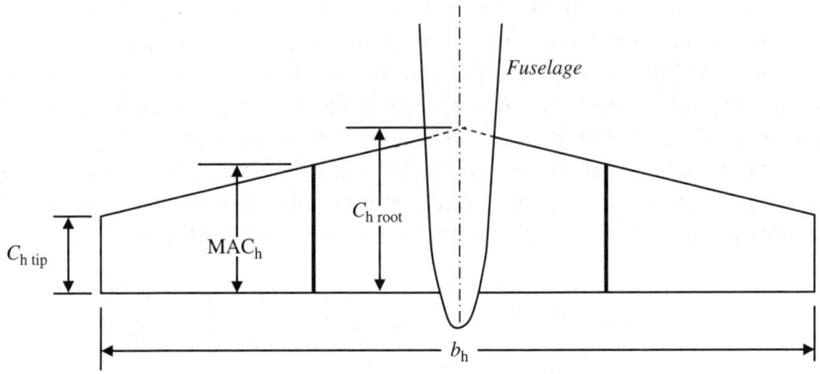

Figure 6.24 Horizontal tail geometry

$$\lambda_h = \frac{C_{h_{tip}}}{C_{h_{root}}} \tag{6.64}$$

$$\overline{C}_h = \frac{2}{3} C_{h_{root}} \left(\frac{1 + \lambda_h + \lambda_h^2}{1 + \lambda_h} \right) \tag{6.65}$$

$$S_h = b_h \cdot \overline{C}_h \tag{6.66}$$

The first two equations have been introduced previously in this section, but the last two equations are reproduced from wing geometry governing equations (see Chapter 5). The required data to solve these equations are the tail planform area, tail aspect ratio, and tail taper ratio.

6.7.11 Control Provision

One of the secondary functions of the horizontal tail is aircraft longitudinal control. The horizontal tail must generate a variety of tail lift forces in various flight conditions to longitudinally trim the aircraft and create the new trim conditions. For this purpose, a fixed and an adjustable horizontal tail have movable sections, which in a conventional aircraft are called elevators. Therefore, in designing the horizontal tail, one must consider some provisions for future control applications. These provisions include insuring sufficient space for the elevator's area, span, and chord as well as the elevator deflection angle to allow for an effective longitudinal control. The design of the aircraft control surfaces, including the elevator design, is examined in Chapter 12.

6.7.12 Final Check

When all horizontal tail parameters have been determined, two design requirements must be examined: (i) aircraft longitudinal trim and (ii) aircraft static and dynamic longitudinal stability. In the analysis of the longitudinal trim, the tail lift coefficient needs to be calculated. The generated horizontal tail lift coefficient should be equal to the required cruise tail lift coefficient. There are several aerodynamic software packages and tools to calculate the horizontal tail lift coefficient. In the early stage of design, it is recommended to employ the lifting line theory as described in Chapter 5. When a whole aircraft is designed, modern CFD software is utilized to determine the aerodynamic features of the aircraft including a horizontal tail. If the longitudinal trim requirements are not satisfied, the horizontal tail parameters such as tail incidence must be adjusted.

The static longitudinal stability is examined through the sign of the longitudinal stability derivative C_{m_α} or the location of the aircraft neutral point. For an aircraft with a fixed aft tail, the aircraft static longitudinal stability derivative is determined [6] as:

$$C_{m_\alpha} = C_{L_{\alpha\_wf}} \left(h - h_o\right) - C_{L_{\alpha\_h}} \eta_h \frac{S_h}{S} \left(\frac{l}{\overline{C}} - h \right) \left(1 - \frac{d\varepsilon}{d\alpha} \right) \tag{6.67}$$

When the derivative C_{m_α} is negative or when the neutral point is behind the aircraft cg, the aircraft is said to be statically longitudinally stable.

The dynamic longitudinal stability analysis is performed after all aircraft components are designed and the roots (λ) of the longitudinal characteristic equation are calculated. A general form of the aircraft longitudinal characteristic equation looks like the following:

$$A_1\lambda^4 + B_1\lambda^3 + C_1\lambda^2 + D_1\lambda + E_1 = 0 \quad (6.68)$$

where the coefficients A_1, B_1, C_1, D_1, and E_1 are functions of several stability derivatives, such as C_{m_α} and C_{m_q}. An aircraft is dynamically longitudinally stable if the real parts of all roots of the longitudinal characteristic equation are negative. Another way to analyze dynamic longitudinal stability is to make sure that longitudinal modes (i.e., short period and long period (phugoid)) are damped.

The reader is encouraged to consult Ref. [1] to see how to derive the aircraft longitudinal characteristic equation. The longitudinal stability derivatives cannot be determined unless all aircraft components, including wing and fuselage, have been designed. This is why we resort to a simplifying criterion that could be a base for the horizontal tail preliminary design. When the horizontal tail volume coefficient ($\overline{V}_H$) is close to the ballpark number (see Table 6.5), we are 90% confident that the longitudinal stability requirements have been satisfied. When the other aircraft components such as fuselage and wing have been designed, the horizontal tail design will be revised and optimized in the longitudinal stability analysis process.

6.8 Vertical Tail Design

6.8.1 Vertical Tail Design Requirements

The third lifting surface in a conventional aircraft is the vertical tail, which is sometimes referred to as a vertical stabilizer or fin. The vertical tail tends to have two primary functions: (i) directional stability and (ii) directional trim. Moreover, the vertical tail is a major contributor in maintaining directional control, which is the primary function of the rudder. These three design requirements are described briefly in this section:

1. The primary function of the vertical tail is to maintain the aircraft directional stability. The static and dynamic directional stability requirements were discussed in Section 6.3. In summary, the stability derivatives C_{n_β} must be positive (to satisfy the static directional stability requirements), but the stability derivatives C_{n_r} must be negative (to satisfy the dynamic directional stability requirements). Two major contributors to the value of these stability derivatives are the vertical tail area (S_v) and the vertical tail moment arm (l_v). If the vertical tail area is large enough and the vertical tail moment arm is long enough, the directional stability requirements could easily be satisfied. The directional stability analysis is performed after all aircraft components have been designed and the roots (λ) of the lateral-directional characteristic equation calculated. A general form of the aircraft lateral-directional characteristic cquation looks like the following:

$$A_2\lambda^4 + B_2\lambda^3 + C_2\lambda^2 + D_2\lambda + E_2 = 0 \quad (6.69)$$

where the coefficients A_2, B_2, C_2, D_2, and E_2 are functions of several stability derivatives such as C_{n_β} and C_{n_r}. An aircraft is dynamically directionally stable if the real parts of all roots of the lateral-directional characteristic equation are negative. Another way to analyze dynamic directional stability is to make sure that directional modes (i.e., dutch-roll and spiral) are damped.

The reader is encouraged to consult Ref. [1] to see how to derive the aircraft lateral-directional characteristic equation. The directional stability derivatives cannot be determined unless all aircraft components, including wing and fuselage, have been designed. Hence, we have to resort to some other simplifying criterion that could be a base for the vertical tail preliminary design. Similar to the horizontal tail volume coefficient, a new parameter that is referred to as the vertical tail volume coefficient (V_v) is defined. If the value of this parameter is close to the ballpark number, we are 90% sure that the directional stability requirements have been satisfied. When other aircraft components have been designed, the vertical tail design will be revised and optimized in the directional stability analysis process. The vertical tail volume coefficient will be introduced in Section 6.8.2.

2. The second function of the vertical tail is to maintain the aircraft directional trim. As discussed in Section 6.3, the summation of all forces along the y-axis and the summation of all moments about the z-axis must be zero:

$$\sum F_y = 0 \tag{6.5}$$

$$\sum N_{cg} = 0 \tag{6.6}$$

An aircraft is normally manufactured symmetrical about the xz plane, so the directional trim is naturally maintained. Although this is an ideal case and is considered in the production of components such as right and left wing sections, in several cases there is a slight asymmetry in the aircraft's xy plane. One source for this asymmetry could be a difference between manufacturing jigs and fixtures of right and left sections (wing and tail). Another reason for directional asymmetry lies in the internal components inside the fuselage, such as the fuel system, electrical wiring, and even the load and cargo inside the load compartment.

However, in a single-engine prop-driven aircraft, the aircraft directional trim is disturbed by the rotation of the engine propeller. In a multi-engine prop-driven aircraft, with odd number of engines, a similar problem exists. Hence, the vertical tail is responsible for maintaining the directional trim by providing an opposing yawing moment about the z-axis. One of the critical parameters influencing the directional trim in such aircraft is the vertical tail incidence angle relative to the xz plane.

Another directional trim case is in multi-engine aircraft, where one engine is inoperative. In such a situation, the operative engines create a disturbing yawing moment and the only way to balance this asymmetric moment is the counteracting yawing moment generated by the vertical tail. A control surface (e.g., rudder) must be deflected to directionally trim the aircraft.

Although the vertical tail is contributing to the aircraft lateral stability and control, this item is not considered as a base for the design of the vertical tail. However, in the analysis of the vertical tail performance, the lateral stability must be studied. This is to make sure that the vertical tail is improving the aircraft lateral stability and not having a negative impact. Recall that the aircraft lateral stability is primarily a function of the wing parameters. The static and dynamic directional trim requirements were discussed in Section 6.2.

3. The third aircraft design requirement in which the vertical tail is a major contributor is directional control. Maneuvering operations such as turning flight and spin recovery are performed successfully using a movable section of the vertical tail which is called a rudder. The design of the rudder is examined in Chapter 12, but the spin recovery requirements will be discussed in Section 6.8.3.

6.8.2 Vertical Tail Parameters

Basically, the vertical tail parameters must be determined initially such that the directional stability requirements are satisfied. In the second and third stages of the vertical tail design process, the directional trim requirements and directional control requirements will be examined.

In the design of the vertical tail, the following parameters must be determined:

1. vertical tail location;
2. planform area (S_v);
3. tail arm (l_{vt});
4. airfoil section;
5. aspect ratio (AR_v);
6. taper ratio (λ_v);
7. tip chord (C_{t_v});
8. root chord (C_{r_v});
9. mean aerodynamic chord (MAC_v or C_v);
10. span (b_v);
11. sweep angle (Λ_v);
12. dihedral angle (Γ_v);
13. incidence (i_v).

Several of these vertical tail parameters are illustrated in Figure 6.25. The vertical tail is a lifting surface, whose aerodynamic force of lift is generated in the direction of the y-axis. In maintaining the directional stability, control and trim, an aerodynamic force along the y-axis needs to be created by the vertical tail (i.e., vertical tail lift L_v):

$$L_v = \frac{1}{2}\rho V^2 S_v C_{L_v} \tag{6.70}$$

where S_v is the vertical tail area and C_{L_v} is the vertical tail lift coefficient. The vertical tail lift generates a yawing moment about the z-axis:

$$N_{cg} = L_v l_v \tag{6.71}$$

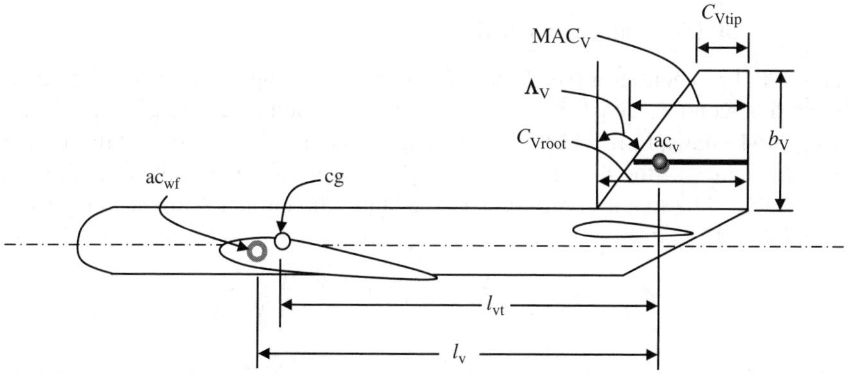

Figure 6.25 The vertical tail parameters

This moment must be large enough to maintain directional trim and must have a positive contribution to the directional stability. As explained in Section 6.8.1, a preliminary evaluation of the directional stability is applied through a parameter called the vertical tail volume coefficient ($\overline{V}_{\mathrm{v}}$):

$$\overline{V}_{\mathrm{v}} = \frac{l_{\mathrm{v}} S_{\mathrm{v}}}{bS} \tag{6.72}$$

where l_{v} is the distance between the vertical tail aerodynamic center (ac_{v}) and the wing/fuselage aerodynamic center (see Figure 6.25), S_{v} is the vertical tail planform area, b is the wing span, and S denotes the wing reference area. The vertical tail aerodynamic center is located at the quarter chord of the vertical tail MAC.

The vertical tail volume coefficient is a non-dimensional parameter which is a direct function of two significant vertical tail parameters: the vertical tail area (S_{v}) and the vertical tail moment arm (l_{v}). The two parameters of l_{v} and l_{vt} are very close, such that one can be determined from the other. The vertical tail volume coefficient is an indirect representative of the aircraft directional stability. A typical value for the vertical tail volume coefficient is between 0.02 and 0.12. Table 6.6 illustrates the vertical tail parameters including the vertical tail volume coefficient for several aircraft. Remember that the vertical tail planform area includes both the fixed section and the movable section (i.e., rudder).

Since the definitions and features of the lifting surface basic parameters – for example the aspect ratio, taper ratio, and airfoil section – have been presented in Chapter 5 and also in the horizontal tail design section (Section 6.7), they are introduced only briefly here.

6.8.2.1 Vertical Tail Location

In order to maintain directional stability, the only location for the vertical tail is aft of the aircraft center of gravity. Three possible candidates are: (i) aft of the fuselage, (ii) wing tips, and (iii) boom(s). If a single aft horizontal tail has been selected, the only place for the vertical tail is on top of the aft fuselage. The vertical tail cannot be placed in front of the fuselage (i.e., forward of the aircraft cg) since it makes the aircraft directionally unstable. The other two options, namely wing tips and boom, are appropriate for some special purposes that have been described earlier in Section 6.4.

6.8.2.2 Vertical Tail Moment Arm (l_{vt})

The vertical tail moment arm (see Figure 6.25) must be long enough to satisfy the directional stability, control, and trim requirements. In a spinnable aircraft, the vertical tail must also satisfy the spin recovery requirements. Increasing the vertical tail moment arm increases the values of the derivatives C_{n_β} and C_{n_r} and thus makes the aircraft directionally more stable. The major contributor to the static directional stability derivative (C_{n_β}) is the vertical tail [1]:

$$C_{n_\beta} \approx C_{n_{\beta\_v}} = K_{f1} C_{L_{\alpha\_v}} \left(1 - \frac{d\sigma}{d\beta}\right) \eta_{\mathrm{v}} \frac{l_{\mathrm{vt}} S_{\mathrm{v}}}{bS} \tag{6.73}$$

where $C_{L_{\alpha\_v}}$ denotes the vertical tail lift curve slope, $\frac{d\sigma}{d\beta}$ is the vertical tail sidewash gradient, and η_{v} is the dynamic pressure ratio at the vertical tail. The parameter K_{f1}

Table 6.6 Vertical tail characteristics for several aircraft

No.	Aircraft	Type	m_{TO} (kg)	Airfoil	$(t/C)_{max}$ (%)	$\overline{V}_V$	S_v/S	AR_v	Λ_v (deg)
1	Wright Flyer	First aircraft in history	420	Flat plate	Low	0.013	0.045	6.3	0
2	Cessna 177	GA single-prop engine	1 100	NACA 0009/0006	7.5	0.14	0.107	1.41	35
3	C-130 Hercules	Large turboprop cargo	70 305	NACA 64A-015	15	0.06	0.18	1.84	18.8
4	DC-9/10	Large jet transport	41 100	DSMA	11	0.08	0.19	0.95	43.5
5	Cessna Citation I	Business jet	5 375	NACA 0012/0008	10	0.0806	0.191	1.58	33
6	Fokker F-27	Turboprop transport	19 773	Modified NACA	15	0.07	0.203	1.55	33
7	Boeing 737-100	Large jet transport	50 300	–	12	0.11	0.27	1.88	35
8	Beechjet 400A	Business jet transport	7 303	–	12	0.123	0.263	1	55
9	DC-8-10	Large jet transport	141 000	DSMA-111/-112	9.85	0.05	0.122	1.91	35
10	Airbus 300B	Large jet transport	165 000	–	12.5	0.102	0.204	1.623	40
11	C-17A	Heavy jet cargo	265 352	–	9	0.08	0.195	1.36	36
12	Eurofighter 2000	Fighter	21 000	–	7	0.035	0.096	1.3	45
13	F-15 Eagle	Fighter	36 741	–	7	0.06	0.346[a]	1.3	35

[a] The aircraft has a twin vertical tail, so the areas of both vertical tails are included in the calculation.

represents the contribution of the fuselage to the aircraft C_{n_β} and depends strongly on the shape of the fuselage and its projected side area. The fuselage contribution to directional static stability tends to be strongly negative. The typical value of K_{f1} for a conventional aircraft is about 0.65–0.85. The value of C_{n_β} for a statically directionally stable aircraft is positive. A higher value for C_{n_β} implies a more directionally statically stable aircraft. The parameter l_{vt} in Equation (6.65) is in the numerator, which implies that a longer moment arm is desirable.

In addition, an increase in the vertical tail moment arm improves the directional and lateral control. In the early stage of the vertical tail design, where other aircraft components have not been designed, the vertical tail moment arm is selected to be equal to the horizontal tail moment arm (l). This assumption means that the vertical tail is located at the same distance from the wing as the horizontal tail. The assumption will be modified in the later design stage, when other aircraft components are designed and the aircraft directional and lateral stability, control, and trim are analyzed.

Another phenomenon that influences the vertical tail moment arm is spin. When an aircraft is spinnable, the aircraft is required to be able to recover from spin safely. Spin is a dangerous flight if the aircraft is not designed to recover safely from it. Some aircraft, however, are not spinnable by design. Most transport aircraft are not spinnable (i.e., spin resistant), while most fighters and maneuverable aircraft are spinnable.

A spin is an aggravated stall resulting in autorotation about the spin axis wherein the aircraft follows a screw path. Spins are characterized by a high angle of attack, low airspeed, high sideslip angle, and high rate of descent. In a spin, both wings are in a stalled condition; however, one wing will be in a deeper stall than the other. This causes the aircraft to autorotate due to the non-symmetric lift and drag. Spins can be entered unintentionally or intentionally. In either case, a specific and often counterintuitive set of actions are needed to influence recovery. If the aircraft exceeds published limitations regarding spin, or is loaded improperly, or if the pilot uses an incorrect technique to recover, the spin may lead to a crash.

The following is reproduced from Section 221 of PAR 23 of FAR [4], which concerns spinning of GA aircraft:

(a) Normal category airplanes. A single-engine, normal category airplane must be able to recover from a one-turn spin or a three-second spin, whichever takes longer, in not more than one additional turn after initiation of the first control action for recovery, or demonstrate compliance with the optional spin resistant requirements of this section.

(b) Utility category airplanes. A utility category airplane must meet the requirements of paragraph (a) of this section. In addition, the requirements of paragraph (c) of this section and §23.807(b)(7) must be met if approval for spinning is requested.

(c) Acrobatic category airplanes. An acrobatic category airplane must meet the spin requirements of paragraph (a) of this section and §23.807(b)(6). In addition, the following requirements must be met in each configuration for which approval for spinning is requested:

(1) The airplane must recover from any point in a spin up to and including six turns, or any greater number of turns for which certification is requested, in not more than one and one-half additional turns after initiation of the first control action for recovery. However, beyond three turns, the spin may be discontinued if spiral characteristics appear.

(2) The applicable airspeed limits and limit maneuvering load factors must not be exceeded. For flaps-extended configurations for which approval is requested, the flaps must not be retracted during the recovery.

(3) It must be impossible to obtain unrecoverable spins with any use of the flight or engine power controls either at the entry into or during the spin.

(4) There must be no characteristics during the spin (such as excessive rates of rotation or extreme oscillatory motion) that might prevent a successful recovery due to disorientation or incapacitation of the pilot.

When a spin occurs, all that is required mainly is a sufficient yaw rate while an aircraft is stalled. Hence, the vertical tail must be able to generate the yawing moment to stop autorotation. Thus, the vertical tail plays a vital role in spin recovery. The vertical tail may have a long moment arm, but there is a situation that could negatively influence the effectiveness of the vertical tail. If the vertical tail is in the horizontal tail wake region, it will lose its effectiveness. Therefore, the vertical tail moment arm needs to be determined so as to provide a wake-free region for the vertical tail.

An experimental rule for the vertical tail effectiveness to achieve a recoverable spin is as follows: at least 50% of the vertical tail planform area must be out of the horizontal tail wake region to be effective in the case of a spin. The horizontal tail wake region is considered to lie between two lines. The first line is drawn at the horizontal tail trailing edge with an orientation of 30 deg. The second line is drawn at the horizontal tail leading edge with an orientation of 60 deg.

So, even if the vertical tail moment arm is theoretically calculated to be sufficient, if the vertical tail is graphically located to be inside the horizontal tail wake region, the moment arm needs to be adjusted. It is clear that if the moment arm needs to be decreased, the vertical tail area must be increased. However, if the adjustment of the vertical tail arm leads to a larger arm, the vertical tail area could be decreased. Another technique to move the vertical tail out of the horizontal tail wake region is to employ a dorsal fin. A graphical method is illustrated in Figure 6.26. Figure 6.26(a) shows a vertical tail that is completely inside the wake region. This configuration does not satisfy spin recovery requirements. Figure 6.26(b) demonstrates a vertical tail that is completely out of the wake region. This configuration does satisfy spin recovery requirements. Figure 6.26(c) depicts a vertical tail that is partly inside the wake region. Although the moment arm of the vertical tail (l_v) in Figure 6.26(c) is shorter than that of the other two vertical tails, the advantage is that it is wake-free.

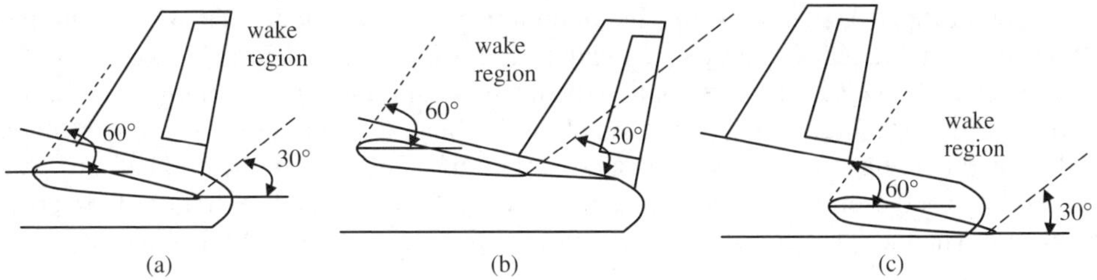

Figure 6.26 The vertical tail effectiveness and the wake region of the horizontal tail. (a) VT is in the wake region; (b) Part of the VT is in the wake region; (c) VT is out of the wake region

6.8.2.3 Planform Area (S_v)

The parameter S_v in Equation (6.65) is in the numerator, which implies a larger vertical tail area is desirable. The vertical tail area must be large enough to satisfy lateral-directional stability, control, and trim requirements. Increasing the vertical tail area increases the values of the derivatives C_{n_β} and C_{n_r} and thus makes the aircraft lateral-directionally more stable. In addition, an increase in the vertical tail area improves the directional and lateral control ($C_{n_{\delta R}}$, $C_{l_{\delta R}}$). If the vertical tail area is too small, the lateral-directional stability requirements will not be satisfied. In contrast, when the vertical tail area is too large, the aircraft will be lateral-directionally too stable, but the directional control requirements are not satisfied. Thus, the middle value is very hard to determine. For this reason, the vertical tail design utilizes a backward design technique. This means that we select a combination of vertical tail area and vertical tail moment arm in a ballpark area through a parameter called the vertical tail volume coefficient. Another criterion for the vertical tail area is that it must be small so as to minimize the manufacturing cost and the aircraft weight.

It is interesting to note that a typical value for the ratio between the vertical tail area and the wing area for a conventional GA aircraft is about 0.1–0.15. The vertical tail planform area is preliminarily determined based on the selection of the vertical tail volume coefficient ($\overline{V}_v$). A typical value for the vertical tail volume coefficient for several aircraft types is introduced in Table 6.4. Hence, the vertical tail area is determined as:

$$S_v = \frac{b \cdot S \cdot \overline{V}_v}{l_v} \tag{6.74}$$

where it is initially assumed that the parameter l_v is equal to the vertical tail moment arm (l_{vt}). This area will be adjusted in the later design stage after other aircraft components are designed and the aircraft directional and lateral stability, control, and trim are analyzed. The design of the vertical tail is one of the most difficult tasks for aircraft designers, since theoretical and experimental results may not match concerning the features of the vertical tail. It is often the case for several aircraft that the vertical tail area is found, in flight tests, insufficient to satisfy lateral-directional stability requirements.

If the aircraft is at the manufacturing stage, and the initial vertical tail design may not be changed, one solution to increase the vertical tail area is to employ a dorsal fin. A dorsal fin[4] (see Figure 6.27(a,b)) is generally a flat plate (i.e., no airfoil section) installed in front of the original vertical tail with a greater sweep angle. The other benefit of a dorsal fin is to reduce the minimum control speed (V_{mc}) during take-off operation (as employed in Piper Arapaho PA-40). In addition, it provides a hidden antenna feature that allows the com antennas to be located under the fin for further drag reduction.

Another approach to solve the small vertical tail area problem is to employ a ventral fin. A ventral fin[5] (see Figure 6.27(c)) is simply a flat plate (i.e., no airfoil section) installed under the aft fuselage (almost in the same longitudinal location as the vertical tail). It is also possible and useful to consider the airfoil section for dorsal and ventral fins to improve their aerodynamic characteristics. These two techniques improve the lateral-directional

[4] This term has been borrowed from fish anatomy. A *dorsal fin* is a polyphyletic fin located on the backs of some fish, whales, and dolphins.

[5] This term has been borrowed from fish anatomy.

(a)

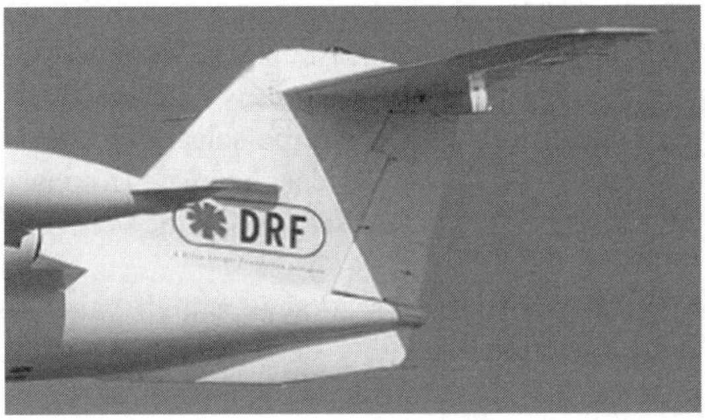

(b) (c)

Figure 6.27 Dorsal fin and ventral fin in three aircraft: (a) Beech 200 Super King Air (dorsal and ventral fin); (b) Gates Learjet 35A (ventral fin); (c) General Atomics Predator (ventral fin). Reproduced from permission of (a) Jenny Coffey; (b) Antony Osborne.

stability of an aircraft, while they do not touch the original vertical tail geometry. Table 6.6 shows the value for the ratio between the vertical tail area and the wing area for several aircraft. Figure 6.27 illustrates the dorsal and ventral fin of a Beech 200 Super King Air, the ventral fin of a Gates Learjet 35A, and the ventral fin of a General Atomics Predator.

The wing and horizontal tail have two right and left sections. But, unlike the wing and horizontal tail, the vertical tail has normally only one section. Thus, the vertical tail span (b_V) is the distance between the vertical tail tip chord and the root chord (see Figure 6.25). For this reason, the vertical tail aerodynamic center in a conventional aircraft is normally above the fuselage center line (and most of the time above the aircraft center of gravity).

6.8.2.4 Airfoil Section

The vertical tail airfoil section is responsible for the generation of the vertical tail lift coefficient (C_{L_V}). The airfoil must generate the required lift coefficient with a minimum drag coefficient. Recall that a non-symmetrical airfoil section creates an aerodynamic pitching moment. One of the basic aircraft design requirements is symmetricity about the xz plane. Therefore, to insure the symmetricity of the aircraft about the xz plane, the vertical airfoil section must be symmetric. Moreover, if the engines, wing, horizontal tail, and fuselage are designed to be symmetric about the *xz* plane, the vertical tail is not required to produce any lift to maintain directional trim under normal flight conditions.

Recall from Chapter 5 that the indication of a symmetric airfoil is that the second digit in a four-digit and the third digit in a five-digit and 6-series NACA airfoil section is zero. This denotes that the airfoil design lift coefficient and zero-lift angle of attack are both zero. NACA airfoil sections such as 0009, 0010, 0012, 63-006, 63-009, 63-012, 63-015, 63-018, 64-006, 64-012, 64A010, 65-009, 65-015, 66-012, 66-018, and 66-021 are all symmetric airfoils. In several GA aircraft, NACA airfoil sections 0009 or 0012 (with 9% or 12% maximum thickness-to-chord ratio) are employed for a vertical tail. Both of these NACA airfoil sections are symmetric.

In addition, another tail requirement is that the vertical tail must be clean of compressibility effect. To satisfy this requirement, the flow Mach number at the vertical tail must be less than the flow Mach number at the wing. This objective will be realized by selecting a vertical tail airfoil section to be thinner (say about 2% of MAC) than the wing airfoil section. For instance, if the wing airfoil section is NACA 23015 (i.e., $(t/C)_{max} = 0.15$ or 15%), the vertical tail airfoil section can be selected to be NACA 0009 (i.e., $(t/C)_{max} = 0.9$ or 9%). Figure 6.13 shows the characteristic graphs of the NACA 0009 airfoil section. Table 6.5 illustrates the airfoil section for the vertical tail of several aircraft.

The third desired feature for the vertical tail airfoil section is a high value for the lift curve slope ($C_{L\alpha\_v}$), since the static directional stability derivative ($C_{n\beta}$) is a direct function of $C_{L\alpha\_v}$ (Equation (6.72)). Thus, as a general rule, a symmetric airfoil section with a high-lift curve slope is desirable for the vertical tail. Recall that the theoretical value for an airfoil section is about 2π 1/rad. Table 6.6 shows the airfoil section of the vertical tail for several aircraft.

6.8.2.5 Incidence (i_v)

The vertical tail incidence is defined as the angle between the vertical tail chord line and the aircraft *xz* plane (when looking at the aircraft from the top). The vertical tail is responsible for the generation of the vertical tail lift coefficient (C_{L_v}). One of the basic aircraft design objectives is symmetricity about the *xz* plane. Hence, if the engines, wing, horizontal tail, and fuselage are designed to be symmetric about the *xz* plane, the vertical tail is not required to produce any lift to maintain directional trim in normal flight conditions. For this reason, the vertical tail incidence must initially be zero.

However, in a prop-driven aircraft with one single engine (or with an odd number of prop-driven engines), the lateral trim is disturbed by the revolution of the propeller and engine shaft about the x-axis. The aircraft body is going to roll as a reaction to the rotation of the propeller and its shaft (recall the third law of Newton). Although this rolling moment is not large, safety requirements require the trim to be maintained and aircraft roll to be avoided. To nullify this yawing moment, the vertical tail is required to generate a lift and cancel this rolling moment. One solution for this problem is to consider a few degrees of incidence for the vertical tail. The vertical tail in most single-engine prop-driven aircraft has about 1–2 deg of incidence to insure the prevention of aircraft roll in a reaction to propeller revolution. Another solution is to select a non-symmetric airfoil for the vertical tail, but this technique has several disadvantages. The exact value for the vertical tail incidence is determined by calculating the propeller rotation's rolling moment. An experimental approach would be more accurate.

6.8.2.6 Aspect Ratio (AR_v)

The vertical tail aspect ratio is defined as the ratio between the vertical tail span b_v (see Figure 6.25) and the vertical tail MAC ($\overline{C}_v$):

$$AR_v = \frac{b_v}{\overline{C}_v} \tag{6.75}$$

The general characteristics of the aspect ratio are introduced in Chapter 5 (see Section 5.6), so they are not repeated here. The vertical tail aspect ratio has several other features that impact various aircraft characteristics. These must be noticed in determining the vertical tail aspect ratio:[6]

1. First of all, a high aspect ratio results in a tall vertical tail that causes the aircraft's overall height to be increased. Many aircraft, especially large transport aircraft and fighter aircraft, have parking limitations in the hangar space. Thus, an aircraft is not allowed to have an overall height beyond a pre-specified value.
2. A high tail aspect ratio weakens the aircraft lateral control, since the vertical tail mass moment of inertia about the x-axis is increased.
3. A vertical tail with a high aspect ratio has a longer yawing moment arm compared with a low aspect ratio vertical tail. Hence, an aircraft with high aspect ratio has a higher directional control.
4. As the vertical aspect ratio is increased, the bending moment and bending stress at the vertical tail root increase, which causes the aft portion of the aircraft to be heavier.
5. A high aspect ratio vertical tail is prone to fatigue and flutter.
6. A high aspect ratio vertical tail is longitudinally destabilizing, since the vertical tail drag generates a nose-up pitching moment.
7. As the aspect ratio of the vertical tail is increased, the aircraft directional stability is improved, due to an increase in the yawing moment arm.
8. As the aspect ratio of the vertical tail is increased, the vertical tail-induced drag is increased.
9. If the aircraft has a T-tail configuration, the horizontal tail location and efficiency are functions of vertical tail aspect ratio. Thus, if deep stall is a major concern, the vertical aspect ratio must be large enough to keep the horizontal tail out of the wing wake when the wing stalls.
10. A high aspect ratio vertical tail is aerodynamically more efficient (i.e., has a higher $(L/D)_{max}$) than a vertical tail with a low aspect ratio. The reason is the vertical tail tip effect.

The above-mentioned advantages and disadvantages for a high and low aspect ratio are general guidelines for the vertical tail designer. As a starting point, a value between 1 and 2 is recommended for the vertical tail aspect ratio. The final value will be determined in

[6] Reference [10] defines the vertical tail aspect ratio as $1.55(b/C)$.

the overall aircraft directional stability analysis. Table 6.6 shows the value for the aspect ratio of a vertical tail for several aircraft.

6.8.2.7 Taper Ratio (λ_v)

As with other lifting surfaces (e.g., wing and horizontal tail), the vertical tail taper ratio is defined as the ratio between the vertical tail tip chord $C_{V_{tip}}$ (see Figure 6.25) and the vertical tail root chord $C_{V_{root}}$:

$$\lambda_V = \frac{C_{V_{tip}}}{C_{V_{root}}} \tag{6.76}$$

General features of the taper ratio are introduced in Section 5.7, so they are not repeated here. The main purposes of the taper ratio are: (i) to reduce the bending stress on the vertical tail root and also (ii) to allow the vertical tail to have a sweep angle. The application of the taper ratio adds a complexity to the tail manufacturing process and also increases the empennage weight. As the taper ratio of the vertical tail is increased, the yawing moment arm is reduced which reduces the directional control of the aircraft. Moreover, an increase in the taper ratio of the vertical tail would reduce the lateral stability of the aircraft. A compromise between these positive and negative features determines the value for the vertical tail taper ratio.

6.8.2.8 Sweep Angle (Λ_v)

The general features of the sweep angle were introduced in Section 5.9, so they are not repeated here. As the sweep angle of the vertical tail is increased, the yawing moment arm is increased which improves the directional control of the aircraft. Subsequently, an increase in the vertical tail sweep angle weakens the aircraft directional stability, since the mass moment inertia about the z-axis is increased. If the aircraft has a T-tail configuration, an increase in the vertical tail sweep angle increases the horizontal tail moment arm which improves the aircraft longitudinal stability and control.

Another reason for the application of the vertical tail sweep angle is to decrease the wave drag in high-subsonic and supersonic flight regimes. For this reason, it is suggested to initially adopt a sweep angle similar to the sweep angle of the wing. The final value for the vertical tail sweep angle will be the result of a compromise between these positive and negative features. Table 6.6 shows the value for the sweep angle of a vertical tail for several aircraft.

6.8.2.9 Dihedral Angle (Γ_v)

Due to the aircraft symmetry requirement about the xz plane, an aircraft with one vertical tail is not allowed to have any dihedral angle. However, if the aircraft has a twin vertical tail (such as a few fighters), the dihedral angle has a positive contribution to the aircraft lateral control. But it reduces the aerodynamic efficiency of the vertical tails, since two vertical tails will cancel part of their lift forces. In addition, the vertical tail dihedral angle will contribute to detectability features of the aircraft. For instance, McDonnell Douglas F-15 Eagle (Figure 9.14) twin vertical tails canted 15 deg to reduce the radar cross section. The exact value for the dihedral angles of a twin vertical tail is determined in the overall aircraft lateral- directional stability analysis process.

6.8.2.10 Tip Chord (C_{t_v}), Root Chord (C_{r_v}), Mean Aerodynamic Chord ($\mathrm{MAC_v}$ or $\mathrm{C_v}$), and Span ($\mathrm{b_v}$)

The other vertical tail geometries include vertical tail span (b_v), vertical tail tip chord ($C_{V_{tip}}$), vertical tail root chord ($C_{V_{root}}$), and vertical tail MAC ($\overline{C}_V$ or $\mathrm{MAC_v}$). These unknown parameters (see Figure 6.25) are determined by solving the following four equations simultaneously:

$$\mathrm{AR_v} = \frac{b_V}{\overline{C}_V} = \frac{b_V^2}{S_V} \tag{6.77}$$

$$\lambda_V = \frac{C_{V_{tip}}}{C_{V_{root}}} \tag{6.78}$$

$$\overline{C}_V = \frac{2}{3} C_{V_{root}} \left(\frac{1 + \lambda_V + \lambda_V^2}{1 + \lambda_V} \right) \tag{6.79}$$

$$S_V = b_V \cdot \overline{C}_V \tag{6.80}$$

The first two equations have been introduced previously in this section, but the last two equations are reproduced from wing geometry governing equations (see Chapter 5). The required data to solve these equations are the vertical tail planform area, vertical tail aspect ratio, and vertical tail taper ratio.

6.9 Practical Design Steps

The tail design flowchart was presented in Section 6.1. Fundamentals of the tail primary functions and design requirements were reviewed in Sections 6.2 and 6.3. Sections 6.4–6.8 introduced the various tail configurations, horizontal tail parameters, vertical tail parameters, and the technique to determine each parameter. The purpose of this section is to outline the practical design steps of the tail. The tail design procedure is as follows:

1. Select tail configuration (Sections 6.4 and 6.7).
   Horizontal tail
2. Select horizontal tail location (aft or forward (canard); Section 6.5).
3. Select horizontal tail volume coefficient, $\overline{V}_H$ (Table 6.4).
4. Calculate optimum tail moment arm (l_{opt}) to minimize the aircraft drag and weight (Section 6.6).
5. Calculate horizontal tail planform area, S_h(Equation (6.24)).
6. Calculate wing/fuselage aerodynamic pitching moment coefficient (Equation (6.26)).
7. Calculate cruise lift coefficient, C_{L_C} (Equation (6.27)).
8. Calculate horizontal tail desired lift coefficient at cruise from trim (Equation (6.29)).

9. Select horizontal tail airfoil section (Section 6.7).
10. Select horizontal tail sweep angle and dihedral (Section 6.7).
11. Select horizontal tail aspect ratio and taper ratio (Section 6.7).
12. Determine horizontal tail lift curve slope, $C_{L_{\alpha\_h}}$ (Equation (6.57)).
13. Calculate horizontal tail angle of attack at cruise (Equation (6.51)).
14. Determine downwash angle at the tail (Equation (6.54)).
15. Calculate horizontal tail incidence angle, i_t(Equation (6.53)).
16. Calculate tail span, tail root chord, tail tip chord, and tail MAC (Equations (6.63)–(6.66)).
17. Calculate horizontal tail generated lift coefficient at cruise (e.g., lifting line theory; Chapter 5). Treat the horizontal tail as a small wing.
18. If the horizontal tail generated lift coefficient (step 17) is not equal to the horizontal tail required lift coefficient (step 8), adjust the tail incidence.
19. Check horizontal tail stall.
20. Calculate the horizontal tail contribution to the static longitudinal stability derivative ($C_{m\alpha}$). The value for the $C_{m\alpha}$ derivative must be negative to insure a stabilizing contribution. If the design requirements are not satisfied, redesign the tail.
21. Analyze the dynamic longitudinal stability. If the design requirements are not satisfied, redesign the tail.
22. Optimize the horizontal tail.

Vertical tail

23. Select vertical tail configuration (e.g., conventional, twin vertical tail, vertical tail at swept wing tip, V-tail) (Section 6.8.2.1).
24. Select the vertical tail volume coefficient, $\overline{V}_V$ (Table 6.4).
25. Assume the vertical tail moment arm (l_v) equal to the horizontal tail moment arm (l).
26. Calculate vertical tail planform area, S_v (Equation (6.74)).
27. Select vertical tail airfoil section (Section 6.8.2.4).
28. Select vertical tail aspect ratio, AR_v(Section 6.8.2.6).
29. Select vertical tail taper ratio, λ_v(Section 6.8.2.7).
30. Determine the vertical tail incidence angle (Section 6.8.2.5).
31. Determine the vertical tail sweep angle (Section 6.8.2.8).
32. Determine the vertical tail dihedral angle (Section 6.8.2.9).
33. Calculate vertical tail span (b_v), root chord ($C_{V_{root}}$), and tip chord ($C_{V_{tip}}$), and MAC_v(Equations (6.76)–(6.79)).
34. Check the spin recovery.
35. Adjust the location of the vertical tail relative to the horizontal tail by changing l_v to satisfy the spin recovery requirements (Section 6.8.2.2).
36. Analyze directional trim (Section 6.8.1).
37. Analyze directional stability (Section 6.8.1).
38. Modify to meet the design requirements.
39. Optimize the tail.

Reminder: Tail design is an iterative process. When the other aircraft components (such as fuselage and wing) are designed, the aircraft dynamic longitudinal-directional stability needs to be analyzed, and based on that, the tail design may need some adjustments.

6.10 Tail Design Example

Example 6.2 provides a tail design example.

Example 6.2

Problem statement: Design a horizontal tail for a two-seat motor glider aircraft with the following characteristics:

$$m_{TO} = 850\,\text{kg}, D_{\text{fmax}} = 1.1\,\text{m}, V_c = 95\,\text{knot (at 10 000 ft)}, \alpha_f = 1\,\text{deg(at cruise)}$$

The wing has a reference area of 18 m^2 and the following features:

$\overline{C} = 0.8$ m, AR = 28, $\lambda = 0.8$, $i_w = 3$ deg, $\alpha_{\text{twist}} = -1.1$ deg, $\Lambda_{LE} = 8$ deg,
$\Gamma = 5$ deg, airfoil : NACA 23012, $C_{L_\alpha} = 5.8$ 1/rad

The aircraft has a high wing and an aft conventional tail configuration, and the aerodynamic center of the wing/fuselage combination is located at 23% of MAC. In cruising flight conditions, the aircraft center of gravity is located at 32% of the fuselage length. Assume that the aircraft cg is 7 cm ahead of the wing/fuselage aerodynamic center.

Then the following tail parameters must be determined: airfoil section, S_h, $C_{h_{tip}}$, $C_{h_{root}}$, b_h, i_h, AR_h, λ_h, Λ_h, Γ_h. At the end, draw a top view of the aircraft that shows the fuselage, wing, and horizontal tail (with dimensions).

Solution:

The tail configuration has already been selected and stated, so there is no need to investigate this item. The only parameter that needs to be decided is the type of setting angle. Since the aircraft is not maneuverable and the cost must be low, a fixed tail is selected. Thus, the design begins with the selection of the horizontal tail volume coefficient:

$$\overline{V}_H = 0.6 \qquad \text{(Table 6.4)}$$

To determine the optimum tail moment arm (l_{opt}), we set the goal to minimize the aircraft drag. Hence:

$$l = l_{\text{opt}} = K_c\sqrt{\frac{4\overline{C}S\overline{V}_H}{\pi D_f}} = 1.2\sqrt{\frac{4 \cdot 0.8 \cdot 18 \cdot 0.6}{\pi \cdot 1.1}} = 3.795\ \text{m} \tag{6.47}$$

where the correction factor K_c is selected to be 1.2. Then, the tail planform area is determined as:

$$\overline{V}_H = \frac{lS_h}{\overline{C}S} \Rightarrow S_h = \frac{\overline{C}S\overline{V}_H}{l} = \frac{0.8 \cdot 18 \cdot 0.6}{3.795} = 2.277\ \text{m}^2 \tag{6.24}$$

The aircraft cruise lift coefficient is:

$$C_L = C_{L_c} = \frac{2W_{\text{avg}}}{\rho V_c^2 S} = \frac{2 \cdot 850 \cdot 9.81}{0.905 \cdot (95 \cdot 0.5144)^2 \cdot 18} = 0.428 \tag{6.27}$$

where the air density at 10 000 ft is 0.905 kg/m^3. The wing/fuselage aerodynamic pitching moment coefficient is:

$$C_{m_{o\_wf}} = C_{m_{af}} \frac{\text{AR} \cos^2(\Lambda)}{\text{AR} + 2\cos(\Lambda)} + 0.01\alpha_t = -0.013 \frac{28 \cdot \cos^2(8)}{28 + 2\cos(8)}$$
$$+ 0.01 \cdot (-1.1) = -0.023 \tag{6.26}$$

where the value for the wing airfoil section pitching moment coefficient ($C_{m_{owf}}$) is usually extracted from the airfoil graphs. Based on Table 5.2, the value of $C_{m_{af}}$ for the NACA 23012 airfoil section is −0.013.

In order to use the trim equation, we need to find h and h_o. Referring to Table 5.2, for this type of aircraft, l_{opt}/L_f is 0.65. So the fuselage length is selected to be:

$$L_f = l_{opt}/0.65 = 3.795/0.65 = 5.838 \text{ m}$$

The aerodynamic center of the wing/fuselage combination is located at 23% of MAC, and the aircraft center of gravity is located at 32% of the fuselage length. This cg is 7 cm ahead of the wing/fuselage aerodynamic center. Combining these three items of data, we have the following relationship regarding the wing:

$$X_{apex} + 0.23 \text{ MAC} = 0.32 L_f + 0.07$$

Thus $X_{apex} = -0.23 \text{ MAC} + 0.32 L_f + 0.07 = 1.754 \text{ m}$

This leads us to find the cg location (X_{cg}) in terms of MAC:

$$X_{cg} = 0.23 \text{ MAC} - 0.07 = 0.23(0.8 \text{ m}) - 0.07 = 0.114 \text{ m} \quad \text{(from wing leading edge)}$$

$$\overline{X}_{cg} = h = \frac{0.114}{\text{MAC}} = \frac{0.114}{0.8} = 0.142 = 14.2\% \text{ MAC}$$

So $h = 0.142$. The tail efficiency is assumed to be 0.98. The horizontal tail required lift coefficient at cruise is calculated using the trim equation:

$$C_{m_{o\_wf}} + C_L(h - h_o) - \eta_h \overline{V}_H C_{L_h} = 0 \Rightarrow C_{L_h} = \frac{C_{m_{o\_wf}} + C_L(h - h_o)}{\overline{V}_H}$$
$$= \frac{-0.023 + 0.428 \cdot (0.114 - 0.23)}{0.6} \Rightarrow C_{L_h} = -0.121 \tag{6.29}$$

The horizontal tail airfoil section must have several properties that are described in Section 6.7. Two significant properties are: (i) symmetric and (ii) thinner than wing airfoil. The wing thickness-to-chord ratio is 12%. There are several airfoil sections that can satisfy these requirements, but we are looking for one with a low drag coefficient. A symmetric airfoil section with a minimum drag coefficient ($C_{d_{min}} = 0.005$) and 3% thinner than the wing airfoil section is NACA 0009. Figure 6.19 provides the characteristic graphs for the NACA 0009 airfoil section. From this figure, other features of this airfoil are extracted as follows:

C_{l_i}	$C_{d_{min}}$	C_m	$(C_l/C_d)_{max}$	α_o (deg)	α_s (deg)	$C_{l_{max}}$	C_{l_α} (1/rad)	$(t/c)_{max}$
0	0.005	0	83.3	0	13	1.3	6.7	9%

The initial tail aspect ratio is determined to be:

$$\mathrm{AR_h} = \frac{2}{3}\mathrm{AR_w} = \frac{2}{3} \cdot 28 = 18.6 \tag{6.59}$$

The tail taper ratio is initially determined to be equal to the wing taper ratio: $\lambda_h = \lambda_w = 0.8$.

The tail sweep angle and the tail dihedral angle are tentatively considered to be the same as those of the wing. The reasons are presented in Section 6.7:

$$\Lambda_h = 10\ \mathrm{deg}, \Gamma_h = 5\ \mathrm{deg}$$

Now we need to determine the tail setting angle (i_h) such that it produces a tail coefficient of −0.121. In order to determine this parameter, we not only need to consider all the tail parameters, but also the wing downwash. At the beginning, the tail angle of attack is determined based on the tail lift curve slope. In the next step, the lifting line theory is used to calculate the tail generated lift coefficient. If the tail generated lift coefficient is not equal to the tail required lift coefficient, the tail incidence will be adjusted until these two are equal. In the last, downwash is applied to determine the tail incidence. The tail lift curve slope is:

$$C_{L_\alpha} = \frac{C_{l_{\alpha\_h}}}{1 + \dfrac{C_{l_{\alpha\_h}}}{\pi \cdot \mathrm{AR_h}}} = \frac{6.7}{1 + \dfrac{6.7}{3.14 \cdot 18.6}} = 6.1\ \frac{1}{\mathrm{rad}} \tag{6.57}$$

The tail angle of attack in cruise is:

$$\alpha_h = \frac{C_{L_h}}{C_{L_{\alpha\_h}}} = \frac{-0.121}{6.1} = -0.018\ \mathrm{rad} = -1.02\ \mathrm{deg} \tag{6.51}$$

To calculate the tail created lift coefficient, the lifting line theory is employed as introduced in Section 5.14. The following MATLAB m-file is utilized to calculate the tail lift coefficient with an angle of attack of −1.02 deg.

```
clc
clear
N = 9; % (number of segments-1)
S = 2.277; % m^2
AR = 18.6; % Aspect ratio
lambda = 0.8; % Taper ratio
alpha_twist = 0.00001; % Twist angle (deg)
a_h = -1.02; % tail angle of attack (deg)
a_2d = 6.1; % lift curve slope (1/rad)
alpha_0 = 0.000001; % zero-lift angle of attack (deg)
```

```
b = sqrt(AR*S); % tail span
MAC = S/b; % Mean Aerodynamic Chord
Croot = (1.5*(1+lambda)*MAC)/(1+lambda+lambda^2); % root chord
theta = pi/(2*N):pi/(2*N):pi/2;
alpha=a_h+alpha_twist:-alpha_twist/(N-1):a_h;
% segment's angle of attack
z = (b/2)*cos(theta);
c = Croot * (1 - (1-lambda)*cos(theta)); % Mean
Aerodynamics chord at each segment
mu = c * a_2d / (4 * b);
LHS = mu .* (alpha-alpha_0)/57.3; % Left Hand Side
% Solving N equations to find coefficients A(i):
for i=1:N
    for j=1:N
    B(i,j)=sin((2*j-1) * theta(i)) * (1+(mu(i) *
    (2*j-1))/sin(theta(i)));
    end
end
A=B\transpose(LHS);
for i = 1:N
    sum1(i) = 0;
    sum2(i) = 0;
    for j = 1 : N
        sum1(i) = sum1(i) + (2*j-1) * A(j)*sin((2*j-1)*theta(i));
        sum2(i) = sum2(i) + A(j)*sin((2*j-1)*theta(i));
    end
   end
CL_tail = pi * AR * A(1)
```

The output of this m-file is:

```
CL_tail = -0.0959
```

The tail is expected to generate a C_{L_h} of −0.121, but it generates a C_{L_h} of −0.0959. To increase the tail lift coefficient to the desired value, we need to increase the tail angle of attack. With trial and error and using the same m-file, we find that the tail angle of attack of −1.29 deg generates the desired tail lift coefficient.

Hence:

$$\alpha_h = -1.29 \text{ deg}$$

Now, we need to take into account the downwash. The ε_o (downwash angle at zero angle of attack) and $\frac{\partial \varepsilon}{\partial \alpha}$ (downwash slope) are:

$$\varepsilon_o = \frac{2C_{L_w}}{\pi \cdot AR} = \frac{2 \cdot 0.428}{\pi \cdot 28} = 0.0097 \text{ rad} = 0.558 \text{ deg} \tag{6.55}$$

$$\frac{\partial \varepsilon}{\partial \alpha} = \frac{2C_{L_{\alpha\_w}}}{\pi \cdot AR} = \frac{2 \cdot 5.8}{\pi \cdot 28} = 0.132 \text{ deg/deg} \tag{6.56}$$

Thus:

$$\varepsilon = \varepsilon_o + \frac{\partial \varepsilon}{\partial \alpha}\alpha_w = 0.0097 + 0.132 \cdot \frac{3}{57.3} = 0.017 \text{ rad} = 0.954 \text{ deg} \quad (6.54)$$

Therefore, the tail setting angle would be:

$$\alpha_t = \alpha_f + i_h - \varepsilon \Rightarrow i_h = \alpha_h - \alpha_f + \varepsilon = -1.29 - 1 + 0.954 = -1.33 \text{ deg} \quad (6.53)$$

The other horizontal tail parameters are determined by solving the following four equations simultaneously:

$$\mathrm{AR_h} = \frac{b_h}{\overline{C}_h} \quad (6.63)$$

$$\lambda_h = \frac{C_{h_{tip}}}{C_{h_{root}}} \quad (6.64)$$

$$\overline{C}_h = \frac{2}{3} C_{h_{root}} \left(\frac{1 + \lambda_h + \lambda_h^2}{1 + \lambda_h} \right) \quad (6.65)$$

$$S_h = b_h \cdot \overline{C}_h \quad (6.66)$$

The solution of these four equations simultaneously yields the following results:

$$b_h = 6.52 \text{ m}, \ \overline{C}_h = 0.349 \text{ m}, \ C_{h_{tip}} = 0.309 \text{ m}, \ C_{h_{root}} = 0.386 \text{ m}$$

The last step is to examine the aircraft static longitudinal stability. The aircraft has a fixed tail, so the aircraft static longitudinal stability derivative is determined as follows:

$$C_{m_\alpha} = C_{L_{\alpha\_wf}} \left(h - h_o\right) - C_{L_{\alpha\_h}} \eta_h \frac{S_h}{S} \left(\frac{l}{\overline{C}} - h \right) \left(1 - \frac{d\varepsilon}{d\alpha} \right) \quad (6.67)$$

$$C_{m_\alpha} = 5.7\,(0.114 - 0.23) - 6.1 \cdot 0.98 \frac{2.277}{18} \left(\frac{3.795}{0.8} - 0.114 \right) (1 - 0.132) = -3.7 \frac{1}{\text{rad}} \quad (6.67)$$

where we assume that the wing/fuselage lift curve slope is equal to the wing lift curve slope. Since the derivative C_{m_α} is negative, the aircraft is *statically longitudinally stable*. The aircraft longitudinal dynamic stability analysis requires the information about other aircraft components that are not provided by the problem statement. So this analysis is not performed in this example. Figure 6.28 shows a top view of the aircraft with details of the tail geometries.

It is important to note that this is the first phase of the horizontal tail design. If the characteristics of the other aircraft components are known, the complete analysis for the longitudinal dynamic and static stability may be performed and the tail could be optimized.

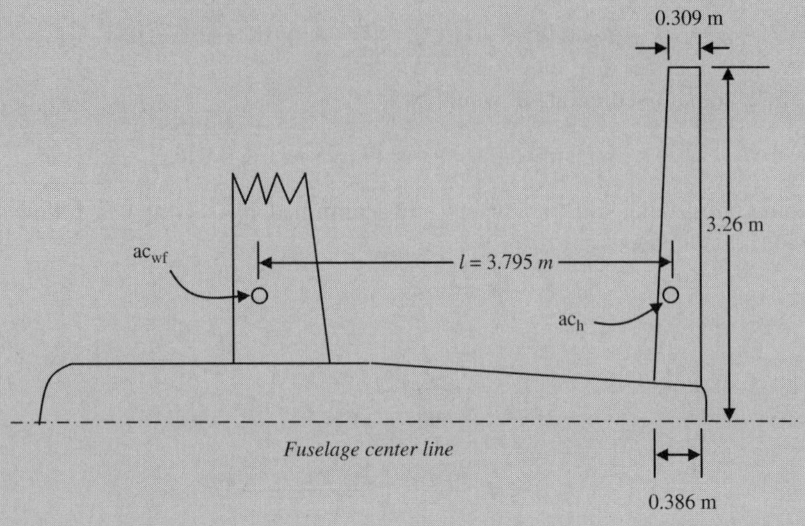

Figure 6.28 Top view of the aircraft in Example 6.2

Problems

1. Using Ref. [5] or other reliable sources, identify the tail configurations of the following aircraft:
   Stemme S10 (Germany), Dassault Falcon 2000 (France), Embraer EMB 145 (Brazil), Canadair CL-415, ATR 42, Aeromacchi MB-339C (Italy), Eagle X-TS (Malaysia), PZL Mielec M-18 Dromader (Poland), Beriev A-50 (Russia), Sukhoi Su-32FN (Russia), Sukhoi S-80, Saab 340B (Sweden), Pilatus PC-12 (Switzerland), An-225 (Ukraine), Jetstream 41 (UK), FLS Optica OA7-300 (UK), Bell/Boeing V-22 Osprey, Boeing E-767 AWACS, Cessna 750 Citation X, Learjet 45, Lockheed F-16 Fighting Falcon, Lockheed F-117A Nighthawk, McDonnell Douglas MD-95, Northrop Grumman B-2 Spirit, Bede BD-10, Hawker 1000, Schweizer SA 2-38, Sino Swearingen SJ30, and Visionaire Vantage

2. Using Ref. [5] or other reliable sources, identify an aircraft for each of the following tail configurations:
   conventional aft tail, V-tail, canard, T-tail, H-tail, non-conventional, cruciform, triplane, boom-mounted, twin vertical tail, and inverted V-tail

3. Using Ref. [5] or other reliable sources, identify an aircraft with a conventional aft tail where the vertical tail is out of the wake region of the horizontal tail.

4. An aircraft has a fuselage with a circular cross-section. Derive an equation for the optimum horizontal tail moment arm such that the aft portion of the aircraft (including aft fuselage and horizontal tail) has the lowest wetted area.

5. An unmanned aircraft has the following features:

$$S = 55\,\text{m}^2,\ \text{AR} = 25,\ S_\text{h} = 9.6\,\text{m}^2,\ l = 6.8\,\text{m}$$

Determine the horizontal tail volume coefficient.

6. The airfoil section of a horizontal tail in a fighter aircraft is NACA 64-006. The tail aspect ratio is 2.3. Using Ref. [8], calculate the tail lift curve slope in 1/rad.

7. The airfoil section of a horizontal tail in a transport aircraft is NACA 64_1-012. The tail aspect ratio is 5.5. Using Ref. [8], calculate the tail lift curve slope in 1/rad.

8. The airfoil section of a horizontal tail in a GA aircraft is NACA 0012. The tail aspect ratio is 4.8. Using Ref. [8], calculate the tail lift curve slope in 1/rad.

9. The wing reference area of an agricultural aircraft is $14.5\,\text{m}^2$ and the wing mean aerodynamic chord is 1.8 m. The longitudinal stability requirements dictate the tail volume coefficient to be 0.9. If the maximum fuselage diameter is 1.6 m, determine the optimum tail arm and then calculate the horizontal tail area. Assume that the aft portion of the fuselage is conical.

10. Consider a single-seat GA aircraft whose wing reference area is $12\,\text{m}^2$ and wing mean aerodynamic chord is 1.3 m. The longitudinal stability requirements dictate the tail volume coefficient to be 0.8. If the maximum fuselage diameter is 1.3 m, determine the optimum tail arm and then calculate the horizontal tail area. Assume that the aft portion of the fuselage is conical.

11. A 19-seat business aircraft with a mass of 6400 kg is cruising with a speed of 240 knot at 26 000 ft. Assume that the aircraft lift coefficient is equal to the wing lift coefficient. The aircraft has the following characteristics:

$$S = 32\,\text{m}^2,\ \text{AR}_\text{w} = 8.7,\ \text{wing airfoil : NACA } 65_1{-}412$$

Determine the downwash angle (in degrees) at the horizontal tail.

12. Suppose that the angle of attack of the fuselage for the aircraft in Problem 11 is 2.3 deg and the horizontal tail has an incidence of −1.5 deg. How much is the horizontal tail angle of attack at this flight condition?

13. The horizontal tail of a transport aircraft has the following features:

$$\text{AR}_\text{h} = 5.4,\ \lambda_\text{h} = 0.7,\ S_\text{h} = 14\,\text{m}^2,\ \Lambda_{\text{h}_\text{LE}} = 30\ \text{deg}$$

Determine the span, root chord, tip chord, and mean aerodynamic chord of the horizontal tail. Then sketch the top view of the tail with dimensions.

14. The horizontal tail of a fighter aircraft has the following features:

$$\text{AR}_\text{h} = 3.1,\ \lambda_\text{h} = 0.6,\ S_\text{h} = 6.4\,\text{m}^2,\ \Lambda_{\text{h}_\text{LE}} = 40\ \text{deg}$$

Determine the span, root chord, tip chord, and mean aerodynamic chord of the horizontal tail. Then sketch the top view of the tail with dimensions.

15. The vertical tail of a transport aircraft has the following features:

$$AR_v = 1.6,\ \lambda_v = 0.4,\ S_v = 35\,m^2,\ \Lambda_{VLE} = 45\ deg$$

Determine the span, root chord, tip chord, and mean aerodynamic chord of the vertical tail. Then sketch the side view of the tail with dimensions.

16. The aircraft in Problem 11 has other features as follows:

$$h = 0.18, h_o = 0.23, \eta_h = 0.97, l = 12\,m, S_h = 8.7\,m^2$$

Determine the aircraft static longitudinal stability derivative (C_{m_α}) and discuss whether the horizontal tail is longitudinally stabilizing or destabilizing.

17. Design a horizontal tail for a twin-jet business aircraft with the following characteristics:

$$m_{TO} = 16\,000\,kg, D_{f_{max}} = 1.8\,m, V_c = 270\,knot\ (at\ 30\,000\ ft),$$
$$\alpha_f = 1.5\ deg\ (at\ cruise)$$

The wing has a reference area of 49 m^2 and the following features:

$$AR = 8, \lambda = 0.6, i_w = 2.4\ deg, \alpha_{twist} =$$
$$-1.3\ deg,\ \Lambda_{LE} = 37\ deg,\ \Gamma = 3\ deg,\ NACA\ 65_2{-}415$$

The aircraft has a low wing and an aft conventional tail configuration, and the aerodynamic center of the wing/fuselage combination is located at 22% of MAC. In cruising flight conditions, the aircraft center of gravity is located at 42% of the fuselage length. Assume that the aircraft cg is 15 cm ahead of the wing/fuselage aerodynamic center.

The following tail parameters must be determined: airfoil section, S_h, $C_{h_{tip}}$, $C_{h_{root}}$, $b_h, i_h, AR_h, \lambda_h, \Lambda_h, \Gamma_h$. At the end, draw a top view of the aircraft that shows the fuselage, wing, and horizontal tail (with dimensions).

18. A large transport aircraft with a mass of 63 000 kg is supposed to cruise with a speed of 510 knots at 42 000 ft. The maximum fuselage diameter is 3.6 m and the fuselage angle of attack at cruise is 3.2 deg. The wing has a reference area of 116 m^2 and the following features:

$$AR = 11.5, \lambda = 0.5, i_w = 2.7\ deg, \alpha_{twist} =$$
$$-1.6\ deg,\ \Lambda_{LE} = 30\ deg,\ \Gamma = 6\ deg,\ NACA\ 64_1{-}412$$

The aircraft has a low wing and a T-tail configuration, and the aerodynamic center of the wing/fuselage combination is located at 20% of MAC. In cruising flight conditions, the aircraft center of gravity is located at 49% of the fuselage length. Assume that the aircraft cg is 18 cm ahead of the wing/fuselage aerodynamic center. Design a horizontal tail to satisfy the longitudinal tail and static longitudinal stability requirements. Then determine the airfoil section, S_h, $C_{h_{tip}}$, $C_{h_{root}}$, $b_h, i_h, AR_h, \lambda_h, \Lambda_h, \Gamma_h$. At the end, draw a top view of the aircraft that shows the fuselage, wing, and horizontal tail (with dimensions).

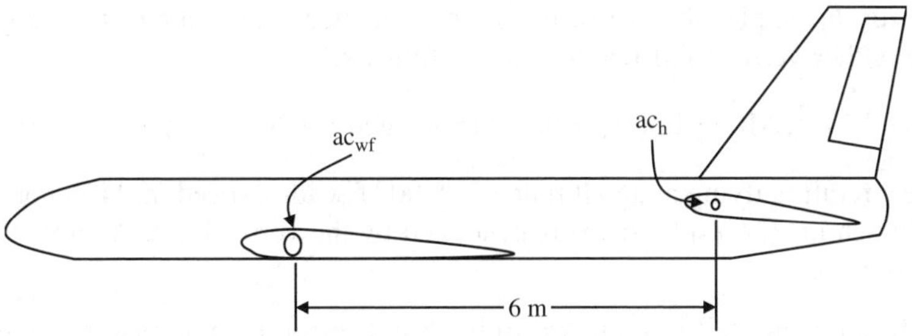

Figure 6.29 Side view of the aircraft in Problem 19

19. Figure 6.29 shows the original design for the empennage of a transport aircraft with a horizontal tail area of $12.3\,\text{m}^2$. The wing reference area is $42\,\text{m}^2$, and the wing aspect ratio is 10.5.

The aircraft is spinnable and the designer found out that the vertical tail is not effective for spin recovery. Move the horizontal tail horizontally such that the vertical tail becomes effective in recovering from spin. Then determine the horizontal tail area such that the horizontal tail volume coefficient remains unchanged. Assume that the sketch in Figure 6.29 is scaled.

20. A fighter aircraft has the following features:

$$S = 57\,\text{m}^2, \text{AR} = 3,\ S_h = 10.3\,\text{m}^2,\ S_v = 8.4\,\text{m}^2,\ l = 6.8\,\text{m},\ l_v = 6.2\,\text{m}$$

Determine the horizontal and vertical tail volume coefficients.

21. Design a vertical tail for the aircraft in Problem 18 to satisfy the static directional stability requirements.

22. The airfoil section of the vertical tail for a twin-jet engine aircraft is NACA 66-009. Other features of the aircraft are as follows:

$$S = 32\,\text{m}^2, \text{AR} = 10.3,\ S_v = 8.1\,\text{m}^2,\ \text{AR}_v = 1.6,\ l = 9.2\,\text{m},\ \frac{d\sigma}{d\beta} = 0.32,\ \eta_v = 0.95$$

Determine the aircraft static directional stability derivative (C_{n_β}). Then analyze the static directional stability of the aircraft.

23. The angle of attack of a horizontal tail for a cargo aircraft is −1.6 deg. Other tail features are as follows:

$$S_h = 12\,\text{m}^2, \text{AR}_h = 5.3,\ \lambda_h = 0.7,\ \text{airfoil section : NACA 64} - 208,\ \eta_h = 0.96$$

If the aircraft is flying at an altitude of 15 000 ft with a speed of 245 knot, determine how much lift is generated by the tail. Assume that the tail has no twist.

24. The sideslip angle of a vertical tail for a maneuverable aircraft during a turn is 4 deg. Other vertical tail features are as follows:

$$S_v = 7.5\,\text{m}^2, \text{AR}_v = 1.4,\ \lambda_v = 0.4,\ \text{airfoil section : NACA 0012},\ \eta_v = 0.92$$

If the aircraft is flying at an altitude of 15 000 ft with a speed of 245 knot, determine how much lift (i.e., side force) is generated by the vertical tail. Assume that the tail has no twist.

25. An aft horizontal tail is supposed to be designed for a single-piston engine aircraft. The aircraft with a mass of 1800 kg is cruising with a speed of 160 knot at an altitude of 22 000 ft. The aircraft center of gravity is at 19% MAC and the wing/fuselage aerodynamic center is located at 24% MAC:

$$S = 12\,\text{m}^2, \text{AR} = 6.4,\ S_h = 2.8\,\text{m}^2,\ l = 3.7\,\text{m},\ C_{m_{o\_wf}} = -0.06,\ \eta_h = 0.1$$

Determine the horizontal tail lift coefficient that must be produced in order to maintain the longitudinal trim.

26. Redo Problem 25 with the assumption that the aircraft has a canard instead of an aft horizontal tail.

References

[1] Roskam, J. (2007) *Airplane Flight Dynamics and Automatic Flight Control Part I*, DAR Corporation.
[2] Nelson, R. (1997) *Flight Stability and Automatic Control*, McGraw-Hill.
[3] Etkin, B. and Reid, L.D. (1995) *Dynamics of Flight Stability and Control*, 3rd edn, John Wiley & Sons, Inc.
[4] Federal Aviation Regulations (2011), Federal Aviation Administration, Department of Transportation, www.faa.gov.
[5] Jackson, P. (1995) *Jane's All the World's Aircraft*, Jane's Information Group, various years.
[6] Hoak, D.E., Ellison, D.E., Fink, R.D. *et al.* (1978) USAF Stability and Control DATCOM, Flight Control Division, Air Force Flight Dynamics Laboratory, Wright-Patterson AFB, Ohio.
[7] Shevell, R.S. (1989) *Fundamentals of Flight*, 2nd edn, Prentice Hall.
[8] Abbott, I.H. and Von Donehoff, A.F. (1959) *Theory of Wing Sections*, Dover.
[9] Lan, E.C.T. (1988) *Applied Airfoil and Wing Theory*, Cheng Chung Book Company.
[10] Lan, E.C.T. and Roskam, J. (2003) *Airplane Aerodynamics and Performance*, DAR Corporation.

7

Fuselage Design

7.1 Introduction

After the wing and tail, the fuselage is the third most important aircraft component. The primary function of the fuselage is to accommodate the payload. This chapter is devoted to the design of the fuselage. It provides the fuselage design requirements, the primary function of the fuselage, the desired characteristics, pilot cockpit design, passenger compartment design, cargo section design, the design of other sections, and design steps. Since the fuselage deals with the human pilot and passengers, human factors must be included in the design considerations. A fully solved example is also provided at the end of the chapter.

A major driving force in the design of the pilot cockpit and passenger cabin is human factors. The human pilots, crew, and passengers are engaged with the aircraft man/machine system mainly through the fuselage. Safety is a theme beneath the surface in all activities connected with flying, including fuselage design. Human factors is an applied technology [1], so human factors in flight is designed primarily for industry and aims to bridge the gap between academic resources of knowledge and the practical operation of aircraft. For an aircraft to have a worldwide sales market, it is necessary that it is designed to meet the appropriate international standards and regulations such as FAR and EASA CS.

It should be noted that the word "man" in this chapter is used in its generic sense. Hence, it includes both sexes unless specifically indicated to the contrary.

7.2 Functional Analysis and Design Flowchart

An early stage in the fuselage design is the functional analysis, which prepares a platform for a systematic approach. Depending upon the aircraft type, desired mission, aircraft configuration, and type of payload, the function of the fuselage may vary a great deal. However, for the majority of aircraft, the fuselage primary function is to accommodate the payload. By definition, the payload is the useful load that the aircraft is intended to carry. Payload does not basically include pilot, crew, or fuel. Therefore, it mainly contains passengers, luggage, and cargo. Therefore, the fuselage is defined as a shell containing the payload which must be carried a certain range at a specified speed. The payload

Aircraft Design: A Systems Engineering Approach, First Edition. Mohammad H. Sadraey.

accommodation must allow for a quick loading before take-off and a rapid unloading after landing.

In addition, in order to reduce aircraft drag, a few other major components and systems – such as landing gear, engine, fuel system, and power transmission system – are highly likely to be enclosed by the fuselage. Therefore, for the fuselage, a couple of secondary functions are defined as listed in Table 7.1.

In general, the fuselage is the most suitable aircraft component for housing the pilot cockpit, where the best location is the nose. In the case of an airliner, the flight crew and other personnel also needs accommodation, which can be considered to be seated in the passenger cabin. The human attendants' (pilot, crew, passengers) accommodation by the fuselage must offer protection against climatic factors such as cold, low pressure, and very high wind speed. In case of a large engine, the fuselage should also keep the flight attendants protected from external noise, such as the engine's loud noise. The extents of comfort which must be provided by the fuselage to the human attendants are specified by the regulations and will be described in Section 7.4.

Another group of secondary functions (Table 7.1) concerns the non-human items such as landing gear, engine, electro-mechanical systems, and fuel tank. For these items, comfort is not a requirement. However, each non-human item that fits inside the fuselage requires specific requirements that will be discussed in later sections. The secondary requirements are not all required for all aircraft; each item is considered if it has been specified in the aircraft configuration design phase.

Table 7.1 Functional analysis of the fuselage

No.	Functions and features	Description
1	Primary function	Accommodate the payload
2	Secondary functions	Accommodate crew members Accommodate flight attendants and other technical personnel Provide space for landing gear (if retracted inside fuselage) Provide space for engine (if inside fuselage) Provide space for fuel tanks (if inside fuselage) Provide sufficient room for systems (electric, hydraulic, mechanical, radio, etc.) Provide structural arm for empennage Keep the integrity of the aircraft structure (e.g., hold the wing)
3	Desired features and expectations	Generate the lowest drag Contribute positively to the lift generation Low weight Provide passenger/pilot/crew comfort Carry structural flight loads External symmetry Loading and unloading effectiveness Safe against environmental hazards (e.g., lightning) Low wetted area

Other than fuselage functions, there are a few expectations that are recommended to be considered during the fuselage design process. The expectations include low weight, low drag, contributing positively to the lift generation, external symmetry, and safety against environmental hazards such as lightning. The fuselage drag usually contributes 30–50% of the aircraft zero-lift drag (C_{D_0}). Furthermore, the fuselage may be aerodynamically designed such that it provides as much as 50% of the total lift. For instance, in the fighter aircraft Mikoyan MIG-29 (Figure 5.56), about 40% of the total lift is created by the lift-generating center fuselage. Furthermore, in the reconnaissance aircraft Lockheed SR-71 Blackbird (Figure 8.21), about 30% of the aircraft lift is generated by the fuselage. It is interesting to note that, in most General Aviation (GA) and transport aircraft, only as much as about 5% of the aircraft lift is produced by the fuselage.

Table 7.1 enumerates the factors which must be given serious attention as they impact most designs. Many of the requirements and expectations which are laid down in relation to the fuselage will limit the designer's range of selections. In order to have an optimum fuselage, the priority of each requirement must be specified in order to enable the designer to measure the contribution of each decision and each selection on the total fuselage effectiveness. For instance, in the case of a subsonic cargo aircraft such as Lockheed C-130 Hercules (Figure 5.4), a nice aerodynamic shape may be sacrificed for easy loading and unloading by means of a readily accessible rear loading door.

Two major fuselage parameters that must be determined during the design process are: (i) fuselage length (L_f) and (ii) maximum diameter (D_f). The fuselage configuration as well as these two parameters are functions of several design requirements. In general, the following are the fuselage design requirements:

1. accommodation requirement;
2. operational and mission requirements;
3. airworthiness requirement;
4. crashworthiness requirement;
5. aerodynamic requirement;
6. aircraft stability requirement;
7. low weight;
8. low wetted area and low side area;
9. symmetry;
10. structural integrity and strength;
11. maintainability;
12. manufacturability;
13. cost;
14. long life;
15. radar detectability.

These design requirements will be described in later sections. Each of the above requirements will influence the fuselage design in a variety of ways. Some fuselage parameters such as the fuselage length are driven more by one design requirement (such as the accommodation requirement) than other requirements (such as the weight requirement). The final decision about each fuselage parameter will be made after careful analysis of the weighted value of the design requirements, as discussed in Chapter 2.

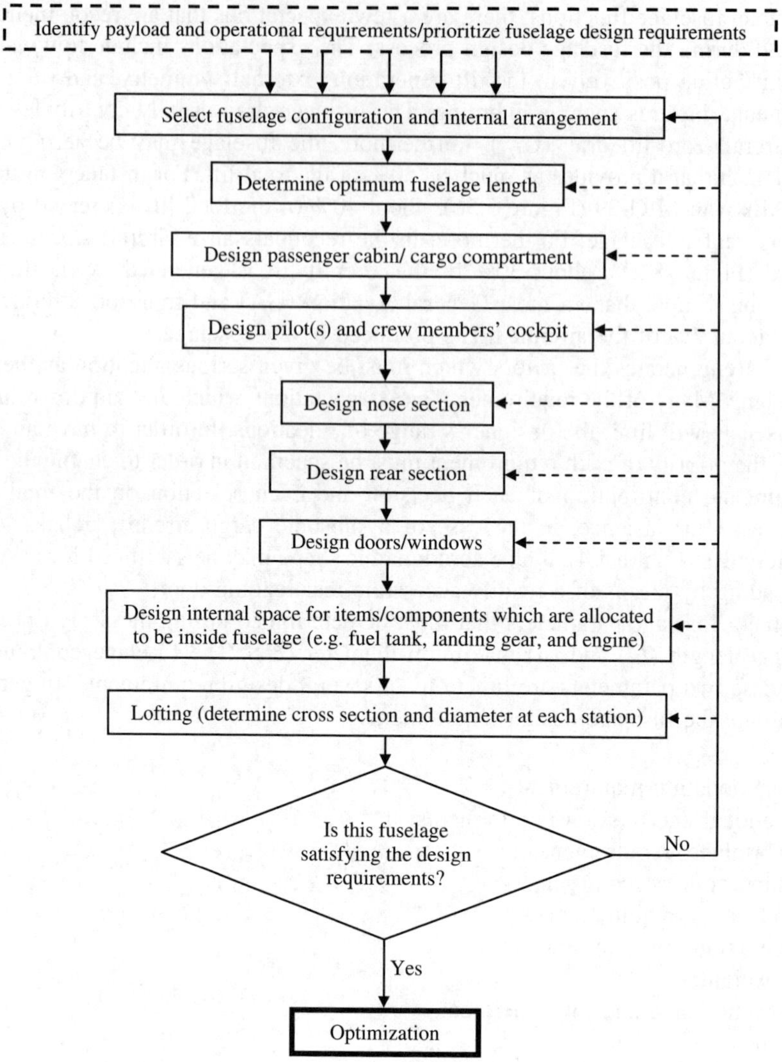

Figure 7.1 Fuselage design flowchart

Figure 7.1 illustrates the fuselage design flowchart including design feedback. As the flowchart shows, the fuselage design is an iterative process and the number of iterations depends on the nature of the design requirements as well as the designer's skills. The fuselage gear design is initiated by identifying the fuselage design requirements and the process ends with optimization. The details of each design box will be presented in the forthcoming sections. As in the case of other aircraft components, there is no unique design solution to satisfy the fuselage requirements. Each acceptable solution will deliver a pool of advantages and disadvantages which must be decided based on the systems engineering technique.

7.3 Fuselage Configuration Design and Internal Arrangement

The first design step, after identification of the payload and design requirements, is to decide on the fuselage configuration and determine the fuselage internal arrangement. This decision is very important and will influence all fuselage parameters. Fuselage configuration design is a conceptual design but at the fuselage level, and does not involve detailed calculations. Indeed, the configuration design of the fuselage requires several skills and long experience. At this point, the external shape as well as the internal arrangement will be determined. Since this is a type of conceptual design, the designer may use a hand drawing to present the selected configuration.

In some cases, a design may look desirable but may not be feasible. Thus, when a designer is deciding about the best seating arrangement, or the best location for cargo, he/she must already be aware of the fundamental solutions. For instance, a short fuselage with low weight and high drag is more desired, or a long fuselage with high weight and low drag. This is a fundamental question of cost versus performance. For a home-build designer the first alternative is the best option, while for a military aircraft designer the second alternative is the most desirable. Therefore, the designer should have the priority list upfront prior to the fuselage configuration design.

A conventional fuselage may consist of the following sections: pilot and crew station (cockpit), passenger compartment (cabin), luggage room, cargo compartment, nose section, doors, windows, rear section, fuel tanks, necessary flight carrying items (e.g., food, water), internal systems (i.e., electrical, mechanical, and hydraulic), and engine(s). Each section needs to be designed separately, since each has a unique design requirement. However, at this stage of design, the locations of these sections relative to each other need to be determined. Figure 7.2 illustrates a side view of four generic fuselage external shapes (note that the figures are not in scale). Although these external shapes have different aerodynamic characteristics, each one is optimum to serve for a particular mission.

The fuselage configuration is also a function of the internal arrangements. In order to specify the location for each internal item, one must first identify and decide what item/component is supposed to be accommodated. Figure 7.3 illustrates a side view of the fuselage for two typical aircraft with their internal arrangements: a civil passenger and a fighter aircraft. The volume and external shape of the fuselage are functions of what is desired to be stored inside.

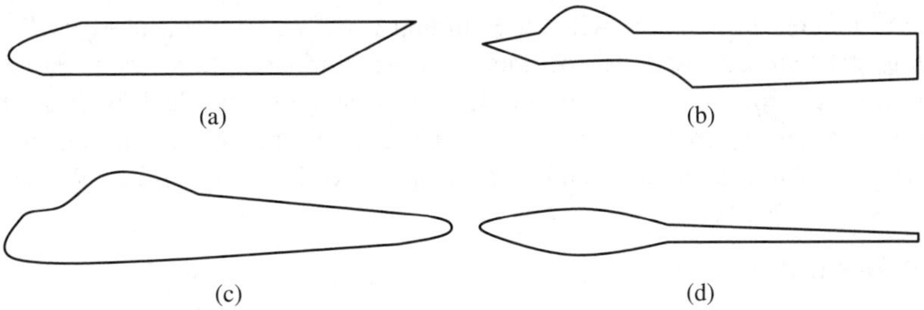

Figure 7.2 Four generic fuselage configurations. (a) Large transport aircraft, (b) Fighter aircraft, (c) Light GA aircraft, and (d) Glider

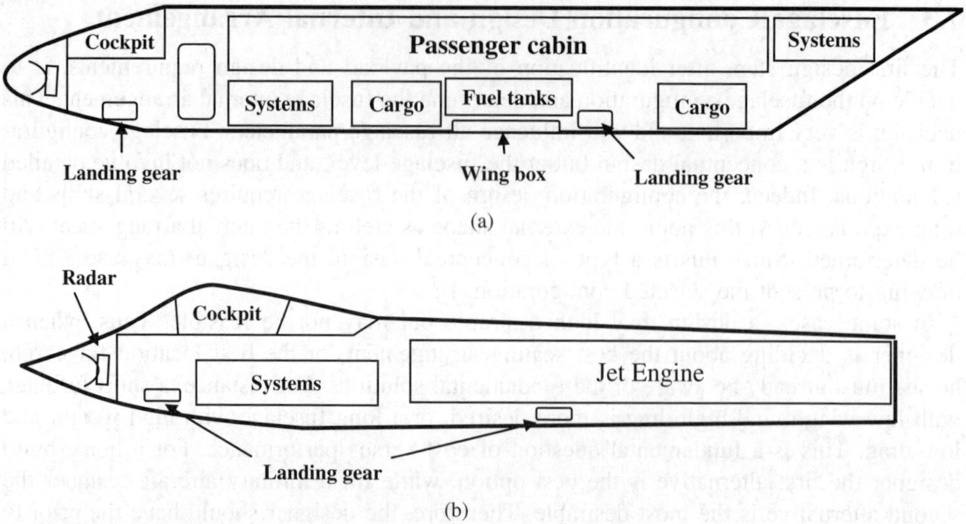

Figure 7.3 Internal arrangement of a civil passenger and a fighter aircraft. (a) Low-wing passenger aircraft, and (b) Fighter aircraft

In general, there are six basic rules for internal arrangement and to locate the allocated items inside the fuselage:

1. Keep the fuselage as small and compact as possible.
2. Arrangement to be symmetric from the top view as far as possible.
3. There must be sufficient space to accommodate all of the items.
4. Usable loads such as fuel must be close to the aircraft center of gravity.
5. The pilot cockpit must be allocated the most forward location of the fuselage, to enable the pilot to view the runway during take-off and landing.
6. Arrangements must be such that the aircraft center of gravity is close to the wing/fuselage aerodynamic center.

Furthermore, the requirements introduced in Section 7.2 must be considered.

Whenever all the fuselage sections are designed, the geometry and dimensions of each section will be finalized. For instance, when the fuel volume is calculated, the exact location will be determined. Or, when the landing gear is designed, the retraction system and landing gear storage will be specified. As a recommendation, try to keep the fuel out of the fuselage for the sake of flight attendants, since it may leak or catch fire in an emergency situation. Figure 7.4 compares the fuselage of three aircraft: a transport (Airbus A-321), a fighter (Sukhoi Su-27), and a GA (Piper PA-28-161 Cherokee Warrior II).

7.4 Ergonomics

7.4.1 Definitions

Whenever an engineering device which deals with humans is planned to be designed, the ergonomic standards need to be considered. Ergonomics (or human factors) is the

(a)

(b)

(c)

Figure 7.4 Three types of aircraft with three different fuselage configurations: (a) Airbus 321; (b) Sukhoi Su-27U; (c) Piper PA-28-161 Cherokee Warrior II. Reproduced from permission of (a) Anne Deus; (b) Antony Osborne; (c) Jenny Coffey

science of designing user interactions with equipment and workplaces to fit the user. The field of human factors engineering uses scientific knowledge about human behavior in specifying the design and use of a human/machine system. The aim is to improve system efficiency by minimizing human error and optimize performance, comfort, and safety. Proper ergonomic design is necessary to prevent repetitive strain injuries, which can develop over time and can lead to long-term disability. The International Ergonomics Association defines ergonomics as follows: "Ergonomics (or human factors) is the scientific discipline concerned with the understanding of interactions among humans and other elements of a system, and the profession that applies theory, principles, data, and methods to design in order to optimize human well-being and overall system performance." References [2–5] are helpful resources in the introduction, principles, fundamentals, and useful data for various aspects of human factors. In this section, a few necessary data is reproduced which will be used in the design of cockpit and cabin.

Because of the health risks posed by poor posture and repetitive stress, proper ergonomics are too important to ignore. Whether it is for a pilot or a passenger, the seat must be designed such that the sitter stays healthy from head to toe when he/she is sitting. The following paragraph contains ergonomic advice for healthy sitting.

Incorrect monitor positioning can cause neck and eye strain, and can lead to poor seat positioning, which creates pressure on the back. The top of the monitor should be positioned just above the eye level when seated. This is the best place for the "vision cone," the most immediate field of vision, which starts at the top of one's eye level and descends at a 30-deg angle. When monitors are too far away, people tend to lean forward to see well. This is increasingly true as people age, since vision almost inevitably declines over time. As a rule of thumb: If you can extend your arm and just touch the screen with your fingertips, then you are in the right position.

To keep wrists and arms at an optimum position, reducing the risk of repetitive-motion injuries, the stick/yoke and switches should be at the same level as the elbows when seated. Since not everybody has a standard size, a simple fix is an adjustable seat. Sitting properly takes 20–30% of the pressure off the lower back. The seat should be between 17 and 19 in. deep, and it should have good lower-back support. The body should be positioned with the back against the seat and the hips open. If you find yourself leaning forward to see the panel, you need to move the seat forward.

Leg positioning contributes to the overall position in the seat, so make sure the legs are bent at about 90-deg angles at the knees. This helps alleviate pressure on the back. Movement is essential for circulation, however, so allow for subtle shifts in positioning and be sure to stand, stretch, and walk a few steps at least every few hours. Feet should be firmly planted on the floor. If the seat positioning required for proper wrist alignment results in the feet not reaching the floor, use some type of footrest to support the feet, such that the height of the support keeps the knees at a right angle.

Matching of working and living areas to human characteristics is one of the primary tasks of those specialized in human factors. Some of the basic characteristics of human beings are those associated with size and shape, and the movements of various parts of the body. Such data is applied in several locations on board an aircraft. On the flight deck, the data are used in the basic geometry, in the provision of adequate inside and outside visibility, in the location and design of controls, in the seat design, and so on. In the cabin, similar basic data are used in the design of galleys, seats, doors, overhead luggage containers, and toilets. Furthermore, maintenance areas take into account human dimensions to assure sufficient access to equipment and working space. In the cargo compartment such data are applied to provide proper access for loaders and adequate room to work. The design of much equipment on board makes use of information about human measurements: life jackets, life rafts, emergency exits, oxygen masks, meal trolleys, wash-basin, seatbelts, and so on.

7.4.2 Human Dimensions and Limits

Photographic techniques have been developed to measure and collect data from a representative sample of people who are to use equipment. However, a caution necessary in applying data concerning physical dimensions is that they are slowly changing. A universal population survey a few years ago indicated a general increase in height over half a century

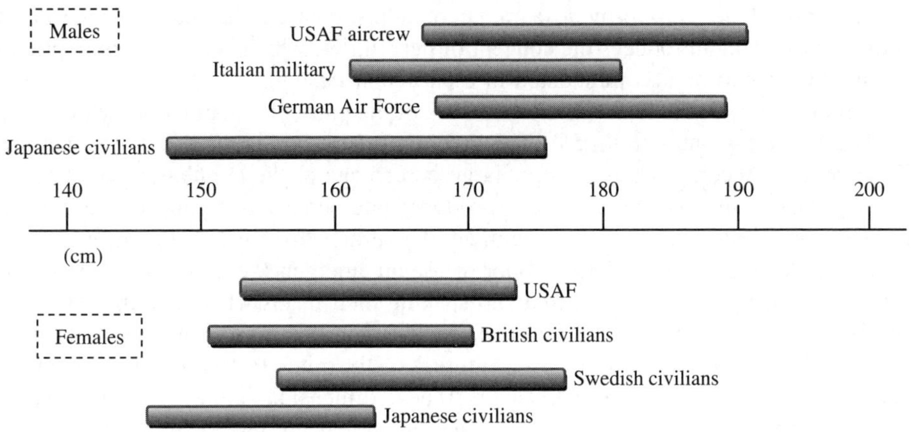

Figure 7.5 Examples of variations in height between males and females and different ethnic groups [6]

of 1.3 mm a year for males and 0.9 mm for females. The most comprehensive source of such data is a three-volume edition published by NASA [6], which covers various groups.

A wide difference in the physical dimensions of people can be expected between ethnic groups. Persons of African descent are relatively long-legged compared with white Caucasians. Asians, on the other hand, are smaller overall but have a relatively long trunk and short legs compared with Europeans. In addition to differences between ethnic groups there are also differences between men and women within one ethnic group (Figure 7.5).

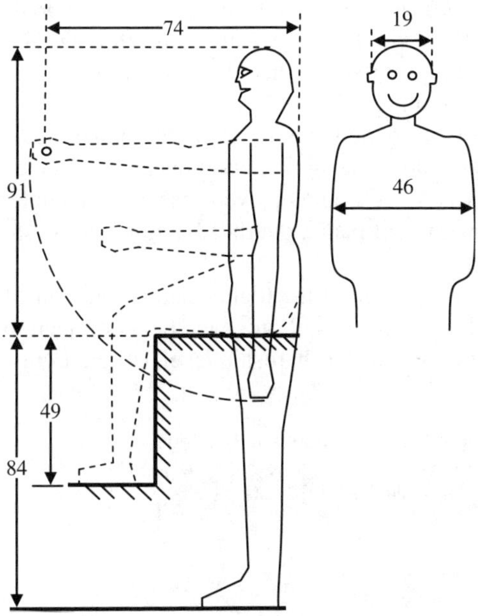

Figure 7.6 Linear body dimensions (cm)

In the USA, the FAA regulations currently require that transport category aircraft be designed for operating by crew members of physical height ranging from 5 ft 2 in. (157 cm) to 6 ft 3 in. (190 cm). For example, at Boeing, all aircraft from the B-747 (Figures 3.7, 3.12, and 9.4) onwards were designed to meet these criteria.

In terms of human weight (for seat design), a study [1] showed that an average male Canadian passenger weighs 76 kg, a male Thai passenger weighs 70 kg, a male Eastern European passenger weighs 84.6 kg, while a female German passenger weighs 68.9 kg, and a female Japanese passenger weighs 57.2 kg.

These variations clearly indicate the need for a compromise in order to make use of these statistics. In many cases, it may not be possible to find a single design solution to suit everybody. One must provide a range of adjustment so that most people can be accommodated. This introduces the concept of percentiles, which is a means of expressing the range of sizes to be accommodated in a particular design.

The recommendation is to accommodate 90% of the particular population, which results in excluding the top and bottom 5%. This is often referred to as the design for the 5th to 95th percentile. Accommodating a very large percentage of the population could be very expensive, since this could involve a disproportionate penalty in trying to accommodate extremes when designing seats or equipment. Therefore, there will be some people who are too wide to enter an aircraft toilet door or use an emergency escape hatch, too short to reach the luggage rack, or too tall to avoid striking their heads when entering the aircraft door. The decision on how many are to be included in this disadvantaged minority is a fundamental one in the design process. Figure 7.6 illustrates the human dimensions that cover 90% of human pilots and passengers. These dimensions are employed in the seat design, cockpit design, and cabin design. The recommended mass/weight for an average passenger and the baggage for a civil transport aircraft are given in Table 7.2.

7.5 Cockpit Design

In a human-piloted aircraft, the cockpit must be designed so as to enable the pilot to control the aircraft through the flight deck. The flight deck is considered as a system with hardware, software, human (i.e., "liveware"), and environment as its components. The flight deck is a workplace with a human as its central component, so the designer must be concerned with matching the other components to the characteristics and limits of humans. Furthermore, some physical constraints will make the job harder. For instance, compare the narrow and cramped environment created by the aerodynamic requirements of Concorde, with a flight deck width of about 148 cm, to the 191 cm of the Boeing 747. In addition, the designer will also face commercial pressure, since the space in the cabin can be sold but the flight deck and galley space cannot.

During the cockpit design process, the following parameters must be determined: seat geometry, seat free space, distance to stick/yoke/side-stick, stick motion distance, distance to pedal, pedal motion range, lower-than-horizon view angle, over-nose vision angle, side-view angle, seatback angle, distance to instrument panel, overhead height, and room behind seat.

The following items must be known prior to the design of the flight deck: (i) number of pilots, (ii) number of crew, (iii) pilot personal equipment (suit, goggles, helmet, ejection seat, pressure system, and parachute), (iv) pilot/crew comfort/hardship level, (v) pilot/crew

Table 7.2 Recommended mass/weight for passengers and baggage

No.	Passenger/baggage	Weight (lb)	Mass (kg)
1	Passenger	180	82
2	Checked baggage – economy	50	23
3	Checked baggage – first class	70	32
4	Carry-on baggage	30	14

mission, (vi) control equipment, (vii) measurement equipment, (viii) level of automation, and (ix) external constraints. Some of these items are under the control of the customer, and some others must comply with airworthiness standards. Indeed, these items are given to the cockpit designer and he/she has no control over them. The following sections will present how to identify/determine these items.

7.5.1 Number of Pilots and Crew Members

The number of pilots and crew members is determined either by the customer order or based on the airworthiness/mission requirements. The number of pilots and crew members for a transport aircraft is regulated by FAR 25 [7], for a GA aircraft by FAR 23 [7], and for a military aircraft by MIL-STD [8]. In the past (say the 1950s), a large transport aircraft tended to have a pilot, a copilot, a flight engineer, a navigator, and a telegraph man. With advances in technology, we could get rid of the telegraph man first. The original Airbus A310 design was for a crew of three and incorporated a lateral panel for the third crew member. By 1979, under pressure from the airline management to reduce costs, the third crew member was dropped and the layout was redesigned.

Throughout the years and by advancing avionics (such as GPS) and computer systems (hardware and software packages), the two jobs of flight engineer and navigation engineer have been omitted. Their functions are distributed between avionic systems and pilots, who are perfectly capable of performing those functions. The removal of the flight engineer created a big dispute between designers, unions and airmen. The argument was that this initiative makes the aircraft unsafe; however, the designers could prove that it is safe to cancel the navigator.

Most GA aircraft (under FAR 23), even GA transport aircraft, tend to have only one pilot. However, most large transport aircraft (under FAR 25) tend to have a pilot and a copilot. The older version of FAR requires an aircraft with a take-off weight of more than 80 000 lb to have a flight engineer in the cockpit. Activity which may demand prolonged head-down operation (excluding instrument flying) in a busy area, precluding visual look-out, will need to be restricted in a two-man operation.

In military fighters, there are fighters with two pilots (e.g., McDonnell Douglas (now Boeing) F-15 Eagle (Figures 3.12 and 9.14)), and also fighters with a single pilot (e.g., General Dynamics F-16 Fighting Falcon (Figure 4.6)). Some combat aircraft, such as the McDonnell Douglas F/A-18 Hornet (Figures 2.11, 6.12, and 12.27), have both versions: one single-seater (e.g., F/A-18A and F/A-18C) and one two-seater (e.g., F/A-18B and F/A-18D). The number of pilots in a fighter aircraft is determined by a special committee comprising military commanders as well as aircraft designers through long analysis. A trainer aircraft (e.g., Pilatus PC-21 Super Tucano (Figure 5.44)), however, as the name implies, has two seats, one for the instructor and one for the students.

FAR Part 25 Section 25.1523 regulates the minimum flight crew for transport aircraft as follows:

> The minimum flight crew must be established so that it is sufficient for safe operation, considering – (i) The workload on individual crew members; (ii) The accessibility and ease of operation of necessary controls by the appropriate crew member; and (iii) The kind of operation authorized.

More details on criteria for determining the minimum flight crew are given in Appendix D to FAR Part 25 as follows:

The following are considered by the Agency in determining the minimum flight crew

1. Basic workload functions. The following basic workload functions are considered:

(a) Flight path control.

(b) Collision avoidance.

(c) Navigation.

(d) Communications.

(e) Operation and monitoring of aircraft engines and systems.

(f) Command decisions.

2. Workload factors. The following workload factors are considered significant when analyzing and demonstrating workload for minimum flight crew determination:

(a) The accessibility, ease, and simplicity of operation of all necessary flight, power, and equipment controls, including emergency fuel shutoff valves, electrical controls, electronic controls, pressurization system controls, and engine controls.

(b) The accessibility and conspicuity of all necessary instruments and failure warning devices such as fire warning, electrical system malfunction, and other failure or caution indicators. The extent to which such instruments or devices direct the proper corrective action is also considered.

(c) The number, urgency, and complexity of operating procedures with particular consideration given to the specific fuel management schedule imposed by center of gravity, structural or other considerations of an airworthiness nature, and to the ability of each engine to operate at all times from a single tank or source which is automatically replenished if fuel is also stored in other tanks.

(d) The degree and duration of concentrated mental and physical effort involved in normal operation and in diagnosing and coping with malfunctions and emergencies.

(e) The extent of required monitoring of the fuel, hydraulic, pressurization, electrical, electronic, deicing, and other systems while en route.

(f) The actions requiring a crew member to be unavailable at his assigned duty station, including: observation of systems, emergency operation of any control, and emergencies in any compartment.

(g) The degree of automation provided in the aircraft systems to afford (after failures or malfunctions) automatic crossover or isolation of difficulties to minimize the need for flight crew action to guard against loss of hydraulic or electric power to flight controls or to other essential systems.

(h) The communications and navigation workload.

(i) The possibility of increased workload associated with any emergency that may lead to other emergencies.

(j) Incapacitation of a flight crew member whenever the applicable operating rule requires a minimum flight crew of at least two pilots.

3. Kind of operation authorized. The determination of the kind of operation authorized requires consideration of the operating rules under which the airplane will be operated.

The number of crew members other than the pilot and copilot is a function of the aircraft mission and number of equipment to employ. For instance, some military aircraft such as combat, reconnaissance, or bomber aircraft may require some technical crew members for aircraft control and mission success to operate particular systems. These are special cases and require special attention via functional analysis and human factors. The Boeing 747-400 (Figures 3.7, 3.12, and 9.4) has [9] a two-crew flight deck, with seats for two observers, and a two-bunk crew rest cabin accessible from the flight deck. The Airbus A-310, A-330 (Figure 5.51), and A-380 (Figure 1.8) all have a crew of two on the flight deck. The twin-turbofan regional airliner EMB-145 with two pilots, flight observer, and cabin attendant accommodates 50 passengers. The twin-turboprop transport EMB-120 aircraft with accommodation for 30 passengers is controlled by two pilots. The flight deck of the business jet aircraft Cessna 560 Citation, to accommodate only seven to eight passengers, is designed for a crew of two.

7.5.2 Pilot/Crew Mission

The pilot/crew mission has a great impact on the design of the cockpit. The main function of the pilot is to control the aircraft and apply command through hands and legs. But, depending upon the mission, the pilot is required to go beyond this basic mission – ranging from civil missions such as training, touring, entertaining, and patrolling to military missions such as commanding, combating, attacking, intercepting, and bombing. In each of these missions, the pilot is expected to perform jobs beyond regular control. These missions must acquire particular equipment, which requires special attention by the pilot. These tasks, along with special devices and levels of comfort (presented in later sections), will shape the configuration of the cockpit. Therefore, the tasks of a pilot must be carefully defined prior to the design of a cockpit.

7.5.3 Pilot/Crew Comfort/Hardship Level

The cockpit provides the internal physical environment for the flight crew. Another parameter which influences the design of the cockpit is the level of comfort/hardship for the pilot and crew members. Humans are constituted to function efficiently under a reasonably narrow set of conditions. Cockpit space, temperature, pressure, humidity, and noise are important crew comfort considerations. The level of comfort/hardship is mainly driven by the space provided for the crew plus the type of seat. Comfortable crew seats are an important item because US flight crews are required to spend all of their flight time in the cockpit with their seatbelt fastened (other than for physiological comfort or scheduled rest periods). The competition for pilot seat and non-pilot-required space makes compromise in the cockpit inevitable.

In most civil missions the level of comfort must be specified, while in most military aircraft the range of hardship needs to be limited. In a civil aircraft such as a firefighter aircraft, the issue of level of hardship should also be considered. The comfort or hardship is a function of the aircraft type and mission. It is obvious that a pilot in a touring aircraft is looking for more comfort and leisure, while a fighter pilot expects that a mission will be accompanied by some kind of hardship. An ear problem is caused by engine noise for most fighter pilots and helicopter pilots, while this problem is seldom found in the latest jet transport pilots.

Another flight variable which impacts the level of comfort is the duration of flight. The late Steve Fossett endured a non-stop flight with the Scaled Composite Model 311 Virgin Atlantic GlobalFlyer (Figure 6.12) of 67 hours from February 28, 2005 to March 3, 2005. This record-breaking mission was very hard, since it required the pilot to not sleep for more than five days. This flight broke the Absolute World Record for the fastest non-stop unrefueled circumnavigation. However, not everybody is capable of tolerating such hardships in a long flight. In contrast, another product of Scaled Composite, the Voyager (Figure 4.20), did fly the same mission 19 years earlier (from December 14 to 23, 1986). This 9-day flight was piloted by Dick Rutan and Jeana Yeager. The cockpit had only one seat for the pilot, and for the second pilot there was a special tube-shaped space for lying and sleeping. A comparison between the two aircraft for the same mission designed by the same designer illustrates that the pilot of the GlobalFlyer had a very hard job, while the pilot of the Voyager had the opportunity (comfort) for sleep/rest during the mission. This explains why the cockpits of these two aircraft are very different.

The Boeing 747-400 has three cockpit seats, in addition to those for the captain and first officer. The flight deck of this long-range aircraft has a dedicated crew rest area, which consists of two bunk-beds in an enclosed area aft of the cockpit. On most modern aircraft the crew seat position can be adjusted horizontally and vertically, with the seat back inclined. The degree of adjustment is driven by the level of comfort. For this reason, a transport aircraft pilot seat is very comfortable, while the seat of a fighter pilot is very tight and uncomfortable. Therefore, the designer must use considerable judgment in reaching a final conclusion.

7.5.4 Pilot Personal Equipment

A safe and comfortable flight by a pilot requires several items of personal equipment. Depending upon the type of aircraft, pilot mission, flight duration, and flight environment, the equipment ranges from seat and seatbelt to suit, goggle, headset, helmet, ejection seat, pressure system, and parachute. Crew members are required to spend hours strapped into their seat; the limited facility for moving the legs on aircraft with a control stick/yoke aggravates blood circulation difficulties.

A hang-glider pilot is not provided with a seat, so he/she has to hang onto the wing structure. The reason pilot comfort is compromised is due to weight and gliding requirements. The seat of an ultralight and most home-built aircraft is very simple and easy to design due to weight and cost priorities. The seat of a fighter pilot is highly complex and very heavy due to various military requirements, the hard pilot task, and harsh environment. When the cockpit noise is too high, such as in a helicopter or some turboprop aircraft (e.g., Lockheed C-130 Hercules (Figure 5.4)), the crew members must be equipped with headsets to talk to other crew members.

Military aircraft are provided with ejection seats to escape when the aircraft crashes, while a civil transport aircraft is not equipped with any ejection seat. Due to combat environments, a fighter pilot must be equipped with a special suit to handle high acceleration (i.e., g), such as 9–12 g during maneuvers. They should also be equipped with special goggles to protect the eyes from the Sun's radiation at high altitude. Fighter pilots need to be provided with a personal pressure system to breathe normally at high altitude, where the air pressure is very low. The pressure system may be incorporated in the pilot helmet. Another piece of equipment that most military crew members must wear is a parachute. The parachute is carefully folded, or packed, to ensure that it will open reliably in case of emergency situations.

When sitting, the main part of the body weight is transferred to the seat. Some weight is also transferred to the floor, back rest, and armrests. Where the weight is transferred is key to a good seat design. When the proper areas are not supported, sitting in a seat for a long trip can put unwanted pressure on the back, causing pain. Seat design aspects which require optimization include the configuration of the seat pan and back structure, armset, headrest, lumbar and thigh support, upholstery characteristics and controls, as well as associated hardware such as seatbelt, shoulder harness, and footrest.

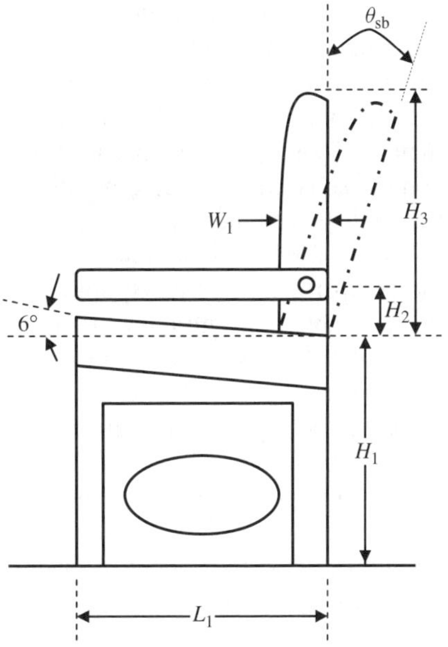

Figure 7.7 Seat geometry (side view)

In order to provide sufficient comfort to the pilot and allow him/her to perform the required tasks during flight operations, the seat must be carefully designed. The geometry of a cockpit seat is illustrated in Figure 7.7. The seat parameters such as cushion angle, armset height, seat height, seat length, seat width, and cushion thickness must be determined. The seatback angle (θ_{sb}) is recommended to be about 13 deg, although angles of up to 30 deg have been used (e.g., F-16) to provide better **g** tolerance for the fighter pilot. Seat back angles of up to 60 deg have been considered for modern fighters to reduce fuselage diameter and so aircraft drag.

Various military specifications and design handbooks (such as Refs [8, 10]) provide detailed requirements for the layout of the seat for different aircraft. All this equipment adds to the pilot volume and weight, hence the cockpit must be synthesized accordingly.

7.5.5 Control Equipment

In order for a pilot to control the aircraft, he/she must move the control surfaces and move the throttle. In a conventional aircraft, the elevator and aileron are controlled through a stick (or yoke, or side-stick), the rudder is controlled via a pedal, and the engine is controlled via an engine throttle. There are a variety of other aircraft components and devices, such as landing gear and flap, which must also be controlled by the pilot. All these devices are utilized by the pilot to control speed, altitude, attitude, and heading. The cockpit must be designed such that all these control levers, sticks, and switches are within reach.

Most light GA aircraft such as the Cessna 172 (Figure 11.15) and military fighters such as the General Dynamics F-16 Fighting Falcon (Figure 4.6) utilize a stick to control the elevator and aileron, while most medium to large transport aircraft such as the Boeing 767 (Figure 5.4) employ a yoke. Ergonomically, the Airbus side-stick controller provides considerable extra flight deck space by replacing the conventional wheel and yoke found in most other airliners. US manufacturers[1] have not gone over to side-stick controls in their fly-by-wire aircraft because of a physiological desire to maintain a high degree of commonality with earlier-generation aircraft, as well as a lack of sufficient feedback from the side-stick to the hand.

Based on FAR 23 Section 23.771 (on pilot compartments), for each pilot compartment, the compartment and its equipment must allow each pilot to perform his duties without unreasonable concentration or fatigue. Based on FAR 23 Section 23.773 (on the pilot compartment view), each pilot compartment must be arranged with a sufficiently extensive, clear, and undistorted view to enable the pilot to safely taxi, take off, approach, land, and perform any maneuvers within the operating limitations of the airplane. Based on FAR 23 Section 23.777 (on cockpit controls), each cockpit control must be located and identified to provide convenient operation and to prevent confusion and inadvertent operation. The controls must be located and arranged so that the pilot, when seated, has full and unrestricted movement of each control without interference from either his clothing or the cockpit structure.

7.5.6 Measurement Equipment

Beside control devices, a pilot also needs to measure and observe the flight parameters to ensure a successful flight. Flight variable measurements are carried out through various avionic instruments by visual cue. For instance, for a transport aircraft, FAR Part 25 Section 25.1303 regulates flight and navigation instruments as follows:

> (a) The following flight and navigation instruments must be installed so that the instrument is visible from each pilot station:
>
> (1) A free air temperature indicator or an air-temperature indicator which provides indications that is convertible to free-air temperature.
>
> (2) A clock displaying hours, minutes, and seconds with a sweep-second pointer or digital presentation.
>
> (3) A direction indicator (nonstabilized magnetic compass).
>
> (b) The following flight and navigation instruments must be installed at each pilot station:
>
> (1) An airspeed indicator. If airspeed limitations vary with altitude, the indicator must have a maximum allowable airspeed indicator showing the variation of V_{MO} with altitude.
>
> (2) An altimeter (sensitive).
>
> (3) A rate-of-climb indicator (vertical speed).
>
> (4) A gyroscopic rate-of-turn indicator combined with an integral slip-skid indicator (turn-and-bank indicator) except that only a slip-skid indicator is required on large airplanes with a third attitude instrument system usable through flight attitudes of 360° of pitch and roll and installed.

[1] Cirrus has recently employed the side-stick yoke in the SR20 and SR22 that handles aileron and elevator.

(5) A bank and pitch indicator (gyroscopically stabilized).

(6) A direction indicator (gyroscopically stabilized, magnetic, or nonmagnetic).

(c) The following flight and navigation instruments are required as prescribed in this paragraph:

(1) A speed warning device is required for turbine engine powered airplanes and for airplanes with V_{MO}/M_{MO} greater than 0.8 V_{FD}/M_D For 0.8 V_D/M_D. The speed warning device must give effective aural warning (differing distinctively from aural warnings used for other purposes) to the pilots, whenever the speed exceeds V_{MO} plus 6 knots or $M_{MO} + 0.01$. The upper limit of the production tolerance for the warning device may not exceed the prescribed warning speed.

(2) A Machmeter is required at each pilot station for airplanes with compressibility limitations not otherwise indicated to the pilot by the airspeed indicating system required under paragraph (b)(1) of this section.

A list of important instruments usually included in a cockpit is as follows: airspeed indicator, altimeter, turn coordinator, bank angle indicator, vertical speed indicator, heading indicator, outside air temperature indicator, GPS, glide slope indicator, transponder, magnetometer, engine instruments (rpm, fuel, exhaust gas temperature, and turbine inlet temperature), compass, electric panel, weather radar, and radio. The basic T panel which forms the core of modern flight instrument panel layout is illustrated in Figure 7.8.

A transport aircraft with two crew members must have duplicated flight controls and instrument panel to allow both pilot and copilot to control the aircraft independently. The work load is defined in Appendix D of FAR Part 25.

7.5.7 Level of Automation

Another issue that can change the cockpit considerably is the decision on the level of automation. Strong competition in the global aviation market is forcing aircraft manufacturers to rethink their approach to design and manufacturing. In order to be competitive, companies need to manufacture a variety of new aircraft models at reasonable cost. This requires aircraft systems to be flexible (to accommodate product variations) and

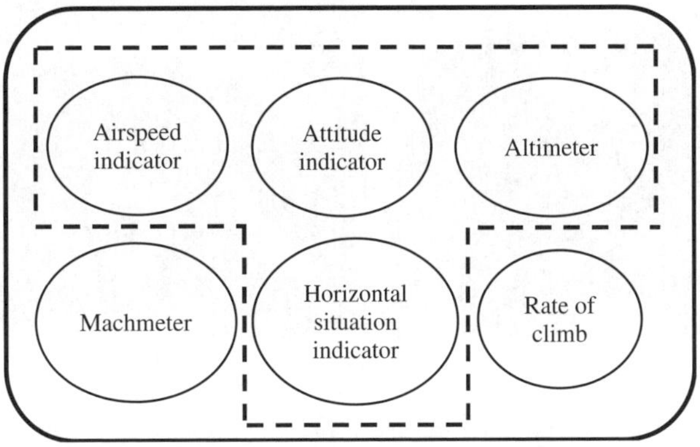

Figure 7.8 Basic T instrument panel

economically viable. Advances in the areas of computer systems and avionics devices have forced cockpit designers to reduce the number of instruments and ease the pilot tasks.

The aviation industry has historically been somewhat automated. The application of an early autopilot goes back to a wing leveler in the 1930s. Nowadays, fully automated unmanned aerial vehicles such as the Global Hawk are able to independently take off, climb, cruise, turn, maneuver, descend, and land successfully. A balanced combination of manual and automated processes increases flexibility, reduces manufacturing costs, improves safety, and provides high quality and throughput. A flight operation control by a team of four crew members in the 1950s is performed in the twenty-first century by only one human pilot, and of course employing various computer systems. All modern transport aircraft, such as the Boeing 767, are equipped with an autopilot to control the segment of cruising flight. These autopilots are usually tasked to keep the altitude and Mach number at a constant value. Figure 7.9 compares the flight decks of four aircraft: (a) GA aircraft Cessna Citation, (b) civil transport Boeing B-757, (c) powered glider Aerotechnik L-13, and (d) two-seat acrobatic aircraft Extra EA-300.

A comparison between the flight deck of a large transport aircraft of the 1950s (such as the Boeing 707 with the flight deck of an advanced very large transport aircraft (such as the Airbus 380 (Figure 1.8)) is clearly an indication of an increase in the level of automation in the cockpit. Current technology also permits the system controls which

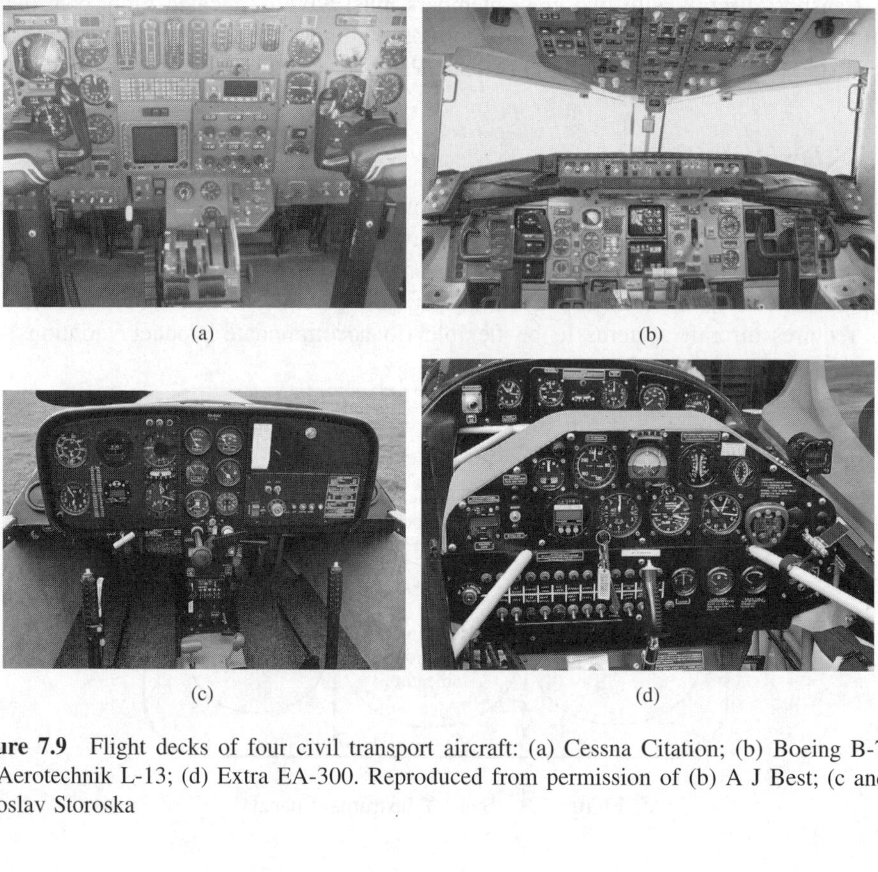

Figure 7.9 Flight decks of four civil transport aircraft: (a) Cessna Citation; (b) Boeing B-757; (c) Aerotechnik L-13; (d) Extra EA-300. Reproduced from permission of (b) A J Best; (c and d) Miloslav Storoska

are displayed on the LCD panel to be operated through a touch-sensitive screen. In view of the above, an optimal level of automation in the cockpit can only be obtained if all relevant aspects of the navigation and control processes are taken into account and optimum levels in terms of cost, productivity, quality, and flexibility are reached. As the level of automation is increased, more mechanical gauges and measurement displays will move into a computer screen or digital display and simplify the cockpit, and therefore the cockpit design gets easier and more efficient.

7.5.8 External Constraints

Even when a cockpit designer is in possession of valid data on human dimensions, one is regrettably required to face a number of external constraints which often limit the extent to which he/she can design as optimum flight deck. The aerodynamic characteristics of the aircraft have a fundamental relationship with the cross-section of the fuselage and the shape of the nose. These sometimes present the cockpit designer with a difficult framework in which to work, and inhibit the generation of an optimum workplace for the pilot. A good example is Concorde, with a narrow and relatively cramped environment (a flight deck width of about 148 cm) imposed by aerodynamic constraints. Compare this cockpit with the wide flight deck of the Boeing 747 (Figures 3.7, 3.12, and 9.4), at 191 cm. Furthermore, the supersonic cruising speed requirement (Mach 2.2) of Concorde presents a severe limitation on the windscreen design. A flush surface, resulting in minimal outside view, is necessary in cruising flight and the complexity of a droop-nose facility is required to provide adequate visibility for approach and landing operations in the terminal area.

Seats in the cockpit area for observers, commonly referred to as "jump-seats," are sometimes an additional requirement. They are required both for observation of flight crew performance by FAA or company and for personal crew movements. Other external constraints which need special attention include structural integrity, operational requirements, security problems, crash survivability, cockpit evacuation, and maintainability considerations. A systems engineering approach will look at the cockpit design as a complex system that should satisfy a variety of design requirements. The final solution will be the output of a trade-off study that yields the optimum design.

7.5.9 Cockpit Integration

So far various cockpit design issues and requirements have been introduced. Now is the time to create the final plan that must meet all design requirements at an optimum level. The detailed design of the cockpit involves the integration of components, assemblies, and instruments that are under the control of various functional disciplines in the systems engineering approach. The cockpit must enable the crew members to easily reach and perform all required flight tasks while seated. When the aircraft has more than one crew member, the flight tasks must be carefully, clearly, and safely distributed among crew members to avoid any confusion during operation. Locating the flight control data on a glare-shield panel both allows pilots to reach it and avoids the disturbing effect on instrument scanning and manual flight controls of having to lean over the control column to the instrument panel. Rain removal on the ground and during low-speed flight is often done by wipers.

The core of the cockpit design is based on ergonomic principles. Incorrect instrument panel positioning can cause neck and eye strain, and can lead to poor seat positioning, which creates pressure on the back. When screens are too far away, the pilot has to lean forward to see well. This is increasingly true as pilots age, since vision almost inevitably declines over time. If the pilot can extend his/her arm and just touch the screen with their fingertips, then the seat is in the right position. The floor is recommended to be horizontal.

To keep wrists and arms at an optimum position, reducing the risk of repetitive-motion injuries, the stick should be at the same level as the elbows when seated. Sitting properly takes 20–30% of the pressure off the lower back. The seat should be between 17 and 19 in. deep, and it should have good lower-back support. The pilot's body should be positioned with the back against the seat and the hips open. Leg positioning contributes to the overall position in the seat, so make sure the legs are bent at about a 90-deg angle at the knees. This helps alleviate pressure on the back. Movement is essential for circulation, so the pilot should allow for subtle shifts in positioning and be sure to stand and stretch at least once an hour. Feet should be firmly planted on the floor. Make sure that the height of the support keeps the knees at a right angle. The pilot must be able to adjust the seat with the seat back in the upright position, so as to locate the midpoint between his/her eyes at the reference eye position.

Pilot inside and outside vision during cruising flight as well as take-off and landing is of critical importance. Over-nose vision is crucial for safety reasons, particularly during landing, and is also important for combat success. Military standards [8] require a 17-deg over-nose angle for transport aircraft, and 10–15 deg for fighters. In a tandem configuration trainer aircraft, a 5–15 deg over-nose angle for the instructor over the student seat is recommended. For civil transport aircraft, 15–25 deg of over-nose is desirable, while an over-nose angle of 10–20 deg for GA aircraft is adequate. Safe landing requires a few degrees of over-nose vision for the pilot during approach; thus the over-nose angle is recommended to be about 5 deg greater than the aircraft approach angle. The over-the-side vision (from 40 to 90 deg) and upward vision also play an important role in the pilot's function. For a fighter aircraft, a complete vision above and all the way to the tail of the aircraft is highly desirable.

Figures 7.10 and 7.11 illustrate a recommended cockpit design for a transport and a fighter aircraft. A general recommendation for the location of the seat and yoke (or wheel) controls for a transport aircraft is presented in Figure 7.10, where the seating arrangement is side-by-side. Reference [7] (FAR 25 Section 25.772) must be consulted for more details. The seating geometry recommendation for a fighter aircraft with a single pilot employing a stick is shown in Figure 7.11. GA cockpits are designed for the range of customer pilots targeted in the market. Note that a fighter pilot is usually wearing a g-suit, helmet, parachute, and other equipment. The McDonnell Douglas F-15 Eagle cockpit is illustrated in Figure 7.12.

7.6 Passenger Cabin Design

When an aircraft is an airliner or is to transport passengers, the passenger compartment or cabin must be designed as part of the fuselage design. A variety of requirements including marketing, economic, and airworthiness regulations must be considered in a cabin design. As the cabin volume is increased, the fuselage volume is increased too

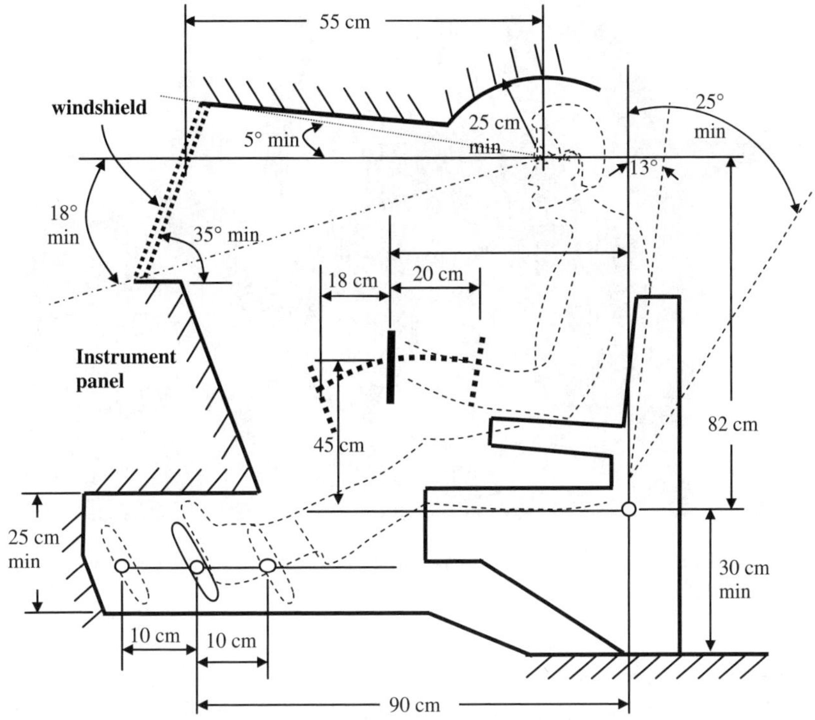

Figure 7.10 Cockpit geometry for a large transport aircraft

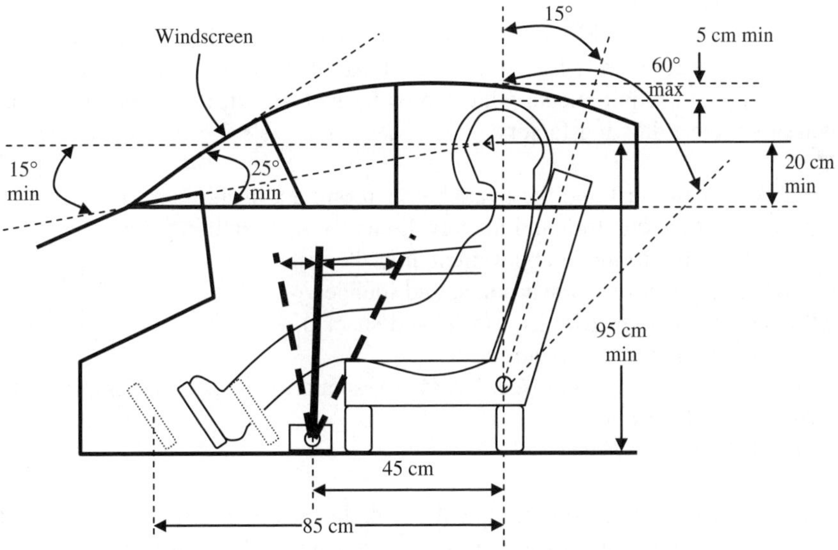

Figure 7.11 Cockpit geometry for a fighter aircraft

Figure 7.12 McDonnell Douglas F-15 Strike Eagle cockpit. Reproduced from permission of Theodore J. Koniares

which is not a desirable outcome. The number of passengers is the only major known parameter that the cabin designer must start with. However, the first step is to determine the number of seats to be placed abreast (n_S). The optimum fuselage length-to-diameter ratio $(L_f/D_f)_{opt}$ or slenderness ratio is a fundamental variable which must be determined by a systems engineering approach. Section 7.8 presents a technique to calculate the optimum slenderness ratio.

The cabin hardware which concerns passengers most is seating. Although passengers evaluate accommodation in a transport aircraft based on level of comfort, there are a number of minimum requirements which must be met. Comfort in a cabin is primarily dependent upon the following factors:

1. Adjustability of the seat and the available legroom, and headroom. It is desirable that each passenger seat has fore and aft travel, swivel, and reclining capability.
2. The room available to move about, including the aisle.
3. The number of lavatories, washrooms, and lounges.
4. Flight attendant services (drinks, meals, and snacks).
5. Air conditioning and pressurization.
6. Interior design including light (e.g., window), sound (or noise), and entertainment.
7. Carry-on bag compartment.
8. Number of flight attendants.

A cabin designer must incorporate the above eight factors into the passenger compartment design to accommodate various types of passengers. Four groups of travelers are defined: (i) very important people (VIP), (ii) first class, (iii) business class, and (iv) tourist or economy class. The highest comfort level is considered for the VIP passenger, but the lowest for the economy one. In fact, VIP individuals such as presidents and heads of state have their own customized aircraft. VIP passengers not only require special seats, but they also have various specific needs so that often a customized cabin needs to be designed. In terms of seat locations, first class is usually considered to be the first few rows of the cabin, then a couple of rows for business seats, and finally the last section is allocated to the economy seats.

The over-wing seats are often considered for economy seats, since the passenger's view is obstructed by the wing. Furthermore, the rear rows are considered for economy class for three reasons: (i) an economy passenger needs to walk longer to reach their seat, (ii) the internal temperature at the rear seats is often a few degrees higher than at the forward seats, and (iii) the rear seats are often noisier due to the engine exit flow. However, in terms of comfort as a function of aircraft spike, the closer to the aircraft cg the less up or down motion there is if a gust hits the aircraft. Since the atmosphere is a dynamic system and the gust is always en route the aircraft course, the gust will force the aircraft to oscillate mostly in the first atmospheric layer. For an aircraft with a cruising altitude above 36 000 ft (which is the case for most large jet transport aircraft), this phenomenon does not happen regularly.

The Boeing 757 (Figure 8.1) has various arrangements; of the 178 seats, 16 are first class and 162 tourist. In the Boeing B-737-800 the first three rows of four seats abreast are assigned for first class, and the remaining 25 rows (six seats abreast) for economy class. The A340-600, the largest A340 (Figure 8.7) variant, is designed for a standard tri-class configuration of 380 seats (12 first, 54 business, and 314 economy seats). The 777-300ER provides 22 first class, 70 business, and 273 economy seats.

Beside the number of seats in each row, the following parameters should also be determined: seat pitch (P_S), seat width (W_S), aisle height, and aisle width (W_A). The seat pitch is defined as the distance between the back of one seat and the back of the next seat (side view). Passenger cabin parameters with six seats abreast are illustrated in Figure 7.13. The FAR 25 regulates various aspects of a cabin. For instance, Section 25.817 limits the number of seats on each side of an aisle to three and also

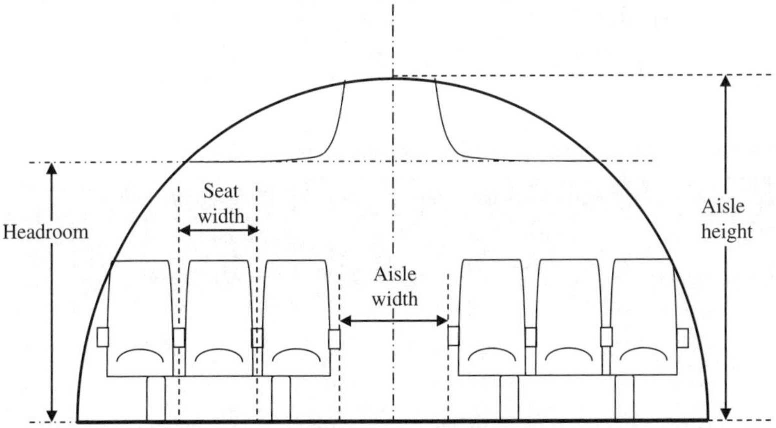

Figure 7.13 Passenger cabin parameters

the minimum permissible width of the aisle. Thus, an aircraft with more than six seats abreast is required to provide two aisles.

FAR Part 25 Section 25.815 has the following requirements on the width of an aisle: The passenger aisle width at any point between the seats must equal or exceed the values in the following table (Table 7.3).

A narrower width, not less than 9 in., may be approved when substantiated by tests found necessary by the Administrator.

Table 7.4 provides some recommended cabin dimensions (based on Refs [7, 11]) for GA and transport aircraft. The price of oil and economic considerations, plus airline competition, is forcing airlines to reduce the seat pitch and seat width for economy seats. The airlines are also reducing the amount of checked luggage and carry-on baggage in order to survive and make a profit. Reference [11] is a helpful resource for cockpit and cabin design.

A twin-turbofan regional airliner with 50 passengers, three abreast at a seat pitch of 79 cm, has an aisle width of 43.2 cm, headroom of 146 cm, and seat width of 44 cm. In the Boeing 757, the first class seats are four abreast, at 96.5 cm pitch, while the tourist seat pitch is 81 or 86 cm, mainly six abreast in mixed arrangements. The typical Boeing 747-400 has 421 seats with a three-class configuration accommodation, with 42 business class seats on the upper deck, 24 first class in the front cabin, 29 business class in the middle cabin, and 326 economy class in the rear cabin on the main deck. The passenger seating of the Airbus A-340 (Figure 8.7), with 295–335 seats, is typically six abreast in first class, six abreast in business class, and eight abreast in economy, all with twin aisles. Figure 7.14 illustrates the seating charts of four transport aircraft.

Table 7.3 Aisle width requirements from FAR 25 for transport aircraft

Passenger seating capacity	Minimum passenger aisle width (in.)	
	Less than 25 in. from floor	25 in. and more from floor
10 or less	12	15
11–19	12	20
20 or more	15	20

Table 7.4 Recommended cabin data (in centimeters)

No.	Cabin parameter	GA aircraft	Transport aircraft		
			Economy		First class
			High density	Tourist	
1	Seat width (W_S)	38–43	42–46	48–55	60–75
2	Seat pitch (P_S)	55–65	65–72	75–86	92–104
3	Headroom	120–130	150–160	160–170	170–185
4	Aisle width (W_A)	35–40	40–50	43–53	60–70
5	Seatback angle (deg)	10–13	13–17	15–20	20–30

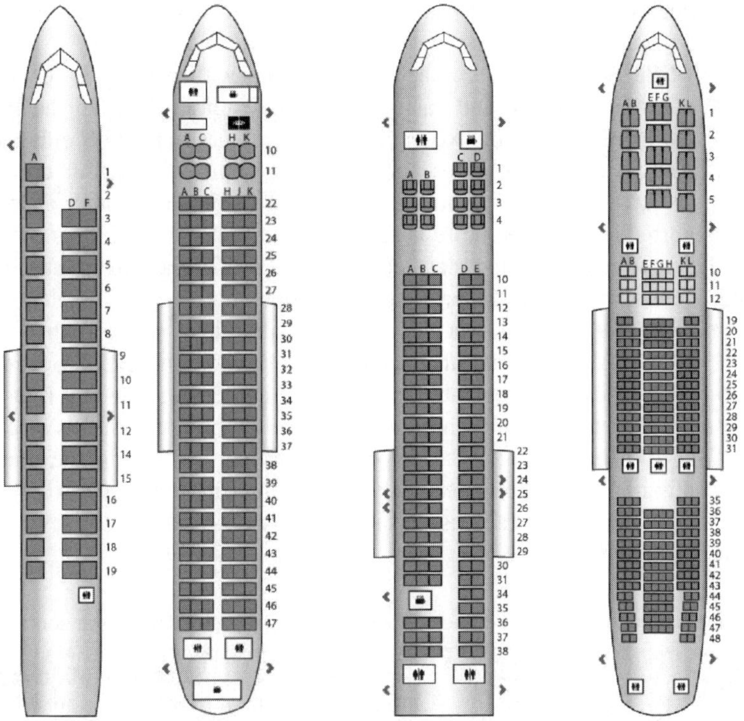

EMB-145
G:Galley,
L: Lavatory,
C: Closet

A-320-200
First two rows: First class

MD-88
First four rows: First class

B-777-200ER
First five rows: First class

Figure 7.14 Seating chart of several transport aircraft (figure not scaled). Reproduced from permission of www.seatplans.com

The business jet aircraft Cessna 560 Citation V, set to accommodate seven to eight passengers, has a customized interior. The accommodation plan includes standard seating for seven passengers with three forward-facing seats and four in club arrangement, or eight passengers in double-club arrangement, on swiveling and fore/aft/inboard-tracking pedestal seats, a refreshment center in the forward cabin area, a lavatory/vanity center with sliding door to the rear, space in the aft section of the cabin for 272 kg of baggage, in addition to an external baggage compartment in the nose and rear fuselage.

The number of flight attendants to serve the passengers, number of doors, emergency exits, galleys, windows, restrooms, and wardrobes are other items that must be determined. References such as [1, 7, 10] must be consulted to determine the minimum requirements. For a maximum of up to 80 passengers, one door is normally sufficient, while two doors are recommended for up to 200 passengers. The window pitch is recommended to match with the seat pitch to provide two windows (one on each side) for each row.

Based on FAR Part 125 Section 125.269, for airplanes having more than 100 passengers, two flight attendants plus one additional flight attendant for each unit of 50 passengers above 100 passengers are required. The Boeing 757-300 has five to seven cabin attendants to serve 289 passengers, while 16 flight attendants in the Boeing 777-300 serve 550 passengers.

Based on FAR 23 Section 23.783, each closed cabin with passenger accommodation must have at least one adequate and easily accessible external door. In the Boeing 757 (Figure 8.17), there are choices of two cabin door configurations, with either three passenger doors and two over-wing emergency exits on each side, or four doors on each side. All versions have a galley at front on the standard side and another at rear, a toilet at front on the port side and three more at rear, or two at rear or amidships. A coat closet is also provided at front of the first class cabins and 214/220 passenger interiors.

According to Section 23.841, if certification for operation over 25 000 ft is requested, the airplane must be able to maintain a cabin pressure altitude of not more than 15 000 ft in event of any probable failure or malfunction in the pressurization system. Sections 25.813 and 25.807 on emergency exits regulate that each required emergency exit must be accessible to the passengers and located where it will afford an effective means of evacuation. The Boeing 747-400 has two modular upper deck toilets and 14 on the main deck, a basic galley configuration on the upper deck, seven centerlines, and two sidewalls on the main deck. The number of seats abreast in the economy class section for several transport aircraft [9] is illustrated in Table 7.5.

Table 7.5 Number of seats abreast in the economy class section for several transport aircraft

No.	Aircraft	Take-off mass (kg)	Cabin width (m)	Total number of passengers	Number of seats abreast
1	Fairchild Metro 23	7 484	1.57	19	1+1
2	Cessna 750 Citation X	16 011	1.7	12	1+1
3	DASH 8 300	19 500	2.51	50	2+2
4	Embraer EMB-145	19 200	2.28	50	1+2
5	Fokker 100	43 090	3.1	107	2+3
6	McDonnell Douglas MD-88	67 800	3.35	172	3+2
7	Boeing 747-400	394 625	6.13	421	3+4+3
8	Boeing B-737-800	78 244	3.53	189	3+3
9	Boeing B-777-200	299 370	6.20	440	2+5+2
10	Airbus A-330-300	235 000	5.64	440	2+4+2
11	Airbus A320-200	78 000	3.7	180	3+3
12	Airbus 380	569 000	Main deck: 6.58 m Upper deck: 5.92 m	525–853	Upper deck: 2+3+2 Main deck: 3+4+3

To provide passengers with a comfortable and seamless travel experience and to improve onboard efficiency, a carry-on baggage policy is established by airlines. A passenger may carry on one bag and one personal item, where all items must fit easily into the overhead stowage or under the seat. Approved personal carry-on items include: one purse, briefcase, camera bag, or diaper bag; or one laptop computer; or one item of a similar or smaller size to those listed above. Additional approved carry-on items often include: a jacket or umbrella; food or drink purchased after clearing the security checkpoint; duty free merchandise; and special items like strollers, child restraint seats, or assistive devices such as wheelchairs or crutches. The size requirements are usually as follows: carry-on baggage may not exceed 115 cm in combined length, width, and height (~56 cm × 36 cm × 23 cm). So, the necessary stowage area (e.g., overhead bin) must be provided in the cabin.

The fuselage cross-section is normally a circle, and the cabin floor is usually flat. Thus the cabin floor is kept level in normal cruising flight to allow for serving food and drink by carts.

When the cabin interior arrangement (e.g., number of economy seats and number of seats abreast) is decided, and the cabin geometry is selected (e.g., seat width, seat pitch), the cabin width and cabin length (Figure 7.15) need to be calculated. The length of the cabin is normally determined by multiplying the number of rows by the seat pitch (P):

$$L_{\mathrm{C}} = \sum_{i=1}^{3} \sum n_{\mathrm{r}_i} \cdot P_{\mathrm{S}_i} \tag{7.1}$$

The summation sign (Σ) is to incorporate three types of seat ($i = 1$ for economy, $i = 2$ for business, $i = 3$ for first class). Then, this number should be revised to include any galley, or lavatory.

The cabin width is determined by multiplying the number of seats abreast (n_{S}) by the seat width (W_{S}), plus the number of aisles (n_{A}) times the width of all aisles (W_{A}):

$$W_{\mathrm{C}} = n_{\mathrm{S}} \cdot W_{\mathrm{S}} + n_{\mathrm{A}} \cdot W_{\mathrm{A}} \tag{7.2}$$

The cabin cross-section dimensions are utilized to determine the external width, D_{f} (diameter, if circular) of the fuselage, by adding the wall thickness. The fuselage overall length (L_{f}) is determined by summing the cabin length, cockpit length, lengths of the nose section, and rear fuselage. Figure 7.15 illustrates the cabin and fuselage for a

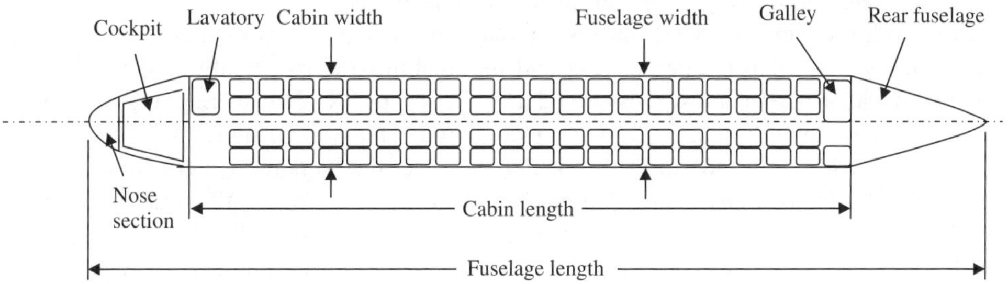

Figure 7.15 Cabin width and cabin length (top view)

(a) (b)

Figure 7.16 Cabins of two transport aircraft: (a) Cessna Citation; (b) Boeing 777. Part (b) reproduced from permission of Toshi Aoki

transport aircraft with 80 passengers. Figure 7.16 demonstrates the cabin arrangement of two transport aircraft.

The cabin of a transport aircraft needs to incorporate a variety of equipment and items to satisfy daily human needs and provide passenger comfort. Galleys and cooking appliances (items such as cooktops, ovens, coffee makers, water heaters, water cooler, and refrigerators) and toilet must be considered in the design of cabin. The bladder problem states that it is incorrect to design a single-engine small aircraft with more than 3 hours of endurance without a restroom. The reason simply is that people should go to the toilet. FAR regulations have several parts, including Part 25, to address these requirements.

7.7 Cargo Section Design

In a transport aircraft, either passenger or pure cargo, a large section of the fuselage must be allocated to the cargo/luggage bay. The fuselage must be designed such that it encompasses a sufficient volume for cargo/luggage. Nowadays, various cargo volumes and configurations are carried by aircraft. Items from mail to vegetables, industrial goods, military equipment, and the Space Shuttle are currently carried by aircraft. Thus, freight presents a wide variety of characteristics, from volume to density. The cargo bay design is based on detailed data concerning the dimensions and weights of the goods to be carried. These data are supplied by potential airlines or customers.

Most airlines regulate that a passenger may check in up to two bags; checked baggage must weigh 70 lb (32 kg) or less and its combined length, width, and height (i.e., length + width + height) must measure 62 in. (158 cm) or less. However, oversize or overweight baggage may be checked in at extra charge. This typical policy is changing due to the high cost of oil and competition. For instance, the baggage weight limit is being reduced to 50 lb for most national flights. A stowage room with sufficient space in the fuselage (usually the lower deck) must be considered to carry all checked baggage. The total volume of passenger cargo (V_C) is primarily equal to the number of travelers (n_t) times the total baggage volume of each traveler (V_b):

$$V_C = n_t \cdot V_b \tag{7.3}$$

Recall that the total number of travelers includes passengers, flight attendants, and pilots. The typical volume of each regular baggage item, based on the combined length of 158 cm, is approximated as:

$$V_b = \frac{158\,\text{cm}}{3} \cdot \frac{158\,\text{cm}}{3} \cdot \frac{158\,\text{cm}}{3} = 146085.6\,\text{cm}^3 = 0.146\,\text{m}^3 \tag{7.4}$$

This is an average baggage volume per passenger for a civil transport aircraft.

To carry cargo and passenger baggage in a secure fashion, and to prevent moving during flight, large passenger aircraft employ cargo containers (Figure 7.17) and pallets. In each container or pallet, several passenger luggage items are placed and tightened securely. This allows a large quantity of cargo to be bundled into a single unit load. Since this leads to fewer units loaded, it saves loading crews time and effort and helps prevent delayed flights. A list of the most common freight container dimensions used today is shown in Table 7.6. While it's true that some companies make custom boxes, these have been designed as the standard for air freight containers by the IATA. The containers are known by their LD number. The shape of most of the containers is rectangular or contoured. The taper of the containers is to allow them to be fitted into the circular shape of the fuselage cross-section (as shown in Figure 7.18).

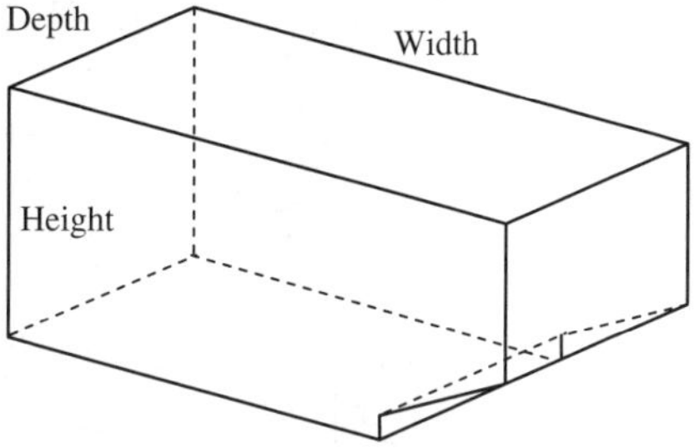

Figure 7.17 Cargo container

Table 7.6 The most common freight containers and pallets

No.	Container	Width (in.)	Height (in.)	Depth (in.)	Volume (ft^3)	Maximum load (lb)
1	LD1	92	64	60	173	3 500
2	LD2	61.5	64	47	120	2 700
3	LD3	79	64	60.4	159	3 500
4	LD4	96	64	60.4	–	5 400
5	LD5	125	64	60.4	–	7 000
6	LD6	160	64	60.4	316	7 000
7	LD7[a]	125	64	80	381	13 300
8	LD8	125	64	60.4	243	5 400
9	LD9	125	64	80	–	13 300
10	LD10	125	64	60.4	–	7 000 (rectangular)
11	LD11	125	64	60.4	253	7 000 (contoured)
12	LD29	186	64	88	–	13 300

[a] LD7 has two pallet variants (type 1 and type 2).

Figure 7.18 Airbus A-300 cross-section (Wikipedia)

The underfloor cargo hold of the Boeing 767-200 can accommodate typically up to 22 LD2 or 11 LD1 containers, while the Boeing 767-300 version can accommodate 30 LD2 or 15 LD1 containers. Three Type 2 pallets are accommodated in the Boeing 767-200 and four in the Boeing 767-300. The Boeing 747-400 cargo bay accommodates up to 32 LD1 containers, or five pallets and 14 LD1 containers.

The Airbus A-300-600 can accommodate 41 LD3 containers and 25 pallets, while the Airbus A-380 can accommodate 38 LD3 containers and 13 pallets. Example 7.1 shows how to determine cargo bay volume, and number of containers, for a large civil transport aircraft.

Example 7.1

A jet transport aircraft is designed to carry 200 passengers.

1. Determine the cargo bay volume to carry the baggage of the travelers.
2. The aircraft is using container LD1. How many containers need to be employed?

Solution:

1. **Cargo bay volume**. Since this is a transport aircraft, the FAR Part 25 regulations must be followed. FAR Part 25 Section 25.1523 and also Appendix D require the

minimum flight crew for a transport aircraft to be two, so a pilot and a copilot are selected. In contrast, based on FAR Part 125 Section 125.269, this aircraft requires four (2 + 2) flight attendants. Hence, a total of 206 people (200 + 2 + 4) are allowed to carry luggage. It is assumed that each traveler is allowed to carry two items of baggage, so the total number of checked baggage items will be considered to be 206 × 2 = 412. Therefore, the total volume of cargo is obtained as:

$$V_C = n_t \cdot V_b = 412 \cdot 0.146 = 60.187\,\mathrm{m}^3 \tag{7.3}$$

This volume must be provided by the fuselage.

2. **Number of containers**. Each LD1 container has a volume of 173 ft^3 or 4.899 m^3. Thus:

$$n = \frac{60.187\,\mathrm{m}^3}{4.899\,\mathrm{m}^3} = 12.28$$

Hence the aircraft needs to carry 13 LD1 containers.

Small transport aircraft do not utilize containers or pallets, but they tend to be equipped with special cargo compartments. The required cargo volume is a function of the number of passengers, and the type of cargo to be transported. The cargo compartment of a military transport aircraft such as the Lockheed C-130 Hercules (Figure 5.4), Lockheed C-5 Galaxy, and McDonnell Douglas C-17 Globemaster (Figure 9.9) is dependent upon the assigned military cargo, some of which may be oversized such as tanks, and large trucks.

A cargo aircraft requires accommodation for load personnel. The number of loadmasters is a function of the cargo type and loading/unloading requirements. It will be determined while assigning tasks to load and unload cargo. The cargo compartment necessitates a cargo handling system including such items as rails and rollers, cranes, and ramps. Loadmasters need a special station to sit and rest during flight. Their seats may be considered in the cockpit or in the cargo bay. However, the station requires a pressure system with air conditioning.

The military transport aircraft C-17A Globemaster III (Figure 9.9) accommodates a pilot and a copilot, side by side, and two observer positions on the flight deck, plus a loadmaster station at the forward end of the main floor. Access to the flight deck is provided via a downward-opening airstair door on the port side of the lower forward fuselage. Crew bunks are immediately aft of the flight deck area, and the crew comfort station is at the forward end of the cargo hold. The main cargo hold is able to accommodate wheeled and tracked vehicles up to an M1 tank, including 5 ton expandable vans in two rows, or up to three AH-64A Apache helicopters.

One of the natural locations for cargo in a large transport aircraft is underneath the cabin. If the cabin is designed to be of circular cross-section, the maximum available space for cargo is:

$$V_{\mathrm{bottom}} = \frac{1}{2}\left(\pi \frac{W_C^2}{4} L_C\right) \tag{7.4a}$$

Please note that not all of this space is available in reality for cargo and luggage. The reason is that a number of other items such as wing box, fuel, and landing gear may be considered to be accommodated by the fuselage. Thus, in case this space is not sufficient for the entire cargo, the rear section of the fuselage may be extended to provide more room.

7.8 Optimum Length-to-Diameter Ratio

Two of the main fuselage design parameters are the fuselage length (L_f) and the maximum diameter (D_f). These two fuselage parameters produce the fuselage volume, wetted area, and weight. The fuselage optimum length-to-diameter ratio (or slenderness ratio) may be determined based on a number of design requirements. The design objectives may be to determine the fuselage length-to-diameter ratio such that it:

1. results in the lowest zero-lift drag;
2. creates the lowest wetted area;
3. delivers the lightest fuselage;
4. provides the maximum internal volume;
5. generates the lowest mass moment of inertia;
6. contributes the most to aircraft stability;
7. requires the lowest cost to fabricate.

The first and second objectives concern the aircraft performance requirements. The third objective aims for weight requirements, and the fourth one satisfies operational requirements. The goal of the fifth one is controllability requirements, while the sixth one satisfies stability requirements. Finally, the last objective targets the lowest cost of manufacturing the fuselage. Depending upon the aircraft mission and design priorities, one of these objectives becomes the most significant.

The length of the fuselage should be enough to provide a sufficient moment arm for both the horizontal and vertical tails. The fuselage length to create the lowest wetted area of the aft aircraft is discussed in Chapter 6. For instance, for a cargo aircraft, the largest fuselage internal volume would be the most desirable objective in the fuselage design. For objectives two to six, the designer is expected to develop a formulation to mathematically express the requirement in terms of fuselage length and diameter. Then, differentiate the formula with respect to fuselage length or diameter. When the result of the differentiation is set equal to zero, the final solution yields the optimum fuselage length and diameter. In this section, a technique to determine the fuselage length-to-diameter ratio is developed to deliver the lowest fuselage zero-lift drag.

7.8.1 Optimum Slenderness Ratio for Lowest f_{LD}

The fuselage drag is proportional to the fuselage slenderness ratio, since the zero-lift drag coefficient of the fuselage is given [12] by the following expression:

$$C_{D_{o\_f}} = C_f f_{LD} f_M \frac{S_{wet_f}}{S_{ref}} \tag{7.5}$$

where C_f is the skin friction coefficient, f_M is a function of aircraft speed, S_{ref} is the wing reference area, and S_{wet_f} is the fuselage wetted area. The second parameter (f_{LD}) in Equation (7.5) is a function of the fuselage length-to-diameter ratio. For a subsonic speed, it is defined [12] as:

$$f_{LD} = 1 + \frac{60}{(L/D)^3} + 0.0025\left(\frac{L}{D}\right) \tag{7.6}$$

where L is the fuselage length and D is its maximum diameter. The variation of this function (f_{LD}) with respect to length-to-diameter ratio is sketched in Figure 7.19, which clearly indicates that this function has a minimum value. To determine the lowest value for this function (f_{LD}), the differentiation of the function with respect to length-to-diameter ratio is set equal to zero:

$$\frac{df_{LD}}{d(L/D)} = 0 \Rightarrow \frac{-180}{(L/D)^4} + 0.0025 = 0 \Rightarrow (L/D)^4 = 72\,000 \tag{7.7}$$

The solution of this equation yields the optimum value for the length-to-diameter ratio as follows:

$$(L/D)_{opt} = 16.3 \tag{7.8}$$

This indicates that when the length of the fuselage is 16.3 greater than the fuselage maximum diameter, the fuselage will generate the lowest zero-lift drag. Therefore, if the goal is to minimize the fuselage zero-lift drag, set the fuselage length to be 16.3 times the fuselage diameter. Another influential parameter in the fuselage drag is the fuselage

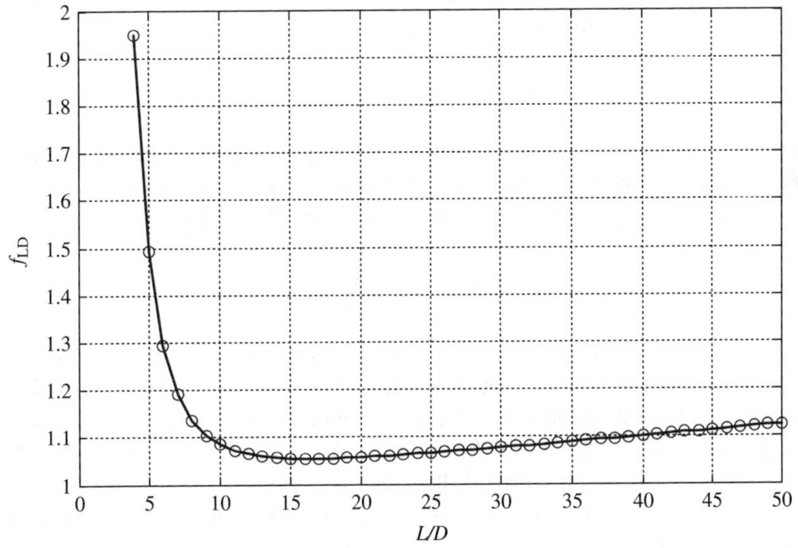

Figure 7.19 The variation of function f_{LD} with respect to slenderness ratio

wetted area (S_{wet_f}). When this parameter is inserted into the differentiation, the results will be different:

$$C_{D_{o\_f}} = C_f f_{LD} f_M \frac{S_{wet_f}}{S_{ref}} = K_1 \left(1 + \frac{60}{(L/D)^3} + 0.0025\left(\frac{L}{D}\right)\right)\left(\frac{L}{D}\right) \tag{7.5a}$$

where K_1 is a constant and does not affect the optimum value. The differentiation of this function with respect to *L*/*D* yields the following solution:

$$(L/D)_{opt} = 5.1 \tag{7.8a}$$

This fuselage length-to-diameter optimum value is not applicable for all aircraft configurations. Since this optimum value does not take into account the overall aircraft wetted area (for instance, this optimum value requires a larger tail), it is recommended to apply the differentiation technique to the combination of fuselage and tail and determine a specific value for your aircraft configuration.

For supersonic speeds, the technique results in a higher length-to-diameter ratio. Table 7.7 presents the fuselage length-to-diameter ratio of several aircraft. Please note that most aircraft do not follow the optimum ratio as stated in Equation (7.8). The reason is that although this ratio yields the lowest zero-lift drag, it does not result in the optimum combination of the above-mentioned seven objectives (e.g., weight, cost, and controllability).

In a transport aircraft, the external fuselage diameter is determined by adding the wall thickness (T_W) (both sides) to the cabin width, which is about 4–10 cm:

$$D_f = W_C + 2T_W \tag{7.9}$$

Similarly, the fuselage length (L_f) in a conventional fuselage is determined by the cockpit length (L_{CP}), the length of the nose section (L_N), and the length of the rear section (L_R):

$$L_f = L_C + L_{CP} + L_N + L_R \tag{7.10}$$

A circular cross-section is recommended for a pressurized cabin (i.e., fuselage) to minimize the hoop stress. In case the fuselage cross-section is not circular, the equivalent diameter is calculated by assuming the cross-section to be a circle:

$$D_{equ} = \sqrt{\frac{4A_{cross}}{\pi}} \tag{7.11}$$

If the pressurized section of the fuselage is limited to the cockpit, a rounded, elliptical, or oval cross-section is feasible. When an aircraft is required to fly above 18 000 ft, the pressurized system must be designed for passengers and flight crew to provide air with sea-level pressure and temperature. For a passenger aircraft, this means that the entire cabin and cockpit must be pressurized. It is also recommended to consider a pressurized system for the luggage/cargo section, since the payload sometimes includes pets and live animals. In most transport aircraft, the fuselage is divided into three subsections: (i) unpressurized nose section, (ii) pressurized flight deck and cabin, and (iii) unpressurized tail section. Example 7.2 illustrates a cabin design to produce low zero-lift drag.

Table 7.7 The fuselage slenderness ratio of several aircraft

No.	Aircraft	Type	Engine	Take-off mass (kg)	L_f/D_f
1	Reims F337F Super Skymaster	Utility	Twin piston	2 000	3.2
2	Cessna 208	Light GA	Piston	3 645	6.8
3	Cessna Citation III	GA light transport	Twin turbofan	9 979	8
4	Pilatus PC-7	Trainer	Turboprop	2 700	7
5	BAE ATP	Transport	Twin turboprop	12 430	9.6
6	STEMME S10	Motor glider	Piston	850	8.4
7	ATR 52C	Cargo	Twin turboprop	22 000	9
8	Firecracker	Trainer	Turboprop	1 830	7.2
9	Embraer Tucano	Trainer	Turboprop	2 250	7.5
10	Dornier 328	Transport	Twin turboprop	11 000	7.5
11	Fairchild Metro VI	Transport	Twin turbofan	7 711	10.7
12	Fokker 100	Airliner	Twin turbofan	23 090	9.85
13	Boeing 737-200	Airliner	Twin turbofan	52 400	8.2
14	Boeing 747-400	Airliner	Four turbofan	394 625	10.5
15	Boeing 757-200	Airliner	Twin turbofan	133 395	12
16	Boeing E-3 Sentry	Relay-communication	Twin turbofan	147 417	11.6
17	Airbus A-330	Airliner	Twin-jet	230 000	11.4
18	Sukhoi SU-27	Fighter	Twin turbofan	25 000	10.3
19	F-16 Fighting Falcon	Fighter	Twin turbofan	27 000	9.5
20	Concorde	Supersonic transport	Four turbojet	141 200	23

Example 7.2

For an aircraft to carry 156 passengers and 4 flight attendants, design a cabin to submit the lowest zero-lift drag in terms of f_{LD}. The length of the nose section (including cockpit) is 3 m and the length of the rear section is 4 m. You are required to determine the cabin length, cabin diameter, and number of seats abreast. Assume that the wall thickness is 4 cm each side. The top view of the fuselage is shown in Figure 7.20. Ignore the galley, lavatories, and assume that all seats are desired to be of economy (tourist) class.

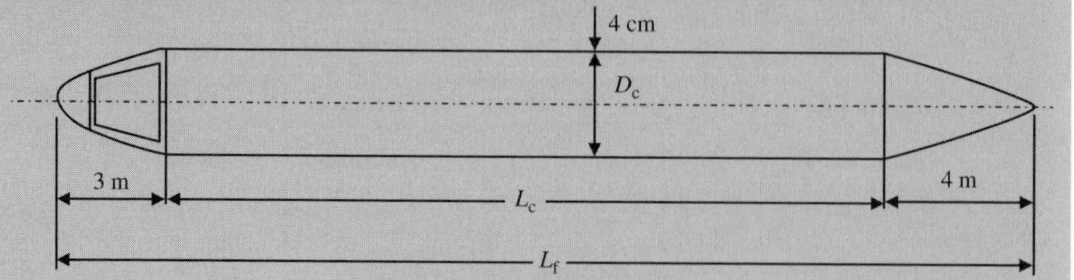

Figure 7.20 Fuselage top view for Example 7.2

Solution:

According to Equation (7.8), the optimum fuselage length-to-diameter to create the lowest zero-lift drag is 13.6. Thus, the number of seats abreast, and number of rows, must satisfy this requirement. The total number of seats is 160. The cabin length and cabin width are determined using Equations (7.1) and (7.2):

$$L_C = \sum_{i=1}^{3} \sum n_{r_i} \cdot P_{s_i} \quad (7.1)$$

$$W_C = n_S \cdot W_S + n_A \cdot W_A \quad (7.2)$$

The seat pitch, seat width, and aisle width for economy class are extracted from Table 7.4 as follows:

- $W_S = 45$ cm
- $P_S = 80$ cm
- $W_A = 45$ cm

In general, the reasonable alternatives are:

1. 160 rows of single seats (plus one aisle);
2. 80 rows (160/2) of dual seats (plus one aisle);
3. 54 rows (160/3) of three seats (plus one aisle);
4. 40 rows (160/4) of four seats (plus one aisle);
5. 32 rows (160/5) of five seats (plus one aisle);
6. 32 rows (160/5) of five seats (plus two aisles);
7. 27 rows (160/6) of six seats (plus one aisle);
8. 27 rows (160/6) of six seats (plus two aisles).

Now we determine the fuselage length-to-diameter for each case by examining these seating alternatives (number of seats abreast) using Equations (7.9) and (7.10). For instance, the calculations for option 6 are as follows:

$$L_C = \sum_{i=1}^{3} \sum n_{r_i} \cdot P_{s_i} = 32 \cdot 80\,\text{cm} = 25.6\,\text{m} \tag{7.1}$$

$$W_C = n_S \cdot W_S + n_A \cdot W_A = (5 \cdot 45) + (2 \cdot 45) = 3.15\,\text{m} \tag{7.2}$$

$$D_f = W_C + 2T_W = 3.15 + (2 \cdot 0.04) = 3.23\,\text{m} \tag{7.9}$$

$$L_f = L_C + L_{CP} + L_N + L_R = 25.6 + 3 + 4 = 32.6\,\text{m} \tag{7.10}$$

Thus, the fuselage length-to-diameter ratio is:

$$\frac{L_f}{D_f} = \frac{32.6}{3.23} = 10.093$$

The calculations for all other seven alternatives are performed in a similar manner. The results are presented in Table 7.8.

As Table 7.8 illustrates, alternative 4 yields the closest fuselage length-to-diameter (i.e., 16.74) to the optimum fuselage length-to-diameter ratio (i.e., 16.3). Therefore the cabin is decided to have four seats abreast and one aisle. This seating in each row includes two seats on one side of the aisle, and two seats on another side. The 40 rows of four seats, a total of 160 seats, are shown in Figure 7.21. It is interesting to note that option 5 yields the smallest fuselage wetted area. Thus, option 5 is also promising.

Table 7.8 Seating alternatives for Example 7.2

No.	Number of rows	Seats abreast	Aisle	W_C (m)	L_C (m)	D_f (m)	L_f (m)	L_f/D_f
1	160	1	1	0.9	128	0.98	135	137.7
2	80	2 (1 + 1)	1	1.35	64	1.43	71	49.6
3	54	3 (2 + 1)	1	1.8	43.2	1.88	50.2	26.7
4	**40**	**4 (2 + 2)**	**1**	**2.25**	**32**	**2.33**	**39**	**16.74**
5	32	5 (2 + 3)	1	2.7	25.6	2.78	32.6	11.7
6	32	5 (2 + 1 + 2)	2	3.15	25.6	3.23	32.6	10.1
7	27	6 (3 + 3)	1	3.15	21.6	3.23	28.6	8.8
8	27	6 (2 + 2 + 2)	2	3.6	21.6	3.68	28.6	7.77

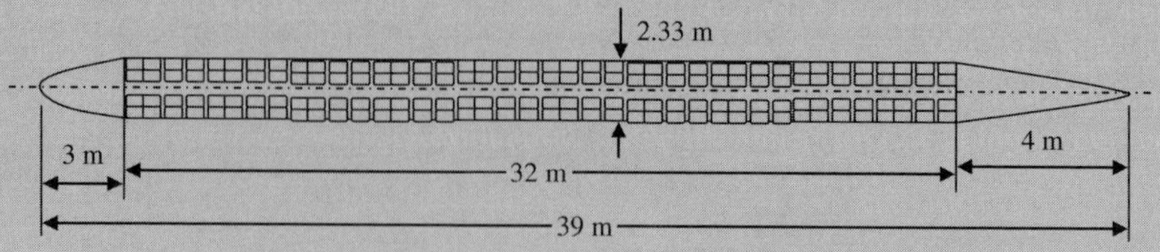

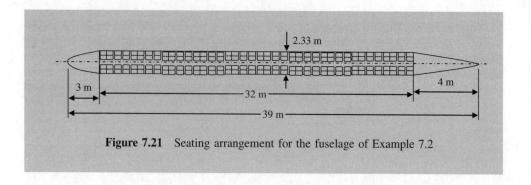

Figure 7.21 Seating arrangement for the fuselage of Example 7.2

7.8.2 *Optimum Slenderness Ratio for Lowest Fuselage Wetted Area*

Another independent variable in the fuselage zero-lift drag coefficient (Equation (7.5)) is the fuselage wetted area (S_{wet_f}). This area is a function of the fuselage geometry. Figure 7.22 demonstrates an example for a transport aircraft which recommends the nose section to have a slenderness ratio of 1.5, and the rear section to have a slenderness ratio of 12. In this section, we examine the fuselage optimum slenderness ratio for a pure cylinder. The interested reader is encouraged to examine other fuselage geometries. The problem statement is to determine the fuselage length-to-diameter ratio to submit the lowest wetted (surface) area of a cylindrical fuselage with a constant volume (due to a constant number of seats and cargo). The volume of a cylinder with radius r and length L is:

$$V = \pi \cdot r^2 L \tag{7.12}$$

So, the length is:

$$\Rightarrow L = \frac{V}{\pi \cdot r^2} \tag{7.13}$$

In contrast, the fuselage wetted area is:

$$S_{\text{wet}} = 2\pi \cdot r^2 + 2\pi \cdot rL \tag{7.14}$$

The fuselage length (L) from Equation (7.13) is plugged into this equation:

$$S_{\text{wet}} = 2\pi \cdot r^2 + 2\pi \cdot r \left(\frac{V}{\pi \cdot r^2} \right) = 2\pi \cdot r^2 + \frac{2V}{r} \tag{7.15}$$

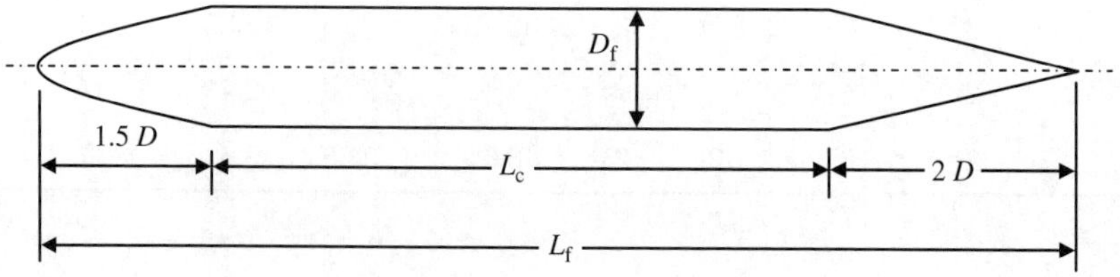

Figure 7.22 A recommended fuselage nose and tail section (top view)

The lowest wetted area is determined by differentiating Equation (7.15) with respect to fuselage radius (r) and setting the equation to zero:

$$\frac{dS_{wet}}{dr} = 2 \cdot 2\pi \cdot r - \frac{2V}{r^2} = 0 \Rightarrow 4\pi \cdot r - \frac{2V}{r^2} = 0 \tag{7.16}$$

which results in the following expression:

$$\frac{dS_{wet}}{dr} = 0 \Rightarrow r^3 = \frac{V}{2\pi} \tag{7.17}$$

Now, the fuselage volume is substituted from Equation (7.12) ($V = \pi r^2 L$) into this expression:

$$r^3 = \frac{V}{2\pi} = \frac{\pi r^2 L}{2\pi} \tag{7.18}$$

However, this expression yields the following interesting results:

$$r = \frac{L}{2} \Rightarrow 2r = D = L \tag{7.19}$$

Thus, for a given volume, a cylindrical fuselage with minimum surface (wetted) area has a length equal to its diameter. In other words, the optimum fuselage slenderness ratio to minimize the fuselage surface area is just 1:

$$\left(\frac{L}{D}\right)_{opt} = 1 \tag{7.20a}$$

It is interesting to compare the optimum fuselage slenderness ratio to minimize the fuselage surface area (i.e., 1) with the optimum fuselage slenderness ratio to minimize the f_{LD} (i.e., 16.3). Since these two do not match, a weighting factor must be applied which comes from the priority of the design requirements. Therefore, it is desirable to have a short fuselage to minimize the fuselage surface area, while it is desirable to have a long fuselage to minimize the fuselage f_{LD}.

Figure 7.23 illustrates the variations in surface area of a cylinder with a volume of 14 m^3. In this figure, the shortest radius is assumed to be 0.5 m^3 and the longest to be 2.93 m^3. The corresponding length-to-diameter ratio is from 0.111 to 17.8.

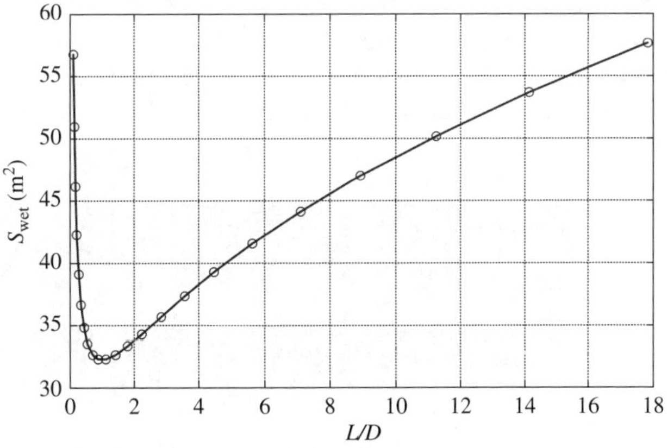

Figure 7.23 Variations of surface area versus L/D for a cylinder with volume 14 m^3

As observed, the smallest surface area (32.194 m^2) happens when the length-to-diameter ratio is one. It is interesting to note that when L/D is 16.3, the corresponding S_{wet} is about twice as large as the minimum S_{wet}.

In this section, we have just derived two optimum fuselage slenderness ratio values. The designer is recommended to derive the optimum slenderness ratio values for other design requirements and compare them in a trade-off study table. Based on the results of Sections 1.8.1 and 7.8.2, the optimum fuselage slenderness ratio has a value between 1 and 16.3. Figure 7.24 illustrates the Aerospatiale-British Aerospace supersonic transport aircraft Concorde with a fuselage length-to-diameter ratio of 23.

7.8.3 Optimum Slenderness Ratio for the Lightest Fuselage

Another interesting goal when designing the fuselage is to minimize its weight. The problem statement is to determine the fuselage length-to-diameter ratio such that the fuselage has the lowest weight. This topic concerns the structural design of the fuselage and to find the weight of the fuselage with regard to skin, longeron, frame, and stiffener. The fuselage structural design is beyond the scope of this text, so it is assumed that the fuselage weight is proportional to the fuselage surface area. The reason is that the weight of a simple cylinder is determined by multiplying the surface area times the thickness times the density of the material. For this reason, the optimum slenderness ratio for the lightest fuselage is the same as the optimum slenderness ratio for the smallest surface area. In other words:

$$\left(\frac{L}{D}\right)_{opt} = 1 \tag{7.20b}$$

Therefore, the optimum slenderness ratio for the lightest fuselage is one. Reference [13] demonstrates the passenger aircraft capacity growth trend, and overall length growth versus gross weight for various transport aircraft.

7.9 Other Fuselage Internal Segments

The primary function of the fuselage is to accommodate the payload. However, there are a number of other items that must be accommodated by the aircraft. The most suitable room for most of these items is in the fuselage. These are items and equipment such as

Figure 7.24 Aerospatiale-British Aerospace Concorde with fuselage length-to-diameter ratio 23. Reproduced from permission of A J Best

fuel tanks, radar, wing box, mechanical systems, hydraulic systems, electrical systems, fuel system, and landing gear retraction bay. When the fuselage configuration is designed (Section 7.3), the optimal spaces for each of these items are allocated. Now is the time to determine the volume needed for each item and to ensure that the fuselage is large enough to encompass these items of equipment.

The following are a few recommendations for the fuselage configuration design and allocating items and instruments.

1. Water condenses in all high-altitude (including transport) aircraft, at cruising flight and during descending; it melts and accumulates on the bottom of the fuselage. If there is no built-in water drainage, some thousands of kilograms of water accumulate in the fuselage. For instance, in a Boeing 777 (Figures 8.6 and 12.27), it is about 10 000 kg after one month of operation. Therefore, every high-altitude aircraft needs built-in drainage. A design solution must also be considered to prevent the water flow into the cabin and cockpit during descending.
2. Water lines should not be very close to the aircraft (e.g., fuselage) skin to avoid water freezing. In the first commercial flight of a Fokker 100 (Figure 10.6), there was a short circuit and power outage in the electrical system [14]. This accident was due to the location of the water line; it was about half an inch from the top skin of the fuselage. In consequence, the water was frozen during cruising flight and melted during descending. Then the water leaked during the approach and flowed into the avionic bay. Thus, the electrical line was short circuited.

7.9.1 Fuel Tanks

An aircraft must carry its own fuel in internal fuel tanks. At the aircraft conceptual design phase (Chapter 3), the best space for storing fuel is determined. The two most conventional locations for the fuel tanks are the wing and the fuselage. For the sake of safety, it is

Figure 7.25 Lockheed CF-104D Starfighter with wing-tip fuel tanks. Reproduced from permission of Antony Osborne

recommended to store the fuel out of the fuselage (e.g., in the wing). Some aircraft, such as the single-engine jet fighter Lockheed CF-104D Starfighter (Figure 7.25), store part of the fuel in wing-tip fuel tanks. This is partly to allow more space in the fuselage for the payload, and partly to enhance passenger safety. Twice a day in the USA a jet aircraft is hit by lighting, typically somewhere at a tip (e.g., wing tip, nose, or tail). Thus, if there is a wing-tip tank, it should be structurally thick enough for safety. However, this may make the aircraft a bit heavy.

There are a number of precautions which must be considered in allocating room for fuel storage in the aircraft, including in the fuselage:

1. The total amount of fuel may be divided into several smaller fuel tanks. This precaution allows for an efficient fuel management system. In this case, the fuel storage must help keep the aircraft symmetric. For instance, if two fuel tanks are considered on the left side of the aircraft, another two fuel tanks (with the same features) need to be considered for the right-hand side.
2. One of the most important precautions is to keep the total fuel center of gravity close to the aircraft center of gravity (along the x-axis). This consideration plays an important role in aircraft longitudinal controllability and stability. More recommendations will be introduced in Chapter 11.
3. The fuel is preferred to be stored near the aircraft rolling axis (x-axis). As the fuel tanks are moved away along the y-axis, the aircraft roll control is degraded. To enhance the aircraft roll control, the aircraft mass moment of inertia about the x-axis must be kept as low as possible.
4. Due to the fire hazard in the event of a leak or aircraft crash, the fuel tank must be away from the passenger cabin, crew cockpit, or engine inlets.
5. In supersonic aircraft, fuel tank locations need special consideration. In subsonic flight, the wing/fuselage aerodynamic center (ac_{wf}) is at about 25% of the wing mean aerodynamic chord (MAC). However, at supersonic speeds, the ac_{wf} moves to about 50% MAC. The longitudinal trim of the aircraft requires the deflection of the elevator. As the distance between the aircraft cg and ac_{wf} is increased, a larger elevator deflection is required. Since the elevator deflection tends to have a limit and also generates trim drag, the goal is to deflect the elevator as little as possible.

   One of the techniques to longitudinally trim the aircraft at supersonic speed is to shift the aircraft cg back closer to the ac_{wf} in order to employ a smaller elevator deflection. This goal is achieved by pumping fuel from front tanks to rear tanks, as was done in the supersonic transport aircraft Concorde. This technique requires the provision of extra rear tanks for pumping fuel from front tanks after the aircraft achieves supersonic speed. This indicates that the fuselage designer should allocate more fuel tanks than the actual fuel volume. The retired Concorde legend will live on forever, since the unique design and supersonic flight service will be remembered by passengers who experienced their first sight of the curvature of the Earth under a deep blue sky and acceleration through Mach 1.
6. If the fuel tanks are considered to be beyond the fuselage and wing (such as wing-tip fuel tanks), the external shape must be such that they aerodynamically generate the

lowest drag. The best aerodynamic shape for a fuel tank is a symmetrical airfoil configuration with a circular cross-section throughout. Almost all Cessna aircraft (except the Citation) have big wings, because the fuel is stored in the wing.

7. In general, the fuel tanks are not recommended to be located at the wing tips, due to the possible hazard when lightning strikes. Furthermore, the fuel should be away from the tips for the same reason.

In case the fuel tanks are considered to be accommodated by the fuselage, the required fuel volume should be calculated and then sections of the fuselage volume must be allocated for fuel tanks. The required fuel volume of the fuel tanks is obtained by dividing the fuel mass (m_f) by the fuel density:

$$V_f = \frac{m_f}{\rho_f} \tag{7.21}$$

The fuel density (ρ_f) is based on the type of fuel. Table 7.9 presents densities for various aviation fuels at 15°C. The fuel mass is a function of the aircraft mission, including range. The fuel mass to fly a specified range (R) is given by the Breguet range equation [15]. Equation (7.22) delivers the range for a jet aircraft (with either turbofan or turbojet engine), while Equation (7.23) specifies the range for a prop-driven aircraft (with either piston or turboprop engine):

$$R = \frac{V\,(L/D)}{C} \ln\left(\frac{1}{1-\left(\frac{m_f}{m_o}\right)}\right) \quad \text{(jet aircraft)} \tag{7.22}$$

$$R = \frac{\eta_P\,(L/D)}{C} \ln\left(\frac{1}{1-\left(\frac{m_f}{m_o}\right)}\right) \quad \text{(prop-driven aircraft)} \tag{7.23}$$

Table 7.9 Density of various fuels at 15°C

No.	Fuel	Density (kg/m^3)	Application
1	Jet A	775–840	Civil jet
2	Jet A-1	775–840	Civil jet
3	JP-4	751–802	Fighter jet
4	JP-5	788–845	Fighter jet
5	JP-7	779–806	Fighter jet
6	JP-8 (military equivalent of Jet A-1)	775–840	Fighter jet
7	Aviation gasoline (100-octane, low lead)	721–740	Piston engine

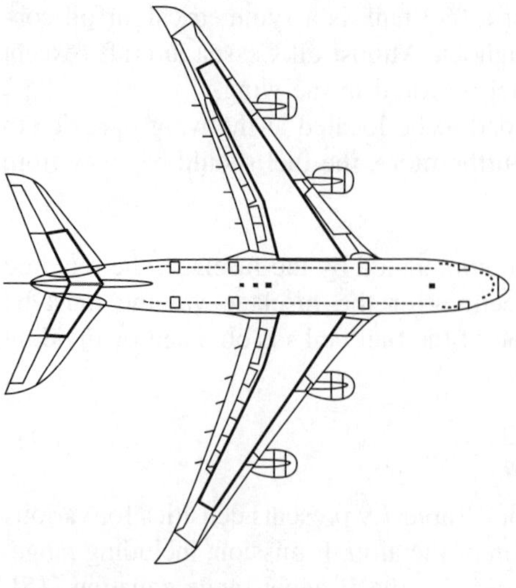

Figure 7.26 Schematic of Airbus A380 fuel tank locations

where V denotes the aircraft airspeed, m_f is the fuel mass, m_o is the initial aircraft mass or simply take-off mass (m_{TO}), C is the engine specific fuel consumption, L/D is the lift-to-drag ratio, and η_P is the propeller efficiency. Typical values of specific fuel consumption for various engines are given in Chapter 3. When the aircraft employs an electric engine (e.g., remote-piloted aircraft), no fuel tank is necessary, since a battery provides the energy for the engine. Figure 7.26 illustrates the Airbus A380 fuel tank locations.

As a rule of thumb, the total fuel mass is about 20% more than the fuel mass obtained by the range equation. This is partly due to take-off, climb, and descent and partly due to safety reserve fuel requirements (about 45 minutes of flight). Combining this rule of thumb with Equations (7.22) and (7.23), after a few mathematical steps, yields:

$$m_f = 1.2 \cdot m_{TO}\left[1 - \exp\left(\frac{-RC}{V\,(L/D)}\right)\right] \quad \text{(jet aircraft)} \tag{7.24}$$

$$m_f = 1.2 \cdot m_{TO}\left[1 - \exp\left(\frac{-RC}{\eta_P\,(L/D)}\right)\right] \quad \text{(prop - driven aircraft)} \tag{7.25}$$

Example 7.3 illustrates how to calculate the total fuel volume based on the specific fuel consumption and range requirements.

Example 7.3

The fuselage design of the following jet transport aircraft is in progress. The fuel tanks are arranged to be accommodated by the fuselage.

$$m_{TO} = 100\,000\,\text{kg}, S = 300\,\text{m}^2, C = 0.7\,\text{lb/h/lb}.$$

The aircraft cruising speed at 30 000 ft is 500 knot with a lift-to-drag ratio of 12.

1. Determine the total fuel volume if the aircraft range is 5000 km. Assume the fuel type is JP-4.
2. If each fuel tank contains 10 m^3 of fuel, how many fuel tanks must be accommodated by the fuselage?

Solution:

The air density at 30 000 ft is 0.459 kg/m^3. The range equation for a jet aircraft (Equation (7.24)) is employed. The specific fuel consumption is equivalent to 0.000194 1/s.

$$m_{\mathrm{f}} = 1.2 \cdot m_{\mathrm{TO}} \left[1 - \exp\left(\frac{-RC}{V\,(L/D)}\right)\right]$$
$$= 1.2 \cdot 100\,000 \cdot \left[1 - \exp\left(\frac{-5000000 \cdot 0.000194}{(500 \cdot 0.5144) \cdot 12}\right)\right] \tag{7.24}$$

So:

$$m_{\mathrm{f}} = 32\,423.1\ \mathrm{kg}$$

The fuel density of JP-4 from Table 7.9 is taken to be 751 kg/m^3, so:

$$V_{\mathrm{f}} = \frac{m_{\mathrm{f}}}{\rho_{\mathrm{f}}} = \frac{32\,423.1}{751} \Rightarrow V_{\mathrm{f}} = 43.173\ \mathrm{m}^3 \tag{7.21}$$

Therefore, five ($43.173/10 = 4.35 = 5$) fuel tanks are required to be accommodated by the fuselage.

7.9.2 Radar Dish

Another piece of equipment that most large and fighter aircraft are equipped with is a communication device such as radar. Radar is an object-detection system which uses electromagnetic waves to determine the range, altitude, direction, or speed of both moving and fixed objects. The radar dish, or antenna, transmits pulses of radiowaves or microwaves which bounce off any object in their path. The object returns a tiny part of the wave's energy to the dish. In modern aircraft, various types of radar are employed including weather radar.

There are several locations to place the radar antenna or dish. The location for an antenna must have an open view to transmit and receive radar signals. One of the best locations is the fuselage nose (see Figure 7.27). The fuselage nose of most modern aircraft (e.g., Boeing B-747 and Airbus A-380) contains vital radar that communicates weather systems to the pilot.

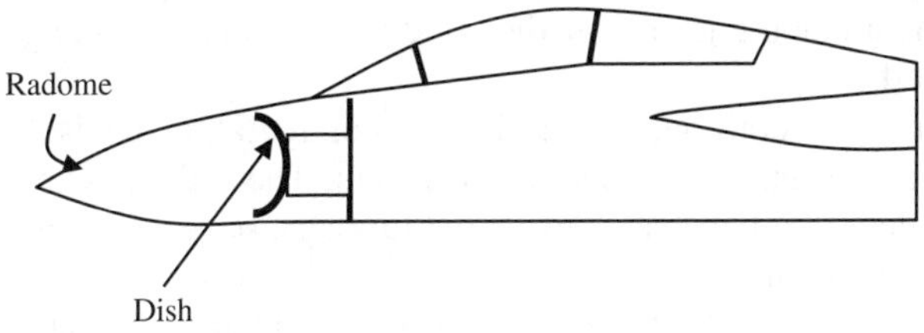

Figure 7.27 Radar dish in the nose section of a fighter aircraft

Most military aircraft also employ a radar dish in the fuselage nose (e.g., C-130 Hercules (Figure 5.4), F-16 Falcon (Figure 4.6), Sukhoi 27 (Figure 6.7), and Global Hawk (Figure 6.12)). The cover of a radar dish is referred to as a radome or radar dome.

A radome is a structural, weatherproof enclosure that protects a radar antenna. The radome is transparent to radar waves. The radome protects the antenna surfaces from the environment (e.g., wind, rain, ice, and sand), and conceals the antenna electronic equipment from public view. Thus, in the fuselage design, the nose may be considered for the radar dish. Furthermore, during landing, the pilot should be able to see below the horizontal when the aircraft is in a tail-down attitude. If the nose section does not have this feature due to other requirements (such as aerodynamic considerations), the nose must be drooped during touch-down. Concorde had these structural characteristics.

7.9.3 Wing Box

Aircraft structural designers prefer to have the wing main spar carried through the fuselage to maintain aircraft structural integrity. This is due to the fact that the wing lift force generates a large bending moment where the wing attaches to the fuselage. The technique by which this moment is carried across the fuselage is a key structural consideration. The wing carry through the structure must be designed to minimize the bending stress and also stress concentration.

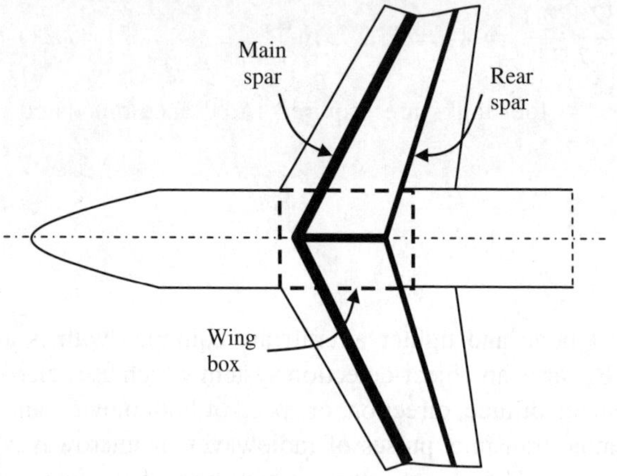

Figure 7.28 Wing box

For a fuselage designer, this consideration is translated into allocating a section of the fuselage volume to the wing spar (i.e., wing box). The wing box of an aircraft, or the section of the fuselage between the wing roots, is the structural component from which the wing extends (see Figure 7.28). This is the strongest structural area of the aircraft, and suffers the most bending and shear stresses. The wing box may be provided mostly for an aircraft with high-wing or low-wing configuration. For instance, in a single-seat light GA aircraft with a low-wing configuration, the wing spar may be considered to pass under the pilot seat. Modern aircraft often locate the main landing gear near the wing roots to take advantage of the structural strength they afford.

The design of the fuselage for an aircraft with a mid-wing configuration tends to generate challenges for the fuselage designer or structural designers. The reason is that either the wing spar must be cut in half to allow space inside the fuselage, or an important amount of space in the fuselage becomes useless.

The first alternative makes the wing spar vulnerable unless it is reinforced structurally, which in turn makes it heavier. The second option obstructs the passenger view in a transport aircraft. Thus, the mid-wing configuration is not a suitable option for a passenger aircraft.

The task for a fuselage designer is to allocate the necessary volume for the wing box and not consider any payload or other component at this location. The wing thickness at the root (t_r) or a fuselage intersection is the wing root maximum thickness-to-chord ratio ($(t/C)_{max}$) times the wing root chord (C_r):

$$t_r = \left(\frac{t}{C}\right)_{max_r} \cdot C_r \quad (7.26)$$

Then the fuselage required volume for the wing box (V_{wb}) is almost equal to:

$$V_{wb} = t_r \cdot C_r \cdot W_f \quad (7.27)$$

where W_f denotes the fuselage width (or diameter if circular) at the wing intersection. The exact volume is determined by taking the wing curvature into account.

Example 7.4 illustrates how to compute the volume of the fuselage that must be allocated for the wing box as a function of wing root airfoil section.

Example 7.4

A large transport aircraft has a low-wing configuration with the following wing characteristics:

$$S = 200\,\text{m}^2, \text{AR} = 10, \lambda = 1, \text{root airfoil: NACA } 64_2\text{-}415$$

Determine what volume of the fuselage must be allocated for the wing box, if the fuselage width at the wing intersection is 4 m.

Solution:

Since the airfoil's last two digits are 15, the wing maximum thickness-to-chord ratio is 15%. Thus, the wing thickness at the root or fuselage intersection is 0.15 times the wing root chord. The wing chord at the root is obtained as follows:

$$b = \sqrt{S \cdot \text{AR}} = \sqrt{200 \cdot 10} = 44.721\,\text{m} \quad (5.19)$$

Since the taper ratio is one, so:

$$C_r = C = \frac{S}{b} = \frac{200}{44.721} = 4.472\,\text{m} \quad (5.18)$$

$$t_r = \left(\frac{t}{C}\right)_{max} \cdot C_r = 0.15 \cdot 4.472 = 0.671\,\text{m} \quad (7.26)$$

The fuselage required volume for the wing box (V_{wb}) is equal to:

$$V_{wb} = t_r \cdot C_r \cdot W_f = 0.671 \cdot 4.472 \cdot 4 = 12\,\text{m}^3 \quad (7.27)$$

7.9.4 Power Transmission Systems

An aircraft is a very complex system in which several subsystems are working constantly to make a successful flight. Other than the aircraft structure, and engine, various systems including the electrical system, hydraulic system, mechanical system, avionic, air conditioning, and fuel system are performing different functions from power transmission to fuel transfer. In the fuselage design, these systems must be accommodated, so the fuselage must be sufficiently large that these systems are capable of working properly without any hazard to passenger safety. References such as [16] can be consulted for more details. There are several interesting stories in Ref. [14] about real experiences and lessons learned in aircraft design. A number of these stories concern fuselage design, and also crashes and mishaps which happened due to the mistakes of the fuselage designers. The reader is encouraged to read these stories and avoid the same mistakes.

7.10 Lofting

When all internal sections of the fuselage are designed and allocated, it is time to look at the fuselage external design, or lofting. Lofting is the process to determine the external geometry of an aircraft, primarily the fuselage. Lofting is performed on the fuselage to improve the overall aerodynamic performance of the fuselage. This means minimizing fuselage drag and potentially producing a fair amount of lift by the fuselage. Sometimes during lofting it is desirable to design a cross-sectional area along the length of the fuselage to minimize any sharp edges. The main location of a sharp change is at the wing/fuselage, tail/fuselage, and pylon/fuselage junctions. Then the fuselage is fabricated to be smoothly lofted for an optimal design.

The fuselage cross-section also contributes to the aircraft spin recovery characteristics. Damping provided by various parts of the aircraft such as the fuselage can counter the yawing moment of the wings during a spin. So, provision of a large amount of damping in yaw for the fuselage is an effective means of preventing a spin. The aerodynamic yawing moment due to rotation of the fuselage about the spin axes is largely dependent on the fuselage shape and its cross-section. Therefore, the aircraft designer can reduce the spin recovery load on the rudder by careful design of the fuselage and a proper aircraft weight distribution.

There are various techniques to cover a sharp edge by adding a curve such as an ellipse, circle, hyperbola, parabola spline, or conic. Engineering drawing software packages such as AutoCAD, CATIA, Unigraphics, Solid Work, and Concepts Unlimited (now Sharks) are equipped with these two-dimensional and three-dimensional graphical tools. Based on the author's information, Lockheed is employing Concepts Unlimited, Boeing and Dassault are utilizing CATIA, and Unigraphics were used by McDonnell Douglas and is used by Pratt and Whitney and Rolls Royce.

In the design of the fuselage external shape, there are several objectives and requirements that drive the design, including aerodynamic considerations, area ruling, radar detectability (or stealth), and operational requirements. These requirements are reviewed in this section.

7.10.1 Aerodynamics Considerations

Aerodynamic considerations for the fuselage external shape introduce very basic requirements such as low drag, low pitching moment, zero rolling moment, low yawing moment, and sometimes generating lift as much as possible. The zero rolling moment requirement

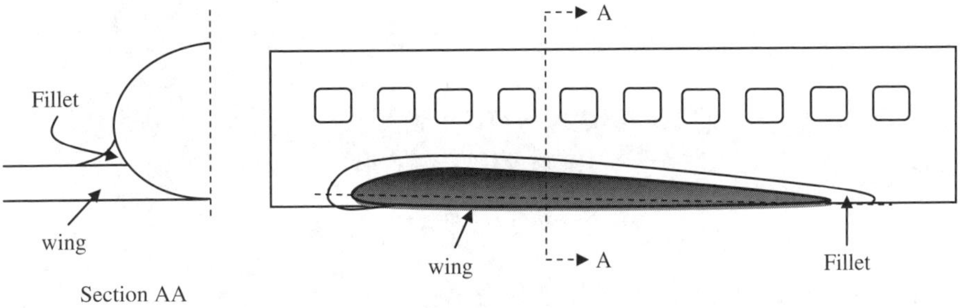

Figure 7.29 The fillet in a low-wing configuration

necessitates a symmetric fuselage about the *xz* plane. Thus, from the top view, the fuselage must be symmetric.

In order to have a lower drag, the fuselage designer needs to select a section close to an airfoil shape. From the side view, other design requirements (e.g., payload) do not usually allow the designer to have a symmetric fuselage. But the streamlining must be considered. So we add a semicircle or semicone to the nose and apply upsweep to the aft fuselage. A nose length-to-diameter ratio of 1.5–2 is recommended. To avoid large regions of boundary layer separation and the associated drag increase in the fuselage, the length of the rear fuselage is often two to three times the diameter of the cylindrical section. Where the wing is connected to the fuselage, some forms of filleting (Figure 7.29) are required to avoid flow separation and turbulence. The exact shape may be determined by wind-tunnel experiments.

In order to avoid any yawing moment by the fuselage, the top view of the fuselage is recommended to follow a symmetrical airfoil (Figure 7.30, top view) such as NACA 0009, NACA 0012, or NACA 0015. However, in order to create lift by the fuselage, the side view of the fuselage needs to be like an airfoil with a positive camber such as NACA 23015 (Figure 7.30, side view). An upsweep in the rear fuselage helps with such an airfoil feature.

In some aircraft, for example supersonic military aircraft such as SR-71 Blackbird (Figure 8.21), the fuselage is modified to produce a fair amount of lift. The extra lift is partly to cover the shortfall in lift generation by the wing. In order to make a fighter more controllable (mainly in roll) and highly maneuverable, the wing area is considered to be small and also the wing span to be short. The solution for lift generation is to have a blended wing/fuselage. Thus, at supersonic speeds, the shock wave at the fuselage edges generates a great deal of lift.

Two unusual fuselage noses are depicted in Figure 7.31. Figure 7.31(a) illustrates the special nose of the transport aircraft Ilyushin IL-76, which features a room for a navigation

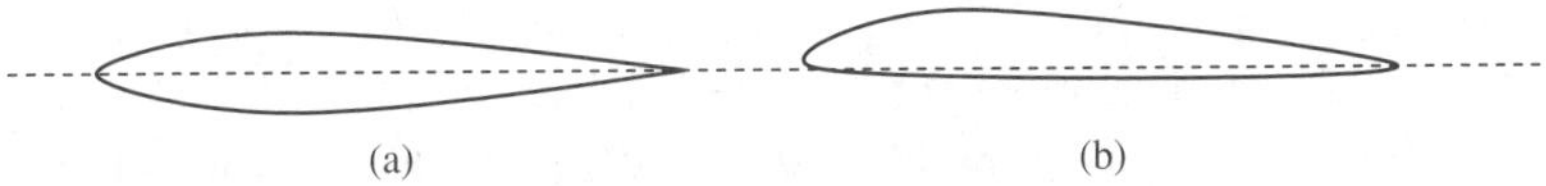

Figure 7.30 Recommended top view and side view for a fuselage (a) Top-view; (b) Side-view

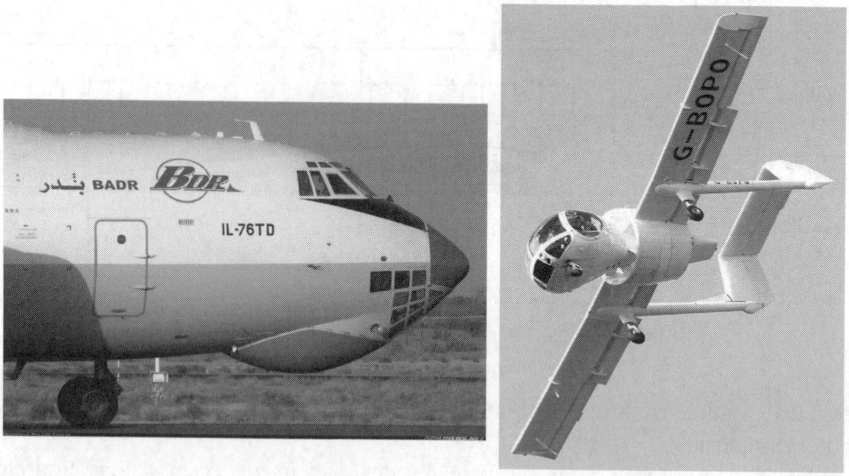

(a) (b)

Figure 7.31 Two aircraft with special fuselage noses: (a) transport aircraft Ilyushin IL-76; (b) Optica OA-7. Part (a) reproduced from permission of Balázs Farkas.

officer to help the pilot in navigation calculations (particularly during landing operations). Figure 7.31(b) illustrates the unusual fuselage of the observation aircraft Optica OA-7. The Optica OA-7, with a crew of one pilot and a capacity of two passengers, has an unusual configuration with a fully glazed forward cabin seating three across. This aircraft nose provides a perfect view for the pilot and passengers. Several Russian transport aircraft have a seat for the navigator in the nose (for landing). With more modern navigation coming up, these are gradually being replaced by radar in later versions (such as newer versions of the Tupolev Tu-134).

7.10.2 Area Ruling

The contribution of the fuselage in aircraft drag at transonic speeds is substantial. In order to reduce the fuselage drag in a transonic flight regime, the fuselage must follow a fashion which is referred to as coke-bottling. This technique to minimize shock wave drag in designing the fuselage base is also called area ruling. It aims to arrange the aircraft components including the fuselage cross-sectional variation so that the total aircraft cross-sectional area, in planes perpendicular to the fuselage center line, has a smooth and prescribed variation in the longitudinal direction (i.e., x-axis). Based on the area rule, all aircraft components that have the same cross-sectional area, including wing and fuselage, will have the same wave drag. The theory was examined and developed by NACA in 1956.

The application of this theory in the fuselage design is to shape the fuselage, and locate other aircraft components relative to the fuselage, so that the shape of the equivalent cross-section is as close as possible to a minimum drag configuration (known as a Sears–Haack body). The area rule is applicable mainly at sonic speed (i.e., Mach 1.0). However, it is very helpful in the entire transonic region from about Mach 0.8 to 1.2.

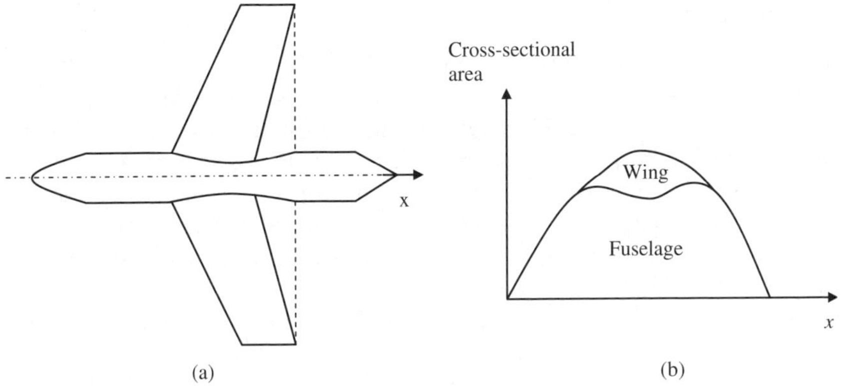

Figure 7.32 Area ruling. (a) Coke bottling, and (b) Contributions of wing and fuselage in cross-sectional area

Hence, for an aircraft which is designed for transonic speeds, such as most large transport aircraft, the fuselage should be narrower at the wing/fuselage intersection than other stations (Figure 7.32(a)). Figure 7.32(b) illustrates the smooth increase in the aircraft cross-sectional area as the area ruling is applied.

The interested reader can find more information in aerodynamics textbooks such as Ref. [17]. Aircraft such as the Boeing B-747 (Figures 3.7, 3.12, and 9.4) and Concorde (Figure 7.33) have employed area ruling in their fuselage design. In the Boeing 747, the fuselage front section (before wing attachment) has a double deck to increase the equivalent diameter (in fact the height is increased). The fuselage has only one deck (smaller height) when the wing is attached to the fuselage. In the case of Concorde, the fuselage has a narrower body in the wing/fuselage attachment area. In the design of the fuselage of the business jet Cessna Citation 10 with a maximum speed of Mach 0.95, the coke-bottle area ruling was considered.

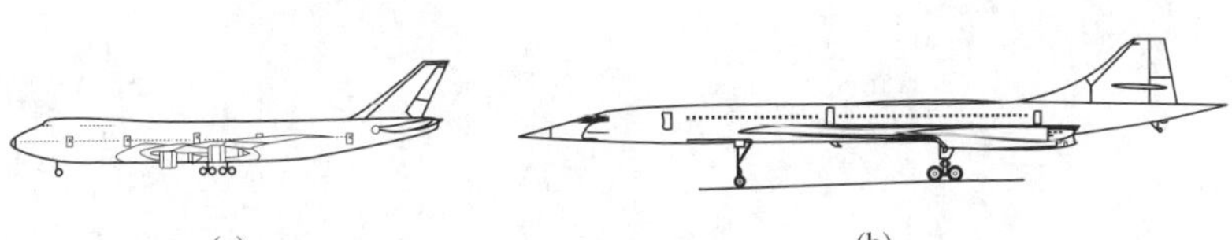

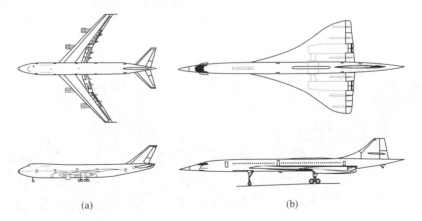

Figure 7.33 The application of area ruling in the Boeing 747 and Concorde. (a) Boeing 747, and (b) Concorde

7.10.3 Radar Detectability

One of the design requirements of modern military aircraft is to be equipped with stealth technology to avoid detection by radar as well as to reduce aircraft visibility in the infrared, visual, and audio spectrum. This capability is obtained by employing a combination of features, such as using composite materials that absorb radar signals. Another technique to improve aircraft detectability is to design the aircraft external shape, including fuselage, such that the aircraft equivalent radar cross-section (RCS) is reduced. RCS is a function of various parameters including the aircraft cross-section perpendicular to the line of radar signal.

Four common techniques to improve the aircraft stealth capability are to: (i) fabricate the aircraft components with radar absorbent materials, (ii) reduce the aircraft size, (iii) remove any surface which is perpendicular to the radar incoming signal and make it inclined, and (iv) hide hot gasses of the engine from direct detection. In the third technique, every component including the fuselage must be shaped such that the incoming radar signal is not reflected to the transmitting source. The stealth technology is employed in several military aircraft such as the F-117 Night Hawk (Figure 7.34), Northrop Grumman B-2 Spirit (Figure 6.8), Lockheed SR-71 Blackbird (Figure 8.21), and Lockheed Martin F-35 Lightning (Figure 2.17). Thus, the fuselage must be designed so that it contributes to the aircraft RCS as little as possible.

7.10.4 Fuselage Rear Section

As recommended in Section 7.10.1, the fuselage is desired to follow an airfoil shape to reduce drag. This implies that the rear fuselage should be tapered to a zero diameter. If the engine is not intended to be accommodated by the rear fuselage, the diameter of the fuselage must be reduced from the cabin diameter to almost zero. If an engine is enclosed by an aft fuselage, the fuselage diameter should be reduced from the diameter at mid-section (e.g., cockpit diameter) to the engine exit nozzle diameter. Caution must be taken not to taper the rear section of the fuselage at too large an angle, or else flow separation will occur. For a subsonic aircraft, the taper angle should be no larger than about 20 deg. For ease of fabrication, part of the rear fuselage may be conical. The transition from

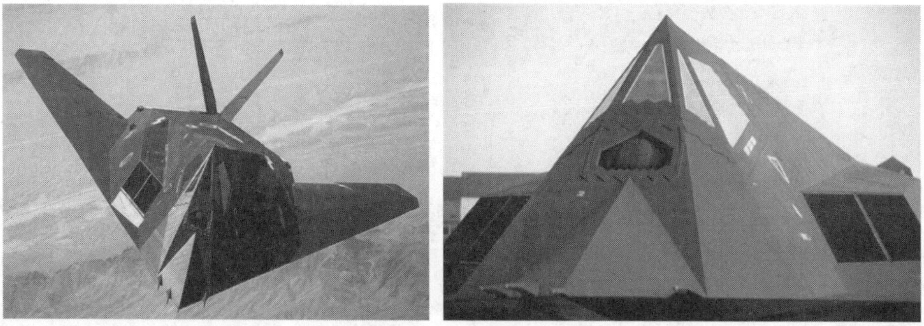

(a) (b)

Figure 7.34 F-117 Night Hawk fuselage is designed to satisfy stealth requirements. (a) Fuselage, engine, and tail design for stealth, and (b) Sharp edges of fuselage (US Air Force) cylinder to cone ought to be smooth with sufficiently large radius of curvature. If the rear fuselage is specious and the aircraft cg limits allow, a portion of the fuel or luggage may be stored in it.

In order to achieve this target, the take-off clearance requirement is simultaneously applied. As discussed in Chapter 9, during take-off and landing, the rear fuselage must clear the ground under regular operating conditions. An aircraft is usually rotating about the main gear in order to increase the lift to prepare for take-off (see Figure 7.35). This is also true for landing operations, in which the aircraft rotates to gain a high angle of attack. In an aircraft with non-tail gear, the upsweep angle must be set such that the tail or rear fuselage does not hit the ground during the take-off rotation or landing with a high angle of attack.

The primary solution to avoid this incident is through an increase in the landing gear height. Another common solution to this problem is to cut the rear fuselage by an upsweep angle (or taper angle) α_{us}. The flow separation avoidance requirement at the rear fuselage has the following recommendation:

$$\alpha_{us} < 20\,\text{deg} \tag{7.28}$$

Thus, there are two requirements which influence the magnitude of the upsweep angle: (i) ground clearance and (ii) smooth transition from a large diameter at the end of the cabin to a zero diameter at the end of the fuselage. These two requirements may be met by determining the three parameters of: (i) upsweep angle, (ii) length of the mid-fuselage after the main gear (without upsweep angle), and (iii) length of the rear fuselage with upsweep applied. Both the length of the rear fuselage and the upsweep angle should be as small as possible. These three parameters must be determined simultaneously to reach an optimum design. If the initial upsweep angle interferes with the cabin, another solution may be sought. One solution is to decrease the length of the rear fuselage; a second solution is to increase the upsweep angle by cutting the rear fuselage (Figure 7.35).

If there is a component such as an engine at the rear fuselage (as in the case for most fighters), this component does not allow for the application of a large upsweep angle (Figure 7.36(c)). Thus, the smallest diameter of the rear fuselage could be equal to the engine nozzle diameter. In case of a small light aircraft, the desire for a large tail moment arm allows for a desirable upsweep angle (Figure 7.36(d)). Figure 7.36 demonstrates the upsweep for two large transport aircraft, a fighter and a light GA aircraft. The cargo aircraft Boeing C-17 Globemaster has an upsweep angle of 16 deg, while the airliner Boeing 777-300 has an upsweep angle of 17 deg. In both aircraft, the upsweep has extended into the main section of the fuselage, so the cargo and passenger compartments are compromised.

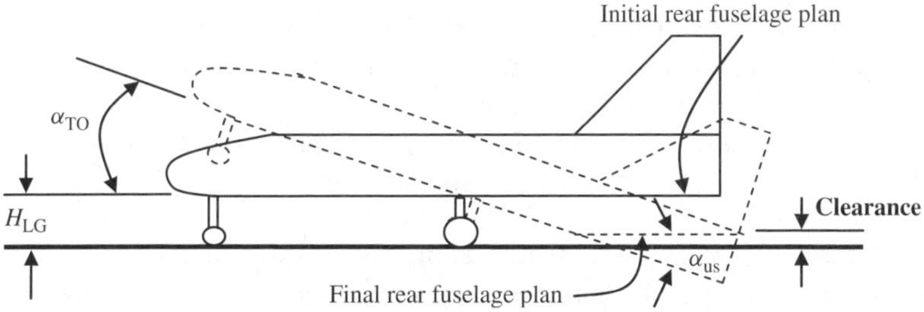

Figure 7.35 Rear fuselage upsweep angle

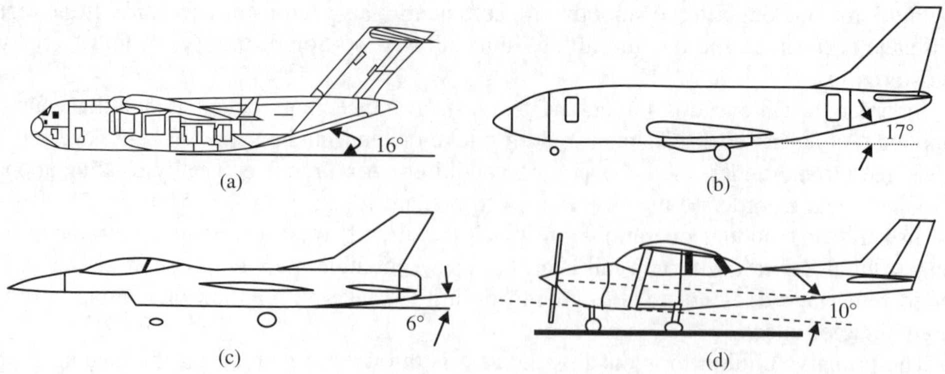

Figure 7.36 Upsweep for four aircraft (drawings not scaled). (a) Boeing C-17 Globemaster, (b) Boeing 777-300, (c) F-16 Fighting Falcon, and (d) Cessna 172

7.11 Fuselage Design Steps

In Sections 7.1–7.10, the fuselage function, configurations, objectives, alternatives, design criteria, parameters, governing rules and equations, formulation, design requirements, as well as how to approach the primary main parameters have been presented in detail. Furthermore, Figure 7.2 illustrates the design flowchart of the fuselage. In this section, the fuselage design procedure is introduced in terms of design steps. It must be noted that there is no unique solution to satisfy the customer and airworthiness requirements in designing a fuselage. Several fuselage designs may satisfy the requirements, but each will have unique advantages and disadvantages.

In order to formulate the design requirements, the designer is encouraged to develop several equations and relations based on the numerical requirements and solve them simultaneously. For instance, for each upsweep, a trigonometric or Pythagorian equation may be built for the corresponding triangle. In this technique, a computer program would allow faster and more accurate design. Based on the systems engineering approach, the fuselage detail design begins with identifying and defining the design requirements and ends with optimization. The following are the fuselage design steps for a conventional aircraft.

Given: payload, aircraft mission, aircraft configuration

1. Identify and list the fuselage design requirements.
2. Determine the number of crew members.
3. Determine the number of flight attendants (for passenger aircraft).
4. Determine the number of technical personnel (for cargo aircraft).
5. Establish human size and target passenger.
6. Select fuselage layout (internal): side view, front view, and top view.
7. Determine the required instruments for the cockpit.
8. Design the cockpit.
9. Establish the fuselage optimum length-to-diameter ratio $(L_f/D_f)_{opt}$.
10. Design the passenger cabin (number of rows and seats abreast) to achieve optimum length-to-diameter ratio.

11. Design the cargo/luggage compartment.
12. Determine the required volume for other components (e.g., fuel, landing gear).
13. Check if the available fuselage space for other components is sufficient.
14. Calculate the fuselage maximum diameter (D_f).
15. Select the number of doors.
16. Design the fuselage nose section.
17. Design the fuselage rear section.
18. Determine the upsweep angle (α_{us}).
19. Calculate the fuselage overall length (L_f).
20. Apply lofting.
21. Check if the fuselage design satisfies the design requirements.
22. If any design requirement is not met, return to the relevant design step and recalculate the corresponding parameter.
23. Optimize.
24. Draw the final design with dimensions.

7.12 Design Example

In this section, a major chapter example (Example 7.5) is presented to design a fuselage for an airliner aircraft. In order to avoid lengthening the chapter, only the major design parameters are determined.

Example 7.5

Problem statement: Design a fuselage for a high-subsonic jet transport aircraft with a low-wing configuration, that can accommodate 120 passengers for a range of 10 000 km at the service ceiling of 35 000 ft. The fuselage must also carry 70% of the total fuel. The landing gear is of tricycle configuration and is retractable into the fuselage. Assume the fuel type is JP-5. The aircraft has a take-off mass of 50 000 kg, and the cruising speed at 35 000 ft is 530 knot with a lift-to-drag ratio of 11. For the necessary crew and luggage, follow the FAA regulations and assume any other necessary parameters. Determine the following:

1. fuselage configuration;
2. fuselage length and fuselage maximum diameter;
3. targeted passenger size;
4. volume of the pressurized part of the fuselage;
5. cabin design (seating arrangement, side view, top view with dimensions);
6. instrument panel (list of instruments);
7. cockpit design (side view, back view with dimensions);
8. cargo and luggage storage design (volume of all cargo, luggage);
9. doors (including cargo and landing gear doors);
10. fuel tanks allocation;
11. systems and equipment locations;

12. upsweep angle (α_{us});
13. lofting;
14. drawing (top view, side view, and front view with dimensions).

Solution:

- **Step 1. Aircraft type, mission, and design requirements**.
  Type: civil high-subsonic jet transport
  Payload: 120 passengers plus luggage
  Range: 10 000 km
  Cruise ceiling: 35 000 ft
  Design requirements: FAR 25 (details are described in the various steps)
- **Step 2. Number of crew members**. Based on FAR Part 25 Section 25.1523, plus Appendix D to FAR 25, two flight crew (a pilot and a copilot) are considered.
- **Steps 3 and 4. Number of flight attendants**. Based on FAR Part 125 Section 125.269 which states that "for airplanes having more than 100 passengers – 2 flight attendants plus 1 additional flight attendant for each unit of 50 passengers above 100 passengers are required," a total of 3 (i.e., 2 + 1) flight attendants is employed. Since the aircraft is an airliner, there is no need for other technical personnel as they are not needed for a cargo aircraft.
- **Step 5. Human size and target passenger**. The human size for crew members, flight attendants, and target passengers is selected to be the same as introduced in Figure 7.6. The size for a male passenger is selected.
- **Step 6. Fuselage configuration**. To minimize weight and drag, a single fuselage with a single-deck configuration is selected. The following items must be accommodated by the fuselage:
  - Passengers (cabin)
  - Flight attendants (cabin)
  - Pilot and copilot (cockpit)
  - Pressurized space (cabin plus cockpit)
  - Cargo
  - Passengers' baggage
  - Fuel
  - Wing spar
  - Measurement and control systems
  - Mechanical, electrical systems
  - Landing gear.

  To locate payloads and all of the above items, we have six basic observations: (i) the fuselage is preferred to be symmetric from a top view, (ii) the fuselage must be as small and compact as possible, (iii) the usable loads (fuel) must be close to the aircraft center of gravity, (iv) a circular cross-section for the pressurized segment is elected to minimize the skin shear stress, (v) the pilot cockpit must be allocated in the most forward location of the fuselage, and (vi) the arrangement must be such that the aircraft center of gravity is close to the wing/fuselage aerodynamic center. Based on these observations, we have allocated every component as shown in Figure 7.37.

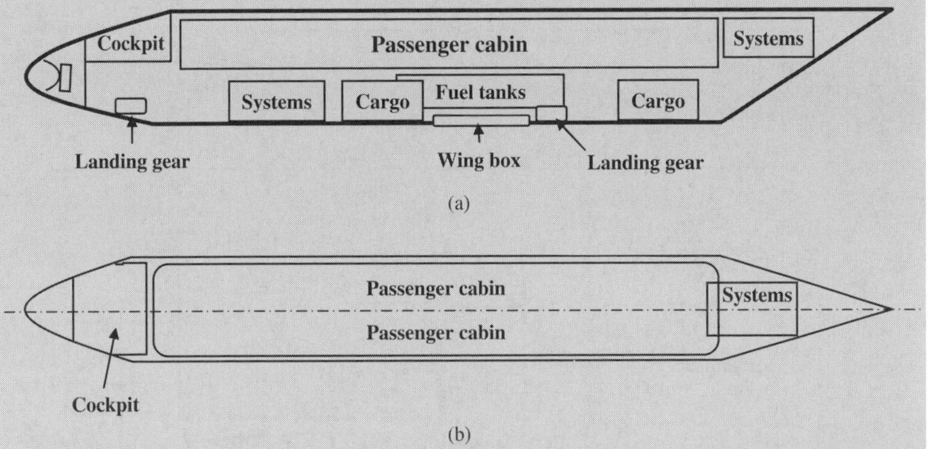

Figure 7.37 Internal arrangement of the fuselage. (a) Side view, and (b) Top view

The fuselage configuration is a function of the internal arrangement. In order to specify the location for each internal item, it is noted that the aircraft is a civil transport aircraft and must carry several items. Figure 7.37 illustrates a side view of the fuselage with the internal arrangements. In terms of front view, the top part of the main section is considered for the cabin, and the lower part for the luggage, cargo, systems, landing gear, and fuel tanks. The volume for each component and the external shape of the fuselage will be determined later.

In the next steps, the volumes of every section and the dimensions for each component are determined.

- **Step 7. Instruments for cockpit**. The flight and navigation instruments as suggested by FAR Part 25 Section 25.1303 are considered for the cockpit (as detailed in Section 7.5.6). In addition, a yoke for roll and pitch control and a pedal for yaw control are employed in the cockpit. The list of important instruments that must be provided in the cockpit is as follows: airspeed indicator, altimeter, altitude indicator, turn coordinator, vertical speed indicator, heading indicator, outside air temperature indicator, inset map, GPS, INS, VORs, ILS glide slope, transponder, magnetometer, engine instruments (rpm, fuel, exhaust gas temperature, turbine temperature), landing gear switch, flap switch, throttle, yoke, pedal, compass, computer monitor, electric panel, weather radar, radio. Since there are two crew members, there must be duplicated flight controls and instrument panel to allow both pilot and copilot to control the aircraft independently. The work load is defined in Appendix D of FAR Part 25.
- **Step 8. Cockpit design**. There are two civil crew members with size as defined in step 5. The 10 000-km flight is a long trip, so both the pilot and copilot must feel comfortable and have the opportunity to move around and take a rest. The list of flight, control, and navigation instruments is specified in step 7. Based on the materials presented in Section 7.5, the cockpit is designed as shown in Figure 7.38.

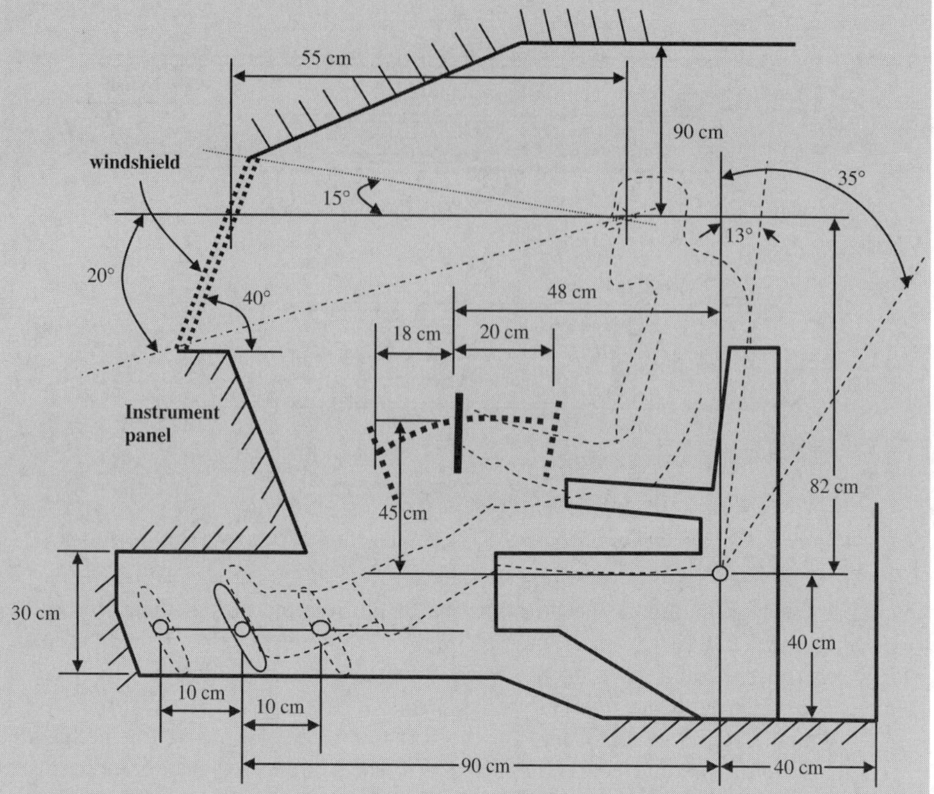

Figure 7.38 Cockpit geometry (side view)

A back view of the cockpit is shown in Figure 7.39, which illustrates the seats for both pilot and copilot. The space between the seats is for the throttle and some other control instruments.

Based on the internal arrangement, the dimensions of the cockpit are 182 cm width, 212 cm height, and 150 cm length. Figures 7.38 and 7.39 show the cockpit design.

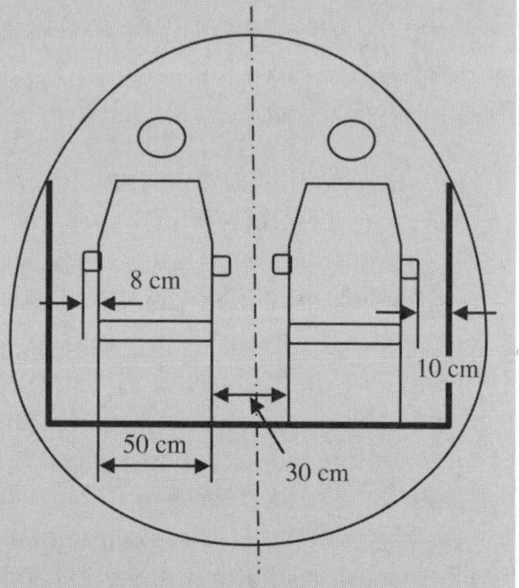

Figure 7.39 Back view of the cockpit

- **Step 9. Determine** $L_f/D_{f_{max}}$. According to the minimum zero-lift drag requirement (in fact f_{LD}), the fuselage length-to-diameter ratio should be 16.3. However, to minimize the surface area and weight, the fuselage length-to-diameter ratio should be 1. To reduce the fuselage surface area and fuselage weight, the theoretical ratio is selected to be 14. Thus, the number of seats abreast, and number of rows, must satisfy this requirement.

- **Step 10. Passenger compartment design**. The cabin is required to accommodate 120 passengers and 3 flight attendants. All seats are considered to be economy (tourist). The wall thickness is assumed to be 6 cm each side. From step 8, the cockpit length is 150 cm. The nose section is assumed to be 1 m. The length of the rear fuselage (behind the cabin) is considered to be 2.5 m. The top view of the fuselage with known values is shown in Figure 7.40. It is required to determine the cabin length, the cabin diameter, and the number of seats abreast, plus services for the passengers.

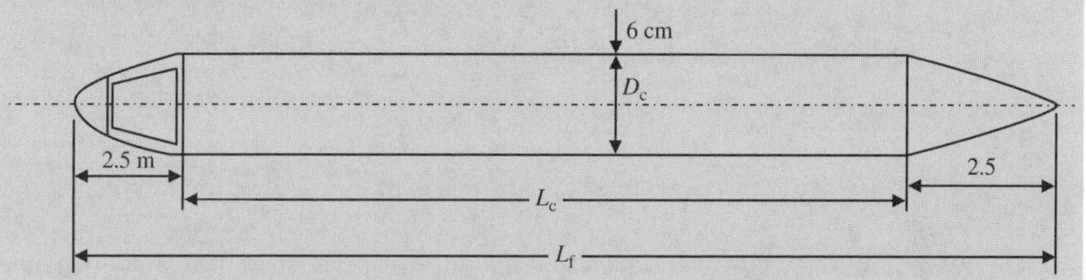

Figure 7.40 Fuselage top view for Example 7.5

Thus, the number of seats abreast and number of rows must satisfy this requirement. The total number of seats is 123. The cabin length and cabin width are determined using Equations (7.1) and (7.2):

$$L_C = \sum_{i=1}^{3} \sum n_{r_i} \cdot P_{s_i} \quad (7.1)$$

$$W_C = n_S \cdot W_S + n_A \cdot W_A \quad (7.2)$$

The seat pitch, seat width, and aisle width for economy class are extracted from Table 7.4 as follows:

- $W_S = 45$ cm
- $P_S = 80$ cm
- $W_A = 45$ cm

In general, the reasonable alternatives are (the numbers are rounded to the nearest number of rows):

- 123 rows of single seats (plus onc aisle)
- 62 rows (123/2) of dual seats (plus one aisle)
- 42 rows (123/3) of three seats (plus one aisle)
- 31 rows (123/4) of four seats (plus one aisle)
- 25 rows (123/5) of five seats (plus one aisle)
- 25 rows (123/5) of five seats (plus two aisles)

– 21 rows (123/6) of six seats (plus one aisle)
– 21 rows (123/6) of six seats (plus two aisles).

Now we determine the fuselage length-to-diameter for each case by examining these seating alternatives (number of seats abreast) using Equations (7.9) and (7.10). For instance, the calculations for option 4 are as follows:

$$L_C = \sum_{i=1}^{3} \sum n_{r_i} \cdot P_{s_i} = 31 \cdot 80\,\text{cm} = 24.8\,\text{m} \tag{7.1}$$

$$W_C = n_S \cdot W_S + n_A \cdot W_A = (4 \cdot 45) + (1 \cdot 45) = 2.25\,\text{m} \tag{7.2}$$

$$D_f = W_C + 2T_W = 2.25 + (2 \cdot 0.06) = 2.37\,\text{m} \tag{7.9}$$

$$L_f = L_C + L_{CP} + L_N + L_R = 24.8 + 1.5 + 1 + 2.5 = 29.8\,\text{m} \tag{7.10}$$

Thus, the fuselage length-to-diameter ratio is:

$$\frac{L_f}{D_f} = \frac{29.8}{2.37} = 12.56$$

The calculations for all other seven alternatives are performed in a similar manner. The results are presented in Table 7.10.

Table 7.10 Seating alternatives for Example 7.5

No.	Number of rows	Seats abreast	Aisle	W_C (m)	L_C (m)	D_f (m)	L_f (m)	L_f/D_f
1	123	1	1	0.9	94.8	01.02	103.4	101.4
2	62	2 (1 + 1)	1	1.35	49.6	1.47	54.6	37.15
3	41	3 (2 + 1)	1	1.8	32.8	1.92	37.8	19.7
4	**31**	**4 (2 + 2)**	**1**	**2.25**	**24.8**	**2.37**	**29.8**	**12.56**
5	25	5 (2 + 3)	1	2.7	20	2.82	25	8.86
6	25	5 (2 + 1 + 2)	2	3.15	20	3.27	25	7.65
7	21	6 (3 + 3)	1	3.15	16.8	3.27	21.8	6.67
8	21	6 (2 + 2 + 2)	2	3.6	16.8	3.72	21.8	5.86

As Table 7.10 illustrates, alternative 4 yields the closest fuselage length-to-diameter ratio (i.e., 12.56), while it is also less than the optimum fuselage length-to-diameter ratio (i.e., 16.3). Therefore, the cabin is decided to have four seats abreast and one aisle. This seating in each row includes two seats on each side of the aisle. The 31 rows of 4 seats, a total of 124 seats, are shown in Figure 7.41.

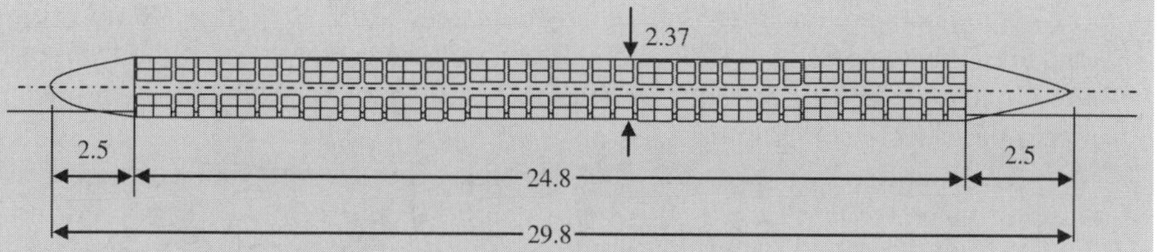

Figure 7.41 Top view of the cabin for Example 7.5 (values in cm)

For this number of passengers, one galley and two toilets are also provided in the cabin (not shown in Figure 7.41). This extends the length of the cabin by about 2.5 m, thus the revised length of the cabin will be:

$$L_C = 24.8 + 2.5 = 27.3\,\text{m}$$

And the fuselage length:

$$L_f = 29.8 + 2.5 = 32.3\,\text{m}$$

The passengers' carry-on baggage is also stowed in the cabin, in overhead containers (see Figure 7.42).

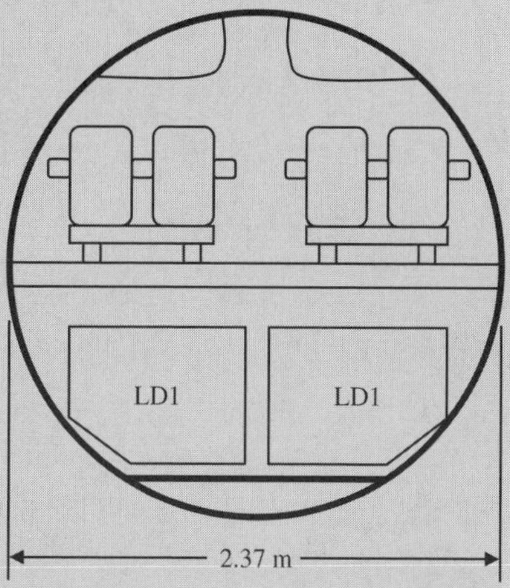

Figure 7.42 Fuselage cross-section

- **Step 11. Cargo/luggage compartment design**. The total volume of passenger cargo (V_C) is primarily equal to the number of travelers (n_t) times the volume of the total bags of each traveler (V_b):

$$V_C = n_t \cdot V_b \tag{7.3}$$

Each passenger may carry two times 60 lb of checked baggage which, according to Equation (7.4), will each have a volume of $0.146\,\text{m}^3$. There are 120 passengers, 3 flight attendants, and 2 flight crew members so the total volume of baggage is:

$$V_C = (120 + 3 + 2) \cdot 2 \cdot 0.146\,\text{m}^3 = 36.5\,\text{m}^3$$

The cargo container LD1 is assumed to be employed. Based on Table 7.6, each LD1 container has a volume of $173\,\text{ft}^3$ (or $4.899\,\text{m}^3$). Thus:

$$\frac{18.25\,\text{m}^3}{4.899\,\text{m}^3} = 7.45$$

This demonstrates that a total of 8 (the closest whole number to 7.45) LD1 containers must be carried. The LD1 containers for cargo and checked baggage will be stored underneath the passenger cabin, as shown in Figure 7.42. The exact locations of LD1 containers relative to the aircraft cg will be determined in the later stages of design.

The cargo doors are under the emergency doors and sized at 70 in. × 70 in. to easily load all items requiring transportation. There are emergency doors at the front of the aircraft, and the galley section. The landing gear doors will accommodate retractable rears in a tricycle configuration.

- **Step 12. Required volume for fuel storage**. The specific fuel consumption is assumed to be 0.6 lb/h/lb, which is equivalent to 0.0001667 1/s. The aircraft cruising speed at 35 000 ft is 530 knot, with a lift-to-drag ratio of 11. The air density at 35 000 ft is 0.38 kg/m^3. The aircraft take-off mass is 50 000 kg. The range equation for a jet aircraft (Equation (7.24)) is employed:

$$m_f = 1.2 \cdot m_{TO}\left[1 - \exp\left(\frac{-RC}{V\,(L/D)}\right)\right]$$

$$= 1.2 \cdot 50\,000 \cdot \left[1 - \exp\left(\frac{-10\,000\,000 \cdot 0.000194}{(530 \cdot 0.5144) \cdot 11}\right)\right] \tag{7.24}$$

So, the total fuel mass is:

$$m_f = 25\,579.8\,\text{kg}$$

According to Table 7.8, the density of fuel JP-5 is 788 kg/m^3 so:

$$V_f = \frac{m_f}{\rho_f} = \frac{25\,579.8}{788} \Rightarrow V_f = 32.46\,\text{m}^3 \tag{7.21}$$

This much fuel may be divided into six fuel tanks, of which four are accommodated in the fuselage. The fuel tanks will be located at the extreme bottom of the fuselage. They are best suited for this location, since the fuel can be stored in very flexible tanks that will fit the curvature of the fuselage. Also, if there is a leak, other cargo will be protected.

- **Step 13. Check if the available fuselage space for other components is sufficient**. The fuselage main section has a circular cross-section. The bottom half of the fuselage in the cabin area (underneath the floor) has the following volume:

$$V_{\text{bottom}} = \frac{1}{2}\left(\pi\frac{W_C^2}{4}L_C\right) = \frac{1}{2}\left(3.14 \cdot \frac{2.25^2}{4} \cdot 27.3\right) = 54.27\,\text{m}^3 \tag{7.4a}$$

So far, the fuselage must accommodate the entire checked luggage, plus 70% of the fuel. The total required volume is:

$$V_{\text{req}} = 0.7V_f + V_C = 0.7 \cdot 32.46 + 36.5 = 59.2\,\text{m}^3$$

The difference between the total required volume and the available space under the cabin is:

$$V_{\text{extr}} = 59.2 - 54.27 = 4.95\,\text{m}^3$$

This space (4.95 m^3) must be provided partly by the rear fuselage and partly by the space under the cockpit. In special circumstances, the tail internal space may also be employed to accommodate fuel tanks. Since the problem statement has not provided the characteristics of other items that are accommodated by the fuselage, they are not discussed at the moment.

- **Step 14. Fuselage maximum diameter**. The fuselage maximum diameter has been calculated in step 10 to be:

$$D_f = 2.37\,\text{m}$$

- **Step 15. Number of doors**. Based on FAR 23 Section 23.783, each closed cabin with passenger accommodation must have at least one adequate and easily accessible external door. Sections 25.813 and 25.807 on emergency exits also regulate that each required emergency exit must be accessible to the passengers and located where it will afford an effective means of evacuation. Thus, three passenger doors and two over-wing emergency exits on each side are selected. Type A doors will be employed. This type is a floor-level exit with a rectangular opening of not less than 42 in. wide by 72 in. high, with corner radii not greater than 7 in.
- **Step 16. Nose section**. The nose section is a perfect location for equipment such as radar dishes. At any rate, the nose of the fuselage must be such that the fuselage produces a minimum amount of drag while being lightweight. So the flat part of the fuselage will be rounded by considering a semicircle or semi-ellipse at the front of the fuselage. A weather radar dish is considered to be accommodated by the nose section. The length of the nose section must be such that it is 1.5 to 2 times the fuselage diameter.

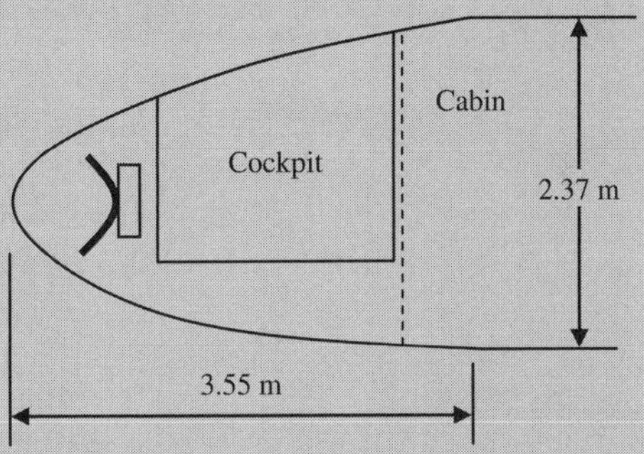

Figure 7.43 Nose section

Thus:

$$L_N = 1.5 \cdot D_f = 1.5 \cdot 2.37 = 3.55\,\text{m}$$

The nose section, which includes the cockpit, is shown in Figure 7.43. Since we already considered 2.5 m for the length of the nose section, we will have to extend the nose curvature into the cabin by about 1.05 m.

- **Step 17. Fuselage rear section**. For this aircraft, the fuselage rear section has the following functions:
  - Provide 4.95 m^3 of space for fuel or luggage, as determined in step 13.
  - Provide ground clearance for the aircraft during take-off rotation.
  - Smoothly reduce the cabin diameter to near zero.

  A conical shape is tentatively considered for the rear fuselage.

  The required length to provide 4.95 m^2 of space is determined by employing the equation for the volume of a cone, as follows:

$$V_{cone} = \frac{1}{3}\pi \cdot r^2 L_R \Rightarrow L_R = \frac{3V_{cone}}{\pi (W_C/2)^2}$$

In practice, not all of the available volume of a cone may be used for storage, so 80% is added to the required volume. Then:

$$L_R = \frac{3 \cdot 1.8 \cdot 4.95}{\pi(2.37/2)^2} \Rightarrow L_R = 1.515\,\text{m}$$

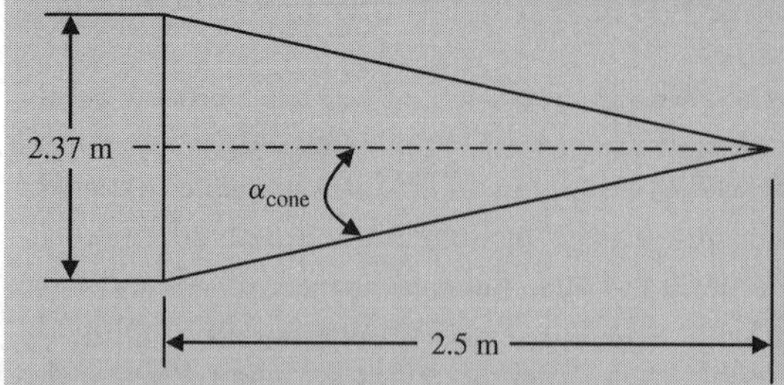

Figure 7.44 Rear section modeled as a cone (top view)

Therefore, the length of the fuselage rear section must be at least 1.515 m. We already considered 2.5 m, which meets this requirement. Moreover, to smoothly reduce the fuselage diameter to zero, a maximum 20 deg cone angle is recommended. Let's see what the current cone angle is (height (i.e., length) 2.5 m and base diameter 2.37 m). The cone (top view of the fuselage rear section) is illustrated in Figure 7.44.

$$\alpha_{cone} = \tan^{-1}\left(\frac{2.37/2}{2.5}\right) = 0.44\,\text{rad}$$

or:

$$\alpha_{cone} = 25.3\,\text{deg}$$

The cone angle is slightly greater than the recommended upsweep angle. The solution is either to extend the length of the fuselage, or to extend the upsweep into the cabin area.

Since we have no extra space to cut, the length of the rear section is extended. Furthermore, a 20-deg cone angle is considered too.

$$\tan(\alpha_{us}) = \frac{D_f/2}{L_R} \Rightarrow L_R = \frac{D_f/2}{\tan(\alpha_{cone})} = \frac{\frac{2.37}{2}}{\tan(20)} \Rightarrow L_R = 3.25\,\text{m}$$

This fuselage extension will increase the length of the fuselage to $2.5 + 27.3 + 3.25 = 33.05$ m and the length-to-diameter ratio will be:

$$\frac{L_f}{D_f} = \frac{33.05}{2.37} = 13.95$$

which is closer to the optimum value for the lowest f_{LD} (i.e., 16.3). If more data was provided by the problem statement, the ground clearance during take-off rotation could be checked.

- **Step 18. Upsweep angle**. Figure 7.45 illustrates the side view of the fuselage rear section featuring the upsweep angle. For the upsweep angle, a triangle is formed.

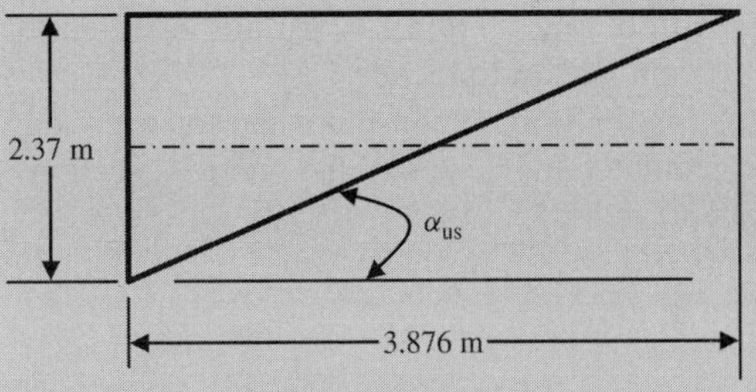

Figure 7.45 Rear section modeled as a cone (side view)

A 17-deg upsweep is considered for the rear section (side view). This angle requires the taper to be applied as follows:

$$\tan(\alpha_{us}) = \frac{D_f/2}{L_{us}} \Rightarrow$$

$$L_{us} = \frac{D_f/2}{\tan(\alpha_{us})} = \frac{2.37/2}{\tan(17)} \Rightarrow L_{us} = 3.876\,\text{m}$$

The tapered length is slightly greater than the length of the rear section (3.876 > 3.25). The solution is either to extend the length of the fuselage, or extend the upsweep into the cabin area. The second alternative is selected.

- **Step 19. Fuselage overall length**. The fuselage overall length is determine in step 17 and is 33.05 m. Please note that the length of the tapered section (3.876 m) due to the upsweep does not change this length.
- **Step 20. Lofting**. Lofting is performed on the fuselage to improve the overall aerodynamic performance of the fuselage. This means minimizing fuselage drag and producing a fair amount of lift by the fuselage. From a top view, the fuselage must be symmetric. In order to have lower drag, we try to have a section close to an airfoil section (Figure 7.46). From the side view, other design requirements do not usually allow us to have a symmetric fuselage. But the streamlining must be considered. So we have added a semicircle to the nose and apply the upsweep to the aft fuselage (Figure 7.46).

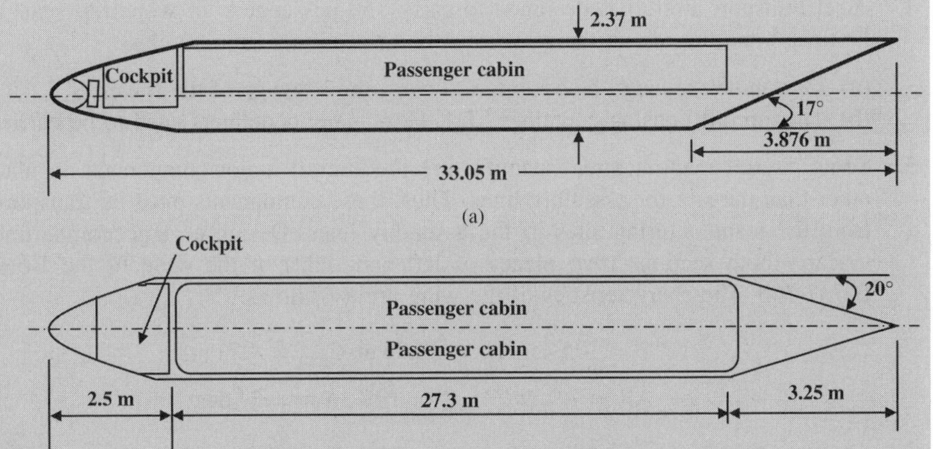

Figure 7.46 Fuselage side view and top view for Example 7.5. (a) Side-view, and (b) Top-view

- **Steps 21–23. Iteration and optimization**. These three steps are open-ended problems. They are left to the reader to continue and practice.
- **Step 24. Drawing of the final design**. Figure 7.46 demonstrates the top view and side view of the fuselage based on the geometry and design of the previous sections. Please note that Figure 7.46 is not scaled.

Problems

1. Using a reference such as [9] or a manufacturer's website, identify the seating chart for the following civil transport aircraft:

Boeing 737-200, Boeing 767-300, Boeing 777-400, Boeing 747SP, Airbus A-380, Airbus A-340-300, Fokker 100, MD-90, Embraer 195, Cessna 750, and Cessna 510.

2. Using a reference such as [9] or a manufacturer's website, identify the personal equipment for the pilot of the following military aircraft:

FA/18 Hornet, F-16 Fighting Falcon, F-15 Eagle, F-117 Night Hawk, SR-75 Blackbird, Panavia Tornado, B-52 Stratofortress, Eurofighter, Dassault Mirage 4000, and Mikoyan MiG-31

3. A jet transport aircraft is designed to carry 400 passengers, of which 30 seats are first class and the rest of the seats must be economy (tourist) class.

(a) Determine the cargo bay volume to carry the baggage of the travelers.
(b) The aircraft is using container LD2. How many containers need to be carried?

4. A jet transport aircraft is designed to carry 150 passengers, of which 20 seats are first class and the rest of the seats must be of economy (tourist) class.

(a) Determine the cargo bay volume to carry the baggage of the travelers.
(b) The aircraft is using container LD1. How many containers need to be carried?

5. Some aircraft manufacturers manufacture the aircraft major components in places other than the site for assembly lines. Thus, these components must be transported from the manufacturing sites to the assembly lines. Design a cargo compartment to carry both sections (two pieces of left and right) of the wing of the Boeing B-737-300. The characteristics of the wing are as follows:

$$S = 105.4\,\mathrm{m}^2, b = 28.88\,\mathrm{m}, C_{\mathrm{root}} = 4.71\,\mathrm{m},$$

$$\mathrm{AR} = 7.9, (t/C)_{\max} = 12\%, \Lambda_{\mathrm{LE}} = 31\,\mathrm{deg}$$

Assume other necessary information.

6. For an aircraft to carry 60 passengers and 2 flight attendants, design a cabin to submit the lowest zero-lift drag. The length of the nose section (including cockpit) is 2 m and the length of the rear section is 2.5 m. You are only required to determine the cabin length, the cabin diameter, and the number of seats abreast. Assume that the wall thickness is 3 cm each side. Ignore the galley, lavatories, and assume that all seats are desired to be of economy (tourist) class.

7. For an aircraft to carry 400 passengers and 10 flight attendants, design a cabin to submit the lowest zero-lift drag. The length of the nose section (including cockpit) is 3 m and the length of the rear section is 5 m. You are required to determine the cabin length, the cabin diameter, and the number of seats abreast. Assume that the wall thickness is 6 cm each side. Ignore the galley, lavatories, and assume that 50 seats are allocated for first class and the rest of the seats are desired to be of economy (tourist) class.

8. The fuselage design of the following jet transport aircraft is in progress. The fuel tanks are arranged so as to be accommodated by the fuselage.

$$m_{TO} = 200\,000\,\text{kg}, S = 450\,\text{m}^2, C = 0.8\,\text{lb/h/lb}$$

The aircraft cruising speed at 35 000 ft is 520 knot with a lift-to-drag ratio of 14.

(a) Determine the total fuel volume if the aircraft range is 7000 km. Assume the fuel type is Jet A.
(b) If each fuel tank contains 10 m^3 of fuel, how many fuel tanks must be accommodated by the fuselage?

9. The fuselage design of the following turboprop-driven transport aircraft is in progress. The fuel tanks are arranged so as to be accommodated by the fuselage.

$$m_{TO} = 40\,000\,\text{kg}, S = 120\,\text{m}^2, C = 0.8\,\text{lb/h/hp}, \eta_P = 0.75$$

The aircraft cruising speed at 25 000 ft is 350 knot with a lift-to-drag ratio of 11.

(a) Determine the total fuel volume if the aircraft range is 3000 km. Assume the fuel type is JP-4.
(b) If each fuel tank contains 8 m^3 of fuel, how many fuel tanks must be accommodated by the fuselage?

10. A cargo aircraft has a high-wing configuration with the following characteristics:

$$S = 100\,\text{m}^2, \text{AR} = 8, \lambda = 0.6, \text{ root airfoil: NACA } 65_3\text{-}418$$

Determine what volume of the fuselage must be allocated for the wing box, if the fuselage width at the wing intersection is 3 m.

11. A business jet aircraft has a low-wing configuration with the following characteristics:

$$S = 60\,\text{m}^2, \text{AR} = 9.4, \lambda = 0.5, \text{ root airfoil: NACA } 64_1\text{-}412$$

Determine what volume of the fuselage must be allocated for the wing box, if the fuselage width at the wing intersection is 2.6 m.

12. Determine the optimum fuselage slenderness ratio for the fuselage shown in Figure 7.47 which is made up of a semisphere, a cylinder, and a cone. The goal is to minimize the fuselage wetted area.

13. Determine the optimum fuselage slenderness ratio for the fuselage shown in Figure 7.48 which is made up of a semisphere, a cylinder, and a cone. The goal is to minimize the fuselage wetted area.

14. Determine the optimum fuselage slenderness ratio for the fuselage shown in Figure 7.49 which is made up of a semisphere, a cylinder, and a long cone. The goal is to minimize the fuselage wetted area.

15. Design a cockpit for a single-engine subsonic trainer aircraft with an instructor pilot and a student in a tandem configuration. You need to first discuss the following items: pilot personal equipment, pilot comfort level, control equipment, measurement equipment, and level of automation. Then the following parameters must be determined: seat geometry, seat free space, distance to stick/yoke/side-stick, stick motion distance, distance to pedal, pedal motion range, lower-than-horizon view angle, over-nose vision angle, side-view angle, seatback angle, distance to instrument panel, overhead height, and room behind seat. Draw the final design with dimensions.

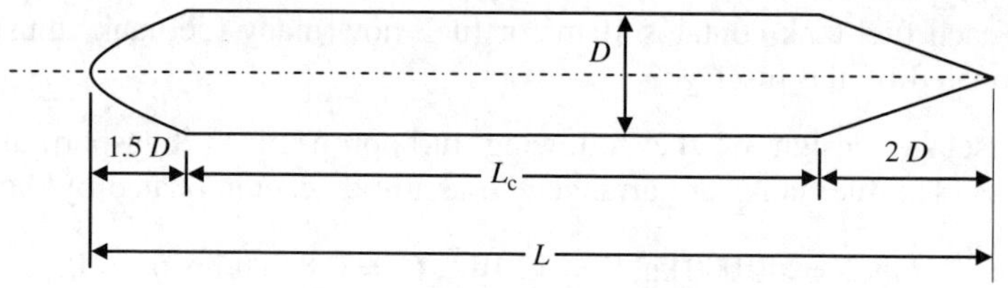

Figure 7.47 A fuselage nose and tail section

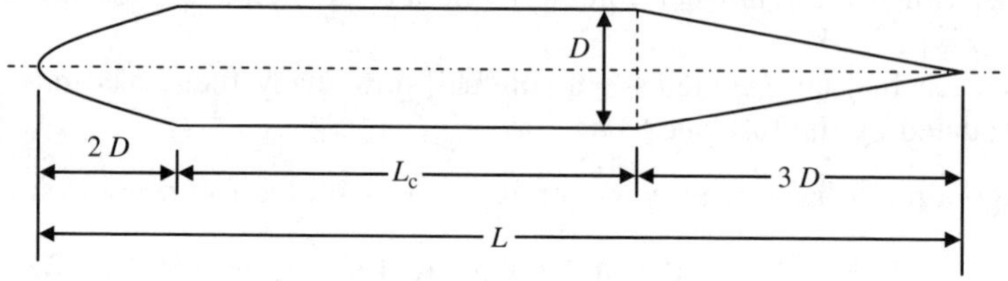

Figure 7.48 A fuselage nose and tail section

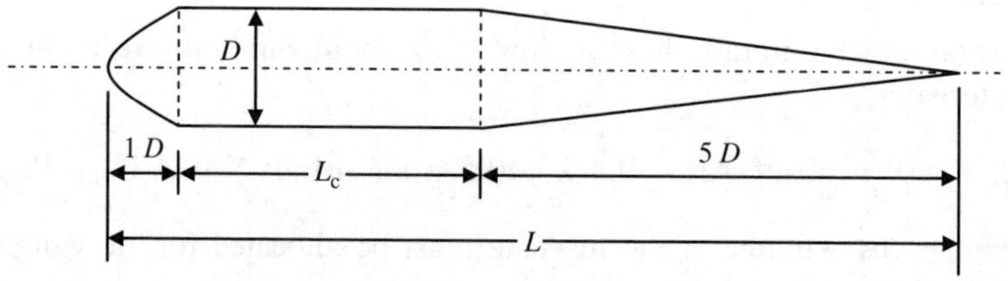

Figure 7.49 A fuselage nose and tail section

16. Design a cockpit for an observation/touring subsonic aircraft with a pilot and a tourist in a side-by-side configuration. You need to first discuss the following items: pilot personal equipment, pilot and tourist comfort level, control equipment, measurement equipment, and level of automation. Then the following parameters must be determined: seat geometry, seat free space, distance to stick/yoke/side-stick, stick motion distance, distance to pedal, pedal motion range, lower-than-horizon view angle, over-nose vision angle, side-view angle, seatback angle, distance to instrument panel, overhead height, and room behind seat. Draw the final design with dimensions.

17. Design a cockpit for a supersonic fighter aircraft with a single pilot. You need to first discuss the following items: pilot personal equipment (suit, goggles, helmet, ejection seat, pressure system, and parachute), pilot hardship level, control equipment, measurement equipment, and level of automation. Then the following parameters must be determined: seat geometry, seat free space, distance to stick/yoke/side-stick, stick motion distance, distance to pedal, pedal motion range, lower-than-horizon view angle, over-nose vision angle, side-view angle, seatback angle, distance to instrument panel, overhead height, and room behind seat. Draw the final design with dimensions.

18. Design a cockpit for a supersonic fighter aircraft with two pilots in a tandem configuration. You need to first discuss the following items: pilot personal equipment (suit, goggles, helmet, ejection seat, pressure system, and parachute), pilot hardship level, control equipment, measurement equipment, and level of automation. Then the following parameters must be determined: seat geometry, seat free space, distance to stick/yoke/side-stick, stick motion distance, distance to pedal, pedal motion range, lower-than-horizon view angle, over-nose vision angle, side-view angle, seatback angle, distance to instrument panel, overhead height, and room behind seat. Draw the final design with dimensions.

19. Design a fuselage for a light single-seat GA aircraft with a high-wing configuration, that can accommodate a pilot and three passengers for a range of 1500 km at a service ceiling of 18 000 ft. The fuselage must also carry 50% of the total fuel. The fuselage must accommodate a piston engine as a tractor aircraft. The aircraft has a take-off mass of 3000 kg, and the cruising speed at 18 000 ft is 220 knot with a lift-to-drag ratio of 8. For necessary crew and luggage, follow the FAA regulations and assume any other necessary parameters. Draw the final design with dimensions.

20. Design a fuselage for a business jet transport aircraft with a low-wing configuration, that can accommodate 8 passengers for a range of 2500 km at a service ceiling of 32 000 ft. The fuselage must also carry 70% of the total fuel. The landing gear is of tricycle configuration and is retractable into the fuselage. The main gear must be at 65% of the fuselage length from the nose. The wing root aerodynamic center is at 60% of the fuselage length from the nose. Assume the fuel type is JP-4. The aircraft has a take-off mass of 5000 kg, and the cruising speed at 32 000 ft is 430 knot with a lift-to-drag ratio of 15. For necessary crew and luggage, follow the FAA regulations and assume any other necessary parameters. Draw the final design with dimensions.

21. Design a fuselage for a high-subsonic jet transport aircraft with a high-wing configuration, that can accommodate 40 passengers for a range of 3000 km at a service ceiling of 25 000 ft. The fuselage must also carry 80% of the total fuel. The landing gear is of tricycle configuration and is retractable into the fuselage. The main gear must be at 62% of the fuselage length from the nose. The wing root aerodynamic center is at 58% of the fuselage length from the nose. Assume the fuel type is JP-5. The aircraft has a take-off mass of 22 000 kg, and the cruising speed at 25 000 ft is 280 knot with a lift-to-drag ratio of 9. For necessary crew and luggage, follow the FAA regulations and assume any other necessary parameters. Draw the final design with dimensions.

22. Design a fuselage for a high-subsonic jet transport aircraft with a high-wing configuration, that can accommodate 18 passengers for a range of 2000 km at a service ceiling of 28 000 ft. The fuselage must also carry 60% of the total fuel. The landing gear is of tricycle configuration and is retractable into the fuselage. The main gear must be at 62% of the fuselage length from the nose. The wing root aerodynamic center is at 58% of the fuselage length from the nose. Assume the fuel type is JP-4. The aircraft has a take-off mass of 13 000 kg, and the cruising speed at 28 000 ft is 280 knot with a lift-to-drag ratio of 9. For necessary crew and luggage, follow the FAA regulations and assume any other necessary parameters. Draw the final design with dimensions.

23. Design a fuselage for a high-subsonic jet transport aircraft with a low-wing configuration, that can accommodate 300 passengers for a range of 12 000 km at a service ceiling of 34 000 ft. The fuselage must also carry 65% of the total fuel. The landing gear is of tricycle configuration and is retractable into the fuselage. The main gear must be at 62% of the fuselage length from the nose. The wing root aerodynamic center is at 58% of the fuselage length from the nose. Assume the fuel type is JP-4. The aircraft has a take-off mass of 200 000 kg and the cruising speed at 34 000 ft is 550 knot with a lift-to-drag ratio of 13. For necessary crew and luggage, follow the FAA regulations and assume any other necessary parameters. Draw the final design with dimensions.

24. Design a fuselage for a fighter jet aircraft with a mid-wing configuration, that can accommodate two pilots for a range of 2000 km at a service ceiling of 45 000 ft. The fuselage must also carry 80% of the total fuel. The landing gear is of tricycle configuration, but is retractable into the wing. The main gear must be at 61% of the fuselage length from the nose. The wing root aerodynamic center is at 59% of the fuselage length from the nose. The aircraft has a take-off mass of 35 000 kg and the cruising speed at 45 000 ft is Mach 1.8 with a lift-to-drag ratio of 7. For necessary crew and luggage, follow the military standards and assume any other necessary parameters. Assume the fuel type is JP-4. Draw the final design with dimensions.

References

[1] Hawkins, F.H. (1998) *Human Factors in Flight*, 2nd edn, Ashgate.
[2] Bridger, R.S. (2008) *Introduction to Ergonomics*, 3rd edn, CRC Press.
[3] Kroemer, K.H.E., Kroemer, H.B., and Kroemer-Elbert, K.E. (2000) *Ergonomics: How to Design for Ease and Efficiency*, 2nd edn, Prentice Hall.
[4] Salyendy, G. (2006) *Handbook of Human Factors and Ergonomics*, 3rd edn, John Wiley & Sons, Inc.
[5] Vink, P. (2011) *Aircraft Interior Comfort and Design; Ergonomics Design Management: Theory and Applications*, CRC Press.

[6] NASA (1978) *Anthropometry Source Book*, vol. **3**, NASA/RP/1024, NASA.
[7] US Department of Transportation, Federal Aviation Administration (2011), FAR 23, FAR 25, www.faa.gov.
[8] US Department of Defense (2008) Military Standards, Performance Specifications and Defense Handbooks.
[9] Jackson, P. *Jane's All the World's Aircraft*, Jane's Information Group, 1996 to 2011.
[10] Garland, D.J., Wise, J.A., and Hopkin, V.D. (1999) *Handbook of Aviation Human Factors*, Lawrence Erlbaum Associates.
[11] Torenbeek, E. (1996) *Synthesis of Subsonic Airplane Design*, Delft University Press.
[12] Hoak, D.E. (1978) USAF Stability and Control DATCOM, Air Force Flight Dynamics Laboratory, Wright-Patterson Air Force Base, Ohio.
[13] Coordinating Research Council Inc. (2007) Commercial Aircraft Design Characteristics – Trends and Growth Projections, 5th edn, International Industry Working Group.
[14] Roskam, J. (2007) *Lessons Learned in Aircraft Design; The Devil is in the Details*, DAR Corporation.
[15] Anderson, J.D. (1999) *Aircraft Performance and Design*, McGraw-Hill.
[16] Ian, M. and Allan, S. (2008) *Aircraft Systems: Mechanical, Electrical and Avionics Subsystems Integration*, 3rd edn, John Wiley & Sons, Inc.
[17] Anderson, J.D. (2011) *Fundamentals of Aerodynamics*, 5th edn, McGraw-Hill.

8

Propulsion System Design

8.1 Introduction

A heavier-than-air craft requires a propulsion system in order to have a sustained flight. Without a proper aero-engine or powerplant, a heavier-than-air vehicle can only glide for a short time; the most flight that gliders and sailplanes are capable of. The contribution of a powerplant to an aircraft is to generate the most influential force in aircraft performance; that is, the propulsive force or thrust. An aircraft engine produces thrust based on Newton's third law. This third law of Newton states that "for every action, there is a reaction equal to but in the opposite direction to the action force." An aircraft engine usually generates a backward force to displace (accelerate) the air flow, thus the aircraft, in reaction, is pushed forward.

The magnitude of engine thrust needed to fulfill a desired aircraft performance is calculated by the technique of matching charts, as discussed in Chapter 4 during the preliminary design phase. In this chapter, other aspects of the propulsion system such as engine type selection, number of engines, and engine locations will be examined. Since the design of the aero-engine is recognized as a distinct discipline, it will not be discussed in this chapter. The reader is forwarded to references such as [1, 2] for more details on aero-engine design. In practice, an aircraft designer does not design the aero-engine, rather the engine is selected to match the design requirements.

This chapter is devoted to the design of a propulsion system. The chapter begins with powerplant functional analysis and a design flowchart, and ends with the design procedure. Various types of engines with their features and technical characteristics, including limitations, advantages, and disadvantages, are briefly introduced. The design considerations and constraints, design requirements, design methodology, engine installation factors, and some engine performance calculations are also presented. A fully solved example to demonstrate the application of the technique is also provided at the end of the chapter.

As expressed throughout the book, the systems engineering approach is followed in the propulsion system design. Although this practice is part of the aircraft detail design phase, the design of engine components such as the inlet, turbine, combustion chamber, and nozzle (in the case of a jet engine) and piston and cylinder, crank, carburetor, propeller, and fuel system (in the case of a piston engine) is not discussed in this chapter. The reason is that these topics are beyond the scope of this textbook.

8.2 Functional Analysis and Design Requirements

In order to design the propulsion system, one needs to perform the engine functional analysis. This allows the designer to select the right engine and determine the most suitable engine parameters. The propulsion design requirements could be satisfied if the designer is able to connect them to the functional analysis. Table 8.1 presents a summary of functions for the aircraft propulsion system. In general, functions are grouped into primary, secondary, and contributing functions. All these functions must be considered in the propulsion system design process, since they impact the design requirements in various ways.

The primary function of an aero-engine is to generate propulsive force. This force is necessary to overcome the aircraft drag and provide the means (airspeed) for the wing to produce lift force. In a jet engine the thrust is created directly by the engine, while in a prop-driven aircraft the thrust is produced by employing a propeller. The engine type selection technique (e.g., jet, prop-driven) will be discussed in Section 8.3.

The secondary function of the propulsion system is to provide power/energy to other subsystems such as the hydraulic system, electric system, pressure system, air conditioning system, and avionics. These subsystems rely on the engine power to operate. In most General Aviation (GA) as well as light transport aircraft, the power for internal consumption is extracted from the propulsion system. However, in large jet transport aircraft, a separate device such as an auxiliary power unit (APU) is often considered to produce energy for other systems. In a small twin-engine transport aircraft, a couple of systems such as electric systems receive their energy from the left engine, while the energy of other systems such as hydraulic systems is provided by the right engine. The energy source selection for each system is a function of a number of factors, such as aircraft mission, handling quality, cost, safety considerations, and operational requirements.

Table 8.1 Propulsion system functions

No.	Category of function	Function
1	Primary function	Generate propulsive force
2	Secondary function	Generate power/energy for various aircraft subsystems such as hydraulic and electric systems
3	Contributing function	Either stabilizing or destabilizing Reduces the comfort of the passengers, crew, and flight attendants due to engine noise Reduces the comfort of the passengers, crew, and flight attendants due to heat exchange to cabin/cockpit Safety contributions in case of one engine inoperative Operating cost due to fuel consumption Structural impact due to engine vibrations

The primary purpose of an aircraft APU is to provide power to start the main engines. An APU is normally a small jet engine running an electric generator to produce electric energy. The APU receives the necessary fuel directly from the fuel tanks and is usually started by an electric motor. Since the APU is independent of the propulsion system, the aircraft tends to have a higher safety rate under an all-engine inoperative condition. For more details on the performance of APUs, the reader is referred to resources such as [3].

There are a number of contributing functions, of which the majority are undesirable. Depending upon the engine locations, it may play a stabilizing or a destabilizing role in aircraft dynamics. The relationship between engine locations and aircraft stability will be addressed in Section 8.5. An engine has lots of moving parts, including rotating shaft, plus rotating fan or rotating prop. The high-speed motion of the engine parts will create a lot of noise, which interferes with passenger comfort. The noise level comparison among various aero-engines will be reviewed in Section 8.3.

Since the engine is burning fuel in a combustion chamber, a great deal of heat is generated, part of which will be exchanged with the cabin and cockpit. The heat transfer from engine to fuselage will heat up the cabin and cockpit and must be accounted for in the engine location selection. Passenger comfort will be degraded if the engine is too close to the cabin and if the engine isolation system is not efficient enough. Cooling provision must be included as part of the engine design operation. The influence of engine type and location on comfort levels will be examined in Sections 8.3 and 8.5.

Another contributing function of the propulsion system relates to safety concerns. In the case of an engine fire, firefighting initiatives must be predicted and provided. In multi-engine aircraft in the case of one-engine inoperative (OEI) conditions, aircraft controllability and passenger safety are of great concern. Federal Aviation Regulations, particularly Parts 23 and 25 [4], have mandated a number of design requirements to provide airworthiness. The aircraft designer should follow the relevant requirements in the propulsion system design process.

Moreover, fuel is expensive and all means must be established to reduce the operating cost. For instance, a Boeing 737-700 (Figure 6.12) burns about 4000 gal of fuel in a flight from Boston to Los Angeles. The significance will be understood when it is noted that in the year 2011, oil cost above \$100 a barrel and the price of a gallon of jet fuel was about \$6 (about 50% more expensive than car gas). This is a huge cost for an airliner; hence, the propulsion system must be designed to minimize fuel costs.

Finally, the aircraft structure is impacted by engine installation as well as engine operation. The engine is installed on an aircraft and mounted by means of a bulkhead or pylon. An engine has several impacts on an aircraft structure through forces, moments, and items such as engine weight, engine thrust, heat exchange, and vibration. The structural integrity must be analyzed as part of the engine design methodology to ensure that the structure is not degraded by the engine design. In general, the following items are considered as the propulsion system design requirements and constraints:

- aircraft performance;
- engine cost;
- operating cost;
- engine weight constraints;
- size constraints;

- flight safety;
- engine efficiency;
- aircraft stability;
- heat exchange;
- structural requirements;
- installation constraints;
- integration;
- noise constraints;
- passenger comfort;
- passenger appeal;
- stealth constraints;
- engine frontal area constraints;
- maintainability;
- manufacturability;
- disposability.

These items are not numbered to demonstrate that their importance is not unique for various aircraft. Depending upon the aircraft mission, available budget, and design priorities, the above-listed requirements and constraints must be evaluated and weighted. Then, a trade-off study should be carried out to determine and finalize the engine parameters along the design process. The relationship between these requirements and constraints will be addressed in Sections 8.3–8.6. In general, the job of an aircraft designer is to determine/design/select the following items: (1) select engine type, (2) select number of engines, (3) determine engine location, (4) select engine from manufacturers' catalogs, (5a) size propeller (if prop-driven engine), (5b) design inlet (if jet engine), (6) design engine installation, (7) iterate and optimize. Items (5a) and (5b) are design activities for an aircraft designer which must be carried out in parallel with the engine design team. This is to emphasize that the aircraft designer has the final say in these two propulsion system parameters.

Figure 8.1 illustrates a propulsion system design flowchart including design feedback. As the flowchart presents, the propulsion system design is an iterative process and the number of iterations depends on the nature of the design requirements as well as the designer's skills. The propulsion system design is initiated by identifying the design requirements and the process ends with optimization. The details of each design box will be presented in the forthcoming sections. As in the case of other aircraft components, there is no unique design solution to satisfy the design requirements. Each acceptable solution will deliver a pool of advantages and disadvantages which must be decided based on the systems engineering technique.

8.3 Engine Type Selection

Soon after the design requirements and constraints have been identified and prioritized, the propulsion system designer will begin to select the type of engine. There are a number of engine types available in the market for flight operations. They include: human-powered, electric (battery), solar-powered, piston-prop, turbojet, turbofan, turboprop, turboshaft, ramjet, and rocket engines. In this section, aero-engine classification including features and characteristics of each type, as well as the engine type selection process, are introduced.

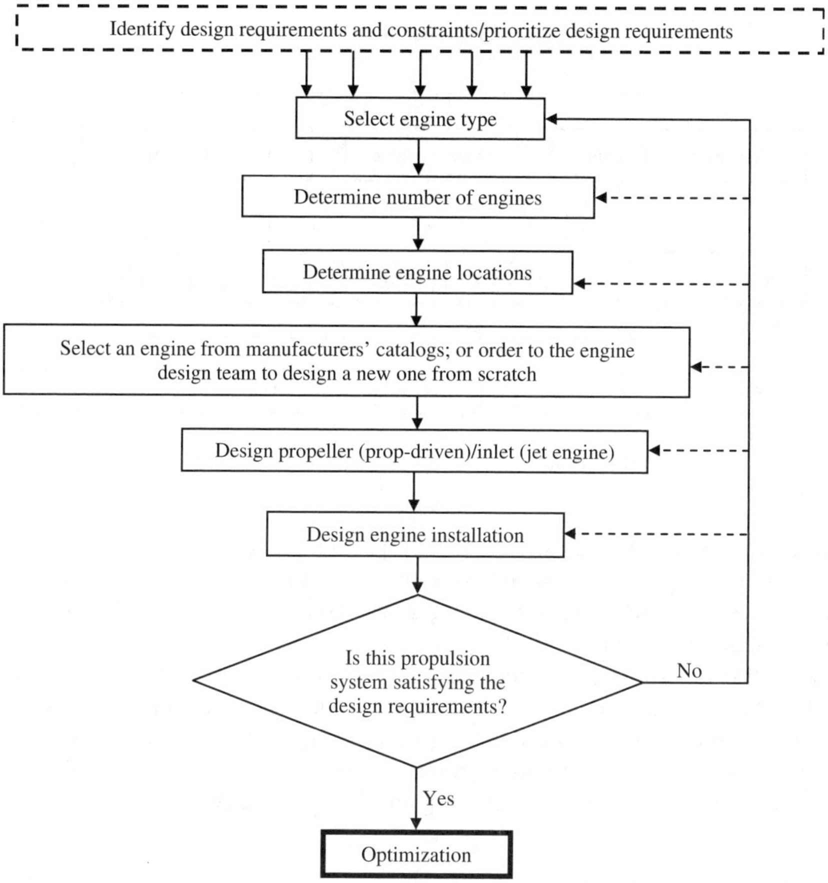

Figure 8.1 Propulsion system design flowchart

8.3.1 Aircraft Engine Classification

The Wright brothers made the first powered flight by Flyer aircraft on December 17, 1903. Nowadays various types of engines are designed, manufactured, and employed in aircraft. Aircraft engines may basically be classified into three major categories: (i) air-breathing engines, (ii) non-air-breathing engines or rocket engines, and (iii) unconventional engines. Figure 8.2 illustrates this classification. In this section, features and characteristics of each type are briefly reviewed. For more details, the reader is referred to references in the area of propulsion such as [5–7].

Three important unconventional engines are: (i) man-powered engine, (ii) Sun-powered engine, and (iii) electrically powered engine. Indeed, the aircraft that uses a man-powered engine has no engine other than the muscular power of the pilot. The Sun-powered or solar-powered aircraft uses solar energy that is absorbed through its solar cells. An electrically powered aircraft uses an electric battery that is powerful enough for its entire flight. In fact, an electric engine is not really a non-conventional engine, since it is very popular and employed in various small remote-controlled aircraft.

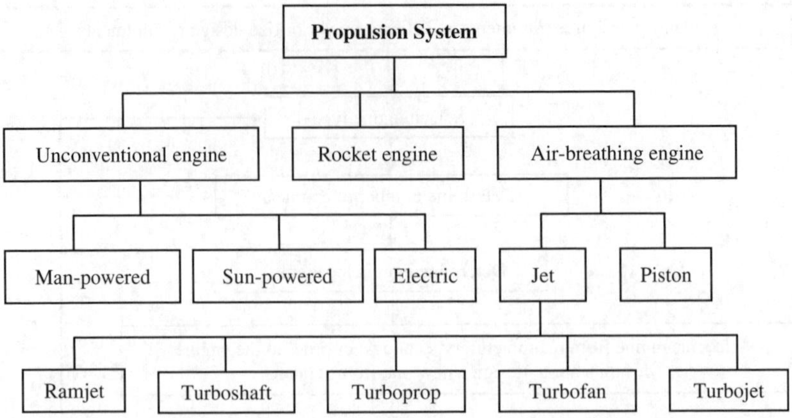

Figure 8.2 Air vehicle engines classification

Another classification is to divide the propulsion system into two groups of non-prop-driven and prop-driven systems. The group of non-prop-driven propulsion systems includes rocket, turbojet, turbofan, and ramjet. The group of prop-driven propulsion systems includes piston-prop, electric, human-powered, solar-powered, turboprop, and turboshaft. The main difference between these two groups is that the non-prop-driven propulsion systems generate thrust directly through a nozzle, while the prop-driven propulsion systems produce thrust with the application of a propeller. The classification in Figure 8.2 is one of many possible propulsion system classifications in the literature. Table 8.2 presents the features of a powerplant for several aircraft.

8.3.1.1 Human-Powered Propulsion System

A human-powered aircraft employs the power of a human (pilot) to generate thrust via a propeller. Hence, the human is assumed as part of the propulsion system. The first successful human-powered aircraft was the Gossamer Albatross in 1979. The aircraft was powered using pedals to drive a large two-bladed propeller and completed a 35.8-km crossing in 2 hours and 49 minutes, achieving a top speed of 29 km/h and an average altitude of 1.5 m. Afterward, several successful human-powered aircraft were designed and flown. A human-powered aircraft built by MIT as a test-bed for the NASA Dryden Flight Research Center is shown in Figure 8.3(a).

The main advantages of a human-powered propulsion system are its independence from fuel, and from a mechanical engine, as well as its low cost. However, the main disadvantages originate from human weaknesses, which include a very low cruising speed (less than 15 knot), a very low ceiling (less than 8000 ft), a very low rate-of-climb (less than 10 m/min), and a low range (a few kilometers). A human-powered engine is surely quieter than a piston-prop engine. A comparison between the features of a human-powered engine and other types of engine is outlined in Table 8.3.

Table 8.2 Engine features for several aircraft [8, 9]

No.	Aircraft	Aircraft type	m_{TO} (kg)	Engine type	Number of engines	Engines	Power or thrust	Engine locations
1	Wright Flyer	First flight in history	338	Piston	1	No brand	89 hp	Loc 1
2	Cessna 182	Light GA	1 406	Piston	1	Lycoming IO-540-AB1A5	230 hp	Loc 1
3	Mooney M20J	Touring	1 315	Piston	1	Continental TSIO-550-G	200 hp	Loc 1
4	Spitfire	British WWII fighter	3 071	Piston	1	Rolls-Royce Merlin 45	1470 hp	Loc 1
5	STEMME S10	Motor glider	850	Piston	1	Rotax 914	93 hp	Above fuselage
6	Voyager	Fly around the globe without refueling	4 398	Piston	2	Continental O-240+IOL-200	100 + 81 kW	Along FCL
7	C-130 Hercules	Military cargo	70 300	Turboprop	4	Allison T56-A-15	$4 \cdot 4590$ hp	Under-wing
8	SkYSpark	100% Electric Airplane	1 497	Electric	1	Valentino synchronous	75 kW	Fuselage nose
9	Bombardier Dash 8 Q400	Regional transport	27 986	Turboprop	2	P&W PW150A	$2 \cdot 5071$ hp	On the wing
10	Bell AH-1Cobra	Attack helicopter	4 310	Turboshaft	1	Lycoming T53-L-703	1100 hp	Loc 2
11	Kawasaki MD500	Light helicopter	1 157	Turboshaft	1	Allison 250-C20R	278 hp	Loc 2
12	Eurocopter EC 135	Light helicopter	2 500	Turboshaft	2	TurbomecaArrius 2B2	$2 \cdot 583$ hp	Loc 2

(continued overleaf)

Table 8.2 (*continued*)

No.	Aircraft	Aircraft type	m_{TO} (kg)	Engine type	Number of engines	Engines	Power or thrust	Engine locations
13	Global hawk	Unmanned surveillance	10 387	Turbofan	1	Allison Rolls-Royce AE3007H	31.4 kN	Over aft fuselage
14	Gulfstream V	Business jet	41 136	Turbofan	2	Rolls-Royce BR710A1-10	2 · 65 kN	Loc 4
15	Cessna Citation X	Business jet	16 374	Turbofan	2	Rolls-Royce AE 3007C-1	2 · 30 kN	Loc 4
16	Boeing 767-300	Airliner	158 760	Turbofan	2	P&W JT9D-7R4	2 · 220 kN	Loc 3
17	Airbus 380-800	Airliner	569 000	Turbofan	4	GP 7270	4 · 311 kN	Loc 3
18	BAe Harrier II	V/STOL close support	8 142	Turbofan	1	Rolls-Royce Pegasus Mk. 105	105.9 kN	Loc 2
19	Eurofighter Typhoon	Fighter	23 500	Turbofan	2	Eurojet EJ200	2 · 60 kN	Loc 2
20	F-16 Eagle	Multirole fighter	19 200	Turbofan	1	F110-GE-100	76.3 kN	Loc 2
21	Concorde	Supersonic airliner	187 000	Turbojet	4	Rolls-Royce/ SNECMA 593	4 · 140 kN	Under-wing
22	SR-71 BlackBird	Reconnaissance	78 000	Turbojet	2	Pratt & Whitney J58-1	2 · 151 kN	Over-wing
23	SpaceShipOne	Suborbital air-launched spaceplane	3 600	Rocket	1	N2O/HTPB SpaceDev	74 kN	Loc 2

FCL: fuselage center line, Loc 1: podded in the fuselage nose, Loc 2: podded inside the fuselage, Loc 3: under the wing, Loc 4: beside the aft fuselage.

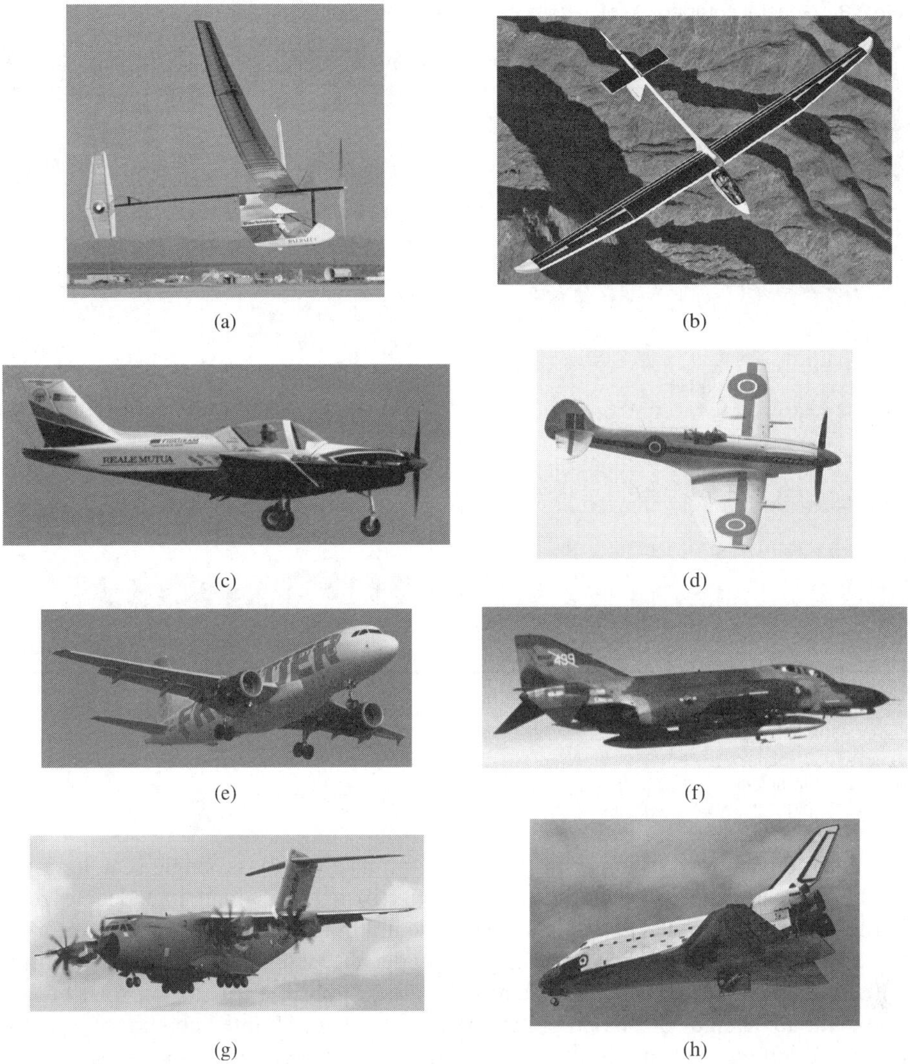

Figure 8.3 Eight aircraft with different engines: (a) Light Eagle, human-powered; (b) Solar Flight's Sunseeker, solar-powered; (c) SkySpark, electric engine; (d) Supermarine 379 Spitfire, piston engine; (e) Airbus A-319, turbofan; (f) McDonnell Douglas F-4 Phantom, turbojet; (g) Airbus A400 Grizzly, turboprop; (h) Space Shuttle, rocket. Reproduced from permission of (d, g) Jenny Coffey, (e) Anne Deus

8.3.1.2 Sun-Powered Propulsion System

A Sun-powered (or solar-powered) aircraft employs a propeller and electric motors which are powered by the solar rays. An example of a Sun-powered aircraft is the unmanned vehicle Pathfinder, with a 98.4-ft wing span and a weight of 560 lb. However, the aircraft structure broke at high altitude and crashed into the ocean due to structural problems.

Table 8.3 General comparison of various parameters for 10 different engines

No.	Engine	SFC	Engine cost	Noise	Specific weight	Propulsive efficiency	Maintainability	Ceiling	Aircraft speed
1	Human-powered	0	1[a]	1	1	8	10	1	1
2	Electric	1	2	3	2[b]	10	9	8	2
3	Solar-powered	0	2[c]	2	2	9	8	9	3
4	Piston-prop	2	3	4	5	3	5	2	5
5	Turbojet	6	6	5	8	6	1	6	8
6	Turbofan	5	9	6	9	7	2	5	7
7	Turboprop	4	7	7	6	4	4	4	6
8	Turboshaft	3	8	8	7	5	3	3	4
9	Ramjet	8	4	6	4	2	6	7	9
10	Rocket	10	5–8	10	3[d]	1	7	10	10

[a] This does not imply that human is cheap, but it means that the pilot does not need to purchase an engine.
[b] Without battery.
[c] Excluding solar panels.
[d] Excluding internal fuel.
1: lowest, 10: highest

A currently developing example is the Solar Impulse, a Swiss single-seat long-range aircraft which is capable of taking off under its own power, and intended to remain airborne for up to 36 hours. The plane has a maximum take-off mass of 2000 kg, a wing span of 80 m, and four 10-hp electric engines under the wing, each with a set of lithium polymer batteries. A total of 11 600 solar cells are responsible for storing solar energy.

The main advantages of this propulsion system are the unlimited endurance, unlimited range, high ceiling, and independence from fuel. The major disadvantages are the low speed (less than 30 knot), low rate-of-climb (less than 50 m/min), and dependence on sunlight. Since the Sun is always available (above the clouds), the aircraft theoretically has an unlimited endurance and an unlimited range. However, the main disadvantages originate from the low storage of solar power, including a very low cruising speed (less than 30 knot) and a very low rate-of-climb (less than 10 m/min). A Sun-powered engine is surely quieter than a piston-prop engine. A comparison between the features of a Sun-powered engine and other types of engine is outlined in Table 8.3. The Solar Flight Sunseeker (Figure 8.3(b)) and Solar Impulse employ a solar-powered engine.

8.3.1.3 Electric Propulsion System

An electric propulsion system includes an electric motor, battery, and propeller. So, in electric-powered aircraft, the powerplant is a battery-powered electric motor. Most model or remote-controlled mini aircraft (wing span of less than 2 m) utilize electric propulsion systems. Since batteries have limited power and a limited life, this type of propulsion system is not widely used in GA and transport aircraft. The main feature of electric propulsion systems is that they are most appropriate for aircraft with a mass of less than about 30 kg. The highest feasible power that a battery or group of batteries can provide is typically less than about 100 hp for less than an hour. Other features of a typical electric-powered aircraft are low speed (less than 60 knot), low range (less than 50 km), low endurance (less than 1 hour), low cost (from a few hundred to a few thousand dollars), compact size, constant center of gravity, and quietness.

The main advantages of an electric-powered propulsion system are its independence from fuel, and from a mechanical engine, as well as its low cost. However, the main disadvantages originate from a limit in the electric energy storage, including a very low cruising speed (less than 100 knot), a low ceiling (less than 40 000 ft), a very low rate-of-climb (less than 15 m/min), and a low range (less than 400 km). An electric engine is generally lighter than a piston engine. However, when the weight of the battery is added, the total weight is heavier than a piston engine, when the fuel weight is considered. As an example, the mass of a typical electric engine to generate 2 hp is about 300 g. But to operate for 15 minutes, it needs a battery which has a mass of about 400 g. A comparison between the features of an electric engine and other types of engine is outlined in Table 8.3. Figure 8.3(c) shows a two-seat SkySpark which uses a 75-kW (101-hp) electric engine powered by lithium polymer batteries. The aircraft achieved a world record of 250 km/h for a human-carrying electric aircraft on June 12, 2009.

8.3.1.4 Piston-Prop Propulsion System

A piston engine, also known as a reciprocating engine, or internal-combustion engine, is a heat engine that uses one or more pistons to convert fuel energy into a rotating mechanical motion. Each piston is inside a cylinder (Figure 8.4), into which the fuel is supplied, heated inside the cylinder either by ignition of a fuel/air mixture. The hot gases expand, pushing the piston to the bottom of the cylinder. The piston is returned to the cylinder top either by a flywheel or the power from other pistons connected to the same shaft. In most types of piston engine, the expanded hot gases are exhausted from the cylinder by this process or stroke.

The rotational motion is then converted into thrust force through the application of a propeller. An aircraft piston engine is similar to an automobile engine with a few slight differences. Commonly used piston engine configurations include: (i) radial, (ii) in-line, (iii) V-type, and (iv) horizontally opposed or flat. The most common configuration is the opposed cylinder engine. The currently available power for the piston-prop engine in the market ranges from 0.5 to 2000 hp, although in the past (1940s and 1950s) more powerful piston engines were manufactured. Currently, the majority of home-built and light GA aircraft employ a piston engine. Due to the application of a propeller, the aircraft airspeed with a piston-prop engine may not exceed about Mach 0.5 due to the occurrence of a shock wave at the prop tip. GA aircraft such as the Cessna 182 (Figure 3.7), Beech Baron 58, Piper Super Cub (Figure 5.56), and Lake LA-270 Turbo Renegade (Figure 8.21) all employ a piston engine.

The altitude performance of a piston engine can be improved by a process called supercharging. This involves compressing the air entering the intake manifold by means of a compressor. In earlier supercharged engines, this compressor was driven by a gear train from the engine crankshaft. A supercharger is often an air compressor (or sometimes a small gas turbine) used to force more air into the combustion chamber(s) of a piston engine than can be achieved at ambient atmospheric pressure or natural aspiration. Today's most supercharged engines employ a turbine-driven compressor powered by the engine's exhaust. Today, most GA aircraft are naturally aspirated.

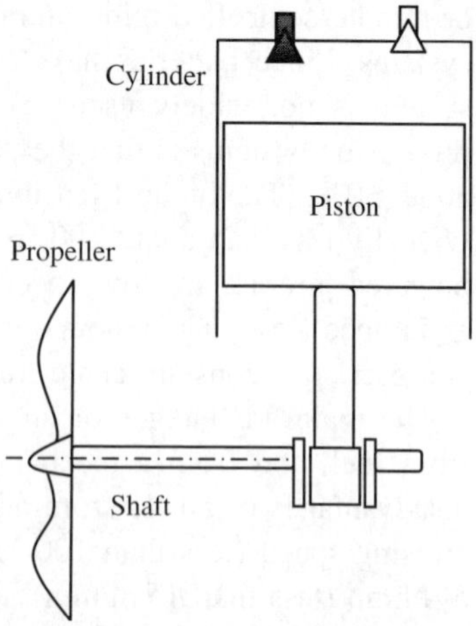

Figure 8.4 Simplified piston engine

The small number of modern piston engines designed to run at high altitudes generally use a turbocharger or turbo-normalizer system rather than a supercharger. Most lighter-than-air craft that are designed to carry a payload with a pilot to control the speed are also equipped with a piston engine. For instance, the semi-rigid helium airship AEROS-50 has one 59.7-kW Rotax 912 flat-four piston engine. A comparison between the features of a piston-prop engine and other types of engine is outlined in Table 8.3. Figure 8.3(d) demonstrates the WWII fighter aircraft Supermarine 379 Spitfire which employs a Rolls-Royce Merlin 45 supercharged V12 piston engine generating 1470 hp (1096 kW).

8.3.1.5 Turbojet Engine

The first gas turbine engine was invented in 1939 and simultaneously developed in Germany and England. A gas turbine engine is a device in which free-stream air is taken in through a carefully designed inlet, compressed in a rotating compressor, heated in a combustion chamber, and expanded through a turbine. The gas then leaves through a nozzle at a velocity much greater than the free stream. The reaction to the ejection of this mass of gas is a forward force on the engine and aircraft: thrust. Thus, a jet engine is an aero-mechanical device which produces forward thrust by forcing the movement of a mass of gases rearward. The gas turbine engine is at the heart of the turbojet engine, turbofan engine, turboprop engine, and turboshaft engine. A pure gas turbine engine which is generating thrust through its nozzle is called a turbojet engine. All the mechanical energy that is produced by the turbine is transferred into the compressor via a shaft to increase the incoming air pressure (Figure 8.5). The remaining energy of the high-temperature, high-pressure air is transferred into the nozzle.

The currently available thrust for turbojet engines in the market ranges from about 10 N to about 100 kN. Small turbojet engines are still used in educational equipment for training purposes. A comparison between the features of a turbojet engine and other types of engine is outlined in Table 8.3. Turbojet engines may operate at a variety of flight regimes including subsonic, transonic, and supersonic speeds. The main disadvantage of a turbojet engine is the high specific fuel consumption, and low propulsive efficiency, which results in a high cost of operation. The first jet airliner De Havilland Comet, supersonic

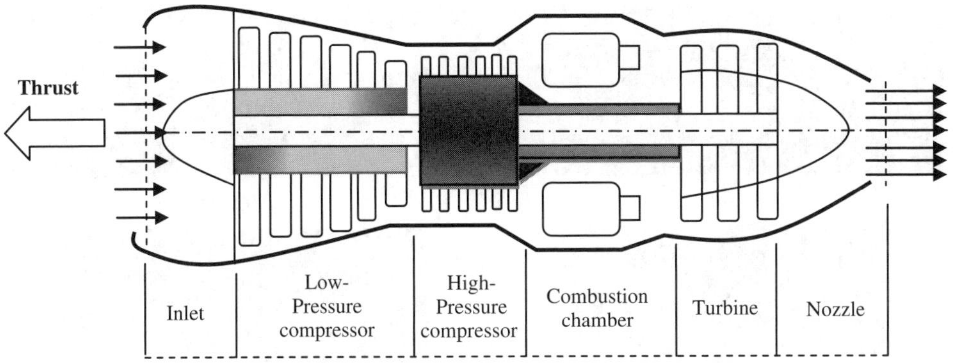

Figure 8.5 Schematic diagram of a turbojet engine

transport aircraft Concorde (Figure 7.24), and military aircraft McDonnell Douglas F-4 Phantom (Figure 8.3(f)) all employed a turbojet engine. Nowadays, the application of a turbojet engine is limited to a few military aircraft.

8.3.1.6 Turbofan Engine

A turbofan engine is a modified version of a turbojet engine to improve the propulsive efficiency and specific fuel consumption. Both of these are improved by adding a fan or a series of fans in front of the compressor with its own exit. A turbofan engine is a gas turbine in which the turbine absorbs power in excess of that required to drive a fan or low-pressure compressor in an auxiliary duct, usually annular around the primary duct. The turbofan engine imparts momentum to greater volumes of air than a turbojet, but the velocity added is less. A turbofan engine is able to operate efficiently at both subsonic and supersonic speeds. An *afterburner*[1] is an additional component added to some jet engines, primarily those on supersonic aircraft. The objective is to provide a temporary increase in thrust, both for supersonic flight and for take-off. The bypass ratio is the ratio between the mass flow rate of air drawn in by the fan bypassing the engine core and the mass flow rate passing through the engine core.

Most light and large transport aircraft plus the majority of military fighters employ a turbofan engine. The currently available thrust for turbojet engines in the market ranges from about 1000 N to about 500 kN. Light transport aircraft such as the Gulfstream 550 (Figure 11.15), Cessna 750, large transport aircraft such as the Boeing B-737 (Figure 6.12), B-767 (Figure 5.4), and B-787 (Figure 1.10), Embraer 195, Airbus A-340 (Figure 8.6), A-380 (Figure 1.8), and military aircraft such as the McDonnell Douglas (now Boeing) F-15 Eagle (Figures 3.12 and 9.14), General Dynamics F-16 Fighting Falcon (Figure 4.6), McDonnell Douglas (now Boeing) F/A-18 Hornet (Figures 2.11, 6.12, and 12.27), Lockheed F-117 Nighthawk (Figure 6.8), and Global hawk (Figure 6.12) all use a turbofan engine. A comparison between the features of a turbofan engine and other types of engine is outlined in Table 8.3. Figure 8.3(e) shows the transport aircraft Airbus A-319 which is powered by two turbofan engines.

[1] In the British literature, afterburning is called reheat.

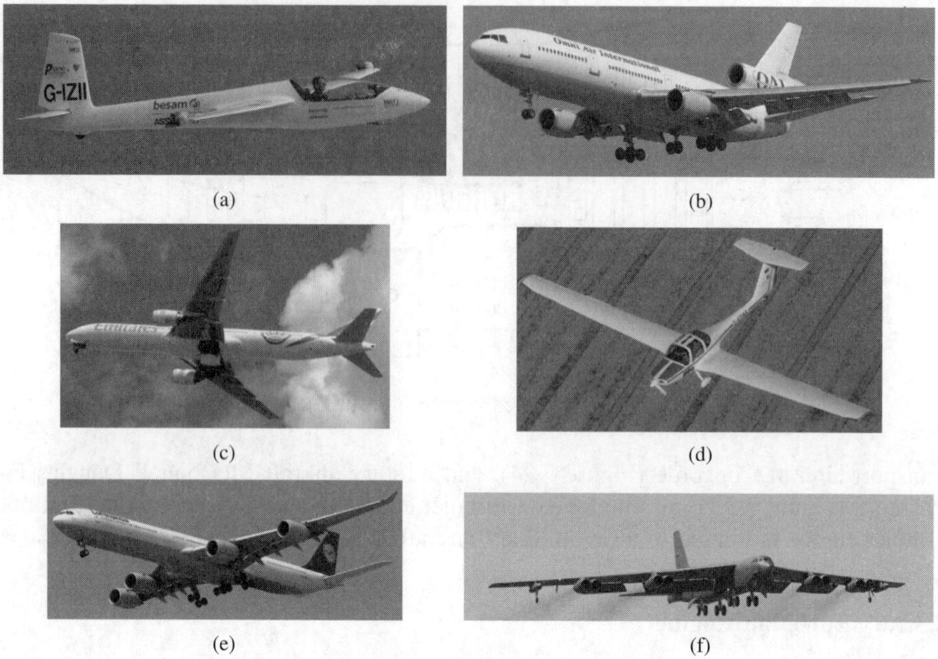

Figure 8.6 Six aircraft with different number of engines: (a) Glider Marganski Swift S-1, no engine; (b) McDonnell Douglas DC-10-30, tri-engine; (c) Boeing B-777, twin engine; (d) Grob G-109B, single engine; (e) Airbus A340, four-engine; (f) Boeing B-52 Stratofortress, multi-engine. Reproduced from permission of (a) Jenny Coffey; (b, c, e) Anne Deus; (d) Rainer Bexten; (f) Antony Osborne

8.3.1.7 Turboprop Engine

A turboprop engine is a gas turbine engine in which the turbine absorbs power in excess of that required to drive the compressor. The excess power is used to drive a propeller. Although most of the energy in the hot gases is absorbed by the turbine, turboprops still have appreciable jet. Thus, most of the gas energy is extracted by the turbine to drive the propeller shaft. A turboprop engine is essentially a propeller driven by a gas turbine engine. By design, most of the available work in the flow is extracted by the turbine, leaving little available for exit nozzle thrust. The propelling nozzle therefore provides a relatively small proportion of the thrust generated by a turboprop. For most turboprop engines, only about 10% of the total thrust is associated with the jet exhaust, and the remaining 90% is generated by the propeller. The large diameter of a propeller requires a reduction gearbox that adds to the propulsion system weight and complexity with its relevant maintainability issues.

With regard to the thrust and efficiency trade-off, the turboprop falls between the piston-prop engine and the turbofan engine. In contrast, the turboprop engine has a specific fuel consumption higher than that of the piston-prop engine, but lower than that of a turbofan or turbojet. A major drawback for a turboprop engine is the high engine noise. Furthermore, the maximum flight speed of a turboprop-powered aircraft is limited to that at which the propeller efficiency becomes seriously degraded by shock wave formation on the propeller tip; usually around Mach 0.6. In terms of power, the maximum available power by a turboprop engine in the market ranges from 100 to 7000 hp.

In the past decade, a few institutions and companies have tried to combine the positive aspects of both the turbofan engine and the turboprop engine and manufactured a new kind of engine, called the turbo propfan or simply propfan engine. This engine has a unique prop that has a smaller diameter compared with a regular turboprop engine and a larger chord compared with a turbofan engine. It also possesses a carefully designed airfoil to reduce the prop noise. So far, it has been installed in only one transport aircraft. Although the performance results of this engine are satisfactory, the new propfan engine is still not popular. Several cargo aircraft such as the C-130 Hercules (Figure 5.4) and some light transport aircraft such as the ATR 72 (Figure 12.42), Embraer EMB 120 Brasilia employ a turboprop engine. A comparison between the features of a turboprop engine and other types of engine is outlined in Table 8.3. Figure 8.3(g) shows the transport aircraft Aibus A400 Grizzly which is powered by four Europrop TP400-D6 turboprop engines each generating 8250 kW (11 060 hp).

8.3.1.8 Turboshaft Engine

A gas-turbine engine that delivers power through a shaft to operate something other than a propeller is referred to as a turboshaft engine. Turboshaft engines are very similar to turboprop engines. The gas turbine may produce some kinds of thrust, but it is primarily designed to produce shaft horsepower. The turboshaft engine has the same basic components found in a turbojet engine, with the addition of a turbine shaft to absorb the power of the escaping gases of combustion. Another use of turboshaft engines is the APU. These small gas-turbine engines are mostly used on large transport aircraft to provide auxiliary power either on the ground or in flight if needed. Turboshaft engines are primarily utilized in helicopters. The Chinook helicopter and Super Cobra AH-1F are two air vehicles which are equipped with a turboshaft engine. A comparison between the features of a turboshaft engine and other types of engine is outlined in Table 8.3.

8.3.1.9 Rocket Engine

A rocket engine is a reaction engine that can be used for aircraft and spacecraft propulsion as well as terrestrial uses, such as missiles. Rocket engines take all their reaction mass from within tanks and form it into a high-speed flow, obtaining thrust in accordance with Newton's third law. Rocket engines produce thrust by the expulsion of a high-temperature, high-speed gaseous exhaust. This is typically created by high-pressure (10–200 bar), high-temperature (2000–4000 K) combustion of solid or liquid propellants, consisting of fuel and oxidizer components, within a combustion chamber. The rocket fuel could be of liquid type or solid type. Liquid-fuel rockets typically pump separate fuel and oxidizer components into the combustion chamber, where they mix and burn. Solid propellants are prepared as a mixture of fuel and oxidizing components and the propellant storage chamber becomes the combustion chamber.

Most military rockets and missiles utilize rocket engines. In non-military applications, two vehicles are named here as examples. The Space Shuttle is equipped with various engines, of which the main engine includes three Rocketdyne Block IIs, each with a sea-level thrust of 1.752 MN. The engines are so powerful that the Space Shuttle reaches its orbit of 220 miles in about 8 minutes. The spacecraft was scheduled to be retired after its last launch on July 8, 2011. Since then, the USA has been paying 50 million dollars to Russia to send one astronaut to the space station at a time. SpaceShipOne is a suborbital air-launched space plane that completed the first manned private space flight in 2004 using a rocket engine. Its mothership aircraft, White Knight, had two turbojet engines. Some military aircraft such as the C-130 sometimes employ rockets, in conjunction with normal engines, to boost the total thrust at take-off in order to reduce the take-off run.

The major aero-engine manufacturers are General Electric, Pratt & Whitney, Rolls Royce, BMW, Alison, Eurojet, Turbomeca, Rocketdyne, SNECMA, Teledyne Continental, Textron Lycoming, and Rotax. Teledyne Continental, Rotax, and Textron Lycoming mainly produce piston engines, while the other manufacturers in the list above mostly fabricate various jet engines. A comparison between the features of a rocket engine and other types of engine is outlined in Table 8.3. Figure 8.3(h) illustrates the Space Shuttle, which employs several rocket engines.

8.3.2 Selection of Engine Type

When the design requirements and constraints are known to a designer, the first step in propulsion system design is to determine the type of engine. In general, the type of engine suitable for a particular aircraft design is mainly determined by the following considerations:

- aircraft performance;
- manufacturing cost;
- operating cost;
- engine weight;
- safety;
- engine propulsive efficiency;
- aircraft stability;
- maintainability;
- heat exchange (cooling provision);
- structural requirements;
- installation requirements;
- integration;
- noise and vibration;
- stealth;
- engine frontal area;
- manufacturability;
- disposability;
- size constraints;
- passenger comfort;
- passenger appeal.

These criteria, which do not tend to have equal importance, are reviewed briefly in this section. Table 8.3 presents a general comparison of various parameters for 10 different engines.

8.3.2.1 Absolute Ceiling and Flight Mach Number

The first and most important criterion to select the engine type relates to the aircraft performance. Two aircraft performance parameters which are most influential in engine selection are absolute ceiling and maximum speed. These parameters form the aircraft flight envelope within which an aircraft will operate. In general, a prop-driven engine operates up to about Mach 0.6. For higher speeds, only a jet engine may be regarded as a suitable means of propulsion. The speed limit plus ceiling limit of each engine type is illustrated in Figure 8.7. Table 8.3 presents a relative comparison of the flight envelope for 10 different engines.

8.3.2.2 Propulsive Efficiency

Figure 8.8 shows the propulsive efficiency by some representative examples in various categories. The figures are for cruising flight at a given thrust. As the figure illustrates, prop-driven engines (e.g., piston-prop, turboprop, electric, and Sun-powered) have the highest propulsive efficiency. Furthermore, the turbofan engine has a slightly higher propulsive efficiency than the turbojet engine, due to the bypass process. Table 8.3 presents a relative comparison of the propulsive efficiency for 10 different engines.

8.3.2.3 Specific Fuel Consumption

In Figure 8.9 the variations of the specific fuel consumption versus flight Mach number are shown for four different engines. The figures are for cruising flight at a given thrust. As the figure illustrates, prop-driven engines (e.g., piston-prop and turboprop) have the lowest specific fuel consumption up to about Mach 0.4. In general, the turboprop engine

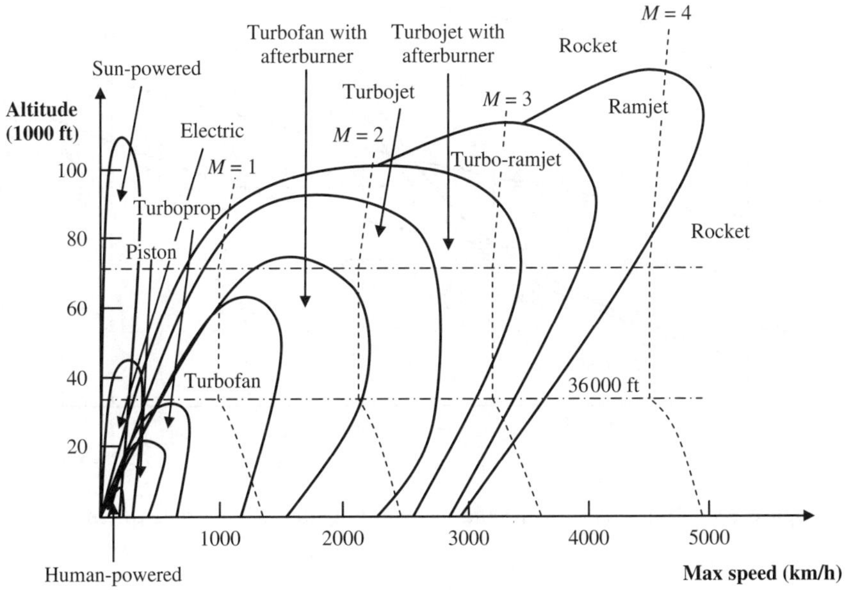

Figure 8.7 A comparison between operating limits of various engines

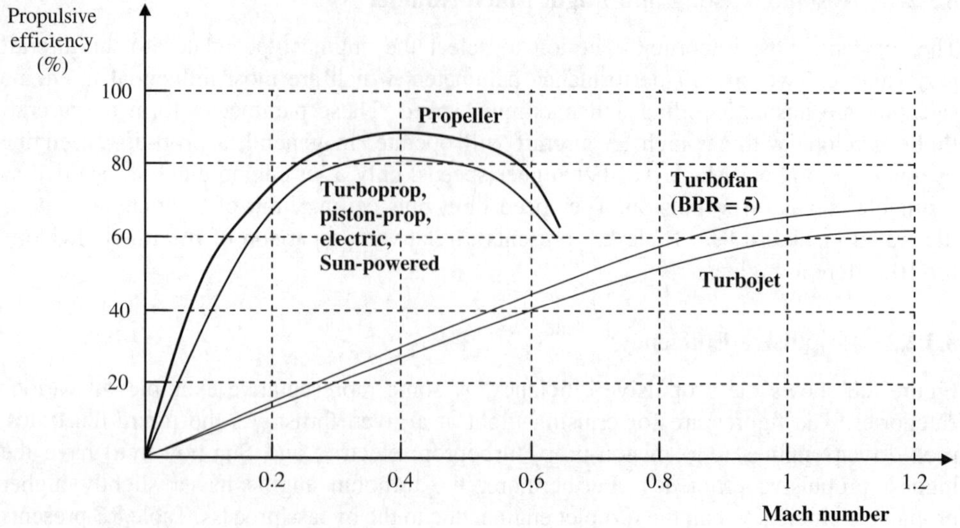

Figure 8.8 A comparison between propulsive efficiency of various engines

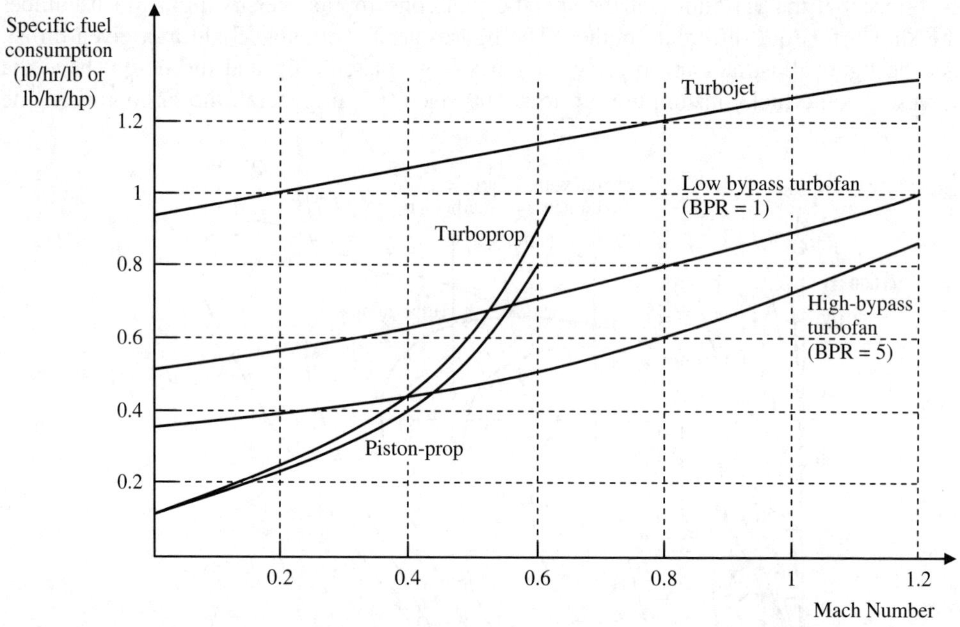

Figure 8.9 Comparison of specific fuel consumption for four types of engine

has a slightly higher fuel consumption than the piston engine. Furthermore, the turbofan engine has a lower specific fuel consumption than the turbojet engine, due to the bypass process. As the bypass ratio is increased, the specific fuel consumption is decreased. Hence, the turbojet engine is the thirstiest engine. However, it burns kerosene which is cheaper than gasoline. The most economical engine with regard to specific fuel consumption at high subsonic speeds is the turbofan engine. Table 8.3 presents a relative comparison of the specific fuel consumption for 10 different engines. Table 4.6 shows typical values of SFC for various engines. The power in the denominator of the unit of SFC for the prop-driven engine is the maximum power. In addition, the lb in the denominator of the unit of SFC for the jet engine is the engine thrust.

8.3.2.4 Engine Weight

In terms of engine weight, Table 8.4 demonstrates a comparison between the average specific weight of various engine types. As the name implies, the lb in the numerator represents the engine weight.

Table 8.4 Specific weight for various engines

No.	Engine type	Specific weight
1	Piston engine	1.5 lb/hp
2	Turboprop	0.4 lb/hp
3	Turbofan	0.2 lb/lb
4	Turbojet	0.3 lb/lb

A direct conclusion from Table 8.4 is that the piston engine is at a disadvantage in terms of weight when compared with the turboprop engine. In addition, the turbojet engine is at a disadvantage in terms of weight when compared with the turbofan engine. To compare an electric propulsion system with a piston-prop engine, in general, the electric engine is lighter than the piston-prop engine. However, when the electric engine plus its battery weight is compared with an equivalent piston-prop engine plus its fuel weight, the overall weight of the electric propulsion system is higher. For instance, the mass of a typical electric engine to generate 2 hp is about 300 g. But to operate for 15 minutes, it needs a battery which has a mass of about 400 g. However, a 2-hp one-cylinder piston engine has a mass of 400 g, but it needs about 250 g of fuel to operate for 15 minutes. So, if a designer is looking for a lighter weight for a remote-controlled model aircraft, a piston-prop engine is recommended.

The combination of propulsive efficiency and engine specific weight implies that any required engine power above say 500 hp is better to be produced by a turboprop engine rather than a piston engine. In fact, a piston engine with more than about 500 hp is inefficient and almost obsolete. Thus, a prop-driven aircraft with a required engine power of more than 500 hp is recommended to have a turboprop engine.

The eight-cylinder, horizontally opposed, four-stroke, air-cooled Textron Lycoming piston engine IO-720-A with a dry mass of 258 kg has a maximum power of 400 hp. In contrast, the Allison turboshaft engine 250-C20B generates 420 hp of power but weighs only 701 N. It is interesting to note that the Allison turboprop engine 250-B17 generates 420 hp of power, but its mass is only 88.4 kg. The SNECMA afterburning turbojet engine Atar 9K50 with a dry mass of 1582 kg generates 70.6 kN of thrust. The Pratt & Whitney turbofan engine F100-220P with a dry mass of 1526 kg generates 74.3 kN of thrust without afterburner, and 120.1 kN of thrust with afterburner. Table 8.3 presents a relative comparison of the specific weight for 10 different engines.

8.3.2.5 Passenger Appeal

One criterion in the marketing department of an airliner manufacturing company is passenger appeal and stylish design. The external shape of an aircraft, including type of engine, will impact a passenger's decision to purchase a flight ticket. It is often believed that jet aircraft possess more passenger appeal than prop-driven aircraft based on human psychology. This is just a psychological judgment by passengers and it is recommended to be included in the design process as an important factor in making a final decision on the type of engine. Therefore, the use of a jet engine enhances the marketing of an aircraft compared with a prop-driven engine. It is also generally accepted that an electric engine possesses more customer appeal than a piston-prop engine. In terms of the environment, a Sun-powered engine is much more attractive than any fuel-consuming engine since it does not pollute.

8.3.2.6 Noise and Vibration

The aviation environment is characterized by multiple sources of noise and vibration, both on the ground and in the air. Exposure of flight crews and passengers to noise is a prevalent issue in aviation. Reciprocating movements of pistons, rotation of propellers and fans, rotating shafts, and engine nozzles are five sources of noise and vibration. In a piston engine, the pistons produce vibrations while the propeller creates noise. The decibel (dB) is the unit used to measure sound and noise intensity. These noises not only make the flight environment more stressful but can, over time, cause permanent hearing impairment, particularly for jet fighter pilots. Ear discomfort may occur during exposure to a 120-dB noise, so the combination of earplugs with earmuffs or communication headsets is recommended when ambient noise levels are above 115 dB.

In a turboprop engine, the rotating shaft is generating relatively less noisy, while the prop produces a noticeable noise. In a turbofan engine, the rotating shaft is the main source of noise, while the fan and nozzle are noise generators. For the occupants of a jet airliner, the cabin is very silent, but an observer on the ground will regard the jet aircraft as a very annoying vehicle. The higher the bypass ratio, the less noise a turbofan engine will generate. The noise of the fan is suppressed by various measures including optimum airfoil design, while the propeller noise is reduced by slowing the revolutions per minute (in fact the tip speed).

Vibration, on the other hand, influences the performance of some measurement devices such as angle of attack and pitch angle meters. Therefore, in flight test operations, special filters must be utilized to filter out the spikes from the measured signals. Although a rocket engine may not have any moving parts, the exhaust flow creates a rather loud noise due to the very high Mach number and the oblique shock waves in the nozzle. An electric engine has considerably less noise and less vibration than a piston-prop engine. So, if a designer is looking for a quieter propulsion system for a remote-controlled model aircraft, a piston-prop engine is recommended.

Both noise and vibration impact the aircraft occupants' comfort level; hence, one of the areas of competition between jet aero-engine manufacturing companies is engine noise level. The operational noise levels of the airliner Airbus A340 (Figure 8.6) are 95 dB during take-off and 97.2 dB in approach, while they are 92.2 dB during take-off and 101 dB in approach [8] for the Airbus A300-600. For the airliner Boeing 757 (Figure 8.16), the noise level is 82.2 dB during take-off but 95 dB in approach. Table 8.3 presents a relative comparison of the noise level for 10 different engines.

8.3.2.7 Engine Maintainability

Engine maintainability and mean time between two overhauls (TBO) are other elements of concern in the engine selection. The systems engineering approach greatly emphasizes engine maintainability, since it will impact on the aircraft as a system. In general, in the case of a turboprop, the TBO is about three times more than that for a piston engine. Moreover, the TBO for a high-bypass ratio turbofan is about twice that for a turbojet engine. Generally speaking, the electric engine is the most maintainable engine, while the turbojet engine is the least maintainable engine. The reason is that a jet engine is very compact and has thousands of parts and elements, while an electric engine constitutes the least amount of moving mechanical parts. Table 8.3 presents a relative comparison of the maintainability for 10 different engines.

8.3.2.8 Engine Size

Whether the engine is podded or installed in a separate nacelle, its dimensions are significant with respect to aircraft performance and configuration design. In the case of a separate nacelle, the nacelle surface area and engine frontal area influence the aircraft drag, which in turn impacts aircraft cost and performance. As the nacelle surface area and the engine frontal area are increased, the aircraft performance is degraded and the flight cost is increased. In the case of a podded engine, the larger the engine, the less useful space is available for payload and fuel. In general, for low engine power (less than about 100 hp) the piston engine has the advantage of lower dimensions, while for higher power the turboprop engine has smaller size. In comparison between turbofan and turbojet engines, the turbofan engine definitely has a larger diameter due to its fan, while it has a smaller length.

An electric engine requires less space than a piston-prop engine, even when the battery is added. As an example, an electric engine to produce 2 hp of power has a length of about 4 cm and a diameter of about 4.5 cm. However, an equivalent piston engine has a length of about 9 cm, a height of about 9 cm, and a width of about 4 cm. So, if a designer is looking for smaller dimensions for a remote-controlled model aircraft, an electric engine is recommended. The Allison turboprop engine 250-B17 with 420 hp of power has a length of 1.143 m and a width of 0.438 m. The General Electric turbofan engine GE90-76B with a length of 5.182 m and a diameter of 3.404 m generates 340 kN of thrust. The SNECMA afterburning turbojet engine Atar 9K50 with a take-off thrust of 70.6 kN has a diameter of 1.02 m and a length of 5.944 m.

8.3.2.9 Engine Production Cost

The cheapest type of engine is the electric engine, after which is the solar-powered engine. However, a human-powered engine may be assumed to have the lowest price, if the human is not included in the cost calculation. In fuel-burning engines, the order from the cheapest to the most expensive engines is as follows: (i) piston engine, (ii) turboprop engine, (iii) turbofan engine, (iv) turbojet engine, (v) turbojet with afterburner engine,

(vi) turboramjet engine, and (vii) rocket engine. This comparison is relative and may change from decade to decade, based on the progress in technology. For instance, in 2010, the turboprop engine cost about twice as much as the piston engine in the aviation market. As the engine power increases, the price gap decreases. Table 8.3 presents a relative comparison of the production cost for 10 different engines.

Now that several engine types have been introduced, it is time to sum up the arguments in this section and present the conclusion. Thus, the technique to select the engine to best satisfy the design requirements is described. The selection of engine type depends upon a number of factors that are listed in this section, and is based on the systems engineering approach. In order to select the best engine, the designer must perform a trade-off study using a comparison table such as Table 8.5. When a rank (R_i) is assigned to each engine relative to each figure of merit (FOM), and a weight (K_i) is designated to each FOM, the total figure of merit (FOM_j) of each engine is determined by summing all the values together as follows:

$$FOM_j = \sum_{i=1}^{n} K_i \cdot R_i \tag{8.1}$$

where the subscript j represents the jth engine and n denotes the number of engines which are capable of performing inside the aircraft flight envelope. A weight is assigned to each FOM based on the priorities. The engine which obtains the highest FOM will be assumed to be the most suitable engine for a given aircraft. Hence, based on aircraft mission and design requirements, one engine is usually the best alternative. It must be noted that, in some design cases, the limited availability of suitable engine types is a decisive factor, forcing the aircraft designer to select a configuration which is feasibly not an ideal one. In such a case, the best engine type for a particular mission can only be arrived at by long and close collaboration between aircraft designer and engine designer. It must be emphasized that the choice of the engine type and the design of the aircraft are so interrelated that it is sometimes very difficult to make the final decision.

Table 8.5 An example for evaluation of engine type figures of merit

		Figure of merit (FOM)								
		SFC	Engine cost	Noise	Specific weight	Propulsive efficiency	Maintainability	Passenger appeal	Size	Total
Weight →		K_1	K_2	K_3	K_4	K_5	K_6	K_7	K_8	
No.	Engine ↓									
1	Piston- prop	R_1	R_1	R_1	R_1	R_1	R_1	R_1	R_1	FOM_1
2	Turbojet	R_2	R_2	R_2	R_2	R_2	R_2	R_2	R_2	FOM_2
3	Turbofan	R_3	R_3	R_3	R_3	R_3	R_3	R_3	R_3	FOM_3
4	Turboprop	R_4	R_4	R_4	R_4	R_4	R_4	R_4	R_4	FOM_4
5	Sun-powered	R_5	R_5	R_5	R_5	R_5	R_5	R_5	R_5	FOM_5
6	Human-powered	R_6	R_6	R_6	R_6	R_6	R_6	R_6	R_6	FOM_6
7	Rocket	R_7	R_7	R_7	R_7	R_7	R_7	R_7	R_7	FOM_7
8	Electric	R_8	R_8	R_8	R_8	R_8	R_8	R_8	R_8	FOM_8

Example 8.1

For a transport aircraft to carry eight passengers, the four engine types of piston-prop, turboprop, turbofan, and turbojet are found to be capable of satisfying the performance requirements. The relevant design requirements are selected to be SFC, engine cost, noise and vibrations, engine weight, propulsive efficiency, maintainability, passenger appeal, and engine size and dimensions. The design priorities for these requirements are considered as follows:

- **Case 1.** The passenger appeal and engine size are more important than the SFC and engine cost.

SFC	Engine cost	Noise	Specific weight	Propulsive efficiency	Maintainability	Passenger appeal	Engine size	Total
13%	6%	8%	10%	11%	17%	20%	15%	100%

- **Case 2.** The SFC and engine cost are more important than the passenger appeal and engine size.

SFC	Engine cost	Noise	Specific weight	Propulsive efficiency	Maintainability	Passenger appeal	Engine size	Total
23%	16%	8%	9%	13%	20%	5%	6%	100%

Determine the most suitable engine among these four alternatives for each case.

Solution:

- **Case 1**. The FOM for four types of engine is evaluated in a table similar to Table 8.5, and the relevant figures are inserted. The result is demonstrated in Table 8.6. Please note that since the order of the ranks must be the same to make for consistent results, the order of ranks for the items where the lowest number is the most desired are reversed. So the ranks for SFC, engine cost, noise, and specific weight are reversed. Thus, "1" represents the worst option and "10" denotes the best alternative. Based on the results of Table 8.5, the most suitable is a turboprop engine (FOM = 624) and the least suitable is a turbojet engine (FOM = 539). The FOM calculations in Table 8.6 explain how the best design is determined. As an example, the calculation of FOM for the piston-prop engine is shown as follows:

$$\mathrm{FOM}_1 = \sum_{i=1}^{4} K_i \cdot R_i = 13 \cdot 10 + 6 \cdot 10 + 8 \cdot 2 + 10 \cdot 5 + 11 \cdot 8$$
$$+17 \cdot 10 + 20 \cdot 2 + 15 \cdot 3 \tag{8.1}$$

$$\mathrm{FOM}_1 = 599$$

Table 8.6 The evaluation of engine type figures of merit for case 1

No.	Engine ↓ / Weight →	SFC	Engine cost	Noise	Specific weight	Propulsive efficiency	Maintainability	Passenger appeal	Size	Total ↓
		13	6	8	10	11	17	20	15	
		Figure of merit (FOM)								
1	Piston-prop	10	10	2	5	8	10	2	3	599
2	Turboprop	8	7	6	8	7	4	5	7	624
3	Turbojet	4	2	7	3	2	1	10	10	539
4	Turbofan	5	4	10	4	5	2	9	8	598

1: worst, 10: best.

- **Case 2**. In this case, the same technique as in case 1 is applied; just new values are inserted. Based on the results in Table 8.7, the most suitable engine is a piston-prop engine (FOM = 783) and the least suitable one is a turbojet engine (FOM = 363). The FOM calculations in Table 8.7 explain how the best design is determined.

Table 8.7 The evaluation of engine type figures of merit for case 2

No.	Engine ↓ / Weight →	SFC	Engine cost	Noise	Specific weight	Propulsive efficiency	Maintainability	Passenger appeal	Size	Total ↓
		23	16	8	9	13	20	5	6	
		Figure of merit (FOM)								
1	Piston-prop	10	10	2	5	8	10	2	3	783
2	Turboprop	8	7	6	8	7	4	5	7	654
3	Turbojet	4	2	7	3	2	1	10	10	363
4	Turbofan	5	4	10	4	5	2	9	8	493

1: worst, 10: best.

Therefore, a comparison of cases 1 and 2 demonstrates that as the order of priorities is changed, the most suitable engine type will be changed.

8.4 Number of Engines

The choice of number of engines has a far-reaching effect on the propulsion system design, in view of the complicated nature of the problem. Accordingly, we shall confine ourselves to summarizing some of the more important aspects. In general, the following items will influence the decision on the number of engines: engine failure rate, safety, aircraft configuration, fuselage design, maximum available power or thrust, engine weight, engine size and dimensions, engine installation, engine location, aircraft controllability, direct operating cost, and additional necessary changes due to multi-engine option. These items, starting with safety precautions, will be discussed in brief in this section.

8.4.1 Flight Safety

Although modern engines (either piston, or turbine, or electric) are very reliable, the possibility of an engine malfunction must never be ignored. Statistics clearly indicates that there have been and will be unfavorable circumstances where an engine may become inoperative during a flight operation. It is always possible that one engine will fail during a flight operation. One of the reasons behind a great number of mishaps and accidents in the past century of flight history was "OEI." Nevertheless, the number of mishaps, accidents, and crashes in aviation is far less than for road (i.e., automobile, bus, and train) transportation. The crash rate in the aviation industry is on the order of one per million flight operations. Due to the low possibility of flight mishaps, it is generally accepted that the aircraft is the safest means of travel.

The occurrence of OEI will not only lead to a considerable decrement in power/thrust, but also disturb aircraft control and equilibrium. Furthermore, a dead engine will increase aircraft drag, so the aircraft performance is also degraded. For this reason, the propulsion system must be designed to provide an acceptable level of safety. The multi-engine propulsion system configuration is one of the best solutions for the OEI case issues. As the number of aircraft occupants (mainly passengers) is increased, it is recommended to employ a higher number of engines.

The airworthiness authorities have established a number of airworthiness standards to regulate aircraft design, including number of engines. For instance, Subpart E of Part 25 of FAR has laid down a variety of regulations concerning powerplants for transport aircraft. Item c in Section 25.901 requires that:

> for each powerplant installation, it must be established that no single failure or malfunction or probable combination of failures will jeopardize the safe operation of the airplane except that the failure of structural elements need not be considered if the probability of such failure is extremely remote.

When an aircraft has only one engine, the only option for a pilot (on the occurrence of engine failure) is to glide and land on the closest available runway. A typical glide angle for most GA and transport aircraft is about 5–7 deg. The typical cruise ceiling for a GA aircraft is about 20 000–30 000 ft, and for a transport aircraft about 30 000–40 000 ft. Thus, if a GA or transport aircraft encounters a dead propulsion system during a cruising flight, the pilot has only about 61 km to 122 km of ground distance to land safely. If there is no safe place to land in this range, such as during a trans-Atlantic flight, or over a mountainous area such as Greenland or parts of Alaska, the aircraft will crash. For this reason, if an aircraft is designed to fly over an area without a safe alternative runway or flat surface, the propulsion system must have at least two engines. Most large transport aircraft, such as the Boeing 737 and Airbus A320, are equipped with more than one engine to enable them to continue on one engine in an OEI during a trans-Atlantic flight.

The likelihood of failure of two engines out of three, or three engines out of four, is very low. Therefore, a twin-engine propulsion system is recommended for long-range transport aircraft. However, there are other issues such as engine installation that force an aircraft designer to select more than two engines. The decision over the number of engines for a fighter aircraft is driven by military issues that sometimes overshadow the safety of the pilot. Therefore, safety in military standards may be sacrificed for mission success or air-power superiority. The availability of a parachute for a fighter pilot is one of the solutions to mitigate this shortcoming. It is interesting to observe that the fighter aircraft General Dynamics F-16 Fighting Falcon (Figure 4.6) has a single engine, but fighters such as McDonnell Douglas F-15 Eagle (Figure 3.12) and Boeing F/A-18 Hornet (Figures 2.11 and 6.12) are equipped with twin engines. Both configurations have advantages and disadvantages, and both are considered to satisfy their different design military requirements.

8.4.2 Other Influential Parameters

Although flight safety is a dominant factor in determining the number of engines, there are requirements and constraints which influence this decision. Progress in technology necessitates a rise in transport aircraft capacity, which in turn requires engine growth. The growth of a propulsion system translates into an increase in engine thrust/power.[2] This increase in engine thrust/power is achieved by raising the turbine blade temperature, increasing the compressor pressure ratio, improving the propulsive efficiency, as well as increasing the diameter of the inlet/piston. Although the aero-engine manufacturing corporations have tried to keep up with market needs, in some instances the aircraft requires a thrust/power that no one single engine can provide. In such circumstances, the designer has to order at least two engines.

There are several design circumstances where the aircraft designer has to select a specific number of engines to comply with some configuration constraints. For instance, if the aircraft configuration has been selected such that there is no way to install the engine along the fuselage center line, the number of engines must be even (e.g., 2 or 4). This is to satisfy symmetry requirements. Flight operation requirements such as a surveillance mission or a particular shape of large payload may force the designer to go with a non-single-engine configuration.

In contrast, there are a number of reasons to recommend a single-engine propulsion system. The negative side of a multi-engine configuration includes factors such as a heavier engine weight, larger engine size, and more direct operating cost. As the number of employed engines increases, the aircraft propulsion system tends to become heavier. Moreover, as the number of employed engines is increased, the aircraft propulsion system tends to get larger in size. If this leads to a larger frontal area the aircraft drag is increased too, which in turn degrades the aircraft performance.

As the number of employed engines increases, the direct operating cost is increased too. Two prop-driven engines with x hp of power each will cost slightly more for a flight operation than one prop-driven engine with $2x$ hp. Similarly, two jet engines with x N of thrust each will cost slightly more for a flight operation than one jet engine with $2x$ N of thrust. Although the engines may have the same power/thrust and the same specific fuel consumption, the direct operating cost is a function of a number of factors such as maintenance cost.

And finally, there is a propulsion system installation requirement that must be considered in the selection of the number of engines. The engine installation must be such that the net yawing moment generated from engine thrust is zero. If the aircraft configuration allows for such a provision, the decision on the number of engines is a practical

[2] Thrust is for a jet aircraft, and power is for a prop-driven engine.

one. This installation requirement restricts the decision for a multi-engine configuration. This implies that for a twin-engine configuration, there are only two engine locations: (i) both engines installed along the fuselage center line, as is the case for the record-breaker Voyager aircraft with two piston engines; (ii) both engines installed at the same distance from the fuselage center line. Moreover, the requirement implies that for a three-engine propulsion system, one must be along the fuselage center line.

In conclusion, safety considerations favor the employment of more than one engine, but almost all other influential parameters recommend utilizing a single engine. The final decision on the number of engines will be determined in a long analysis by taking into account all effective factors with their weights. As stated in Section 8.4.1, a large transport aircraft with a long range is required by airworthiness standards (e.g., FAR 25) to have at least two engines, whether prop-driven or jet engines. Table 8.2 illustrates features of a propulsion system, including engine locations for several aircraft. Six aircraft with different number of engines are illustrated in Figure 8.6: (a) acrobatic glider Marganski Swift S-1 (without engine); (b) GA aircraft Grob G-109B (single engine); (c) transport aircraft Boeing B-777 (twin engine); (d) transport aircraft McDonnell Douglas DC-10-30 (tri-engine); (e) transport aircraft Airbus A340 (four-engine); (f) military bomber Boeing B-52 Stratofortress (multi-engine).

8.5 Engine Location

This section is devoted to the issue of engine(s) location selection and installation. The design requirements, design alternatives, placement classifications, general guidelines, and comparisons between various options emphasizing the advantages and disadvantages of each design will be presented in this section.

8.5.1 Design Requirements

In general, factors affecting the engine location selection are as follows: flight safety, cost, drag, frontal area, inlet design, exit nozzle, stability, structural considerations, maintainability, operational requirements (e.g., radar detectability), aircraft cg, engine maintenance, cabin noise, cockpit noise, foreign object ingestion, stall, fuel system, longitudinal equilibrium, engine-out control, fire hazard, aerodynamic interferences, heat exchange, and special considerations. Most of these requirements may be addressed and met by a combination of engine location and other aircraft configuration parameters such as landing gear height. Thus the decision on the engine location selection should be finalized in meetings with members present from each design group. The engine may be placed in a variety of aircraft locations, but these locations may be classified into the following configurations:

- buried inside fuselage nose;
- buried inside fuselage main section;
- buried inside rear fuselage nose;
- buried inside wing;
- podded over the wing;
- podded on the wing tip;

- podded under the wing;
- podded over the fuselage;
- podded under the fuselage;
- podded beside the rear section of the fuselage;
- behind the pilot seat without pod.

In the majority of light single-engine prop-driven GA aircraft, the engine is buried in the fuselage nose (e.g., Cessna 172 (Figure 11.15), Beech Bonanza, Piper PA-28R-201 Arrow, and Pilatus PC-9), while a few aircraft bury the engine in the aft fuselage (e.g., Rutan Long-EZ, RutanVariEze (Figure 3.12), RutanVariViggen, Grob G 850 Strato 2C, and AFI Prescott II). In contrast, most ultralight aircraft such as the two-seat Merlin E-Z, CFM Starstreak, AirBorne XT912, and Pegasus Quantum 145-912 have their single engine installed (without cover) behind the pilot seat.

It is interesting to know that in the range record-breaker aircraft Voyager (Figure 4.5) with two piston engines, one engine was buried in the fuselage nose and the other in the aft fuselage. Both engines were installed along the fuselage center line. In most jet fighter aircraft, either single-engine or twin-engine, the engine is buried in the rear fuselage (e.g., General Dynamics F-16 Fighting Falcon (Figure 4.6), McDonnell Douglas (now Boeing) F/A-18 Hornet (Figures 2.11, 6.12, and 12.27), Dassault Rafale (Figure 6.8), and Eurofighter (Figure 3.7)).

The engines of the great majority of medium and large jet transport aircraft are podded under the wing (e.g., Boeing 777 (Figures 8.6 and 12.27), McDonnell Douglas MD-11 (Figure 3.16), and Airbus A380 (Figure 1.8)), while there are a number of twin-engine medium and business jet transport aircraft with podded engines beside the rear fuselage (e.g., Fokker 100 (Figure 10.6), Dassault Falcon 2000, Gulfstream 550 (Figure 11.15), Hawker 1000, Bombardier CRJ1000, Embraer EMB-145, and Cessna 750 Citation). Furthermore, the majority of twin-engine prop-driven aircraft such as the Fairchild Metro 23 (Figure 6.18), Beech Super King B200, ATR 72-200 (Figure 12.42), Dornier 328, and Jetstream 41 have their engines podded on the wing.

All four turbofan engines of the stealth aircraft strategic bomber Northrop Grumman B-2 Spirit are mounted and buried in the wing structure by an inlet and exhaust nozzle at the top of the wing to satisfy low radar observability requirements. A similar propulsion system configuration is employed in the ground attack aircraft Lockheed F-117 Nighthawk (Figure 6.8), where both engines are covered and integrated into the wing/fuselage structure.

Some GA and almost all amphibian aircraft have their engines podded way above the fuselage (e.g., Thurston TA16 Seafire, Lake LA-270 Turbo Renegade (Figure 8.21), Piper PA-47, Seawind 300C, and A-40 Albatross) or above the wing (e.g., side-by-side kitplane four-seat Creative Flight Aerocat with two Jbiru 3300 piston engines, and Bombardier 415 water-bomber with two PW123AF turboprops) to protect the propulsion system from sea water being splashed into the engine inlet. A similar propulsion system configuration but for a different purpose is selected in some twin-jet engine close-support military aircraft (such as the Fairchild Republic A-10 Thunderbolt II (Figure 6.12)) and STOL transport aircraft (e.g., Antonov An-74). The purpose is either to protect the engine from enemy fire (in military aircraft) or to improve the take-off performance (in the civil case). The four turbofan engines of the Hawker Siddeley Nimrod are located inside the wing

(Figure 9.12). Each configuration has a number of advantages and disadvantages, with their features briefly discussed in this section.

8.5.2 General Guidelines

In general, there are various considerations and requirements which the following guidelines address. It must be emphasized that some features are inherently conflicting, and only a systematic approach will identify the most suitable engine for a particular mission.

1. The engine(s) in a civil aircraft must be located such that the aircraft center of gravity is a small percentage of the wing MAC (say 10% MAC) in front of the aircraft neutral point. This provision guarantees the aircraft will be longitudinally statically stable. In a fighter aircraft which is often unstable, the engine location should help the aircraft cg to be at the desired location.
2. An aircraft with a buried engine tends to have a smaller wetted area than that with a podded engine.
3. A propulsion system with a buried engine is usually lighter than that with a podded engine.
4. The engine is recommended to be located in a place such that a probable fire in the engine does not create a fire in the fuel tanks. This recommendation implies that the engine had better be at a considerable distance from the fuel tanks.
5. As the engine is placed farther from the cabin or cockpit, the passengers and flight crew will experience a quieter environment. So the farther the engine is from the cabin/cockpit, the more comfortable the aircraft occupants will be.
6. As the height of an engine increases from the ground, the possibility of foreign object (e.g., dirt, sand) ingestion during take-off is reduced. This will increase the TBO of the engine, and helps the engine to serve longer.
7. In a twin-engine configuration, positioning the prop-driven engine on the wing (with propeller in front of the wing) often results in the most attractive design from a structural and aerodynamic point of view. Figure 8.10 demonstrates the location options for a multi-engine podded configuration with respect to the wing.
8. The engine must be located in a place resulting in the least negative aerodynamic interference with the wing and tails and the most favorable aerodynamic interferences. This guideline is applied for both buried and podded configurations. In contrast, an engine ahead of the horizontal tail will decrease the downwash behind the wing, so the longitudinal trim and stability will be influenced.

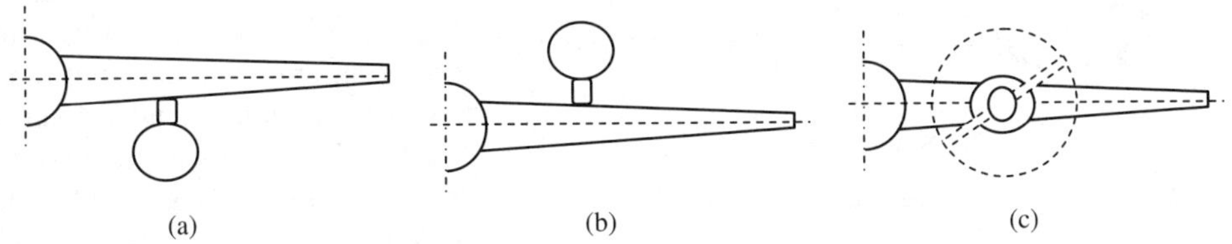

Figure 8.10 The location options for a multi-engine podded configuration with respect to the wing: (a) Engine under wing; (b) Engine over wing; and (c) Engine on the wing (prop-driven)

9. The engine location must be selected such that the engine installation incurs the least weight burden on the structure. For instance, the slipstream of a propeller in front of the wing reinforces the wing lift and induces a positive effect on wing stall.
10. The engine should be located such that the hot gas flow from a prop-driven engine exhaust or jet engine nozzle does not impinge on any aircraft structure. Otherwise, the affected skin will gradually be heated and lose its effectiveness. This guideline is to safeguard the aircraft structural integrity.
11. In the case of a prop-driven engine, the engine location must provide a reasonable propeller clearance. For instance, a low-wing configuration with a twin engine on the wing requires a long landing gear, which is not favorable.
12. When an engine is considered to be placed on a wing (either under-wing or over-wing configuration), it has a favorable effect on the wing structure during flight (Figure 8.10(c)). The reason is understood when the wing lift bending moment is summed up with the engine weight bending moment. The lift bending moment is counteracted by the engine weight bending moment (see Figure 8.11). However, when an aircraft is on the ground (since no lift is produced), the engine weight purely imposes a bending moment on the wing root. Therefore, an engine on a wing supports the wing structurally during flight operations.
13. When an engine is considered to be placed on a wing, an under-wing configuration (see Figure 8.10(a)) is structurally favored over an over-wing configuration. To understand why this is true, just compare the dynamic of a hanging pendulum with an inverted pendulum. A regular pendulum is naturally dynamically stable, while an inverted pendulum is inherently dynamically unstable. Hence, an over-wing engine configuration will induce a flutter to the wing structure, while an under-wing engine will not have such a negative impact.
14. When three engines are used, there is always a problem of locating the third engine. The symmetry requirement forces the designer to place it in the aircraft plane of symmetry (e.g., in the vertical tail). It is clear that the other two engines are generally placed on the two sides of the plane of symmetry (either on the left and right sides of the wing or the fuselage).
15. With regard to the wing structure, a buried engine in the fuselage is favored over the case of a podded engine on the wing. An engine on a wing will create a bending moment on the wing root, while an engine buried in the fuselage does not generate such unfavorable moment. The wing structure must be reinforced to hold a podded engine, which in turn makes it considerably heavier.
16. In the case of an aircraft with stealth requirements, the inlet and particularly the exhaust nozzle must be hidden from ground radars by positioning and burying inside the wing (e.g., B-2 Spirit (Figure 6.8)) or fuselage (e.g., F-117 Nighthawk (Figure 6.8)).

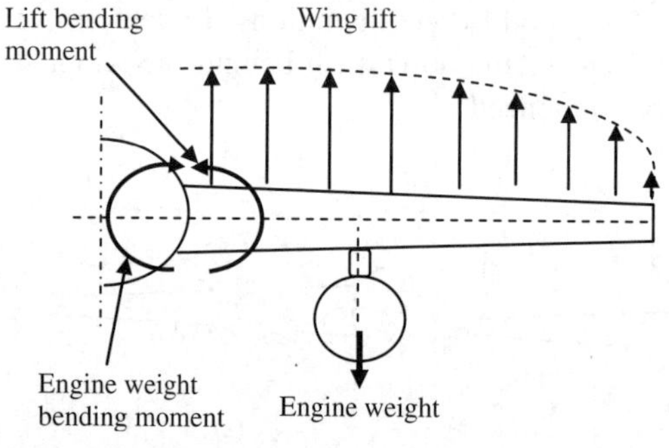

Figure 8.11 Lift bending moment and engine weight bending moment

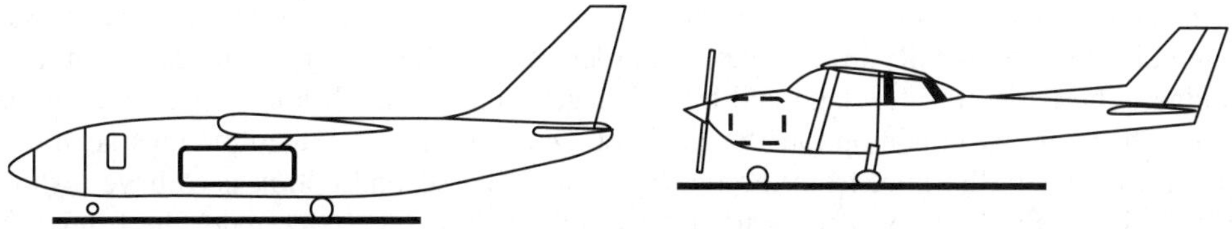

Figure 8.12 An example of a podded engine and an example of a buried engine: (a) Podded under the wing; and (b) Buried in the fuselage nose

17. When an engine is installed on a wing, the wing spar will be cut in half. This engine/wing configuration creates a design challenge for the wing structural engineer. In such circumstance, the spar on both sides of the engine is attached to a frame around the engine, which makes the wing structure heavier.

There are other design guidelines which are introduced in the forthcoming sections. Some more details, applications of these guidelines, and the relationship between design alternatives and design requirements will also be provided in Sections 8.5.3–8.5.6.

8.5.3 Podded versus Buried

One of the fundamental choices for the engine location is to either bury it inside an aircraft component (e.g., fuselage) or leave it out of the aircraft component by housing it inside a pod or nacelle. The third option is to leave the engine without any cover. This decision is applicable for both jet and piston-prop engines. Figure 8.12 illustrates an example of a jet podded engine under a high wing for a transport aircraft and an example of a piston engine buried in the fuselage nose.

The simplest design is to install the engine in the aircraft structure without any cover. This plan keeps the manufacturing cost low, but will increase the aircraft drag which in turn downgrades the aircraft performance. Some home-built and most ultralight aircraft have such a configuration (e.g., the two-seat aircraft Merlin E-Z, CFM Starstreak, AirBorne XT912, and Pegasus Quantum 145-912). In this case, the engine is mounted without any cover behind the pilot seat or above the wing. However, engine cooling is provided naturally by the external air flow.

A propulsion system with a podded engine appears to be heavier than a propulsion system with a buried engine. This is due to the structural requirements of a podded engine. As the name implies, a podded engine requires a nacelle and in most cases a pylon. A podded jet engine in a nacelle under a wing is depicted in Figure 8.13. The nacelle is attached to the wing via a pylon.

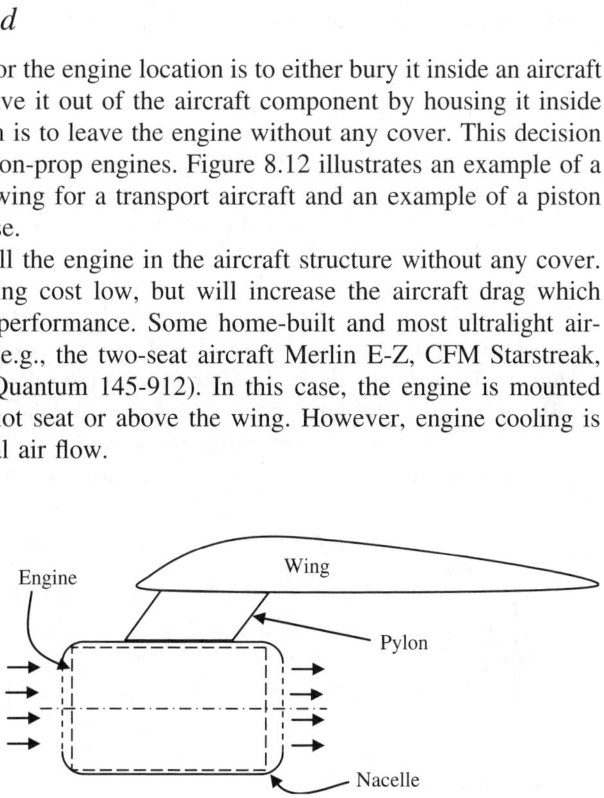

Figure 8.13 Engine nacelle and pylon

In the case of a jet engine (either turbojet or turbofan), a buried engine creates/adds some kinds of complexity in the inlet design and nozzle design. For instance, when a jet engine is buried in the fuselage and the fuselage length is much longer than the engine length, there are two main options for the inlet. One option is to extend the inlet such that the airflow enters the inlet from the fuselage nose. The second option is to have a short inlet and turn the flow by placing the inlet beside the fuselage (say under the fuselage). Both options have one main advantage and one main disadvantage. The first option makes the inlet longer, which degrades the inlet efficiency. However, this type of inlet is easier to design. The short intake and exhaust ducts enable the engine to operate under optimal conditions. To minimize the negative effects of a pylon on the wing, the pylon must not extend above and around the wing leading edge.

In general, an aircraft with a buried engine tends to have a smaller wetted area than one with a podded area. In this regard, a buried engine is recommended over a podded engine. In contrast, a buried engine tends to limit the available space to the payload and fuel tanks. For this reason, a podded engine is recommended over a buried engine. Furthermore, in the case of a podded engine, the designer has more freedom to locate the engines, while in the case of a buried engine, there is not much freedom. Moreover, from a safety point of view, the likelihood of spreading a possible fire to the fuel in a podded engine is lower than that for a buried engine. The final design will be determined after considering all features and weighing the priorities.

8.5.4 Pusher versus Tractor

When the engine location, particularly in the case of a single-engine configuration, is compared with respect to the aircraft center of gravity, two categories of propulsion systems are identified: (i) pusher, where the engine is located behind the aircraft center of gravity and (ii) tractor, where the engine is located ahead of the aircraft center of gravity. A typical prop-driven tractor aircraft and a prop-driven pusher aircraft are depicted in Figure 8.14.

This definition is broad and needs more clarification, particularly for a multi-engine configuration. For the case of an aircraft with a multi-engine propulsion system, the pusher versus tractor is defined with regard to the relationship between the overall thrust of all engines and the aircraft cg. For instance, when an aircraft has three engines at different locations, the resultant thrust is considered to determine if the propulsion system is a pusher or a tractor. In the case of a twin-engine prop-driven propulsion system,

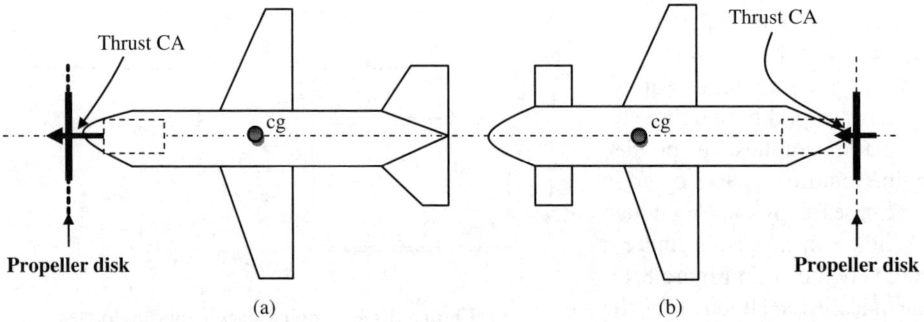

Figure 8.14 Pusher and tractor configurations: (a) Tractor; and (b) Pusher

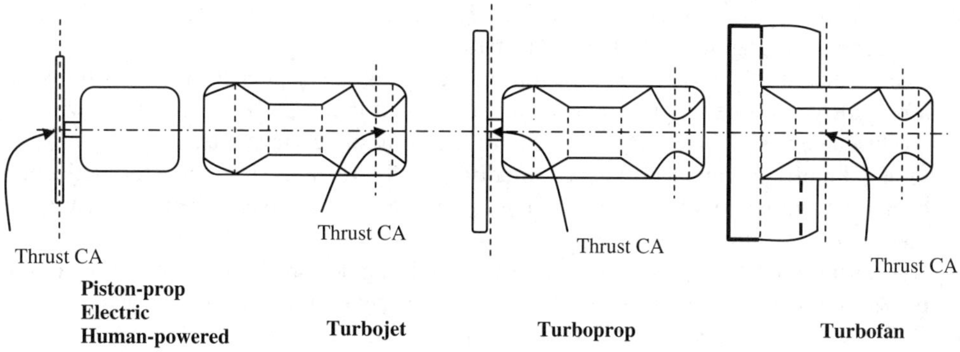

Figure 8.15 Estimation of thrust center of action for various engines

where the prop is behind the engine (e.g., Piaggio Aero P180 or Beechcraft Starship), the propulsion system may be considered a pusher. In contrast, when the propeller is located ahead of the engine (e.g., C-130 Hercules or Beechcraft King Air 350), the propulsion system may be considered a tractor. The reason why the word "may" is used is that the relationship between the engine thrust center of effect (or center of action (CA)) and the aircraft cg must be known to judge the type of system.

The CA of the engine thrust along the fuselage center line is one point. It is very hard to determine the location of this point exactly, since the engine is a long device and every piece is involved in thrust production. However, the following gives some hints to help the reader identify the thrust CA for each type of engine.

For a prop-driven engine, the propeller disk can be considered as the thrust center of effect. However, for a jet aircraft, the issue is somewhat complicated. In a turbojet engine, a great deal of jet thrust is generated by the exhaust nozzle. Hence, a point slightly ahead of the nozzle exit area might be considered as the thrust CA (see Figure 8.15). For the case of a turbofan engine, a location between the fan duct and the exhaust nozzle is the equivalent thrust CA. In a turbofan engine, the thrust is generated partly through the core exhaust nozzle, and partly through the fan nozzle. Therefore, in a high-bypass ratio turbofan engine, the thrust CA is closer to the fan duct, while in a low-bypass ratio turbofan engine, the thrust CA is closer to the exhaust nozzle. In a turboprop engine, the thrust CA is slightly behind the propeller disk. For an electric engine (including a solar-powered engine) and a human-powered engine, the thrust CA is at the propeller disk. The equivalent thrust CA is compared with the aircraft cg to determine whether the propulsion system is a pusher or a tractor. The exact location of thrust CA is determined through an engine ground test.

The primary factors influencing the decision for a pusher or a tractor configuration are: aircraft longitudinal stability, aircraft longitudinal controllability, aircraft longitudinal trim, aerodynamic interferences, structural integration, and aircraft performance. In general, the following factors and facts will influence the decision on pusher or tractor configuration:

1. The pusher engine moves the aircraft cg rearward. Recall that from the point of view of static longitudinal stability, there is an aft limit for the aircraft cg. So other aircraft components must be located so as to achieve the desirable cg. In a severe case, the aircraft has to carry ballast to move the cg forward and inside the acceptable region.

2. The tractor engine moves the aircraft cg forward. As the aircraft cg moves forward, the aircraft is longitudinally less controllable.
3. The net thrust of a prop-driven tractor engine is slightly greater than the net thrust of a prop-driven pusher engine. This is due to the fact that a propeller aft of the wing or the fuselage has a lower efficiency due to the wing or fuselage wake.
4. In a take-off operation, a pusher aircraft usually requires less elevator deflection compared with a tractor aircraft. Thus, a pusher aircraft has a better longitudinal controllability during take-off operation.
5. Due to the relationship between aircraft cg and engine thrust, a tractor configuration provides better directional stability than a pusher configuration.
6. Both pusher and tractor configurations impose some limits on other component configurations. For instance, the installation of an aft tail is problematic when the engine is installed at the fuselage rear section, so a canard is more convenient for this type of pusher. On the other hand, a canard is hard to install when the engine is mounted on the fuselage nose.

In conclusion, the final decision must be made only after considering all these facts, the features of each option, and weighing the priorities.

8.5.5 Twin-Jet Engine: Under-Wing versus Rear Fuselage

Two attractive alternatives for the engine locations of a podded twin-engine jet low-wing aircraft are: (i) under the wing and (ii) beside the rear fuselage. In Section 8.5.2, it was stated that an under-wing podded engine is often favored over an over-wing podded engine. In this section, the under-wing configuration is compared with the case of a rear fuselage podded configuration (Figure 8.16), with more details. Both configurations tend to have a number of advantages and disadvantages. The final decision will be made after weighing all features with respect to the design priorities. The following is a general comparison between the under-wing podded engines versus the rear fuselage podded ones.

1. Both configurations have almost the same wetted area (nacelle plus pylon). Thus, in this regard, both configurations are evaluated as being the same.
2. Due to the locations of the engine relative to the aircraft x-axis: the under-wing engine configuration is usually located below the x-axis, while the beside-fuselage engine configuration is often above the x-axis. Accordingly, an under-wing engine creates a nose-up pitching moment, while a beside-fuselage engine creates a nose-down pitching moment. Therefore, an aircraft with under-wing engines requires less elevator deflection compared with an aircraft with a beside-fuselage engines configuration.
3. Based on the logic presented in item 2, an aircraft with under-wing engines is more prone to pitch-up than an aircraft with a beside-fuselage engines configuration.
4. An under-wing configuration adds complexity to the wing structural design, while a rear-fuselage configuration creates complexity in the fuselage structural design.
5. An under-wing configuration improves the wing aerodynamic performance (i.e., wing lift, drag, and pitching moment) while a rear-fuselage configuration does not induce such an effect.

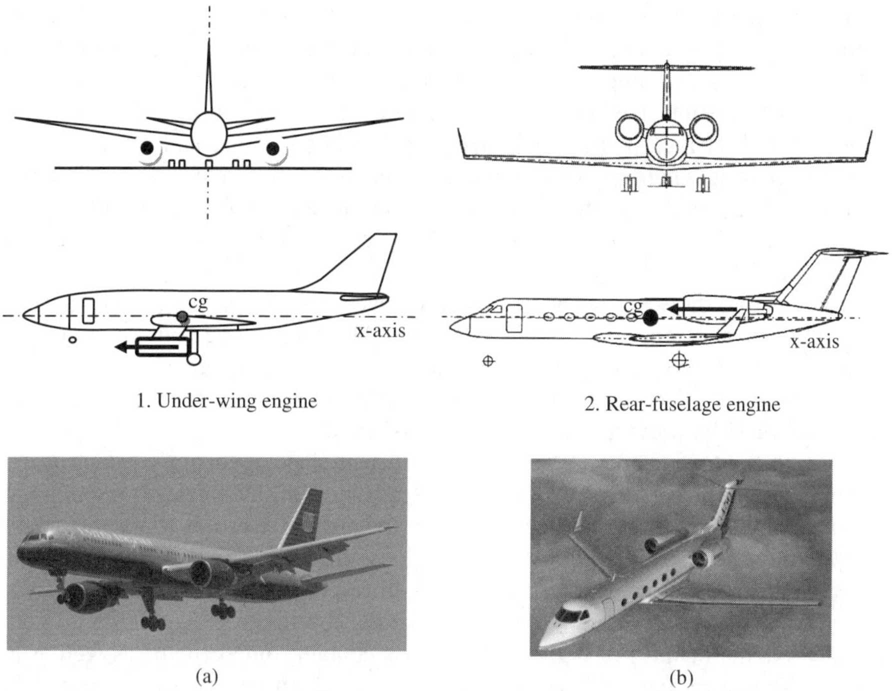

Figure 8.16 Under-wing engine configuration versus rear-fuselage engine configuration: (a) Boeing B-757; (b) Gulfstream G-450. Reproduced from permission of (a) Anne Deus; (b) Gulfstream

6. The mass of the engines and the pylons for the case of a rear-fuselage configuration leads to a reduction in the bending moment compared with that for an under-wing configuration, thus lightening the overall aircraft weight.
7. A rear fuselage podded engine will impose a very hot gas over the horizontal tail, which in turn interferes with the longitudinal trim and dynamic stability. Thus, if a rear-fuselage engine configuration is considered, it is beneficial to employ a T-tail rather than a conventional tail (as in the HondaJet).
8. For the case of a rear-fuselage configuration, the temperature at the cabin near the engine installation area will be slightly higher (say about 5°C) than in the front section of the cabin.
9. The moment arm of the engine thrust of an under-wing configuration is usually longer than that for a rear-fuselage configuration. Therefore, the vertical tail and the rudder for the case of an aircraft with an under-wing engine configuration must be considerably larger than that for an aircraft with rear-fuselage engine configuration.
10. The landing gear height will be affected by an under-wing configuration. Due to a need for ground clearance provision, the landing gear height will be longer than that for an aircraft with rear-fuselage engines. The complexity of the problem is better understood when observing that the typical maximum diameter of a turbofan engine of a large transport aircraft is on the order of 2 m. The solution to avoid long landing gear is to select a high-wing configuration.

11. Since the fuel tanks are usually located close to the aircraft cg, an aircraft with an under-wing engine configuration is less safe with regard to fire hazard in the cabin. The chance of spreading a fire in the engine into the fuel tanks is higher in an aircraft with an under-wing engine configuration. For instance, a Boeing 737-700 stores 8600 lb of fuel in each wing, and 28 800 lb in the center tank (in the fuselage).
12. The under-wing engine configuration appears to provide a positive effect on the airflow at large angles of attack, so tends to counteract the pitch-up of a swept wing. This is due to the presence of a pylon, which acts as a fence in that it avoids flow in the y-direction.
13. Engines below the wing are generally more accessible from the ground than in the rear-up fuselage. Hence, an aircraft with an under-wing engine configuration is more maintainable compared with an aircraft with rear-fuselage engines.
14. Rear-fuselage engines cause the aircraft cg to move backward, so a greater aircraft cg travel must be provided during load distribution.
15. The rear-fuselage nacelle causes the aircraft neutral point to move backward, so it has a positive role in aircraft longitudinal stability.
16. At large angles of attack, particularly when the airflow over the wing has separated, the wake created by the nacelles and the pylons will greatly reduce the horizontal tail effectiveness. This is an important factor with respect to the deep stall problem.

As observed, the features of both configurations contain some benefits as well as some drawbacks. Therefore, the final design and the choice between the two layouts will be determined after considering all features and weighing the priorities. In fact, it is suggested to employ wind-tunnel experiments to investigate the aerodynamic characteristics of each configuration. Table 8.8 provides some recommendations on engine location for several aircraft configurations. Due to the fact that the advantages of an under-wing engine configuration considerably outweigh the disadvantages, almost all low-wing large transport aircraft such as the Boeing 737 (Figure 6.12), Boeing 767 (Figure 5.4), Airbus 340 (Figure 8.6), and MD-80 employ such a configuration. However, there are aircraft such as the Fokker 100 (Figure 10.5) which benefit from a rear-fuselage engine configuration. A Boeing B-757 with two under-wing engines and a Gulfstream G-450 with two rear-fuselage engines are depicted in Figure 8.16.

8.6 Engine Installation

After the engine type and engine locations are determined, engine installations must be investigated. The installation challenges primarily include the engine cooling provision, cabin and cockpit isolation against engine heat, intake duct, exhaust nozzle design, fire safety precautions, and mechanical attachments. Engines, like aircraft, are subject to a variety of airworthiness standards that the propulsion system designer must follow. When an engine is buried in the fuselage, these items are more critical than for a podded engine.

An aero-engine responsible for thrust production generates heat, contains fire, and is often so heavy that it needs special handling. The heat generated in the combustion chamber must be transferred to the environment efficiently. The engine heat transfer requires cooling plates and cowl for an air-cooled system, and pipes and pumps for an

Table 8.8 Engine location recommendations for several aircraft configurations

No.	Aircraft	Engine	Location recommendation
1	Single-engine light GA with aft tail	Prop-driven	Buried in the fuselage nose (tractor)
2	Single-engine light GA with canard	Prop-driven	Buried in the rear fuselage (pusher)
3	Multi-engine GA with canard	Prop-driven	Podded on the wing (pusher)
4	Twin-engine light GA	Prop-driven	Podded on the wing
5	Agriculture	Prop-driven	Buried in the fuselage nose
6	Twin-engine medium and large transport	Jet	Podded under the wing
7	Twin-engine light transport	Jet	Podded beside rear fuselage
8	Multi-engine large transport	Jet	Under-wing
9	Single-engine amphibian	Jet/prop-driven	Podded over the fuselage
10	Multi-engine amphibian	Jet/prop-driven	Podded over the wing
11	Multi-engine cargo	Turboprop	Podded on the wing
12	Motor-glider	Prop-driven	Over the fuselage behind pilot seat
13	Ultralight with aft tail	Prop-driven	Over the fuselage behind pilot seat
15	Fighter	Jet	Buried in the rear section of the fuselage and the inlet under the wing
16	Military aircraft with stealth requirements	Jet	Buried inside the wing or fuselage; inlet and exhaust above wing and fuselage

oil-cooled system. The engine casing tends to have a temperature limit which must not be surpassed in any circumstances. This temperature is a target for the nacelle design in a podded engine configuration. Therefore, special measures must be taken to keep the engine cool, particularly at low altitude and in hot seasons.

Furthermore, heat shielding is another concern in the engine installation. Heat must be stopped from transferring into the passenger cabin and crew cockpit. The severity of the challenge is understood when we note that metals are very heat conductive. Isolation of the engine combustion chamber from the rest of the aircraft is of great importance.

The structural integrity will be endangered if a great deal of heat is transferred to frames and spars. Therefore, a suitable space between hot areas of the engine and the fuselage, or special isolation materials, must be employed to maintain the aircraft structure from fracture and creep.

The uninstalled thrust for a jet engine is provided by the engine manufacturer, but the installed thrust will be determined after the engine is installed. The installation usually reduces the maximum engine thrust/power by a small percentage (sometimes as high as 10%) based on the inlet design and engine installation. Therefore, the engine must be installed such that the engine installed maximum thrust/power is as close as possible to the engine maximum uninstalled thrust/power. The engine installation must provide the necessary clean and undisturbed airflow to allow for an engine to generate the maximum uninstalled rated thrust/power. The installed engine thrust/power is obtained by correcting the uninstalled thrust/power for the actual losses. The correction is to account for installation-related issues such as pressure recovery, shock-induced boundary layer separation, flow distortion, inlet spillage drag, and starting process of a supersonic convergent/divergent inlet to swallow the starting normal shock wave.

8.6.1 Prop-Driven Engine

In the case of an engine installed along the fuselage center line, the engine mounting requires special attention. A piece of aircraft structure for mounting the engine is usually called a firewall. As the name implies, the firewall provides support for the engine as well as maintaining a safe distance between any engine fire and the occupants in the cockpit/cabin and fuel tanks. In several single-piston engine aircraft with the engine in the fuselage nose, the nose gear is attached to the firewall to save aircraft weight. Figure 8.17 illustrates a firewall which is employed for a piston engine installation in a single-engine light GA aircraft.

According to FAR 25 Section 25.1191, each engine and the combustion, turbine, and tailpipe sections of turbine engines must be isolated from the rest of the airplane by firewalls, shrouds, or equivalent means. Each firewall and shroud must be fireproof, constructed so that no hazardous quantity of air, fluid, or flame can pass from the compartment to other parts of the airplane and constructed so that each opening is sealed with close-fitting fireproof grommets, bushings, or firewall fittings.

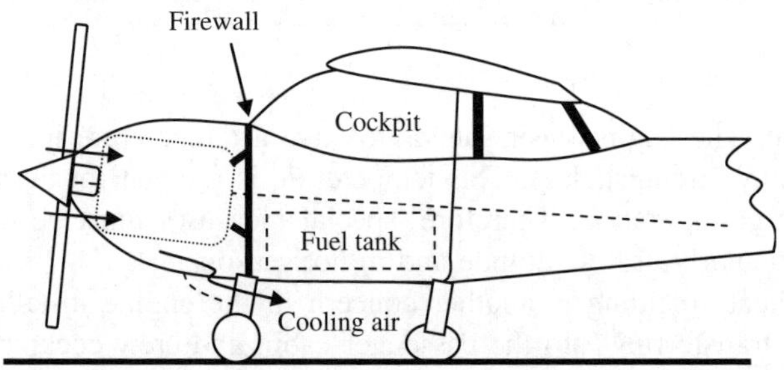

Figure 8.17 Firewall for a piston engine installation

The vertical location of a single tractor prop-driven engine is dictated by the below-the-horizon pilot view as well as prop ground clearance. In case of an air-cooled piston engine, a proper exit for engine cooling, and a proper opening for the supply of air, must be provided.

Section 23.925 of FAR Part 23 on propeller clearance requires that propeller clearances, with the airplane at the most adverse combination of weight and center of gravity, and with the propeller in the most adverse pitch position, may not be less than the following:

1. **Ground clearance**. There must be a clearance of at least 7 in. (for each airplane with nose wheel landing gear) or 9 in. (for each airplane with tail wheel landing gear) between each propeller and the ground with the landing gear statically deflected and in the level, normal take-off, or taxiing attitude, whichever is most critical. In addition, for each airplane with conventional landing gear struts using fluid or mechanical means for absorbing landing shocks, there must be positive clearance between the propeller and the ground in the level take-off attitude with the critical tire completely deflated and the corresponding landing gear strut bottomed. Positive clearance for airplanes using leaf spring struts is shown with a deflection corresponding to 1.5 **g**.
2. **Aft-mounted propellers**. In addition to the clearances specified in paragraph (1) of this section, an airplane with an aft-mounted propeller must be designed such that the propeller will not contact the runway surface when the airplane is in the maximum pitch attitude attainable during normal take-offs and landings.
3. **Water clearance**. There must be a clearance of at least 18 in. between each propeller and the water.
4. **Structural clearance**. There must be (i) at least 1 in. radial clearance between the blade tips and the airplane structure, plus any additional radial clearance necessary to prevent harmful vibration; (ii) at least 0.5 in. longitudinal clearance between the propeller blades or cuffs and stationary parts of the airplane; and (iii) positive clearance between other rotating parts of the propeller or spinner and stationary parts of the airplane.

Furthermore, FAR Part 23 Section 23.771 requires that for each pilot compartment, the aerodynamic controls (excluding cables and control rods) must be located with respect to the propellers so that no part of the pilot or controls lies in the region between the plane of rotation of any inboard propeller and the surface generated by a line passing through the center of the propeller hub making an angle of 5° forward or aft of the plane of rotation of the propeller. It is also recommended to avoid positioning passenger seats in this region. This requirement is shown in Figure 8.18.

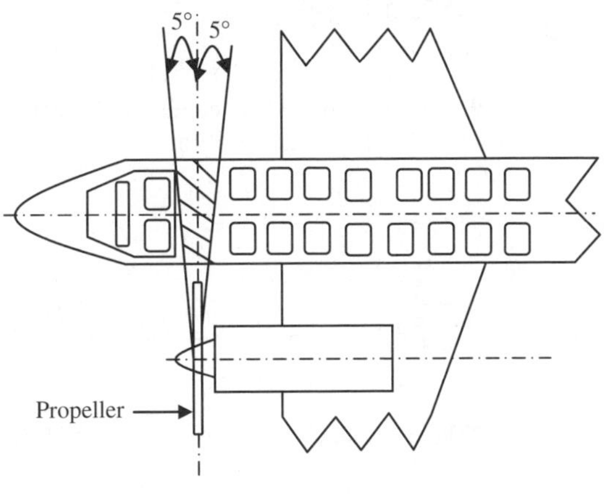

Figure 8.18 Region in which flight crew may not be located

One of the reasons behind this precaution and requirement is the possibility of lumps of ice being thrown from the propeller. For this reason the fuselage structure needs to be locally reinforced to protect it from such an incident. The engine exhaust also needs special consideration in order to positively contribute to the aircraft performance. For instance, the Cessna 150 has a great deal of cooling drag, since the direction of the engine exhaust is perpendicular to the flight direction.

8.6.2 Jet Engine

For a single- or twin-engine jet aircraft with engines mounted inside the fuselage, the intake and exhaust ducts usually present a problem. The inlet duct has to efficiently supply a constant airflow in different flight conditions and at different engine settings. In general, the options for the inlet in a fuselage-buried jet engine lie in two main groups: (i) non-split and (ii) split. The non-split inlets are divided mainly into three variations: (i) under-fuselage (chin), (ii) over-fuselage, and (iii) pitot. Figure 8.19 illustrates various alternatives for non-split-type inlets of a fuselage-buried jet engine. The split-type inlet tends to have mainly three variants: (i) under-wing, (ii) over-wing, and (iii) beside-fuselage. Figure 8.20 illustrates various variants for non-split-type inlets of a fuselage-buried jet engine. The features of these inlets will be discussed briefly here.

In terms of manufacturability and ease of production, a long non-split inlet all the way to the fuselage nose (Figure 8.19(c)) originally seems an attractive design. This inlet configuration was used in the past as in the ex-USAF North American F-86A Sabre, and also in some Soviet jet fighters such as the MiG-21. This design, which is also known

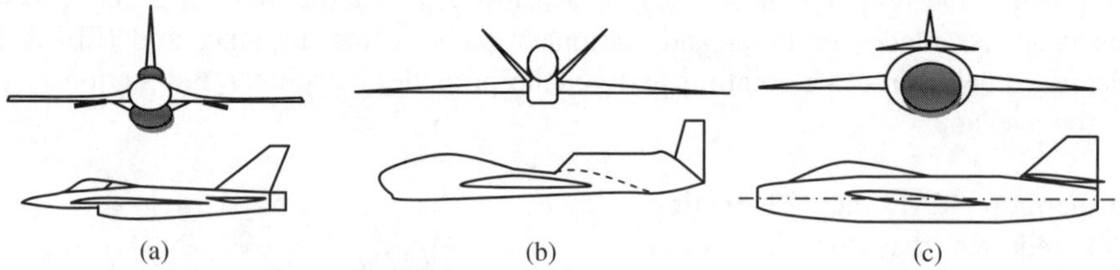

Figure 8.19 Fuselage-buried jet engine inlet location for non-split-type inlet: (a) Under fuselage (chin); (b) Over fuselage; and (c) Pitot

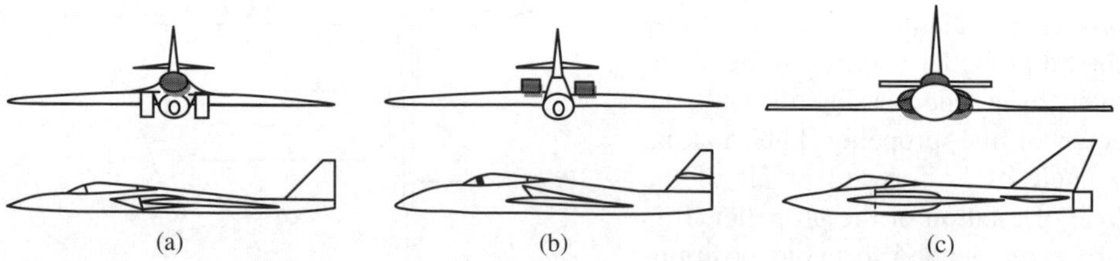

Figure 8.20 Fuselage-buried jet engine inlet location for split-type inlet: (a) Under wing; (b) Over wing; and (c) Beside fuselage

as the pitot-type inlet, may reduce the inlet curvature of a split-type inlet. The pitot-type inlet is not desirable in a few respects. On the one hand, a long inlet will cost weight and space. Furthermore, a long inlet will negatively influence the inlet efficiency. In a regular inlet, the inlet efficiency is about 0.96–0.98, but in a long inlet, the efficiency may be reduced to less than 0.9. A low-efficiency inlet will reduce the engine overall thrust, due to pressure loss. Nevertheless, the pitot inlet carries the lowest challenge to the aircraft structural designer.

The other two non-split inlets tend to have more desirable features. An inlet in the top or bottom of the fuselage stands somehow between a pitot inlet and a split inlet. Since the inlet length is short, both save cost and weight, so they provide higher efficiency. If ingestion of debris during take-off is avoided by some means, the inlet in the bottom of the fuselage (Figure 8.19(a)) is suitable for high-wing and low-wing aircraft. Moreover, this design has very low flow interference between inlet and wing. However, pronounced curvature in the inlet must be avoided to limit flow distortion and turbulence. The fighter aircraft F-16 Fighting Falcon has adopted such an inlet.

The debris ingestion problem is solved by using an inlet on top of the fuselage (Figure 8.19(b)). Furthermore, an inlet above the fuselage favors stealth requirements, since it is beyond reach of ground radars. However, to avoid fuselage boundary layer and wake ingestion at large angles of attack, the inlet opening has to be raised sufficiently above the fuselage. The unmanned aircraft Global hawk and the piloted aircraft Paragon Spirit have employed such an inlet configuration.

A split inlet requires a fairly long and curved inlet, which causes loss of inlet efficiency and extra weight. This also applies to the third engine of a three-engine jet aircraft, when the engine is installed (such as Boeing 727) in the plane of symmetry.

When split intakes (Figure 8.20) are employed, a sideslip angle in a turning flight will cause a dissimilar flow pattern, which may result in unstable flow. In extreme cases, the airflow may oscillate instead of entering the inlet. This is true especially for a split type beside the fuselage (close to the wing root) since the inlet duct must supply the required flow of air (kg/s) at different velocities, angles of attack, and sideslip angles. Furthermore, a split inlet beside the fuselage on either side forms a scoop, and produces additional drag. To keep the scoop drag low, the inlet must be carefully designed and faired. Moreover, the interference between fuselage boundary layer and this type of inlet may also create a new drag. The opening of the inlet must be located sufficiently ahead of the wing leading edge to minimize the negative effects of the inlet flow and wing flow.

In a fuselage-buried jet engine, if the designer decides to continue the wing spar without interruption, the wing thickness must be sufficient to allow for the inlet to go through the spar web. Supersonic fighter aircraft often tend to have a thin airfoil which does not allow for such installation. A solution in such circumstances is to lead the inlet either under or over the wing. The McDonnell Douglas (now Boeing) AV-8B Harrier II (Figure 4.4) has adopted a split inlet beside the fuselage.

A common challenge to the split-type inlet is the symmetry of the inlet flow in the presence of sideslip angle. In an aircraft turning maneuver with an extreme sideslip angle, the loss of sufficient airflow may cause the engine to shut down. So, at different yaw rate, the direction of the incoming air must not be excessive. This is also true for various angles of attack, particularly at a high angle of attack. The optimum design of an inlet will be determined by wind-tunnel experiments and a thorough analysis of all pros and cons.

The McDonnell Douglas (now Boeing) F-15 Eagle (Figure 4.6) with twin-engine propulsion system has utilized a split engine under the wing. The Lockheed F-117 Nighthawk (Figure 6.8) single-seat, twin-engine stealth ground attack aircraft has located both inlets over the wing to improve radar detectability. Although this design causes interference between the wing and inlet, the stealth requirement for this military aircraft has led to such a decision.

A precaution for the installation of a twin-jet engine is to account for the wing/fuselage interference with the inlet airflow. To align the inlet to the local airflow for the wing-mounted engine, the nacelle is recommended to be tilted nose inward about 2–3° and nose down about 2–4°. To align the inlet to the local airflow for the rear fuselage-mounted engine, the nacelle is recommended to be installed with a nose-up pitch of about 2–5° and a nose-outward angle of about 1–3°.

Two parameters which must be determined in the propulsion system design are the engine (in fact, inlet) ground clearance and engine span. The engine ground clearance is to avoid debris ingestion, and also to provide the inlet with the highest efficiency. For these objectives, there is a minimum height for the engine and inlet locations that must be determined. The inlet height is a function of engine maximum thrust/power, wing and landing gear configurations, inlet type, and safety considerations.

In a multi-engine configuration, the engine distance from the fuselage center line should also be determined. Each pair of engines must have the same distance from the fuselage center line to neutralize the thrust yawing moment. The distance between each two corresponding engines is referred to as the engine span. The engine span is a function of a couple of factors, including the inlet efficiency, wing aerodynamic considerations, yaw control in the event of asymmetric thrust, number of engines, and engine weight. As an engine is moved to the outer wing sections, the wing spar will experience a higher bending moment which in turn makes the wing heavier. Furthermore, as the engine span is increased, the rudder must be larger to insure aircraft directional control, when an engine is out. There is a minimum distance between an inlet and the fuselage, and also between two neighboring inlets, which provides undisturbed airflow. This distance is a function of the engine maximum thrust/power, wing features, and wheel track.

Reference [10] provides the statistics for engine span versus wing span for several jet transport aircraft. Based on this reference, the maximum engine spread is about 5% of wing span for two-engined aircraft and 24% of wing span (Airbus A-380 (Figure 1.8)) for four-engined aircraft from the outside tire edge. The optimum vertical distance between an inlet and the wing, as well as the distance between each pair of neighboring inlets, must be determined by wind-tunnel experiments or CFD software packages. Reference [10] also provides the statistics for engine ground clearance versus overhang from main gear.

Figure 8.21 demonstrates six aircraft with different propulsion systems and engine locations. The Lockheed SR-71 Blackbird (Figure 8.21(a)) was an advanced, long-range, Mach 3+ strategic reconnaissance aircraft, equipped with two Pratt & Whitney J58-1 afterburning turbojet engines, each generating 151 kN of thrust. Engines are installed on the wing with special inlets. In addition, two vertical tails are placed over the engines. The Bell Boeing MV-22 Osprey (Figure 8.21(b)) is a multi-mission, military, tilt-rotor aircraft with a vertical take-off and landing capability (VTOL). It is equipped with two Rolls-Royce Allison turboshaft engines of 6150 hp (4590 kW) each. The engines are installed at the wing tip and are allowed to rotate 90°.

Figure 8.21 Six aircraft with different propulsion systems and engine locations: (a) Lockheed SR-71 Blackbird; (b) Bell-Boeing MV-22B Osprey; (c) Lockheed Martin F-22A Raptor; (d) Aermacchi MB-339; (e) Lake LA-270 Turbo Renegade; (f) Saab 340B. Reproduced from permission of (b, c) Antony Osborne; (d, e) Jenny Coffey; (f) Anne Deus

The Lockheed Martin F-22 Raptor (Figure 8.21(c)) is a single-seat, twin-engine super-maneuverable fighter aircraft that employs stealth technology. The engines are Pratt & Whitney F119-PW-100 pitch thrust vectoring turbofans installed inside the rear fuselage with under-wing inlets. Each engine is capable of generating a dry thrust of 104 kN and more than 156 kN of thrust with afterburner. Due to the introduction of the advanced fighter F-35, only about 110 Raptors were manufactured and in the Air Force inventory. The Aermacchi MB-339 (Figure 8.21(d)) is an Italian military trainer and light attack aircraft (Figure 8.21(a)) with one Rolls-Royce Viper Mk. 632 turbojet engine which generates 4000 lb (17.8 kN) of thrust. The engine is installed inside the rear fuselage, while there are two inlets beside the fuselage in front of the wing leading edge.

The Lake Turbo 270 Renegade (Figure 8.21(e)) is a five-passenger amphibious utility aircraft which is equipped with one turbocharged 201-kW piston engine. The engine is located far above the fuselage, and the aircraft is assumed to be a pusher since the engine thrust line (prop) is behind the aircraft cg. The Saab 340 (Figure 8.21(f)) is a Swedish twin-engine turboprop transport aircraft. Each General Electric CT7-9B engine generates 1305 kW of power. The engines are placed over the wing, with propellers in front of the wing leading edge.

8.7 Propeller Sizing

The propeller design to determine parameters such as the blade airfoil section and twist angle is beyond the scope of this text. However, in order to determine variables such as engine location and landing gear height for a prop-driven aircraft, the propeller diameter must be known. Thus, a prop-driven aircraft designer needs to have a rough estimate of the propeller diameter. If the engine is selected to be prop-driven (e.g., piston prop, turboprop), the prop must be sized for each engine. To provide rapid initial analysis and trade studies, a ballpark estimate of the propeller diameter is presented in this section.

The propeller is a means to convert the engine power to engine thrust. The aerodynamic equations and principles that govern the performance of a wing are generally applied to a propeller. Hence, the propeller may be called a rotating wing. It simply creates the lift (i.e., thrust) with the cost of drag. For this reason, the prop efficiency could never reach 100%. In a cruising flight and with the optimum angle of twist (the best propeller pitch), the propeller efficiency (η_P) may be around 75–85%. With this in mind, a method to estimate the propeller diameter is presented.

The propulsion system of a prop-driven aircraft in a steady-state cruising flight with an airspeed of V_C and a prop efficiency of η_P will generate a thrust which is a function of engine power as follows:

$$T = \frac{P \cdot \eta_P}{V_C} \tag{8.2}$$

where P is the engine power. In contrast, a rotating propeller with an aerodynamic three-dimensional finite wing will produce a lift force in the direction of flight as follows:

$$L_P = \frac{1}{2}\rho V_{av}^2 S_P C_{L_P} \tag{8.3}$$

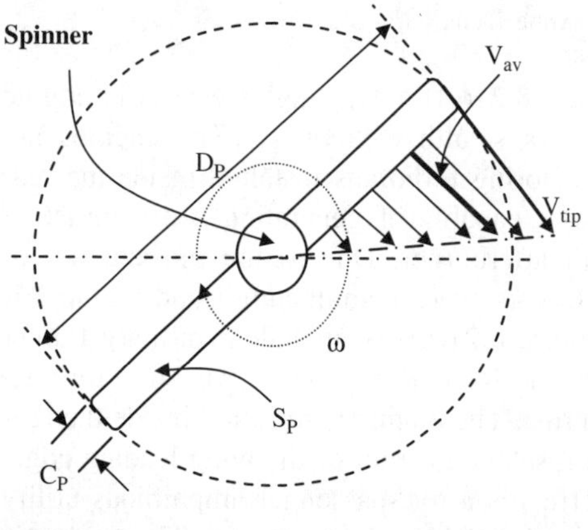

Figure 8.22 Propeller airspeed along prop span

where ρ is the air density at cruising altitude, S_P is the propeller planform area, and C_{L_P} is the propeller lift coefficient. The parameter V_{av} is the average airspeed at the propeller, which may be assumed to be about 70% of the propeller tip speed ($V_{tip_{cruise}}$). The reason is that the airspeed at the propeller center is zero and is increased as we move toward the tip (see Figure 8.22). Since the central part of the propeller does not contribute considerably to thrust generation, a spinner or cone-shaped fairing is often employed at the hub.

It is interesting to note that this propeller-generated lift (L_P) is also the engine thrust (T). Hence, we can write:

$$L_P = T \Rightarrow \frac{1}{2}\rho V_{av}^2 S_P C_{L_P} = \frac{P\eta_P}{V_C} \tag{8.4}$$

Therefore, the required propeller planform area to generate such a lift (i.e., thrust), when the engine power P is supplied, is:

$$S_P = \frac{2P\eta_P}{\rho V_{av}^2 C_{L_P} V_C} \tag{8.5}$$

In contrast, the geometry of a typical propeller may be modeled as a rectangle; so the planform area is:

$$S_P = D_P C_P \tag{8.6}$$

where C_P is the average propeller chord. Furthermore, the propeller aspect ratio (AR_P) is the ratio between the propeller span (i.e., diameter) and propeller chord:

$$AR_P = \frac{D_P}{C_P} \tag{8.7}$$

Inserting Equations (8.6) and (8.7) into Equation (8.5) would allow us to derive the following expression for the propeller diameter:

$$D_P = \sqrt{\frac{2P\eta_P AR_P}{\rho V_{av}^2 C_{L_P} V_C}} \tag{8.8}$$

The typical value for the propeller aspect ratio is between 7 and 15; and a typical propeller lift coefficient is between 0.2 and 0.4.

It must be noted that the propeller is moving forward (V_C) as it spins ($V_{tip_{static}}$) due to the aircraft cruising velocity (see Figure 8.23). Therefore, the propeller tip speed during a cruising flight ($V_{tip_{static}}$) is simply determined by employing the Pythagorean equation, which yields:

$$V_{tip_{cruise}} = \sqrt{V_{tip_{static}}^2 + V_C^2} \tag{8.9}$$

where the static prop tip speed is:

$$V_{tip_{static}} = \frac{D_P}{2}\omega \tag{8.10}$$

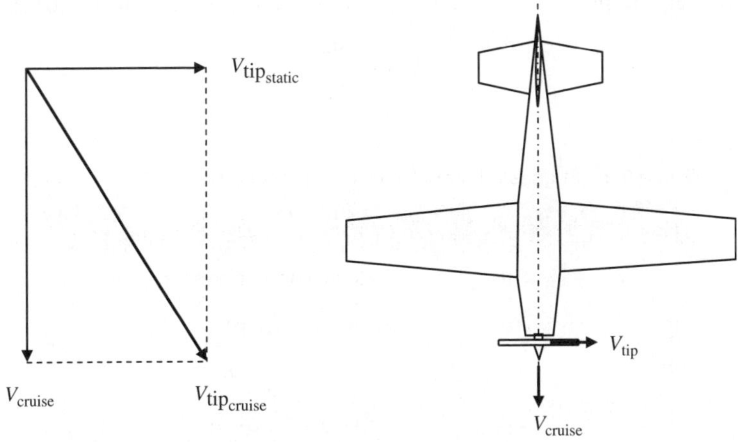

Figure 8.23 Propeller rotational speed and forward speed

The prop rotational speed is usually given in terms of revolutions per minute (n, rpm), so the angular speed of the prop (in rad/s) is:

$$\omega = \frac{2\pi \cdot n}{60} (\text{rad/s}) \tag{8.11}$$

The average airspeed for lift generation at the propeller is typically about 70% of the prop tip speed, so:

$$V_{\text{av}} = 0.7 V_{\text{tip}_{\text{cruise}}} \tag{8.12}$$

This is to account for the inactivity of the hub segment and the linear reduction of the propeller airspeed from tip to root. The prop tip speed must be less than a value such that it does not pass the speed of sound and the prop does not vibrate. To insure the prop has the optimum performance, it is recommended to keep the propeller tip speed at cruise below a certain value as suggested in Table 8.9. The recommended values are to prevent shock wave occurrence at the prop tip, as well as to avoid prop vibration and noise.

Equation (8.8) was derived for a two-blade propeller. For a prop with a higher number of blades, a correction factor (K_{np}) is applied as follows:

$$D_{\text{P}} = K_{\text{np}} \sqrt{\frac{2P \eta_{\text{P}} \text{AR}_{\text{P}}}{\rho V_{\text{av}}^2 C_{L_{\text{P}}} V_{\text{C}}}} \tag{8.13}$$

where K_{np} is 1 for a two-blade propeller, and 0.72 for a six-bladed and beyond. For any other number of blades, you can use linear interpolation to find the appropriate correction factor. Generally speaking, as the engine power increases, the number of blades must increase to reduce the prop diameter in order to avoid the tip speed exceeding the speed of sound.

Equation 8.13 provides an estimate of the propeller diameter as a function of engine power, aircraft speed, prop rotational speed, and prop aerodynamic characteristics. If the prop diameter seems too long, there are two solutions: (i) increase the number of blades or (ii) reduce the prop angular speed using a proper gearbox. The gearbox ratio (GR) is the ratio between the propeller rotational speed and the engine shaft rotational speed:

$$\text{GR} = \frac{n_{\text{P}}}{n_{\text{S}}} \tag{8.14}$$

Table 8.9 Suggested propeller cruise tip speed limit

No.	Tip speed limit (m/s)	Propeller type
1	310	Metal high-performance prop
2	270	Metal regular prop
3	250	Composite prop
4	210	Wooden prop
5	150	Plastic prop for RC model aircraft

Table 8.10 Features of several propellers

No.	Manufacturer	Designation	Number of blades	Diameter	Features	Aircraft
1	Smiths	R381	6	12 ft 6 in.	Constant speed	Saab 2000
2	Smiths	R391	6	13 ft 6 in.	Constant speed, composite	Lockheed C-130J, Alenia C-27J
3	Smiths	R408	6	13 ft 6 in.	Constant speed, composite	Dash 8 Q400, Y-8F600
4	Sensenich	W72CK-42	2	70 in.	Wood	Aeronca 7AC
5	Hamilton Sundstrand	14SF-5	4	13 ft	Aluminum and composite	ATR 72, Canadair 215T, Dash 8 Q300
6	Hamilton Sundstrand	568F-1	6	13 ft	Constant speed, composite	Casa c-295, Ilyushin Il-114
7	Hartzell	HC-E4A-2/E9612	4	77 in.	Metal, fully feathering	Beechcraft T-6
8	Hartzell	–	5	77 in.	Composite, constant speed, reversible pitch	Embraer EMB-314 Super Tucano
9	Hartzell	HC-D4N-ZA/09512A	4	77 in.	Metal, constant speed, fully feathering	Pilatuc PC-9 M
10	Hartzell	HC-B3TN-3	4	77 in.	Metal, constant speed, fully feathering	Beechcraft King Air 90
11	Hartzell	HC-E5N	5	77 in.	Aluminum, constant speed, fully feathering, reversible pitch	Piaggio P-180 Avanti II
12	Hartzell	HC-E4N-3Q	4	77 in.	Metal	Piper PA-46-500TP Meridian
13	Hartzell	BHC-J2YF-1BF	2	72 in.	Metal	Cirus SR20
14	Hartzell	HC-B5MP-3F	5	72 in.	Metal, constant speed, feathering, reversible pitch	Air tractor AT-802
15	Hartzell	HC-C2YK-1BF	2	74 in.	Metal, constant speed	Aermacchi SF-260EA

The single-piston engine light GA aircraft Cessna 172 (Figure 11.15) employs a gearbox ratio of 2 : 1 to reduce the engine shaft rotational speed of 4200 rpm down to 2100 rpm at the two-bladed propeller which has a diameter of 1.95 m. The two-seat lightplane Gobosh G-800 aircraft is equipped with a 79.9-hp Rotax 912 ULS flat-four piston engine and employs an Elprop three-blade, composite propeller. The business aircraft Piper PA-46-500TP Malibu Meridian uses one 1029-hp Pratt & Whitney PT56A-42A turboprop engine which employs a Hartzell HC-E4N-3Q four-bladed constant-speed reversible propeller. Each turboprop engine of the Lockheed Martin 382U/V Super Hercules (C-130J) four-engine turboprop military transport aircraft generates 4591 shaft hp and uses a six-bladed metal propeller. The gearbox reduces the shaft rotational speed of 3820 rpm to the lower speed needed by the 4.11 m propeller (1020 rpm).

In general, there are five common families of propeller: (i) fixed pitch, (ii) ground adjustable, (iii) in-flight adjustable, (iv) constant speed, and (v) folding. The last two are both examples of variable-pitch propellers. The earliest and most common type of propeller in GA aviation is the fixed-pitch propeller. A primitive definition of pitch is the distance an aircraft travels by one revolution of the propeller. A variable-pitch propeller may be utilized to improve the efficiency across a broad speed range. A constant-speed propeller is automatically controlled in pitch by a governor to maintain the engine at its rated revolutions per minute. An in-flight adjustable propeller has its pitch changed directly by the pilot as needed. The use of variable-pitch and constant-speed propellers greatly enhances the rate-of-climb for aircraft, compared with that for a fixed-pitch propeller. The folding propeller is a means to reduce drag for special applications such as motor gliders in engine-off flight modes.

Propellers must be carefully matched with the characteristics of the aircraft structure, engine, and reduction gearbox to which they are fitted. At best a mismatch could make the engine, and aircraft, incapable of delivering its designed performance, or create a situation where the engine cannot be opened up to full throttle. At worst, a mismatch could lead to torsional vibration or centrifugal force-induced propeller blade destruction.

The major propeller manufacturers are Hartzell, Ivoprop, Powrefin, Sensenich, Hamilton Sundstrand, and Whirl Wind in the USA; Avia, Kasparaero, VZLU, and Woodcomp in the Czech Republic; DCU, e-Props, EVRA, Halter, Ratier-Figeac, and Valex in France; Falter, Helix, Hoffmann, MT-Propeller, and Neuform in Germany; and Smiths in the UK. Table 8.10 shows the features of several propellers which are currently produced and employed.

Example 8.2

A light GA aircraft with a cruising speed of 130 knot at 15 000 ft employs a 180-hp piston engine. A regular two-blade metal prop is going to be used. Assume that the engine power is kept constant up to the cruising altitude by using a turbocharger.

1. Estimate the propeller diameter for this engine.
2. What would be the propeller rotational speed (in rpm) for this cruising flight?
3. The engine shaft rotational speed is 4600 rpm. What gearbox ratio must be employed?

Solution:
The air density at 15 000 ft is 0.592 kg/m^3. A propeller aspect ratio of 10, a prop lift coefficient of 0.3, and a prop efficiency of 0.75 are selected.

1. Based on Table 8.9, the prop tip speed must not exceed 270 m/s. Using Equations (8.12) and (8.13), we can obtain the prop diameter as follows:

$$D_P = K_{np}\sqrt{\frac{2P\eta_P AR_P}{\rho\left(0.7V_{tip_{cruise}}\right)^2 C_{L_P} V_C}}$$

$$= 1 \cdot \sqrt{\frac{2\cdot 180\cdot 745.7\cdot 0.75\cdot 10}{0.592\cdot(0.7\cdot 270)^2\cdot 0.3\cdot(130\cdot 0.514)}} \Rightarrow D_P = 2.178\,\text{m} \tag{8.13}$$

2. Propeller rotational speed:

$$V_{tip_{cruise}} = \sqrt{V_{tip_{static}}^2 + V_C^2} \Rightarrow V_{tip_{static}} = \sqrt{V_{tip_{cruise}}^2 - V_C^2} = \sqrt{270^2 - (130\cdot 0.514)^2} \tag{8.9}$$

$$\Rightarrow V_{tip_{static}} = 261.6\,\text{m/s}$$

$$V_{tip_{static}} = \frac{D_P}{2}\omega \Rightarrow \omega = \frac{2V_{tip_{static}}}{D_P} = \frac{2\cdot 261.6}{2.178} = 240.2\,\text{rad/s} \tag{8.10}$$

$$\omega = \frac{2\pi\cdot n}{60} \Rightarrow n = \frac{60\omega}{2\pi} = \frac{60\cdot 240.2}{2\cdot 3.14} = 2293.7\,\text{rpm} \tag{8.11}$$

3. Gearbox ratio:

$$GR = \frac{n_P}{n_S} = \frac{2293.7\,\text{rpm}}{4500\,\text{rpm}} = 0.51 \approx \frac{1}{2} \tag{8.14}$$

8.8 Engine Performance

Aircraft engines are very complex machines, and at the same time they are not efficient. They waste most of the energy released by the fuel during the combustion process. The overall efficiency of most aero-engines is around 20–30%. Engine performance is based on many factors, including altitude and aircraft speed. The best source of analysis for engine performance is the catalogs published by the engine manufacturers. The basic tools for modeling engine performance are presented briefly in this section.

8.8.1 Prop-Driven Engine

The propulsive efficiency, η_P, of a propulsion system is a measure of how effectively the engine power is used to push/pull the aircraft. Propulsive efficiency is the ratio of the aircraft required power (thrust (T) times aircraft velocity (V)) to the available power (P_{in}) out of the engine:

$$\eta_P = \frac{TV}{P_{in}} \tag{8.15}$$

This equation is valid for all types of prop-driven engine, such as piston, turboprop, solar-powered, and electric engines. As the aircraft climbs, the power and thrust of an air-breathing engine is decreased, since the available air is dropping. This is true for both jet and prop-driven aircraft. The rate of loss of power in terms of altitude depends on several parameters, including manufacturing technology and configuration. There is no unique expression for power modeling in terms of altitude; hence, we need to resort to empirical relationships as follows:

$$P_{max} = P_{max_{SL}} \left(\frac{\rho}{\rho_o}\right)^m \tag{8.16}$$

where P_{max} and ρ represent the maximum shaft power output and air density, respectively, at a given altitude and $P_{max_{SL}}$ and ρ_o are the corresponding values at sea level. The value of m changes as the technology advances. It is suggested to assume 0.9 for a piston engine and 1.2 for a turboprop engine.

8.8.2 Jet Engine

In a turbojet engine, Newton's second and third laws govern the relationship between forces and motion. The output thrust is obtained as:

$$T = \dot{m}\left(V_e - V_i\right) + A_e\left(P_e - P_a\right) \tag{8.17}$$

where $\dot{m}$ represents the air mass flow rate into the engine, V_e is the gas exit velocity from the nozzle, V_i is the velocity of the incoming air to the inlet, A_e is the cross-sectional area of the engine nozzle exit, P_e is the static pressure of the gas exiting the nozzle, and P_a is the ambient pressure at which the aircraft is flying. The incoming air to the inlet depends on the aircraft speed and configuration, but at ideal conditions, this speed is close to the aircraft speed. For a turbofan engine, Equation (4.17) must be modified by inclusion of a bypass ratio. It is left to the interested reader to further develop such a relationship. References [6, 7] may be consulted for more details. The variation of engine thrust with altitude is approximated by:

$$T_{max} = T_{max_{SL}} \left(\frac{\rho}{\rho_o}\right)^n \tag{8.18}$$

where n is taken to be 1 for both turbofan and turbojet engines. The variable T_{max} denotes the maximum engine thrust and $T_{max_{SL}}$ is the corresponding value at sea level. More accurate data may be extracted from manufacturers' catalogs.

8.9 Engine Selection

One of the last tasks of the propulsion system designer is to order the engine; either from an engine design team or a manufacturer (i.e., off-the-shelf engine). It is often more practical to select an engine from a manufacturer's list which is closest to the aircraft need. In situations where a very special engine is needed, such as in the reconnaissance aircraft SR-71 Blackbird, the engine must be internally designed. After preliminary calculations have been carried out, the thrust/power per engine will be determined. The next step is to select the best engine from a list of engines satisfying the design requirements.

Every engine manufacturer publishes engine charts and specifications that help the aircraft designer and performance engineer. Tables 8.11–8.14 present the most important specifications of some manufactured aero-engines. These tables, plus the criteria introduced in this section, may help the designer to make up his/her mind to make the final decision. The most suitable engine for a particular aircraft can only be selected by long and close collaboration between aircraft and engine designers. The performance of several similar engines may not be the same, but a case study will identify the most appropriate one.

The propulsion system design requires the dimensions of the engine, as well as the SFC, weight, noise level, maximum thrust/power, shaft rotational speed, maintainability, installation data, engine cost, and operating cost. Since there are some ways to play with the engine thrust/power, some variables such as engine power-to-weight ratio, engine thrust-to-weight ratio, engine power-to-volume ratio, and engine thrust-to-volume ratio are also employed to judge and compare various engines. Furthermore, engine charts and curves such as thrust/power versus altitude, thrust/power versus Mach number, SFC versus Mach number, propulsive efficiency versus Mach number, and SFC versus thrust/power are also subject to investigation.

Table 8.11 Primary specifications for several electric engines [8]

No.	Manufacturer	Designation	Length (mm)	Diameter (mm)	Mass (kg)	Maximum current (A)	Max power (kW)
1	Hacker	A20-26M EVO	28	28	0.042	12 A; 1130 rpm/V	0.150
2	Raiden	T30A	42.7	60	0.271	58	0.400
3	Applied Motion	M1500-232-7-000	190	100	5.7	9.5	1.5
4	Leopard	LBP4074	40	38	0.347	120 A; 2000 rpm/V	2.6
5	Yuneec	Power drive 10	–	160	4.54	180	10
6	Electroavia	GMPE 102 Devoluy	200	210	11.57	250	19.4
7	Electroavia	GMPE 201 Arambre	200	210	12	275	32
8	Yuneec	Power drive 40	–	240	17	285	40

Table 8.12 Primary specifications for several piston engines [8]

No.	Manufacturer	Designation	Arrangement	Number of cylinders	Cooling	Mass (kg)	Max power (hp)
1	Hirth	F33B	–	1	Air	13	24
2	Rotax	447 UL-1V	In-line	2	Air	26.8	39.6
3	BMW	R115ORS	Opposed	2	Air + oil	76.3	96.6
4	Subaru	EA81-140	Opposed	4	Liquid	100	140
5	Wilksch	WAM 160	In-line	4	Liquid	120	160
6	Textron-Lycoming	O-320-H	Opposed	4	Air	128	160
7	PZL	F 6A6350-C1	Opposed	6	Air	150	205
8	TCM	Tsio-360-RB	Opposed	6	Air	148.6	220
9	Textron-Lycoming	IO-540-C	Opposed	6	Air	170	250
10	TCM	IO-470-D	Opposed	6	Air	193.3	260
11	Bombardier	V300	Vee		Liquid	210	300
12	TCM	TSIOL-550-C	Opposed	6	Liquid	188.4	350
13	Textron-Lycoming	IO-270-A	Opposed	8	Air	258	400
14	VOKBM	M-9F	Radial	9	Air	214	420
15	Orenda	OE600 Turbo	Vee	8	Liquid	–	750
16	PZL	K-9	Radial	9	Air	580	1 170

8.10 Propulsion System Design Steps

In Sections 8.1–8.9, the propulsion system function, engine types, objectives, alternatives, design criteria, parameters, governing rules and equations, formulation, design requirements, as well as how to approach the primary parameters have been presented in detail. Table 8.1 presents a summary of the functions for the aircraft propulsion system. Furthermore, Figure 8.1 illustrates the design flowchart of the propulsion system. In this section, the propulsion system design procedure in terms of design steps is introduced. The data given to the propulsion system design team are: (i) total required power/thrust, (ii) aircraft mission, and (iii) aircraft configuration.

It must be noted that there is no unique solution to meet the customer and airworthiness requirements in designing a fuselage. Several propulsion system designs may satisfy the requirements, but each will have unique advantages and disadvantages. A main task of the propulsion system designer is to select one alternative from the variety of available

Table 8.13 Primary specifications for several turboprop engines [8]

No.	Manufacturer	Designation	Arrangement	Airflow (kg/s)	Length (mm)	Width (mm)	Mass (kg)	Max power (hp)
1	Innodyn	255TE	C	–	762	360	85.3	255
2	Rolls Royce	250-B17	6A + C	1.56	1143	483	88.4	420
3	Turbomeca	Arrius 2F	C	–	945	459	103	504
4	P&WC	PT6A-27	3A + C	3.08	1 575	483	149	680
5	Honeywell	TPE331-3	C + C	3.54	1 092	533	161	840
6	PZL	TWD-10B	6A + C	4.58	2 060	555	230	1 011
7	P&WC	PT6A-65b	4A + C	4.31	1 880	483	225	1 100
8	P&WC	PT6A-69	4A + C	–	1 930	483	259.5	1 600
9	GE	CT7-9	5A + C	5.2	2 438	737	365	1 940
10	P&WC	PW123C	C, C	–	2 143	635	450	2 150
11	Klimov	TV3-113VMA-SB2	12A	9	2 860	880	570	2 500
12	DEMC	WJ5E	10A	14.6	2 381	770	720	2 856
13	Rolls Royce	AE 2100C	14A	16.33	2 743	1151	715.8	3 600
14	Progress	AI-20M	10A	20.7	3 096	842	1 040	3 943
15	P&WC	PW150A	3A + C	–	2 423	767	690	5 071
16	EPI	TP400-D6	5A	26.31	3 500	924.5	1 795	11 000

A: axial stage, C: centrifugal stage, C, C: two stages on different shafts.

choices, such as engine type, engine location, number of engines, and engine installation. In such areas, a trade-off study (as introduced in Chapter 2) must be followed, which involves weighing each option with regard to design requirements and priorities.

Based on the systems engineering approach, the propulsion system detail design begins with identifying and defining design requirements and ends with optimization. The following lists the propulsion system design steps for a heavier-than-air craft:

1. Identify and list the propulsion system design requirements.
2. Determine the engine type.
3. Determine the number of engines.
4. Determine the engine locations.
5. Select an engine from the manufacturers' catalogs, or order an engine design team to design a new engine from scratch.
6. Design the propeller (if a prop-driven engine).

Table 8.14 Primary specifications for several turbofan and turbojet engines [8]

No.	Manufacturer	Designation	Arrangement	Airflow (kg/s)	BPR	Length (mm)	Diameter (mm)	Mass (kg)	Max thrust (kN)
1	GE Honda	HF120	1A + 2A + 2A	–	2.9	1118	538	181	9.12
2	Honeywell	TFE731-20	1F, 4A + C	66.2	3.1	1 547	716	406	15.57
3	Rolls Royce	Viper 680	8A	27.2	0	10 806	740	379	19.39
4	P&W	J-52-408	5A, 7A	64.9	0	3 020	814	1 052	49.8
5	Rolls Royce	Spey 512	5A, 12A	94.3	0.71	2 911	942	1 168	55.8
6	SNECMA	Atar 9K50	9A, a/b	73	0	5 944	1 020	1 582	70.6
7	Volvo/GE	RM12	3F, 7A, a/b	68	0.28	4 100	880	1 050	80.5
8	P&W	JT8D-219	1F + 6A, 7A	221	1.77	3 911	1 250	2 092	93.4
9	SNECMA	M53 P2	3F, 5A, a/b	86	0.35	5 070	1 055	1 500	95
10	CFM	CFM56-2B	1F + 3A, 9A	370	6	2 430	1 735	2 119	97.9
11	P&W	F-100-220P	3F, 10A, a/b	112.5	0.6	5 280	1 181	1 526	120.1
12	Saturn	AL-31FM	4F, 9A, a/b	112	0.57	4 950	1 277	1 488	122.6
13	Soyuz	R-79	5F, 6A, a/b	120	1	5 229	1 100	2 750	152
14	Rolls-Royce	RB211-524B	1F, 7A, 6A	671	4.4	3 106	2 180	4 452	222
15	P&W	JT9D-7R4H	1F + 4A, 11A	769	4.8	3 371	2 463	4 029	249
16	GE	GE90-76B	1F + 3A, 10A	1 361	9	5 182	3 404	7 559	340
17	Rolls-Royce	Trent 895	1F, 8A, 6A	1217	5.79	4 369	2 794	5981	425
18	GE	GE90-115B	1F + 4A, 9A	1641	8.9	7 290	3 442	8 761	511.6

BPR: bypass ratio, a/b: afterburner, A: axial, C: centrifugal, F: fan.

7. Design the inlet (if a jet engine).
8. Design the engine installation.
9. Check if the propulsion system design satisfies the design requirements.
10. If any design requirement is not met, return to the relevant design step and reselect/recalculate the corresponding parameter.
11. Optimize.

It must be emphasized again that the best design for a particular mission can only be arrived at by a long and close collaboration between the aircraft designer and propulsion system designer. Furthermore, the choice of the engine type and the design of the aircraft are so interrelated that it is sometimes very difficult to make the final decision.

8.11 Design Example

In this section, a major chapter example is presented to design a propulsion system for a transport aircraft. In order to avoid lengthening the chapter, only the major design parameters are determined.

Example 8.3

Design a propulsion system for a low-wing, T-tail, transport aircraft to carry eight passengers for a range of 4000 km with the following characteristics:

$$m_{\mathrm{TO}} = 7000\,\mathrm{kg},\ S = 29\,\mathrm{m}^2,\ C_{D_o} = 0.028,\ \mathrm{AR} = 8,\ e = 0.92$$

The aircraft must be capable of cruising with a maximum cruising speed of 320 KTAS at 20 000 ft altitude. For this problem, you need to discuss and determine the following:

1. engine thrust and engine power at cruise;
2. engine type;
3. number of engine(s);
4. engine(s) location;
5. engine selection;
6. prop diameter and number of blades (if prop-driven engine).

The propulsion system must be of low manufacturing cost, low operating cost, with high efficiency, and airworthiness requirements must be met. Then, sketch a front view and a top view of the aircraft to show the propulsion system installation. Assume any other parameters as needed.

Solution:

The solution is presented in six sections. The optimization is left to the interested reader.

1. **Design Requirements**. The following design requirements are identified and listed in order of importance: aircraft performance (maximum speed), engine

manufacturing cost, engine operating cost, flight safety, engine efficiency, maintainability, and manufacturability. Other general requirements (such as structural requirements, installation constraints, and integration) are important, but not considered at this moment. Other performance items such as ceiling, rate of climb, and take-off run are not given by the problem statement, so they are not discussed here.

2. **Engine type**. It is observed that the first three requirements are low manufacturing cost, low operating cost, and high efficiency. Since the vehicle is a transport aircraft carrying human passengers, the noise pollution cabin must be addressed and mitigated. Noise level can be controlled using dynamic vibration absorbers mounted throughout the cabin flight deck plus bagged glass fiber insulation. For these reasons, the prop-driven engine is the most suitable engine for this design problem. Due to the high speed and high altitude, only a turboprop engine will meet these required performance elements. According to Figure 8.7, piston-prop, electric, or solar-powered engines are not capable of meeting a maximum speed of 320 KTAS at 20 000 ft altitude. Although turbofan and turbojet engines are capable of this performance mission, they are costlier (both in manufacturing cost and in operating cost) compared with a turboprop engine.
3. **Number of Engines**. The aircraft is carrying eight passengers. The flight safety of passengers is of prime importance in a civil transport aircraft. To have greater safety, a multi-engine propulsion system is adopted. The more engines, the safer the flight. But as we increase the number of engines, the flight cost plus maintenance cost is increased. We start with two engines; if a suitable rate of safety can be achieved with these we stick with it, otherwise we go for a higher number of engines.

   Furthermore, the aircraft range is required to be 4000 km, so it may fly over the ocean such as in a flight from Los Angeles to Hawaii. Statistics clearly indicates that there have been and there will be unfavorable circumstances where an engine may become inoperative during a flight operation. The multi-engine propulsion system configuration is one of the best solutions to the OEI case issue. More details on the importance of flight safety regarding a multi-engine configuration are provided in Section 8.4.1.

   First we need to determine the required engine power for this mission. The air density at 20 000 ft is 0.653 kg/m^3. The cruise lift coefficient is:

$$C_{L_C} = \frac{2\,\mathrm{mg}}{\rho S \left(V_C\right)^2} = \frac{2 \cdot 7000 \cdot 9.81}{0.653 \cdot 29 \cdot (320 \cdot 0.514)^2} = 0.267 \qquad (5.1)$$

   Aircraft drag at cruise:

$$K = \frac{1}{\pi \cdot e \cdot \mathrm{AR}} = \frac{1}{3.14 \cdot 0.92 \cdot 8} \Rightarrow K = 0.043 \qquad (5.22)$$

$$C_D = C_{D_o} + K{C_L}^2 = 0.028 + 0.043 \cdot 0.267^2 = 0.031 \qquad (4.40)$$

$$D = \frac{1}{2}\rho V^2 S C_D = \frac{1}{2} \cdot 0.653 \cdot (320 \cdot 0.514)^2 \cdot 29 \cdot 0.031 \Rightarrow D = 7980.4\ \mathrm{N} \qquad (5.2)$$

Required engine thrust at cruise:

$$T = D = 7980.4\ \text{N} \tag{8.19}$$

Required engine power at 20 000 ft:

$$P_{20\,000} = \frac{TV_C}{\eta_P} = \frac{7980.4 \cdot (320 \cdot 0.514)}{0.75} \Rightarrow P_{20\,000} = 1\,751\,675\ \text{W}$$

$$= 1751.675\ \text{kW} = 2349\ \text{hp} \tag{8.15}$$

Required engine power at sea level:

$$P_{\max} = P_{\max_{SL}} \left(\frac{\rho}{\rho_o}\right)^{1.2} \Rightarrow 1751.6 = P_{\max_{SL}} \left(\frac{0.653}{1.225}\right)^{1.2} = P_{\max_{SL}}(0.47)$$

$$\Rightarrow P_{\max_{SL}} = \frac{1751.6}{0.47} \Rightarrow P_{\max_{SL}} = 3725.73\ \text{kW} = 4996.3\ \text{hp} \tag{8.16}$$

Referring to engine manufacturers' catalogs (e.g., Table 8.12), it is observed that there is no piston engine available in the market that delivers this much power even with three engines. This is another reason that we have chosen the turboprop engine. Moreover, there are very few turboprop engines (e.g., Table 8.13) that generate about 5000 hp; hence, this is another reason for the decision of a multi-engine configuration. There are a number of turboprop engines which deliver about 2500 hp, therefore two turboprop engines are selected for this aircraft.

4. **Engine locations**. For the case of two engines, both engines must be placed such that the locations satisfy the symmetry requirement. To satisfy this requirement, there are mainly two options. One option is to place an engine in the fuselage nose and another option is to place an engine in the rear fuselage, both along the fuselage center line. The one pusher and one tractor configuration is not a practical and viable alternative for a civil transport aircraft, since one engine obstructs the pilot view and the other engine interferes with the conventional tail. Furthermore, in a twin-engine configuration, positioning the prop-driven engine on the wing (with propeller in front of the wing) often results in the most attractive design from a structural and aerodynamic point of view. Based on these reasons and advantages, both engines are placed on the wing with propellers in front of the wing. Other advantages of locating twin engines on the wing are introduced in Section 8.5. The exact distance between each engine and the fuselage will be determined based on the prop clearance and aerodynamic interference considerations.
5. **Engine selection from manufacturers' catalogs**. Most turboprop engine manufacturers fabricate engines with a rated power of 2500 hp. Two turboprop engines from Pratt & Whitney are selected with the following features:

   designation PW 127; arrangement C, C; prop drive free turbine; length 2134 mm; width 600 mm; dry mass 481 kg; and TO rating 2750 shp

As observed, the engines deliver slightly higher shaft powers. We intentionally selected a more powerful engine for future design precautions and considerations.

6. **Propeller design**. The engine power at 20 000 ft is:

$$P_{\max} = P_{\max_{SL}} \left(\frac{\rho}{\rho_o}\right)^{1.2} = 2750 \left(\frac{0.653}{1.225}\right)^{1.2} = 1293\ \text{hp} = 964137\ \text{W} \tag{8.16}$$

A two-blade propeller with a lift coefficient of 0.3, a prop efficiency of 0.75, and a prop aspect ratio of 9 is adopted. Two regular metal propellers are selected. According to Table 8.9, the suggested tip speed is less than 270 m/s. Using Equations (8.12) and (8.13), we can obtain the prop diameter as follows:

$$D_P = K_{np} \sqrt{\frac{2P\eta_P \text{AR}_P}{\rho \left(0.7V_{\text{tip}_{\text{cruise}}}\right)^2 C_{L_P} V_C}}$$

$$= 1 \cdot \sqrt{\frac{2 \cdot 964137 \cdot 0.75 \cdot 9}{0.653 \cdot (0.7 \cdot 270)^2 \cdot 0.3 \cdot (320 \cdot 0.514)}} \Rightarrow D_P = 3.361\ \text{m} \tag{8.13}$$

So two propellers with a diameter of 3.361 m are needed.

Propeller rotational speed:

$$V_{\text{tip}_{\text{cruise}}} = \sqrt{V_{\text{tip}_{\text{static}}}^2 + V_C^2} \Rightarrow V_{\text{tip}_{\text{static}}} = \sqrt{V_{\text{tip}_{\text{cruise}}}^2 - V_C^2} = \sqrt{270^2 - (320 \cdot 0.514)^2} \tag{8.9}$$

$$\Rightarrow V_{\text{tip}_{\text{static}}} = 214\ \text{m/s}$$

$$V_{\text{tip}_{\text{static}}} = \frac{D_P}{2}\omega \Rightarrow \omega = \frac{2V_{\text{tip}_{\text{static}}}}{D_P} = \frac{2 \cdot 214}{3.361} = 127.35\ \text{rad/s} \tag{8.10}$$

$$\omega = \frac{2\pi \cdot n}{60} \Rightarrow n = \frac{60\omega}{2\pi} = \frac{60 \cdot 127.35}{2 \cdot 3.14} = 1216.1\ \text{rpm} \tag{8.11}$$

So a gearbox must reduce the engine shaft revolution to 1216.1 rpm.

7. **Engine Installation**. The aircraft weight and type place the aircraft under FAR 23 airworthiness regulations. FAR Part 23 Section 23.771 requires that each pilot compartment must be located with respect to the propellers so that no part of the pilot or the controls lies in the region between the plane of rotation of any inboard propeller and the surface generated by a line passing through the center of the propeller hub making an angle of 5° forward or aft of the plane of rotation of the propeller.

Thus, the engines are placed far ahead of the wing such that the propeller planes are ahead of the cabin. This provision will provide sufficient distance between the propellers and the wing to minimize the negative effect of the prop wake on the wing aerodynamics. Moreover, a clearance of about one-quarter of the diameter of the props between the prop tip and the fuselage is considered to minimize the interference between fuselage and propellers. Figure 8.24 illustrates more details on the engine installation. The wing inboard section of the structure is further reinforced to make it strong enough to carry the wing weight and the corresponding bending moments at the wing root.

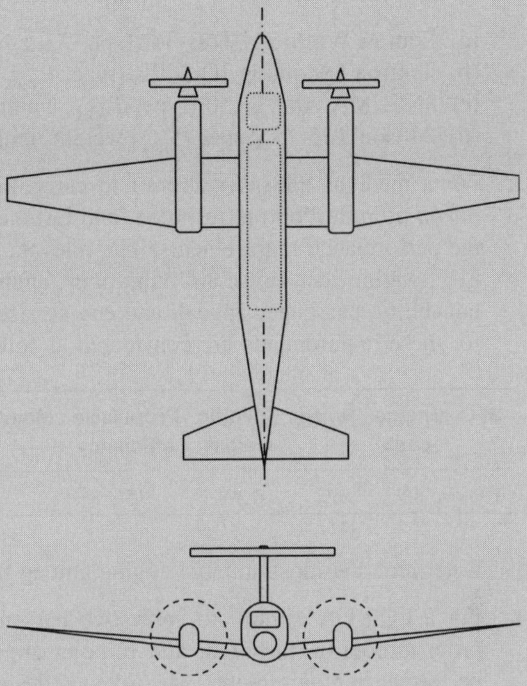

Figure 8.24 The engine installation for the aircraft in Example 8.3

Problems

1. Using a reference such as [8], identify and introduce one aircraft with the following engines:

   (a) human-powered
   (b) electric (battery) engine
   (c) solar-powered
   (d) piston prop
   (e) turbojet
   (f) turbofan
   (g) turboprop
   (h) turboshaft
   (i) rocket engine (non-air-breathing).

   For each aircraft, provide the name of the aircraft, type of the aircraft, and its picture or three-view.

2. Using a reference such as [8], introduce features of the following aero-engines:
   (a) Pratt & Whitney PT6B-36 (type, T_{max}, weight, length, air flow)
   (b) Textron Lycoming IO-540B (type, P_{max}, weight, cooling, capacity)
   (c) SNECMA Atar 9K50 (type, T_{max}, weight, length, air flow)
   (d) Allison T56-15 (type, P_{max}, weight, length, air flow).

3. For a medium transport aircraft to carry 12 passengers, the four engine types of piston-prop, turboprop, turbofan, and turbojet are found to be capable of satisfying the performance requirements. The relevant design requirements are selected to be SFC, engine cost, noise and vibrations, engine weight, propulsive efficiency, maintainability, passenger appeal, and engine size and dimensions. The design priorities for these requirements are considered as follows:

SFC	Engine cost	Noise	Specific weight	Propulsive efficiency	Maintainability	Passenger appeal	Engine size	Total
19%	4%	6%	12%	15%	13%	22%	9%	100%

Determine the most suitable engine among these four alternatives.

4. For a light GA aircraft to carry two passengers, the four engine types of piston-prop, turboprop, turbofan, and turbojet are found to be capable of satisfying the performance requirements. The relevant design requirements are selected to be SFC, engine cost, noise and vibrations, engine weight, propulsive efficiency, maintainability, passenger appeal, and engine size and dimensions. The design priorities for these requirements are considered as follows:

SFC	Engine cost	Noise	Specific weight	Propulsive efficiency	Maintainability	Occupant appeal	Engine size	Total
12%	7%	20%	2%	12%	8%	30%	9%	100%

Determine the most suitable engine among these four alternatives.

5. For a model aircraft to carry 1 kg of payload, the three engine types of piston-prop, turbojet, and electric are found to be capable of satisfying the performance requirements. The relevant design requirements are selected to be operating cost, engine cost, noise and vibrations, engine weight, propulsive efficiency, maintainability, customer appeal, and engine size and dimensions. The design priorities for these requirements are considered as follows:

Operating cost	Engine cost	Noise	Specific weight	Propulsive efficiency	Maintainability	Customer appeal	Engine size	Total
20%	25%	2%	2%	14%	16%	15%	6%	100%

Determine the most suitable engine among these three alternatives.

6. For a motor-glider aircraft to carry one pilot, the six engine types of piston-prop, turboprop, turbofan, turbojet, electric, and solar-powered are found to be capable of satisfying the performance requirements. The relevant design requirements are selected to be SFC, engine cost, noise and vibrations, engine weight, propulsive efficiency, maintainability, pilot appeal, and engine size and dimensions. The design priorities for these requirements are considered as follows:

Operating cost	Engine cost	Noise	Specific weight	Propulsive efficiency	Maintainability	Pilot appeal	Engine size	Total
4%	7%	18%	20%	7%	8%	17%	19%	100%

Determine the most suitable engine among these six alternatives.

7. A light GA aircraft with a cruising speed of 150 knot at 18 000 ft employs a 210-hp piston engine. A regular two-blade wooden prop is going to be used. Assume that the engine power is kept constant up to the cruising altitude by using a turbocharger.

(a) Estimate the propeller diameter for this engine.
(b) What would be the propeller rotational speed (in rpm) for this cruising flight?
(c) The engine shaft rotational speed is 4400 rpm. What gearbox ratio must be employed?

8. A medium transport aircraft with a cruising speed of 300 knot at 28 000 ft employs two 2100-hp turboprop engines. A high-performance three-blade composite prop is going to be used for each engine.

(a) Estimate the propeller diameter for each engine.
(b) What would be the propeller rotational speed (in rpm) for this cruising flight?
(c) The engine shaft rotational speed is 8000 rpm. What gearbox ratio must be employed?

9. A large transport aircraft with a cruising speed of 360 knot at 28 000 ft employs four 4200-hp turboprop engines. A high-performance six-blade metal prop is going to be used for each engine. Assume $C_{L_P} = 0.25$, $AR_P = 12$.

(a) Estimate the propeller diameter for each engine.
(b) What would be the propeller rotational speed (in rpm) for this cruising flight?
(c) The engine shaft rotational speed is 10 000 rpm. What gearbox ratio must be employed?

10. A small RC model aircraft with a cruising speed of 40 knot at 3000 ft employs a 2-hp electric engine. A plastic two-blade prop is going to be used for each engine.

(a) Estimate the propeller diameter for this engine.
(b) What would be the propeller rotational speed (in rpm) for this cruising flight?

11. A business transport aircraft with a cruising speed of 300 knot at 26 000 ft employs two 1200-hp turboprop engines. A regular four-blade composite prop is going to be used for each engine. Assume $C_{L_P} = 0.35$, $AR_P = 9$.

(a) Estimate the propeller diameter for this engine.
(b) What would be the propeller rotational speed (in rpm) for this cruising flight?
(c) The engine shaft rotational speed is 5200 rpm. What gearbox ratio must be employed?

12. A very large cargo aircraft with a cruising speed of 400 knot at 31 000 ft employs four 13 000-hp turboprop engines. A high-performance eight-blade metal prop is going to be used for each engine. Assume $C_{L_P} = 0.28$, $AR_P = 8$.
(a) Estimate the propeller diameter for this engine.
(b) What would be the propeller rotational speed (in rpm) for this cruising flight?
(c) The engine shaft rotational speed is 9000 rpm. What gearbox ratio must be employed?

13. Design a propulsion system for a low-wing, conventional-tail, six-seat GA aircraft for a range of 2000 km with the following characteristics:

$$m_{TO} = 1700\,\text{kg},\ S = 17\,\text{m}^2,\ C_{D_o} = 0.026,\ AR = 9,\ e = 0.88$$

The aircraft must be capable of cruising with a maximum cruising speed of 190 KTAS at 15 000 ft altitude. For this problem, you need to discuss and determine the following: (i) engine thrust/power at cruise, (ii) engine type, (iii) number of engine(s), (iv) engine(s) location, (v) engine selection, and (vi) prop diameter and number of blades (if prop-driven engine). The propulsion system must be of low manufacturing cost, low operating cost, with high efficiency, and airworthiness requirements must be met. Then, sketch a front view and a top view of the aircraft to show the propulsion system installation. Assume any other parameters as needed.

14. Design a propulsion system for a high-wing, T-tail, 12-seat utility aircraft for a range of 2500 km with the following characteristics:

$$m_{TO} = 5000\,\text{kg},\ S = 22\,\text{m}^2,\ C_{D_o} = 0.022,\ AR = 7.5,\ e = 0.9$$

The aircraft must be capable of cruising with a maximum cruising speed of 220 KTAS at 10 000 ft altitude. For this problem, you need to discuss and determine the following: (i) engine thrust/power at cruise, (ii) engine type, (iii) number of engine(s), (iv) engine(s) location, (v) engine selection, and (vi) prop diameter and number of blades (if prop-driven engine). The propulsion system must be of low manufacturing cost, low operating cost, with high efficiency, and airworthiness requirements must be met. Then, sketch a front view and a top view of the aircraft to show the propulsion system installation. Assume any other parameters as needed.

15. Design a propulsion system for a low-wing, conventional-tail, single-seat ultralight aircraft for a range of 1000 km with the following characteristics:

$$m_{TO} = 300\,\text{kg},\ S = 7\,\text{m}^2,\ C_{D_o} = 0.024,\ AR = 7.5,\ e = 0.9$$

The aircraft must be capable of cruising with a maximum cruising speed of 110 KTAS at sea-level altitude. For this problem, you need to discuss and determine the following: (i) engine thrust/power at cruise, (ii) engine type, (iii) number of engine(s), (iv) engine(s) location, (v) engine selection, and (vi) prop diameter and number of blades (if prop-driven engine). The propulsion system must be of low manufacturing cost, low operating cost, with high efficiency, and airworthiness requirements must be met. Then, sketch a front view and a top view of the aircraft to show the propulsion system installation. Assume any other parameters as needed.

16. Design a propulsion system for a low-wing, cruciform-tail, 19-seat business aircraft for a range of 8000 km with the following characteristics:

$$m_{TO} = 22\,000\,\text{kg},\ S = 51\,\text{m}^2,\ C_{D_o} = 0.021,\ \text{AR} = 10.4,\ e = 0.93$$

The aircraft must be capable of cruising with a maximum cruising speed of 350 KTAS at 25 000 ft altitude. For this problem, you need to discuss and determine the following: (i) engine thrust/power at cruise, (ii) engine type, (iii) number of engine(s), (iv) engine(s) location, (v) engine selection, and (vi) prop diameter and number of blades (if prop-driven engine). The propulsion system must have low operating cost, be comfortable, have high efficiency, and airworthiness requirements must be met. Then, sketch a front view and a top view of the aircraft to show the propulsion system installation. Assume any other parameters as needed.

17. Design a propulsion system for a high-wing, T-tail, cargo aircraft to carry 40 000 kg of payload and 12 flight crew and staff members for a range of 9000 km with the following characteristics:

$$m_{TO} = 240\,000\,\text{kg},\ S = 355\,\text{m}^2,\ C_{D_o} = 0.025,\ \text{AR} = 9.3,\ e = 0.9$$

The aircraft must be capable of cruising with a maximum cruising speed of 480 KTAS at 30 000 ft altitude. For this problem, you need to discuss and determine the following: (i) engine thrust/power at cruise, (ii) engine type, (iii) number of engine(s), (iv) engine(s) location, (v) engine selection, and (vi) prop diameter and number of blades (if prop-driven engine). The propulsion system must have low operating cost and high efficiency, and airworthiness requirements must be met. Then, sketch a front view and a top view of the aircraft to show the propulsion system installation. Assume any other parameters as needed.

18. Design a propulsion system for a low-wing, conventional airliner to carry 160 passengers for a range of 5000 km with the following characteristics:

$$m_{TO} = 74\,000\,\text{kg},\ S = 125\,\text{m}^2,\ C_{D_o} = 0.018,\ \text{AR} = 10,\ e = 0.93$$

The aircraft must be capable of cruising with a maximum cruising speed of 350 KTAS at 35 000 ft altitude. For this problem, you need to discuss and determine the following: (i) engine thrust/power at cruise, (ii) engine type, (iii) number of engine(s), (iv) engine(s) location, (v) engine selection, and (vi) inlet considerations. The propulsion system must have low operating cost and high efficiency, be comfortable, and airworthiness requirements must be met. Then, sketch a front view and a top view of the aircraft to show the propulsion system installation. Assume any other parameters as needed.

19. Design a propulsion system for a single-seat supersonic fighter aircraft for a combat radius of 700 km with the following characteristics:

$$m_{TO} = 15\,000\,\text{kg},\ S = 48\,\text{m}^2,\ C_{D_o} = 0.017,\ \text{AR} = 3.4,\ e = 0.83$$

The aircraft must be capable of cruising with a maximum speed of Mach 2 at 50 000 ft altitude. For this problem, you need to discuss and determine the following: (i) engine thrust/power at cruise, (ii) engine type, (iii) number of engine(s), (iv) engine(s) location, (v) engine selection, and (vi) inlet considerations. The propulsion system must have low observation and make the aircraft agile. Then, sketch a front view and a top view of the aircraft to show the propulsion system installation. Assume any other parameters as needed.

20. Design a propulsion system for a high-wing, four-seat amphibian aircraft for a range of 2700 km with the following characteristics:

$$m_{TO} = 2400\,\text{kg},\ S = 18\,\text{m}^2,\ C_{D_o} = 0.022,\ \text{AR} = 11,\ e = 0.83$$

The aircraft must be capable of cruising with a maximum speed of 160 KTAS at 15 000 ft altitude. For this problem, you need to discuss and determine the following: (i) engine thrust/power at cruise, (ii) engine type, (iii) number of engine(s), (iv) engine(s) location, (v) engine selection, and (vi) prop diameter and number of blades (if prop-driven engine); inlet considerations (if jet engine). The propulsion system must have low operating cost and high efficiency, be comfortable, and airworthiness requirements must be met. Then, sketch a front view and a top view of the aircraft to show the propulsion system installation. Assume any other parameters as needed.

21. Design a propulsion system for a low-wing, conventional-tail airliner to carry 1000 passengers for a range of 10 000 km with the following characteristics:

$$m_{TO} = 750\,000\,\text{kg},\ S = 930\,\text{m}^2,\ C_{D_o} = 0.019,\ \text{AR} = 8.2,\ e = 0.94$$

The aircraft must be capable of cruising with a maximum cruising speed of Mach 0.93 at 40 000 ft altitude. For this problem, you need to discuss and determine the following: (i) engine thrust/power at cruise, (ii) engine type, (iii) number of engine(s), (iv) engine(s) location, (v) engine selection, and (vi) inlet considerations. The propulsion system must have low operating cost and high efficiency, be comfortable, and airworthiness requirements must be met. Then, sketch a front view and a top view of the aircraft to show the propulsion system installation. Assume any other parameters as needed.

22. Design a propulsion system for a high-wing, conventional-tail model remote-controlled aircraft to carry 2 kg of payload for an endurance of 45 minutes with the following characteristics:

$$m_{TO} = 10\,\text{kg},\ S = 0.6\,\text{m}^2,\ C_{D_o} = 0.029,\ \text{AR} = 6.5,\ e = 0.88$$

The aircraft must be capable of cruising with a maximum cruising speed of 35 knot at sea-level altitude. For this problem, you need to discuss and determine the following: (i) engine thrust/power at cruise, (ii) engine type, (iii) number of engine(s), (iv) engine(s) location, (v) engine selection, and (vi) prop diameter and number of blades (if prop-driven engine); inlet considerations (if jet engine). The propulsion system must have low operating cost, low manufacturing cost, and high efficiency. Then, sketch a front view and a top view of the aircraft to show the propulsion system installation. Assume any other parameters as needed.

23. Design a propulsion system for a high-wing, conventional-tail reconnaissance unmanned aircraft to carry 300 kg of payload for an endurance of 150 hours and a range of 100 000 km with the following characteristics:

$$m_{\mathrm{TO}} = 800\,\mathrm{kg},\ S = 12\,\mathrm{m}^2,\ C_{D_o} = 0.026,\ \mathrm{AR} = 13.6,\ e = 0.94$$

The aircraft must be capable of cruising with a maximum cruising speed of 200 knot at 70 000 ft altitude. For this problem, you need to discuss and determine the following: (i) engine thrust/power at cruise, (ii) engine type, (iii) number of engine(s), (iv) engine(s) location, (v) engine selection, and (vi) prop diameter and number of blades (if prop-driven engine); inlet considerations (if jet engine). The propulsion system must have low operating cost, low manufacturing cost, and high efficiency. Then, sketch a front view and a top view of the aircraft to show the propulsion system installation. Assume any other parameters as needed.

24. Design a propulsion system for a high-wing, T-tail, single-seat motor-glider for a range of 700 km with the following characteristics:

$$m_{\mathrm{TO}} = 640\,\mathrm{kg},\ S = 13.2\,\mathrm{m}^2,\ C_{D_o} = 0.015,\ \mathrm{AR} = 32,\ e = 0.96$$

The aircraft must be capable of cruising with a maximum cruising speed of 90 knot at 15 000 ft altitude. For this problem, you need to discuss and determine the following: (i) engine thrust/power at cruise, (ii) engine type, (iii) number of engine(s), (iv) engine(s) location, (v) engine selection, and (vi) prop diameter and number of blades (if prop-driven engine); inlet considerations (if jet engine). The propulsion system must have low operating cost, low manufacturing cost, and high efficiency. Then, sketch a front view and a top view of the aircraft to show the propulsion system installation. Assume any other parameters as needed.

25. Design a propulsion system for a high-wing, single-seat close support military aircraft for a range of 3200 km with the following characteristics:

$$m_{\mathrm{TO}} = 15\,000\,\mathrm{kg},\ S = 24\,\mathrm{m}^2,\ C_{D_o} = 0.022,\ \mathrm{AR} = 5,\ e = 0.87$$

The aircraft must be capable of a vertical take-off and landing at 5000 ft altitude and cruising with a speed of 500 knot at 20 000 ft altitude. For this problem, you need to discuss and determine the following: (i) engine thrust/power at cruise, (ii) engine type, (iii) number of engine(s), (iv) engine(s) location, (v) engine selection, and (vi) prop diameter and number of blades (if prop-driven engine); inlet considerations (if jet engine). The propulsion system must have low manufacturing cost and satisfy the VTOL requirement. Then, sketch a front view and a top view of the aircraft to show the propulsion system installation. Assume any other parameters as needed.

References

[1] Mattingly, J., Heiser, W., and Pratt, D. (2003) *Aircraft Engine Design*, 2nd edn, AIAA.
[2] Stone, R. (1999) *Introduction to Internal Combustion Engines*, 3rd edn, Society of Automotive Engineers Inc.
[3] Kroes, M.J. and Wild, T.W. (1995) *Aircraft Powerplants*, 7th edn, Glencoe.
[4] Federal Aviation Regulations, FAR 23 and FAR 25 (2011), Federal Aviation Administration, US Department of Transportation, www.faa.org.
[5] Taylor, C.F. (1985) *Internal Combustion Engine in Theory and Practice*, 2nd edn, MIT Press.
[6] Farokhi, S. (2008) *Aircraft Propulsion*, John Wiley & Sons, Inc.
[7] Mattingly, J. (2006) *Elements of Propulsion: Gas Turbines and Rockets*, AIAA.
[8] Jackson, P., Munson K., Peacock, L. *Jane's All the World's Aircraft*, Jane's Information Group, various years, 1996 to 2011.
[9] www.wikipedia.org.
[10] International Industry Working Group (2007) Commercial Aircraft Design Characteristics – Trends and Growth Projections, 5th edn.

9

Landing Gear Design

9.1 Introduction

Another aircraft major component that needs to be designed is the landing gear (undercarriage). The landing gear is the structure that supports an aircraft on the ground and allows it to taxi, take off, and land. In fact, landing gear design tends to have several overlaps with aircraft structural design. In this book, the structural design aspects of landing gear are not addressed, but those design parameters that strongly impact the aircraft configuration design and aircraft aerodynamics will be discussed. In addition, some aspects of landing gear – such as shock absorber, retraction mechanism, and brakes – are assumed to be non-aeronautical issues and may be determined by a mechanical engineer. Thus, these purely mechanical parameters will not be considered in this chapter either. In general, the following are the landing gear parameters to be determined here:

1. type (e.g., nose gear (tricycle), tail gear, bicycle);
2. fixed (faired, or un-faired), or retractable, partially retractable;
3. height;
4. wheel base;
5. wheel track;
6. distance between main gear and aircraft center of gravity (cg);
7. strut diameter;
8. tire sizing (diameter, width);
9. landing gear compartment if retracted;
10. load on each strut.

The landing gear usually includes wheels, but some aircraft are equipped with skis for snow or floats for water. In case of a vertical take-off and landing aircraft such as a helicopter, wheels may be replaced with skids. Figure 9.1 illustrates landing gear primary parameters. The descriptions of primary parameters are as follows. Landing gear height is the distance between the lowest point of the landing gear (i.e., bottom of the tire) and the attachment point to the aircraft. Since the landing gear may be attached to the fuselage or to the wing, the term *height* has a different meaning. Furthermore, the landing gear

Aircraft Design: A Systems Engineering Approach, First Edition. Mohammad H. Sadraey.

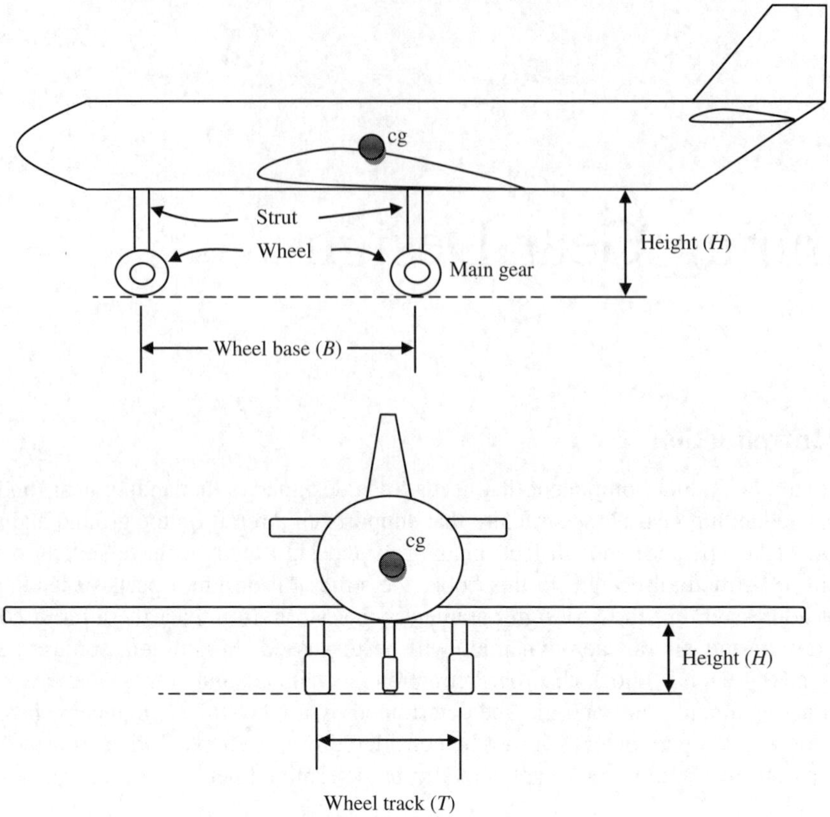

Figure 9.1 Landing gear primary parameters

height is a function of the shock absorber and the landing gear deflection. The height is usually measured when the aircraft is on the ground; it has maximum take-off weight and the landing gear has maximum deflection (i.e., lowest height).

Thus, the landing gear when it has the maximum extension is still at height, but is less important in applications. The distance between the lowest point of the landing gear (i.e., ground) and the aircraft cg is also of significance and will be employed during calculations. The wheel base is the distance between the main gear and any other gear (from a side view). The landing gear is divided into two sections: (i) main gear or main wheel[1] and (ii) secondary gear or secondary wheel. The main gear is the gear which is the closest to the aircraft cg. During landing operation, the main wheel touches first the point of contact with the ground. Furthermore, during take-off, the main wheel leaves the ground last. In contrast, the main gear carries a great portion of the aircraft load on the ground. The wheel track is the distance between two main gears (left and right) from a front view. If a gear is expected to carry a high load, it may have more than one wheel. In general, the landing gear weight is about 3–5% of the aircraft take-off weight.

[1] The term "wheel" is often used to mean the entire wheel/brake/tire assembly.

For instance, in the case of a Boeing 747 (Figures 3.7, 3.12, and 9.4), the landing gear assembly weighs about 16 000 lb.

This chapter is organized as follows. Section 9.2 addresses the landing gear functional analysis and design requirements. The landing gear configuration and its selection process are examined in Section 9.3. In Section 9.4, the decision on fixed, retractable, or separable landing gear is discussed. Section 9.5 deals with landing gear geometry, including wheel height, wheel base, and wheel track. In this section, a number of significant design requirements which influence the determination of the landing gear parameters (e.g., aircraft general ground clearance requirement, and take-off rotation clearance requirement) are examined. Section 9.6 deals with the landing gear and aircraft center of gravity and three design requirements (tipback, tipforward angles and take-off rotation requirements) are introduced. Landing gear mechanical subsystems/parameters including tire sizing, shock absorber, strut sizing, steering, and retraction subsystems are presented in Section 9.7. The landing gear design steps and procedure are introduced in Section 9.8. Finally, a fully solved design example is presented in Section 9.9.

9.2 Functional Analysis and Design Requirements

In terms of design procedure, the landing gear is the last aircraft major component which is designed. In other words, all the major components (such as wing, tail, fuselage, and propulsion system) must be designed prior to the design of the landing gear. Furthermore, the aircraft most aft cg and most forward cg must be known for landing gear design. In some instances, the landing gear design may drive the aircraft designer to change the aircraft configuration to satisfy the landing gear design requirements.

The primary functions of the landing gear are as follows:

1. to keep the aircraft stable on the ground and during loading, unloading, and taxi;
2. to allow the aircraft to move freely and maneuver during taxiing;
3. to provide a safe distance between other aircraft components such as the wing and fuselage while the aircraft is on the ground to prevent any damage by the ground contact;
4. to absorb the landing shocks during landing operations;
5. to facilitate take-off by allowing aircraft acceleration and rotation with the lowest friction.

In order to allow for a landing gear to function effectively, the following design requirements are established:

1. ground clearance requirement;
2. steering requirement;
3. take-off rotation requirement;
4. tipback prevention requirement;
5. overturn prevention requirement;
6. touch-down requirement;
7. landing requirement;
8. static and dynamic load requirement;
9. aircraft structural integrity;

10. ground lateral stability;
11. low cost;
12. low weight;
13. maintainability;
14. manufacturability.

Table 9.1 shows more details of the design requirements, plus the relationship between the requirements and landing gear parameters. Technical aspects of these requirements are described in Sections 9.5 and 9.6. In the next sections, techniques to determine landing gear parameters to satisfy all requirements will be presented.

Table 9.1 Relationship between landing gear design requirements and landing gear parameters

No.	Requirements and constraints	Requirement	Parameter affected
1	Ground clearance	Wing, engine, fuselage, prop clearance must be reasonable	Height
2	Controllability (steering)	Load on nose wheel must be limited	Wheel base, X_n to X_{cg}
3	Take-off rotation	Aircraft must be able to rotate about the main gear with a desired angular rate	Height, X_m to X_{cg}
4	Take-off rotation clearance	Aft fuselage and tail during take-off rotation must not have a strike	Height, wheel base
5	Tipback/forward prevention	Prevent tipback on its tail during take-off, prevent nose hit during loading	Height
6	Overturn prevention	Lateral angle must be such as to prevent overturn when taxied around sharp corner	Wheel track
7	Touch-down	Shock absorber must absorb and mitigate dynamic loads	Shock absorber, tire
8	Landing	Landing speed must be brought to zero before end of runway	Brake
9	Static and dynamic loading	Tires and struts must be able to function in static and dynamic loading	Strut
10	Structural integrity	The wing structural deflection at the center line on the ground due to the aircraft weight must be minimal	Wheel track
11	Ground lateral stability	The aircraft should not roll over due to a cross-wind	Wheel track, height

While the aircraft landing gear is a crucial component for take-off and landing, it is a dead weight during airborne flight operations. For this reason, it is recommended to retract the landing gear inside the aircraft to reduce aircraft drag and improve aircraft performance. Figure 9.2 illustrates a landing gear design flowchart including design feedback. As the flowchart shows, the landing gear design is an iterative process and the number of iterations depends on the nature of the design requirements as well as the designer's skills. Furthermore, the design of mechanical subsystems and parameters is grouped into one box and should be performed by the mechanical design group. We initiate the landing gear design by defining the landing gear design requirements and the process ends

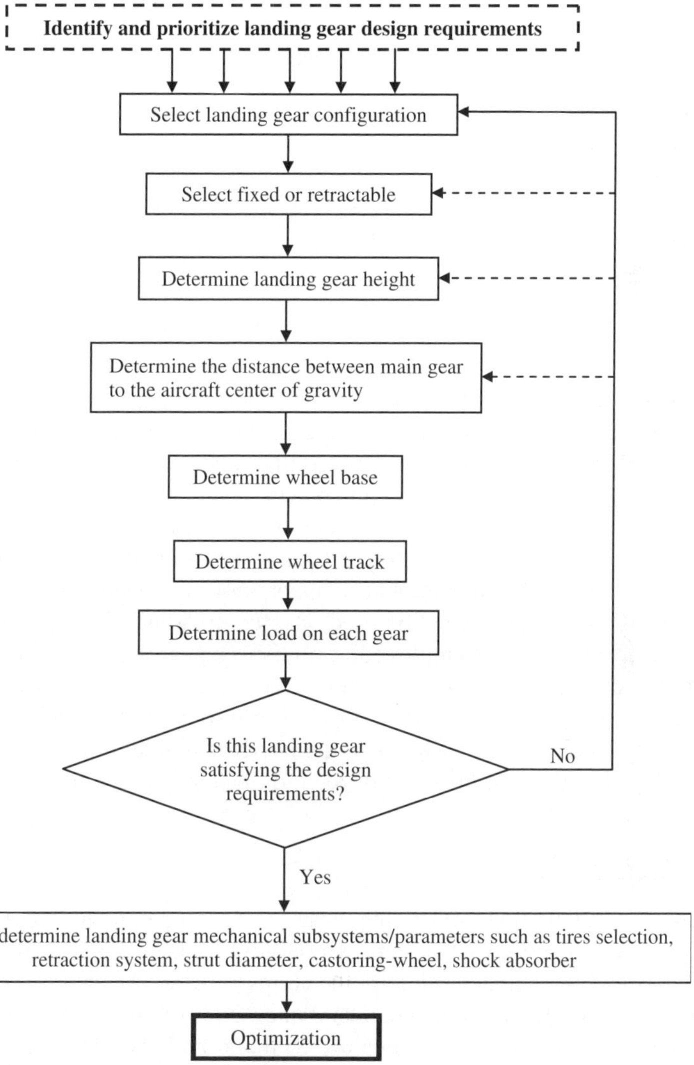

Figure 9.2 Landing gear design flowchart

with optimization. The details of the design of such items are beyond the scope of this textbook, and the reader is referred to other references such as [1].

9.3 Landing Gear Configuration

The first job of an aircraft designer in the landing gear design process is to select the landing gear configuration. Landing gear functions may be performed through the application of various landing gear types and configurations. Landing gear design requirements are part of the aircraft general design requirements, including cost, aircraft performance, aircraft stability, aircraft control, maintainability, producibility, and operational considerations. In general, there are 10 configurations for a landing gear as follows:

1. single main;
2. bicycle;
3. tail-gear;
4. tricycle or nose-gear;
5. quadricycle;
6. multi-bogey;
7. releasable rail;
8. skid;
9. seaplane landing device;
10. human leg.

The features and technical descriptions of each landing gear configuration will be presented in this section. The common alternatives for landing gear configurations are illustrated in Figure 9.3. The landing gear configuration selection process includes setting up a table of features that can be compared in a numerical fashion. The details of the process were covered in Chapter 2. It needs to be clarified that for simplicity, the term "gear" or "wheel" is sometimes employed for a single strut and whatever that is connected to – comprising such items as tire, wheel, shock absorber, actuators, and brake assembly. Hence, when the term "nose-gear" is used, it refers to a landing gear configuration while when the term "nose gear" is employed, it refers to a gear that is attached under the fuselage nose. In general, most General Aviation (GA), transport, and fighter aircraft employ tricycle landing gear, while some heavyweight transport (cargo) aircraft use quadricycle or multi-bogey landing gear. Nowadays, the tail gear is seldom used by GA aircraft, but it was employed in the first 50 years of aviation history by the majority of aircraft.

9.3.1 Single Main

The simplest configuration of landing gear is the single main (see Figure 9.3(a)). It includes one large main gear that carries a large portion of the aircraft weight and load plus a very small gear under the nose. In terms of size, the main gear is much larger (both strut and wheel) than the secondary one. Both of these gears are in the aircraft symmetrical plane. The main gear is close to the aircraft cg, while the other gear is far from it. In the majority of cases, the main gear is located in front of the aircraft cg and the other one is behind the cg (under the tail section). In case the main gear is aft of the

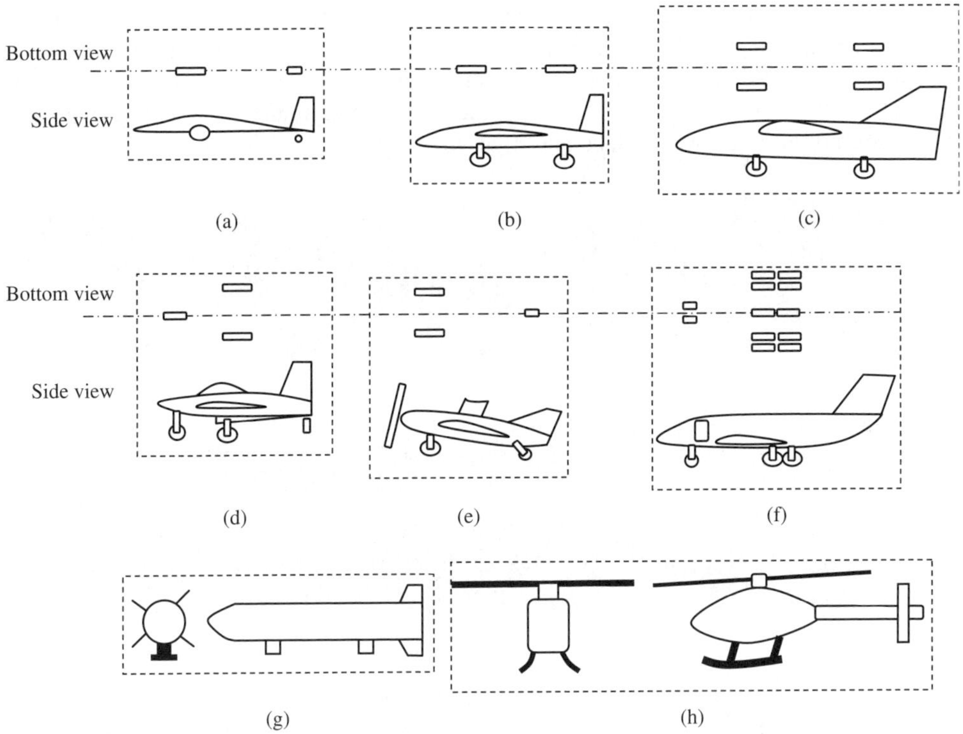

Figure 9.3 Landing gear types: (a) Single main; (b) Bicycle; (c) Quadricycle; (d) Tricycle; (e) Tail-gear; (f) Multi-bogey; (g) Releasable rail; (h) Skid

aircraft cg, the secondary gear is usually converted to a skid under the fuselage nose. The majority of sailplanes employ a single main landing gear because of its simplicity.

The single main landing gear is not usually retracted, so it is very short in height. An aircraft with a single main landing gear is not stable on the ground, so the aircraft will tip over one side (usually on the wing tips) while staying on the ground. In such a landing gear configuration, an operator must hold the wing level when the aircraft is stationary and prior to take-off. To prevent sideways tipping, some aircraft are equipped with two auxiliary small gears under two wing sections. In an aircraft without auxiliary wheels, the wing tips must be repaired on a regular basis, since the wing tips are damaged during each tipping. Two advantages of this arrangement are the simplicity and low weight of the landing gear. In contrast, beside the ground instability, a disadvantage of this configuration is the longer take-off run, since the take-off rotation is limited.

9.3.2 Bicycle

Bicycle landing gear, as the name implies, has two main gears (Figure 9.3(b)), one aft and one forward of the aircraft cg, and both wheels have a similar size. To prevent the aircraft from tipping sideways, two auxiliary small wheels are employed on the wings. The distance between two gears to the aircraft cg is almost the same, thus both gears carry a similar load. The bicycle landing gear has some similar features with a single main, and is in fact an extension to the single main. This arrangement is not popular among aircraft designers due to its ground instability. The main advantages of this configuration are its design simplicity and low weight. This landing gear configuration is a cheap candidate for an aircraft with narrow fuselage and high-wing configuration. Figure 9.4(a) illustrates

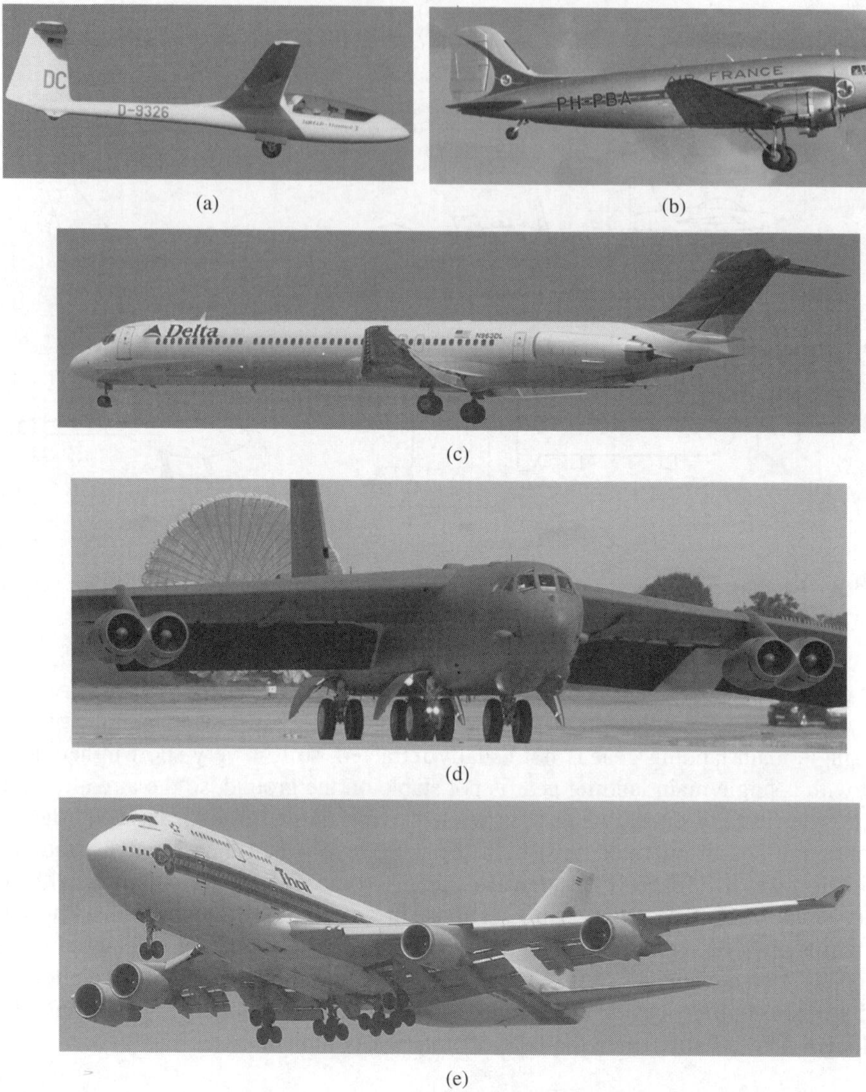

Figure 9.4 Five example aircraft with various landing gear configurations: (a) glider PZL-Bielsko SZD-48 Jantar Standard 3 with bicycle landing gear; (b) Douglas C-47A Skytrain; (c) transport aircraft McDonnell Douglas MD-88 with tricycle landing gear; (d) bomber aircraft B-52 Stratofortress with quadricycle landing gear using a parachute during landing operation; (e) transport aircraft Boeing 747 with multi-bogey landing gear. Reproduced from permission of (a) Miloslav Storoska; (b) Jenny Coffey; (c, e) Anne Deus; (d) Antony Osborne

the glider aircraft ASK21with bicycle landing gear. The Lockheed U-2, the McDonnell Douglas (now Boeing) AV-8B Harrier II (Figure 4.4), and the British Aerospace Sea Harrier (Figure 5.51) employ a bicycle landing gear configuration with two outrigger units under the wing.

9.3.3 Tail-Gear

Tail-gear landing gear has two main wheels forward of the aircraft cg and a small wheel under the tail. Figure 9.3(e) illustrates the side and top views of the gear in a typical aircraft. The wheels in front of the aircraft cg are very close to it (compared with the aft wheel) and carry much of the aircraft weight and load; thus they are referred to as the main wheel. Two main gears are at the same distance from the cg in the x-axis and the same distance in the y-axis (in fact left and right sides); thus both carry the same load. The aft wheel is far from the cg (compared with the main gear); hence it carries a much smaller load and thence is called an auxiliary gear. The share of the main gear from the total load is about 80–90%, so the tail gear carries about 10–20%.

This configuration of landing gear is referred to as a conventional landing gear, since it was the primary landing gear during the first 50 years of aviation history. But currently, only about 10% of the aircraft produced employing tail gear. In order to reduce drag, in some aircraft, a skid (vertical flat plate) is used instead of the tail wheel. Such landing gear is referred to as a tail-dragger. Most agricultural and some GA aircraft are equipped with tail gear. The aircraft is not level on the ground, due to the fact that the main gear is much larger and taller than the tail gear. Thus the passengers must climb the floor on such aircraft as the passenger aircraft Boeing 80 during the 1940s in order to get aboard. Since the aircraft has a high angle of attack during ground roll, the tail will be lifted up during take-off. This attitude makes the take-off run longer compared with a tricycle landing gear. Another consequence of high angle of attack during take-off is the low pilot view of the runway.

Since the aircraft has three wheels (supporting points), the aircraft is stable on the ground. However, it is inherently directionally unstable during ground maneuver (turn). The reason is that when an aircraft with a tail gear starts to turn on the ground around the main gear, the cg behind the main gear generates a centrifugal force. If the aircraft ground speed is high enough, the moment of the centrifugal force will be larger than the moment of the friction force on the tail gear, so it causes the aircraft to yaw around the main gear. Thus, the aircraft will roll and tip on its outer wing tip, or will skid off the side of the runway. This aircraft behavior can easily be controlled by lowering the speed during taxi. However, it is potentially possible to go out of control during landing and touch-down, due to cross-winds. To prevent this, the pilot needs to dance on the rudder pedals until the aircraft slows down. The World War II aircraft Spitfire (Figure 8.3) and Tiger Moth, as well as the GA aircraft Piper Super Cub (Figure 5.56) and Cessna 185, all had tail gear. Figure 9.4(b) shows the old transport aircraft Douglas C-47A Skytrain (DC-3) with its tail-gear configuration.

9.3.4 Tricycle

Tricycle is the most widely used landing gear configuration. Figure 9.3(d) shows the side and top views of the gear in a typical aircraft. The wheels aft of the aircraft cg are very close to it (compared with forward gear) and carry much of the aircraft weight and load;

thus they are referred to as the main wheel. Two main gears are at the same distance from the cg in the x-axis and the same distance in the y-axis (left and right sides); thus both carry the same load. The forward gear is far from the cg (compared with the main gear); hence it carries a much smaller load. The share of the main gear from the total load is about 80–90%, so the nose gear carries about 10–20%. This arrangement is sometimes called nose-gear.

GA, transport, and fighter aircraft are frequently equipped with a tricycle configuration. Both main and nose gears have the same height, so the aircraft is level on the ground, although the main gears often tend to have larger wheels. This allows the floor to be flat for passengers and cargo loading. Unlike tail gear, a nose-gear configuration aircraft is directionally stable on the ground as well as during taxiing. The reason is that if the aircraft yaws slightly while taxiing, the rolling and skidding resistance of the main gear, acting behind the cg, tends to straighten the aircraft out. This feature enables the aircraft to have a fairly large crab angle during cross-wind landing. The pilot view during take-off and landing is much better compared with tail gear. Aircraft such as the Boeing 737 (Figure 6.12), Airbus 320, General Dynamics F-16 Fighting Falcon (Figure 4.6), Pilatus PC-9, Piper Cherokee (Figure 7.4), Cessna 208, Embraer EMB 314 Super Tucano (Figure 10.5), and Mikoyan Mig-29 (Figure 5.56) all have nose-gear configuration. Figure 9.4(c) shows the transport aircraft McDonnell Douglas MD-88 (Figure 5.51) with tricycle configuration.

Most large transport aircraft (e.g., Fokker 100 (Figure 10.6)), fighters (e.g., McDonnell Douglas F/A-18 Hornet (Figures 2.11 and 6.12)), and some military aircraft (e.g., Northrop Grumman B-2 Spirit (Figure 6.8)) employ two wheels on the nose gear to increase the safety during take-off and landing in case of a flat tire. This is also the case for an aircraft with large load on the nose gear. For such a case, instead of one large wheel, two small wheels are utilized to decrease the gear frontal area and also aircraft drag. Carrier-based aircraft such as the F-14 Tomcat (Figure 5.44) and F/A-18 Hornet (Figures 2.11 and 6.12) need to employ two wheels for the nose gear in order to be capable of using a catapult launch mechanism.

As the number of wheels is increased, the manufacturing, maintaining, and operating costs will be increased too while the safety is improved. Furthermore, as the number of wheels is increased, the wheel frontal area is reduced, so the aircraft performance (especially during take-off) will be improved. Another reason for having multiple wheels is to tailor the wheel's overall volume to match the retraction bay geometry inside the wing or fuselage. Typically, when the aircraft weight is between 70 000 and 200 000 lb, two wheels per main strut are employed. The cargo aircraft Lockheed C-5 Galaxy with a very heavy weight (maximum take-off weight of 840 000 lb) employs four nose wheels to spread out the gear load among the tires.

9.3.5 *Quadricycle*

As the name implies, a quadricycle landing gear (see Figure 9.3(c)) utilizes four gears, similar to a conventional car wheel system: two wheels at each side, with two wheels in front of the aircraft cg and the other two aft. The load on each gear depends on its distance from the cg. If aft and forward wheels have the same distance from the cg, they will have to carry the same load. In this case, it is very hard to rotate the aircraft during take-off and landing so the aircraft will perform a flat take-off and landing. This characteristic causes the aircraft to have a longer take-off run, compared with a tricycle configuration.

This feature enables the aircraft to have a very low floor, which permits easier loading and unloading. The quadricycle landing gear configuration is usually used in a very heavy cargo or bomber aircraft. The bomber aircraft Boeing B-52 Stratofortress (Figure 9.4(d)) utilizes quadricycle landing gear and also has two outrigger units under the wing tips to divide the very heavy weight of the aircraft. An aircraft with quadricycle landing gear is very stable on the ground and during taxiing.

9.3.6 Multi-Bogey

As the aircraft gets heavier, the number of gears needs to be increased. A landing gear configuration with multiple gears of more than four wheels also improves take-off and landing safety. When multiple wheels are employed in tandem, they are attached to a structural component (see Figure 9.3(f)) referred to as a "bogey" that is connected to the end of the strut. An aircraft with multi-bogey landing gear is very stable on the ground and also during taxiing. Among various landing gear arrangements, a multi-bogey is the most expensive and most complex to manufacture. When the aircraft weight is beyond 200 000 lb, multiple bogeys each with four to six wheels are used. Large transport aircraft such as the Boeing B-747 (Figures 3.7, 3.12, and 9.4) and Airbus A-380 (Figure 1.8) utilize multi-bogey landing gear. The Boeing B-747 (Figure 9.4(e)) is equipped with four four-wheel bogies on the main gear and a twin-wheel nose unit.

9.3.7 Releasable Rail

For those aircraft designed to take off while airborne and not expected to land on the ground or sea, there is a special type of gear. Rockets and missiles (see Figure 9.3(g)) are in the same category in terms of landing gear configuration. These air vehicles are either launched, or released to get airborne. Take-off or launch gear usually consists of two or three fixed pieces. One piece is a flat plate T-shape part (Figure 9.5) that is attached to the mother vehicle (e.g., fighter) or launcher. The main function of this attachment is to hold the vehicle while launched.

9.3.8 Skid

Some vertical take-off and landing aircraft and helicopters do not need to taxi on the ground, so they are equipped with a beam-type structure referred to as skids (see Figure 9.3(h)) instead of regular landing gear. The configuration of skids mainly

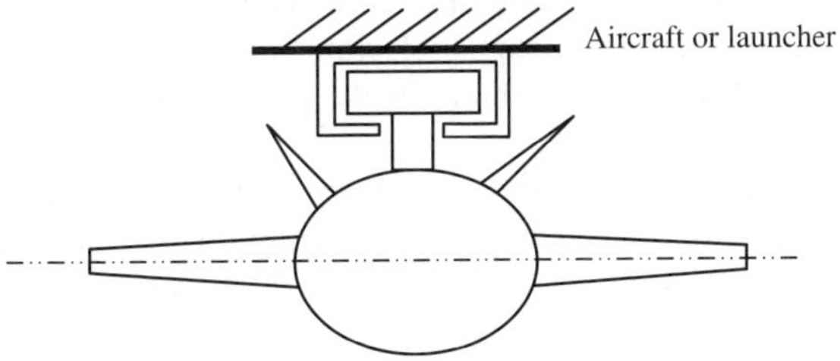

Figure 9.5 Missile attachment

comprises three or four fixed cantilever beams which are deflected outward when a load (i.e., aircraft weight) is applied. The deflection of skids plays the role of a shock absorber during landing operations. However, due to the nature of the beams, they are not as efficient as oleo shock absorbers. The design of skids compared with regular landing gear equipped with wheels is much simpler. Basic equations for beam deflection and bending stress might be employed in the design and analysis of skids. In addition, fatigue loading and fatigue life must be taken into account to predict the skid endurance.

9.3.9 Seaplane Landing Device

Take-off and landing on the sea requires a special landing gear configuration. The technical features of the water runway are totally different from a hard surface tarmac. Thus, a seaplane is not able to employ the advantages of wheels on water. The seaplane landing gear and shape of the hull are governed by the following design requirements:

1. slipping;
2. water-impact load reduction;
3. floating;
4. lateral static stability.

A seaplane usually lands on the water first by its fuselage and then by utilizing a special skid to remain stable. The fuselage (or hull) bottom shape constitutes the primary part of a seaplane landing gear. The fuselage shape must be designed to satisfy the above requirements as well the fuselage original design requirements for accommodating payload. The slipping and reduction of the water-impact load requirements often influence the design of the fuselage bottom shape, while the floating requirement affects the fuselage height. Lateral static stability on the water is usually provided by wing-mounted skids. These skids must be located such that they contact the water when the seaplane tips sideways by about less than 10 deg.

One of the important variables in designing the fuselage bottom shape is the water line (see Figure 9.6), which is borrowed from ship dynamics. The purpose of a "load line" is to ensure that a ship (as in the seaplane) has sufficient freeboard (i.e., the height from the water line to the main deck) and thus sufficient reserve buoyancy. The freeboard of sea vessels is measured between the lowest point of the uppermost continuous deck at the side and the water line, and this must not be less than the allowable freeboard. The water line or load line indicates the legal limit to which a ship may be loaded. Any section of the aircraft under the water line will submerge. The aircraft take-off/landing speed is determined by, amongst other parameters, the water line length. The length of the water line can change significantly as the vehicle heels, and can dynamically affect the speed of the vehicle.

A body in a fluid is buoyed up by a force equal to the weight of the fluid displaced. The buoyancy force (F_b) acts vertically upward through the centroid of the displaced volume. Thus the exact location of the load line is calculated using the Archimedes principle as follows:

$$F_b = \rho_f g V_d \tag{9.1}$$

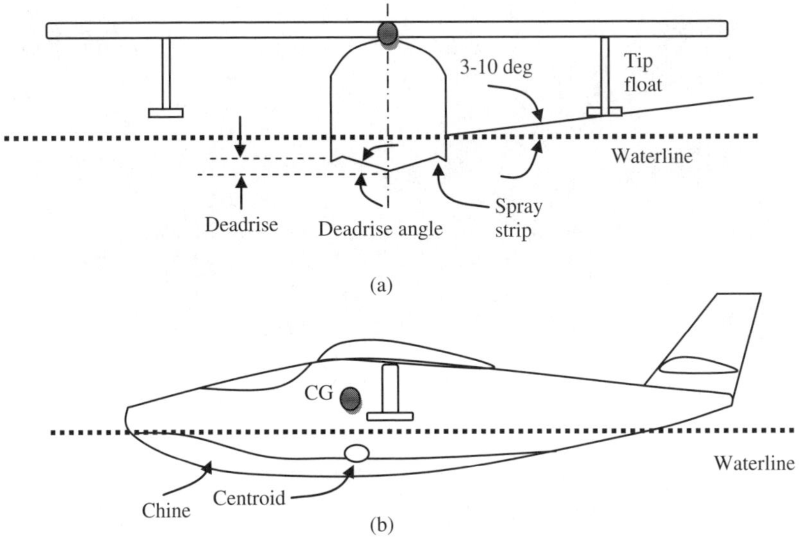

Figure 9.6 Seaplane landing provision geometry: (a) Front view; (b) Side view

where ρ_f is the density of the fluid (water has a density of $1000\,kg/m^3$), g is the gravity, and V_d is the displaced volume of the fluid. The centroid of the area on the submerged volume should be close to the aircraft center of gravity.

The reduction of the water-impact load requirement may be satisfied using a V-shaped bottom. The height of the V is referred to as the dead rise, and the angle is the dead-rise angle. The dead-rise angle needs to be increased for higher landing speeds. The dead-rise angle is also increased toward the fuselage nose to about 40 deg to better cut through water waves. To reduce water spray, spray strips may be applied to the edges of the bottom. The spray strips are usually angled about 40 deg below the horizon.

An important parameter that strongly influences seaplane performance during landing and take-off is the ratio between the water line length (L_W) and the fuselage width (W_f). The landing impact as well as the water-dynamic resistance are functions of this ratio (L_W/W_f). A wide fuselage has a lower water resistance, but suffers a higher landing impact. Figure 9.7 shows the amphibious (or flying boat) aircraft Canadian Vickers PBV-1A Canso.

9.3.10 Human Leg

When an aircraft is very light and the cost is supposed to be as low as possible, the human leg can function as the landing gear. This is the case for both hang glider and paraglider. The pilot must use his/her leg during take-off and landing operations. Due to human physical weaknesses, the landing speed must be very low (e.g., less than 10 knot) in order to have a safe landing. Pilot skill and nimbleness are a requirement, besides the leg, for successful landing. In such a case, there is no need for landing gear design; just assume that it has been designed and fabricated and is ready for flight. Figure 9.8 illustrates a pilot during take-off aboard a hang glider (note his leg as a landing gear).

Figure 9.7 Amphibious aircraft Canadian Vickers PBV-1A Canso. Reproduced from permission of Jenny Coffey

Figure 9.8 A pilot running to launch himself off the top of a hill aboard a hang glider. Reproduced from permission of Christopher Huber

9.3.11 Landing Gear Configuration Selection Process

Now that several configurations of landing gear arrangements have been introduced, it is time to describe how to select one to satisfy the design requirements. Choice of landing gear depends upon a number of factors, and one should not automatically assume that a nose gear (i.e., tricycle) is necessarily the best. There are several design requirements which affect the decision on selection of landing gear configuration. These include: cost, weight, performance, take-off run, landing run, ground static stability, ground taxi stability, and maintainability. In order to select the best landing gear configuration, the designer must perform a trade-off study using a comparison table such as Table 9.2. The candidate which gains the highest point is often the most appropriate landing gear for the aircraft. Hence, based on aircraft mission and design requirements, one arrangement is usually the best alternative.

In the USA, landing certification is only based on brake while in Europe, thrust reverse is also considered. The main reason is that runways in the USA are often dry, while in Europe they are frequently wet. However, in Russia, parachutes are still used in some parts of the country, due to snow and bad weather.

Table 9.2 A comparison among various landing gear configurations

No.		Single main	Bicycle	Tail-gear	Nose-gear	Quadricycle	Multi-bogey	Human leg
1	Cost	9	7	6	4	2	1	10
2	Aircraft weight	3	4	6	7	9	10	1
3	manufacturability	3	4	5	7	9	1	10
4	Take-off/landing run	3	4	6	10	5	8	2
5	Stability on the ground	1	2	7	9	10	8	5
6	Stability during taxi	2	3	1	8	10	9	–

10: best, 1: worst.

9.3.12 Landing Gear Attachment

When the configuration of the landing gear has been selected, the landing gear attachment must also be decided. Two primary options for attachment are the fuselage and the wing. The attachment between landing gear and aircraft will influence several design requirements, such as weight, take-off and landing performance, cost, and ground stability. A few main alternatives for the attachment between landing gear and aircraft are usually as follows:

1. All struts/wheels are attached to the fuselage (e.g., F/A-18 (Figures 2.11 and 6.12) and Boeing 747 (Figure 9.4(e))).
2. The main gear is attached to the wing, but the nose gear is attached to the fuselage (e.g., long-range British airliner Vickers VC10 (Figure 9.9(c))).
3. The main gears are attached to the wing, but the tail gear is attached to the fuselage (in a tail-wheel configuration). An example is the WWII fighter aircraft P-51 Mustang (Figure 3.14) and GA aircraft Van's RV-7 (Figure 9.9(a)).
4. The main gears are attached to the nacelle, but the nose gear is attached to the fuselage (in a nose-wheel configuration). A few examples are the Boeing B-47 Stratojet, Cessna 340, and Ilyushin IL-18 (Figure 9.9(b)).

A natural option for the attachment is to attach the landing gear to the fuselage. However, there are cases where the designer should consider other alternatives. For instance, when the fuselage is not wide enough to allow for a long wheel track, attachment to the wing will provide a solution. However, in the case of a high-wing configuration, the attachment of the landing gear to the wing makes the landing gear very long and heavy, as well as the retraction system hard to design. Another solution for an aircraft with a narrow fuselage is to accommodate a special bay for the landing gear retraction storage. This technique has been employed in the military cargo aircraft C-17 Globemaster (Figure 9.9(d)).

Figure 9.9 Example aircraft for landing gear attachments: (a) Van's RV-7; (b) Ilyushin IL-18; (c) Vickers VC10; (d) McDonnell Douglas C-17A Globemaster. Reproduced from permission of (a) Jenny Coffey; (b, c) A J Best; (d) Anne Deus

As a safety precaution, it is recommended not to attach the strut such that it is under the fuel tank. In case the touchdown is mistakenly very rapid, the high sink rate may cause the fuel tanks to explode. Two Boeing 727 aircraft crashed in the past [2], due to pilots' mistake in high rate touchdown, and so the fuel tanks exploded. The design of landing gear was corrected in future production.

In order to decide on the landing gear attachment, the designer must perform a trade-off study using a comparison table. The fundamentals of the technique are introduced in Section 9.3.11. In summary, the purpose of this Section 9.3 is to give the designer an overall understanding of the fundamental trade-offs associated with different landing gear configurations. This understanding is helpful for discussions about landing gear design. In subsequent sections, various aspects and parameters of landing gear will be examined and the relationships between landing gear parameters and design requirements will be discussed.

9.4 Fixed, Retractable, or Separable Landing Gear

Another design aspect of the landing gear is to decide what to do with it after take-off. In general, there are four alternatives:

1. Landing gear is released after take-off.
2. Landing gear hangs underneath the aircraft (i.e., fixed).
3. Landing gear is fully retracted inside the aircraft (e.g., inside the wing or fuselage).
4. Landing gear is partially retracted inside the aircraft.

Each of these four alternatives has various advantages and disadvantages which must be evaluated prior to decision making. In the first case, the landing gear is released after take-off so the aircraft does not have to carry it during flight mission. Hence the aircraft weight will be reduced after take-off and this is assumed to be an advantage. However, this alternative does not have anything to do with landing. It means that the aircraft is not supposed to land, which is the case for drones that are used as targets for missile tests. Or, the aircraft must use another landing gear to land safely. Such wheels are sometimes mounted onto axles that are part of a separate dolly (for main wheels only) or trolley (for a three-wheel set with a nose wheel) chassis. The major advantage of such an arrangement is the weight reduction, which results in a higher performance. If the aircraft is planned to land at the end of its mission, this option is not recommended since landing on a moving cart is not a safe operation. There are very few aircraft with such a landing gear configuration.

One of the longest-established jet target drones with a separable landing gear is the *Jindivik*, developed in Australia and used for decades in Britain and Australia. Over 400 were built, and small numbers were also supplied to the US Navy and to Sweden. The name is Aborigine for "that which is hunted." Development was begun in 1948 by the Australian Government Aircraft Factory. Figure 9.10 shows a Jindivik right after take-off. The aircraft utilizes a cart during take-off operation, which is released after lift-off.

In the second, third, and fourth cases, the landing gear will be a deadweight and has no positive function while the aircraft in onboard. However, it is saved and employed during landing operations. Advantages and disadvantages of these two arrangements are compared in Table 9.3. In general, two major criteria are cost versus performance. If the primary design objective is higher performance, the retractable landing gear is the best design. However, if the designer's main concern is to reduce the aircraft cost, one way is to select a fixed landing gear. Currently all transport aircraft (such as the Boeing 777 (Figure 8.7) and Airbus 340 (Figure 8.7)), most military aircraft (such as the Lockheed C-5, F/A-18 Hornet, and F-16 Falcon), and a great portion of GA aircraft (e.g., Cessna 550 and Gulfstream 550 (Figure 11.15)) employ retractable landing gear. But most home-built aircraft and some GA aircraft (e.g., Cessna 182 (Figure 3.7)) have fixed landing gear. Figure 9.2 illustrates a few examples. If a retractable landing gear needs to be

Figure 9.10 Aircraft Jindivik releases landing gear after take-off. Reproduced from permission of http://www.militaryimages.net

Table 9.3 Fixed and retractable landing gear comparison

No.	Item	Fixed (non-retractable) landing gear	Retractable landing gear
1	Cost	Cheaper	Expensive
2	Weight	Lighter	Heavier
3	Design	Easier to design	Harder to design
4	Manufacturing	Easier to manufacture	Harder to manufacture
5	Maintenance	Easier to maintain	Harder to maintain
6	Drag	More drag	Less drag
7	Aircraft performance	Lower aircraft performance (e.g., maximum speed)	Higher aircraft performance (e.g., maximum speed)
8	Longitudinal stability	More stable (stabilizing)	Less stable (destabilizing)
9	Storing bay	Does not require a bay	Bay must be provided
10	Retraction system	Does not require a retraction system	Requires a retraction system
11	Fuel volume	More available internal fuel volume	Less available internal fuel volume
12	Aircraft structure	Structure is uninterrupted	Structural elements need reinforcement due to cutout

compromised to provide internal volume for other components such as fuel, a partially retractable landing gear is the solution. For instance, the close support military aircraft Fairchild A-10 Thunderbolt (Figure 6.12) feature a partially retractable landing gear in order to provide more room for stores.

In case of retractable landing gear, it folds after take-off into the fuselage where it is stored during flight until shortly before landing. Related features of a retractable landing gear are: (i) retracting system design and (ii) provision of sufficient room for landing gear after retraction. Most mechanisms for landing gear retraction are based upon a four-bar linkage, using three members connected by pivots. The fourth bar is the aircraft structure. A retraction mechanism clearly increases the aircraft weight, design complexity, and maintenance, and reduces the internal fuel volume.

The major options for main landing gear home (see Figure 9.11) are: (i) in the wing, (ii) in the fuselage, (iii) wing-podded, (iv) fuselage-podded, (v) wing-fuselage junction, and (vi) in the nacelle. In a high-wing configuration, retracting and locating landing gear in the fuselage makes the strut shorter. In general, a retracted position inside the aircraft will chop up the aircraft structure, which consequently increases the aircraft weight. Examples are locating the landing gear in the wing, in the fuselage, or in the wing/fuselage. In contrast, a podded bay configuration tends to increase the aircraft frontal area, causing

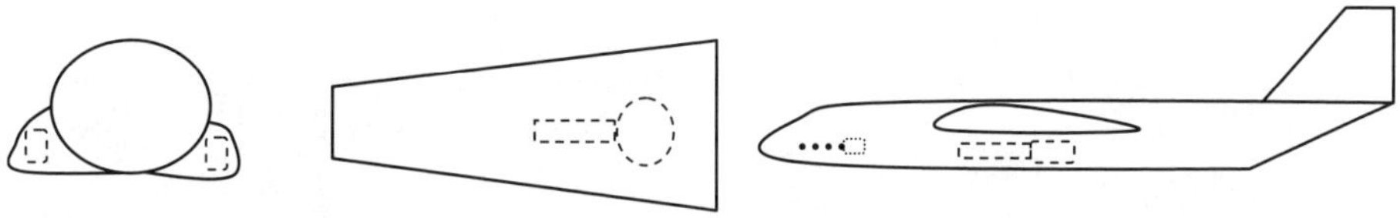

(a) (b) (c)

Figure 9.11 Landing gear storage bay: (a) Fuselage podded (front view); (b) In the wing (top view); (c) In the fuselage (side view)

additional aerodynamic drag. The example is locating the landing gear in a pod beside the fuselage. In terms of aircraft structural design complexity, a landing gear bay in the wing requires a wing cutout that leads to stronger spars. The best candidate for a bay in the wing is the space between the main spar and the rear spar. A landing gear bay in the fuselage also requires a fuselage cutout that leads to stronger frames and longerons. The aerodynamic benefits of the wing or fuselage bay arrangements outweigh the drawbacks for high-speed aircraft.

Most low-wing transport aircraft (such as the Boeing 767 (Figure 5.4) and Airbus 320) retract the main gear into the wing/fuselage junction, while most high-wing transport (cargo) aircraft retract the main gear into the fuselage. Most fighters (such as the F-16 Falcon and F/A-18 Hornet) with low-wing configuration retract the main gear and also the nose wheel into the fuselage. Some GA aircraft retract the main gear into the wing (e.g., Cessna 525), while some GA aircraft (e.g., Learjet 85) into the wing/fuselage junction. The fuselage-podded or wing-podded landing gear bay reduces the aircraft weight significantly, since the fuselage and wing structure is uncut. The close support aircraft A-10 Thunderbolt (Figure 6.12) has a wing-podded landing gear configuration due to its military mission requirements.

In case of a twin propeller-driven engine aircraft with main gears underneath the engines (e.g., P-38 Lightning), a typical location for the main gear bay is the nacelle behind the engines. A retractable landing gear bay normally requires a couple of doors to be closed after retraction in order to reduce the drag. In some aircraft such as the Boeing 737-700, the main wheels are retracted into the fuselage bay without any landing gear door to save weight. Figure 9.12 illustrates a Dassault Mirage 2000 (Figure 9.12(a)) and a Hawker Siddeley Nimrod (Figure 9.12(c)) with retractable landing gear, and a Robin DR-400-120 Dauphin (Figure 9.12(b)) with faired fixed landing gear.

A technique to reduce the fixed landing gear drag is to employ fairing. Fairing is a special airfoil-shaped cover which mainly covers the wheel. As a rule of thumb, a well-designed fairing will reduce the wheel drag by as much as 1000%. Thus an unfaired wheel (see Figure 9.12(d)) will generate about 10 times more drag than a faired wheel. However, the landing gear wheels generate ~5% of the aircraft total drag; hence the application of wheel fairing will reduce the aircraft total drag by as much as 4.5%. Figure 9.12(b) shows a Robin DR-400-120 Dauphin with faired fixed landing gear.

9.5 Landing Gear Geometry

At this point, the landing gear configuration is selected and the retraction configuration is decided. Now, the designer needs to perform mathematical calculations to determine

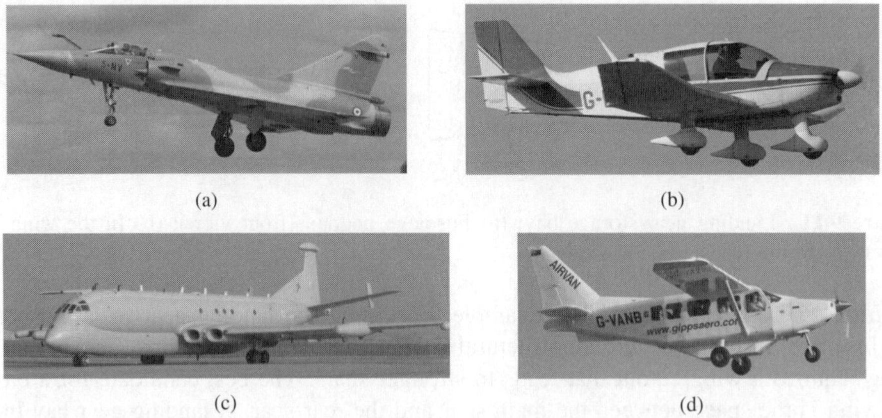

Figure 9.12 Four aircraft with various types of landing gear: (a) Dassault Mirage 2000 with retractable landing gear; (b) Robin DR-400-120 Dauphin with faired fixed landing gear; (c) Hawker Siddeley Nimrod with retractable landing gear; (d) Gippsland GA-8 Airvan with unfaired fixed landing gear. Reproduced from permission of (a, b, d) Jenny Coffey; (c) Antony Osborne

a few parameters such as height, wheel base, wheel track, and the distance between the main gear and the aircraft center of gravity. These parameters are interrelated through geometrical relations and several mathematical principles. These relationships are described in this section. The guidelines for determining these parameters are presented in the following sections.

9.5.1 Landing Gear Height

9.5.1.1 Definition of Height

Landing gear height (H_{LG}) is defined as the distance between the ground and the conjunction between the main gear strut and the aircraft structure. Figure 9.13 illustrates several aircraft with different landing gear height cases. The main gear may be attached to the fuselage (Figure 9.13(a)), wing (Figure 9.13(b)), or nacelle (Figure 9.13(d)). The connection might be through a variety of ways, including a strut (Figure 9.13(b)), solid spring (Figure 9.13(a)), solid axle (Figure 9.13(f)), rubber bungee (Figure 9.13(e)), hinge, or oleo (Figure 9.13(d)). Hence, the landing gear height could be shorter when the aircraft is on the ground due to the spring deflection or oleo compression because of the aircraft weight. In order to have a uniform definition, the landing gear height is measured when the aircraft is on the ground and the fuselage is horizontal.

The tires themselves provide some kind of shock-absorbing ability by deflection when a bump is encountered. An aircraft with rigid axle relies solely upon the tires for shock absorbing. There are five main design requirements in which the landing gear height plays an important role. They are:

1. Landing gear height provides aircraft clearance during taxi.
2. Landing gear height provides rear fuselage clearance during take-off rotation.

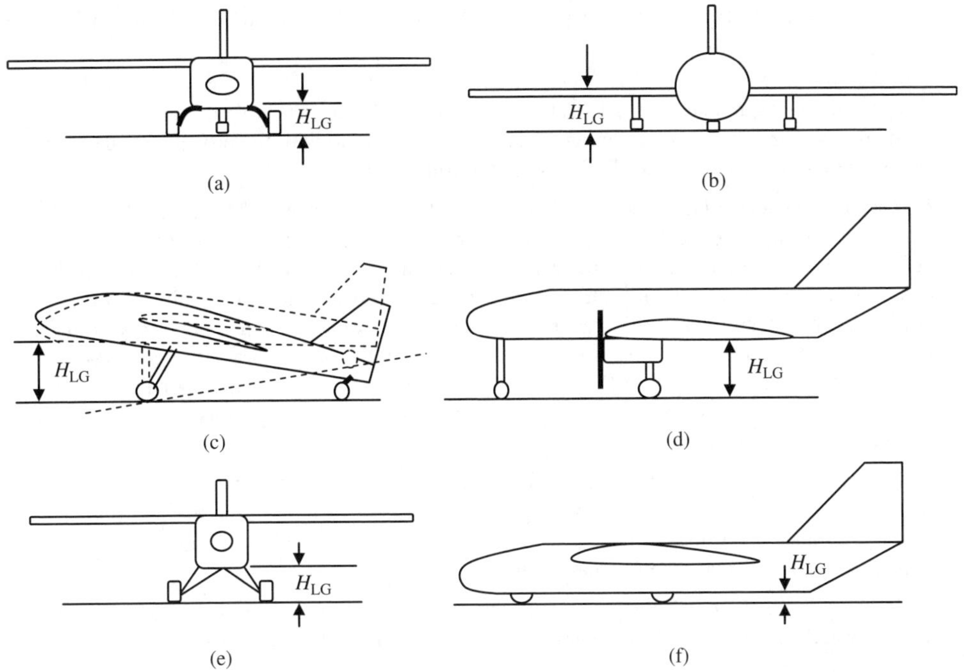

Figure 9.13 Landing gear height in various aircraft configurations: (a) LG is attached to the fuselage (solid spring); (b) Main gear is attached to the wing; (c) LG is attached to the fuselage (with outriggers); (d) Main gear is attached to the nacelle; (e) LG is attached to the fuselage (rubber bungee); (f) LG is attached into the fuselage (no strut)

3. Landing gear height contributes to tipback prevention.
4. Landing gear height contributes to overturn prevention.
5. Landing gear height satisfies loading and unloading requirements.

In the early stages of design, it is not clear which of the above requirements is the most critical. Thus, the designer should examine all five requirements to make sure that the landing gear height does not violate any of these requirements.

9.5.1.2 Aircraft General Ground Clearance Requirement

One of the primary functions of the landing gear is to protect the aircraft structure from the ground. This job is performed by providing a clearance with the ground. The clearance is measured from the lowest point of the aircraft to the ground. In some aircraft the lowest component is the wing (e.g., low wing), while in some aircraft it is the fuselage (e.g., high wing), and in some other aircraft the jet engine has the lowest height from the ground (e.g., a transport aircraft with engines hanging underneath the low wing). In the case of an aircraft with prop-driven engine(s), the prop tip is often the lowest point. In any case, clearance needs to be provided via the landing gear height. The minimum magnitude of the clearance is a function of several design parameters, including cost, safety, performance, weight, stability, engine inlet, loading, and operational requirements.

The following is reproduced from FAR [3] Part 23 Section 23.925 on propeller clearance:

> Unless smaller clearances are substantiated, propeller clearances, with the airplane at the most adverse combination of weight and center of gravity, and with the propeller in the most adverse pitch position, may not be less than the following:
>
> (a) Ground clearance. There must be a clearance of at least seven inches (for each airplane with nose wheel landing gear) or nine inches (for each airplane with tail wheel landing gear) between each propeller and the ground with the landing gear statically deflected and in the level, normal takeoff, or taxing attitude, whichever is most critical. In addition, for each airplane with conventional landing gear struts using fluid or mechanical means for absorbing landing shocks, there must be positive clearance between the propeller and the ground in the level takeoff attitude with the critical tire completely deflated and the corresponding landing gear strut bottomed. Positive clearance for airplanes using leaf spring struts is shown with a deflection corresponding to 1.5 g.
>
> (b) Aft-mounted propellers. In addition to the clearances specified in paragraph (a) of this section, an airplane with an aft mounted propeller must be designed such that the propeller will not contact the runway surface when the airplane is in the maximum pitch attitude attainable during normal takeoffs and landings.
>
> (c) Water clearance. There must be a clearance of at least 18 inches between each propeller and the water, unless compliance with §23.239 can be shown with a lesser clearance.

For an aircraft with one piston-prop engine, the typical value for the prop ground clearance is about 20 cm. For an aircraft with jet engine(s), the inlet must be high enough such that sand or debris is not pulled into the engine inlet during take-off. The inlet height is a function of aircraft speed and engine thrust. A typical value for the inlet height of a jet engine with 50 kN of thrust is about 70 cm. Figure 9.13 illustrates several aircraft configurations with various clearances. For a transport aircraft with prop-driven engine(s), the recommendation is to provide a prop clearance of as much as the height of a human (about 180 cm). This safety initiative avoids the possible accident of a human being hit by a rotating prop while moving around the aircraft.

The clearance for various aircraft components is as recommended in Table 9.4. The recommended clearance has a range of values due to the fact that aircraft type, aircraft mission, aircraft speed, type of runway, and cost dictates other constraints. For instance, a very light remote-controlled aircraft requires a much smaller clearance (say 20 cm) compared with a very large civil transport aircraft (say 1 m). Furthermore, a large military cargo aircraft, such as the McDonnell Douglas C-17A Globemaster (Figure 9.9(d)), requires much smaller clearance (say 30 m) due to the loading requirements.

9.5.1.3 Take-Off Rotation Ground Clearance Requirement

An aircraft is usually rotating about the main gear in order to increase the lift to prepare for take-off (see Figure 9.14). This is also true for landing operation, in which the aircraft rotates to gain a high angle of attack. In an aircraft with non-tail gear, the height of the landing gear must be set so that the tail or rear fuselage does not strike the ground during the take-off rotation or landing with a high angle of attack. However, in practice, transport

Table 9.4 Recommended clearance for various aircraft components

No.	Aircraft components	Clearance (m)	Remarks
1	Fuselage	0.2–1.2	
2	Rear fuselage	0.2–0.5	During take-off rotation
3	Wing	0.2–1.5	Includes flap clearance
4	Turbofan/turbojet engine	0.5–1.5	Inlet clearance
5	Propeller (piston or turboprop) – landplane	0.2–1	Tip clearance
6	Propeller (piston or turboprop) – seaplane	1–2	Tip clearance
7	Store/fuel tank/pitot tube/antenna/probe	0.2–0.6	

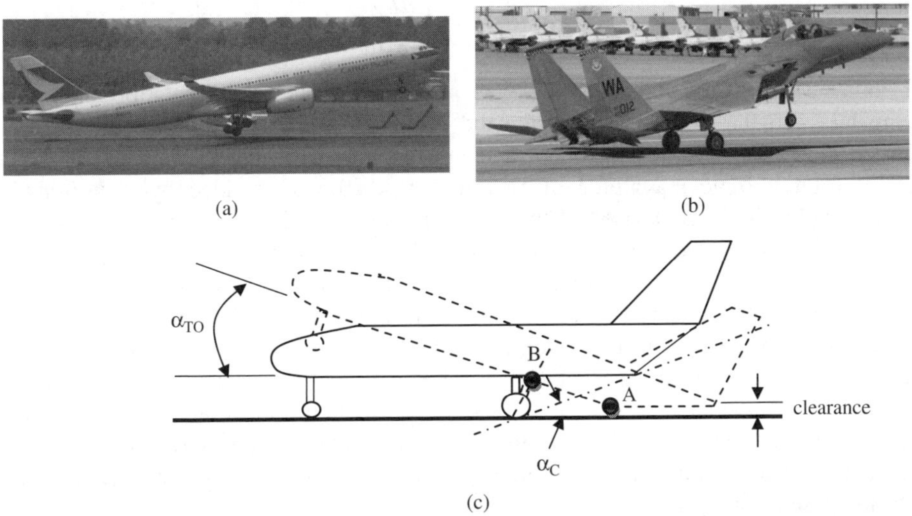

Figure 9.14 Take-off rotation and rear fuselage clearance: (a) Airbus A330; (b) McDonnell Douglas F-15C Eagle. Reproduced from permission of (a) Anne Deus; (b) Antony Osborne

aircraft are provided with removable shields that protect the fuselage from striking the ground, due to the fact that some unskilled pilots rotate the aircraft so fast that the rear fuselage strikes the ground. These rear fuselage protective shields are replaced on a regular base. The same is true for landing operation, where the aircraft angle of attack and wheel height must be such that there is no danger of a tail-strike and the crew members have a good view of the runway. In spite of including ground clearance in landing gear design, each year there are several tail strike reports by transport aircraft.

Tail-strike accidents must be prevented through an increase in the landing gear height. Another common solution to this problem is to cut the rear fuselage by an upsweep angle. The occurrence of hit is examined by looking at the angle between the ground and the line passing from the main gear contact with the ground to the beginning of the upsweep

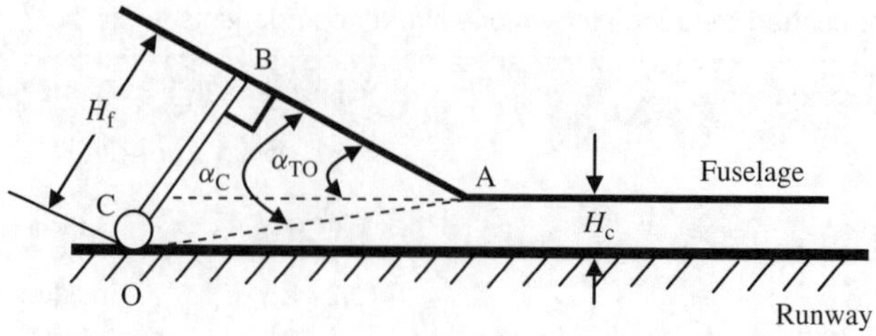

Figure 9.15 Examination of rear fuselage clearance during take-off rotation

angle at the fuselage (i.e., α_C). The take-off rotation ground clearance requirement to prevent a fuselage hit is as follows:

$$\alpha_C \geq \alpha_{TO} \tag{9.2}$$

where the clearance angle is:

$$\alpha_C = \tan^{-1}\left(\frac{H_f}{AB}\right) \tag{9.3}$$

In other words, if the clearance angle (α_C) is less than the aircraft rotation angle (α_{TO}) during take-off, the fuselage will strike the ground. Otherwise, there will be clearance between the fuselage and the ground, and the fuselage will not be damaged during the take-off rotation. The magnitude of clearance could be determined by examining the triangle (Figure 9.15) comprised of the following sides: (i) distance aft of the main gear to the beginning of the upsweep angle (i.e., AB); (ii) fuselage height (H_f); and (iii) take-off rotation angle (α_{TO}). Figure 9.15 shows the triangle ABC (part of the aircraft in Figure 9.14) that is formed between the fuselage lower surface and the main gear. The aircraft is rotated about the main gear (O or C) by the amount of take-off rotation angle. The minimum clearance between the fuselage and the ground (H_C) during take-off rotation is about 30 cm. Example 9.1 illustrates the application of this triangle to determine the acceptability of the main gear height regarding this requirement. If the clearance H_C is determined to be negative or below the limit, the main gear height needs to be increased accordingly.

Example 9.1

A pilot of the jet aircraft shown in Figure 9.16 is going to take off with 12 deg of fuselage angle of attack. Determine if the aircraft rear fuselage will hit the ground during take-off rotation. If yes, what must be the main gear height to achieve a clearance of 30 cm?

Solution:

First, we need to determine the clearance angle:

$$\alpha_C = \tan^{-1}\left(\frac{H_f}{AB}\right) = \tan^{-1}\left(\frac{1}{5}\right) = 0.197\,\text{rad} = 11.31\,\text{deg} \tag{9.3}$$

Since the clearance angle is less than the fuselage rotation angle (12 deg), the fuselage will hit the ground during take-off rotation. Next, a new value for the main gear height must be determined to prevent the occurrence of the fuselage hit. Since the aircraft take-off rotation angle is 12 deg, we tentatively consider an angle of 12 deg to prevent the hit:

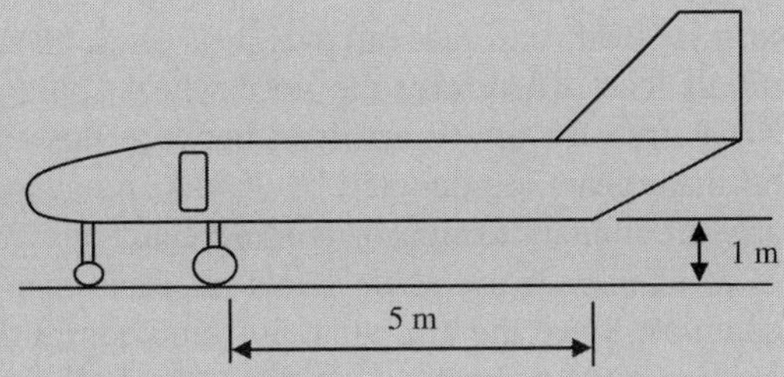

Figure 9.16 Figure for Example 9.1

$$\alpha_C = \tan^{-1}\left(\frac{H_f}{AB}\right) \Rightarrow 12\,\text{deg} = \tan^{-1}\left(\frac{H_f}{5}\right) \Rightarrow H_f = 5 \cdot \tan(12\,\text{deg}) = 1.063\,\text{m} \tag{9.3}$$

When the landing gear height is 1.063 m, the fuselage is about to have contact with the ground. A landing gear height of 1.369 m (1.063 + (30/(cos(12)))) provides a 30 cm clearance during a 12 deg take-off rotation.

9.5.2 *Wheel Base*

Wheel base (B) plays an important role in the load distribution between primary (i.e., main) gear and secondary (e.g., nose or tail) gear. This parameter also influences the ground controllability and ground stability. Thus, the wheel base must be carefully determined and an optimum value needs to be calculated to ensure it meets all relevant design requirements. In this section, the load distribution between main and nose gear is examined. The effect of wheel base on the ground controllability and ground stability will be discussed in subsequent sections.

Figure 9.17 shows a stationary aircraft with a tricycle landing gear on the ground. The aircraft weight (W) is carried by three wheels (i.e., two main and one nose gear). Due to the ground mobility (i.e., steering) requirement, typically the nose gear must not carry

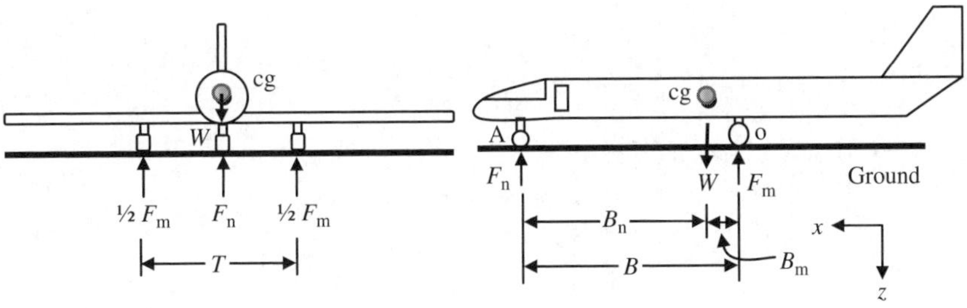

Figure 9.17 Wheel load geometry

less than about 5% of the total load and also must not carry more than about 20% of the total load (e.g., aircraft weight). Thus, the main gear carries about 80–95% of the aircraft load. Therefore, the nose wheel could be much smaller than the main wheels. This is true for the comparison between nose strut and main struts. The loads on nose and main gears are denoted by F_n and F_m respectively. These data are employed in the early preliminary design of landing gear.

Calculation of the static loads on each gear is performed by employing equilibrium equations. Since the aircraft is in static equilibrium, the summation of all forces in the z direction must be zero:

$$\sum F_z = 0 \Rightarrow F_n + F_m = W \tag{9.4}$$

Furthermore, the summation of all moments about o is zero:

$$\sum M_o = 0 \Rightarrow F_n B - W B_m = 0 \tag{9.5}$$

Thus, the percentage of the static load (i.e., aircraft weight) which is carried by the nose gear is:

$$F_n = \frac{B_m}{B} W \tag{9.6}$$

In addition, the percentage of the static load which is carried by the main gear is:

$$F_m = \frac{B_n}{B} W \tag{9.7}$$

In the case of a tricycle landing gear, the load on the main gear is divided between the left and right gear, so each wheel will carry one-half of the main gear load (i.e., $\frac{1}{2} F_m$). The wheel bases for several aircraft are tabulated in Table 9.5. Example 9.2 illustrates how to calculate the static loads that are carried by the nose gear and main gear based on the aircraft weight.

Example 9.2

A GA aircraft with a mass of 5000 kg has a tricycle landing gear configuration. The wheel base and wheel track are 10.2 m and 1.8 m respectively, and the distance between the main gear and the aircraft cg is 0.84 m. Determine the static load on each gear. What percentage of the aircraft weight is carried by the nose gear?

Solution:

$$\sum M_o = 0 \Rightarrow W_{TO}(0.84) - F_{nose}(10.2) = 0 \Rightarrow F_{nose} = \frac{5000 \cdot 9.81 \cdot 0.84}{10.2} = 4038\,N$$

$$\sum F_z = 0 \Rightarrow F_{main} + F_{nose} = W_{TO} \Rightarrow F_{main} = W_{TO} - F_{nose} = 5000 \cdot 9.81 - 4038 = 44\,995.2\,\text{N}$$

$$\frac{F_{nose}}{W_{TO}} = \frac{4038}{5000 \cdot 9.81} = 0.0824 = 8.24\%$$

Thus, 8.24% of the aircraft weight is carried by the nose gear.

Table 9.5 Wheel base and wheel track for several aircraft

No.	Aircraft	Type	Take-off mass (kg)	Overall length (m)	Wheel base (m)	Wheel track (m)
1	Airbus A-380	Airliner	590 000	72.73	30.4	14.3
2	Airbus A-300-600	Airliner	170 500	54.08	18.62	9.60
3	Airbus A-319	Airliner	75 500	33.84	11.04	7.59
4	Airbus A-340-500	Airliner	372 000	67.9	27.59	10.69
5	MD-11	Airliner	237 289	61.24	24.61	10.56
6	Boeing B-767-200	Airliner	136 080	48.81	19.69	9.30
7	Boeing B-747-400	Airliner	362 875	70.66	25.6	11
8	Boeing B-737-300	Airliner	56 470	33.40	12.45	5.23
9	Northrop Grumman B-2 Spirit	Bomber	170 550	21.03	9.76	12.2
10	Mooney M20J MSE	Touring	1 315	7.52	1.82	2.79
11	Piper PA-44-180 Malibu	Trainer	1 723	8.41	2.56	3.20
12	Beech super king 200	Transport	5 670	13.34	4.56	5.23
13	Beechjet 400A	Trainer	7 303	14.75	5.86	2.84
14	Cessna 208	Light GA	3 629	11.46	3.54	3.56
15	Cessna 650	Business	10 183	16.9	6.5	2.84
16	Gulfstream IV-SP	Transport	33 838	26.92	11.61	4.17
17	Lockheed C-130J Hercules	Tactical transport	70 305	29.79	12.3	4.43
18	C-17A Globemaster III	Transport	265 352	53.04	20.05	10.27
19	F-15E Eagle	Fighter	36 741	19.43	5.42	2.75
20	F/A-18 Hornet	Attack	16 651	17.07	5.42	3.11

The above-mentioned relationships are applicable only in static situations. There are two other interesting conditions that cause the landing gear to experience different loadings: (i) change in the aircraft center of gravity location and (ii) dynamic loading. Due to the possibility of a change in the load distribution, or having different combinations of cargo, or number of passengers, the gears must carry a load other than the nominal static load. In the x-axis, an aircraft center of gravity is allowed to move between two extreme limits: (i) most aft location ($X_{cg_{aft}}$) and (ii) most forward location ($X_{cg_{for}}$).

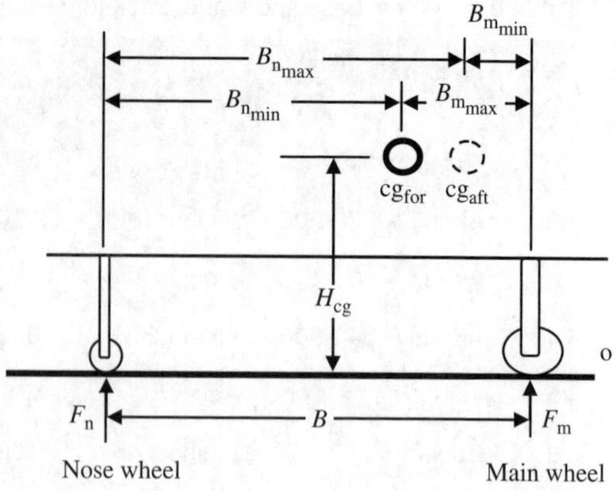

Figure 9.18 Wheel load geometry

Figure 9.18 illustrates a tricycle configuration with most aft and most forward cg locations. The following equations govern the minimum and maximum static loads on each gear:

$$F_{m_{max}} = \frac{B_{n_{max}}}{B} W \tag{9.8}$$

$$F_{n_{max}} = \frac{B_{m_{max}}}{B} W \tag{9.9}$$

$$F_{m_{min}} = \frac{B_{n_{min}}}{B} W \tag{9.10}$$

$$F_{n_{min}} = \frac{B_{m_{min}}}{B} W \tag{9.11}$$

Furthermore, the landing gear tends to experience a dynamic loading due to aircraft acceleration and deceleration during take-off and landing. The nose gear will have to carry a dynamic loading during the landing operation when the aircraft is braking. During the braking segment of the landing operation, the following equilibrium equation may be written (see Figure 9.17):

$$\sum M_o = 0 \Rightarrow F_n B - W B_m - \frac{W}{g} |a_L| H_{cg} = 0 \tag{9.12}$$

where a_L is the braking deceleration and g is the gravitational acceleration. Therefore, the nose gear load is:

$$F_n = W \frac{B_m}{B} + \frac{W |a_L| H_{cg}}{gB} \tag{9.13}$$

The first term of Equation (9.13) is the static load, but the second term is referred to as the dynamic load:

$$F_{n_{dyn}} = \frac{|a_L| W H_{cg}}{gB} \tag{9.14}$$

Hence, the total load on the nose gear during landing will be:

$$F_{\mathrm{n}} = F_{\mathrm{n_{max}}} + F_{\mathrm{n_{dyn}}} \tag{9.15}$$

To insure ground controllability in a tricycle landing gear configuration, the parameter $B_{\mathrm{m_{min}}}$ should be greater than 5% of the wheel base and the parameter $B_{\mathrm{m_{max}}}$ should be less than 20% of the wheel base. These equations and requirements are employed to determine the wheel base plus the distance between cg and the nose gear, and cg and the main gear. With a similar approach, the dynamic loading on the main gear during take-off acceleration with an acceleration of a_{T} will be determined as follows:

$$F_{\mathrm{m_{dyn}}} = \frac{a_{\mathrm{T}} W H_{\mathrm{cg}}}{gB} \tag{9.16}$$

Thus, the total load on the main gear is:

$$F_{\mathrm{m}} = F_{\mathrm{m_{max}}} + F_{\mathrm{m_{dyn}}} = W \frac{B_{\mathrm{n_{max}}}}{B} + \frac{W a_{\mathrm{T}} H_{\mathrm{cg}}}{gB} \tag{9.17}$$

These static and dynamic loadings are utilized to determine the nose and main gear locations, strut load, and wheel and tire design. It must be noted that the main gear usually carries a total load which is greater than the aircraft weight.

Although an aircraft during landing tends to have a landing weight which is much less than the take-off weight, the landing gear must be designed based on the aircraft maximum take-off weight, not the landing weight. This is the current FAR regulation. The aircraft weight at landing is frequently about 20–30% less than the take-off weight. In the 1960s, about once a month, a Boeing 747 (Figures 3.7, 3.12, and 9.4) was dumping its fuel in the sky due to an aborted landing. This was because the landing gear was designed based on the aircraft normal landing weight to save weight and cost. Due to this design policy, the aircraft was not able to land with its take-off weight, and the pilot had to pour fuel into the sky to reduce the weight. The landing gear was designed based on $W_{\mathrm{L}}/W_{\mathrm{TO}} = 0.65$ at that time. When the environmentalists discovered that this flight policy was polluting the environment, they marched against it and lobbied in US Congress. After a few years, Congress passed a law and FAR 36 forced the Boeing Company to redesign the landing gear. This true story reveals the fact that law and regulations must be in place; otherwise, some designers and companies are willing to sacrifice the environment to get more profit. Example 9.3 illustrates how to calculate the dynamic loads that are carried by the nose gear and main gear based on the landing deceleration.

Example 9.3

A small business jet aircraft with a mass of 6500 kg has a tricycle landing gear configuration. The aircraft cg is allowed to move between 7.1 and 6.5 m from the nose gear.

1. The nose gear is desired to carry a maximum of 15% of the aircraft weight in static equilibrium, determine the wheel base.

2. The deceleration during landing braking is $-3\,\mathrm{m/s^2}$ and the acceleration during take-off is $4\,\mathrm{m/s^2}$. The distance between the aircraft cg and the ground is 2 m. Determine the maximum load on each gear.

Solution:

1.

$$F_{n_{max}} = \frac{B_{m_{max}}}{B} W \Rightarrow B = B_{m_{max}} \frac{W}{F_{n_{max}}} = (B - 6.5) \frac{W}{0.15W} = \frac{B}{0.15} - \frac{6.5}{0.15}$$

$$= 6.667B - 43.333 \qquad (9.9)$$

$$\Rightarrow B = 7.647\,\mathrm{m}$$

2. The maximum load on the nose gear will be during landing braking:

$$B_{m_{max}} = B - B_{n_{min}} = 7.647 - 6.5 = 1.147\,\mathrm{m}$$

$$F_n = F_{n_{max}} + F_{n_{dyn}} = W \frac{B_{m_{max}}}{B} + \frac{W\,|a_L|\,H}{gB} = 6500 \cdot 9.81 \cdot \frac{1.147}{7.647} + \frac{6500 \cdot 9.81 \cdot 3 \cdot 2}{9.81 \cdot 7.647} \Rightarrow$$

$$F_n = 14661.5\,\mathrm{N} \qquad (9.13)$$

It is interesting to note that this load is 23% of the aircraft weight. The maximum load on the main gear will be during take-off acceleration:

$$F_m = F_{m_{max}} + F_{m_{dyn}} = W \frac{B_{n_{max}}}{B} + \frac{W a_T H_{cg}}{gB} = 6500 \cdot 9.81 \cdot \frac{7.1}{7.647} + \frac{6500 \cdot 9.81 \cdot 4 \cdot 2}{9.81 \cdot 7.647}$$

$$\Rightarrow F_m = 65983.1\,\mathrm{N} \qquad (9.17)$$

It is interesting to note that this load is 103.5% of the aircraft weight. This implies that the main gear during take-off has to carry a total load which is 3.5% greater than the aircraft weight.

9.5.3 Wheel Track

Wheel track (T) is defined as the distance between the most left and the most right gears (when looking at a front view) and is measured at the ground (Figure 9.1). Three main design requirements which drive the magnitude of this parameter are: (i) ground lateral control, (ii) ground lateral stability, and (iii) structural integrity. The wheel track of the main wheel should be arranged so that the aircraft cannot roll over too easily due to wind or during a ground turn. Some aircraft, such as the British WWII single-seat fighter aircraft Supermarine Spitfire (Figure 8.3), were critical in this regard. To determine the wheel track, the overturn angle is introduced. The overturn angle is the angle which is critical to the aircraft overturn. There are two overturn angles (Figure 9.19), and the smaller one is considered in this technique.

1. When looking at the aircraft front view, the angle between the vertical line passing through the aircraft cg and the line between the aircraft cg and that of the main wheels is the overturn angle (Figure 9.19(b)). In this figure, the parameter H_{cg} is the height of the aircraft cg from the ground.
2. When looking at the aircraft top view, first draw a line passing through the main gears (say the left one) and the nose gear. Then, draw a line parallel to this line passing through the aircraft cg. The next step is to form a triangle by selecting a distance on this line equal to the length of *Hcg* (see Figure 9.19(a)), and draw a line perpendicular to this point. The last step is to pass a line from the intersection of the last line with the aircraft cg. The overturn angle is formed by this line, as shown.

As a rule of thumb, the wheel track must be such that the overturn angle Φ_{ot} is inside the following recommended limit:

$$\Phi_{ot} \geq 25^\circ \tag{9.18}$$

For an accurate determination of the wheel track, the three design requirements of (i) ground lateral control, (ii) ground lateral stability, and (iii) structural integrity must be examined, as explained in the following subsections. The minimum allowable value for the wheel track must satisfy the overturn angle requirements (Section 9.5.3.1). The maximum allowable value for the wheel track must satisfy the structural integrity requirements (Section 9.5.3.2). The wheel tracks for several aircraft are tabulated in Table 9.5.

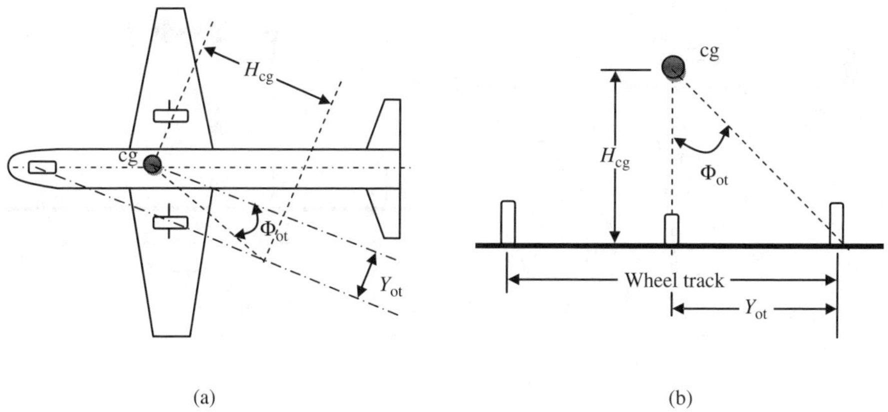

Figure 9.19 Overturn angle: (a) Φ_{ot} based on top view; (b) Φ_{ot} based on front view

9.5.3.1 Overturn Angles Requirement

One of the influencing requirements on the design of the landing gear is the overturn angle requirement. This requirement sets minimum and maximum limits for the wheel track. In general, there are two disturbing moments which are able to overturn an aircraft: (i) centrifugal force in a ground turn and (ii) cross-wind force. The first force is addressed in the ground controllability requirement, while the second one is examined in the ground stability requirement. The wheel track, or overturn angle, contributes to meeting these two design requirements in two separate ways.

A. Ground Controllability

The wheel track must be large enough such that the aircraft is not rolled over during a ground turn taxi. The force that may roll over the aircraft is the centrifugal force (F_C) which is created during a turn due to centripetal acceleration:

$$F_C = m\frac{V^2}{R} \tag{9.19}$$

where m represents the aircraft mass, V is the aircraft ground speed, and R is the radius of turn (see Figure 9.20(a)). The force to prevent the overturn is the aircraft weight. The two moments contributing to an overturn are the moment of the centrifugal force and the moment of the aircraft weight (Figure 9.20(b)). The restoring moment of the aircraft weight is a function of the wheel track. The summation of the two contributing moments about the outer main gear is as follows:

$$\sum M_O = 0 \Rightarrow W \cdot Y_{ot} + F_C \cdot H_{cg} = 0 \tag{9.20}$$

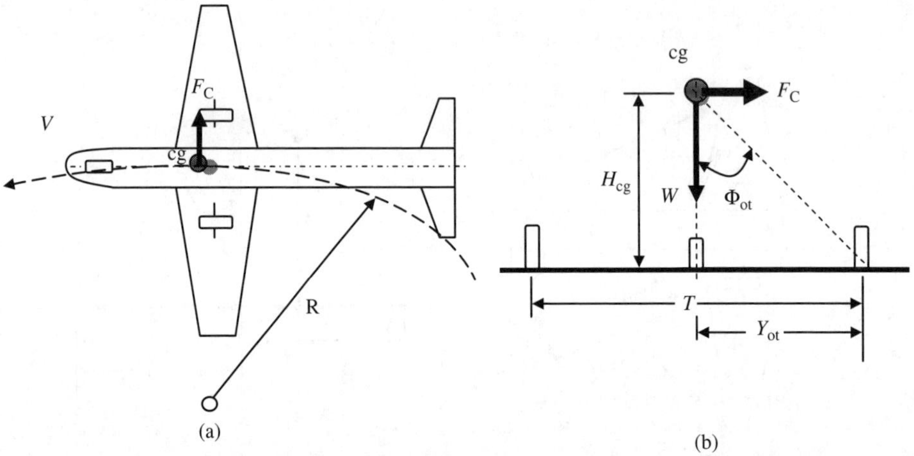

Figure 9.20 An aircraft in ground turn and overturn contributing factors: (a) Top view; (b) Front view

Thus:

$$Y_{ot} = \frac{F_C \cdot H_{cg}}{mg} \tag{9.21}$$

Therefore the wheel track must be:

$$T > 2\frac{F_C \cdot H_{cg}}{mg} \tag{9.22}$$

For the triangle in Figure 9.20(b), we can write:

$$\tan\left(\Phi_{ot}\right) = \frac{Y_{ot}}{H_{cg}} \tag{9.23}$$

Hence, the overturn angle must be:

$$\Phi_{ot} > \tan^{-1}\left(\frac{\dfrac{F_C \cdot H_{cg}}{mg}}{H_{cg}}\right) \Rightarrow \Phi_{ot} > \tan^{-1}\left(\frac{F_C}{mg}\right) \tag{9.24}$$

Thus, the wheel track (T) plays an important role in the aircraft ground controllability. It must be large enough to prevent the aircraft rolling over during a ground turn. The critical condition is when the aircraft has the lowest possible weight. Example 9.4 illustrates how to determine the minimum overturn angle and the wheel track to prevent an overturn during taxi.

Example 9.4

A twin-engine jet transport aircraft with a take-off mass of 60 000 kg and a wing area of 100 m^2 is turning on a runway. The ground speed is 20 knot and the turn radius is 30 m. The height of the aircraft center of gravity from the ground is 3.5 m.

1. Determine the minimum overturn angle to prevent an overturn in this taxi maneuver.
2. Determine the wheel track corresponding to this overturn angle.

Solution:

1.

$$F_C = m\frac{V^2}{R} = 60\,000 \cdot \frac{(20 \cdot 0.5144)^2}{30} = 211\,722.6\,\text{N} \tag{9.19}$$

$$\Phi_{ot} = \tan^{-1}\left(\frac{F_C}{mg}\right) = \tan^{-1}\left(\frac{211\,722.5}{60\,000 \cdot 9.81}\right) = 0.345\,\text{rad} = 23\,\text{deg} \tag{9.24}$$

Thus, any overturn angle greater than 23 deg will prevent the aircraft overturning in this taxi maneuver.

2.

$$T = 2\frac{F_C \cdot H_{cg}}{mg} = 2 \cdot \frac{211\,722.5 \cdot 3.5}{60\,000 \cdot 9.81} = 2.52\,\text{m} \tag{9.22}$$

The wheel track corresponding to this overturn angle is 2.52 m.

B. Ground Stability

One of the atmospheric phenomena affecting the aircraft ground stability is the wind. The most noticeable wind affecting an aircraft on the ground is the cross-wind, where it is perpendicular to the aircraft ground path or fuselage center line. A cross-wind creates a force on an aircraft at the ground, which in turn generates a moment that is capable of overturning the aircraft. The restoring moment is the aircraft weight times its corresponding arm (half of the wheel track). Thus, the wheel track (T) plays an important role in the aircraft ground stability. It must be large enough to prevent the aircraft rolling over when on the ground due to a cross-wind.

Figure 9.21 illustrates an aircraft on the ground with a cross-wind. Whenever a cross-wind is blowing, it will create a force (F_W) which is applied on the aircraft side area (Figure 9.22). The centroid of the aircraft side area (CA) may be obtained by integrating over the projected side area from nose to tail. The details of this technique are introduced in any statics textbook (e.g., Ref. [4]). In Figure 9.21 or 9.22, H_C is the height of the centroid from the ground.

The cross-wind force (F_W) on an aircraft may be modeled as a drag force and is calculated as follows:

$$F_W = \frac{1}{2}\rho V_W^2 A_S C_{D_S} \tag{9.25}$$

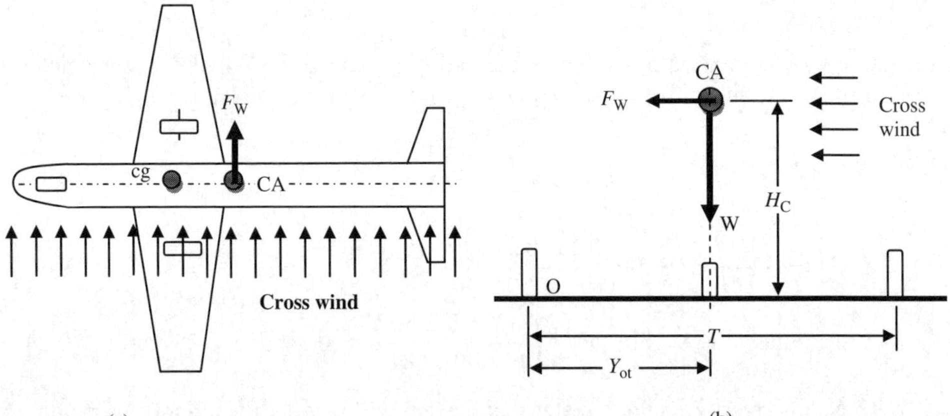

Figure 9.21 An aircraft on ground when a cross-wind is blowing. (a) Top view; (b) Front view

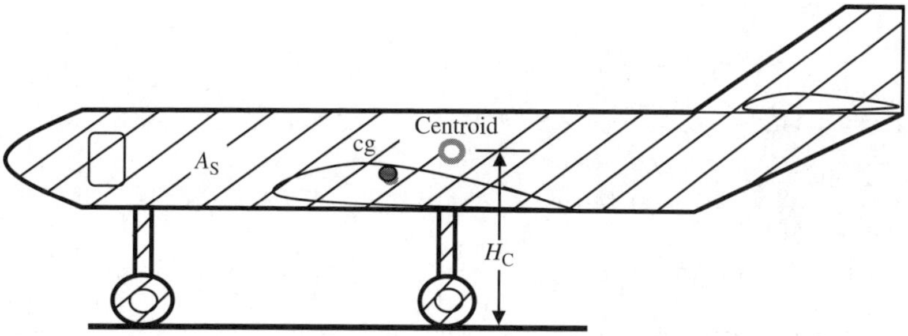

Figure 9.22 Aircraft side area and its centroid

where V_W represents the wind speed, and A_S represents the aircraft side area (hatched area in Figure 9.22). The parameter C_{D_S} is called the aircraft side drag coefficient and its value varies between 0.3 and 0.8. For the exact value of C_{D_S} you may consult any fluid mechanics textbook.

To prevent an aircraft from overturning under a cross-wind, the moment of the aircraft weight must be greater than the moment of the wind force (see Figure 9.21(b)). Taking the moment about the left main gear yields:

$$\sum M_O = 0 \Rightarrow W \cdot Y_{ot} + F_W \cdot H_C = 0 \tag{9.26}$$

Thus:

$$Y_{ot} = \frac{F_W \cdot H_C}{W} \tag{9.27}$$

Hence, the wheel track must be greater than twice the value of this Y_{ot} in order for an aircraft to be stable on the ground in case of a cross-wind:

$$T > 2Y_{ot} \tag{9.28}$$

Please note that the critical condition is when the aircraft has the lowest possible weight and the runway is located at sea-level altitude. In the majority of aircraft cases, satisfaction of the ground controllability automatically meets the ground stability requirement. Example 9.5 illustrates how to determine the minimum wheel track to prevent an overturn due to a cross-wind.

Example 9.5

Problem statement: Consider the aircraft in Example 9.4, on a runway at sea-level altitude. The aircraft side area is 150 m^2, and the height of the aircraft centroid of side area from the ground is 3.6 m. A cross-wind with a speed of 50 knot is blowing. Assume the aircraft side drag coefficient is 0.8. Determine the minimum wheel track to prevent an overturn due to this cross-wind. The lowest possible mass is 40 000 kg when there is no passenger onboard and zero fuel.

Solution:

$$F_W = \frac{1}{2}\rho V_W^2 A_S C_{D_S} = \frac{1}{2} \cdot 1.225 \cdot (50 \cdot 0.5144)^2 \cdot 150 \cdot 0.8 = 48\,630\,\text{N} \quad (9.25)$$

$$Y_{ot} = \frac{48\,630 \cdot 3.6}{40\,000 \cdot 9.81} = 0.446\,\text{m} \quad (9.27)$$

$$T > 2Y_{ot} = 2 \cdot 0.446 = 0.893\,\text{m} \quad (9.28)$$

Therefore, the minimum wheel track for this aircraft to avoid a rollover due to this cross-wind is 0.9 m.

9.5.3.2 Structural Integrity

The previous section introduced the technique to obtain a minimum value for the wheel track to avoid a rollover. Another limit for the wheel track is the maximum value, which is presented in this section. The maximum value for the wheel track is limited by the aircraft structural integrity requirement. When looking at an aircraft from a front view, the aircraft structure may be viewed as a beam with a few simple supports (Figure 9.23). In an aircraft with a tricycle configuration, at the main gear station the beam is the wing and two simple supports are the two main gears. Thus, the wheel track is another name for the distance between two supports.

Based on the basic theory of structural engineering, a beam with two simple supports will deflect. The maximum deflection (y_{max}) will be at the middle of the beam. As the distance between two supports is increased (i.e., the wheel track increases), the beam deflection will increase too. The limiting factors for this deflection (i.e., wheel track) are as follows:

1. An increase in the wheel track will be translated as an increase in the wing dihedral, which in turn degrades the aircraft lateral stability and roll control.
2. An increase in the wheel track will cause the fuselage to deflect down, and in the worst case the fuselage may touch the ground.
3. An increase in the wheel track may degrade the aircraft structural integrity, aerodynamic integrity, and in the worst case, the structure may break.

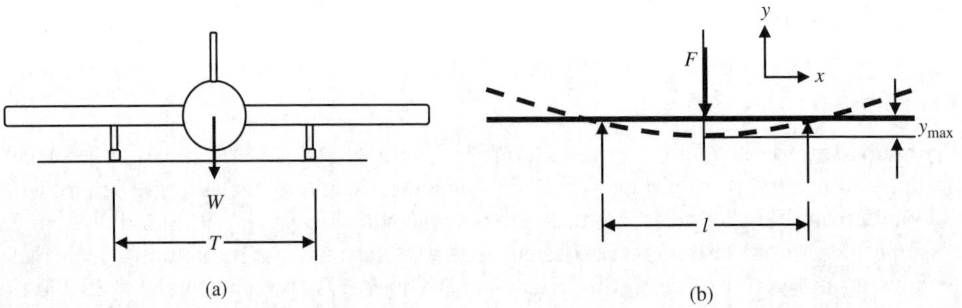

Figure 9.23 The aircraft structure at front view may be modeled as a beam with two simple supports: (a) Aircraft structure; (b) Beam with two simple supports

As soon as we know the allowable deflection for the structure at the main wheel station, the wheel track is obtained.

The maximum deflection (y_{max}) in a beam with a force F at the middle of the beam (Figure 9.23(b)) is determined [5] as follows:

$$y_{max} = -\frac{Fl^3}{48EI} \tag{9.29}$$

where E is the modulus of elasticity and I is the second moment of the area of the beam. This equation may be applied to the aircraft (Figure 9.23(a)) as follows:

$$y_{max} = -\frac{F_{m_{max}} T^3}{48EI} \tag{9.30}$$

where $F_{m_{max}}$ is the maximum load on the main gear which was obtained earlier in this chapter:

$$F_{m_{max}} = \frac{B_{n_{max}}}{B} W \tag{9.8}$$

where B denotes the wheel base and $B_{n_{max}}$ denotes the maximum distance between the aircraft cg and the nose gear in a tricycle configuration. Substituting Equation (9.8) into Equation (9.30) yields:

$$y_{max} = -\frac{B_{n_{max}} W T^3}{48EIB} \tag{9.31}$$

Now we can write the wheel track in terms of the maximum allowable deflection and other parameters:

$$T = \left[\frac{48EIBy_{max}}{WB_{n_{max}}}\right]^{\frac{1}{3}} \tag{9.32}$$

Using this equation, one can determine the maximum limit for the wheel track in terms of aircraft weight, aircraft geometry, and structural coefficients. Since the wheel track is inversely proportional to the aircraft weight, the critical condition is with maximum take-off weight. This technique can easily be revised for other landing gear configurations. Example 9.6 illustrates how to determine the maximum allowable wheel track to satisfy a structural integrity requirement.

Example 9.6

Problem statement: An aircraft with a mass of 30 000 kg and a wing span of 42 m has a tricycle landing gear configuration. The wheel base is 15 m, and the maximum distance between the aircraft cg and the nose gear is 13 m. The wing is made of aluminum with a modulus of elasticity of 70 GPa. Assume that the wing can be modeled with a beam of I-section with a second moment of area 0.003 m^4. If the maximum allowable wing deflection is 3 cm, determine the maximum allowable wheel track.

Solution:

$$T = \left[\frac{48EIBy_{max}}{WB_{n_{max}}}\right]^{\frac{1}{3}} = \left[\frac{48 \cdot 70 \cdot 10^9 \cdot 0.003 \cdot 15 \cdot 0.03}{30\,000 \cdot 9.81 \cdot 13}\right]^{\frac{1}{3}} \Rightarrow T = 10.58\ \text{m} \tag{9.32}$$

9.6 Landing Gear and Aircraft Center of Gravity

An important factor in the landing gear design process is to determine the location of the main gear relative to the aircraft center of gravity. An aircraft usually has two extreme cg locations:

1. most forward cg;
2. most aft cg.

In an aircraft with a tricycle landing gear, the location of the main gear with respect to the most aft cg is governed by the tipback angle requirement. Furthermore, the location of the main gear with respect to the most forward cg is governed by the take-off rotation requirement. The tipback angle requirement is described in Section 9.6.1, while the details of the take-off rotation requirements will be presented in Section 9.6.2. For other landing gear configurations, the reader is asked to identify and develop the requirements with respect to cg locations.

In contrast, in an aircraft with a tail-wheel landing gear, the location of the main gear with respect to the most forward cg is governed by the tipback angle requirement. But the location of the main gear with respect to the most aft cg is governed by the take-off rotation requirement. For other configurations, the reader is asked to identify the requirements with respect to the cg locations. The tipforward angle requirement is described in Section 9.6.1, while the take-off rotation requirements are presented briefly in Section 9.6.2. For other landing gear configurations, the reader is asked to identify and develop the requirements with respect to cg locations.

The significance of relating the landing gear design to the aircraft center of gravity is to make sure that the major landing gear variables – such as the wheel base, wheel track, and wheel height – satisfy all requirements. When the above-mentioned requirements are satisfied, one or more changes in the design must be applied. In the majority of cases, the designer at this point needs to iterate the landing gear design and revise the values. In rather noticeable cases, the designer is forced to redesign other aircraft components (e.g., wing, tail, and fuselage). Even, in some cases, the designer has to switch to a new aircraft configuration. Thus the satisfaction of these three requirements is very crucial in the entire aircraft design process.

9.6.1 Tipback and Tipforward Angle Requirements

The tipback and tipforward angle requirements are defined to prevent the aircraft from tipping back on its tail or tipping forward on its nose. The tipback angle requirement regulates the distance between the aircraft most aft cg and the main gear in a tricycle configuration. In contrast, the tipforward angle requirement regulates the distance between the aircraft most forward cg and the main gear in a tail-gear configuration.

In an aircraft with a tricycle landing gear, if during a take-off rotation the aircraft cg moves aft of the main gear, the aircraft will fall back onto the ground. Similarly, in an aircraft with a tail-wheel landing gear, if during a take-off rotation the aircraft cg moves forward of the main gear, the aircraft nose will fall forward onto the ground. To prevent such accidents as tipback and tipforward, two requirements are defined. These two requirements are examined in this section. For other landing gear configurations, the fundamentals of these two requirements need to be applied accordingly.

9.6.1.1 Tipback Angle Requirement

The tipback angle is the maximum aircraft nose-up attitude with the tail touching the ground and the strut fully extended. To prevent a tipback in a tricycle configuration, the tipback angle (α_{tb}) must always be greater than the take-off rotation angle (α_{TO}) (see Figure 9.24(a)):

$$\alpha_{tb} \geq \alpha_{TO} + 5\,\deg \tag{9.33a}$$

According to Figure 9.24, the tipback angle is:

$$\alpha_{tb} = \tan^{-1}\left(\frac{x_{mg}}{h_{cg}}\right) \tag{9.34}$$

In Equation (9.33), the angular difference of 5 deg is selected as a safety assurance to cover uncertainties. The typical take-off rotation angle is about 10–15 deg, so the tipback angle must be equal to or greater than 15–20 deg. Furthermore, the tipback angle must be less than the angle measured from the vertical (at the main gear location) to the aircraft most aft center of gravity. One of the techniques to increase the tipback angle is to reduce the landing gear height. The second way is to move back the main gear.

9.6.1.2 Tipforward Angle Requirement

For the case of an aircraft with a tail-wheel landing gear, the term tipforward angle (α_{tf}) is employed (see Figure 9.24(b)). The tipforward angle is the angle between the vertical and the line passing through the aircraft most forward cg and the contact point between tire and ground. The tipforward angle must be greater than the fuselage incline angle (α_{fi}). The angle is measured when the aircraft is in the horizontal position:

$$\alpha_{tf} \geq \alpha_{fi} + 5\,\deg \tag{9.33b}$$

In Equation (9.33b), the angular difference of 5 deg is selected as a safety assurance to cover uncertainties. An aircraft with a tail gear during take-off is normally rotated about its main gear due to a local increase in the tail lift. Thus, if the cg during take-off rotation

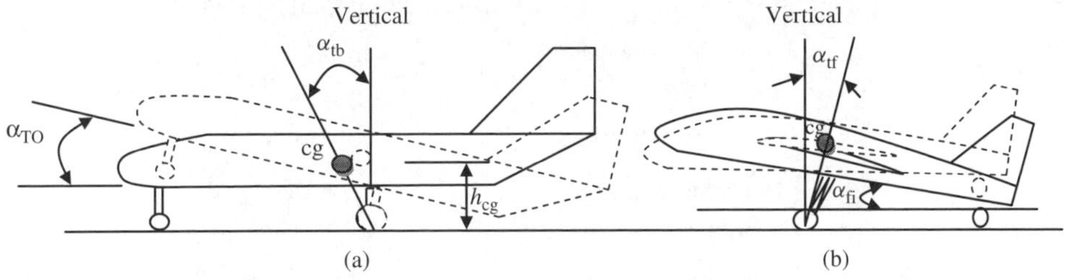

Figure 9.24 Tipback angle, tipforward angle, and take-off rotation: (a) Aircraft with tricycle landing gear; (b) Aircraft with tail-wheel landing gear

passes the vertical limit, the nose will fall forward onto the ground. To avoid this accident, the landing gear height (i.e., main gear height) must be increased or its location must be moved forward. As a rule of thumb, the tipforward angle is usually between 12 and 20 deg.

9.6.2 Take-Off Rotation Requirement

For an aircraft with a landing gear configuration in which the main gear is behind the aircraft cg (e.g., tricycle landing gear) , the take-off rotation requirement is defined to regulate the distance between the main gear and the most forward cg. Most aircraft, to become airborne, must be rotated about the main gear to achieve the angle of attack required for lift-off. Exceptions to this are aircraft like the military bomber aircraft Boeing B-52 Stratofortress (Figures 8.6 and 9.4). The take-off rotation requirement requires the distance between the main gear and the most forward cg to be such that the pitch angular acceleration ($\ddot{\theta}$) is greater than a desired value. In this section, the requirement is mathematically developed and we specifically focus on the relationship with landing gear design.

The angular acceleration about the main gear rotation point, $\ddot{\theta}$, is a function of a couple of parameters including the horizontal tail area, horizontal tail arm, elevator control power, aircraft weight, rotation speed, and finally the distance between the main gear and the aircraft cg. Typical rotational acceleration is given in Table 9.6 for various types of aircraft. For acceleration requirements for military aircraft, the reader is recommended to refer to military standards such as [12]. The rotation acceleration is the aircraft acceleration at the time the aircraft begins to rotate about the main gear. This speed must be slightly more than the stall speed (V_s). During the landing gear design process, it may be assumed that the airplane rotation speed is:

$$V_R = 1.1 \text{ to } 1.3\, V_s \tag{9.35}$$

In this section, an analysis of the distance between the main gear and the aircraft cg required to generate a given level of pitch angular acceleration about the main gear contact point is presented. Consider the aircraft with a tricycle landing gear in Figure 9.25, which is at the onset of a rotation about the main gear in a take-off operation. The figure illustrates all forces and moments contributing to this moment of take-off. Contributing forces include the wing/fuselage lift (L_{wf}), horizontal tail lift (L_h), aircraft drag (D),

Table 9.6 Take-off rotational acceleration for various aircraft

No.	Aircraft type	Take-off pitch angular acceleration $\ddot{\theta}$ (deg/s^2)
1	Highly maneuverable (e.g., acrobatic, fighter)	10–20
2	Utility, semi-acrobatic	10–15
3	Normal light general aviation	8–10
4	Small transport	6–8
5	Large transport	4–6

friction force between tires and the ground (F_f), aircraft weight (W), engine thrust (T), and acceleration force (ma). Please note that the latter force (ma) is acting backwards due to Newton's third law (as a reaction to the acceleration). Furthermore, the contributing moments are the wing/fuselage aerodynamic pitching moment ($M_{o_{wf}}$) plus the moments of preceding forces about the rotation point. The distance between these forces is measured with respect to both the x reference line (i.e., fuselage nose) and the z reference line (i.e., ground), as shown in Figure 9.25.

For a conventional aircraft with tricycle landing gear, the horizontal tail lift is negative during rotation. It is recommended to consider the ground effect on the lift and drag. The friction coefficient μ depends on the type of terrain. Table 9.7 introduces the friction coefficients for different terrains.

There are three governing equations of motion that govern the aircraft equilibrium at the instant of rotation, two force equations and one moment equation:

$$\sum F_x = m\frac{dV}{dt} \Rightarrow T - D - F_f = ma \Rightarrow T - D - \mu N = ma \tag{9.36}$$

$$\sum F_z = 0 \Rightarrow L + N = W \Rightarrow L_{wf} - L_h + N = W \Rightarrow N = W - \left(L_{wf} - L_h\right) \tag{9.37}$$

$$\sum M_{cg} = I_{yy_{mg}}\ddot{\theta} \Rightarrow -M_W + M_D - M_T + M_{L_{wf}} + M_{ac_{wf}} + M_{L_h} + M_a = I_{yy_{mg}}\ddot{\theta} \tag{9.38}$$

In Equation (9.36), the force N is the normal force on the ground which is obtained from

$$N = W - L_{TO} \tag{9.39}$$

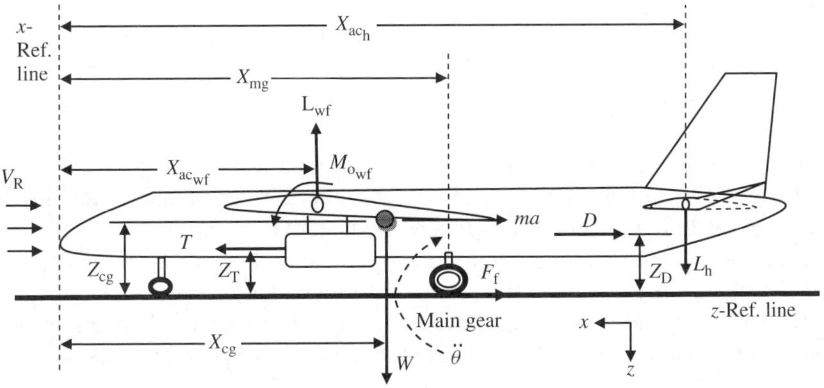

Figure 9.25 Forces and moments during take-off rotation

Table 9.7 Friction coefficient for various runways

Type of terrain	Concrete	Asphalt	Hard turf	Short grass	Long grass	Firm dirt
Wheel-to-ground friction coefficient	0.03–0.04	0.04–0.05	0.05	0.05–0.07	0.07–0.1	0.04–0.06

So, the friction force (F_f) is:

$$F_f = \mu N = \mu \left(W - L_{TO} \right) \quad (9.40)$$

The aircraft take-off lift is obtained by the following expression:

$$L_{TO} = \frac{1}{2} \rho V_R^2 C_{L_{TO}} S_{ref} \quad (9.41)$$

where the aircraft lift is equal to the sum of the wing/fuselage lift (L_{wf}) plus the horizontal tail lift (L_h):

$$L_{TO} = L_{wf} + L_h \Rightarrow L_{wf} = L_{TO} - L_h \quad (9.42)$$

where

$$L_h = \frac{1}{2} \rho V_R^2 C_{L_h} S_h \quad (9.43a)$$

$$L_{wf} = \frac{1}{2} \rho V_R^2 C_{L_{wf}} S_{ref} \quad (9.43b)$$

A negative sign for the horizontal tail in Equation (9.37) indicates that this force acts downward. This force is generated by upward deflection of the elevator. The other aerodynamic forces and pitching moments are obtained from the following expressions:

$$D = \frac{1}{2} \rho V_R^2 C_D S_{ref} \quad (9.44)$$

$$M_{ac_{wf}} = \frac{1}{2} \rho V_R^2 C_{m_{ac\_wf}} S_{ref} \overline{C} \quad (9.45)$$

where V_R denotes the aircraft linear forward speed at the instant of rotation, S_{ref} represents the wing planform area, S_h is the horizontal tail planform area, ρ is the air density, and $\overline{C}$ is the wing mean aerodynamic chord. Furthermore, the four coefficients of C_D, $C_{L_{wf}}$, C_{L_h}, and $C_{m_{ac\_wf}}$ denote drag, wing/fuselage lift, horizontal lift, and wing/fuselage pitching moment coefficients respectively.

In Equation (9.38), the clockwise rotation is assumed to be a positive rotation. Thus, the aircraft weight and engine thrust both create negative moments. Recall that the wing/fuselage pitching moment is also inherently negative, so its sign is already included. In Equation (9.38), the contributing moments are aircraft weight moment (M_W), aircraft drag moment (M_D), engine thrust moment (M_T), wing/fuselage lift moment ($M_{L_{wf}}$), wing/fuselage aerodynamic pitching moment ($M_{ac_{wf}}$), horizontal tail lift moment (M_{L_h}), and linear acceleration moment (M_a). These moments are obtained as follows:

$$M_W = W \left(x_{mg} - x_{cg} \right) \quad (9.46)$$

$$M_D = D \left(z_D - z_{mg} \right) \quad (9.47)$$

$$M_T = T \left(z_T - z_{mg} \right) \quad (9.48)$$

$$M_{L_{wf}} = L_{wf} \left(x_{mg} - x_{ac_{wf}} \right) \quad (9.49)$$

$$M_{L_h} = L_h \left(x_{ac_h} - x_{mg} \right) \quad (9.50)$$

$$M_a = ma \left(z_{cg} - z_{mg} \right) \quad (9.51)$$

In Equations (9.46)–(9.51), the subscript "mg" denotes main gear, since the distances are measured from the main gear. The inclusion of the moment generated by the aircraft acceleration (Equation (9.51)) is due to the fact that based on Newton's third law, any action creates a reaction (ma). This reaction force produces a moment when its corresponding arm is taken into account. Substituting these moments into Equation (9.38) yields:

$$\sum M_{cg} = I_{yy}\ddot{\theta} \Rightarrow -W\left(x_{mg} - x_{cg}\right) + D\left(z_D - z_{mg}\right) - T\left(z_T - z_{mg}\right) + L_{wf}\left(x_{mg} - x_{ac_{wf}}\right)$$
$$+ M_{ac_{wf}} - L_h\left(x_{ac_h} - x_{mg}\right) + ma\left(z_{cg} - z_{mg}\right) = I_{yy_{mg}}\ddot{\theta} \quad (9.52)$$

where $I_{yy_{mg}}$ represents the aircraft mass moment of inertia about the y-axis at the main gear. Thus, the aircraft mass moment of inertia about the cg (y-axis) must be transferred to the main gear contact point ($I_{yy_{mg}}$) by employing the parallel axis theorem:

$$I_{yy_{mg}} = I_{yy_{cg}} + m\left(d_{cg\text{-}mg}\right)^2 \quad (9.53)$$

where $d_{cg\text{-}mg}$ is the distance between the aircraft cg and the main gear contact point, and m is the aircraft mass. Please note that for a tricycle landing gear, the tail lift moment, wing/fuselage moment, drag moment, and acceleration moment are all clockwise, while the weight moment, thrust moment, and wing/fuselage aerodynamic pitching moment are counterclockwise. These directions must be considered when assigning a sign to each one. Equation (9.52) is only a function of one unknown (x_{mg}), the distance between the main gear and a reference line, which can be obtained from Equation (9.52). The result is as follows:

$$x_{mg} = \frac{I_{yy_{mg}}\ddot{\theta} - D\left(z_D - z_{mg}\right) + T\left(z_T - z_{mg}\right) - M_{ac_{wf}} - ma\left(z_{cg} - z_{mg}\right) - Wx_{cg} + L_{wf}x_{ac_{wf}} + L_h x_{ac_h}}{L_{wf} + L_h - W} \quad (9.54)$$

Then, this distance will be used to determine the main gear location with respect to the aircraft most forward cg ($x_{mg} - x_{cg}$) in order to satisfy the take-off rotation requirement. The magnitude of the linear acceleration is determined by employing Equation (9.36). It is interesting to note that this distance ($x_{mg} - x_{cg}$) is the maximum allowable distance for the main gear location. You may reduce this distance to account for other design requirements.

Another important landing gear design in determining the main gear location is to avoid auto-rotation (pitch-up) at lift-off right after rotation. A few passenger aircraft are notorious in this regard. This phenomenon will occur when the distance between the wing/fuselage aerodynamic center and the main gear is too large. In such an aircraft, the pilot must immediately return the stick, after pulling it back.

Example 9.7

Problem statement: A small subsonic business aircraft (Figure 9.26) with a take-off mass of 13 000 kg and a wing area of 45 m^2 has two turbofan engines, each generating 20 000 N of thrust. The overall length of the aircraft is 15 m, it has a tricycle landing gear, and the runway is concrete. Assume that the forward cg is at 20% MAC, and the wing/fuselage ac is at 24% MAC. The aircraft is equipped with a single-slotted flap which is set to generate an extra lift coefficient of 0.6 during take-off. The elevator deflection during take-off rotation generates a tail lift coefficient of −1.1.

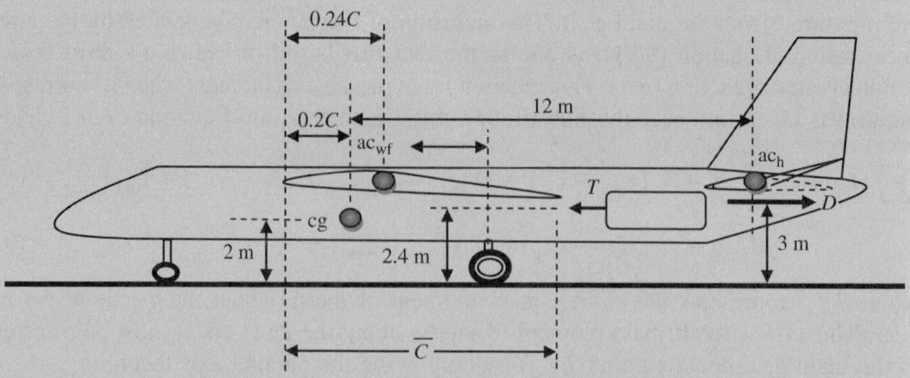

Figure 9.26 Aircraft in Example 9.7

Some dimensions of the aircraft are shown in Figure 9.26, and other characteristics of the aircraft are as follows:

$$V_c = 400 \text{ KTAS (at 20 000 ft)}, \; V_s = 80 \text{ KEAS}, \; C_{D_o} = 0.025, \; C_{D_{o_{TO}}} = 0.035,$$
$$I_{yy_{mg}} = 20\,000\,\text{kg m}^2, \; \text{AR} = 10, \; C_{m_o} = -0.04, \; e = 0.92, \; S_h = 9\,\text{m}^2$$

The aircraft is required to rotate about the main gear with an angular acceleration of 7 deg/s^2 during the take-off operation at sea-level altitude. Determine the distance between the main wheel and the aircraft forward cg.

Solution:
From Figure 9.26, we can extract the following dimensions:

$$h_{cg} = 2\,\text{m}, \; h_D = 3\,\text{m}, \; h_T = 2.4\,\text{m}, \; l_h = 12\,\text{m}, \; x_{L_{wf}} = x_{mg} - (0.24 - 0.2)\overline{C}$$

The air density at sea level is 1.225 kg/m^3, and at 20 000 ft is 0.653 kg/m^3. To obtain the wing mean aerodynamic chord:

$$b = \sqrt{S.\text{AR}} = \sqrt{45 \cdot 10} = 21.213\,\text{m} \tag{5.19}$$

$$\overline{C} = \frac{S}{b} = \frac{45}{21.213} = 2.121\,\text{m} \tag{5.18}$$

To find aircraft drag:

$$K = \frac{1}{\pi.e.\text{AR}} = \frac{1}{3.14 \cdot 0.92 \cdot 10} = 0.035 \tag{5.22}$$

$$C_{L_C} = \frac{2W}{\rho V_C^2 S} = \frac{2 \cdot 13\,000 \cdot 9.81}{0.653 \cdot (400 \cdot 0.5144)^2 \cdot 45} = 0.205 \tag{5.1}$$

$$C_{L_{\rm TO}} = C_{L_C} + \Delta C_{L_{\rm flap}} = 0.205 + 0.6 = 0.805 \quad (4.69c)$$

$$C_{D_{\rm TO}} = C_{D_{\rm o_{TO}}} + KC_{L_{\rm TO}}^2 = 0.035 + 0.035 \cdot 0.805^2 = 0.057 \quad (4.68)$$

$$V_{\rm R} = 1.1V_{\rm s} = 1.1 \cdot 80 = 88\,\text{knot} \quad (9.35)$$

$$D_{\rm TO} = \frac{1}{2}\rho V_{\rm R}^2 S C_{D_{\rm TO}} = \frac{1}{2} \cdot 1.225 \cdot (88 \cdot 0.5144)^2 \cdot 45 \cdot 0.057 = 3244.9\,\text{N} \quad (9.44)$$

Other aerodynamic forces and moments:

$$L_{\rm TO} = \frac{1}{2}\rho V_{\rm R}^2 S_{\rm ref} C_{L_{\rm TO_f}} = \frac{1}{2} \cdot 1.225 \cdot (88 \cdot 0.5144)^2 \cdot 45 \cdot 0.805 = 45\,490\,\text{N} \quad (9.41)$$

$$L_{\rm h} = \frac{1}{2}\rho V_{\rm R}^2 S_{\rm h} C_{L_{\rm h}} = \frac{1}{2} \cdot 1.225 \cdot (88 \cdot 0.5144)^2 \cdot 9 \cdot (-1.1) = -12\,433\,\text{N} \quad (9.43a)$$

$$M_{\rm ac_{wf}} = \frac{1}{2}\rho V_{\rm R}^2 C_{\rm m_{ac\_wf}} S_{\rm ref} \overline{C} = \frac{1}{2} \cdot 1.225 \cdot (88 \cdot 0.5144)^2 \cdot 45 \cdot (-0.04) \cdot 2.121$$

$$= -4795.4\,\text{Nm} \quad (9.45)$$

$$L_{\rm wf} = L_{\rm TO} - L_{\rm h} = 45\,490 - (-12\,433) = 57\,923\,\text{N} \quad (9.42)$$

Friction force:

$$F_{\rm f} = \mu\left(W - L_{\rm TO}\right) = 0.02\,(13\,000 \cdot 9.81 - 45\,490) = 1640\,\text{N} \quad (9.40)$$

Aircraft linear acceleration at the time of take-off rotation:

$$a = \frac{T - D - F_{\rm R}}{m} = \frac{20\,000 \cdot 2 - 3244.9 - 1640}{13\,000} \Rightarrow a = 2.701\,\text{m/s}^2 \quad (9.36)$$

Contributing moments are:

$$M_W = W\left(x_{\rm mg} - x_{\rm cg}\right) = W\left(x_{\rm mg}\right) \quad (9.46)$$

$$M_D = D\left(z_D - z_{\rm mg}\right) = 3244.9 \cdot 3 = 9734.6\,\text{Nm} \quad (9.47)$$

$$M_T = T\left(z_T - z_{\rm mg}\right) = 20\,000 \cdot 2 \cdot 2.4 = 96\,000\,\text{Nm} \quad (9.48)$$

$$M_{L_{\rm wf}} = L_{\rm wf}\left(x_{\rm mg} - x_{\rm ac_{wf}\,to\,cg}\right) = 57\,923 \cdot \left(x_{\rm mg} - 0.04 \cdot 2.121\right) \quad (9.49)$$

$$M_{L_{\rm h}} = L_{\rm h}\left(x_{\rm ac_h} - x_{\rm mg}\right) = -12\,433.3 \cdot \left(12 - x_{\rm mg}\right) \quad (9.50)$$

$$M_{\rm a} = ma\left(z_{\rm cg} - z_{\rm mg}\right) = 13\,000 \cdot 2.701 \cdot 2 = 70\,230.4\,\text{Nm} \quad (9.51)$$

Please note that in this example, the x reference line is assumed to be the aircraft cg, thus $x_{cg} = 0$. Furthermore, for all moment arms, the absolute value is utilized. Now, all the moments are substituted into Equation (9.54):

$$x_{mg} = \frac{I_{yymg}\ddot{\theta} - D\left(z_D - z_{mg}\right) + T\left(z_T - z_{mg}\right) - M_{ac_{wf}} - ma\left(z_{cg} - z_{mg}\right) - Wx_{cg} + L_{wf}x_{ac_{wf}} + L_h x_{ac_h}}{L_{wf} + L_h - W} \quad (9.54)$$

$$x_{mg} = \frac{20\,000 \cdot \frac{7}{57.3} - 9734.6 + 96\,000 - (-4795.4) - 70\,230.4 + (57\,923 \cdot 0.04 \cdot 2.121) + (-12\,433.3 \cdot 12)}{57\,923 - 12\,433.3 - (13\,000 \cdot 9.81)} \quad (9.54)$$

which yields:

$$x_{mg} = 1.476\,\mathrm{m}$$

This distance indicates (according to Figure 9.15) that the aircraft has the following tipback angle:

$$\alpha_{tb} = \tan^{-1}\left(\frac{x_{mg}}{h_{cg}}\right) = \tan^{-1}\left(\frac{1.476}{2}\right) = 0.636 \Rightarrow \alpha_{tf} = 36.4\,\mathrm{deg} \quad (9.34)$$

9.7 Landing Gear Mechanical Subsystems/Parameters

The scope of this book concerns the aeronautical engineering aspects of landing gear design, including parameters such as landing gear configuration, fixed or retractable, landing gear height, wheel base, wheel track, and distance from the main wheel to the aircraft center of gravity. The mechanical engineering aspects/subsystems of landing gear design have been left to other references which discuss these items in more detail. In other words, these parameters would often be left for mechanical engineers to deal with. The design of the landing gear subsystems/parameters – such as retraction system, steering subsystem, shock absorber, tire sizing, braking subsystem, and strut sizing – are reviewed in brief in this section.

9.7.1 Tire Sizing

Technically, the term "wheel" refers to a circular metal/plastic object around which the rubber "tire" is mounted. The brake system is mounted inside the wheel to slow the aircraft during landing. However, in the majority of cases, the entire wheel, tire, and brake system is also referred to as the wheel. The fundamental materials of modern tires are synthetic or natural rubber, fabric, and wire, along with other compound chemicals. Today, most tires are pneumatic inflatable and include a doughnut-shaped body of cords and wires encased in rubber. So they consist of a tread and a body (Figure 9.27). Tires perform four important functions with the assistance of the air contained within them:

1. Tires support the aircraft structure off the ground.
2. They help absorb shocks from the runway surface.
3. They help transmit acceleration and braking forces to the runway surface.
4. They help change and maintain the direction of motion.

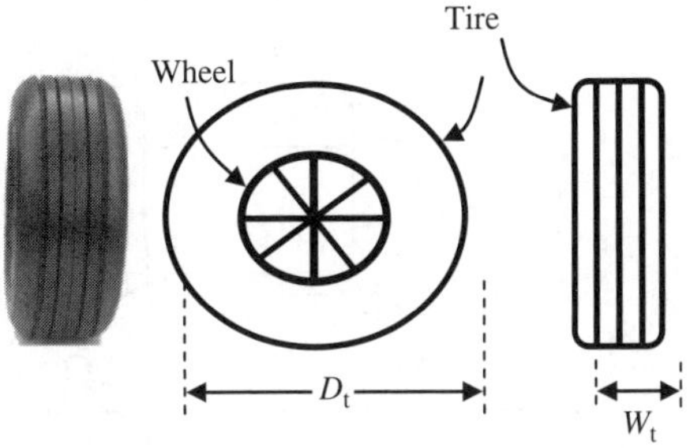

Figure 9.27 Tire geometry

A tire carries the load almost entirely by its internal pressure. Tire sizing includes the calculation of the tire outer diameter (D_t) and the tire width (W_t), then selecting the closest tire in the market from a manufacturer's catalog (e.g., Refs [6, 7]). Tire selection should be based on the smallest diameter rated to carry the desired dynamic and static loads.

As a guideline, the following is the information about tires for a civil transport, a military fighter, and a GA aircraft .The transport aircraft Boeing 777-200 employs [8] Goodyear main tires H49 · 19-22, and Michelin radial nose-wheel tires 44 · 18-18. The fighter aircraft McDonnell Douglas F-15 Eagle (Figures 4.6 and 9.14) utilizes [8] Bendix wheels and Michelin AIR X with nose-wheel tires of size 22 · 7.75–9, and main wheel tires of size 36 · 11–18 where the tire pressure is 305 psi. The main-wheel tire of the business jet Cessna 650 Citation VII [8] is of size 22 · 5.75 (pressure of 168 psi), while the nose-wheel tire is of size 188 · 4.4 (140 psi).

Generally speaking, for a tricycle configuration, nose tires may be assumed to be about 50–100% the size of the main tires. For quadricycle and bicycle configurations, the front tires are often the same size as the main tires.

9.7.2 *Shock Absorber*

The landing gear must be able to absorb the shocks exerted on the structure during the landing operation (mainly at touchdown phase). Some light, ultralight, small, and home-built aeroplanes, most helicopters, plus sailplanes are built with rigid axles or solid springs, relying solely on the tires and solid springs for absorbing shocks. Although the tires themselves provide some shock-absorbing abilities by deflection, for medium/large aircraft, the requirements for absorbing shock are higher than what the tires are offering. The solid spring (Figure 9.28(a)), which tends to be fairly simple in design, is employed in many GA light aircraft (e.g., Cessna 172 (Figure 11.15), Cessna Caravan (Figure 3.7), Beech 77 Skipper, AkroTech Aviation Giles G-200). However, almost all modern transport aircraft and military fighters (e.g., Boeing 737 (Figure 6.12), Boeing 767 (Figure 5.4), Airbus 330 (Figures 5.51 and 9.14), F/A-18 (Figures 2.11 and 6.12), C-130 Hercules (Figure 5.4), and F-16 Falcon (Figure 4.6)) are equipped with oleo-pneumatic shock absorbers or "oleo" (Figure 9.28(b)). The oleo combines a mechanical coil spring (in air) with a hydraulic damper (piston/oil/cylinder/orifice).

In general, if the landing gear is selected to be fixed, a solid spring (i.e., bar), a rigid axle, or a rubber bungee would be suitable options. However, if the landing gear is decided

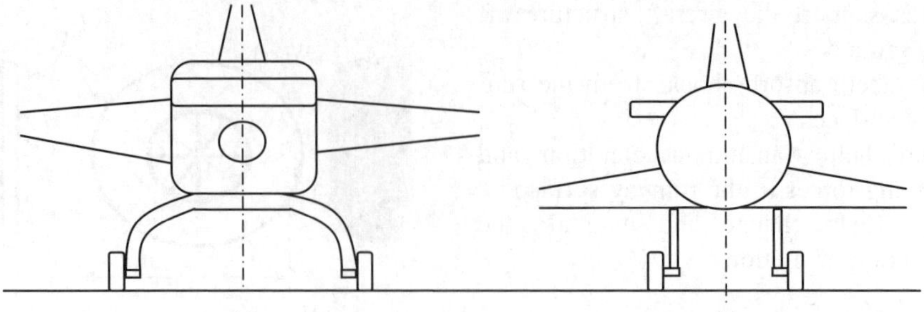

Figure 9.28 Landing gear with shock absorber: (a) Solid spring; (b) Oleo shock strut

to be retracted, the hydraulic shock absorber (preferably oleo-pneumatic shock strut) is an appropriate option. In terms of cost, an oleo shock absorber is much more expensive than a solid spring. Furthermore, the maintenance of an oleo shock absorber is very much more labor-extensive than a solid spring.

In both cases, the deflection of the solid spring or oleo will change the length of the strut, the parameter which must be taken into account during the landing gear design process. The desired deflection of the shock absorbing system (i.e., stroke) is a function of aircraft landing speed during touchdown, as well as the damping requirements. A smoother landing requires a longer deflection, which in turn applies less "g" on the structure. The total aircraft energy that must be absorbed during touchdown is a kinetic energy which is derived by the aircraft mass as well as the aircraft vertical velocity at the instance of touchdown. In determining the ground loads on the nose wheels and tail wheel, and affected supporting structures, it must be assumed that the shock absorbers and tires are in their static positions.

When a solid spring is chosen, the main parameter for the design is the geometry and cross-section of the beam. For more information on the solid spring (i.e., beam) design, the reader is referred to references such as [5]. In case a hydraulic shock absorber is selected for the landing gear, the typical parameters which must be determined include stroke, orifice, outer and inner diameter, and internal spring sizing. References such as [9] may be consulted for more information.

9.7.3 Strut Sizing

The wheel strut must be sized, in that the cross-section and its area need to be determined. The cross-section is primarily a function of aircraft mass, load per wheel, landing gear height, safety factor, strut deflection, strut material, and "g" load during touchdown. There are various mechanical engineering references in the literature such as [5], which the reader is referred to for more details. Two typical strut cross-sections are circular and rectangular. If the landing gear is non-retractable, it is recommended to use fairing for the struts such that the cross-sectional area resembles a symmetric airfoil. This technique will considerably reduce the strut drag.

Most aircraft are designed to be able to land safely while there is a cross-wind. One of the techniques in such conditions is referred to as crabbed landing. An impact of crabbed landing is on the landing gear design, due to the lateral force on touchdown. As the crab angle is increased, the banding moment on the struts of the main gear is increased. The landing gear of the Boeing 747 can tolerate about 15 deg crabbed landing, while the bomber Boeing B-52 (Figures 8.6 and 9.4) is designed for 15-deg crabbed landing.

9.7.4 Steering Subsystem

An aircraft must be able to taxi on the ground in an airport, including turning maneuvers. For instance, the minimum ground turning radius of the transport aircraft Boeing 757 (Figure 8.17) is 71 ft at the nose wheel, and 98 ft at the wing tip. For the purpose of ground steering, a nose wheel, the main wheel, or a tail wheel must be capable of being turned (castored). For an aircraft with tricycle landing gear, a steerable nose wheel is usually employed, while for an aircraft with tail-wheel landing gear, a steerable tail wheel is often utilized. However, the steering capability may be augmented by the use of differential braking on the main gear. For a multi-engine aircraft, the use of differential thrust is another technique to steer the aircraft. The steering mechanism is frequently connected to the rudder pedal, providing direct control of the turning angle. Most modern and large aircraft are equipped with hydraulic-type steering systems.

A castoring wheel may cause wheel shimmy, a rapid side-to-side motion of the wheel which can break the landing gear. A typical solution to wheel shimmy is to employ the rake angle (Figure 9.29) and trail (i.e., offset), or frictional shimmy damper. If the castoring wheel is free to swivel, as is the case for most tail wheels, shimmy could be prevented by utilizing a small angle of rake, as well as an appropriate trail.

The twin-turbofan business transport aircraft Gulfstream IV-SP employs a steerable nose wheel forward. The Learjet 60 nose wheel is equipped with a twin dual-chine tire, size 18 · 4.4, with steer-by-wire. The Boeing 777 employs a twin-wheel steerable nose gear: two main legs carrying six-wheel bogies with steering rear axles automatically engaged by nose-gear steering angle. The Beech Super King Air 200 has a single wheel on a steerable nose unit.

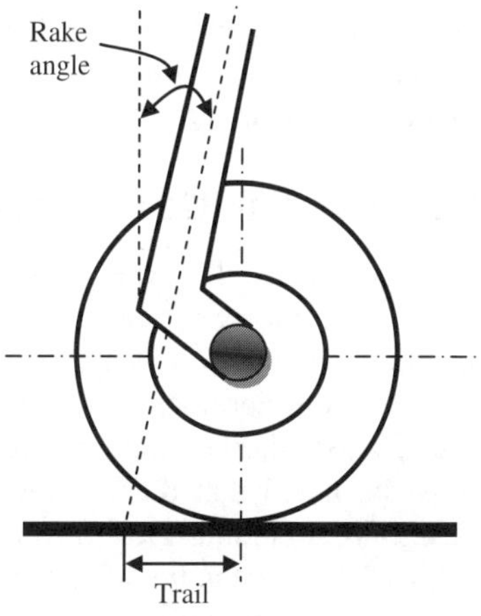

Figure 9.29 Steering wheel geometry

9.7.5 Landing Gear Retraction System

One of the very last landing gear subsystems which must be designed in a retractable landing gear is the retraction subsystem. At this point, the geometry of the landing gear plus the home for the retraction (Figure 9.11) must be known. The retraction subsystem is another mechanical engineering topic that is covered in brief in this section. References [10, 11] contain basic principles and a comprehensive introduction to modern mechanism design with a focus on theoretical foundations.

The landing gear retraction mechanism typically includes a couple of mechanical members and/or a piston-cylinder. The direction of retraction (inward, outward, forward, backward) is another decision which must be made prior to considering more details. The criteria for the selection of type of landing gear retraction mechanism include mechanism weight, volume, cost, maintenance, landing gear/aerostructure integration, and power transmission system.

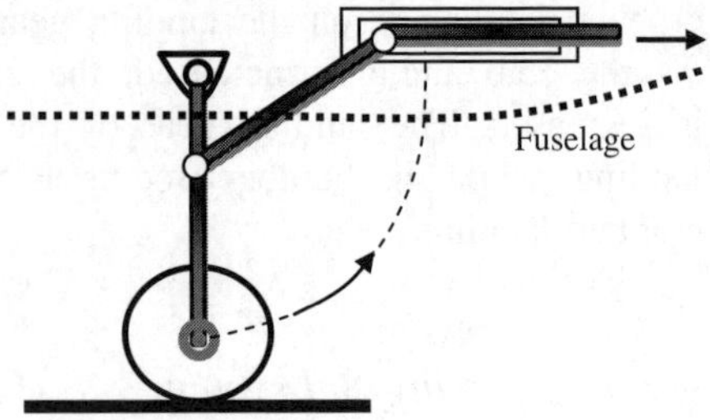

Figure 9.30 Landing gear retraction subsystem

There are a variety of design options, but two convenient retraction systems are hydraulic and mechanical linkage. In general, a hydraulic system is more expensive and heavier than a mechanical linkage. An example of a retraction system is illustrated in Figure 9.30. The following are a couple of real-world applications. In the commuter aircraft Fairchild SA227, all wheels retract forward, but the main gear into the engine nacelles and the nose wheel into the fuselage. In Gulfstream IV-SP (Figure 11.15), the main wheels retract inward while in a Learjet 60 the main legs retract inward, but the nose leg forward. In the light transport aircraft Beech Super King Air 200 hydraulically retractable tricycle landing gear, the main unit retracts forward, the nose wheel rearward. And finally, in the world of fighters: in the F-15 Eagle all units retract forward, while in the F/A-18 Hornet (Figures 2.11, 6.12, and 12.27) the nose unit retracts forward, but the main wheel rearward.

9.8 Landing Gear Design Steps

In Sections 9.1–9.7, the landing gear function, configurations, objectives, alternatives, design criteria, parameters, governing rules and equations, formulation, design requirements, as well as how to approach the primary aero-related parameters have been presented in detail. Furthermore, Figure 9.2 illustrates the design flowchart of the landing gear. In this section, the landing gear design procedure is introduced in terms of design steps. It must be noted that there is no unique solution to satisfy the customer requirements in designing a landing gear. Several landing gear designs may satisfy the requirements, but each will have unique advantages and disadvantages.

In order to formulate the design requirements, the designer is encouraged to develop several equations and relations based on the numerical requirements and solve them simultaneously. For instance, for each angle requirement, a trigonometric or Pythogorian equation may be built for each triangle. With this technique, a computer program would allow a faster and more accurate design. Based on the systems engineering approach, the landing gear detail design begins with identifying and defining design requirements and ends with optimization. The following are the landing gear design steps for a land-based aircraft:

1. Identify and list the landing gear design requirements. It is recommended to consult references such as [12] and [3].
2. Select the landing gear configuration (e.g., tricycle, tail gear, bicycle, quadricycle, multi-bogey).

3. Select fixed, or retractable, or partially retractable.
4. If fixed, select faired or un-faired.
5. Determine the aircraft forward and aft center of gravity (assume no landing gear at this moment).
6. Calculate the landing gear height, based on ground clearance requirements.
7. Determine the distance between the main gear and the aircraft most forward center of gravity.
8. Determine the distance between the main gear and the aft center of gravity.
9. Check the tipback (or tipforward if tail gear) requirement.
10. Check the take-off rotation clearance requirement.
11. Calculate the wheel base.
12. Determine the wheel track (distance between left and right wheels of main gear) in the lateral axis.
13. Determine the landing gear attachments.
14. If retractable, determine where the landing gear is going to be retracted (e.g., inside wing, inside fuselage).
15. Determine the aircraft forward and aft center of gravity when the landing gear weight is added to the aircraft weight.
16. Check the overturn angle requirement.
17. Investigate the structural integrity.
18. Investigate the aircraft ground clearance requirement.
19. Investigate the aircraft ground stability.
20. Investigate the aircraft ground controllability.
21. Check other design requirements (e.g., cost, maintainability, and weight).
22. If any of the design requirements are not satisfied, return to the relevant design step and recalculate the corresponding parameter.
23. If any landing gear parameters are changed, the entire landing gear needs to be revisited and revised.
24. Determine the load on each gear.
25. Size the wheels and tires.
26. Design the struts.
27. Design the shock absorber.
28. Design the gear retracting mechanism.
29. Optimize.
30. Draw the final design for the landing gear.

For other aircraft configurations (e.g., seaplane) or other landing gear configurations (e.g., bicycle), the reader needs to revise the above-mentioned steps, and establish a revised design procedure.

9.9 Landing Gear Design Example

In this section, a major chapter example is presented to design landing gear for a transport aircraft. In order to avoid lengthening the chapter, it only covers the major design parameters.

Example 9.8

Problem statement: Design a landing gear for the following subsonic civil transport aircraft to carry 18 passengers. The aircraft has two turboprop engines, and is equipped with a split flap which is deflected 30 deg during the take-off operation on a concrete runway. Assume that the aircraft forward cg is at 18% MAC, aft cg is at 30% of MAC, and wing/fuselage aerodynamic center is located at 22% MAC. The distance between the horizontal tail aerodynamic center and the wing/fuselage aerodynamic center is 13 m:.

$m_{TO} = 18\,000\,\text{kg}$, $D_{f_{max}} = 2.4\,\text{m}$, $V_{max} = 370$ KTAS (at 27 000 ft), $V_s = 85$ KEAS,

$D_{prop} = 3.8\,\text{m}$, $C_{D_{o\_clean}} = 0.02$, $C_{D_{o\_TO}} = 0.03$, $I_{yy} = 23\,000\,\text{kg}\,\text{m}^2$,

$P_{max} = 8\,000\,\text{hp}$, $C_{mo} = -0.03$, $\eta_{P_{TO}} = 0.5$, $\alpha_{TO} = 14\,\text{deg}$

Wing: airfoil, $S = 60\,\text{m}^2$, NACA 64_1-112, $AR = 12$, $e = 0.9$, $\Delta C_{L_{flap}} = 0.9$, $\lambda = 1$

Horizontal tail: $S_h = 14\,\text{m}^2$, NACA 0009, $AR_t = 5$, $C_{L_{h\_TO}} = -0.8$

The aircraft configuration and other geometry variables are illustrated in Figure 9.31. The following parameters must be determined: landing gear configuration, fixed or retractable, height, wheel track, wheel base, distance between main wheel and aircraft cg.

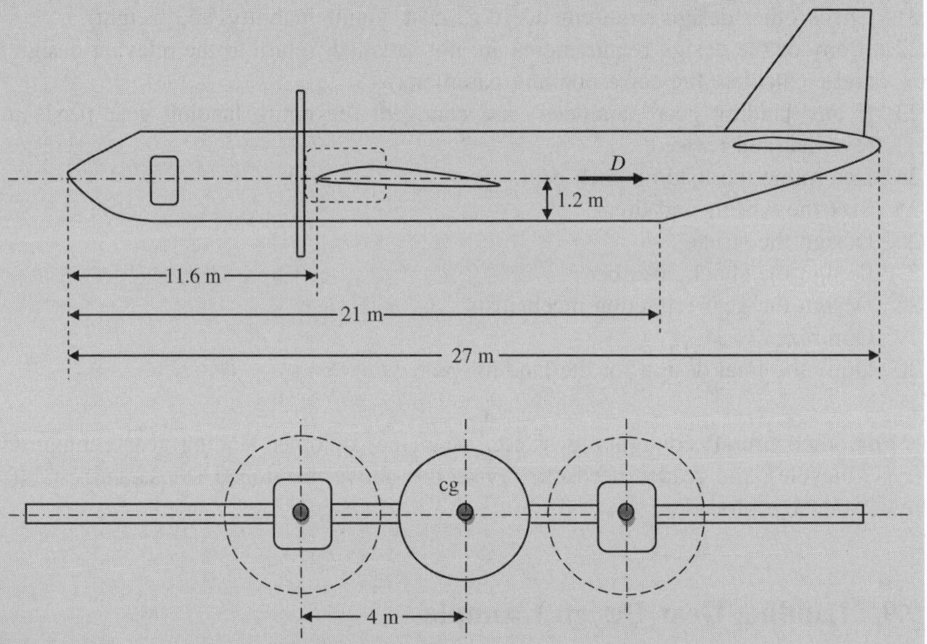

Figure 9.31 Aircraft in Example 9.8

Solution:

- **Step 1. Landing gear design requirements**. The following design requirements must be satisfied: ground clearance requirement, tipback (or tipforward angle if tail gear) angle requirement, take-off rotation requirement, overturn angle requirement, structural integrity, aircraft ground stability, aircraft ground controllability, low cost, maintainability, and manufacturability.
- **Step 2. Landing gear configuration**. This is a transport aircraft, and the passenger's comfort is an important requirement. So, the tail gear, bicycle, and single main configurations would not satisfy this requirement. Three viable configurations are: (i) tricycle or nose gear, (ii) quadricycle, and (iii) multi-bogey. Since the aircraft weight is not very high, both the quadricycle and multi-bogey configurations are set aside due to their cost and weight. Therefore, the best landing gear configuration for this aircraft is nose gear or tricycle. An attractive feature of this configuration is that the aircraft will be horizontal on the ground. The passengers do not have to climb during the boarding period. The nose gear also decreases the take-off run, and at the same time the aircraft will take off sooner.
- **Step 3. Fixed or retractable**. The aircraft must compete with other transport aircraft in the market, and it must have a fairly high performance, so a retractable landing gear is the best option. The cost of this configuration is covered by the customers (passengers). Then, this will reduce the aircraft drag during flight and therefore the aircraft will feature a higher performance. The higher landing gear weight due to the retraction system will be paid off compared with the other advantages of a retractable landing gear.
- **Step 4. Landing gear height**. Based on Figure 9.31, the lowest point of the aircraft is the propeller tip. There must be a reasonable clearance between the prop and the ground. Due to the fact that the aircraft engine is turboprop, and for the sake of passenger safety considerations, a 1.2 m ground clearance for the propeller is considered necessary. This distance may be revised in later design phases (i.e., $\Delta H_{\text{clear}} = 1.2\,\text{m}$). Hence, the distance between the aircraft center of gravity and the ground would be:

$$H_{\text{cg}} = \Delta H_{\text{clear}} + \frac{D_{\text{prop}}}{2} = 1.2 + \frac{3.8}{2} = 3.1\,\text{m}$$

This clearance is shown in Figure 9.32. Please note that as Figure 9.31 illustrates, the aircraft cg is at the same height as the wing mid-plane. The landing gear height is a function of its attachment location. The nose gear will naturally be attached to the fuselage. But the main gear attachment tends to have two main alternatives: (i) attach to the fuselage and (ii) attach to the wing. As soon as the wheel track is determined, we are able to decide about the landing gear attachment, and then the landing gear height may be determined.

  - **Case 1. Attach main gear to the fuselage**. In this case, the landing gear height will be:

$$H_{\text{LG}} = H_{\text{cg}} - \frac{D_{\text{fuse}}}{2} = 3.1 - \frac{2.4}{2} = 1.9\,\text{m}$$

– **Case 2. Attach main gear to the wing**. Wing mean aerodynamic cord is:

$$b = \sqrt{S.\mathrm{AR}} = \sqrt{60 \cdot 12} = 26.833\,\mathrm{m} \tag{5.19}$$

$$\overline{C} = \frac{S}{b} = \frac{60}{26.833} = 2.236\,\mathrm{m} \tag{5.18}$$

The wing airfoil is NACA 64_1-112, so the wing thickness-to-chord ratio is 12%. Thus, the wing thickness is:

$$t_{\mathrm{w}} = \left(\frac{t}{C}\right)_{\max} \overline{C} = 0.12 \cdot 2.236 = 0.268\,\mathrm{m}$$

In this case, the landing gear height will be:

$$H_{\mathrm{LG}} = H_{\mathrm{cg}} - \frac{t_{\mathrm{w}}}{2} = 3.1 - \frac{0.268}{2} = 2.966\,\mathrm{m}$$

These two landing gear heights are shown in Figure 9.32. When the wheel track and wheel base are determined, the main gear attachment will be finalized. Furthermore, in the later steps, other landing gear requirements will be checked to make sure this clearance does not violate any other design requirements.

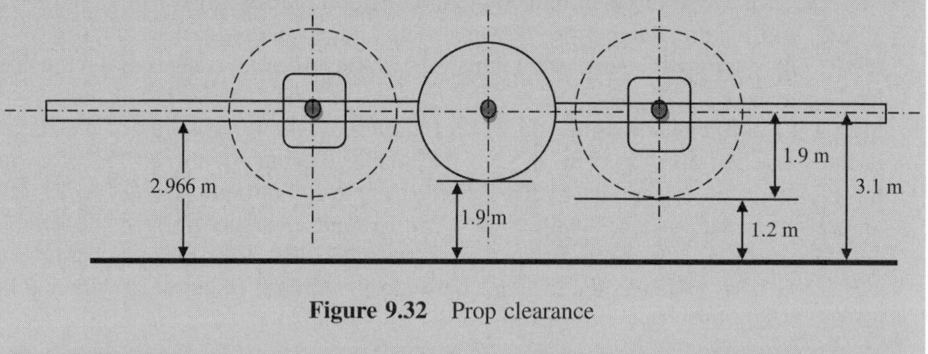

Figure 9.32 Prop clearance

- **Step 5. The distance between the main gear and the aircraft forward cg**. Now we determine the location of the main landing gear. The take-off rotation requirement will be employed to obtain this distance. The aircraft is required to be able to rotate during the transition segment of the take-off operation by an amount of 9 deg/s^2. This requirement must be examined for the aircraft critical cg location, which is the most forward cg.

  Since the aircraft forward cg is at 18% MAC, and the wing/fuselage aerodynamic center is located at 22% MAC, we can write the following relationship for the wing/fuselage lift moment arm:

$$x_{L_{\mathrm{wf}}} = x_{\mathrm{mg}} - \left(\overline{X}_{\mathrm{ac}} - \overline{X}_{\mathrm{cg_{fwd}}}\right)\overline{C} = x_{\mathrm{mg}} - (0.22 - 0.18) \cdot 2.236 = x_{\mathrm{mg}} - 0.089$$

  Furthermore, the distance between the horizontal tail aerodynamic center and the wing/fuselage aerodynamic center is 12 m, hence the tail moment arm would be:

$$x_{\mathrm{h}} = x_{\mathrm{ac_h}} - x_{\mathrm{mg}} = l_{\mathrm{h}} + \left(\overline{X}_{\mathrm{ac}} - \overline{X}_{\mathrm{cg_{fwd}}}\right)\overline{C} - x_{\mathrm{mg}}$$
$$= 13 + (0.22 - 0.18) \cdot 2.236 - x_{\mathrm{mg}} = 13.089 - x_{\mathrm{mg}}$$

where x_{mg} is measured from the main gear to the forward cg in meters.

From Figures 9.31 and 9.32, we can extract the following dimensions: $h_D = H_{cg} = h_T = 3.1\,\mathrm{m}$.

The air density at sea level is $1.225\,\mathrm{kg/m^3}$, and at 27 000 ft is $0.512\,\mathrm{kg/m^3}$. To obtain the wing mean aerodynamic chord, proceed as follows.

To find the aircraft drag:

$$K = \frac{1}{\pi.e.\mathrm{AR}} = \frac{1}{3.14 \cdot 0.9 \cdot 12} = 0.029 \tag{5.22}$$

$$C_{L_C} = \frac{2W}{\rho V_C^2 S} = \frac{2 \cdot 18\,000 \cdot 9.81}{0.512 \cdot (370 \cdot 0.5144)^2 \cdot 60} = 0.317 \tag{5.1}$$

$$C_{L_{\mathrm{TO}}} = C_{L_C} + \Delta C_{L_{\mathrm{flap}}} = 0.317 + 0.9 = 1.217 \tag{4.69c}$$

$$C_{D_{\mathrm{TO}}} = C_{D_{\mathrm{o\_TO}}} + KC_{L_{\mathrm{TO}}}^2 = 0.03 + 0.029 \cdot 1.217^2 = 0.074 \tag{4.68}$$

$$V_{\mathrm{R}} = 1.1V_{\mathrm{s}} = 1.1 \cdot 85 = 93.5\ \mathrm{knot} = 48.1\,\mathrm{m/s} \tag{9.35}$$

$$D_{\mathrm{TO}} = \frac{1}{2}\rho V_{\mathrm{R}}^2 S C_{D_{\mathrm{TO}}} = \frac{1}{2} \cdot 1.225 \cdot (48.1)^2 \cdot 60 \cdot 0.074 = 6267.4\,\mathrm{N} \tag{9.44}$$

Other aerodynamic forces and moments:

$$L_{\mathrm{TO}} = \frac{1}{2}\rho V_{\mathrm{R}}^2 S_{\mathrm{ref}} C_{L_{\mathrm{TO}}} = \frac{1}{2} \cdot 1.225 \cdot (48.1)^2 \cdot 60 \cdot 1.217 = 103\,554.6\,\mathrm{N} \tag{9.41}$$

$$L_{\mathrm{h}} = \frac{1}{2}\rho V_{\mathrm{R}}^2 S_{\mathrm{h}} C_{L_{\mathrm{h}}} = \frac{1}{2} \cdot 1.225 \cdot (48.1)^2 \cdot 14 \cdot (-0.8) = -15\,879\,\mathrm{N} \tag{9.43}$$

$$M_{\mathrm{ac_{wf}}} = \frac{1}{2}\rho V_{\mathrm{R}}^2 C_{m_{\mathrm{ac\_wf}}} S_{\mathrm{ref}}\overline{C} = \frac{1}{2} \cdot 1.225 \cdot (48.1)^2 \cdot 60 \cdot (-0.03) \cdot 2.236$$
$$= -5706\,\mathrm{N} \tag{9.45}$$

$$L_{\mathrm{wf}} = L_{\mathrm{TO}} - L_{\mathrm{h}} = 103\,554.6 - (-15\,879) = 119\,434\,\mathrm{N} \tag{9.42}$$

Friction force:

$$F_{\mathrm{f}} = \mu\left(W - L_{\mathrm{TO}}\right) = 0.04\,(18\,000 \cdot 9.81 - 103\,554.6) = 2918.6\,\mathrm{N} \tag{9.40}$$

The engine total power is 8 000 hp, which is equivalent to 5 965 599 W. The engine thrust at the instance of rotation is:

$$T = \frac{P\eta_{\mathrm{P}}}{V_{\mathrm{R}}} = \frac{5\,965\,599 \cdot 0.5}{48.1} \Rightarrow T = 62\,011.7\,\mathrm{N} \tag{8.15}$$

Aircraft linear acceleration at the time of take-off rotation:

$$a = \frac{T - D - F_R}{m} = \frac{62\,011.7 - 6267.4 - 2918.6}{18\,000} \Rightarrow a = 2.935\,\text{m/s}^2 \quad (9.36)$$

Contributing moments:

$$M_W = W\left(x_{mg} - x_{cg}\right) = 18\,000 \cdot 9.81\left(x_{mg}\right) \quad (9.46)$$

$$M_D = D\left(z_D - z_{mg}\right) = 6267.4 \cdot 3.1 = 19\,429\,\text{Nm} \quad (9.47)$$

$$M_T = T\left(z_T - z_{mg}\right) = 62\,011.7 \cdot 3.1 = 192\,236.4\,\text{Nm} \quad (9.48)$$

$$M_{L_{wf}} = L_{wf}\left(x_{mg} - x_{ac_{wf\ to\ cg}}\right) = 119\,433.7 \cdot \left(x_{mg} - 0.089\right) \quad (9.49)$$

$$M_{L_h} = L_h\left(x_{ac_h} - x_{mg}\right) = -15\,879 \cdot \left(13.089 - x_{mg}\right) \quad (9.50)$$

$$M_a = ma\left(z_{cg} - z_{mg}\right) = 18\,000 \cdot 2.935 \cdot 3.1 = 163\,760\,\text{Nm} \quad (9.51)$$

where, for the sake of simplicity, the x reference line is considered to be the aircraft forward cg. Now all moments are substituted into Equation (9.54):

$$x_{mg} = \frac{I_{yy_{mg}}\ddot{\theta} - D\left(z_D - z_{mg}\right) + T\left(z_T - z_{mg}\right) - M_{ac_{wf}} - ma\left(z_{cg} - z_{mg}\right) - Wx_{cg} + L_{wf}x_{ac_{wf}} + L_h x_{ac_h}}{L_{wf} + L_h - W} \quad (9.54)$$

By substituting moments and forces, we have:

$$x_{mg} = \frac{23\,000 \cdot \frac{9}{57.3} - 19\,429 + 192\,236.4 - (-5706.4) - 163\,760 + 0 + (119\,433.7 \cdot 0.089) + (-15\,879 \cdot 13)}{119\,433.7 - 15\,879 - (18\,000 \cdot 9.81)} \quad (9.54)$$

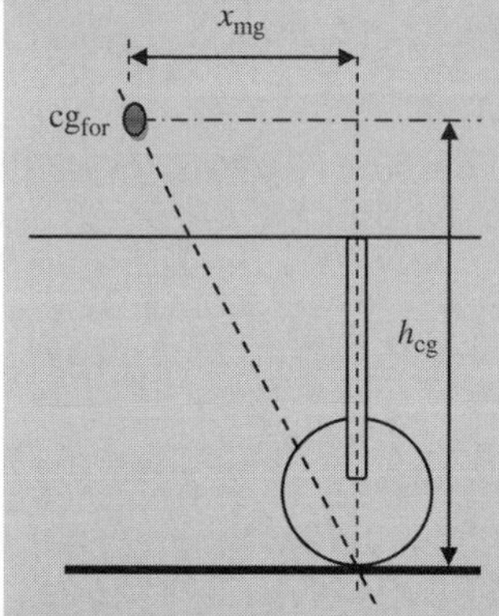

Figure 9.33 Main gear and forward cg

The solution is:

$$x_{mg} = 2.431\,\text{m}$$

So far, the fact that the given mass moment of inertia is about the aircraft cg has been ignored. Hence, the calculation must be repeated with the revised equation to include the parallel axis theorem. Thus, the aircraft mass moment of inertia about cg (y-axis) must be transferred to the main gear contact point ($I_{yy_{mg}}$) by employing the parallel axis theorem (Equation (9.53)):

$$I_{yy_{mg}} = I_{yy_{cg}} + m\left(\sqrt{x_{mg}^2 + h_{cg}^2}\right)^2 = I_{yy_{cg}} + m\left(x_{mg}^2 + h_{cg}^2\right) \quad (9.53a)$$

where x_{mg} and h_{cg} are shown in Figure 9.33.

Now, the $I_{yy_{mg}}$ from Equation (9.53a) is inserted into Equation (9.54). The result would be a non-linear equation with only one unknown parameter (x_{mg}) as follows:

$$x_{mg} = \frac{\left(I_{yy_{cg}} + m\left(x_{mg}^2 + h_{cg}^2\right)\right)\ddot{\theta} - D\left(z_D - z_{mg}\right) + T\left(z_T - z_{mg}\right) - M_{ac_{wf}} - ma\left(z_{cg} - z_{mg}\right) + L_{wf}x_{ac_{wf}} + L_h x_{ac_h}}{L_{wf} + L_h - W} \tag{9.54a}$$

The solution of this revised equation would be:

$$x_{mg} = 1.916\,\text{m}$$

- **Step 6. Check tipback requirement**. In order to check the tipback angle, we have to obtain the distance between the aft cg and the main gear. Based on the problem statement, the forward cg is located at 18% MAC, while the aft cg is at $0.3C$. Thus, the distance between the aircraft aft cg and forward cg is:

$$\Delta x_{cg} = x_{cg_{for}} - x_{cg_{aft}} = (0.30 - 0.18)\,\overline{C} = 0.12 \cdot 2.236 = 0.268\,\text{m} \tag{11.16}$$

So, the distance between the main gear and the aft cg based on Figure 9.18 is:

$$x_{mg_{aft}} = x_{mg} - \Delta x_{cg} = 1.916 - 0.268 = 1.648\,\text{m}$$

This distance indicates (according to Figure 9.15) that the aircraft has the following tipback angle:

$$\alpha_{tb} = \tan^{-1}\left(\frac{x_{mg}}{h_{cg}}\right) = \tan^{-1}\left(\frac{1.648}{3.1}\right) = 0.489\,\text{rad} \Rightarrow \alpha_{tf} = 28\,\text{deg} \tag{9.34}$$

This tipback angle is greater than the aircraft take-off rotation angle (14 deg):

$$28 > 14 + 5 \tag{9.33a}$$

Therefore, the distance between the main gear and the aft cg satisfies the tipback angle requirement. This x_{mg} is the distance between the main gear and the aircraft forward cg just to satisfy the take-off rotation requirement as well as the tipback angle requirement. In the forthcoming steps, this value must be examined again to ensure it meets the other design requirements.

- **Step 7. Check the take-off rotation clearance requirement**. The take-off rotation ground clearance requirement to prevent a fuselage hit is as follows:

$$\alpha_C \geq \alpha_{TO} \tag{9.2}$$

In order to determine the clearance angle (α_C), two distances should be obtained: (i) height between lowest point of the fuselage to the ground and (ii) distance between the main gear to the fuselage upsweep point. Figure 9.32 illustrates that the fuselage height (H_f) is 1.9 m. In contrast, from Figure 9.31, the length between the nose and the fuselage upsweep point is 21 m, and the distance between the wing leading edge and the fuselage nose in 11.6 m. Thus, the distance between the fuselage upsweep point and the wing leading edge is:

$$21\,\text{m} - 11.6\,\text{m} = 9.4\,\text{m}$$

Furthermore, the distance between the main gear and the wing leading edge is:

$$X_{\text{mg-LE}} = x_{\text{mg}_{\text{for}}} + 0.18\overline{C} = 1.916 + (0.18 \cdot 2.236) = 2.319\,\text{m}$$

Thus, the distance between the main gear and the fuselage upsweep point (Figure 9.34) is:

$$9.4 - 2.319 = 7.081\,\text{m}$$

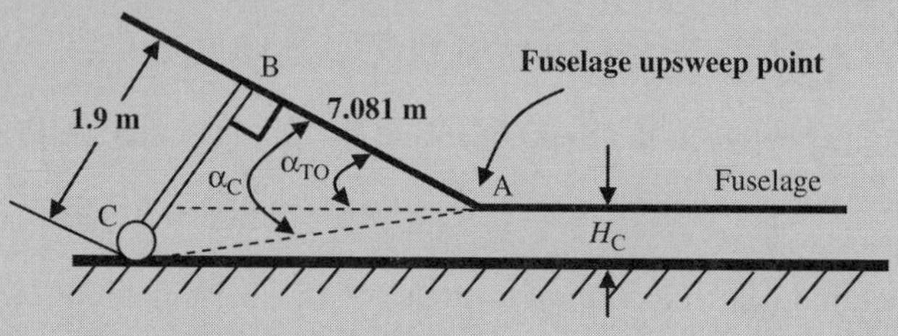

Figure 9.34 Examination of rear fuselage clearance during take-off rotation

Therefore, the clearance angle is:

$$\alpha_C = \tan^{-1}\left(\frac{H_f}{AB}\right) = \tan^{-1}\left(\frac{1.9}{7.081}\right) = 0.262\,\text{rad} = 15.02\,\text{deg} \tag{9.3}$$

Since the clearance angle (α_C, 15 deg) is greater than the aircraft rotation angle (α_{TO}, 12 deg), the fuselage will not hit the ground during take-off operations.

- **Step 8. Wheel base**. Due to the ground controllability requirement, the nose gear must not carry less than about 5% of the total load and also must not carry more than about 20% of the total load (e.g., aircraft weight). Thus, the main gear carries about 80–95% of the aircraft load. To meet this requirement, it is decided that the nose gear should carry 15% of the total load and the main gear 85% of the total load. The wheel base is determined using Equation (9.6):

$$F_n = \frac{B_m}{B} W \tag{9.6}$$

where F_n is selected to be 10% of the total weight, so:

$$B = \frac{B_m}{F_n} W = \frac{B_m}{0.15W} W = \frac{B_m}{0.15} = 6.667B_m \tag{9.6}$$

where B_m is obtained previously as 1.916 m. Thus, the wheel base (B) is:

$$B = 6.667 \cdot 1.916 = 12.775\,\text{m}$$

When the cg is at the aft location, the nose wheel will carry less than 15% of the aircraft weight. This value for the wheel base could be revised later for optimization when examining other requirements.

- **Step 9. Wheel track**. The three main design requirements which drive the wheel track (T) are: (i) ground lateral control, (ii) ground lateral stability, and (iii) structural integrity. The overturn angle is the angle which is critical to the aircraft overturn. There are two overturn angles (Figure 9.19), and the smaller one is considered in this technique.

The minimum allowable value for the wheel track must satisfy the overturn angle requirements (Section 9.5.3.1). The maximum allowable value for the wheel track must satisfy the structural integrity requirements (Section 9.5.3.2).

In the first place, to determine the wheel track, we use the criterion of overturning prevention. The lateral distance between each main gear to the cg must be greater than 25 deg (Equation (9.18)). Here, we consider 30 deg. Figure 9.35 illustrates the front view of the aircraft, showing one of the main wheels relative to the aircraft cg.

In step 4, the height of the cg from the ground was determined to be 3.1 m. Using the triangle shown in Figure 9.35, the wheel track is determined as follows:

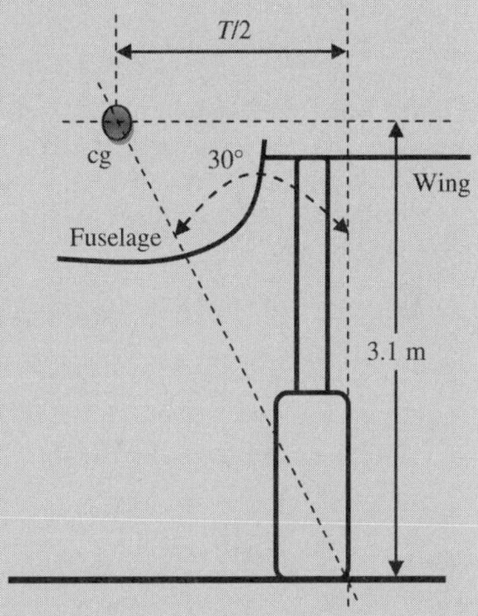

Figure 9.35 Wheel track (front view)

$$\tan(30) = \frac{T/2}{H_{cg}} \Rightarrow T = 2\tan(30)\,H_{cg} = 2\tan(30)\cdot 3.1 \Rightarrow T = 3.58\,\text{m} \qquad (9.23)$$

Now, we need to examine the overturn angle based on the top view. Figure 9.36 shows the top view of the aircraft and a triangle to determine Φ_{ot} based on the top view.

In order to determine the overturn angle (Φ_{ot}) for this aircraft, we first need to determine the parameter Y_{ot} as shown in Figure 9.36(a). This parameter is calculated by using the sine law in the triangle ADE (Figure 9.36(b)) using the angle $\phi 1$. However, this angle is obtained via the triangle ACF. In the triangle ABC, the side AC is the wheel base, and the side FC is one-half of the wheel track. Thus, in triangle ACF:

$$\tan(\phi 1) = \frac{\text{BC}}{\text{AC}} = \frac{T/2}{B} = \frac{3.58/2}{12.775} \Rightarrow \phi 1 = \tan^{-1}\left(\frac{3.58/2}{12.775}\right)$$

$$= 0.273\,\text{rad} = 15.65\,\text{deg}$$

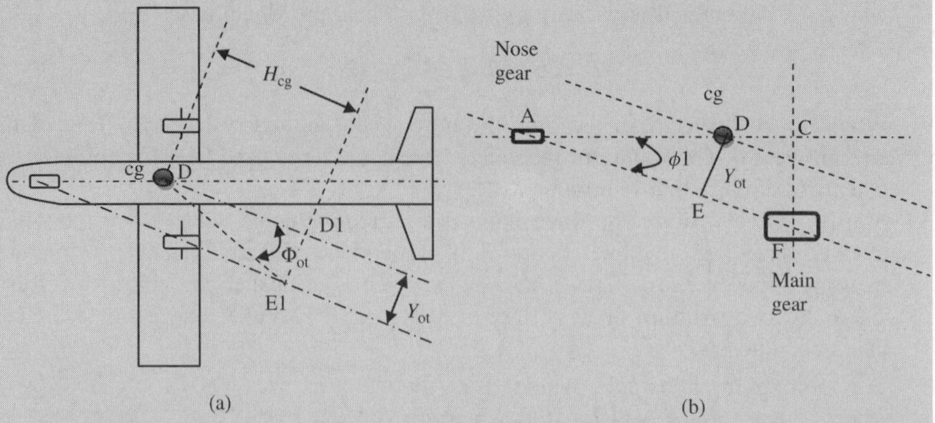

Figure 9.36 Calculation of the overturn angle for the aircraft in Example 9.8: (a) Aircraft top view; (b) Top view (main gear and nose gear)

Similarly, in the triangle ADE:

$$\tan(\phi 1) = \frac{Y_{ot}}{AD} = \frac{Y_{ot}}{B_{n_{min}}} \Rightarrow Y_{ot} = B_{n_{min}} \tan(\phi 1)$$

$$= (12.775 - 1.916)\tan(15.65\,\text{deg}) \Rightarrow Y_{ot} = 3.042\,\text{m}$$

Finally, in the triangle DE1D1, we can write:

$$\tan\left(\Phi_{ot}\right) = \frac{E1D1}{DD1} = \frac{Y_{ot}}{H_{cg}} = \frac{3.042}{3.1} \Rightarrow \Phi_{ot} = \tan^{-1}\left(\frac{3.042}{3.1}\right)$$

$$= 0.776\,\text{rad} = 44.5\,\text{deg}$$

The overturn angle is greater than 25 deg, so the wheel track satisfies the rule of thumb for overturn prevention requirement. In the later steps, ground lateral control, ground lateral stability, and structural integrity must be examined to validate the wheel track.

- **Step 10. Landing gear attachment**. As a natural selection, the nose gear is attached to the fuselage nose. However, for the main gear, we need to compare the fuselage diameter with the wheel track. It is observed that the fuselage diameter (2.4 m) is smaller than the wheel track (3.58 m). Hence, the main gear cannot be attached to the fuselage. Thus, the main gear may be either attached directly to the wing or attached under the nacelle. In order to determine the best location, several design requirements must be examined, which is beyond the scope of this example. For the time being, it is decided to attach the landing gear to the wing. Thus, the landing gear height will be:

$$H_{LG} = 2.996\,\text{m (as shown in Figure 9.32)}$$

- **Step 11 through step 29. Mechanical parameters of the landing gear**. Although the landing gear designed so far has satisfied several requirements, there are still other design requirements which have not been examined. During the design process, several iterations will take place until we have the optimum design. The rest of the landing gear design – including examining mechanical parameters such as tire, shock absorber, and strut – is left to the reader for practice.

- **Step 30. Drawing**. The calculated dimensions for the wheel base, wheel track, wheel height, and distance between main gear and aircraft cg are illustrated in Figure 9.37.

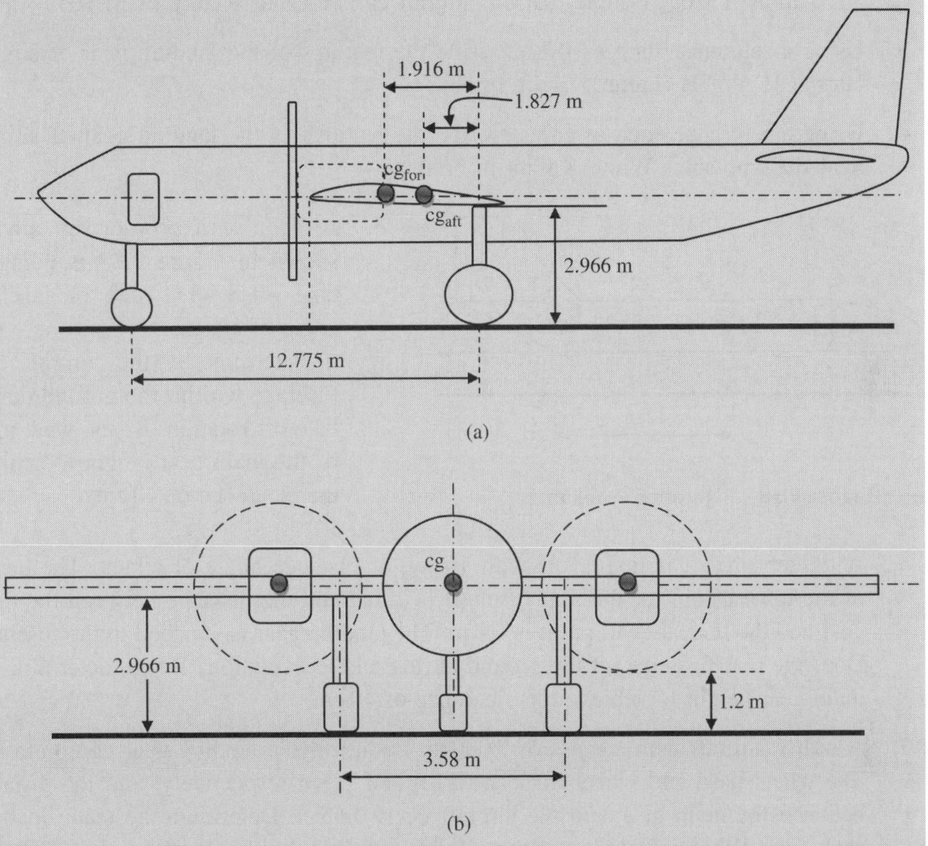

Figure 9.37 The aircraft in Example 9.8 with the designed landing gear (figure not scaled): (a) Wheel base, landing gear height, and main gear to cg; (b) Wheel track, clearance, and wheel height

Problems

1. Using a reference such as [8], identify one aircraft with fixed tricycle landing gear, one aircraft with retractable tricycle landing gear, one aircraft with tail gear, one aircraft with quadricycle landing gear, and one aircraft with partially retractable landing gear (either main or nose gear is retracted). For each aircraft, provide the name of the aircraft, the type of the aircraft, and its picture or three-view.

2. Using a reference such as [8], determine the following:
   (a) The ratio between the wheel track and fuselage length, and the ratio between the wheel base and wing span for twin-turboprop regional transport ATR 42 (Figure 3.8).
   (b) The lateral angle between the main wheels off the cg (front view) for fighter F-16 Falcon (Figure 4.6).
   (c) What percentage of aircraft weight is carried by the nose gear of jet transport Airbus A310? Assume that the aircraft cg is located at 20% of MAC.

3. Using a reference such as [8], describe the features of the landing gear of aircraft Harrier II AV-8B (Figure 4.4) in brief.

4. Using a reference such as [8], describe the features of the landing gear of aircraft Scaled Composites White Knight in brief.

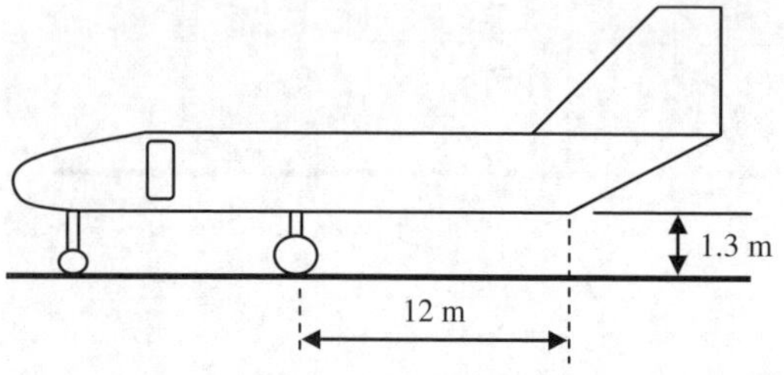

Figure 9.38 Figure for Problem 5

5. A pilot of a prop-driven aircraft shown in Figure 9.38 is going to take off with 14 deg of fuselage angle of attack.

   Determine if the aircraft rear fuselage will hit the ground during take-off rotation. If yes, what must be the main gear height to achieve the clearance of 20 cm?

6. A fighter aircraft is taking off with 16 deg of fuselage angle of attack. The height of the lowest point of the rear fuselage is 1.4 m and the distance between the main gear and the fuselage tail point is 6.8 m. The landing gear is attached to the fuselage. Does the rear fuselage hit the ground during take-off rotation? If yes, determine the main gear height to achieve the clearance of 40 cm.

7. A utility aircraft with a mass of 7 000 kg has a tricycle landing gear configuration. The wheel base and wheel track are 11.6 and 1.9 m respectively, and the distance between the main gear and the aircraft cg is 0.65 m. Determine the static load on each gear. What percentage of the aircraft weight is carried by the main gear?

8. A large transport aircraft with a mass of 70 000 kg has a tricycle landing gear configuration. The wheel base and wheel track are 25 and 4.2 m respectively, and the distance between the main gear and the aircraft cg is 1.2 m. Determine the static load on each gear. What percentage of the aircraft weight is carried by the nose gear?

9. A twin-turboprop aircraft with a take-off mass of 20 000 kg has a tricycle landing gear configuration. The aircraft cg is allowed to move between 0.8 and 1.2 m from the main gear.

   (a) The nose gear is desired to carry a maximum of 10% of the aircraft weight in static equilibrium. Determine the wheel base.

(b) The deceleration during landing braking is $-5\,\text{m/s}^2$ and the acceleration during take-off is $7\,\text{m/s}^2$. The distance between the aircraft cg and the ground is 2.4 m. Determine the maximum dynamic load on each wheel.

10. A large transport aircraft with a take-off mass of 300 000 kg has a tricycle landing gear configuration. The aircraft cg is allowed to move between 1.2 and 1.8 m from the main gear.

(a) The nose gear is desired to carry a maximum of 18% of the aircraft weight in static equilibrium, determine the wheel base.
(b) The deceleration during landing braking is $-7\,\text{m/s}^2$ and the acceleration during take-off is $10\,\text{m/s}^2$. The distance between the aircraft cg and the ground is 4 m. Determine the maximum load on each gear.

11. A jet transport aircraft with a mass of 40 000 kg and a wing area of $85\,\text{m}^2$ is turning on a runway. The ground speed is 15 knot and the turn radius is 25 m. The height of the aircraft center of gravity from the ground is 2.7 m.

(a) Determine the minimum overturn angle to prevent an overturn in this taxi maneuver.
(b) Determine the wheel track corresponding to this overturn angle.

12. A single-engine prop-driven aircraft with a mass of 4000 kg and a wing area of $14\,\text{m}^2$ is turning on a runway. The ground speed is 18 knot and the turn radius is 15 m. The height of the aircraft center of gravity from the ground is 0.8 m.

(a) Determine the minimum overturn angle to prevent an overturn in this taxi maneuver.
(b) Determine the wheel track corresponding to this overturn angle.

13. Consider the aircraft in Problem 11 is on a runway at 5 000 ft altitude. The aircraft side area is $120\,\text{m}^2$, and the height of the aircraft centroid of side area from the ground is 2.6 m. A cross-wind with a speed of 35 knot is blowing. Assume the aircraft side drag coefficient is 1.1. Determine the minimum wheel track to prevent an overturn due to this cross-wind. The lowest possible mass is 25 000 kg when there is no passenger onboard and zero fuel.

14. Consider the aircraft in Problem 12 is on a runway at 3000 ft altitude. The aircraft side area is $16\,\text{m}^2$, and the height of the aircraft centroid of side area from the ground is 1.2 m. A cross-wind with a speed of 30 knot is blowing. Assume the aircraft side drag coefficient is 0.7. Determine the minimum wheel track to prevent an overturn due to this cross-wind. The lowest possible mass is 2000 kg when there is no passenger onboard and zero fuel.

15. An aircraft with a mass of 20 000 kg and wing span of 28 m has a tricycle landing gear configuration. The wheel base is 12 m, and the maximum distance between the aircraft cg and the nose gear is 11 m. The wing is made of aluminum with a modulus of elasticity of 74 GPa. Assume that the wing can be modeled with a beam of I-section with a second moment of area $0.0025\,\text{m}^4$. If the maximum allowable wing deflection is 2 cm, determine the maximum allowable wheel track.

16. An aircraft with a mass of 100 000 kg and wing span of 38 m has a tricycle landing gear configuration. The wheel base is 20 m, and the minimum distance between the aircraft cg and the main gear is 1.3 m. The wing is made of aluminum with a modulus of elasticity of 70 GPa. Assume that the wing can be modeled with a beam of I-section with a second moment of area 0.008 m^4. If the maximum allowable wing deflection is 3 cm, determine the maximum allowable wheel track.

17. A business aircraft (Figure 9.39) with a take-off mass of 20 000 kg and a wing area of 60 m^2 has two turbofan engines, each generating 25 000 N of thrust. The overall length of the aircraft is 25 m, it has a tricycle landing gear, and the runway is concrete. Assume that the forward cg is at 15% MAC, and the wing/fuselage ac is at 22% MAC. The aircraft is equipped with a double-slotted flap which is set to generate an extra lift coefficient of 0.9 during take-off. The elevator deflection during take-off rotation generates a tail lift coefficient of −1.3.

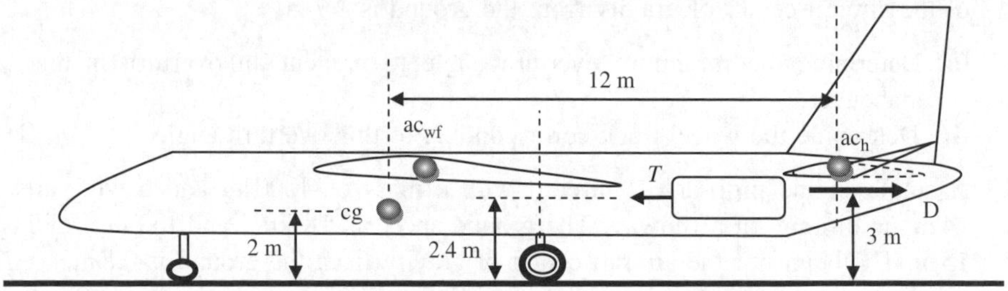

Figure 9.39 Aircraft in Problem 17

Some dimensions of the aircraft are shown in Figure 9.39, and other characteristics of the aircraft are as follows:

$$V_c = 350 \text{ KTAS (at 25 000 ft)},\ V_s = 82 \text{ KEAS},\ C_{D_o} = 0.022,\ C_{D_{o\_TO}} = 0.031,$$

$$I_{yy_{mg}} = 30\,000 \text{ kg m}^2,\ \text{AR} = 10,\ C_{m_o} = -0.05,\ e = 0.87,\ S_h = 13 \text{ m}^2$$

The aircraft is required to rotate about the main gear with an angular acceleration of 6 deg/s^2 during the take-off operation at sea-level altitude. Determine the distance between the main wheel and the aircraft forward cg.

18. A transport aircraft with a take-off mass of 15 000 kg and a wing area of 52 m^2 has two turbofan engines, each generating 24 000 N of thrust. The overall length of the aircraft is 17 m, it has a tricycle landing gear, and the runway is concrete. Assume that the forward cg is at 18% MAC, and the wing/fuselage ac is at 26% MAC. The aircraft is equipped with a single-slotted flap which is set to generate an extra lift coefficient of 0.8 during take-off. The elevator deflection during take-off rotation generates a tail lift coefficient of −1.3. Other characteristics of the aircraft are as follows:

$$V_c = 440 \text{ KTAS (at 27 000 ft)},\ V_s = 85 \text{ KEAS},\ C_{D_o} = 0.023,\ C_{D_{o\_TO}} = 0.032,$$

$$I_{yy_{mg}} = 22\,800\,\text{kg m}^2,\ C_{m_o} = -0.06,\ \text{AR} = 12,\ e = 0.87,\ S_h = 12\,\text{m}^2,$$

$$h_{cg} = 2.2\,\text{m},\ h_D = 3.1\,\text{m},\ h_T = 1.7\,\text{m},\ l_h = 11\,\text{m}$$

The aircraft is required to rotate about the main gear with an angular acceleration of 9 deg/s^2 during the take-off operation at 5 000 ft altitude. Determine the distance between the main wheel and the aircraft forward cg.

19. Design a landing gear for the following transport aircraft to carry 25 passengers. The aircraft has two turboprop engines, and is equipped with a single-slotted flap which is deflected 20 deg during the take-off operation on a concrete runway. Assume that the aircraft forward cg is at 14% MAC, the aft cg is at 34% of MAC, and the wing/fuselage aerodynamic center is located at 23% MAC. The distance between the horizontal tail aerodynamic center and the wing/fuselage aerodynamic center is 18 m.

$$m_{TO} = 40\,000\,\text{kg},\ D_{f_{max}} = 2.8\,\text{m},\ V_{max} = 420\ \text{KTAS (at 30 000 ft)},\ V_s = 75\ \text{KEAS},$$

$$D_{prop} = 3.4\,\text{m},\ C_{D_{o\_clean}} = 0.018,\ C_{D_{o\_TO}} = 0.032,\ I_{yy} = 30\,000\,\text{kg m}^2,$$

$$P_{max} = 12\,000\,\text{hp},\ C_{m_o} = -0.02,\ \eta_{P_{TO}} = 0.5,\ \alpha_{TO} = 15\,\text{deg}$$

Wing: airfoil, $S = 100\,\text{m}^2$, NACA 64_2-215, $\text{AR} = 14$, $e = 0.93$,

$$\Delta C_{L_{flap}} = 0.9,\ \lambda = 1$$

Horizontal tail: $S_h = 25\,\text{m}^2$, NACA 0009, $\text{AR}_t = 6$, $C_{L_{h\_TO}} = -0.9$

The aircraft configuration and other geometry variables are illustrated in Figure 9.40. The following parameters must be determined: landing gear configuration, fixed or retractable, height, wheel track, wheel base, distance between main wheel and aircraft cg, and applied load on each wheel.

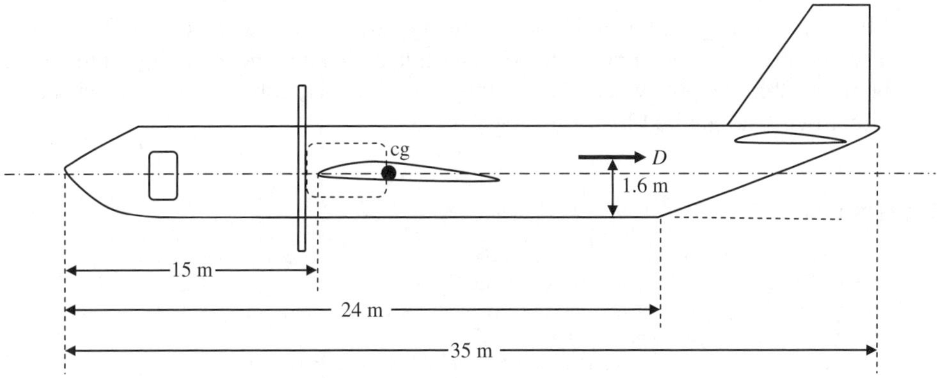

Figure 9.40 Aircraft in Problem 19

20. Design a landing gear for the following early warning jet aircraft. The aircraft has two jet engines, and is equipped with a single-slotted flap which is deflected 25 deg during the take-off operation on a concrete runway. Assume that the aircraft forward cg is at 15% MAC, the aft cg is at 30% of MAC, and the wing/fuselage aerodynamic center is located at 24% MAC. The distance between the horizontal tail aerodynamic center and the wing/fuselage aerodynamic center is 26 m.

$m_{TO} = 180\,000\,\text{kg},\ D_{\text{fmax}} = 3.5\,\text{m},\ V_{\text{max}} = 460\ \text{KTAS (at } 35\,000\,\text{ft)},$

$V_s = 110\ \text{KEAS},\ C_{D_{o\_clean}} = 0.019,\ C_{D_{o\_TO}} = 0.028,\ I_{yy} = 3 \cdot 10^7\,\text{kg m}^2,$

$T_{\text{max}} = 2 \cdot 270\,\text{kN},\ C_{m_o} = -0.06,\ \alpha_{TO} = 13\,\text{deg}$

Wing: airfoil, $S = 320\,\text{m}^2$, NACA 65_2-415, AR $= 10$, $e = 0.85$,

$\Delta C_{L_{\text{flap}}} = 1.4,\ \lambda = 1$

Horizontal tail: $S_h = 75\,\text{m}^2$, NACA 0012, $AR_t = 4$, $C_{L_{h\_TO}} = -1.3$

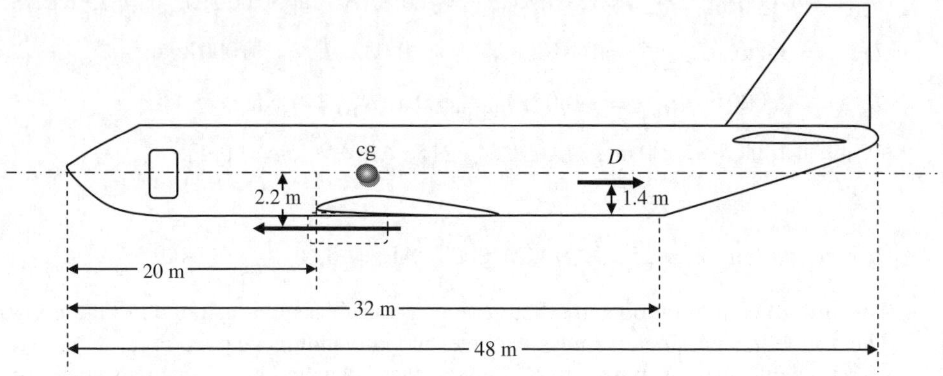

Figure 9.41 Aircraft in Problem 20

The aircraft configuration and other geometry variables are illustrated in Figure 9.41. The following parameters must be determined: landing gear configuration, fixed or retractable, height, wheel track, wheel base, distance between main wheel and aircraft cg, and applied load on each wheel.

References

[1] Currey, N.S. (1988) *Aircraft Landing Gear Design: Principles and Practices*, AIAA.
[2] Roskam, R.J. (2006) *Airplanes War Stories*, DAR Corporation.
[3] Federal Aviation Regulations, Airworthiness Standards for GA Aircraft, FAR 23 (2011), Federal Aviation Administration, US Department of Transportation.
[4] Hibbeler, R.C. (2009) *Engineering Mechanics: Statics*, 12th edn, Prentice Hall.
[5] Budynas, R.G. and Nisbett, J.K. (2011) *Shigley's Mechanical Engineering Design*, 9th edn, McGraw-Hill.
[6] Aircraft Tire Data (2002) The Goodyear Tire & Rubber Company.
[7] *Aircraft Tire Data* (2009) Bridgestone Corporation.
[8] Jackson, P., Munson, K., Peacock, L. *Jane's All the World's Aircraft*, Jane's Information Group, various years 1996 to 2011.

[9] Green, W.L. (1986) *Aircraft Hydraulic Systems: An Introduction to the Analysis of Systems and Components*, John Wiley & Sons, Inc.
[10] Norton, R.L. (2008) *Design of Machinery: An Introduction to the Synthesis and Analysis of Mechanisms and Machines*, McGraw-Hill.
[11] Erdman, A.G., Sandor, G.N., and Kota, S. (2001) *Mechanism Design: Analysis and Synthesis*, 4th edn, Prentice Hall.
[12] Anonymous (1990) Flying Qualities of Piloted Airplanes, Air Force Flight Dynamic Laboratory, MIL-F-1797C, WPAFB, Dayton, OH.

10

Weight of Components

10.1 Introduction

In Chapters 5–9, the detail design of the aircraft major components (wing, tail, fuselage, propulsion system, and landing gear) is presented. These designs were based on an initial weight estimation which was performed at the preliminary design phase. Now that the components have been designed, the aircraft weight can be calculated. Basically, there are three types of analyses regarding the aircraft weight as follows:

1. aircraft weight estimation;
2. aircraft weight calculation;
3. aircraft weight measurement.

The first design analysis regarding aircraft weight is covered in Chapter 4, and the second is introduced in this chapter. The first two analyses are performed prior to the aircraft fabrication, but the third analysis is in fact initiated after the aircraft is manufactured. The accuracy of the aircraft weight estimation during preliminary design phase is about 70–90%, since it is based on a crude statistical technique. However, the aircraft weight calculation is about 85–95% accurate, since it employs a rather more sophisticated empirical approach. It is clear that only the aircraft weight measurement delivers an accuracy of 100%. This comparison indicates that there is a need to modify some of the aircraft parameters, such as aircraft center of gravity (cg) limit, once the aircraft exact weight is obtained.

The aircraft weight calculation is based on the components geometry, dimensions, and the density of the materials they will be manufactured from. But the measurement is based on the components which are designed according to the detail design phase. The aircraft weight measurement simply weighs all components one by one, and then adding them all to find the overall aircraft weight. By comparing these three types of analyses, one could readily conclude that only the aircraft weight measurement is reliable.

The design evolution passes through the weight calculation a number of times until the accuracy is within an acceptable range. The aircraft weight estimation is a useful and necessary basis for the aircraft component design. In contrast, the aircraft weight

Aircraft Design: A Systems Engineering Approach, First Edition. Mohammad H. Sadraey.

calculation is a very significant tool for the balance analysis. This technique is based on industrial experiences and utilizes the detailed statistical equations. The weight analysis process is illustrated in Figure 10.1, which includes two feedback loops. The loop between the aircraft weight measurement and the aircraft balance and cg calculation is often performed once. However, the loop between the aircraft weight calculation and the aircraft components design is repeated several times. The iterations are necessary to minimize the difference between the aircraft weight and the base weight for which the components are designed.

In the first loop, the aircraft major components are initially designed with details based on the aircraft weight which is estimated at the aircraft preliminary design phase. Then, the weight of each component is calculated using the technique offered in this chapter. The aircraft new weight is then determined as the sum of the weight of components. In the next step, the calculated weight is compared with the estimated weight. If any considerable difference is reported, the components must be redesigned according to the new weight. This loop is traveled by the design team a number of times until the difference is acceptable (less than 3% is suggested).

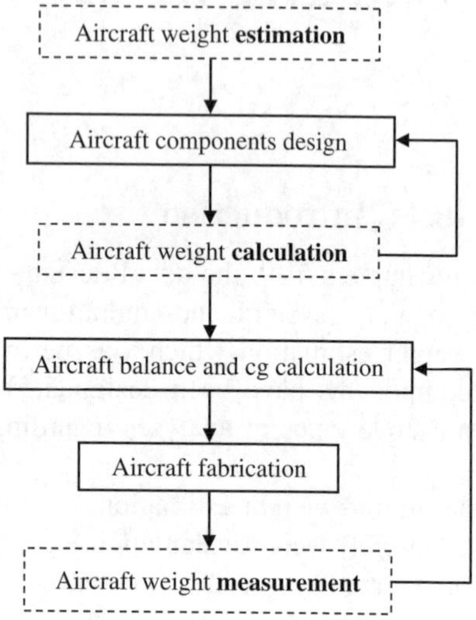

Figure 10.1 Aircraft weight analysis process

Since the manufacturing technology and engineering materials are advancing each year, these equations, and particularly the empirical coefficients, must be updated accordingly. Every year, several new engineering materials are produced that not only are lighter, but much stronger. The new materials include advanced aluminum alloys and advanced composite materials. These new productions make the aircraft lighter in weight and longer in life.

In general, modern aircraft structures are usually manufactured either from aluminum alloys or composite materials. Parts of the landing gears, engine shafts, propellers, and turbine blades are often fabricated from steel alloys. Nowadays, about 85% of aircraft structures are made of aluminum alloys, while about 15% are manufactured from composite materials. Although the Boeing Company used composite materials for its newest production, the long-range, mid-size, wide-body, twin-engine jet airliner Boeing 787 Dreamliner (Figure 1.10), to make it super efficient, most civil transport aircraft are still manufactured from aerospace aluminum alloys (e.g., aluminum alloy 7075, 2024, or 6061). Aluminum alloy plate is used in a large number of aircraft, ranging in complexity and performance requirements from simple components to primary load-bearing structures in aircraft such as the Boeing 777 (Figures 8.6 and 12.27) and Airbus A-340 (Figure 8.7).

The same is true for current military aircraft, where the majority of fighters and bombers are made mainly of aluminum alloys. Stealth aircraft such as the B-2 Spirit (Figure 6.8), F-117 Night Hawk (Figure 6.8), and SR-71 Blackbird (Figure 8.21) are primarily fabricated from advanced composites to satisfy radar detectability requirements. However, the trend is to utilize composite materials in future aircraft. The structure of the majority of model remote-controlled airplanes is made of composite materials such as foam, wood, and plastics (e.g., Monokote). In the General Aviation (GA) aircraft arena, again the majority of aircraft such as Cessna, Beech, and Piper aircraft are made of aluminum alloys. The engineering materials which the aero-structures are made of play a significant role in the aircraft weight calculation. Hence, the designer should be aware of the materials for aircraft production in advance in order to calculate the weight of various components.

The home-built aircraft BD-5J Microjet is the world's lightest jet (please note: turboprop engine) aircraft, with an empty mass of 196 kg and a take-off mass of 390 kg. However, the world's heaviest aircraft is the Antonov An-225 Mriya with six turbofan engines and a mass of 640 000 kg. These two numbers demonstrate two extreme limits of aircraft weight, and provide a vision for the aircraft designer of the range the numbers may occupy.

The sensitivity of weight calculation, including the role of aircraft weight and center of gravity in aircraft control and stability, will be introduced in Section 10.2. The aircraft is often divided into several component groups for the purpose of weight calculation. Section 8.3 is devoted to the aircraft weight breakup for various aircraft families. The main section of the chapter is Section 8.4, which presents the empirical technique to determine the weight of components. Finally, an example is solved to demonstrate the application of the technique.

10.2 Sensitivity of Weight Calculation

One of the primary differences between an aircraft and other types of structures and vehicles such as buildings, bridges, automobiles, trains, and ships lies in the weight sensitivity. In a non-aero-structure, the weight is not as sensitive as in an aero-structure. In non-aero-structures, the weight could be compromised to gain a higher strength. However, in aircraft structures, the weight has a limit and may not be increased even to improve the stiffness or strength. The semi-monocoque structures are very effective and are employed in an aircraft due to a need for stressed skin. Although a thicker spar web or fuselage skin may result in a stronger structure, it degrades the mission success due to its heavy weight. Hence, the aircraft designer should try his/her best to reduce the aircraft weight and establish an optimal weight.

The logic of weight sensitivity must be considered in the aircraft weight calculations. So, the calculation of the weight of an aircraft is one of the critical parts of the design process. A mistake in the weight calculation may result in a catastrophe for a design program. Three main reasons for an accurate weight calculation are: (i) aircraft manufacturing cost, (ii) aircraft performance, and (iii) aircraft center of gravity. These parameters are directly influenced by aircraft weight.

Aircraft center of gravity and corresponding parameters such as aircraft stability and control will be addressed in Chapter 11. The aircraft performance is a strong function of the aircraft weight. When the aircraft weight is increased, while the aircraft geometry and engine setting remain the same, the aircraft performance variables such as maximum speed, ceiling, and rate of climb will decline. For instance, consider a GA aircraft when the aircraft weight increases by 10%. The consequences of such an increase are typically as follows.

- Increment in stall speed: 5%
- Reduction in maximum speed: 4%
- Reduction in range: 8%
- Reduction in endurance: 9%
- Increment in take-off run: 8%
- Reduction in rate-of-climb: 16%
- Reduction in ceiling: 7%.

These performance variation numbers indicate that the aircraft weight must be minimized as much as possible. Furthermore, it demonstrates the crucial role of aircraft weight calculation during the detail design phase.

The relationship between aircraft weight and three major design phases is discussed in Chapter 2. Here, a review on this topic seems helpful. Figure 10.2 shows the conceptual relationship between the ease of change in aircraft weight and aircraft design phases. Experience indicates that there may be a large commitment in terms of technology applications, the establishment of an aircraft configuration and its characteristics, the obligation or resources, and potential lifecycle cost at the early stages of the design process. It is at this point when aircraft-specific knowledge (e.g., aerodynamics, flight dynamics, and propulsion) is limited, but when major decisions are made pertaining to items such as selection of technologies, selection of materials and potential sources of supply, equipment packaging scheme and levels of diagnostics, selection of manufacturing process, and establishment of maintenance approach. Reference [1] may be consulted for more details on the interrelationship between the systems engineering discipline and the aircraft weight calculation.

Aircraft manufacturing cost is influenced directly by aircraft weight. However, the relationship is not linear, as shown in Figure 10.3. It is clear that the aircraft manufacturing cost is largely based on the weight. However, there is only one optimal weight that incurs

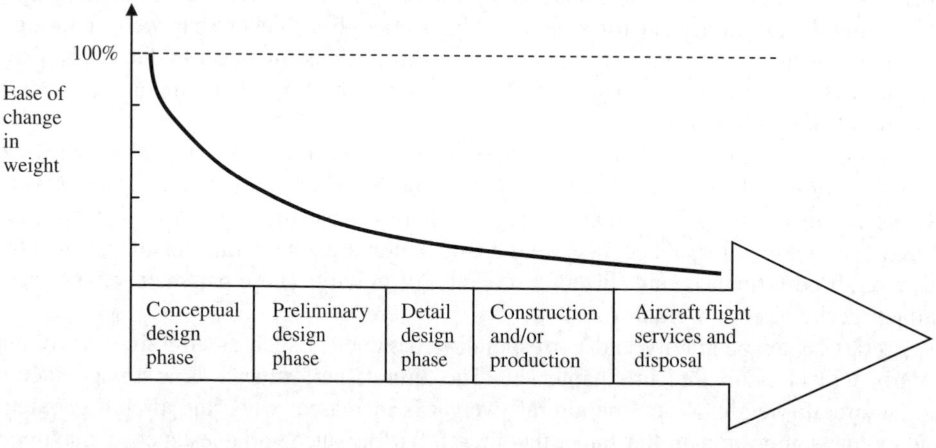

Figure 10.2 Lifecycle commitment and ease of change of aircraft weight

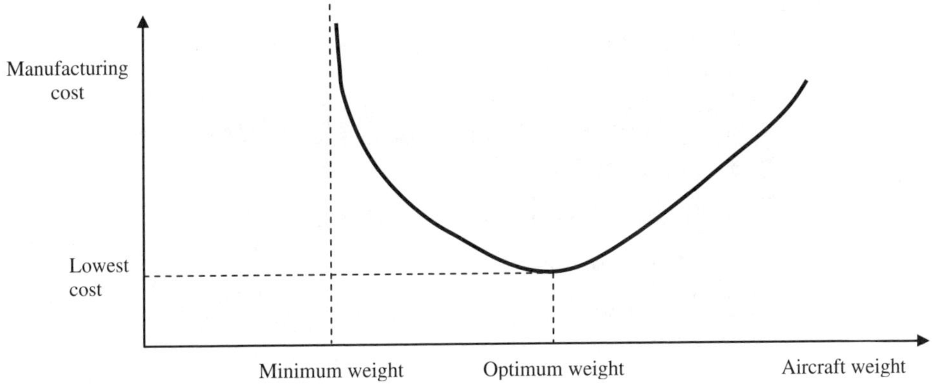

Figure 10.3 Non-linear relationship between manufacturing cost and aircraft weight

the lowest fabrication cost. Any other aircraft weight, either lighter or heavier, will cost more to produce. The heavier weight requires more materials, so the aircraft manufacturer should spend more money to purchase more materials. However, lighter engineering materials are often more expensive. For instance, 1 kg of carbon fiber is about 10 times more expensive than 1 kg of fiberglass. Or, 1 kg of aerospace aluminum alloy is about 1/20th cheaper than some advanced aluminum alloys.

In some cases, particularly in military applications, to reduce the aircraft weight by 10% requires an increase of about 10 000% in manufacturing cost. The huge jump in production cost is due to the fact that new materials and new fabrication techniques require thousands of man-hours in research and development (R&D) to develop advanced materials and techniques. Handling some light and advanced materials requires precise manufacturing techniques, which are highly expensive. The stories of the stealth aircraft which were designed, manufactured, and assembled by Lockheed Skunk Work [2] reveal the amount of money spent on advancing the new materials.

The maximum take-off weight (MTOW) of a conventional aircraft is a function of payload weight. The payload is the useful load which must be carried by the aircraft as part of its mission. It does not include the weight of flight crew and fuel weight. For each type of aircraft, the ratio between MTOW and payload weight has a specific range. The maximum take-off weight-to-payload weight ratio for several aircraft is presented in Table 10.1. The data has been extracted by the author from various references, such as [3]. This table may be employed to compare the calculated aircraft weight with current values to obtain the position of your aircraft with respect to the statistics. The Society of Allied Weight Engineers (www.sawe.org) has data for aircraft weight and balance.

Three major feedback loops in the entire design process are illustrated in Figure 10.4. The aircraft weight calculation lies in the second loop, which provides feedback on aircraft stability analysis. This feedback initially uses the weight of components to balance the aircraft by allocating them at the right place to submit the desired center of gravity location. As Figure 10.4 shows, the calculated aircraft weight is employed to modify all components that were previously designed based on an estimated aircraft weight. The sensitivity of the weight calculation is clearly depicted in Figure 10.4, since the feedback will cause all components to be redesigned.

Table 10.1 Maximum take-off weight-to-payload weight ratio for several aircraft

No.	Aircraft	MTOW/payload weight
1	Transport	3–4
2	Single-engine light GA	3–6
3	Twin-engine GA	2–4
4	Glider/sailplane	3–6
5	Hang glider	1.2–1.4
6	Motor glider	4–9
7	Supersonic fighter	10–18
8	Remote-controlled model	1.5–2.5
9	Human-powered	1.3–1.6
10	Agriculture	2–3
11	Basic trainer	6–15
12	Ultralight	2–3

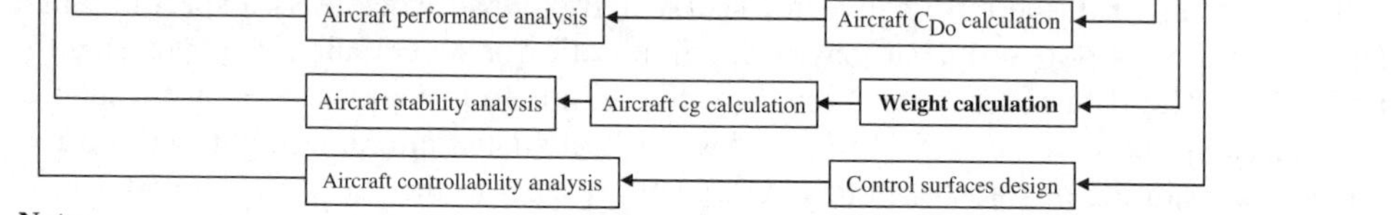

Note:

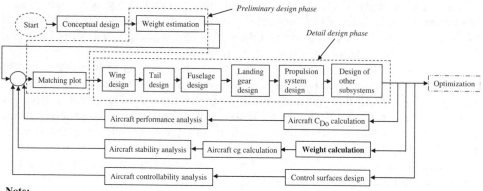

(1) Loop 1: Aircraft performance analysis.
(2) Loop 2: Aircraft stability analysis.
(3) Loop 3: Aircraft controllability analysis.
(4) The relationship between three design phases (conceptual, preliminary, and detail) is briefly depicted.
(5) Aircraft center of gravity (cg) calculation and aircraft balance are examined in Chapter 11.
(6) Control surfaces design is addressed in Chapter 12.
(7) Aircraft performance analysis, stability analysis, and controllability analysis are beyond the scope of this book.

Figure 10.4 Three major feedback loops in the design process

10.3 Aircraft Major Components

An aircraft is manufactured and assembled from a large number of parts and elements which may be combined into several groups according to the group function.

The aircraft maximum take-off weight (W_{TO}) may be divided into four major weight groups: (i) empty weight (W_E), (ii) payload (W_P), (iii) crew members (W_C), and (iv) fuel (W_F). The summation of these four elements makes up the aircraft overall weight:

$$W_{TO} = W_E + W_P + W_C + W_F \tag{10.1}$$

The weight of the payload is easily determined based on the given features of what is supposed to be carried (e.g., cargo, passengers, baggage, and store) by the aircraft. The weight of flight crew members is a function of the number of people necessary for the mission success. The technique to determine the weight of flight crew members, flight attendants, as well as passengers was examined in Chapter 7 on fuselage design. The weight of fuel is determined based on the performance mission parameters such as range and endurance. The details of the fuel weight calculation technique are addressed in Chapter 4. This chapter is devoted to the technique to determine the aircraft empty weight. The sum of weights of payload, flight crew members, and fuel is sometimes referred to as removable weight (W_R):

$$W_R = W_P + W_C + W_F \tag{10.2}$$

These three elements may be removed without disintegrating the aircraft integrity. The aircraft empty weight is divided into three main weight groups:

1. structure
2. engine (including nacelle and pylon), and
3. systems and equipment.

Furthermore, the aircraft structure is subdivided into four/five major weight groups:

1. wing;
2. fuselage;
3. tail (horizontal and vertical);
4. landing gear;
5. nacelle (if engine is podded).

In this classification, the control surfaces are included in the corresponding lifting surfaces. For example, the elevator and rudder weight is included in the tail group, while the aileron weight is accounted for in the wing group. Table 10.2 illustrates the weight percentage of aircraft empty weight and payload for several aircraft types. Table 10.3 also presents the average aircraft weight breakup (engine, fuel, payload, structure, and systems and equipment) for several aircraft types. In defining the payload, there is a subtle point that must be explained. In a human-powered aircraft, the pilot is assumed as part of the propulsion system. In a twin-seat ultralight aircraft, one of the occupants is assumed as part of the payload. In a single-seat GA aircraft, the pilot is the main part of the payload, while in a multi-seat GA aircraft, the pilot is not assumed as part of the payload.

Table 10.2 Average weight breakdown for several aircraft

No.	Aircraft	Empty weight (%)	Removable weight (%)	Payload weight (%)	Fuel weight (%)
1	Hang glider	25	75	75	0
2	Glider/sailplane	60	40	40	0
3	Human-powered	30	70	70	0
4	Model RC	40	60	53	7
5	Ultralight	55	42	42	3
6	Single-engine GA	60–82	18–40	8–30	10
7	Twin-engine GA	62	38	16	20
8	Agriculture	50	50	40	10
9	Subsonic transport	48	52	26	25
10	Supersonic fighter	43	57	40	16
11	Rutan Voyager	23	77	5	72

Table 10.3 Average group weight breakdown for several aircraft

No.	Aircraft	Fuel weight (%)	Payload weight (%)	Crew (%)	Engine (%)	Structure (%)	Equipment (%)
1	Hang glider	0	75	75	0	25	0
2	Glider/sailplane	0	40	40	0	58	2
3	Human-powered	0	70	70	75[a]	23	2
4	Model RC	5	55	0	6	32	2
5	Ultralight	3	42[b]	21	20	32	3
6	Single-engine GA	10	8–30[c]	8	23	30–52	2
7	Twin-engine GA	14	24	1	24	31	3
8	Agriculture	10	40	1	20	25	4
9	Subsonic transport	25	26.5	0.5	12	24	12
10	Supersonic fighter	16	40	1	13	20	10

[a] The pilot is assumed as part of the propulsion system.
[b] One of the crew members is part of the payload.
[c] In a single-seat GA aircraft, the pilot is the main part of the payload, while in a multi-seat GA aircraft, the pilot is not assumed as part of the payload.

According to Table 10.2, a hang glider tends to have the highest percentage of the payload weight (i.e., 75%), which means it has the lightest structure. However, a single-seat GA aircraft has the lowest percentage of removable weight, which implies the relatively heaviest structure. Hang gliders, gliders/sailplanes, and human-powered aircraft carry no fuel and have no fuel tank. However, in the case of the human-powered aircraft, the pilot is part of the propulsion system. Subsonic transport aircraft historically carry the highest fuel weight ratio, due to the long endurance and reserve fuel requirements.

Based on Table 10.3, hang gliders tend to have the highest percentage of crew weight, while subsonic transport aircraft have the lowest percentage of crew weight. In general, transport aircraft are the most economical in that they are designed to be the most profitable. Furthermore, subsonic transport aircraft carry the maximum number of equipment due to the safety and airworthiness requirements. Although gliders have the highest percentage of structure weight, this is because of lack of engine and fuel.

The last row in Table 10.2 demonstrates the weight breakdown for the aircraft Rutan Model 76 Voyager (Figure 4.5), which was designed and built for a special mission. The mission of this aircraft, with a maximum take-off mass of 4397 kg, was to fly around the globe without stopping or refueling. The mission was accomplished successfully on a nine-day flight in December 1986. This record-breaking composite aircraft carried two crew members, and 72% of the aircraft weight was fuel weight. This aircraft weight breakdown is an exception and is not the norm in conventional aircraft.

Table 10.4 illustrates a typical aircraft structural weight breakdown for several aircraft. In practice, every value has a range which is based on a number of factors including the manufacturer approach, structural materials, load factor, and aircraft configuration. For instance, as the load factor is increased (particularly in fighter aircraft), the percentage weight of the wing is increased. Moreover, the landing gear weight is varied if the tricycle configuration is changed to a tail-gear. A T-tail is considerably heavier than a conventional

Table 10.4 Structural weight breakdown for several aircraft

No.	Aircraft	Wing (%)	Fuselage (%)	Tail (horizontal and vertical) (%)	Landing gear (%)	Structure (%)
1	Hang glider	20	5	0	0	25
2	Glider/sailplane	30	23	3	2	58
3	Human-powered	9	10	2	3	23
4	Model RC	14	11	2	4	32
5	Ultralight	15	9	3	5	32
6	Single-engine GA	13	11	2	4	30
7	Twin-engine GA	14	11	2	4	31
8	Agriculture	10	9	2	4	25
9	Subsonic transport	10	8	2	4	24
10	Supersonic fighter	8	7	2	3	20

tail to carry the banding moment of the vertical tail by the horizontal tail. Please note that all numbers are quoted relative to the aircraft weight in that specific group.

Generally speaking, gliders or sailplanes have relatively the heaviest wing due to a very high aspect ratio (AR) (usually on the order of 30). In contrast, fighters often tend to have relatively the lightest wing due to a very short AR and small wing area. A hang glider uses no mechanical landing gear (0%), since the pilot lands using his/her legs. A hang glider does not have a regular tube-shaped fuselage, but a mechanical linkage that the pilot hangs on to is assumed to be the fuselage (5%). In the majority of aircraft, the wing group has the largest contribution to the weight of the structure. In Table 10.4, the weight of the nacelle for a podded engine configuration is included in the wing or fuselage. Figure 10.5 illustrates the medium-haul airliner Fokker 100 with MTOW/payload weigh ratio of 1.75, and the two-seat turboprop trainer Embraer EMB-314 Super Tucano with MTOW/payload weigh ratio of 1.318.

Table 10.5 [4–6] demonstrates the actual weight of major structural components for several prop-driven and jet aircraft. It contains the weight of components for a home-built, two single-piston engine, four twin-prop engine, an agricultural, two business jet, three jet transport, three fighter, and one large cargo aircraft. This historical tabulated data may be used in conjunction with the empirical technique presented in the next section, to calculate the aircraft component weights.

10.4 Weight Calculation Technique

The purpose of this section is to present a technique for calculating weights of aircraft components, which is the basis for calculating aircraft empty weight. Almost all aircraft manufacturers have developed their own technique for calculating the weight of components. Most of these techniques are not published and are proprietary. There are

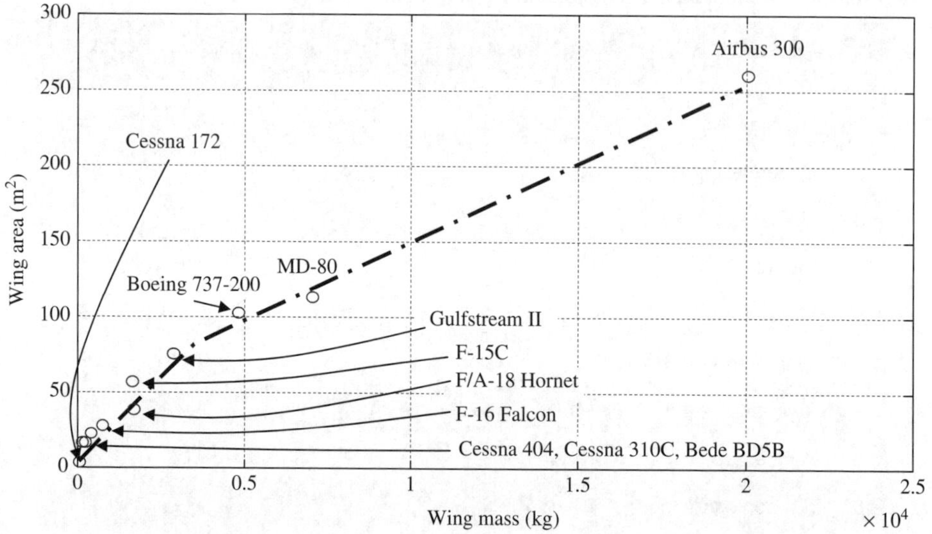

Figure 10.5 Wing mass versus wing area for several aircraft

Table 10.5 Actual weight (in lb) of major components for several aircraft

No.	Aircraft	Type	Wing	Tail	Fuselage	Nacelle	Landing gear	Engine	Equipment	Fuel	Payload
1	Cessna 172	Single-engine prop GA	226	57	353	27	111	267	159	252	702
2	Cessna 182	Single-engine prop GA	235	62	400	34	132	545	173	390	715
3	Cessna 310C	Twin-engine prop GA	453	118	319	129	263	1 250	498	612	1 186
4	Cessna 404 Titan	Twin-engine prop GA	860	181	610	284	316	1 626	1 129	1 379	1 900
5	PZL M21 Dromader	Agriculture	1 455	181	540	106	322	1 471	719	1 984	2 425
6	Learjet 28	Business twin-jet	1 939	361	1 624	214	584	1 284	2 605	4 684	1 962
7	Gulfstream II	Business twin-jet	6 372	1 965	5 944	1 239	2 011	6 886	11 203	23 300	5 380
8	Fokker F-27-100	Twin-engine turboprop	4 408	977	4 122	628	1 940	4 347	5 673	9 198	12 500
9	Short skyvan	Twin-engine turboprop	1 220	374	2 154	254	466	1 524	1 703		4 924
10	MD-80	Jet transport	15 560	3 320	16 150	2 120	5 340	11 000	25 460	39 362	43 050
11	Boeing 737-200	Jet transport	10 613	2 718	12 108	1 392	4 354	8 177	14 887	34 718	34 790
12	Airbus 300	Jet transport	44 131	5 941	35 820	7 039	13 611	22 897	35 053	76 512	69 865
13	F-15C Eagle	Fighter	3 642	1 104	6 245	102	1 393	9 205	5 734	13 455	2 571
14	F-16 Falcon	Fighter	1 699	650	3 069	598	867	3 651	4 191	–	–
15	F/A-18	Fighter	3 798	945	4 685	143	1 992	6 277	5 134	17 592	7 684
16	Lockheed C-5A	Large cargo	100 015	12 461	118 193	9 528	38 353	40 575	44 059	332 500	200 000
17	Bede BD5B	Home-built prop-driven	87	17	89	0	32	189	125	340	170

Table 10.6 Density of various aerospace materials

No.	Engineering materials	Density (kg/m^3)
1	Aerospace aluminum	2711
2	Fiberglass/epoxy	1800–1850
3	Graphite/epoxy	1520–1630
4	Low-density foam	16–30
5	High-density foam	50–80
6	Steel alloys	7747
7	Titanium alloys	4428
8	Balsa wood	160
9	Plastics (including Monokote)	900–1400

some exceptions, such as the one in Ref. [7]. The components weight calculation is a mixture of rational analysis and statistical methods. The equations presented in this section are developed based on four sources:

1. direct relationship between weight of an object and its average density (Table 10.6);
2. actual published data on weight of various components (e.g., Table 10.5);
3. derived empirical factors by the author;
4. published empirical equations [6–9].

The first and second sources make the technique very accurate and reliable. However, the third source indicates that there must be a calibration of the results to adjust the empirical factors. Empirical factors are due to the fact that aero-structure elements are generally hollow and not solid objects, and include skin, spar, frame, rib, stiffener, and longeron.

Statistical equations for many components are presented in exponential form, where the constant of proportionality and the exponents of the design parameters are determined by the standard regression analysis approach. The equations employ weight and geometric data of actual components, and are subjected to the condition of minimum standard deviation. For each contributing design factor, a relevant parameter is selected based on a rational technique. A curve fit approach is utilized and a linear function is obtained to model the contribution of each design parameter to the weight of a component.

Table 10.7 provides the empirical factors readily derived from data in Refs [3, 5] to rapidly estimate the weight of the wing. Figure 10.6 illustrates the wing mass versus wing area for several aircraft. As this figure demonstrates, there is a linear relationship between wing weight and wing planform area. The relationship is clearly a function of several factors, including the engine attachment to the wing, the fuel tank in the wing, the number of stores (e.g., missiles and rockets), their locations relative to the wing root, and maximum dynamic pressure.

Table 10.7 Wing mass versus wing area for various aircraft

No.	Aircraft	Type	Wing area (m^2)	Wing mass (kg)	Wing mass (kg) / Wing area (m^2)
1	Bede BD5B	Home-built prop-driven	3.5	39.5	11.3
2	Cessna 172	Single-engine prop GA	16.2	102.5	6.3
3	Cessna 310C	Twin-engine prop GA	16.6	205.5	12.4
4	Cessna 404	Twin-engine prop GA	22.24	390	17.5
5	Gulfstream II	Business twin-jet	75.21	2 890	38.5
6	MD-80	Jet transport	112.3	7 058	62.8
7	Boeing 737-200	Jet transport	102	4 814	48
8	Airbus 300	Jet transport	260	20 017	77
9	Lockheed C-5A	Large cargo	576	45 366	76.8
10	F-15CEagle	Fighter	56.5	1 652	29.2
11	F-16 Falcon	Fighter	27.9	771	27.6
12	F/A-18 Hornet	Fighter	38	1 723	45.3

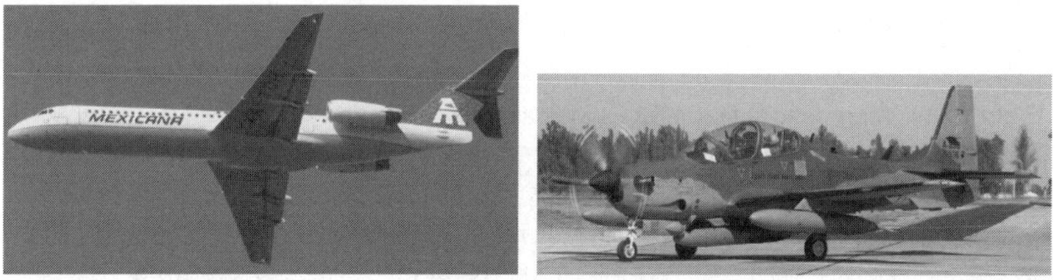

(a) (b)

Figure 10.6 Two aircraft with different MTOW/payload weigh ratios: (a) Fokker 100 with MTOW/payload weigh ratio of 1.75; (b) Embraer EMB-314 Super Tucano with MTOW/payload weigh ratio of 1.318. Reproduced from permission of (a) Anne Deus; (b) Antony Osborne

The equations introduced in this section are valid in both metric and British units. If metric units are employed for the right-hand-side variables, the weight will be obtained in terms of Newton (N). However, when British units are employed for the right-hand-side variables, the weight will be obtained in terms of pound (lb).

10.4.1 *Wing Weight*

The weight of the wing (W_W) is a function of wing planform area (S_W), the items (e.g., fuel and engine) carried by the wing, wing maximum thickness-to-chord ratio, aircraft maximum load factor, wing structural configuration (e.g., single-spar, twin-spar), construction material (e.g., aluminum, composite), and other wing geometry (AR, taper ratio, and sweep angle). The wing weight is calculated as follows:

$$W_W = S_W \cdot \text{MAC} \cdot \left(\frac{t}{C}\right)_{\max} \cdot \rho_{\text{mat}} \cdot K_\rho \cdot \left(\frac{\text{AR} \cdot n_{\text{ult}}}{\cos\left(\Lambda_{0.25}\right)}\right)^{0.6} \cdot \lambda^{0.04} \cdot g \qquad (10.3)$$

where S_W denotes the wing planform area, MAC the wing mean aerodynamic chord, $(t/C)_{\max}$ the maximum thickness-to-chord ratio, ρ_{mat} the density of construction material (Table 10.6), AR the aspect ratio, n_{ult} the ultimate load factor, $\Lambda_{0.25}$ the quarter chord sweep angle, λ the taper ratio, and g the gravitational constant (9.81 m/s^2 or 32.17 ft/s^2). The parameter K_ρ is the wing density factor and is obtained from Table 10.8. This parameter has no unit; hence, Equation (10.3) may be employed in both SI and British units.

For the purpose of structural safety considerations, the ultimate load factor (n_{ult}) is usually 1.5 times the maximum load factor (i.e., a safety factor of 1.5):

$$n_{\text{ult}} = 1.5 \cdot n_{\max} \qquad (10.4)$$

Typical maximum load factors ($n_{\max}$) for various aircraft are shown in Table 10.9. The range for the value of K_ρ in Table 10.8 is due to the fact that as the distance between the fuel tank/engine and the wing root increases, the wing will be heavier in order to handle the larger root bending moment.

Table 10.8 Wing density factor for various aircraft

No.	Aircraft – wing structural installation condition	K_ρ
1	GA, no engine, no fuel tank in the wing	0.0011–0.0013
2	GA, no engine on the wing, fuel tank in the wing	0.0014–0.0018
3	GA, engine installed on the wing, no fuel tank in the wing	0.0025–0.003
4	GA, engine installed on the wing, fuel tank in the wing	0.003–0.0035
5	Home-built	0.0012–0.002
6	Transport, cargo, airliner (engines attached to the wing)	0.0035–0.004
7	Transport, cargo, airliner (engines not attached to the wing)	0.0025–0.003
8	Supersonic fighter, few light stores under wing	0.004–0.006
9	Supersonic fighter, several heavy stores under wing	0.009–0.012
10	Remotely controlled model	0.001–0.0015

Table 10.9 Maximum positive load factor for various aircraft

No.	Aircraft	Maximum load factor (n_{max})
1	GA normal	2.5–3.8
2	GA utility	4.4
3	GA acrobatic	6
4	Home-built	2.5–5
5	Remote-controlled model	1.5–2
6	Transport	3–4
7	Supersonic fighter	7–10

10.4.2 Horizontal Tail Weight

The weight of the horizontal tail (W_{HT}) is a function of horizontal tail planform area (S_{HT}), tail maximum thickness-to-chord ratio, tail configuration (e.g., conventional, T-tail, or V-tail), construction material, elevator chord, and other horizontal tail geometry (AR, taper ratio, and sweep angle). The horizontal tail weight is calculated as follows:

$$W_{HT} = S_{HT} \cdot \mathrm{MAC}_{HT} \cdot \left(\frac{t}{C}\right)_{\max_{HT}} \cdot \rho_{mat} \cdot K_{\rho HT} \cdot \left(\frac{\mathrm{AR}_{HT}}{\cos\left(\Lambda_{0.25_{HT}}\right)}\right)^{0.6} \cdot \lambda_{HT}^{0.04} \cdot \bar{V}_H^{0.3} \cdot \left(\frac{C_e}{C_T}\right)^{0.4} \cdot g \qquad (10.5)$$

where S_{HT} denotes the horizontal tail exposed (i.e., net) planform area, MAC_{HT} the horizontal tail mean aerodynamic chord, $(t/C)_{max\_HT}$ the horizontal tail maximum thickness-to-chord ratio, ρ_{mat} the density of construction material (Table 10.6), AR_{HT} the horizontal tail aspect ratio, $\Lambda_{0.25\_HT}$ the horizontal tail quarter chord sweep angle, λ_{HT} the horizontal tail taper ratio, C_e/C_T the elevator-to-tail chord ratio, and $\bar{V}_H$ the horizontal tail volume ratio (Chapter 6). The parameter $K_{\rho HT}$ is the horizontal tail density factor and is obtained from Table 10.10. This parameter has no unit; hence, Equation (10.5) may be utilized in both metric and British units. For other types of tail in Table 10.10, an interpolation yields a reasonable result.

10.4.3 Vertical Tail Weight

The weight of the vertical tail (W_{VT}) is a function of the vertical tail planform area (S_{VT}), vertical tail maximum thickness-to-chord ratio, construction material, vertical tail volume

Table 10.10 Tail density factors for various aircraft

No.	Aircraft – tail configuration	$K_{\rho HT}$	$K_{\rho VT}$
1	GA, home-built – conventional		
	tail/canard	0.022–0.028	0.067–0.076
2	GA, home-built – T-tail/H-tail	0.03–0.037	0.078–0.11
3	Transport – conventional tail	0.02–0.03	0.035–0.045
4	Transport – T-tail	0.022–0.033	0.04–0.05
5	Remotely controlled model	0.015–0.02	0.044–0.06
6	Supersonic fighter	0.06–0.08	0.12–0.15

ratio, and other vertical tail geometry (AR, taper ratio, and sweep angle). The vertical tail weight is calculated as follows:

$$W_{\mathrm{VT}} = S_{\mathrm{VT}} \cdot \mathrm{MAC}_{\mathrm{VT}} \cdot \left(\frac{t}{C}\right)_{\max_{\mathrm{VT}}} \cdot \rho_{\mathrm{mat}} \cdot K_{\rho\mathrm{VT}} \cdot \left(\frac{\mathrm{AR}_{\mathrm{VT}}}{\cos\left(\Lambda_{0.25_{\mathrm{VT}}}\right)}\right)^{0.6}$$

$$\cdot \lambda_{\mathrm{VT}}^{0.04} \cdot \bar{V}_{\mathrm{V}}^{0.2} \left(\frac{C_{\mathrm{r}}}{C_{\mathrm{V}}}\right)^{0.4} \cdot g \tag{10.6}$$

where S_{VT} denotes the vertical tail planform area, $\mathrm{MAC}_{\mathrm{VT}}$ the vertical tail mean aerodynamic chord, $(t/C)_{\max\_\mathrm{VT}}$ the vertical tail maximum thickness-to-chord ratio, ρ_{mat} the density of construction material (Table 10.6), $\mathrm{AR}_{\mathrm{VT}}$ the aspect ratio, $\Lambda_{0.25\_\mathrm{VT}}$ the vertical tail quarter chord sweep angle, λ_{VT} the vertical tail taper ratio, $C_{\mathrm{r}}/C_{\mathrm{V}}$ the rudder-to-vertical tail chord ratio, and $\bar{V}_{\mathrm{V}}$ the vertical tail volume ratio (Chapter 6). The parameter K_{ρ_VT} represents the vertical tail density factor and is obtained from Table 10.10. This parameter has no unit; hence, Equation (10.6) may be employed in both SI and British units.

10.4.4 Fuselage Weight

The weight of the fuselage (W_{F}) is a function of the fuselage volume, fuselage configuration, construction material, fuselage structural arrangement, and aircraft mission. The fuselage weight is calculated as follows:

$$W_{\mathrm{F}} = L_{\mathrm{f}} \cdot D_{\mathrm{f_{max}}}^2 \cdot \rho_{\mathrm{mat}} \cdot K_{\rho\mathrm{f}} \cdot n_{\mathrm{ult}}^{0.25} \cdot K_{\mathrm{inlet}} \cdot g \tag{10.7}$$

where L_{f} denotes the fuselage length, $D_{\mathrm{f_{max}}}$ the fuselage maximum diameter of the equivalent circular cross-section, ρ_{mat} the density of construction material (Table 10.6). The

Table 10.11 Fuselage density factor for various aircraft

No.	Aircraft	$K_{\rho f}$
1	General aviation, home-built	0.002–0.003
2	Unmanned aerial vehicle	0.0021–0.0026
3	Transport, cargo, airliner	0.0025–0.0032
4	Remotely controlled model	0.0015–0.0025
5	Supersonic fighter	0.006–0.009

parameter K_{ρ_f} represents the fuselage density factor and is obtained from Table 10.11. This parameter has no unit; hence, Equation (10.7) may be used in both SI and British units. The parameter K_{inlet} is 1.25 for the case of inlets on the fuselage, and 1 for inlets elsewhere. Since unmanned and model aircraft do not carry humans, they are not designed to have pressurized cabins. Thus, these two types of aircraft tend to have a relatively lighter fuselage.

10.4.5 Landing Gear Weight

The landing gear is composed mainly of strut, wheel, tire, shock absorber (e.g., hydraulic system), retraction system (if any), and braking system. The weight of the landing gear is largely a function of aircraft weight at landing (W_L), and also affected by landing gear height (H_{LG}), landing gear configuration, landing speed, landing run, retraction system, construction material, and landing ultimate load factor ($n_{ult_{land}}$). The landing gear weight (W_{LG}) is calculated as follows:

$$W_{LG} = K_L \cdot K_{ret} \cdot K_{LG} \cdot W_L \cdot \left(\frac{H_{LG}}{b}\right) \cdot n_{ult_{land}}^{0.2} \tag{10.8}$$

where b is the wing span and K_{ret} is 1 for fixed landing gear and 1.07 for retractable landing gear. The parameter K_{LG} is the landing gear weight factor and is presented in Table 10.12 for various aircraft. The parameter K_L is the landing place factor, and is 1.8 for Navy aircraft and 1 otherwise. This factor indicates that, since Navy aircraft land on an aircraft carrier and employ arresting gear (tail hook), the landing gear is about 80% heavier.

The ratio between main and nose/tail gear is almost proportional to the percentage of the aircraft weight carried by each set of gears. For instance, if 80% of the aircraft weight is supposed to be carried by the main gear, the weight of the main gear is equal to 0.8 multiplied by W_{LG} from Equation (10.8). Equation (10.8) may be employed in both SI and British units.

Table 10.12 Landing gear weight factor for various aircraft

No.	Aircraft	K_{LG}
1	General aviation, home-built	0.48–0.62
2	Transport, cargo, airliner	0.28–0.35
3	Supersonic fighter	0.31–0.36
4	Remotely controlled model	0.35–0.52

10.4.6 Installed Engine Weight

The engine itself is selected by the aircraft designer, and then ordered from a manufacturer for production. Thus, the weight of each engine (W_E) is readily available based upon the engine manufacturer's data and scaling factor. However, the engine installation is another story that must be dealt with. Engine installation may require a firewall, engine mount, cowl, nacelle, pylon, inlet provision, and starting system. The installed engine weight ($W_{E_{ins}}$) including propeller(s) for GA aircraft is calculated as follows:

$$W_{E_{ins}} = K_E \cdot N_E \cdot (W_E)^{0.9} \tag{10.9}$$

where N_E is the number of engines. The parameter K_E is the engine weight factor and is 2.6 when using British units (i.e., lb) and 3 for metric units (i.e., N). If the engine is not selected yet, the engine weight may be estimated using Table 8.3. For installed engine weight of other types of aircraft (e.g., fighter, transport), the interested reader is referred to Ref. [5]. For fighter and transport aircraft, the propulsion system weight includes weight of engine cooling, weight of starter, weight of inlet system, weight of firewall, weight of nacelle, weight of engine control, and weight of auxiliary power unit.

10.4.7 Fuel System Weight

The fuel system includes items such as pipes, hoses, pumps, tanks, and valves. The weight of the fuel system is mainly a function of the total fuel weight (W_{fuel}), and also affected by the type of fuel tank(s), fuel tank location(s), pumps, valves, pipes, number of fuel tanks, and number of engines. The three equations in this subsection are reproduced from Ref. [6].

1. **GA aircraft**

$$W_{FS} = K_{fs} \cdot \left(\frac{W_{fuel}}{\rho_f}\right)^{n_{fs}} \tag{10.10}$$

where K_{fs} is 2 for single-engine aircraft and 4.5 for multi-engine aircraft, and n_{fs} is 0.667 for single-engine aircraft and 0.60 for multi-engine aircraft. The weight W_{fuel}

must be in lb and ρ_f is the fuel density in lb/gal. Recall that aviation gasoline has a density of 5.87 lb/gal and JP-4 has a density of 6.55 lb/gal. The technique to determine the total aircraft fuel weight is introduced in Section 7.1. The resultant fuel system weight (W_{FS}) will be in lb.

2. **Transport and fighter aircraft**
   a. For transport and fighter aircraft equipped with non-self-sealing bladder tanks:

$$W_{FS} = K_{fs} \cdot \left(\frac{W_{fuel}}{\rho_f}\right)^{n_{fs}} \tag{10.11}$$

   where K_{fs} is 1.6 and n_{fs} is 0.727. The fuel weight (W_{fuel}) must be in lb and ρ_f is the fuel density in lb/gal. The resultant fuel system weight (W_{FS}) will be in lb.

   b. For transport and fighter aircraft equipped with integral fuel tanks (i.e., wet wing, such as in F-111):

$$W_{FS} = 15\left(N_t\right)^{0.5} \cdot \left(\frac{W_{fuel}}{\rho_f}\right)^{0.333} + 80\left(N_E + N_t - 1\right) \tag{10.12}$$

   where N_t denotes the number of separate fuel tanks, and N_E is the number of engines. The resultant fuel system weight (W_{FS}) will be in lb.

   It must be noted that an aircraft with electric engine(s) (e.g., remote-controlled model) must carry a battery (and maybe a solar cell). The application of fuel tank terminology does not seem appropriate, so a fuel cell is recommended. For this type of aircraft, the weight of the battery and fuel cells must be known to determine the weight of the fuel system.

10.4.8 Weight of Other Equipment and Subsystems

There is a variety of other subsystems that an aircraft is often equipped with. For instance, power transmission system to control surfaces, hydraulic system, electric system, avionic system, instruments, air conditioning system, anti-ice system, and furnishing are part of most modern aircraft. The sum of the weight of all these subsystems could add up to about 3–8% of the aircraft MTOW. References [5, 6] introduce some techniques to estimate the weight of these subsystems. Table 10.13 presents the weight of some typical civil and military components and instruments. The equipment weight breakdown for several aircraft is shown in Table 10.14.

10.5 Chapter Examples

In this section, three solved examples are presented to demonstrate the application of the technique. Examples 10.1, 10.2, and 10.3 show the application of the components weight calculation technique with regard to a wing, a fuselage, and a vertical tail respectively

Table 10.13 Mass of some miscellaneous components [3]

No.	Component	Type, description, details	Mass (kg)
1	Seat	Flight deck – civil	24–28
2		Fighter pilot (ejection seat)	95–110
3		Passenger – economy	13–16
4		Passenger – tourist	20–28
5		Troop	4–6
6	Missile and bomb	ACM, AGM-129	1250
7		AGM-130	1323
8		HARM, AGM-88	254
9		Harpoon, AGM-84A	530
10		Hellfire, AGM-114A	46
11		Maverick, AGM-65A	210
12		Penguin 2, AGM-119B	385
13		Sea Eagle	600
14		Sidewinder, AIM-9J	87
15		Sparrow, AIM-7F	227
16		Stinger, FIM-92	16
17		TOW, BGM-71A/B	19
18		Standard, AGM-78	615
19		SLAM, AGM-84E	630
20	Stick, yoke, wheel	Side-stick	0.1–0.2
21		Stick	0.5–1
22		Yoke, wheel	1–2
23	Parachute	Civil	4–6
24		Military	8–20
25	Instruments	Compass, tachometer, altimeter, airspeed indicator, clock, rate of climb, bank angle indicator, accelerometer, GPS, etc.	0.3–0.7 each
26		Gyroscope (x, y, z)	0.5–2
27		Display	1–4
28	Lavatories	Short-range aircraft	$0.13N_{\text{seat}}^{1.3}$
29		Long-range aircraft	$0.5N_{\text{seat}}^{1.3}$
30		Business jet	$1.7N_{\text{seat}}^{1.3}$

Table 10.14 Equipment weight (in lb) breakdown for several aircraft

No.	Aircraft	MTOW	Equipment group						
			AUP	Instruments/ navigation	Avionic	Hydraulic/ pneumatic	Electric	Air conditioning/ anti-ice	Furnishings
1	Cessna 172	2 200	–	7	–	3	41	4	99
2	Cessna 210	2 900	–	16	–	4	60	12	116
3	Cessna 310	4 830	–	46	–	–	121	46	154
4	Beech 760	7 650	–	70	158	–	284	49	169
5	Cessna T-37	6 436	–	132	86	56	194	69	256
6	Lockheed C-130	151 522	466	665	2 432	671	2 300	2 126	4 765
7	Gulfstream I	33 600	355	97	99	235	966	755	415
8	Fokker F-27	39 000	–	81	386	242	835	1 225	2 291
9	MD DC-8	328 000	–	1 271	1 551	2 196	2 398	3 144	14 335
10	MD DC-9/10	91 500	818	719	914	714	1 663	1 476	7 408
11	Fokker F-28	65 000	346	302	869	364	1 023	1 074	4 030
12	Airbus A-300	302 000	983	377	1 726	3 701	4 923	3 642	13 161
13	Boeing 727-100	160 000	60	756	1 591	1 418	2 142	1 976	10 257
14	Boeing 737-200	100 400	836	625	956	837	1 066	1 416	6 643
15	Boeing 747-100	710 000	1 130	1 909	4 429	4 471	3 348	3 969	37 245

Example 10.1

The wing of a four-seat single-piston engine GA normal aircraft with a maximum take-off mass of 1400 kg has the following characteristics:

$$\text{AR} = 8,\ \lambda = 0.8,\ (t/C)_{\max} = 0.12,\ \Lambda_{0.25} = 15\,\text{deg},\ C_{L_{\max}} = 1.6$$

The wing is constructed with aerospace aluminum alloy, and two fuel tanks are located in the wing left and right sections. The engine is placed in the fuselage nose. The aircraft stall speed at sea level is 58 knot. Calculate the weight of the wing.

Solution:

The aircraft type is normal GA, so from Table 10.9 a maximum load factor of 3 is selected. Based on the engine and fuel tank locations, a K_ρ of 0.0016 is taken from Table 10.8. The density of aerospace aluminum alloy from Table 10.6 is 2711 kg/m^3. The air density at sea level is 1.225 kg/m^3.

$$n_{\text{ult}} = 1.5 \cdot n_{\max} = 1.5 \cdot 3 = 4.5 \tag{10.4}$$

$$S_{\text{W}} = \frac{2W_{\text{TO}}}{\rho V_s^2 C_{L_{\max}}} = \frac{2 \cdot 1400 \cdot 9.81}{1.225 \cdot (58 \cdot 0.544)^2 \cdot 1.5} = 16.785\,\text{m}^2 \tag{5.13}$$

$$\text{AR} = \frac{b^2}{S_{\text{W}}} \Rightarrow b = \sqrt{\text{AR} \cdot S_{\text{W}}} = \sqrt{8 \cdot 16.785} = 11.59\,\text{m} \tag{5.19}$$

$$S_{\text{W}} = b \cdot \text{MAC} \Rightarrow \text{MAC} = \frac{S_{\text{W}}}{b} = \frac{16.785}{11.59} = 1.448\,\text{m} \tag{5.18}$$

$$W_{\text{W}} = S_{\text{W}} \cdot \text{MAC} \cdot \left(\frac{t}{C}\right)_{\max} \cdot \rho_{\text{mat}} \cdot K_\rho \cdot \left(\frac{\text{AR} \cdot n_{\text{ult}}}{\cos\left(\Lambda_{0.25}\right)}\right)^{0.6} \cdot \lambda^{0.04} \cdot g \tag{10.3}$$

$$W_{\text{W}} = 16.785 \cdot 1.448 \cdot 0.12 \cdot 2711 \cdot 0.0016 \cdot \left(\frac{8 \cdot 4.5}{\cos(15)}\right)^{0.6}$$

$$\cdot (0.8)^{0.04} \cdot 9.81 = 1078.2\,\text{N} \tag{10.3}$$

The mass of each object is obtained just by dividing its weight by the gravitational constant:

$$m_{\text{W}} = \frac{W_{\text{W}}}{g} = \frac{1078.2}{9.81} = 109.9\,\text{kg}$$

Example 10.2

The fuselage of a 170-seat twin-jet engine transport aircraft with a maximum take-off mass of 63 000 kg has a length of 31 m and a maximum diameter of 3.7 m. The fuselage is constructed with aerospace aluminum alloy. Calculate the weight of the fuselage. Both jet engines are attached under the wing.

Solution:
In Table 10.11, the fuselage density factor is 0.0025–0.0032. Tentatively, a value of 0.0028 is selected. The aircraft type is transport, so from Table 10.9 a maximum load factor of 4 is selected. The density of aerospace aluminum alloy from Table 10.6 is 2711 kg/m^3.

$$n_{\text{ult}} = 1.5 \cdot n_{\max} = 1.5 \cdot 4 = 6 \quad (10.4)$$

Since the engine inlets are not on the fuselage, the parameter K_{inlet} is 1.

$$W_{\text{F}} = L_{\text{f}} \cdot D_{\text{f}_{\max}}^{2} \cdot \rho_{\text{mat}} \cdot K_{\rho_{\text{f}}} \cdot n_{\text{ult}}^{0.25} \cdot K_{\text{inlet}} \cdot \text{g} \quad (10.7)$$

$$W_{\text{F}} = 31 \cdot (3.7)^2 \cdot 2711 \cdot 0.0028 \cdot (6)^{0.25} \cdot 1 \cdot 9.81 = 49443.7\,\text{N}$$

Thus, the fuselage mass is:

$$m_{\text{F}} = \frac{W_{\text{F}}}{g} = \frac{49443.7}{9.81} = 5041.8\,\text{kg}$$

Example 10.3

The vertical tail of a remotely controlled model aircraft with a maximum take-off mass of 8 kg has the following characteristics:

$$S_{\text{VT}} = 0.4\ \text{m}^2,\ \text{AR}_{\text{VT}} = 1.5,\ \Lambda_{0.25\text{VT}} = 20\,\text{deg},\ \lambda_{\text{VT}} = 0.6,\ \bar{V}_{\text{V}} = 0.04,\ C_{\text{r}}/C_{\text{V}} = 0.2$$

The vertical tail has a thickness-to-chord ratio of 12% and its structure is constructed with balsa wood and a skin of Monokote. Determine the weight of the vertical tail. Tail has a conventional configuration.

Solution:
The densities of balsa wood and Monokote from Table 10.6 are 160 kg/m^3 and 900–1400 kg/m^3 respectively. It is assumed that the average density is 500 kg/m^3. From Table 10.10, a vertical tail density factor of 0.05 is obtained.

$$W_{\text{VT}} = S_{\text{VT}} \cdot \text{MAC}_{\text{VT}} \cdot \left(\frac{t}{C}\right)_{\max_{\text{VT}}} \cdot \rho_{\text{mat}} \cdot K_{\rho\text{VT}} \cdot \left(\frac{\text{AR}_{\text{VT}}}{\cos\left(\Lambda_{0.25_{\text{VT}}}\right)}\right)^{0.6}$$

$$\cdot \lambda_{\text{VT}}^{0.04} \cdot \bar{V}_{\text{V}}^{0.2} \left(\frac{C_{\text{r}}}{C_{\text{V}}}\right)^{0.4} \cdot \text{g} \quad (10.6)$$

$$\text{AR}_{\text{VT}} = \frac{b_{\text{VT}}^2}{S_{\text{VT}}} \Rightarrow b_{\text{VT}} = \sqrt{\text{AR}_{\text{VT}} \cdot S_{\text{VT}}} = \sqrt{1.5 \cdot 0.4} = 0.775\,\text{m} \quad (6.77)$$

$$S_{\text{VT}} = b_{\text{VT}} \cdot \text{MAC}_{\text{VT}} \Rightarrow \text{MAC}_{\text{VT}} = \frac{S_{\text{VT}}}{b_{\text{VT}}} = \frac{0.4}{0.775} = 0.516\,\text{m} \quad (6.80)$$

$$W_{VT} = 0.4 \cdot 0.516 \cdot 0.12 \cdot 500 \cdot 0.05 \cdot \left(\frac{1.5}{\cos(20)}\right)^{0.6}$$

$$\cdot (0.6)^{0.04} \cdot 0.04^{0.2} (0.2)^{0.4} \cdot 9.81 = 2.14\text{N}$$

The mass of the vertical tail is:

$$m_{VT} = \frac{2.14}{9.81} = 0.218\,\text{kg} = 218\,\text{g}$$

Problems

1. The wing of a four-seat single-piston engine GA normal aircraft with a maximum take-off mass of 2500 kg has the following characteristics:

$$AR = 10, \lambda = 0.7, (t/C)_{max} = 0.12, \Lambda_{0.25} = 10\,\text{deg}, C_{L_{max}} = 1.9$$

The wing is constructed with aerospace aluminum alloy, and two fuel tanks are located in the wing left and right sections. The engine is placed in the fuselage nose. The aircraft stall speed at sea level is 58 knot. Calculate the weight of the wing.

2. The wing of a transport aircraft with a maximum take-off mass of 30 000 kg has the following characteristics:

$$AR = 11,\ \lambda = 0.6,\ (t/C)_{max} = 0.15,\ \Lambda_{0.25} = 30\,\text{deg},\ C_{L_{max}} = 2.4$$

The wing is constructed with aerospace aluminum alloy, and two fuel tanks are located in the wing left and right sections. The engine is placed on the wing. The aircraft stall speed at sea level is 85 knot. Calculate the weight of the wing.

3. The wing of a supersonic fighter aircraft with a maximum take-off mass of 6000 kg has the following characteristics:

$$S = 17\ \text{m}^2,\ AR = 4,\ \lambda = 0.5,\ (t/C)_{max} = 0.07,\ \Lambda_{0.25} = 35\,\text{deg}$$

The wing is constructed with aerospace aluminum alloy, and few light stores are under wing. The engine is placed in the fuselage rear section. Calculate the weight of the wing.

4. The wing of a single-engine prop-driven RC model aircraft with a maximum take-off mass of 10 kg has the following characteristics:

$$S = 2\ \text{m}^2,\ AR = 6,\ \lambda = 0.5,\ (t/C)_{max} = 0.15,\ \Lambda_{0.25} = 0\,\text{deg}$$

The wing is constructed with graphite/epoxy, and the battery is located in the fuselage. The engine is placed in the fuselage nose section. Calculate the weight of the wing.

5. The fuselage of a 60-seat twin-jet engine transport aircraft with a maximum take-off mass of 35 000 kg has a length of 27 m and a maximum diameter of 3.1 m. The fuselage is constructed with aerospace aluminum alloy. Calculate the weight of the fuselage. Both jet engines are attached over the wing.

6. The fuselage of a 200-seat twin-turbofan engine transport aircraft with a maximum take-off mass of 80 000 kg has a length of 35 m and a maximum diameter of 3.8 m. The fuselage is constructed with aerospace aluminum alloy. Calculate the weight of the fuselage. Both jet engines are attached under the wing.

7. The fuselage of a twin-engine supersonic fighter aircraft with a maximum take-off mass of 25 000 kg has a length of 16 m and an average maximum diameter of 1.4 m. The fuselage is constructed with aerospace aluminum alloy. Calculate the weight of the fuselage. The engine is located in the rear fuselage with the inlet under the wing.

8. The vertical tail of a GA aircraft with a maximum take-off mass of 1200 kg has the following characteristics:

$$S_{VT} = 3.7\ \text{m}^2,\ \text{AR}_{VT} = 1.6,\ \Lambda_{0.25VT} = 30\,\text{deg},\ \lambda_{VT} = 0.5,$$

$$\bar{V}_V = 0.05,\ C_r/C_V = 0.25$$

The vertical tail has a thickness-to-chord ratio of 9% and its structure is constructed with aerospace aluminum alloy. Determine the weight of the vertical tail. Tail has a conventional configuration.

9. The vertical tail of a large transport aircraft with a maximum take-off mass of 70 000 kg has the following characteristics:

$$S_{VT} = 35\ \text{m}^2, \text{AR}_{VT} = 1.4,\ \Lambda_{0.25VT} = 25\,\text{deg},\ \lambda_{VT} = 0.3,\ \bar{V}_V = 0.08,\ C_r/C_V = 0.22$$

The vertical tail has a thickness-to-chord ratio of 10% and its structure is constructed with graphite/epoxy. Determine the weight of the vertical tail. Tail has a T-tail configuration.

10. Consider a twin-turboprop aircraft with a maximum take-off mass of 6000 kg and a horizontal tail with the following characteristics:

$$S_{HT} = 7.3\ \text{m}^2,\ \text{AR}_{HT} = 4.6,\ \Lambda_{0.25HT} = 24\,\text{deg},\ \lambda_{HT} = 0.4,\ \bar{V}_H = 0.9,$$

$$C_e/C_t = 0.18,\ (t/C)_{max} = 0.09$$

The horizontal tail is constructed with aerospace aluminum. Determine the weight of the horizontal tail.

11. Consider a sailplane with a maximum take-off mass of 640 kg and a horizontal tail with the following characteristics:

$$S_{HT} = 3.2\ \text{m}^2,\ \text{AR}_{HT} = 36,\ \Lambda_{0.25HT} = 10\,\text{deg},\ \lambda_{HT} = 0.4,\ \bar{V}_H = 0.7,$$

$$C_e/C_t = 0.24,\ (t/C)_{max} = 0.09$$

The horizontal tail is constructed with fiberglass/epoxy. Determine the weight of the horizontal tail. Tail has an H-tail configuration.

12. Consider a Navy fighter with a maximum take-off mass of 17 000 kg and a wing span of 7.6 m. The landing gear is retractable and its height is 0.92 m. Determine the weight of the landing gear. Assume that the landing weight is 80% of the maximum take-off weight.

13. A twin-turbofan regional airliner aircraft with a maximum take-off weight of 42 000 lb and a wing span of 100 ft has a retractable landing gear with a height of 4.2 ft. Determine the weight of the landing gear. Assume that the landing weight is 90% of the maximum take-off weight.

14. A two-seat home-built aircraft with a maximum take-off weight of 1800 lb and a wing span of 50 ft has a fixed landing gear with a height of 1.7 ft. Determine the weight of the landing gear. Assume that the landing weight is 70% of the maximum take-off weight.

15. Consider a four-turboprop cargo aircraft with a maximum take-off mass of 36 000 kg. Each engine has a dry mass of 530 kg. Calculate the total installed engine weight.

16. Consider a twin-turbofan airliner with a maximum take-off weight of 250 000 lb. Each engine has a dry weight of 4000 lb. Calculate the total installed engine weight.

17. Consider an eight-seat twin-turbofan business/military trainer with a maximum take-off weight of 16 000 lb. The aircraft carries 502 US gal of JP-4. Calculate the fuel system weight. The aircraft is equipped with non-self-sealing bladder tanks.

18. Consider a two-seat twin-engine supersonic fighter with a maximum take-off weight of 60 000 lb. The aircraft is equipped with three integral fuel tanks with a total fuel capacity of 2000 US gal of JP-4. Calculate the fuel system weight.

19. Consider a four-turbofan airliner with a maximum take-off weight of 800 000 lb. The aircraft is equipped with six non-self-sealing bladder tanks with a total fuel capacity of 50 000 US gal of JP-4. Calculate the fuel system weight.

20. A five-seat single-piston engine light GA aircraft with a maximum take-off weight of 3000 lb has a fuel capacity of 75 US gal of aviation gasoline. Determine the fuel system weight.

21. Consider a ten-seat single-turboprop light utility GA aircraft with maximum take-off mass 3600 kg and wing area 27 m^2 having the following characteristics:

$$\text{Wing: } AR = 9,\ \lambda = 0.8,\ (t/C)_{max} = 0.15,\ \Lambda_{0.25} = 10\,\text{deg}$$

$$\text{Horizontal tail: } S_{HT} = 5.8\ \text{m}^2,\ AR_{HT} = 4.2,\ \Lambda_{0.25HT} = 0\,\text{deg},\ \lambda_{HT} = 0.7,$$

$$\bar{V}_H = 0.85,\ C_e/C_t = 0.23$$

$$\text{Vertical tail: } S_{VT} = 4.1\ \text{m}^2,\ AR_{VT} = 1.3,\ \Lambda_{0.25VT} = 25\,\text{deg},\ \lambda_{VT} = 0.6,$$

$$\bar{V}_V = 0.035,\ C_r/C_V = 0.3$$

The engine is in the fuselage nose and has a mass 170 kg and the fuel capacity is 350 US gal of aviation gasoline. Aircraft has a conventional tail configuration and two fuel tanks are located in the wing left and right sections. Each occupant may carry up to 20 kg of luggage. The landing gear is fixed with a height of 0.632 m. The landing weight is 80% of the maximum take-off weight. Both horizontal tail and vertical tail airfoils have a thickness-to-chord ratio of 9%. The fuselage has a length of 10.6 m and a maximum diameter of 1.6 m. The entire aircraft structure is constructed with aerospace aluminum. Calculate the weight of the following components: wing, horizontal tail, vertical tail, fuselage, installed engine, fuel system, and landing gear. What is the weight of other equipment and instruments? Assume the weight of each occupant is 190 lb.

22. Consider a 50-passenger twin-turbofan transport aircraft with maximum take-off mass 20 000 kg and wing area 52 m^2 having the following characteristics:

$$\text{Wing: AR} = 11,\ \lambda = 0.4,\ (t/C)_{\max} = 0.12,\ \Lambda_{0.25} = 26\,\text{deg}$$

$$\text{Horizontal tail: } S_{\text{HT}} = 13.6\ \text{m}^2,\ \text{AR}_{\text{HT}} = 5.1,\ \Lambda_{0.25\text{HT}} = 18\,\text{deg},$$

$$\lambda_{\text{HT}} = 0.8,\ \bar{V}_{\text{H}} = 0.93,\ C_{\text{e}}/C_{\text{t}} = 0.21$$

$$\text{Vertical tail: } S_{\text{VT}} = 10.2\ \text{m}^2,\ \text{AR}_{\text{VT}} = 1.4,\ \Lambda_{0.25\text{VT}} = 50\,\text{deg},$$

$$\lambda_{\text{VT}} = 0.5,\ \bar{V}_{\text{V}} = 0.05,\ C_{\text{r}}/C_{\text{V}} = 0.24$$

The engines are installed beside the rear fuselage and each engine has a mass of 730 kg and a fuel capacity of 4200 kg of jet fuel. The landing gear is retractable with a height of 1.8 m. The landing weight is 85% of the maximum take-off weight. Both horizontal tail and vertical tail airfoils have a thickness-to-chord ratio of 9%. The fuselage has a length of 28.2 m and a maximum diameter of 2.28 m. The entire aircraft structure is constructed with aerospace aluminum. Calculate the weight of the following components: wing, horizontal tail, vertical tail, fuselage, installed engine, fuel system, and landing gear. What is the weight of other equipment and instruments? Assume the weight of each occupant is 190 lb and each one is allowed to carry up to 50 lb of luggage. Aircraft has a T-tail configuration and is equipped with non-self-sealing bladder tanks.

References

[1] Blanchard, B.S. and Fabrycky, W.J. (2006) *Systems Engineering and Analysis*, 4th edn, Prentice Hall.

[2] Rich, B.R. and Janos, L. (1994) *Skunk Works*, Little-Brown Company.

[3] Jackson, P., Munson, K., Peacock, L. *Jane's All the World's Aircraft, Jane's Information Group*, various years 1996 to 2011.

[4] Airworthiness Standards for GA and Transport Aircraft, FAR 23 and 25 (2011) US Department of Transportation, Federal Aviation Administration.

[5] Roskam, J. (2003) *Airplane Design, Part V: Component Weight Estimation*, DAR Corporation.

[6] Torenbeek, E. (1996) *Synthesis of Subsonic Airplane Design*, Delft University Press.

[7] Schmitt, R.L., Foreman, K.C., Gertsen, W.M., and Johnson, P.H. (1959) *Weight Estimation Handbook for Light Aircraft*, Cessna Aircraft Company.

[8] Staton, R. (1968) Statistical Weight Estimation Methods for Fighter/Attack Aircraft. Vought aircraft, Report 2-59320/8R-50475, Vought Corporation.

[9] Staton, R. (1969) Cargo/Transport Weight Estimation Methods. Vought aircraft, Report 2-59320/9R-50549, Vought Corporation.

11

Aircraft Weight Distribution

11.1 Introduction

One of the primary concerns during the aircraft design process, even during the conceptual design phase, is the aircraft weight distribution. The distribution of aircraft weight (sometimes referred to as weight and balance) will greatly influence airworthiness as well as aircraft performance. Hence, the aircraft designer must always take into account the effect of each design decision/selection on the aircraft weight distribution. The distribution of aircraft weight will influence the airworthiness and performance via two aircraft parameters: (i) aircraft center of gravity (cg) and (ii) aircraft mass moment of inertia. The technique to distribute aircraft weight in order to achieve an ideal cg location and ideal mass moment of inertia is the objective of this chapter.

The aircraft center of gravity is the cornerstone for aircraft stability, controllability, and trim analysis, as well as handling qualities evaluation. All the analyses and evaluations are aimed at determining airworthiness aspects of the aircraft. In addition, the aircraft cg is the center of the coordinate axis system that all calculations are based on. All non-aerodynamic moments are measured with respect to the aircraft cg. Therefore, aircraft cg determination is a vital task in the aircraft design process. The main objective of aircraft weight distribution is to achieve an ideal cg location and ideal cg range. By definition, the center of gravity is the point at which an aircraft would balance when suspended. Its distance from the reference datum is determined by dividing the total moment by the total weight of the aircraft.

The center of mass or center of gravity of a complex system is the mean location of all the mass in the system. The term *center of mass* is often used interchangeably with center of gravity, but they are physically different concepts. They happen to coincide in a uniform gravitational field, but where gravity is not uniform the center of gravity refers to the mean location of the gravitational force acting on an object. For a rigid body, the position of the center of mass is fixed in relation to the body. The center of mass of a body does not often coincide with its geometric center. In the case of a movable distribution of masses in a compound, such as the passengers from a transport aircraft, the position of the center of mass is a point in space among them that may

Aircraft Design: A Systems Engineering Approach, First Edition. Mohammad H. Sadraey.

not correspond to the position of any individual mass. The application of the center of gravity often allows the use of simplified (e.g., linear) governing equations of motion to analyze the movement of a dynamic system. The center of gravity is also a convenient reference point for many other calculations in dynamics, such as the mass moment of inertia. In many applications, such as aircraft design, components can be replaced by point mass located at their centers of gravity for the purposes of analysis.

The distance between the forward and aft center of gravity (or center of mass) limits is called the center of gravity range or limit along the x-axis. One of the crucial tasks in aircraft design is to balance the components and loads such that the center of gravity lies in an acceptable region. The cg must remain within the specified limits as fuel is burned, and as the stores are expended. For the case of fuel, an automated fuel-management system may be used, but it will impose additional cost and complexity. The allowable limits on the cg vary with Mach number, since at supersonic speeds the wing aerodynamic center moves considerably rearward. So at supersonic speeds, the forward-cg limit has to move forward to allow for longitudinal trim. The aft cg influences primarily the aircraft longitudinal and directional stability, while the forward cg will influence mainly the aircraft longitudinal and directional controllability. The aft and forward cg will determine several parameters, including the size of the horizontal and vertical tails as well as elevator and rudder design.

The main goals in aircraft design are frequently to obtain adequate stability in all phases of flight, high performance, low dispersion, and large payload mass and capacity. The aircraft weight distribution has primarily two aspects; (i) internal, such as seating arrangement; (ii) external, such as wing or engine locations. In practice, it is difficult to achieve all of these objectives due to the complicated nature of various requirements. These conflicting design requirements generate a complex situation where an optimal solution is the only resort.

The distribution of weight in flight operation is also of vital importance, since the position of the center of gravity affects the performance, stability, and controllability of the aircraft. In loading an aircraft, the cg must be within the permissible range and remain so during the flight to ensure the stability and maneuverability of the aircraft during flight. Aircraft manufacturers publish weight and balance limits for their aircraft. This information can be found in two sources: (i) The Aircraft Weight and Balance Report and (ii) The Airplane Flight Manual (e.g., Refs [1, 2]). The aircraft with all equipment installed is weighed and the cg limits calculated, and this information is tabulated on the report that accompanies the aircraft logbook.

Aircraft must be stable, controllable, and safe for all allowable aircraft cg locations during the flight envelope. The safety of an aircraft is influenced by several factors, including the cg location, and overloading it will cause serious problems. If the cg is too forward, the take-off run necessary to become airborne will be longer. In some cases, the required take-off run may be greater than the available runway. The angle of climb and the rate of climb will be reduced. The maximum ceiling will be lowered and the range shortened. The landing speed will be higher and the landing roll will be longer. In addition, the extra weight may cause bending moment and structural stresses during maneuvers, and turbulence that could lead to damage. In an aircraft on the ground with a tricycle landing gear configuration, if the loads are placed such that the aircraft cg moves aft of the main gear, the aircraft will pitch up and the tail will hit the ground. Two

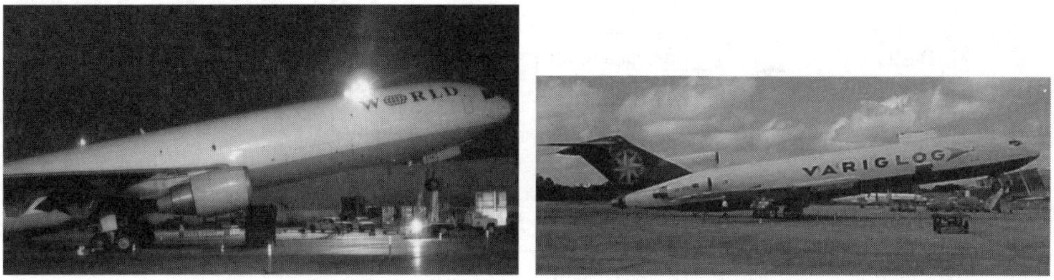

(a) (b)

Figure 11.1 Too much load beyond the allowed aft cg creates an accident. (a) A McDonnell Douglas MD-11 nose up due to the cg aft of main gear; (b) A Boeing 727 nose up due to loading error in Brazil, June 9, 2001. Reproduced from permission of (a) Randy Crew; (b) Orlando J. Junior

examples of such incidents for a McDonnell Douglas MD-11 and for a Boeing 727 are shown in Figure 11.1.

In the 1990s, one of the famous long-haul transport aircraft at Paris Charles de Gaulle Airport had a nose-up during boarding passengers, because all the passengers went to the back of the cabin. Thus, the passengers were asked to move forward to bring the nose down. Two bomber B-52Gs were lost during take-off acceleration, because fuel moved aft of their cells (in fact, aft of the allowable cg), so they crashed. Such accidents indicate the importance of weight distribution during the design process.

As a rule of thumb, the best aircraft cg location is around the wing/fuselage aerodynamic center (ac_{wf}). The reason is that the aerodynamic forces (lift and drag) are produced at the ac_{wf}. As the distance between the aircraft cg and the wing/fuselage aerodynamic center increases, the need for a balancing moment to trim the aircraft is increased. The balancing moment (either longitudinal or directional) has cost and controllability consequences. A careful design will make sure that the cost is low and controllability is at an acceptable level.

Modern transport aircraft are equipped with an onboard weight and balance system [3]. In this system, dynamic aircraft/vehicle-specific load planning and weight balancing systems and methods are used to automate the process of weighing passengers and their baggage, including carry-on baggage, to accurately and quickly determine the passenger and baggage location on a specific aircraft to generate an efficient and precise aircraft load plan and provide data to determine the loaded aircraft weight and balance. Digital scales and passenger boarding pass scanners are used to acquire the weight and location data, which can be fed to a processor to process the data to determine the appropriate weight and balance for each flight and/or transmit data to other systems to be included in their determination of vehicle weight and weight distribution. These types of systems are helpful only during flight operations; during the aircraft design process and configuration selection, a systems engineering approach must be followed to locate components optimally.

In some rare design cases, the designer has to change the aircraft configuration in order to achieve the desired cg locations. For instance, the location of the engines may be moved from a tractor configuration to a pusher configuration to move the cg rearward. Or the

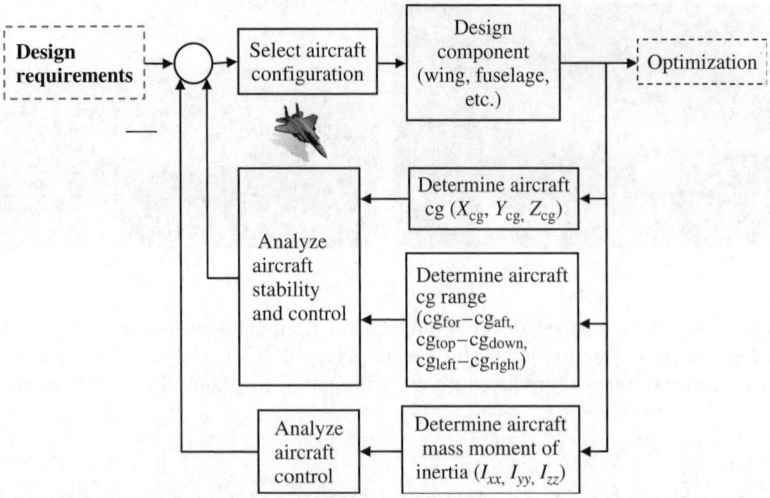

Figure 11.2 Feedback loops provided by cg and moment of inertia calculations

aft tail may be replaced with a canard to move the cg forward. The aircraft cg and mass moment of inertia calculations provide two feedback loops in the overall design process, as depicted in Figure 11.2. The aircraft cg, aircraft cg range, and aircraft mass moment of inertia about the three axes of x, y, and z are required groups of data to analyze aircraft stability and control.

This chapter is organized as follows. Section 11.1 presents the fundamental techniques to determine the aircraft center of gravity along three axes. The aircraft cg range and the reasons and incidents that cause the cg to move will be introduced in Section 11.3. In this section, the features of the ideal cg location and ideal cg range will be examined, and some general remarks on how to define the ideal cg location will be offered. The technique to distribute aircraft components and to determine the aircraft cg range will be presented in Section 11.4. Section 11.5 is devoted to a build-up technique to obtain aircraft mass moment of inertia about three aircraft axes. In the last section, Section 11.6, a design example for aircraft weight distribution will be fully solved and a step-by-step solution outlined.

11.2 Aircraft Center of Gravity Calculation

An aircraft is composed of several components, such as wing, fuselage, tail, engine, and landing gear, plus payload, fuel, and crew. Each component has a unique mass (weight), and a unique center of gravity, and thus contributes to the aircraft overall center of gravity. In order to determine the aircraft cg, a coordinate axis system must be defined.

There are mainly four coordinate axis systems; namely (i) earth fixed axis system, (ii) body axis system, (iii) wind axis system, and (iv) stability axis system. The body, wind, and stability axis systems are moving and rotating with the aircraft motion and centered at the aircraft cg. Each of these four axis systems has applications and advantages; here, a body axis system as depicted in Figure 11.3 is selected.

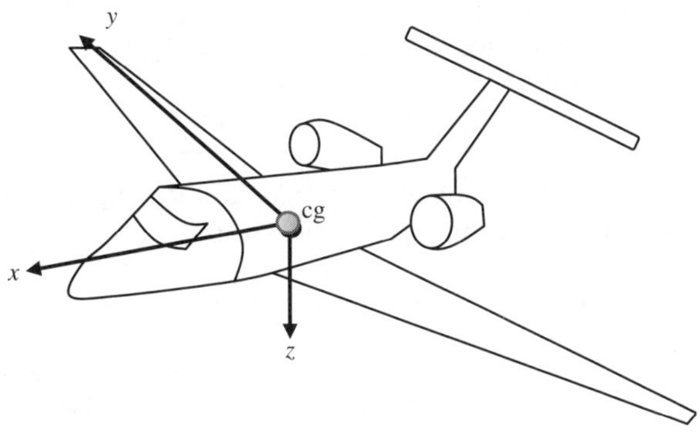

Figure 11.3 Definition of body-axis coordinate system

In a body axis system, an orthogonal axis is defined where the x-axis is along the fuselage center line and the y-axis is defined using the right-hand rule. Thus, the y-axis is to the right, and the z-axis is downward. The coordinate of the aircraft cg along the x-axis is represented by X_{cg}, along the y-axis by Y_{cg}, and along the z-axis by Z_{cg}. The aircraft cg coordinates (X_{cg}, Y_{cg}, and Z_{cg}) with n components are determined using the following formulas:

$$X_{\text{cg}} = \frac{\sum_{i=1}^{n} W_i x_{\text{cg}_i}}{\sum_{i=1}^{n} W_i} = \frac{\sum_{i=1}^{n} m_i x_{\text{cg}_i}}{\sum_{i=1}^{n} m_i} \tag{11.1}$$

$$Y_{\text{cg}} = \frac{\sum_{i=1}^{n} W_i y_{\text{cg}_i}}{\sum_{i=1}^{n} W_i} = \frac{\sum_{i=1}^{n} m_i y_{\text{cg}_i}}{\sum_{i=1}^{n} m_i} \tag{11.2}$$

$$Z_{\text{cg}} = \frac{\sum_{i=1}^{n} W_i z_{\text{cg}_i}}{\sum_{i=1}^{n} W_i} = \frac{\sum_{i=1}^{n} m_i z_{\text{cg}_i}}{\sum_{i=1}^{n} m_i} \tag{11.3}$$

where W_i denotes the weight of each aircraft component, m_i denotes the mass of each aircraft component, and x_{cg_i}, y_{cg_i}, and z_{cg_i} are the coordinates of each individual component. The coordinates are measured with respect to a particular reference line. The selection of reference lines is arbitrary, and does not affect the final result. However, it is recommended to select a vertical line passing through the foremost point of the aircraft (e.g., the fuselage nose) as the reference line for the x coordinate. The reference line for the y coordinate is recommended to be the fuselage center line; that is, the xz plane. The reference line for the z coordinate is recommended to be the ground level (i.e., the contact surface between the landing wheels and the ground).

When an aircraft carries the maximum allowable payload, maximum fuel, and maximum crew members, the summation of the weights of all components is equal to the maximum take-off weight:

$$\sum W_i = W_{\mathrm{TO}} \tag{11.4}$$

Otherwise, the summation is equal to the weight of the components which are present in that particular configuration and condition:

$$\sum W_i = W_{\mathrm{w}} + W_{\mathrm{F}} + W_{\mathrm{HT}} + W_{\mathrm{VT}} + W_{\mathrm{E}} + W_{\mathrm{LG}} + W_{\mathrm{PL}} + W_{\mathrm{fuel}} + W_{\mathrm{C}} + \ldots \tag{11.5}$$

where W_{w}, W_{F}, W_{HT}, W_{VT}, W_{E}, W_{LG}, W_{PL}, W_{fuel}, W_{C} represent the wing, fuselage, horizontal tail, vertical tail, engine, landing gear, payload, fuel, and crew members respectively. Thus, Equations (11.1)–(11.3) are revised into the following forms:

$$X_{\mathrm{cg}} = \frac{\left[\begin{array}{c}\sum W_{\mathrm{w}}x_{\mathrm{w}} + W_{\mathrm{F}}x_{\mathrm{F}} + W_{\mathrm{HT}}x_{\mathrm{HT}} + W_{\mathrm{VT}}x_{\mathrm{VT}} + W_{\mathrm{E}}x_{\mathrm{E}} \\ +W_{\mathrm{LG}}x_{\mathrm{LG}} + W_{\mathrm{PL}}x_{\mathrm{PL}} + W_{\mathrm{fuel}}x_{\mathrm{fuel}} + W_{\mathrm{C}}x_{\mathrm{C}} + \ldots\end{array}\right]}{\sum W_i} \tag{11.6}$$

$$Y_{\mathrm{cg}} = \frac{\left[\begin{array}{c}\sum W_{\mathrm{w}}y_{\mathrm{w}} + W_{\mathrm{F}}y_{\mathrm{F}} + W_{\mathrm{HT}}y_{\mathrm{HT}} + W_{\mathrm{VT}}y_{\mathrm{VT}} + W_{\mathrm{E}}y_{\mathrm{E}} \\ +W_{\mathrm{LG}}y_{\mathrm{LG}} + W_{\mathrm{PL}}y_{\mathrm{PL}} + W_{\mathrm{fuel}}y_{\mathrm{fuel}} + W_{\mathrm{C}}y_{\mathrm{C}} + \ldots\end{array}\right]}{\sum W_i} \tag{11.7}$$

$$Z_{\mathrm{cg}} = \frac{\left[\begin{array}{c}\sum W_{\mathrm{w}}z_{\mathrm{w}} + W_{\mathrm{F}}z_{\mathrm{F}} + W_{\mathrm{HT}}z_{\mathrm{HT}} + W_{\mathrm{VT}}z_{\mathrm{VT}} + W_{\mathrm{E}}z_{\mathrm{E}} \\ +W_{\mathrm{LG}}z_{\mathrm{LG}} + W_{\mathrm{PL}}z_{\mathrm{PL}} + W_{\mathrm{fuel}}z_{\mathrm{fuel}} + W_{\mathrm{C}}z_{\mathrm{C}} + \ldots\end{array}\right]}{\sum W_i} \tag{11.8}$$

These equations must be modified based on the components of an aircraft. For instance, if an aircraft has a special system (such as automatic flight control system or autopilot in an unmanned aerial vehicle), this system must be included in these equations. Table 11.1 demonstrates a tabulated technique to obtain the aircraft cg. The table illustrates elements, subsystems, components, and various loads which influence the aircraft center of gravity location. Each component group is broken down into elements, subsystems, and items which make the calculation straightforward.

The center of gravity of homogeneous objects such as a cylinder, a rod, a sphere, a plate, a disk, a solid cube, and a rectangular prism are simple to determine. However, it is very hard to accurately obtain the center of gravity of non-homogenous components such as the wing, tail, and fuselage. A rough cg estimate based on statistical data is presented in Table 11.2. The range of values in the table is due to the fact that the structural design (e.g., rib, spar, and frame) may vary from aircraft to aircraft.

Among the three centers of gravity, the cg along the x-axis is the most significant one, since it varies considerably. The aircraft center of gravity concerns designers, flight crews, and load masters due to the fact that it significantly influences aircraft longitudinal stability and control. Hence, it tends to have more applications and calculations. In order to simplify the calculation and bookkeeping, the cg along the x-axis is frequently expressed in terms of wing mean aerodynamic chord (MAC). Another reason why this technique is selected is that the aircraft cg along the x-axis is often close to the wing aerodynamic center, which is at quarter chord of MAC.

Table 11.1 Weight and center of gravity statement

No.	Component group	Elements	Weight	X_{cg}	Y_{cg}	Z_{cg}
1	Wing	1.1. Wing main structure				
		1.2. Ailerons				
		1.3. Flaps				
		1.4. Aileron controls				
		1.5. Flap controls				
		1.6. Spoilers				
		1.7. Fairing				
		1.8. Strut (if any)				
		1.9. Miscellaneous				
		1.10. Wing overall				
2	Fuselage	2.1. Fuselage main structure				
		2.2. Seats				
		2.3. Furnishing				
		2.4. Doors				
		2.5. Windows				
		2.6. Fillets				
		2.7. Toilets				
		2.8. Galleys				
		2.9. Stowage				
		2.10. Miscellaneous				
		2.11. **Fuselage overall**				
3	Empennage	3.1. Horizontal tail				
		3.2. Vertical tail				
		3.3. Elevator				
		3.4. Rudder				
		3.5. Elevator tab				
		3.6. Rudder tab				
		3.7. Elevator control				
		3.8. Rudder control				
		3.9. Mass balance (if any)				
		3.10. **Empennage overall**				
4	Powerplant	4.1. Engine				
		4.2. Nacelle				
		4.3. Pylon				

(*continued overleaf*)

Table 11.1 (*continued*)

No.	Component group	Elements	Weight	X_{cg}	Y_{cg}	Z_{cg}
		4.4. Propeller				
		4.5. Gear box				
		4.6. Inlet				
		4.7. Exhaust				
		4.8. Fuel tanks				
		4.9. Fuel system				
		4.10. Oil				
		4.11. Oil system				
		4.12. Miscellaneous				
		4.13. **Powerplant overall**				
5	Landing gear	5.1. Tires				
		5.2. Wheels				
		5.3. Struts				
		5.4. Brake system				
		5.5. Shock absorbers				
		5.6. Retraction system				
		5.7. Fairing (if any)				
		5.8. Wheel control				
		5.9. Landing gear door				
		5.10. **Landing gear overall**				
6	Equipment and instruments	6.1. Lights				
		6.2. Batteries				
		6.3. Auxiliary power unit				
		6.4. Transmitter				
		6.5. Antenna				
		6.6. Paint				
		6.7. Stick/wheel/yoke				
		6.8. Cockpit instruments panel				
		6.9. First aid				
		6.10. Wiring/cable				
		6.11. Air conditioning system				
		6.12. Pedals				

Table 11.1 (*continued*)

No.	Component group	Elements	Weight	X_{cg}	Y_{cg}	Z_{cg}
		6.13. Electric wiring				
		6.14. Avionic system				
		6.15. Weather radar				
		6.16. Store				
		6.17. Pressurization system				
		6.18. Radome				
		6.19. Emergency escape hatch				
		6.20. Hydraulic system				
		6.21. Fire extinguisher				
		6.22. De-ice system				
		6.23. Autopilot				
		6.24. Lightning protection				
		6.25. Communication system				
		6.26. INS/IRS/GPS/radio[a]				
		6.27. Miscellaneous				
		6.28. **Equipment overall**				
7	Payload, unpaid load	7.1. Flight crew members				
		7.2. Flight attendants				
		7.3. Passengers				
		7.4. Technical crew				
		7.5. Luggage				
		7.6. Carry-on baggage				
		7.7. Cargo				
		7.8. Food, refreshment				
		7.9. Water				
		7.10. Fuel				
		7.11. **Load overall**				

[a] INS: Inertial navigation system; IRS: Inertial reference system; GPS: Global positioning system.

Employing this approach yields a non-dimensional value for x_{cg} with the symbol h. For instance, when the aircraft cg is located at 20% of MAC or simply C, the value of x_{cg} is written as $h = 0.2$. Therefore, the reference line for the aircraft cg along the x-axis is the wing leading edge at MAC. To extend this approach, the symbol h_o is also utilized for the location of the wing aerodynamic center along MAC. Figure 11.4 illustrates the relationship between wing aerodynamic center and aircraft center of gravity.

Table 11.2 Center of gravity locations for various non-homogenous components

No.	Component	Center of gravity location		
		X_{cg}	Y_{cg}	Z_{cg}
1	Wing	35–42% MAC	Along FCL[a]	5–10% thickness above mid-plane
2	Horizontal tail	30–40% MAC_{HT}	Along FCL	Mid-thickness
3	Vertical tail	30–40% MAC_{VT}	Along FCL	30–40% vertical tail span
4	Fuselage	40–48% length	Along FCL[b]	1–5% diameter above FCL
5	Landing gear	15–35% wheel base (from main gear)	Along FCL	30–40% of the gear height from ground
6	Turbine engine	30–45% engine length from inlet	Along shaft	Along the shaft center line
7	Piston engine	Treat it as a rectangular prism		
8	Human while seating	See Chapter 7	Along mid-plane	See Chapter 7

[a] Fuselage center line.
[b] In the case where the seating arrangement is not symmetric, the cg is shifted toward the side with more seats.

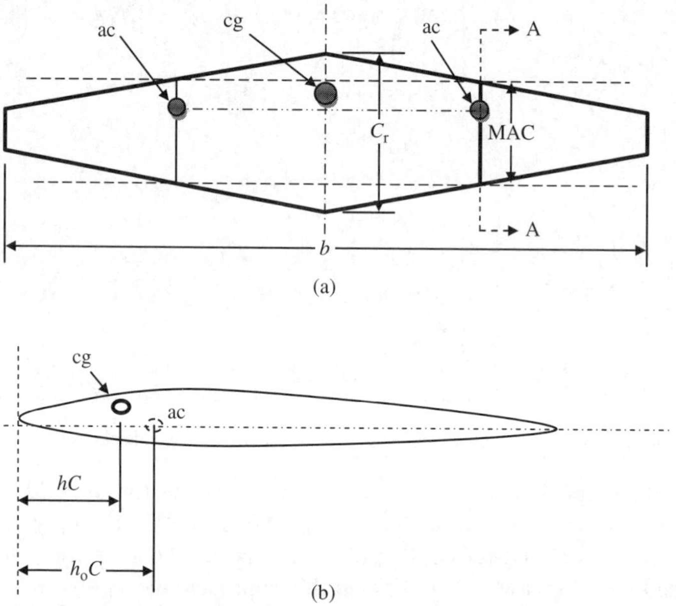

Figure 11.4 Wing mean aerodynamic chord, wing aerodynamic center, and aircraft center of gravity. (a) Wing top view; (b) Wing side-view at Section A-A

11.3 Center of Gravity Range

11.3.1 Fixed or Variable Center of Gravity

One of the pieces of information an aircraft designer should have for aircraft weight distribution is whether the cg is fixed or variable. The weight distribution for an aircraft with a fixed cg is much easier than for a variable cg aircraft. For an aircraft with a fixed cg there is no cg range (i.e., the cg range is zero), while for an aircraft with a variable cg the cg range must be determined. In an aircraft where the aircraft weight may vary from time to time due to payload variations, or burning fuel, the center of gravity will not be fixed. As the flight of the aircraft progresses and fuel is consumed, the weight of the airplane decreases. As a consequence, the aircraft weight distribution changes and hence the cg moves to new positions. Thus, in most extreme cases, there are such cg locations as most forward, most aft, most top, most down, most left, and most right.

In some aircraft such as remote-controlled model aircraft with an electric engine, solar-powered, and human-powered, the cg may remain at a relatively fixed position during flight operations, since no fuel is burned. However, if the payload varies from mission to mission, the cg location will change too. There are a variety of reasons and causes for the aircraft cg movement during flight operations. The following are some examples:

1. Fuel is burned and consumed by an air-breathing engine during flight.
2. The payload weight may vary from one flight to another in a transport/cargo aircraft.
3. Human pilots have different sizes in nature, and for different flight operations a lighter or heavier pilot may be onboard.
4. Human passengers have different sizes in nature, and for different flight operations a combination of different passengers may be onboard. On one flight there may be lots of obese passengers, while on another flight there may be lots of petite passengers.
5. All passengers may be adult during one flight operation, while on another flight several youths and a few babies may be flying with the aircraft.
6. An airliner may not fly with full capacity on all flights, so the number of passengers on a passenger aircraft may change from flight to flight.
7. On an airliner during cruising, the passengers may walk and change their seats.
8. On an airliner during cruising, flight attendants need to walk and serve the passengers by distributing foods and refreshments.
9. On a fighter aircraft, stores (e.g., missile, rocket, and bomb) may be fired or dropped.
10. On a military aircraft, troops may jump from the aircraft for military exercises and operations using a parachute.
11. On a firefighter aircraft, a huge amount of water or dirt may be released to contain a fire.
12. In a humanitarian flight operation, an aircraft may drop boxes of food or containers of water to affected areas.
13. On a sport aircraft, students may learn/practice jumping using a parachute.
14. Some military aircraft are equipped with a refueling system. Thus, a refueling operation will increase the weight of the aircraft while it will decrease the weight of the tanker. In both cases, the aircraft cg will move.

If the aircraft/payload/mission situations such as those mentioned above are not predicted for an aircraft, the designer could proceed with the weight distribution process on the assumption that the aircraft cg remains at a rather constant location. Any combination of passenger weight, fuel weight, passenger location, state of store, number of passengers, and payload feature results in a unique cg location. Hence, in practice, there are an infinite number of cg positions that the aircraft may have. Therefore, in the aircraft weight distribution, all the probable cg locations must be determined and the airworthiness analyzed.

11.3.2 Center of Gravity Range Definition

The distance between the most forward and most aft center of gravity limits is called the center of gravity range or limit along the x-axis. The distance between the most left and most right center of gravity limits is called the center of gravity range or limit along the y-axis. The distance between the most top and most bottom center of gravity limits is called the center of gravity range or limit along the z-axis (see Figure 11.5). The designer must take into account these situations and calculate the weight and balance not only for the beginning of the flight, and at the end of it, but also for every possible weight scenario.

The position of the center of gravity along the x-axis greatly affects the longitudinal stability and longitudinal controllability of the aircraft. There are forward and aft limits which must be established by the aircraft designer beyond which the cg should not be located during flight operations. These limits are set to assure that sufficient elevator deflection is available in all phases of flight for longitudinal trim and control. If the cg is too far forward, the aircraft is referred to by the pilot community as nose heavy; if too far aft, tail heavy. An aircraft whose center of gravity is too far aft may be dangerously unstable and may enter abnormal stall and possess spin characteristics. Recovery may be difficult if not impossible because the pilot will run out of elevator/rudder deflection. It is, therefore, the pilot's (or technical crew's) responsibility when loading an aircraft to see that the cg lies within the recommended limits.

The position of the center of gravity along the y-axis largely affects the lateral stability and lateral controllability of the aircraft. There are left and right limits which must be established by the aircraft designer beyond which the cg should not be located during

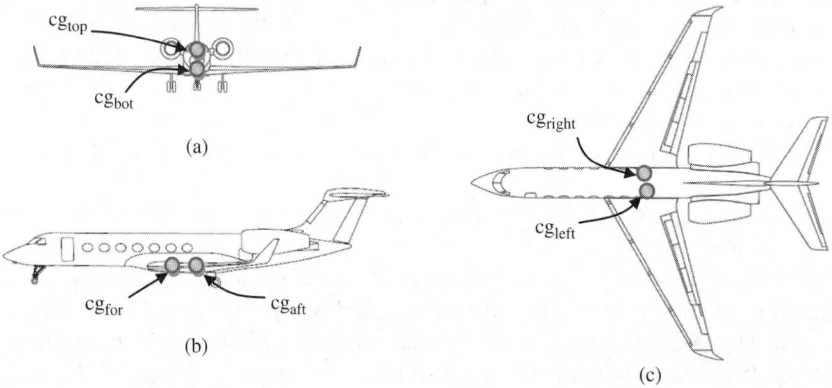

Figure 11.5 Most extreme aircraft cg locations. (a) Most top and bottom cg; (b) Most aft and forward cg; (c) Most left and right cg

flight operations. These limits are set to assure that sufficient aileron deflection is available in all phases of flight for lateral trim and control. The position of the center of gravity along the z-axis greatly influences the directional stability and directional controllability of the aircraft. There are top and bottom limits which must be established by the aircraft designer beyond which the cg should not be located during flight operations. These limits are set to assure that sufficient rudder deflection is available in all phases of flight for lateral trim and control.

In contrast, different maximum weights may be defined for different situations. For example, a large transport aircraft may have a maximum landing weight that is lower than the maximum take-off weight (because some weight is expected to be lost as fuel is burned during the flight). When the center of gravity or weight of an aircraft is outside the acceptable range, the aircraft may not be able to sustain flight or it may be impossible to maintain the aircraft in level flight under some or all circumstances. Placing the cg or weight of an aircraft outside the allowed range can lead to an unavoidable crash of the aircraft.

Few aircraft impose a minimum weight for flight, but all impose a maximum weight. If the maximum weight is exceeded, the aircraft may not be able to achieve or sustain trimmed and controlled flight. Excessive take-off weight may make it impossible to take off within available runway lengths, or it may completely prevent take-off. Excessive weight in flight may make climbing beyond a certain altitude difficult or impossible, or it may make it impossible to maintain an altitude. It is necessary that a cg envelope versus aircraft weight is plotted and available in the pilot flight manual. One such envelope is illustrated in Figure 11.6.

11.3.3 Ideal Center of Gravity Location

An aircraft with a given weight and a fixed configuration has a unique cg location. One of the objectives in the aircraft configuration design is to achieve the best cg location. This goal will limit several alternatives in aircraft component allocation. For instance, to achieve a desired cg location, a business jet designer may have to switch from a twin-jet

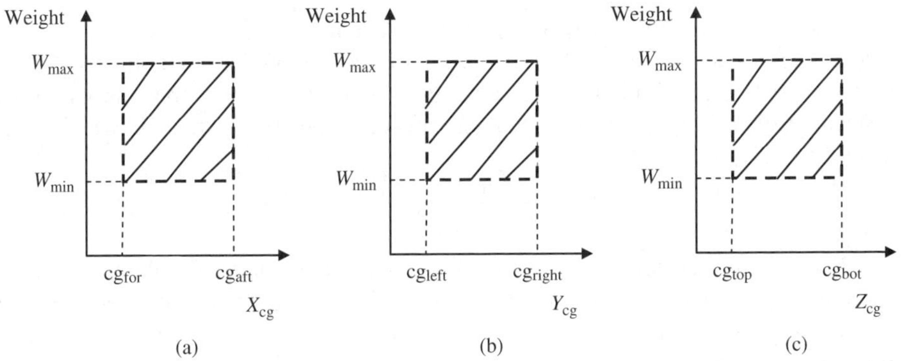

Figure 11.6 Weight and cg range for a typical flight. (a) The cg range along x-axis; (b) The cg range along y-axis; (c) The cg range along z-axis

engine attached to the rear fuselage to an under-wing attachment. This approach will definitely move the cg backward, so another provision must be decided on to maintain stability and controllability. A fundamental question in the weight distribution phase is where the ideal cg location is. To find an answer to this question, the features of an ideal cg location must be identified, prioritized, and listed. In this section, characteristics for an ideal cg location along the x-, y-, and z-axis are reviewed.

11.3.3.1 Ideal Longitudinal cg Location

The determination of features of an ideal cg location is mainly a function of aircraft mission and type, longitudinal stability, longitudinal control, longitudinal trim, operating cost, and aircraft performance. In this section, the ideal longitudinal cg location for civil transport where the cost is assumed to be the top priority is introduced. The ideal longitudinal cg location is where the aircraft does not require any horizontal tail lift to longitudinally trim the aircraft. This initiative results in a reduction of the aircraft trim drag to zero and thus a reduction in the flight cost. Since the wing/fuselage lift longitudinal moment about the y-axis (i.e., cg) must be cancelled, the easiest scenario is to configure the aircraft component and locate the aircraft cg almost at the location of the wing/fuselage aerodynamic center (ac_{wf}). In addition, any lifting surface including a wing generates a zero-lift pitching moment (M_o) about the wing aerodynamic center that contributes to the aircraft longitudinal trim. Therefore the ideal aircraft cg location is moved slightly since, in fact, the summation of the wing zero-lift pitching moment and wing/fuselage lift longitudinal moment about the y-axis must be zero. This is true only for an aircraft cg that is fixed.

However, the aircraft cg usually moves during cruising flight operation aft or forward due to burning fuel by the engine. This fact requires the aircraft designer to provide a precaution in order to trim the aircraft throughout flight. This precaution must be such that the overall aircraft trim drag during cruising flight is minimized. Hence, the ideal location of the aircraft cg must be such that the overall absolute elevator deflection ($|\delta_E|$) during cruise is minimized. Since the elevator deflection could be both positive (down) and negative (up), a performance index (I_P) is defined as follows:

$$I_P = \int \sum (\delta_E)^2 dt \qquad (11.9)$$

where t represents the duration of flight. The cg locations may be found by minimizing the performance index ($dI_P/dt = 0$). The elevator deflection under any flight conditions is obtained by casting longitudinal trim equations. For instance, for a conventional aircraft in a steady cruising flight, the longitudinal trim equation yields the following elevator deflection:

$$\delta_E = \frac{-C_{L_\alpha} C_{m_o} - C_{m_\alpha}\left(C_L - C_{L_o}\right)}{C_{L_\alpha} C_{m_{\delta E}}} \qquad (11.10)$$

where $C_{m_{\delta E}}$ is an aircraft longitudinal control derivative and denotes the slope of variations of pitching moment with respect to elevator deflection. For details of the derivation, and also other flight conditions, the interested reader is referred to flight dynamics textbooks such as Ref. [4].

This criterion implies that it is beneficial to employ positive elevator deflections during half of the cruising flight and negative elevator deflections during the other half. This implication further implies that it is desired to locate the aircraft cg in front of the wing/fuselage aerodynamic center during one half of the cruising flight and forward of the wing/fuselage aerodynamic center during the other half. It should be remembered that this cg location recommendation is based on the cost minimization objective. Other design requirements, such as control and stability, may impose other limits on the cg location. This criterion may readily be utilized to allocate the fuel tanks.

The influence of cg location along the x-axis on the rudder design was intentionally left unaddressed, since it has a slight effect on the rudder design. The reason is that the rudder deflection is minimal during a turn. For other types of aircraft, the designer must establish priorities and define the appropriate performance index (similar to Equation (11.9)). Minimizing the new performance index results in the best location of the cg along the x-axis.

11.3.3.2 Ideal Lateral cg Location

The ideal or optimum lateral cg location is a function of a number of factors, including lateral stability, lateral control, lateral trim, operating cost, and aircraft performance. One must prioritize these design requirements and establish a lateral performance index. Minimizing the lateral performance index yields the best location of cg along the y-axis. In general, the ideal lateral cg location is where the aircraft does not require any aileron deflection to hold the aircraft lateral trim (i.e., wing level). This criterion implies that the aircraft must be symmetrical about the xz plane. Thus, the cg is preferred to be along the fuselage center line. For this basic reason, the wing and horizontal tail have two similar left and right sections, and engines are placed such that the aircraft symmetry is maintained. This result is true for almost all types of aircraft. However, in some cases, the designer may select a configuration that moves the aircraft cg away from the xz plane. For instance, in some transport aircraft, the number of seats on one side of the aisle is more than the number of seats on the other side of the aisle. Examples are the regional transport aircraft Bombardier CRJ900 that has one seat on one side and two seats on the other side of the aisle and the Fokker 100 (Figure 10.6) with three seats on one side and two seats on the other side of the aisle. This type of seating arrangement moves the aircraft cg away from the fuselage center line.

11.3.3.3 Ideal Directional cg Location

The ideal or optimum directional cg location is a function of a number of factors, including directional stability, directional control, directional trim, operating cost, and aircraft performance. One must prioritize these design requirements and establish a directional performance index. Minimizing the lateral performance index yields the best location of cg along the z-axis. In general, the ideal directional cg location is where the aircraft has the lowest mass moment of inertia about the x-axis. This provides the best lateral control. The motion of cg along the z-axis determines the distance between the cg and the x-axis. This implies that the components must be distributed such that the cg along the z-axis is close to the x-axis as far as possible. The technique to determine aircraft mass moment of inertia is presented in Section 11.7.

11.4 Longitudinal Center of Gravity Location

We have already specified the cg range as longitudinal (forward and aft), lateral (left and right), or directional (up and down) limits within which the aircraft's center of gravity must be located during flight operations. Among these three cg limits, the cg range along the x-axis is the most important and critical one. When the weight of the aircraft is at or within the allowable limits for its configuration (e.g., parked, taxi, take-off, climb, cruise, and landing) and its center of gravity is within the allowable range, and will remain so for the duration of the flight, the aircraft is said to be within weight and balance.

The longitudinal cg location not only influences the longitudinal control, longitudinal stability, longitudinal trim, longitudinal handling qualities, and take-off and landing performance, but also largely affects the elevator design. The location of the aircraft cg along the x-axis will also impact directly the operating cost, since the engine's required thrust is a function of aircraft drag. The aircraft drag is directly a function of the aircraft angle of attack and elevator deflection. As indicated by Equation (11.10), the elevator deflection is largely dominated by the location of the aircraft cg. For a long duration aircraft such as an airliner (transport aircraft), the cg variation along the x-axis during flight is inevitable, but the cost component must be minimized.

When the fore or aft center of gravity is out of the allowable range, the aircraft may pitch uncontrollably down or up. This pitch tendency may exceed the control authority available to the pilot, causing a loss of control. The excessive pitch may be apparent in all phases of flight, or only during certain phases, such as take-off or climb. Because the burning of fuel gradually produces a loss of weight and usually a shift in the center of gravity, it is possible for an aircraft to take off with the center of gravity in a position that allows full control, and yet later develop an imbalance that exceeds control authority. The aircraft weight distribution must take this issue into account. The bulk of this job must often be calculated in advance by the designer and incorporated into cg limits.

A lightly loaded aircraft at the end of a flight when the fuel is almost all consumed may experience a situation where the cg moves forward beyond the allowable cg range. When an aircraft cg location is not achievable by a regular weight distribution technique, a special lump mass which is referred to as ballast will be employed. *Ballast* is a removable or permanently installed weight in an aircraft used to bring the center of gravity into the allowable range. In some aircraft, when flying with only one pilot onboard and no passengers or baggage, it is necessary to carry some suitable type of ballast to compensate for a too-far-forward or too-far-aft cg. The flight characteristics of an aircraft at maximum take-off weight with the cg very near its most aft limits are very different from those of the same aircraft lightly loaded. Typical cg variations along the x-axis for a large transport aircraft with under-wing engines and a business jet with engines mounted beside the rear fuselage are shown in Figure 11.7.

It is convenient to express the aircraft cg in terms of percentage MAC. To obtain this non-dimensional parameter (h or $\overline{X}_{cg}$), one must divide the distance between the aircraft cg and the wing leading edge at the MAC by the wing MAC:

$$h = \overline{X}_{cg} = \frac{x_{cg} - x_{LE_{MAC}}}{\overline{C}} \tag{11.11}$$

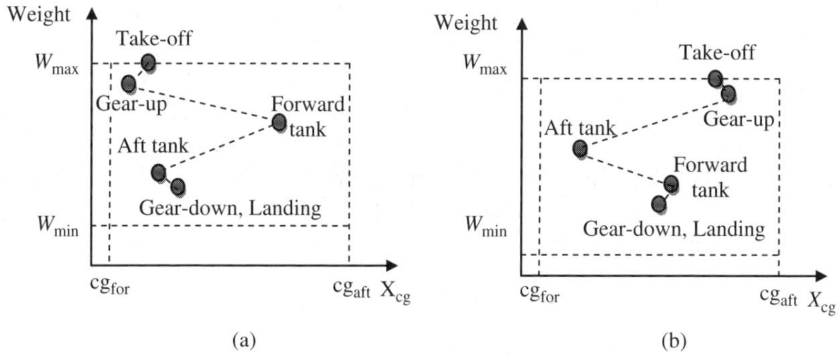

Figure 11.7 Typical center of gravity longitudinal variations during flight operations. (a) A transport aircraft with a forward cg at maximum take-off weight (underwing engines); (b) A business jet aircraft with a rear cg at maximum take- off weight (engines beside rear fuselage)

The same technique is applied to the most forward and the most aft cg:

$$h_{for} = \overline{X}_{cg_{for}} = \frac{x_{cg_{for}} - x_{LE_{MAC}}}{\overline{C}} \tag{11.12}$$

$$h_{aft} = \overline{X}_{cg_{aft}} = \frac{x_{cg_{aft}} - x_{LE_{MAC}}}{\overline{C}} \tag{11.13}$$

As the aircraft cg moves aft during a flight operation, the aircraft longitudinal stability is downgraded, until the cg passes the neutral point at which the aircraft becomes statically longitudinally unstable. A conventional aircraft will usually be dynamically longitudinally unstable when the aircraft cg is closer than a small percentage MAC (about 2–3%) to the aircraft neutral point. In contrast, as the aircraft cg moves forward during a flight operation, the aircraft longitudinal controllability is downgraded, until the cg passes a certain point at which the aircraft becomes uncontrollable. Both of these cg scenarios are undesirable, and must be prevented during the design process and avoided during the load distribution period by load masters. The majority of cg-related aircraft crashes are due to cg locations beyond the allowable cg range along the x-axis. Thus, the aircraft designer must be careful during weight distribution and even include some safety factor for crew members during load allocation.

Although the aircraft neutral point is almost fixed for a fixed configuration, the wing aerodynamic center is not so for a supersonic aircraft. The aircraft neutral point is often at somewhere about 40–50% MAC. However, the wing aerodynamic center at subsonic speed is at about wing quarter chord (25% MAC), while it shifts to about 50% MAC at supersonic speed. This aerodynamic phenomenon creates a unique situation for a supersonic aircraft designer and makes it very hard to distribute weight during the design process. To maintain the aircraft longitudinal equilibrium, the aircraft cg must be moved rearward at supersonic speeds. In addition, as the cg moves aft, the rudder becomes less effective. Since the rearward motion of cg downgrades the aircraft directional control, there is a narrow band for allowable cg limits. Therefore, a compromise is necessary to maintain a balance between longitudinal control and directional control.

A technique to shift the aircraft cg rearward during flight is fuel transfer using a pump from the front tanks to the rear tanks. This technique was employed [1] in the supersonic transport aircraft Concorde (Figures 7.24 and 11.15). In the majority of transport aircraft, the cg location varies as the fuel burns. However, in a few aircraft such as the McDonnell Douglas DC-10 (Figure 8.7), the fuel tanks are located such that the aircraft cg remains the same while in cruise. In some aircraft, such as the McDonnell Douglas MD-11 (Figure 3.16), there is a fuel tank in the horizontal tail into which fuel is pumped during cruise to keep the cg at the most aft limit.

Table 11.3 Features of forward and aft cg positions

No.	Criterion	Forward cg	Aft cg
1	Stability	Aircraft is longitudinally/directionally more stable	Aircraft is longitudinally/directionally less stable
2	Controllability	Aircraft is longitudinally/directionally less controllable	Aircraft is longitudinally/directionally more controllable
3	Elevator design	Aircraft requires more elevator deflection during take-off rotation	Aircraft requires less elevator deflection during take-off rotation
4	Rudder design	Aircraft requires more rudder deflection during asymmetric thrust	Aircraft requires less rudder deflection during asymmetric thrust
5	Load on wheel	There will be more load on the nose wheel (in a tricycle configuration)	There will be more load on the main wheel (in a tricycle configuration)
6	Taxi	It is easier for aircraft to turn during taxi	It is harder for aircraft to turn during taxi
7	Fuel cost	Cruising flight will often burn more fuel	Cruising flight will often burn less fuel
8	Stall	Aircraft is safer (to enter stall)	Aircraft is more prone to stall
9	Spin	Aircraft is safer (to enter spin)	Aircraft is more prone to spin
10	Spin recovery	Recovery is slower (if aircraft is spinnable)	Recovery is faster (if aircraft is spinnable)
11	Crash	Aircraft is safer and there is less possibility of crash	Aircraft is more prone to crash
12	Mishap	Aircraft is safer during taxi	Aircraft is prone to tip back during take-off
13	Gust	It takes more oscillations to recover when a gust hits and disturbs the longitudinal trim	It takes fewer oscillations to recover when a gust hits and disturbs the longitudinal trim

In general, forward and aft center of gravity positions tend to have different and even almost opposite features. For some design requirements, an aft cg is preferred while for other design requirements, a forward cg is favored. Table 11.3 tabulates the relationship between cg locations and various design requirements. The table demonstrates the consequences for an aircraft when the cg is at the most forward or most aft cg location. The table can be regarded as a guide to establish the cg range along the x-axis. The aircraft cg must remain within the specified limits as fuel is burned, as passengers are walking, as loads are dropped, as landing gear is retracted, and as stores are released. It is recommended to sequence the fuel tanks, selecting to burn fuel from different tanks at different times to keep the aircraft cg within the permissible range. It is also suggested to design and implement an automated fuel management system to distribute fuel from various tanks symmetrically.

Another topic which is directly related to the aircraft weight distribution is to decide whether to widen the cg range or shorten it along the x-axis. The answer to this challenge by two groups of aircraft-related people (i.e., designer and customer) is almost opposite. The customer in this case does not mean the passenger, but refers to the person who deals with load handling. In this specific issue, the interest or wish of the designer does not match with that of the customer. Generally speaking, an aircraft customer often seeks an aircraft with a wider cg range, while the designer will try to shorten the cg range. Table 11.4 illustrates some aspects of short and wide cg range along the x-axis. The decision about how wide the cg range needs to be must be made after consultation with the marketing department. In summary, a load-relating customer would rather purchase an aircraft with a large cg range, while an aircraft designer tries to limit the cg range to the shortest distance. Table 11.5 demonstrates the aft cg, forward cg, and cg range for several aircraft. Based on this historical data, the most forward cg location was at 5% MAC, and the most aft cg at 41% MAC.

Figure 11.8 shows the recommended location for an aircraft cg about the wing/fuselage aerodynamic center. The aerodynamic center for a wing alone is located at the quarter chord MAC (25% MAC). However, when the fuselage is attached to a wing, the wing/fuselage combination aerodynamic center will be at a new place. In other words,

Table 11.4 Features of cg range

No.	Criterion	Wider cg range	Shorter cg range
1	Load handling (customer)	Aircraft can carry more diverse combinations of load/cargo (in terms of both size and volume)	Aircraft can carry less diverse combinations of load/cargo (both size and volume)
2	Accident/crash	Aircraft is less prone to accident/crash due to loading issues	Aircraft is more prone to accident/crash due to loading issues
3	Configuration design	Aircraft configuration design is more challenging	Aircraft configuration design is less challenging
4	Alternatives	There are fewer aircraft configuration alternatives	There are more aircraft configuration alternatives

Table 11.5 Aft cg, forward cg, and cg range for several aircraft in terms of percentage MAC

No.	Aircraft	Engine	m_{TO} (kg)	Forward cg	Aft cg	cg Range
1	Cessna 172	Single-piston	1 111	15.6	36.5	20.9
2	Cessna 177-Utility	Single-piston	1 100	5	18.5	13.5
3	Cessna 206 Skywagon	Single-piston	1 632	12.2	39.4	27.2
4	Cessna Skymaster	Twin-piston	2 000	17.3	30.9	13.6
5	Air Tractor AT-602	Single-turboprop	7 257	23	35	12
6	Piper PA-30 Comanche	Twin-piston	1 690	12	27.8	15.8
7	Beechcraft Queen Air	Twin-piston	3 992	16	29.3	13.3
8	Dornier Do 28	Twin-piston	2 720	10.7	30.8	20.1
9	Douglas DC-6	Four-radial	44 129	12	35	23
10	Pilatus PC-12	Single-turboprop	4 740	13	46	33
11	Beechcraft B-45	Single-turboprop	1 950	19	28	9
12	Pilatus PC-6	Single-turboprop	6 108	11	34	23
13	Fokker F-27	Twin-turboprop	19 773	18.7	40.7	22
14	Lockheed C-130E	Four-turboprop	70 300	15	30	15
15	Learjet 25	Twin-turbojet	6 802	9	30	21
16	Gulfstream G200	Twin-turbofan	16 080	22	40	18
17	Cessna Citation III	Twin-turbofan	9 527	14	31	17
18	Fokker F-28	Twin-turbofan	29 000	17	37	20
19	DC-9-10	Twin-turbofan	41 100	15	40	25
20	Gulfstream G550	Twin-turbofan	41 277	21	45	24
21	Boeing 737-100	Twin-turbofan	50 300	11	31	20
22	Boeing 707-120	Four-turbofan	116 570	16	34	18
23	Boeing 747-200	Four-turbofan	377 842	12.5	32	19.5
24	Douglas DC-8	Four-turbofan	140 600	16.5	32	15.5
25	Lockheed C-141	Four-turbofan	147 000	19	32	13
26	Lockheed C-5A	Four-turbofan	381 000	19	41	22
27	Concorde	Four-turbojet	185 700	20	59	39

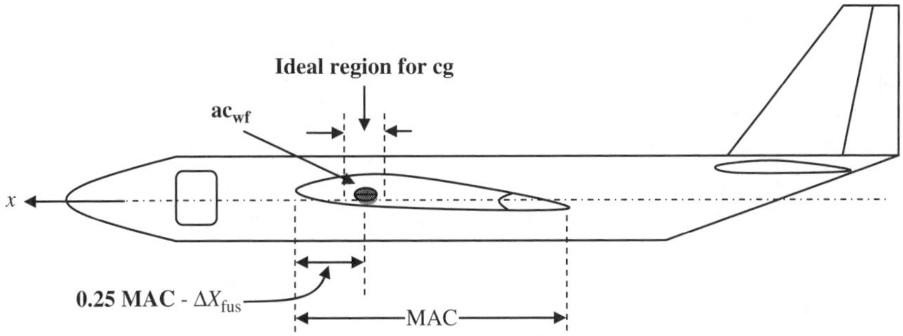

Figure 11.8 Ideal region for cg location along the x-axis

the aerodynamic center is shifted (ΔX_{fus}) by the fuselage (usually forward) due to the ability of the fuselage to generate lift. This shift is referred to as the Munk shift. It is interesting to note that the aerodynamic center of a wing + fuselage tends to shift aft with Mach number more or less like that of a wing alone. Multhopp [5] developed a technique to predict the shift of the wing ac due to the fuselage attachment. Nacelles and stores, when mounted under a wing such that they protrude forward from the wing leading edge, also cause a shift in the wing aerodynamic center. The fuselage/store/nacelle-induced shift in aerodynamic center location is very significant and must be accounted for in the weight distribution process. Table 11.6 demonstrates the position of the wing/fuselage combination aerodynamic center for several aircraft.

In general, in large transport aircraft, the most forward cg is located as forward as 5% MAC, while the most aft location as aft as 40% MAC. The average cg limits in large transport aircraft are between 20 and 30% MAC. In General Aviation (GA) aircraft, the most forward cg location is at about 10–20% MAC, while the most aft location is at about 20–30% MAC. The average cg range in GA aircraft is between 10 and 20% MAC. Large cargo aircraft have unique conditions, since they are designed to carry various combinations of payloads. For instance, the most aft cg of the Tupolev 154 (Figure 11.14) with three turbofan engines is located at about 50% MAC (at empty weight). The center of

Table 11.6 Position of the wing/fuselage combination aerodynamic center for several aircraft

No.	Aircraft/wing	Configuration	Wing/fuselage ac (% MAC)	Muck shift (% MAC)
1	Wing alone	Lifting surface	25	0
2	Cessna 172	Single-engine light GA	21	–4
3	Learjet 24	Six-seat twin-jet engine	11	–14
4	Piaggio P-180	Nine-seat twin-turboprop pusher	–7	–32

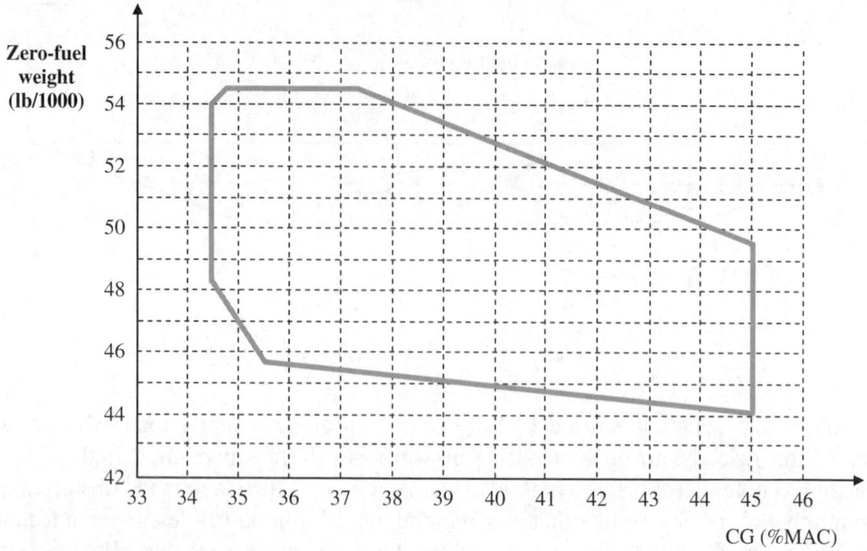

Figure 11.9 Zero-fuel gross weight cg envelope for business jet Gulfstream G550. Reproduced from permission of Gulfstream

gravity envelope [6, 10] for the twin-engine business jet Gulfstream G 550 (Figure 11.15) is shown in Figure 11.9. Example 11.1 illustrates the determination of aircraft cg in terms of percentage MAC.

Example 11.1

Consider a two-seat (side-by-side) light aircraft with a maximum take-off mass of 1200 kg, wing area of 16 m^2, and overall length of 8 m (Figure 11.10). The mass of each major component and their corresponding centers of gravity from the propeller spinner (reference line) are as follows:

No.	Component	Mass (kg)	X_{cg} (m)
1	Wing	180	3.2
2	Tails	50	7.3
3	Engine	170	0.6
4	Fuselage	150	3.5
5	Pilots	160	2.9
6	Landing gear	40	2.8
7	Fuel	140	3.2
8	Systems	310	3.4

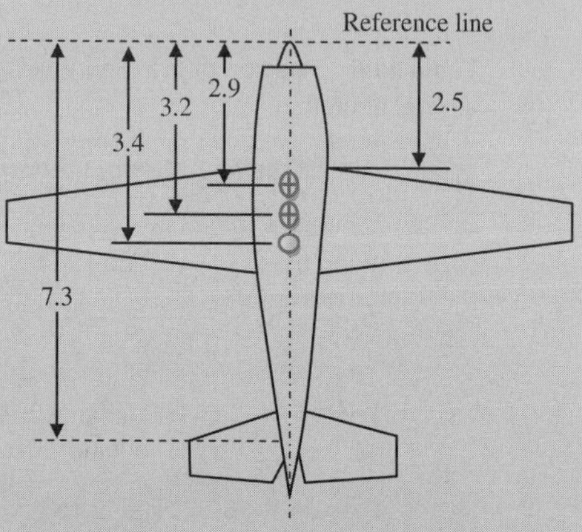

Figure 11.10 Aircraft of Example 11.1 (numbers are in meters) (US Government)

Determine the aircraft cg in terms of percentage MAC. The wing is straight-tapered with an aspect ratio of 6, a taper ratio of 0.6, and the wing apex is 2.5 m behind the reference line.

Solution:
We first need to determine the wing MAC, root (in fact center line) chord, and tip chord:

$$AR = \frac{b^2}{S} \Rightarrow b = \sqrt{S \cdot AR} = \sqrt{16 \cdot 6} \Rightarrow b = 9.8 \text{ m} \tag{5.19}$$

$$AR = \frac{b}{\overline{C}} \Rightarrow \overline{C} = \frac{b}{AR} = \frac{9.8}{6} \Rightarrow \overline{C} = 1.633 \text{ m} \tag{5.17}$$

$$\overline{C} = \frac{2}{3}C_r\left(\frac{1+\lambda+\lambda^2}{1+\lambda}\right) \Rightarrow$$

$$1.633 = \frac{2}{3}C_r\left(\frac{1+0.6+0.6^2}{1+0.6}\right) \Rightarrow C_r = 2 \text{ m} \tag{5.26}$$

$$\lambda = \frac{C_t}{C_r} \Rightarrow 0.6 = \frac{C_t}{2} \Rightarrow C_t = 2 \cdot 0.6 = 1.2 \text{ m} \tag{5.24}$$

Next, the aircraft center of gravity must be determined. There are eight components, so $n = 8$:

$$X_{cg} = \frac{\sum_{i=1}^{8} m_i x_{cg_i}}{\sum_{i=1}^{8} m_i} = \frac{m_w x_w + m_t x_t + m_e x_e + m_b x_b + m_p x_p + m_{lg} x_{lg} + m_f x_f + m_s x_s}{m_w + m_t + m_e + m_b + m_p + m_{lg} + m_f + m_s}$$

$$= \frac{\left[\begin{array}{c}(180 \cdot 3.2) + (50 \cdot 7.3) + (170 \cdot 0.6) + (150 \cdot 3.5) + (160 \cdot 2.9) \\ + (40 \cdot 2.8) + (140 \cdot 3.2) + (310 \cdot 3.4)\end{array}\right]}{180 + 50 + 170 + 150 + 160 + 40 + 140 + 310}$$

$$= \frac{3646}{1200} \Rightarrow X_{cg} = 3.038 \text{ m} \tag{11.1}$$

Based on the geometry shown in Figure 11.11, the distance between the aircraft cg and the wing apex is:

$$X_{cg} - X_{LE} = 3.038 - 2.5 = 0.538 \text{ m}$$

The distance between the wing apex and the wing leading edge at the MAC (Figure 11.11) is:

$$\frac{C_r - MAC}{2} = \frac{2 - 1.633}{2} = 0.183 \text{ m}$$

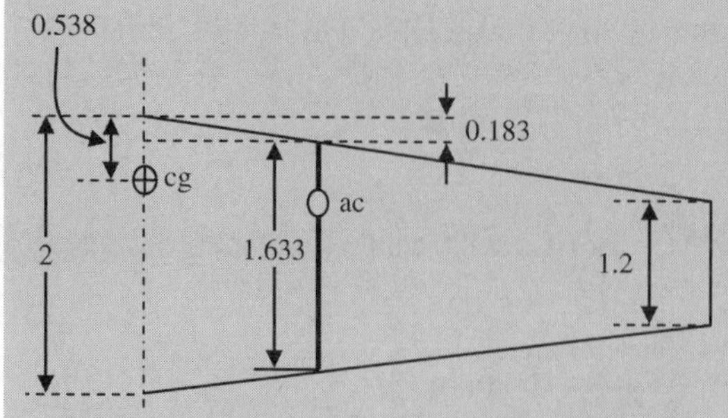

Figure 11.11 Half wing (numbers are in meters)

The distance between the aircraft cg and the wing leading edge at the MAC location (see Figure 11.11) is:

$$X_{cg\text{-}LE} = X_{cg} - X_{LE} - \frac{C_r - MAC}{2}$$
$$= 3.038 - 2.5 - \frac{2 - 1.633}{2}$$
$$= 0.538 - 0.183 = 0.355\,\text{m}$$

Ultimately, the aircraft cg in terms of MAC is:

$$\overline{X}_{cg} = \frac{X_{cg\text{-}LE}}{MAC} = \frac{0.355}{1.633} = 0.217 \quad (11.11)$$

Thus, the aircraft cg is located at 21.7% of wing MAC.

11.5 Technique to Determine the Aircraft Forward and Aft Center of Gravity

In earlier sections, it was emphasized that the aircraft center of gravity is a key point and a major concern for an aircraft designer. This is reflected in several design requirements, such as stability requirements, controllability requirements, and handling or flying quality requirements. However, the aircraft cg is moving throughout the flight operation due to fuel burning. At any rate, in extreme cases, there will be a most forward cg and a most aft cg. The aircraft cg range has two extreme locations, namely most forward cg and most aft cg.

The aircraft longitudinal cg range is defined as the distance between the most forward and the most aft cg locations. The aircraft center of gravity is a function of two major elements: (i) center of gravity of aircraft components such as wing, fuselage, tail, fuel, engine, passengers, luggage, systems, cargo, etc. and (ii) rate of change of location of movable or removable components such as passengers and fuel. There are basically two ways to determine the most forward and most aft cg locations. One method is trial and error, and the second is a systematic approach. In this section, a systems engineering technique will be presented. It considers all removable elements such as passengers, cargo, fuel, and store.

An aircraft may experience an infinite number of loading scenarios throughout its flight operations. A few examples of loading scenarios are: no store at one side of a fighter, a few empty seats in a passenger airplane, passengers with different weights and seating arrangements, an empty aft fuel tank, various cargo packages, and the addition of external fuel tanks to a fighter for a long-range mission. It may initially seem that if all probable load scenarios are known, by using Equation (11.1) the aircraft cg for all cg cases could be obtained and accordingly the extreme locations could be found. However, there is an infinite number of loading variants (i.e., cg cases) that an aircraft may experience during flight operations. Therefore it is impossible to determine the number of possible loading scenarios and include them in the calculation. A technique based on the systems engineering approach is introduced in this section to determine the aircraft most forward and most aft center of gravity locations. The technique is developed based on the laws which govern the aircraft cg motion due to various factors. These laws are as follows:

1. The aircraft cg will move with the motion of moving elements (e.g., passengers walking), but at a smaller rate.
2. The aircraft cg will move farther from an absent element which already exists (e.g., an empty seat when, in the first place, there was a passenger).
3. The aircraft cg will move farther from a lighter load/item (petite size teenage passenger compared with a large size adult).
4. The aircraft cg will move farther from the cg of burning fuel as long as it is consumed. Therefore, as fuel is burned during a flight operation, the aircraft cg will also move either forward or aft depending upon the fuel tank location relative to the original aircraft cg.

Equations (11.1)–(11.3) are general equations that apply at any point in the flight operations of any aircraft. When either the weight or the location of any aircraft component is changed, the aircraft cg will follow suit. Thus, the aircraft stability, controllability, handling quality, and aircraft performance vary throughout a flight operation. When the aircraft cg at maximum take-off weight is determined, the next step is to determine the aircraft cg range for various loading scenarios. In this technique a term "removable load" is employed and needs to be defined here. A removable load is any payload or unpaid load that can be removed from an aircraft with the aircraft still able to fly safely. It includes fuel, payload, and all other unpaid load except one pilot. Payload generally includes cargo, passengers, stores, and luggage (both checked and carry-on).

The following is the procedure to determine the most aft and most forward cg of an aircraft:

1. Determine the aircraft maximum take-off weight (or mass).
2. Determine the aircraft center of gravity (with maximum take-off weight) in the longitudinal axis using Equation (11.1), and call it X_{cg_1}.
3. Identify all removable loads and their centers of gravity. In the case of an airliner, the cg of each individual passenger and each luggage/cargo package must be known.
4. Remove any removable load whose cg is in front of the aircraft cg calculated in step 2. Now, determine the aircraft center of gravity in the longitudinal axis excluding these removable elements using the following equation:

$$x_{\mathrm{cg}_2} = \frac{\sum_{j=1}^{n-k1} x_{\mathrm{cg}_j} m_j}{\sum_{j=1}^{n} m_j - \sum_{j=1}^{k1} m_j} \tag{11.14}$$

This cg (X_{cg_2}) will be assumed as the most aft cg of the aircraft ($X_{\mathrm{cg}_{\mathrm{aft}}}$) up to this moment. The parameter $k1$ denotes the number of removable loads which are located ahead of X_{cg_1}.

5. Investigate if there is any removable load in front of $X_{cg_{aft}}$ of step 4. If yes, repeat the calculations by removing it and determine the aircraft center of gravity in the longitudinal axis excluding this new removable load, by using Equation (11.14) again. In this case, the new cg will be assumed as the most aft cg of the aircraft ($X_{cg_{aft}}$). This process must be continued until no removable element is observed ahead of this $X_{cg_{aft}}$.
6. Remove any removable load whose cg is aft of the aircraft cg calculated in step 2. Now, determine the aircraft center of gravity in the longitudinal axis excluding these removable elements using the following equation:

$$x_{cg_3} = \frac{\sum_{j=1}^{n-k2} x_{cg_j} m_j}{\sum_{j=1}^{n} m_j - \sum_{j=1}^{k2} m_j} \tag{11.15}$$

This cg (X_{cg_3}) will be assumed as the most forward cg of the aircraft ($X_{cg_{for}}$) up to this moment. The parameter $k2$ denotes the number of removable loads which are located aft of X_{cg_1}.
7. Investigate if there is any removable load in front of $X_{cg_{for}}$ of step 6. If yes, repeat the calculations by removing it and determine the aircraft center of gravity in the longitudinal axis excluding this new removable load, by using Equation (11.15) again. In this case, the new cg will be assumed as the most forward cg of the aircraft ($X_{cg_{for}}$). This process must be continued until no removable element is observed aft of this $X_{cg_{for}}$.
8. Determine the non-dimensional longitudinal cg range (Δx_{cg}), or the non-dimensional distance between the most aft and most forward cg:

$$\Delta x_{cg} = \frac{x_{cg_{aft}} - x_{cg_{for}}}{\overline{C}} \tag{11.16}$$

Many technical references are suggesting to considering tens of possible loading scenarios and determining aircraft cg via employing Equation (11.1), and then comparing all cg locations to see which one is located in the farthest location and which one in the foremost. The systematic technique described above will eliminate such a hectic and repetitive operation and yields reliable results in just seven steps. Example 11.2 gives an example of finding the most aft and most forward cg location of an aircraft.

Example 11.2: Aircraft forward and aft cg

Consider a twin-engine business jet aircraft with 12 passengers and 2 crew members (Figure 11.12). The aircraft has the following mass and wing characteristics:

$$m_{TO} = 30\,000 \text{ kg},\ S = 90 \text{ m}^2,\ AR = 9,\ \lambda = 0.5$$

The seat pitch is 1 m and there is one baggage compartment right above each seat. The mass and cg location (X_{cg}) for each major component are given in Table 11.7.

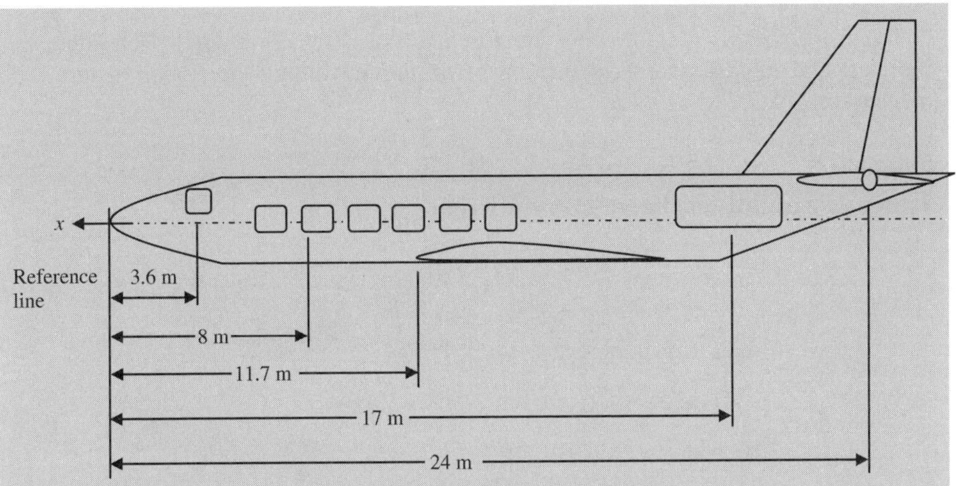

Figure 11.12 Aircraft in Example 11.2

Table 11.7 Mass and cg locations (X_{cg}) of major components for aircraft in Example 11.2

No.	Component	Mass (kg)	X_{cg} (m)
1	Wing	4200	13
2	Tails	600	24
3	Engine	6900	17
4	Fuselage	3300	13
5	Pilots + bag	2·(80 + 10)	3.6
6	Passengers (first row)	2·80	8
7	Flight attendant	80 + 10	5
8	Carry-on baggage (first row)	2·10	8
9	Checked baggage	600	15
10	Landing gear	1 200	9
11	Wing fuel	3050	13
12	Fuselage fuel	6000	9
13	Systems and other equipment	2800	8

The wing apex is at 11 m from the fuselage nose (reference line). Determine the most aft and most forward cg location of the aircraft along the x-axis in terms of percentage MAC.

Solution:

- **Step 1. Aircraft maximum take-off weight.**

$$m_{\mathrm{TO}} = m_{\mathrm{w}} + m_{\mathrm{T}} + m_{\mathrm{e}} + m_{\mathrm{F}} + m_{\mathrm{P}} + m_{\mathrm{pass}} + m_{\mathrm{fa}} + m_{\mathrm{bg1}} + m_{\mathrm{bg2}} + m_{\mathrm{LG}} + m_{\mathrm{fl1}} + m_{\mathrm{fl2}} + m_{\mathrm{s}}$$

$$m_{\mathrm{TO}} = 4200 + 600 + 6900 + 3300 + (2 \cdot 80) + (12 \cdot 80)) + 80 + (15 \cdot 10)) + 600 + 1200 + 4000 + 5050 + 2800$$

$$m_{\mathrm{TO}} = 30\,000\,\mathrm{kg}$$

$$W_{\mathrm{TO}} = m_{\mathrm{TO}} g = 30\,000 \cdot 9.81 = 294\,199\,\mathrm{N}$$

- **Step 2. X_{cg} for maximum take-off weight.** X_{cg} of passengers plus their carry-on bags

  There are six rows, each with two seats having a pitch of 1 m. Each passenger carries 10 kg of carry-on baggage and stows it in the overhead baggage compartment. So each row has 180 kg (i.e., 80 + 10 + 80 + 10) of mass. The cg of the first seat is at 8 m from the reference line. So:

$$x_{\mathrm{pass}} = \frac{\sum_{i=1}^{6} m_i x_{\mathrm{cg}_i}}{\sum_{i=1}^{6} m_i} = \frac{m_1 x_1 + m_2 x_2 + m_3 x_3 + m_3 x_3 + m_4 x_4 + m_5 x_5 + m_6 x_6}{m_{\mathrm{pass}}} \quad (11.1)$$

$$x_{\mathrm{pass}} = \frac{\left[\begin{array}{c}(2 \cdot 90 \cdot 8) + (2 \cdot 90 \cdot 9) + (2 \cdot 90 \cdot 10) + (2 \cdot 90 \cdot 11) \\ + (2 \cdot 90 \cdot 12) + (2 \cdot 90 \cdot 13)\end{array}\right]}{12 \cdot (80 + 10)} = 10.5\,\mathrm{m}$$

X_{cg} of entire aircraft.

There are 13 components, so $n = 13$:

$$X_{\mathrm{cg}_1} = \frac{\sum_{i=1}^{13} m_i x_{\mathrm{cg}_i}}{\sum_{i=1}^{13} m_i} = \quad (11.1)$$

$$\frac{\left[\begin{array}{c} m_{\mathrm{w}} x_{\mathrm{w}} + m_{\mathrm{t}} x_{\mathrm{t}} + m_{\mathrm{e}} x_{\mathrm{e}} + m_{\mathrm{b}} x_{\mathrm{b}} + m_{\mathrm{p}} x_{\mathrm{p}} + m_{\mathrm{pass}} x_{\mathrm{pass}} + m_{\mathrm{fa}} x_{\mathrm{fa}} \\ + m_{\mathrm{bg1}} x_{\mathrm{bg1}} + m_{\mathrm{bg2}} x_{\mathrm{bg2}} + m_{\mathrm{lg}} x_{\mathrm{lg}} + m_{\mathrm{f1}} x_{\mathrm{f1}} + m_{\mathrm{f2}} x_{\mathrm{f2}} + m_{\mathrm{s}} x_{\mathrm{s}} \end{array}\right]}{m_{\mathrm{w}} + m_{\mathrm{t}} + m_{\mathrm{e}} + m_{\mathrm{b}} + m_{\mathrm{p}} + m_{\mathrm{pass}} + m_{\mathrm{fa}} + m_{\mathrm{bg1}} + m_{\mathrm{bg2}} + m_{\mathrm{lg}} + m_{\mathrm{f1}} + m_{\mathrm{f2}} + m_{\mathrm{s}}}$$

$$= \frac{\begin{bmatrix} 4200 \cdot 13 + 600 \cdot 24 + 6900 \cdot 17 + 3300 \cdot 13 + 180 \cdot 3.6 + 1080 \cdot 10.5 \\ +90 \cdot 5 + 600 \cdot 15 + 1200 \cdot 9 + 4000 \cdot 14 + 5050 \cdot 9 + 2800 \cdot 8 \end{bmatrix}}{\begin{bmatrix} 4200 + 600 + 6900 + 3300 + 180 + 1080 + 90 \\ +600 + 1200 + 4000 + 5050 + 2800 \end{bmatrix}}$$

$$= \frac{385\,288}{30\,000} \Rightarrow X_{cg_1} = 12.843\,\mathrm{m}$$

- **Step 3. Identify all removable loads**. In this aircraft, removable loads are all passengers, fuel tanks (in the wing and in the fuselage), checked baggage (in the fuselage), carry-on baggage (overhead in the cabin), flight attendant, and one of the pilots. Removable components and their locations compared with the aircraft cg at maximum take-off weight are tabulated in Table 11.8.

Table 11.8 Removable components and their locations compared with aircraft cg at maximum take-off weight

Component	X_{cg} (m)	Location
Pilots	3.6	Ahead of X_{cg_1}
Flight attendant	5	Ahead of X_{cg_1}
Passengers + carry-on bag (first row)	8	Ahead of X_{cg_1}
Fuselage fuel	9	Ahead of X_{cg_1}
Passengers + carry-on bag (second row)	9	Ahead of X_{cg_1}
Passengers + carry-on bag (third row)	10	Ahead of X_{cg_1}
Passengers + carry-on bag (fourth row)	11	Ahead of X_{cg_1}
Passengers + carry-on bag (fifth row)	12	Ahead of X_{cg_1}
Aircraft (with maximum take-off weight)	**12.843**	$\mathbf{X_{cg_1}}$
Passengers + carry-on bag (sixth row)	13	Aft of X_{cg_1}
Wing fuel	14	Aft of X_{cg_1}
Checked baggage	15	Aft of X_{cg_1}

- **Step 4. Remove any removable load whose cg is in front of the aircraft cg**. By comparing the aircraft X_{cg} in step 2 and the cg locations of removable items in Table 11.8, it is observed that one pilot (i.e., co-pilot), a flight attendant (plus his/her carry-on bag), first, second, third, fourth, and fifth rows of passengers (plus their carry-on baggage) are located ahead of X_{cg_1}. So, a new cg is calculated by removing these eight items:

$$x_{cg_2} = \frac{\sum_{j=1}^{13-8} x_{cg_j} m_j}{\sum_{j=1}^{13} m_j - \sum_{j=1}^{8} m_j} \tag{11.14}$$

$$x_{cg_2} = \frac{\left[\begin{array}{c} 385\,288 - (90 \cdot 3.6) - (90 \cdot 5) - (180 \cdot 8) - (5050 \cdot 9) - (180 \cdot 9) \\ - (180 \cdot 10) - (180 \cdot 11) - (180 \cdot 12) \end{array}\right]}{30\,000 - (12 \cdot 80) - (12 \cdot 10) - 5050}$$

$$x_{cg_2} = 13.836\,\text{m}$$

- **Step 5. The most aft cg**. X_{cg_2} is behind the sixth row of seats (13.836 is greater than 13), so it must be removed too. Therefore a newer X_{cg_2} is calculated by removing nine items:

$$x_{cg_2} = \frac{\sum_{j=1}^{13-9} x_{cg_j} m_j}{\sum_{j=1}^{13} m_j - \sum_{j=1}^{9} m_j} \tag{11.14}$$

$$x_{cg_2} = \frac{\left[\begin{array}{c} 385288 - (90 \cdot 3.6) - (90 \cdot 5) - (180 \cdot 8) - (5050 \cdot 9) - (180 \cdot 9) \\ - (180 \cdot 10) - (180 \cdot 11) - (180 \cdot 12) - (180 \cdot 13) \end{array}\right]}{30\,000 - (14 \cdot 80) - (14 \cdot 10) - 5050}$$

$$x_{cg_2} = 13.843\,\text{m}$$

This is the most aft cg of the aircraft ($X_{cg_{aft}}$).

- **Step 6. Remove any removable load whose cg is aft of the aircraft cg**. By looking at Table 11.8, it is observed that only two elements (wing fuel and checked baggage) are aft of X_{cg_1}. Thus:

$$x_{cg_3} = \frac{\sum_{j=1}^{13-2} x_{cg_j} m_j}{\sum_{j=1}^{13} m_j - \sum_{j=1}^{2} m_j} = \frac{385\,288 - (4000 \cdot 14) - (600 \cdot 15)}{30\,000 - 4000 - 600} \Rightarrow x_{cg_3} = 12.61\,\text{m} \tag{11.15}$$

- **Step 7. The most aft cg**. X_{cg_3} is in front of the sixth row of seats (12.61 is less than 13), so it must be removed too. Therefore a newer X_{cg_3} is calculated by removing three items and recalculating:

$$x_{\mathrm{cg}_3} = \frac{\sum_{j=1}^{13-3} x_{\mathrm{cg}_j} m_j}{\sum_{j=1}^{13} m_j - \sum_{j=1}^{3} m_j} = \frac{385\,288 - (4000 \cdot 14) - (600 \cdot 15) - (2 \cdot 90 \cdot 13)}{30\,000 - 4000 - 600 - (2 \cdot 90)} \quad (11.15)$$

$$\Rightarrow x_{\mathrm{cg}_3} = 12.607\,\mathrm{m}$$

- **Step 8. The cg range and X_{cg} of aircraft in terms of percentage MAC**. The wing MAC, root (in fact center line) chord, and tip chord are obtained as follows:

$$\mathrm{AR} = \frac{b^2}{S} \Rightarrow b = \sqrt{S \cdot \mathrm{AR}} = \sqrt{90 \cdot 9} \Rightarrow b = 28.46\,\mathrm{m} \quad (5.19)$$

$$\mathrm{AR} = \frac{b}{\overline{C}} \Rightarrow \overline{C} = \frac{b}{\mathrm{AR}} = \frac{28.46}{9} \Rightarrow \overline{C} = 3.162\,\mathrm{m} \quad (5.17)$$

$$\overline{C} = \frac{2}{3} C_{\mathrm{r}} \left(\frac{1 + \lambda + \lambda^2}{1 + \lambda} \right) \Rightarrow 3.162 = \frac{2}{3} C_{\mathrm{r}} \left(\frac{1 + 0.5 + 0.5^2}{1 + 0.5} \right) \Rightarrow C_{\mathrm{r}} = 4.066\,\mathrm{m} \quad (5.26)$$

$$\lambda = \frac{C_{\mathrm{t}}}{C_{\mathrm{r}}} \Rightarrow 0.5 = \frac{C_{\mathrm{t}}}{4.066} \Rightarrow C_{\mathrm{t}} = 4.066 \cdot 0.5 = 2.033\,\mathrm{m} \quad (5.24)$$

According to the geometry shown in Figure 11.13, the distance between the wing apex and the wing leading edge at the MAC is:

$$\frac{C_{\mathrm{r}} - \mathrm{MAC}}{2} = \frac{4.066 - 3.162}{2} = 0.452\,\mathrm{m}$$

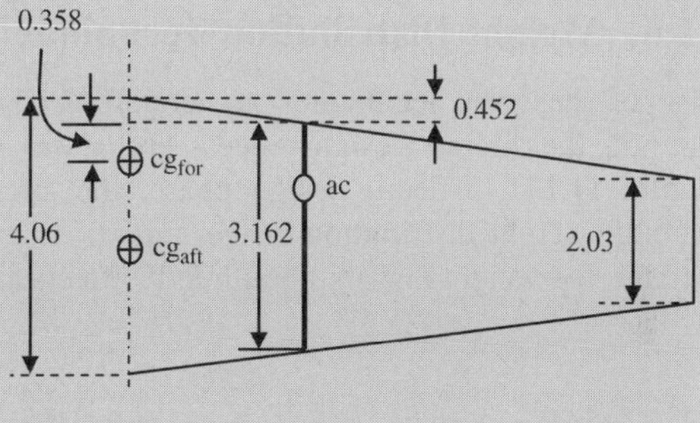

Figure 11.13 Half wing (numbers are in meters)

The distance between the aircraft forward cg and the wing leading edge at the MAC location is:

$$X_{\mathrm{cg}_{\mathrm{for}}} = x_{\mathrm{cg3}} - x_{\mathrm{LE}} - \frac{C_{\mathrm{r}} - \mathrm{MAC}}{2} = 12.607 - 11.8 - 0.452 = 0.358\,\mathrm{m}$$

Ultimately, the aircraft most forward cg in terms of MAC is:

$$\overline{X}_{cg_{for}} = h_{for} = \frac{x_{cg_{for}}}{MAC} = \frac{0.358}{3.162} = 0.112 \tag{11.12}$$

For the most aft cg, the distance between the aircraft aft cg and the wing leading edge at the MAC location is:

$$x_{cg_{aft}} = x_{cg_2} - x_{LE} - \frac{C_r - MAC}{2} = 13.483 - 11.8 - 0.452 = 1.585\,\text{m}$$

Finally, the aircraft most aft cg in terms of MAC is:

$$\overline{X}_{cg_{aft}} = h_{aft} = \frac{x_{cg_{aft}}}{MAC} = \frac{1.585}{3.162} = 0.503 \tag{11.13}$$

Thus, the aircraft most forward cg is located at 11.2% of the wing MAC, while the aircraft most aft cg is located at 50.3% of the wing MAC. The non-dimensional longitudinal cg range, or non-dimensional distance between the most aft and most forward cg, is:

$$\Delta x_{cg} = \frac{x_{cg_{aft}} - x_{cg_{for}}}{\overline{C}} = \frac{13.483 - 12.607}{3.162} = 0.391 \tag{11.16}$$

Therefore, the aircraft cg range is 39.1% MAC.

11.6 Weight Distribution Technique

One of the basic techniques an aircraft designer must master is the weight distribution. As the aircraft design process progresses, from early conceptual design phase (see Figure 11.14) to detail design phase, the aircraft weight is automatically distributed. Aircraft weight distribution is the apportioning of the weight within the aircraft. In an aircraft, the weight distribution directly affects a variety of flight characteristics, including

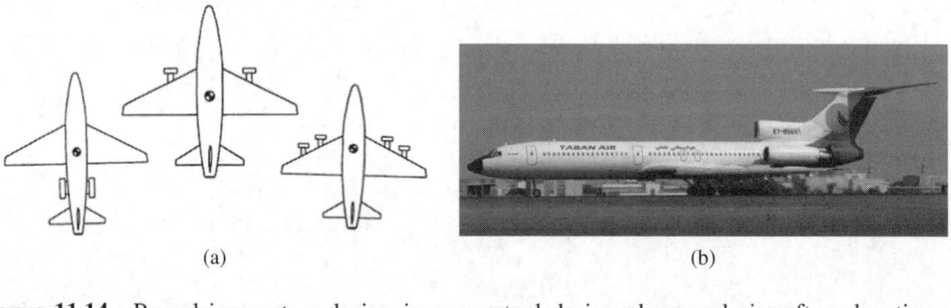

(a) (b)

Figure 11.14 Propulsion system design in conceptual design phase and aircraft cg location. (a) as the engines are moved aft, the aircraft cg will follow suit; (b) Tupolev Tu-154 with the most aft cg at about 50% MAC (empty weight). Part (b) reproduced from permission of A J Best airworthiness, stability, controllability, operating cost, and aircraft life. An ideal weight distribution will vary from aircraft to aircraft and from mission to mission. For example, the weight distribution for a transport aircraft will be different from that of a fighter. As Figure 11.14 shows, the selection of number of engines and engine locations will greatly influence the aircraft cg. Thus, the designer should study the effect of design decisions on the aircraft cg, since it is indirectly part of the aircraft weight distribution.

11.6.1 Fundamentals of Weight Distribution

In the airline industry, load balancing is often used to evenly distribute the weight of passengers, cargo, and fuel throughout an aircraft, so as to keep the aircraft's cg close to its wing/fuselage aerodynamic center to minimize the elevator deflection for longitudinal trim. In military transport aircraft, it is common to have a loadmaster as part of the crew members team; their responsibilities include calculating accurate load information for cg calculations, and ensuring cargo is properly placed and secured to prevent its shifting. In large aircraft, multiple fuel tanks and pumps are often used, so that as fuel is consumed, the remaining fuel can be positioned to keep the aircraft balanced, and to reduce stability problems associated with the free surface effect.

Basically, the term *free surface effect* implies a liquid under the influence of gravity and can cause vehicle instability. The free surface effect is a phenomenon which can cause an aircraft to become unstable and roll over. It refers to the tendency of liquids to slosh about – to move in response to changes in the attitude of an aircraft's cargo holds, decks, or fuel tanks in reaction to pilot-induced motions. In a refueling tanker aircraft, firefighting aircraft, or an aircraft with partially filled fuel tanks, any rolling motion is countered by a moment generated from the increased volume of water displaced by the tank on the lowered side. This assumes the center of gravity of the aircraft is relatively constant. If moving fuel inside the tank moves in the direction of the roll, this counters the righting effect by moving the center of gravity toward the lowered side.

The free surface effect can become a major problem in an aircraft with large partially full fuel or water tanks (e.g., refueling tanker aircraft, and firefighter aircraft). If a fuel tank is either empty or full, there is no change in the aircraft center of gravity as it pitches, rolls or yaws, or turns. However, if the fuel tank or water tank is only partially full, the liquid in the tank will respond to the aircraft's roll, pitch, and yaw. For instance, as an aircraft rolls to the left, fuel will displace to the left side of a tank and this will move the aircraft center of gravity toward the left. This has the effect of slowing the wing return to level. Also, the linear and angular momentum of large volumes of moving fuel causes significant forces, which act against the righting effect. When the wing returns to level, the roll continues and the effect is repeated on the opposite side.

In gusty atmospheric flight conditions, this can become a positive feedback loop, causing each roll to become more and more extreme, eventually overcoming the righting effect leading to an unstable situation (e.g., invert or stall). To mitigate this hazard, tanker aircraft use multiple smaller fuel tanks, instead of fewer larger ones, and possibly baffling within fuel tanks to minimize the free surface effects on the aircraft as a whole. Keeping individual fuel tanks either relatively empty or full is another way to minimize the effect and its attendant problems.

There are various ways to maintain the aircraft cg within allowable limits during flight operations. An in-flight technique is to transfer fuel among various fuel tanks. A nice example is the supersonic transport aircraft Concorde, whose first flight was in 1976 and which was in service for 27 years. The problem with Concorde was that as the aircraft accelerated to supersonic flight, the wing aerodynamic center moved backwards. To counteract this problem, Concorde had to move fuel from the forward compartments to the rearward compartments to keep the aircraft stable and balanced (so they were effectively changing the position of the center of gravity to balance the aircraft).

In general, design requirements that influence the decision on the aircraft weight/load distribution (i.e., cg location and limits) are: (i) controllability requirements, (ii) stability requirements, (iii) flying quality requirements, and (iv) operational requirements and constraints. However, various aircraft have different missions and requirements, so they do not have the same priority and similar constraints. The following is a comparison between three groups of aircraft to express what must be the basis in weight distribution.

11.6.1.1 Fighter Aircraft

For a fighter aircraft, the controllability requirements are much more significant than other design requirements (e.g., stability). The reasons are as follows:

1. The primary mission is to fight, which is fundamentally based on aircraft control (e.g., quick turn, fast pitch, and pull-up).
2. There is only one pilot (or sometimes two pilots).
3. A fighter pilot is normally stronger than a regular passenger and can also handle uncomfortable situations.
4. A fighter pilot has a parachute/ejection seat in case of accident or crash.
5. A fighter pilot is aware that his/her mission may face a hazard or accident.

11.6.1.2 Civil Transport Aircraft (Airliner)

For a passenger aircraft, the stability requirements are much more significant than other design requirements (e.g., controllability). The reasons are as follows:

1. As the name implies, a civil transport aircraft has a civil mission for citizen travel.
2. Airworthiness is the highest priority in an airliner. Even the low cost (i.e., annual profit) is assumed as the number two priority.
3. There are tens or hundreds of passengers and any accident/crash may require compensation and impose a cost to the airline.
4. Passengers are regular humans, and are not expected to handle uncomfortable situations.
5. The pilots/passengers do not have a parachute in case of accident or crash (in fact it is not allowed).
6. The passengers are not expecting to experience any accident and not ready for that situation.

11.6.1.3 General Aviation Aircraft

For a GA aircraft, the preference of stability over controllability or controllability over stability depends on the aircraft mission. However, due to the low number of seats (less than 19), low cost is often the highest priority. For some aircraft such as a business jet luxury aircraft designed and manufactured for Very Important People (VIP and wealthy passengers), the top priority is passenger comfort and super handling qualities.

In weight distribution, the location of some items and components such as the pilot, passengers, and tail is self-evident. Other components – such as the wing, fuel tanks, engines, stores, and cargo – can be shifted around to a certain extent to achieve the desired cg location. Table 11.9 illustrates recommended cg locations for various aircraft in terms of percentage wing MAC. The table also recommends a cg range for various aircraft. These are non-dimensional numbers. As an aircraft gets larger, the cg range in meters will be larger too, since the aircraft MAC gets longer. For instance, the MAC of a light GA aircraft Cessna 182 (Figure 3.7) is 1.5 m, while that of a large transport aircraft Boeing 747 (Figures 3.7, 3.12, and 9.4) is 8.32 m. Thus, a 15% MAC cg range for a Cessna 182 means 0.224 m, while that for the Boeing 747 means 1.248 m. In the following Sections 11.6.2–11.6.4, the details of longitudinal stability requirements, longitudinal controllability requirements, and longitudinal handling quality requirements are briefly reviewed. Figure 11.15 shows three aircraft (Gulfstream G550, Aerospatiale-BAC Concorde, and Cessna 172) with different cg features.

11.6.2 Longitudinal Stability Requirements

For a civil aircraft, the static longitudinal stability criterion requires that the center of gravity is never allowed to be behind the aircraft neutral point or aircraft aerodynamic center (X_{np}). In terms of non-dimensional derivative, the rate of change of pitching moment with respect to angle of attack must be negative for an aircraft to be statically longitudinally stable. This derivative is determined by:

$$C_{m_\alpha} = C_{L_\alpha}\left(\overline{X}_{\text{cg}} - \overline{X}_{\text{np}}\right) \tag{11.17}$$

Table 11.9 Recommended longitudinal cg locations for various aircraft

No.	Aircraft	Forward cg (% MAC)	Aft cg (% MAC)	Range (% MAC)
1	GA-subsonic	15–20	25–30	5–15
2	Subsonic transport	5–20	20–35	10–30
3	Supersonic transport	15–35	40–60	20–40
4	Fighter – subsonic speeds	15–20	35–45	15–30
5	Fighter – supersonic speeds	45–50	50–55	10–30

(a)

(b) (c)

Figure 11.15 Three aircraft with different cg features. (a) twin turbofan Gulfstream G550 with a maximum take-off mass of 41 277 kg, forward cg at 21% MAC, aft cg at 45% MAC, and a cg range of 24% MAC; (b) supersonic transport aircraft Aerospatiale-BAC Concorde with a maximum take-off mass of 185 700 kg, forward cg at 20% MAC, aft cg at 59% MAC, and a cg range of 39% MAC; (c) light GA aircraft Cessna 172 with a maximum take-off mass of 1111 kg, forward cg at 15.6% MAC, aft cg at 36.5% MAC, and a cg range of 20.9% MAC. Reproduced from permission of (a) Gulfstream; (b) A J Best

This criterion dictates that for a statically longitudinally stable aircraft, the most aft location of the cg must be forward of the aircraft neutral point. Equation (11.17) indicates that the location of the aircraft cg will directly influence longitudinal static stability. The static margin (SM) is defined as the non-dimensional difference between the aircraft center of gravity and the aircraft neutral point:

$$\mathrm{SM} = \frac{x_{\mathrm{np}} - x_{\mathrm{cg}}}{\overline{C}} \tag{11.18}$$

The longitudinal dynamic stability criterion requires that the real parts of the roots of the longitudinal characteristic equation are all negative. Although several factors would affect the magnitude of the real part of the longitudinal characteristic equation, in the majority of conventional aircraft the pitch-damping derivative (C_{m_q}) is the most dominating factor. The pitch-damping derivative [7] is proportional to the square of the moment arm of the horizontal tail:

$$C_{m_q} = -2C_{L_{\alpha_h}} \overline{V}_{\mathrm{H}} \left(\overline{x}_{\mathrm{ac_h}} - \overline{x}_{\mathrm{cg}}\right) \tag{11.19}$$

where $C_{L_{\alpha_h}}$ is the horizontal tail lift curve slope, $\overline{V}_{\mathrm{H}}$ is the horizontal tail volume coefficient, $\overline{x}_{\mathrm{ac_h}}$ is the non-dimensional location of the horizontal tail aerodynamic center, and

$\overline{x}_{cg}$ is the non-dimensional location of the aircraft cg. The derivative C_{m_q} must also be negative and much larger than C_{m_α}. The horizontal tail volume coefficient is defined by:

$$\bar{V}_H = \frac{S_h l_h}{S\overline{C}} \tag{11.20}$$

where S_h is the horizontal tail planform area, l_h is the horizontal tail arm to aircraft cg, and S and C are respectively the wing reference area and wing MAC. The combination of Equations (11.19) and (11.20) reflects that the increased tail area would increase the weight of the aircraft as well as moving the cg rearward. The reason is that a conventional tail is located at the rear of the fuselage.

A similar discussion applies for a canard, since it is located forward of the fuselage. Thus, the increased canard area would increase the weight of an aircraft as well as moving the cg forward. Hence, increasing the tail or canard area is twofold. The tail canard size would influence the aircraft neutral point, while the tail or canard size would move the cg negatively. In the majority of aircraft, a tail volume coefficient of more than 0.3, along with a positive SM, yields a dynamically stable aircraft.

Equation (11.19) implies that the location of the aircraft cg will directly influence the longitudinal dynamic stability, since C_{m_q} is a function of the aircraft center of gravity (x_{cg}). This non-dimensional longitudinal dynamic stability derivative is then converted to the dimensional stability derivative M_q. This dimensional derivative plus other longitudinal derivatives such as M_α and $M_{\alpha dot}$ will dominate the dynamic longitudinal stability. Increasing $M_q + M_{\alpha dot}$ will increase the damping ratio of short-period mode, while increasing M_α will increase the frequency of short-period mode. The damping and frequency of both short- and long-period modes could be determined in terms of stability derivatives. In summary, two requirements of longitudinal stability of an aircraft are as follows:

$$\overline{V}_H > 0.3 \tag{11.21}$$

$$x_{np} - x_{cg} > 0 \tag{11.22}$$

These two numerical requirements will be utilized later as a basis for aircraft weight distribution through the longitudinal center of gravity range.

In the case of a fighter aircraft, it is beneficial to make the aircraft longitudinally unstable to improve the longitudinal controllability and maneuverability. Thus, for a fighter, it is desired that the aircraft cg is behind the aircraft neutral point. Since it is very hard for an unstable aircraft to be controlled by a human pilot, an automatic flight control system (i.e., autopilot) is required. The stealth bomber Northrop Grumman B-2 Spirit (Figure 6.8) has an SM of −0.1, which implies it is longitudinally unstable. Therefore, the aircraft is equipped with a stability augmentation system. Another example is the experimental aircraft X-29 with an SM of −0.35 to investigate the aircraft controllability under extreme conditions. The static and dynamic longitudinal stability criteria must be followed in the aircraft weight distribution process.

11.6.3 Longitudinal Controllability Requirements

The longitudinal controllability criterion requires the center of gravity to not be allowed forward of a specific location such that the aircraft is longitudinally controllable. In the majority of cases, it is translated to take-off rotation requirements. This criterion dictates the most forward location of the cg. For an aircraft with a tricycle landing gear, the forward position of the cg must be at a location such that the elevator is able to rotate the aircraft about the main gear and lift the nose when the aircraft has obtained 80% of its take-off speed. The initial angular acceleration about the main gear (rotation point) should have a value of 6–8 deg/s^2 [8]. Then a constant angular velocity of 2–3 deg/s should be maintained such that the take-off rotation process does not take more than 3–4 seconds.

Furthermore, in a conventional aircraft with a tricycle landing gear, when the aircraft is on the ground, the aircraft cg should not be located aft of the main gear. Otherwise, the aircraft will tip on its tail and the rear fuselage/tail will be damaged. A B-1B bomber suffered a mishap at Ellsworth AFB on October 22, 2000, when a faulty fuel pump caused fuel to migrate from the forward section to the aft section of the aircraft. Thus, the cg exceeded the aft cg limit and caused the aircraft to tip on its tail. Damage was reported to be minor.

The elevator deflection (δ_E) to rotate an aircraft during take-off is a function of four longitudinal derivatives (C_{L_α}, C_{m_α}, $C_{L_{\delta E}}$, $C_{m_{\delta E}}$), plus aircraft lift coefficient (C_L), and pitching moment coefficient (C_{m_o}):

$$\delta_E = -\frac{C_{L_\alpha} C_{m_o} + C_{m_\alpha} C_L}{C_{L_\alpha} C_{m_{\delta_E}} - C_{L_{\delta_E}} C_{m_\alpha}} \tag{11.23}$$

The status of C_{m_α} was described in previous sections. Two derivatives ($C_{L_{\delta_E}}$ and $C_{m_{\delta_E}}$) are referred to as control power derivatives. Both $C_{L_{\delta_E}}$ (variation of lift coefficient versus elevator deflection) and $C_{m_{\delta_E}}$ (variation of pitching moment coefficient versus elevator deflection) are direct functions of elevator geometry. Elevator geometry (i.e., elevator area, elevator chord, elevator span, and elevator hinge location) must be such that sufficient control power is generated in extreme cases. Two control power derivatives are determined as follows:

$$C_{L_{\delta_E}} = \frac{S_h}{S} \frac{dC_{L_t}}{d\delta_E} \tag{11.24}$$

$$C_{m_{\delta_E}} = -V_H \frac{dC_{L_t}}{d\delta_E} \tag{11.25}$$

Equations (11.24) and (11.25) indicate that the location of the aircraft cg along with the elevator geometry, wing area, and wing chord will directly influence the aircraft longitudinal controllability. The angular acceleration about the aircraft main gear rotation point, $\ddot{\theta}_{mg}$, should have a value such that the take-off rotation does not take more than 3–4 seconds. Applying the moment equation about the main gear will yield the governing equation that can be used to calculate the angular acceleration. Full governing equations for take-off rotation are introduced in Chapter 9.

Weight distribution contributes to aircraft spin recovery features of a spinnable aircraft. It is very important that the weight is distributed such that the aircraft moments of inertia are anti-spin. The magnitude and sign of the inertia term $\left(\left(I_{xx} - I_{yy}\right)/I_{zz}\right)$ greatly influences the effectiveness of the rudder and consequently spin recovery. When the magnitudes of pitch (I_{yy}) and roll (I_{xx}) inertia are close, the effect of the inertia term is small, and hence the rudder will be the primary control for spin recovery. But whenever the inertia term becomes significant, it has a considerable impact on the spin motion and thus the size of the rudder could be smaller. The controllability requirements will be utilized later as another basis for the aircraft weight distribution process.

11.6.4 Longitudinal Handling Quality Requirements

The handling or flying qualities in the case of an aircraft flown by human pilots requires that the interaction between pilot cockpit control inputs and aircraft response to the cockpit control inputs must be such that the pilot can achieve the mission objectives with reasonable physical and mental effort. In other words, the aircraft is required to have acceptable handling qualities anywhere inside the operational flight envelope. Human factors are significant in aircraft handling qualities. The flying qualities of an aircraft are related to those stability and control characteristics that are important in forming the pilot's impression of the aircraft. One of the parameters that shapes the flight envelope is the aircraft longitudinal cg limit.

Fundamentally, the flying qualities of an aircraft must be such that [4] the following characteristics are present anywhere inside the operational flight envelope:

1. The aircraft must have sufficient control power to maintain steady-state, straight-line flight as well as steady-state maneuvering flight consistent with mission objectives.
2. The aircraft must be maneuverable from one steady-state flight condition to another.
3. The aircraft must have sufficient control power to accomplish the following transitions: (i) transition from ground operations to airborne operations (take-off, lift-off, and initial steady-state climb), (ii) transition from airborne operations to ground operations (steady-state approach, touchdown, and landing).

In the case of military aircraft, these three characteristics must be present with certain asymmetrical weapon and/or store loading as well as under certain conditions of combat damages. Examples of flying quality parameters which deal with physical pilot effort are maximum required stick force, stick force per g, and stick force/speed gradient. The flying quality requirements are presented in the two references MIL-F-8785C [9] and MIL-STD-1797A [8]. Flying qualities depend upon the aircraft classes (I, II, III, or IV), flight phase categories (terminal (C) or non-terminal (A, B)), and levels of acceptability[1] (1, 2, or 3). The longitudinal control (stick) force is a critical parameter in handling qualities and must be less than the comfort limit of the human pilot. FAR 23, FAR 25 [10], and MIL-STD each have different requirements considering the aircraft type and mission objectives.

The longitudinal response of an aircraft to a gust may be modeled by two simultaneous modes of oscillation: short-period and long-period. A short-period oscillation for a dynamically longitudinally stable aircraft takes about a few seconds, while the long period takes about a few minutes. Two parameters that will be impacted by aircraft cg location are the short-period damping ratio (ζ_{sp}) and the short-period undamped natural frequency (ω_{n_sp}). MIL-F-8785C [9] requires that the short-period damping ratio of the short-period longitudinal mode be within the limits presented in Table 11.10. In addition, MIL-F-8785C also requires that the short-period undamped natural frequency of the

[1] Aircraft classes, flight phase categories, and levels of acceptability are presented in Chapter 12.

Table 11.10 Short-period damping ratio limits [5]

Level	Category A and C flight phase	Category B flight phase
Level 1	$0.35 < \zeta_{sp} < 1.3$	$0.3 < \zeta_{sp} < 2.0$
Level 2	$0.25 < \zeta_{sp} < 2.0$	$0.2 < \zeta_{sp} < 2.0$
Level 3	$0.15 < \zeta_{sp}$	$0.15 < \zeta_{sp}$

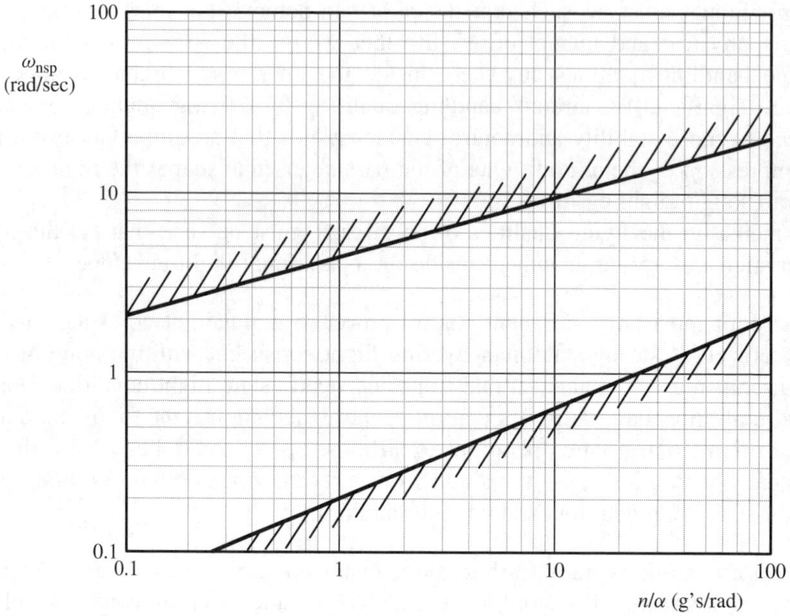

Figure 11.16 Level 1 short-period undamped natural frequency requirements

short-period longitudinal mode be within the limits shown in Figure 11.16. This figure only illustrates the requirements for flight phase B. Other flight phase requirements can be found in Ref. [4].

The exact technique to determine the frequency and damping ratio of the short-period mode can be found in most flight dynamics textbooks. An approximation to the short-period mode is assumed reasonable due to its acceptable accuracy. Based on this approximation [4], the undamped natural frequency of the short-period mode is a function of aircraft cg as follows:

$$\omega_{n_{sp}} = \sqrt{-\frac{\rho V^2 S \overline{C} C_{m_\alpha}}{2 I_{yy}}} \tag{11.26}$$

The two parameters of C_{m_α} and I_{yy} (aircraft mass moment of inertia about the y-axis) are functions of aircraft cg. The handling or flying quality requirements (Table 11.10

and Figure 11.15) must be employed as another basis for the aircraft weight distribution process.

11.7 Aircraft Mass Moment of Inertia

Aircraft controllability and maneuverability is a function of several factors, including aircraft mass moment of inertia. In contrast, the weight distribution will greatly influence the aircraft mass moment of inertia. Thus, the aircraft designer should be careful to select a configuration, and design aircraft components, such that they yield the desirable mass moments of inertia. Since an aircraft has three rotational axes, namely x, y, and z, there are generally three moments of inertia. Weight distribution along the x-axis affects the mass moment of inertia about the y- and z-axes, and consequently influences the aircraft pitch (longitudinal) and yaw (directional) control. Weight distribution along the y-axis affects the mass moment of inertia about the x- and z-axes, and consequently influences the aircraft roll (lateral) and yaw (directional) control. Weight distribution along the z-axis affects the mass moment of inertia about the x- and z-axes, and consequently influences the aircraft pitch (longitudinal) and roll (lateral) control. In this section, the technique to compute aircraft moments of inertia about the x-, y-, and z-axes is addressed.

The mass moment of inertia provides information on how easy or difficult it is (how much inertia there is) to rotate an object around a given axis. The mass moment of inertia is one measure of the distribution of the mass of an object relative to a given axis. The mass moment of inertia is denoted by I and is given for a single rigid object of mass m as:

$$I = mR^2 \tag{11.27}$$

where R is the perpendicular distance between the mass and the axis of rotation. The mass moment of inertia has the unit of mass times length squared. The mass moment of inertia should not be confused with the area moment of inertia, which has the unit of length to the power four. The mass moment of inertia usually appears naturally in the equations of motion, whereas the area moment of inertia appears in the bending stress equation of a beam under a bending load.

To calculate the mass moment of inertia of an object consisting of n particles, each having a mass of dm, integration is used to sum the moment of inertia of each dm to get the mass moment of inertia of the entire body:

$$I = \int_1^n R^2 dm \tag{11.28}$$

Table 11.11 illustrates the mass moment of inertia for several standard geometries. When the axis of rotation is different from the object's center of gravity, the parallel-axis theorem is employed to transfer the moment of inertia from object cg to the axis of rotation. The moment of inertia around any axis can be calculated from the moment of inertia around the parallel axis which passes through the center of mass. The equation to calculate this is called the parallel-axis theorem, and is given as:

$$I_O = I_C + md^2 \tag{11.29}$$

Table 11.11 Mass moment of inertia of aircraft components [11, 12]

No.	Aircraft component	Component model	Geometry, axis	Mass moment of inertia
1	Wing, horizontal tail, vertical tail	Rectangular prism or thin plate of thickness *t*, length *b*, chord C	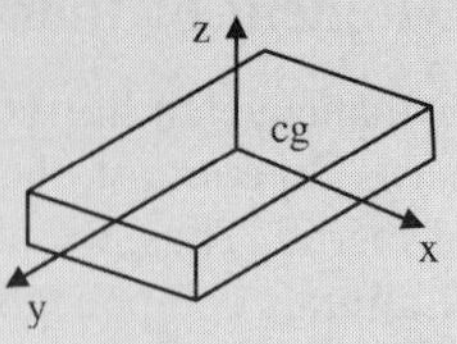	$I_{xx} = \frac{1}{12}m\left(b^2 + t^2\right)$ $I_{yy} = \frac{1}{12}m\left(t^2 + C^2\right)$ $I_{zz} = \frac{1}{12}m\left(b^2 + C^2\right)$
2	Fuselage	Thin cylindrical shell of radius *r* and length *L*	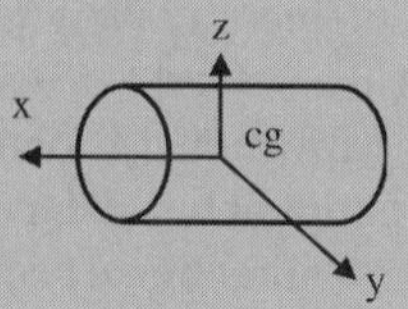	$I_{xx} = mr^2$ $I_{yy} = \frac{m}{12}\left(6r^2 + L^2\right)$ $I_{zz} = \frac{m}{12}\left(6r^2 + L^2\right)$
3	Engine	Solid cylinder of radius *r*, length *L*	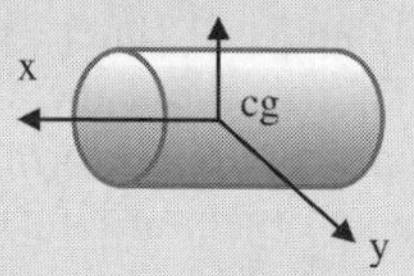	$I_{xx} = \frac{1}{2}mr^2$ $I_{yy} = \frac{1}{12}m\left(3r^2 + L^2\right)$ $I_{zz} = \frac{1}{12}m\left(3r^2 + L^2\right)$
4	Propeller	Slender rod of length *L*	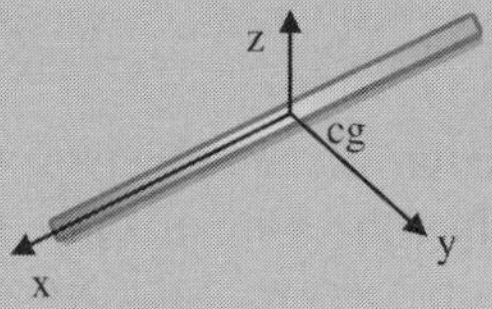	$I_{xx} = 0$ $I_{yy} = \frac{1}{12}mL^2$ $I_{zz} = \frac{1}{12}mL^2$
5	Human pilot, passenger, seat, fuel tank, miscellaneous items	Point mass of mass *m*	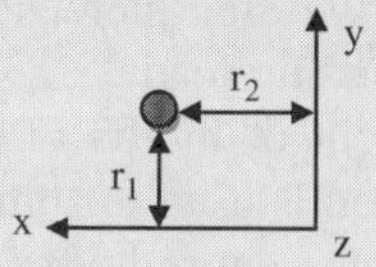	$I_{xx} = mr_1^2$ $I_{yy} = mr_2^2$ $I_{zz} = mr_3^2$ $r_3 = \sqrt{r_1^2 + r_2^2}$

where I_O and I_C are the object mass moments of inertia about the axis of rotation and the object cg respectively. The parameter d is the distance between the axis of rotation and the axis passing through the center of gravity of the object, and m is the mass of the object.

An aircraft has a complex geometry and so it is very hard to employ the integration (Equation (11.28)) to determine the aircraft mass moment of inertia. In order to calculate the aircraft mass moment of inertia, one must model the aircraft shape with a number of standard geometries such as the sphere, rod, cylinder, and prism. For instance, a fuselage could be modeled as a thin cylindrical shell or a hollow circular cylinder, and a wing as a rectangular thin plate or prism. Since the cg of each component does not coincide

Table 11.12 Body-axis mass moments of inertia for several aircraft

No.	Aircraft	m_{TO} (kg)	I_{xx} (kg m^2)	I_{yy} (kg m^2)	I_{zz} (kg m^2)	I_{xz} (kg m^2)
1	Cessna 182	1 200	1 285	1 825	2 667	0
2	Beech 99	4 990	20 593	27 455	46 290	5926
3	Cessna 620	6 800	87 872	23 455	87 508	0
4	McDonnell F-4	15 100	32 132	159 300	181 300	2170
5	Boeing 747-200	288 500	24 675 886	44 877 574	67 384 152	1 315 143

with the aircraft cg, the parallel-axis theorem must also be utilized. Table 11.12 illustrates body-axis moments of inertia for several aircraft. Examples 11.3 and 11.4 demonstrate one application of the technique.

Example 11.3: Fuselage Mass Moment of Inertia

Problem statement: Consider a twin-turbofan medium-haul transport aircraft (Figure 11.17) with a maximum take-off mass of 45 000 kg and wing area of 94 m^2. The fuselage has a length of 32 m, a maximum diameter of 3.3 m, and a mass of 3600 kg. Model the fuselage as a thin cylindrical shell and determine its longitudinal, lateral, and directional mass moment of inertia about the aircraft cg. Assume that the aircraft cg is along the fuselage center line and located 15.2 m from the fuselage nose.

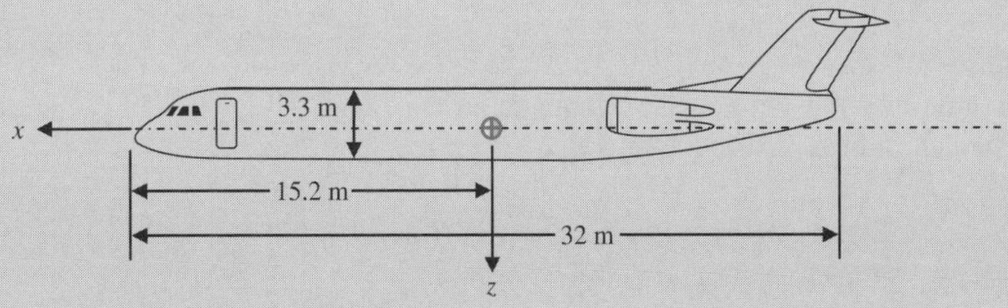

Figure 11.17 The geometry of the aircraft in Example 11.3

Solution:
The fuselage is modeled as a thin cylindrical shell, so using Table 11.11 the following equations are used to obtain the fuselage mass moments of inertia about its own cg:

$$I_{xx_f} = mr^2 = 3600 \cdot \left(\frac{3.3}{2}\right)^2 = 9801\ \text{m}^2\ \text{kg (Table 11.11)}$$

$$I_{yy_f} = \frac{m}{12}\left(6r^2 + L^2\right) = \frac{3600}{12} \cdot \left[6 \cdot \left(\frac{3.3}{2}\right)^2 + 32^2\right] = 312\,100.5\ \text{m}^2\ \text{kg (Table 11.11)}$$

$$I_{zz_f} = I_{yy_f} = 312\,100.5\ \text{m}^2\ \text{kg (Table 11.11)}$$

The fuselage cg is along its center line, so it coincides with the aircraft cg along the x-axis. Hence, the fuselage mass moment of inertia about the x-axis is 9801 m^2 kg. But, since the fuselage cg about the y- and z-axes does not coincide with the aircraft cg, the parallel-axis theorem must be used to transfer the other two mass moments of inertia to those of the aircraft cg. The fuselage cg is located at 32/2 = 16 m from the nose. Thus:

$$d = \frac{32}{2} - 15.2 = 0.8\,\text{m}$$

Thus, the fuselage cg is 0.8 m aft of the aircraft cg:

$$\begin{aligned} I_{yy} &= I_{yy_f} + md^2 = 312100.5 + 3600 \cdot (0.8)^2 = 314404.5\,\text{m}^2\text{kg} \\ I_{zz} &= I_{zz_f} + md^2 = 312100.5 + 3600 \cdot (0.8)^2 = 314404.5\,\text{m}^2\text{kg} \end{aligned} \tag{11.29}$$

The fuselage moment of inertia about the y- and z-axes is about 32 times greater than that about the x-axis. Thus, it is much harder to apply the yaw and pitch control than the roll control on this fuselage.

Example 11.4: Wing Mass Moment of Inertia

The wing (see Figure 11.18) of an aircraft has the following features:

$$S = 12\ \text{m}^2,\ \text{AR} = 8,\ \text{airfoil section: NACA 23015},\ m_W = 100\ \text{kg}$$

Determine the wing mass moment of inertia about three aircraft axes (i.e., I_{xx}, I_{yy}, I_{zz}).

Solution:

We assume that the wing is a rectangular prism (Figure 11.19). The wing geometry must be determined first.

$$\text{Span: AR} = \frac{b^2}{S} \Rightarrow b = \sqrt{\text{AR} \cdot S} = \sqrt{8 \cdot 12} \Rightarrow b = 9.8\,\text{m} \tag{5.19}$$

$$\text{Chord: } S = b \cdot \overline{C} \Rightarrow \overline{C} = \frac{S}{b} = \frac{12}{9.8} = 1.225\,\text{m} \tag{5.18}$$

According to the wing airfoil section notation (23015), the wing maximum thickness-to-chord ratio is 15% (i.e., the last two digits). The wing average thickness is assumed to be half of the wing maximum thickness-to-chord ratio.

Thickness: $(t/C)_{\max} = 15\% \Rightarrow t = 0.5 \cdot 0.15 \cdot C = 0.5 \cdot 0.15 \cdot 1.225 = 0.092\,m$

1. x-axis:

$$I_{xx_G} = \frac{1}{12} m \left(a^2 + b^2\right) = \frac{1}{12} \cdot 50 \cdot \left(9.8^2 + 0.092^2\right) = 800.1\,\text{kg m}^2$$

$$= 590.1\ \text{slug ft}^2\ \text{(Table 11.11)}$$

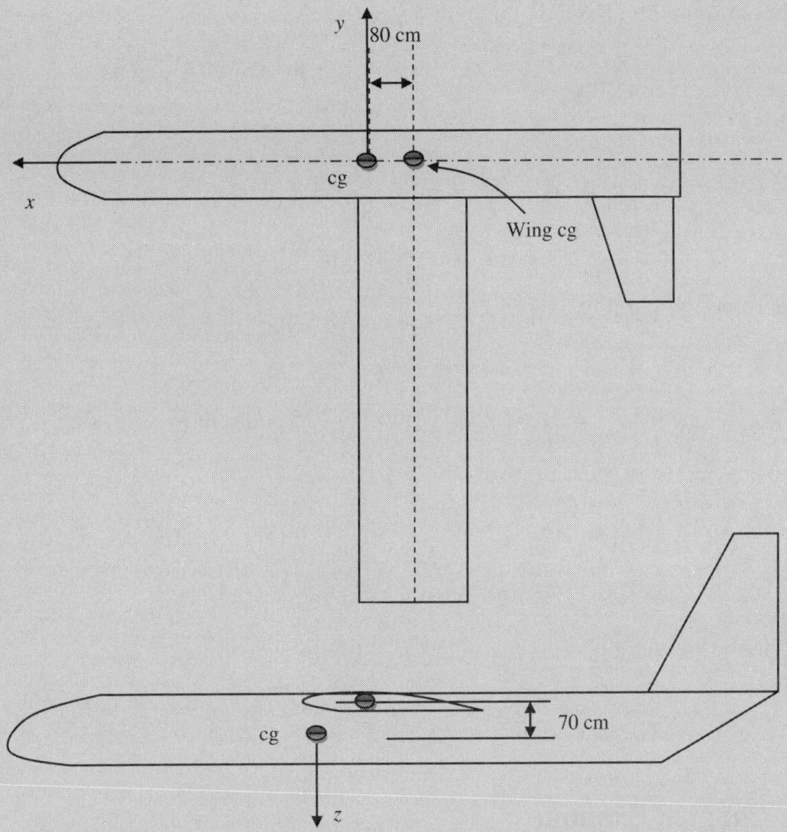

Figure 11.18 The geometry of the aircraft in Example 11.4

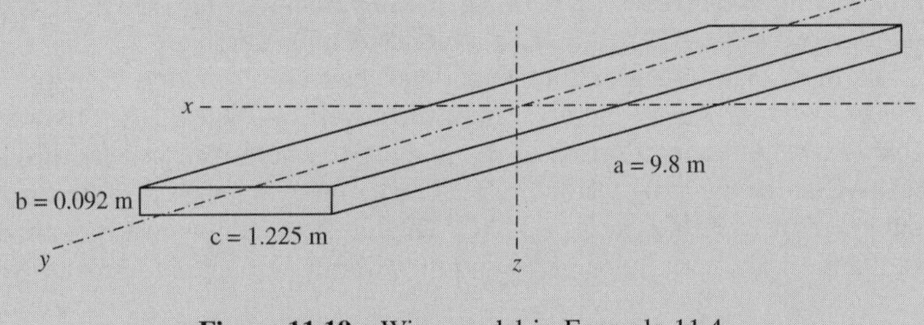

Figure 11.19 Wing model in Example 11.4

Applying the parallel-axis theorem ($d = 0$):

$$I_{xx} = I_{xx_G} + md^2 = 800.1\,\text{kg m}^2 = 590.1\,\text{slug ft}^2$$

2. y-axis:

$$I_{yyG} = \frac{1}{12}m\left(b^2 + c^2\right) = \frac{1}{12}\cdot 50\cdot\left(0.092^2 + 1.225^2\right)$$
$$= 12.57\,\text{kg m}^2 = 9.27\,\text{slug ft}^2\ \text{(Table 11.11)}$$

Applying the parallel-axis theorem:

$$d = \sqrt{0.7^2 + 0.8^2} = 1.06\,\text{m}$$
$$I_{yy} = I_{yy_G} + md^2 = 12.57 + 50\cdot 1.06^2 = 125.57\,\text{kg m}^2 = 92.6\,\text{slug ft}^2 \quad (11.29)$$

3. z-axis:

$$I_{zzG} = \frac{1}{12}m\left(a^2 + c^2\right) = \frac{1}{12}\cdot 50\cdot\left(9.8^2 + 1.225^2\right) = 812.5\,\text{kg m}^2$$
$$= 599.3\,\text{slug ft}^2\ \text{(Table 11.11)}$$

Applying the parallel-axis theorem ($d = 0.8$ m):

$$I_{zz} = I_{zz_G} + md^2 = 812.5 + 50\cdot(0.8)^2 = 844.5\,\text{kg m}^2 = 622.9\,\text{slug ft}^2 \quad (11.29)$$

11.8 Chapter Example

Example 11.5: Weight Distribution

Problem statement: Consider a two-seat (tandem) single-turboprop trainer aircraft with a maximum take-off mass of 2500 kg, a wing area of 17 m^2, and an overall length of 10 m (Figure 11.20). The wing has a straight rectangular shape with an aspect ratio of 6.5. The mass of each major component and their corresponding centers of gravity from the propeller spinner (reference line) are given in Table 11.13. Two fuel tanks are located inside the wing in front of the main spar where their centers of gravity are at 20% MAC aft of the wing leading edge. Locate the wing such that the aircraft cg at maximum take-off weight is at 30% MAC. Then determine the most forward cg of the aircraft, the most aft cg of the aircraft, and the aircraft cg range in terms of percentage MAC.

Solution:

The locations of all components are given except wing and fuel tank. Since fuel tanks are inside the wing, when the wing location is determined, the fuel tanks are automatically obtained.

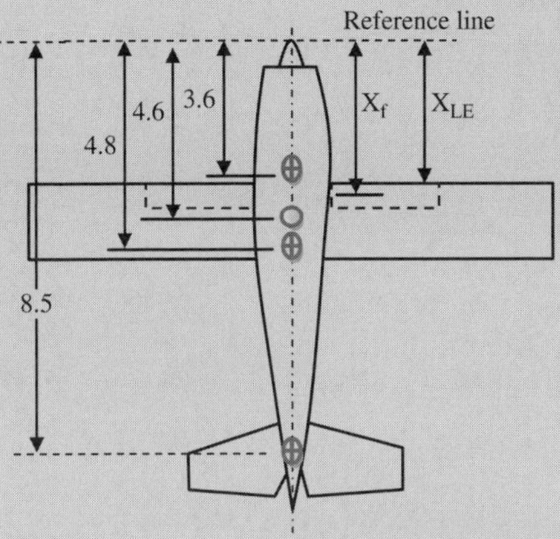

Figure 11.20 Aircraft of Example 11.5 (numbers are in meters)

Table 11.13 Mass and cg locations of various components of the aircraft in Example 11.5

No.	Component	Mass (kg)	Symbol	X_{cg} (m)
1	Engine	220	m_e	1.8
2	Fuselage	320	m_b	4.6
3	Pilot 1	90	m_{P1}	3.6
4	Pilot 2	90	m_{P2}	4.8
5	Landing gear	130	m_{lg}	3.5
6	Tails	70	m_t	8.5
7	Systems	655	m_s	4.1
8	Wing	375	m_w	$X1$
9	Fuel	550	m_f	$X2$

- **Step 1. Wing MAC**. The wing is expressed as a straight rectangle, so the wing root chord, wing tip chord, and wing MAC are all the same:

$$\mathrm{AR} = \frac{b^2}{S} \Rightarrow b = \sqrt{S \cdot \mathrm{AR}} = \sqrt{17 \cdot 6.5} \Rightarrow b = 10.512\,\mathrm{m} \tag{5.19}$$

$$\mathrm{AR} = \frac{b}{\overline{C}} \Rightarrow \overline{C} = \frac{b}{\mathrm{AR}} = \frac{10.512}{6.5} \Rightarrow \overline{C} = 1.617\,\mathrm{m} \tag{5.17}$$

- **Step 2. Aircraft cg**. The aircraft cg is a function of the wing cg and fuel cg:

$$X_{cg} = \frac{\sum_{i=1}^{9} m_i x_{cg_i}}{\sum_{i=1}^{9} m_i} = \frac{\begin{bmatrix} m_e x_e + m_b x_b + m_{P1} x_{P1} + m_{P2} x_{P2} + m_t x_t \\ + m_g x_g + m_s x_s + m_w x_w + m_f x_f \end{bmatrix}}{m_e + m_b + m_{P1} + m_{P2} + m_t + m_g + m_s + m_w + m_f} \tag{11.1}$$

$$x_{cg} = \frac{\begin{bmatrix} (22 \cdot 1.8) + (320 \cdot 1.8) + (90 \cdot 3.6) + (90 \cdot 4.8) + (70 \cdot 8.5) \\ + (130 \cdot 3.5) + (655 \cdot 4.1) + (375 \cdot x_w) + (550 \cdot x_f) \end{bmatrix}}{220 + 320 + 90 + 90 + 70 + 130 + 655 + 375 + 550}$$

$$x_{cg} = \frac{5463.5 + (375 x_w) + (550 x_f)}{2500} \tag{11.30}$$

- **Step 3. Wing and fuel cg relationship**. We need to derive a relationship between fuel cg and wing cg. According to Table 11.2, the wing cg is at 35–42% MAC. A value of 40% MAC is selected.

$$x_w = x_{LE} + 0.4C = x_{LE} + 0.4 \cdot 1.617 = x_{LE} + 0.647 \tag{11.31}$$

Furthermore, the fuel cg is at 20% wing MAC. Thus:

$$x_f = x_{LE} + 0.2C = x_{LE} + 0.2 \cdot 1.617 = x_{LE} + 0.323 \tag{11.32}$$

Equations (11.31) and (11.32) are inserted into Equation (11.30):

$$x_{cg} = \frac{5463.5 + [375(x_{LE} + 0.647)] + [550(x_{LE} + 0.323)]}{2500} \tag{11.33}$$

However, the aircraft cg is required to be at 30% MAC, so:

$$x_w = x_{LE} + 0.3C \tag{11.34}$$

Inserting Equation (11.34) into Equation (11.33) yields:

$$x_{LE} + 0.3C = \frac{5463.5 + [375(x_{LE} + 0.647)] + [550(x_{LE} + 0.323)]}{2500} \tag{11.35}$$

- **Step 4. Wing location**. Equation (11.35) has only one unknown. Solving this equation results in:

$$x_{LE} = 2.966\,\text{m}$$

$$x_w = x_{LE} + 0.647 = 2.966 + 0.647 = 3.613\,\text{m} \tag{11.31}$$

Thus, the wing leading edge must be 3.613 m behind the reference line in order for the aircraft cg to be at 30% MAC at take-off weight.

- **Step 5. Fuel location**.

$$x_f = x_{LE} + 0.323 = 3.613 + 0.323 = 3.289\,\text{m} \tag{11.32}$$

- **Step 6. Aircraft cg location**. The wing and fuel cg locations are inserted into Equation (11.30):

$$x_{cg} = \frac{5463.5 + (375x_w) + (550x_f)}{2500} = \frac{5463.5 + (375 \cdot 3.613) + (550 \cdot 3.289)}{2500} \tag{11.30}$$

$$\Rightarrow x_{cg} = 3.451\,\text{m}$$

- **Step 7. Aircraft most forward cg**. In order to determine the aircraft most forward cg, all removable items aft of the aircraft cg at maximum take-off weight must be removed. Only fuel and one pilot can be removed. The locations of all components are listed in Table 11.14 in order. As can be seen, the pilot in the front seat and the fuel are aft of the aircraft cg and must be removed:

$$x_{cg_3} = \frac{\sum_{j=1}^{n-k2} x_{cg_j} m_j}{\sum_{j=1}^{n} m_j - \sum_{j=1}^{k2} m_j} \tag{11.15}$$

So, $n = 9$ and $k2 = 2$:

$$X_{cg_{for}} = \frac{\sum_{j=1}^{9-2} x_{cg_j} m_j}{\sum_{j=1}^{9} m_j - \sum_{j=1}^{2} m_j} = \frac{m_e x_e + m_b x_b + m_{P2} x_{P2} + m_t x_t + m_g x_g + m_s x_s + m_w x_w}{m_e + m_b + m_{P2} + m_t + m_g + m_s + m_w} \quad (11.15)$$

$$x_{cg_{for}} = \frac{\left[\begin{array}{c}(22 \cdot 1.8) + (320 \cdot 1.8) + (90 \cdot 4.8) + (70 \cdot 8.5) + (130 \cdot 3.5) \\ + (655 \cdot 4.1) + (375 \cdot 3.613)\end{array}\right]}{220 + 320 + 90 + 70 + 130 + 655 + 375}$$

$$\Rightarrow x_{cg_{for}} = 3.191\ \text{m}$$

- **Step 8. Aircraft most aft cg**. By looking at Table 11.14, it is observed that there is no removable item in front of the aircraft cg at maximum take-off weight, thus the cg at maximum take-off weight is the aircraft most aft cg:

$$\Rightarrow x_{cg_{aft}} = 3.451\ \text{m}$$

Table 11.14 The cg of components in Example 11.5

No.	Component	X_{cg} (m)
1	Engine	1.8
	Aircraft	**3.451**
2	Landing gear	3.5
3	Fuel	3.597
4	Pilot 1	3.6
5	Wing	3.921
6	Systems	4.1
7	Fuselage	4.6
8	Pilot 2	4.8
9	Tails	8.5

- **Step 9. Aircraft most forward and aft cg in terms of percentage MAC**.

$$h_{for} = \overline{X}_{cg_{for}} = \frac{x_{cg_{for}} - x_{LE_{MAC}}}{\overline{C}} = \frac{3.191 - 2.966}{1.617} = 0.14 \quad (11.12)$$

$$h_{aft} = \overline{X}_{cg_{aft}} = \frac{x_{cg_{aft}} - x_{LE_{MAC}}}{\overline{C}} = \frac{3.451 - 2.966}{1.617} = 0.3 \quad (11.13)$$

Therefore, the aircraft most aft cg is located at 30% MAC, while the most forward cg is at 14% MAC.

- **Step 10. Aircraft cg range**.

$$\Delta x_{cg} = \frac{x_{cg_{aft}} - x_{cg_{for}}}{\overline{C}} = \frac{3.451 - 3.191}{1.617} = 0.3 - 0.14 = 0.16 \quad (11.16)$$

Thus, the aircraft cg range is 16% MAC.

Problems

1. Consider an acrobatic two-seat (side-by-side) light aircraft with a wing area of $25\,m^2$ and an overall length of 10 m. The mass of each major component and their corresponding centers of gravity from the propeller spinner (reference line) are shown here:

No.	Component	Mass (kg)	X_{cg} (m)
1	Wing	400	4.5
2	Tails	120	9.1
3	Engine	330	0.8
4	Fuselage	310	4.6
5	Pilots	170	3.1
6	Landing gear	75	3.3
7	Fuel	400	3.4
8	Systems	560	4.4

The wing is straight tapered with an aspect ratio of 10, a taper ratio of 0.4, and the wing apex is 3.2 m behind the reference line. Determine the aircraft cg in terms of percentage MAC.

2. Consider a trainer two-seat (tandem) aircraft with a wing area of $9.2\,m^2$ and an overall length of 6.2 m. The mass of each major component and their corresponding centers of gravity from the propeller spinner (reference line) are as follows:

No.	Component	Mass (kg)	X_{cg} (m)
1	Wing	95	2.4
2	Tails	18	5.7
3	Engine	75	0.5
4	Fuselage	87	2.9
5	Instructor	80	2.7
6	Student	75	3.4
7	Landing gear	15	3.1
8	Fuel	82	3.4
9	Systems	68	3.4

The wing is swept back and has a leading edge sweep of 20 deg, an aspect ratio of 9, a taper ratio of 0.72, and the wing apex is 1.9 m behind the reference line. Determine the aircraft cg in terms of percentage MAC.

3. Consider a single-seat fighter aircraft with a wing area of $50\,m^2$ and an overall length of 16 m. The mass of each major component and their corresponding centers of gravity from the fuselage nose (reference line) are shown here:

No.	Component	Mass (kg)	X_{cg} (m)
1	Wing	1680	2.4
2	Tails	420	15.5
3	Engine	2730	12
4	Fuselage	1470	7
5	Pilot	95	2.9
6	Stores	3000	9.5
7	Landing gear	630	7.6
8	Fuel	7000	10.1
9	Systems	4000	5.3

The wing is swept back and has a leading edge sweep of 55 deg, an aspect ratio of 4.5, a taper ratio of 0.2, and the wing apex is 10.1 m behind the reference line. Determine the aircraft cg in terms of percentage MAC.

4. Determine the most forward and most aft cg locations in terms of percentage MAC for the aircraft in Problem 1.

5. Determine the most forward and most aft cg locations in terms of percentage MAC for the aircraft in Problem 2.

6. Determine the most forward and most aft cg locations in terms of percentage MAC for the aircraft in Problem 3.

7. Consider a transport aircraft with a maximum take-off mass of 11 000 kg and wing area of 32 m^2. The fuselage has a length of 17 m, a maximum diameter of 2.7 m, and a mass of 900 kg. Model the fuselage as a thin cylindrical shell and determine its longitudinal, lateral, and directional mass moment of inertia about the aircraft cg. Assume that the aircraft cg is along the fuselage center line and located 8 m from the fuselage nose.

8. Consider a twin-turboprop military aircraft with a maximum take-off mass of 3200 kg and wing area of 28 m^2. The fuselage has a length of 11 m, a maximum diameter of 1.3 m, and a mass of 290 kg. Model the fuselage as a thin cylindrical shell and determine its longitudinal, lateral, and directional mass moment of inertia about the aircraft cg. Assume that the aircraft cg is along the fuselage center line and located at 6 m from the fuselage nose.

9. Consider a high-wing cargo aircraft with a maximum take-off mass of 12 000 kg and overall length of 17 m. The wing has the following characteristics:

$$S = 34\ \text{m}^2,\ \text{AR} = 10,\ \lambda = 0.65,\ (t/C)_{\text{max}} = 0.12,\ \Lambda_{0.5C} = 0\ \text{deg},\ m_{\text{w}} = 1200\ \text{kg}$$

The aircraft cg is along the fuselage center line and located at 8 m from the fuselage nose. The wing cg is at 9 m behind the fuselage nose and 1.3 m above the fuselage center line. Model the wing as a rectangular thin plate of thickness $0.5t$ and determine its longitudinal, lateral, and directional mass moment of inertia about the aircraft cg.

10. Consider a single-seat low-wing sailplane with a maximum take-off mass of 800 kg and overall length of 6 m. The wing has the following characteristics:

$$S = 11\ \text{m}^2,\ \text{AR} = 30,\ \lambda = 0.75,\ (t/C)_{\text{max}} = 0.12,\ \Lambda_{0.5C} = 10\ \text{deg},\ m_{\text{w}} = 110\ \text{kg}$$

The aircraft cg is along the fuselage center line and located at 2.7 m from the fuselage nose. The wing cg is located at 2.6 m behind the fuselage nose and 0.6 m below the fuselage center line. Model the wing as a rectangular thin plate of thickness $0.5t$ and determine its longitudinal, lateral, and directional mass moment of inertia about the aircraft cg.

11. Consider a low-wing large transport aircraft with a maximum take-off mass of 270 000 kg and overall length of 63 m. The wing has the following characteristics:

$$S = 340\ \text{m}^2,\ \text{AR} = 14,\ \lambda = 0.3,\ (t/C)_{\text{max}} = 0.15,\ \Lambda_{0.5C} = 35\ \text{deg},\ m_{\text{w}} = 22\,000\ \text{kg}$$

The aircraft cg is along the fuselage center line and located at 29 m from the fuselage nose. The wing apex is at 25 m behind the fuselage nose and 2.5 m below the fuselage center line. Model the wing as a rectangular thin plate of thickness $0.5t$ and determine its longitudinal, lateral, and directional mass moment of inertia about the aircraft cg.

12. Consider a light aircraft with a maximum take-off mass of 1000 kg and overall length of 6 m. The horizontal tail has the following characteristics:

$$S_{\text{h}} = 2.7\,\text{m}^2, \text{AR}_{\text{h}} = 4, \lambda_{\text{h}} = 0.8, (\text{t/C})_{\text{max}} = 0.09, \Lambda_{\text{h}0.5\text{C}} = 12\ \text{deg}, \text{m}_{\text{h}} = 21\ \text{kg}$$

The aircraft cg is along the fuselage center line and located at 2.8 m from the fuselage nose. The horizontal tail cg lies at 5.4 m behind the fuselage nose and 1.3 m above the fuselage center line. Model the horizontal tail as a rectangular thin plate of thickness $0.5t$ and determine its longitudinal, lateral, and directional mass moment of inertia about the aircraft cg.

13. Consider a canard aircraft with a maximum take-off mass of 2500 kg and overall length of 10 m. The canard has the following characteristics:

$$S_c = 3.2\ \text{m}^2,\ AR_c = 5,\ \lambda_c = 0.75,\ (t/C)_{max} = 0.08,\ \Lambda_{0.5C} = 18\ \text{deg},\ m_c = 45\ \text{kg}$$

The aircraft cg is along the fuselage center line and located at 4.4 m from the fuselage nose. The canard cg lies at 0.6 m behind the fuselage nose and 0.3 m below the fuselage center line. Model the canard as a rectangular thin plate of thickness $0.5t$ and determine its longitudinal, lateral, and directional mass moment of inertia about the aircraft cg.

14. Consider a transport aircraft with a maximum take-off mass of 42 000 kg and overall length of 27 m. The vertical tail has the following characteristics:

$$S_{vt} = 22\ \text{m}^2,\ AR_{vt} = 1.4,\ \lambda_{vt} = 0.3,\ (t/C)_{max} = 0.09,\ \Lambda_{vt0.5C} = 40\ \text{deg},\ m_{vt} = 510\ \text{kg}$$

The aircraft cg is along the fuselage center line and located at 12 m from the fuselage nose. The vertical tail cg lies at 26.5 m behind the fuselage nose and 3.3 m above the fuselage center line. Model the vertical tail as a rectangular thin plate of thickness $0.5t$ and determine its longitudinal, lateral, and directional mass moment of inertia about the aircraft cg.

15. Consider a fighter aircraft with a maximum take-off weight of 20 000 lb and overall length of 50 ft. The aircraft has twin vertical tails, each with the following characteristics:

$$S_{vt} = 90\ \text{ft}^2,\ AR_{vt} = 1.7,\ \lambda_{vt} = 0.4,\ (t/C)_{max} = 0.06,\ \Lambda_{vt0.5C} = 50\ \text{deg},\ m_{vt} = 180\ \text{lb}$$

The aircraft cg is along the fuselage center line and located at 27 ft from the fuselage nose. The vertical tail cg lies at 46 ft behind the fuselage nose and 6 ft above the fuselage center line. Model the vertical tail as a rectangular thin plate of thickness $0.5t$ and determine its longitudinal, lateral, and directional mass moment of inertia of both vertical tails as a group about the aircraft cg.

16. Consider a single-engine jet trainer aircraft with a maximum take-off mass of 7500 kg and overall length of 13 m. The jet engine has the following characteristics:

$$L_e = 3.6\ \text{m}^2,\ D_e = 1.1\ \text{m},\ m_e = 650\ \text{kg}$$

The aircraft cg is along the fuselage center line and located at 6 m from the fuselage nose. The engine cg lies at 10 m behind the fuselage nose and 0.8 m below the fuselage center line. Model the engine as a solid cylinder and determine its longitudinal, lateral, and directional mass moment of inertia about the aircraft cg. Assume that the engine center line coincides with the fuselage center line.

17. Consider a twin-turbofan business jet aircraft with a maximum take-off mass of 18 000 kg and overall length of 17 m. Engines are located beside the rear fuselage and each engine has the following characteristics:

$$L_e = 1.3 \text{ m}^2, D_e = 0.95 \text{ m}, m_e = 340 \text{ kg}$$

The aircraft cg is along the fuselage center line and located at 7 m from the fuselage nose. The engines' centers of gravity are 12 m behind the fuselage nose and 0.1 m above the fuselage center line. The lateral distance between the center lines of engines is 5 m. Model each engine as a solid cylinder and determine the longitudinal, lateral, and directional mass moments of inertia of both engines as a group about the aircraft cg.

18. Consider a twin-turboprop cargo aircraft with a maximum take-off mass of 50 000 kg and overall length of 34 m. The aircraft cg is along the fuselage center line and located at 15 m from the fuselage nose. The aircraft has a tricycle landing gear configuration. The main gear has a mass of 350 kg and its cg is located at 16 m aft of the fuselage nose and 2.5 m below the fuselage center line. The main gear has two wheels with a wheel track of 7 m. Model the main gear as two point masses and determine the longitudinal, lateral, and directional mass moment of inertia of the main gear as a group about the aircraft cg.

19. An aircraft has two cylindrical fuel tanks at the wing tip. Each tip tank has a mass of 280 kg, a diameter of 40 cm, and a length of 1.2 m. The wing span is 14 m. Determine the tip tanks' lateral mass moments of inertia (i.e., I_{xx}) as a group about the aircraft cg. Assume the aircraft cg is along the fuselage center line.

20. Consider a single-seat single-engine GA aircraft with a maximum take-off weight of 6500 lb. The wing is straight rectangular with a planform area of 195 ft^2 and aspect ratio of 9.5. The aircraft has two fuel tanks, located one in each half wing. Each fuel tank has a capacity of 140 US gallons of JP-4 and their centers of gravity are located 1.5 m from the fuselage center line. Determine the most left and most right cg of the aircraft in terms of wing span.

21. Consider a two-seat (tandem) single-engine trainer aircraft with a wing area of 15 m^2 and overall length of 9 m (Figure 11.21). The wing has a straight tapered shape with an aspect ratio of 8 and taper ratio of 0.7. The mass of each major component and their corresponding centers of gravity from the propeller spinner (reference line) are given in Table 11.15. Two fuel tanks are located inside the wing between the main spar and the rear spar, where their centers of gravity are at 50% MAC aft of the wing leading edge. Locate the wing such that the aircraft cg at maximum take-off weight is at 28% MAC. Then determine the most forward cg of the aircraft, the most aft cg of the aircraft, and the aircraft cg range in terms of percentage MAC.

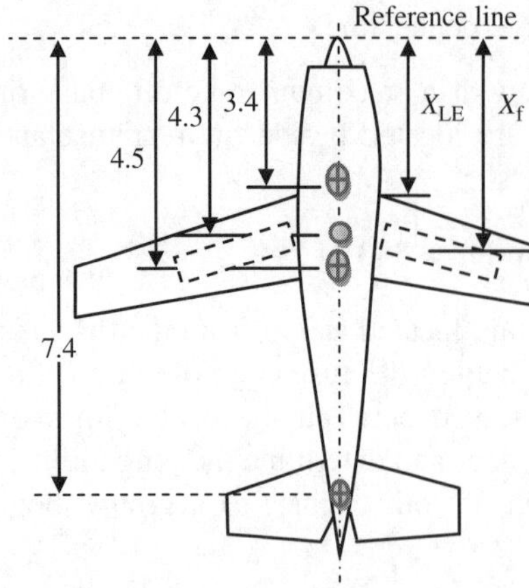

Figure 11.21 Aircraft of Problem 21 (numbers are in meters)

Table 11.15 Mass and cg locations of various components of the aircraft in Problem 21

No.	Component	Mass (kg)	Symbol	X_{cg} (m)
1	Engine	220	m_e	1.75
2	Fuselage	300	m_b	4.3
3	Pilot 1	80	m_{P1}	3.4
4	Pilot 2	80	m_{P2}	4.5
5	Landing gear	130	m_{lg}	3.3
6	Tails	70	m_t	7.4
7	Systems	630	m_s	4
8	Wing	360	m_w	x_w
9	Fuel	50	m_f	x_f

22. For the following 10-seat business jet aircraft:

$$m_{TO} = 7000\,\text{kg},\ S = 31\,\text{m}^2,\ C_{D_o} = 0.022,\ \text{AR} = 9,\ \lambda = 1,$$
$$S_h/S = 0.2,\ S_{vt}/S = 0.17,\ m_{empty} = 4200\,\text{kg}$$

(a) Determine the aircraft center of gravity in terms of percentage MAC for maximum take-off weight.
(b) Locate the wing such that the aircraft cg is at 22% MAC (with maximum take-off weight).
(c) Determine the aircraft aft and forward center of gravity in terms of percentage MAC.

The mass, cg location, and features of each component and item are shown in Table 11.16.

23. Consider a twin-turboprop transport aircraft with two crew members (Figure 11.22). The main cabin accommodates two flight attendants and 30 passengers in three-abreast seating at 80-cm pitch. The aircraft has the following mass and wing characteristics:

$$S = 40\ \text{m}^2,\ \text{AR} = 10,\ \lambda = 0.7,\ L = 20\ \text{m}$$

There are two baggage compartments right above each row. The mass and cg location (X_{cg}) for each major component is given in Table 11.17.

The wing apex is at 7.5 m from the fuselage nose (reference line). Determine the most aft and most forward cg location of the aircraft along the x-axis in terms of percentage MAC.

Table 11.16 Mass and cg locations of various components of the aircraft in Problem 22

No.	Component	Mass	X_{cg} (from fuselage nose)	Features
1	Fuselage	12% of m_{TO}	41% of L_f	$L_f = 15$ m, $D_f = 1.6$ m
2	Wing	14% of m_{TO}	Unknown for designer	$AR = 9$, $\lambda = 1$, $\Lambda_{LE} = 0$
3	Horizontal tail	4% of m_{TO}	91% of L_f	$AR = 5$, $\lambda = 1$, $\Lambda_{LE} = 0$
4	Vertical tail	2% of m_{TO}	93% of L_f	$AR = 2.5$, $\lambda = 1$, $\Lambda_{LE} = 30$ deg
5	Engine	18% of m_{TO}	63% of L_f	Twin-engine beside fuselage
6	Landing gear	5% of m_{TO}	Xcg_main = 48% of L_f Xcg_nose = 10% of L_f	Tricycle configuration $W_{mainLG} = 4\%$ of m_{TO}
7	Fuel	1550 kg	58% of L_f	In three fuel tanks
8	Passengers	10·85 kg	Xcg_front_row = 15% of L_f Xcg_ last_row = 50% of L_f	Two seats in each row
9	Luggage	10·30 kg	64% of L_f	In rear fuselage
10	Pilot	100 kg	10% of L_f	One
11	Other systems	5% of m_{TO}	50% of L_f	–

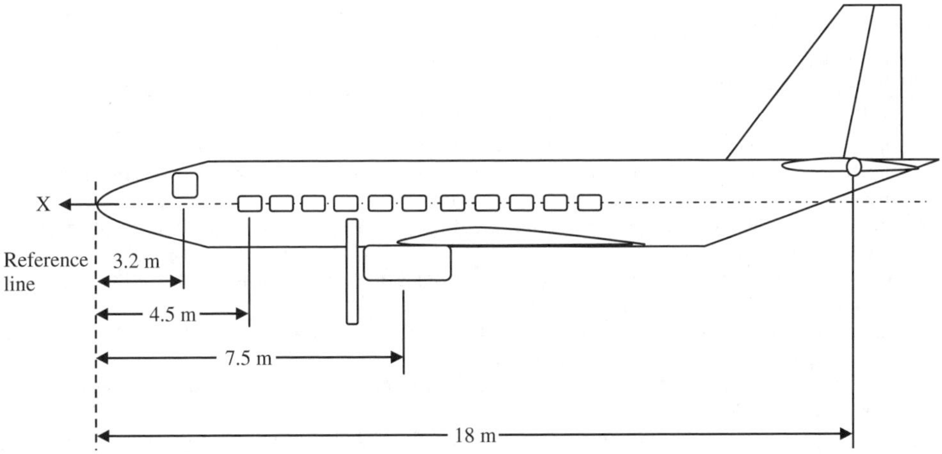

Figure 11.22 Aircraft in Problem 23

Table 11.17 Mass and cg location (X_{cg}) of major components for aircraft in Problem 23

No.	Component	Mass (kg)	X_{cg} (m)
1	Wing	2000	9.5
2	Tails	400	18
3	Engine	2400	7.7
4	Fuselage	1600	9.3
5	Pilots + bag	200	3.2
6	Passengers (first row)	3·85	4.5
7	Flight attendant	2·100	15.7
8	Carry-on baggage (first row)	3·15	4.5
9	Checked baggage	34·40	15
10	Landing gear	800	6.8
11	Wing fuel	1000	8.6
12	Fuselage fuel	2000	6.4
13	Systems and other equipment	7000	9

References

[1] (2004) Aviation Francais Virtuel, Concorde Operating Manual, British Aerospace/Aerospatiale Concorde.

[2] (2003) Gulfstream G 550 Flight Manual, Gulfstream Aerospace.

[3] Vetsch, L.E. and Burgener, L.L. (1988) Design of a certifiable primary on-board aircraft weight and balance system, AIAA/IEEE Digital Avionics Systems Conference, San Jose, October 17–20, AIAA-1988-3919.

[4] Roskam, J. (2007) *Airplane Flight Dynamics and Automatic Flight Control, Part I*, DAR Corporation.

[5] Multhopp, H. (1942) Aerodynamics of the Fuselage, NACA Technical Memorandum No. 1036.

[6] DOT, FAA (2011) Citation 550 Type Certificate Data Sheet.

[7] Anonymous (1976) USAF Stability and Control DATCOM, Flight Control Division, Air Force Flight Dynamics Laboratory, Wright-Patterson Air Force Base, Ohio.

[8] Anonymous (1990) Flying Qualities of Piloted Airplanes, MIL-F-1797C, Air Force Flight Dynamic Laboratory WPAFB, Dayton.

[9] Anonymous (1980) Military Specification Flying Qualities of Piloted Airplanes, MIL-F-8785C, Air Force Flight Dynamics Laboratory WPAFB, Dayton.

[10] Airworthiness Standards for GA and Transport Aircraft, FAR 23 and 25 (2011), US Department of Transportation, Federal Aviation Administration.

[11] Avallone, E., Baumeister, T., and Sadegh, A. (2006) *Marks' Standard Handbook for Mechanical Engineers*, 11th edn, McGraw-Hill.

[12] Tapley, B.D. (1990) *Eshbach's Handbook of Engineering Fundamentals*, 4th edn, Wiley-VCH Verlag GmbH.

12

Design of Control Surfaces

12.1 Introduction

Two primary prerequisites for a safe flight are stability and controllability. In addition, pilot and occupant comfort is of significant importance, which is often referred to as handling qualities. These three aircraft design objectives will influence the design of control surfaces and create a variety of design constraints. Flight stability is defined as the inherent tendency of an aircraft to oppose any input and return to the original trim condition if disturbed. When the summation of all forces along each of the three axes, and the summation of all the moments about each of the three axes are zero, an aircraft is said to be in trim or equilibrium. In this case, the aircraft will have a constant linear speed and/or a constant angular speed. Control is the process of changing the aircraft flight condition from an initial trim point to a final or new trim point. This is performed mainly by the pilot, through moving the control surfaces/engine throttle. The desired change is basically expressed with reference to the time that it takes to move from the initial trim point to the final trim point (e.g., pitch rate and roll rate).

Maneuverability is profoundly significant for fighter aircraft and missiles, and is a branch of controllability. Control systems should be designed with sufficient redundancy to achieve two orders of magnitude more reliability than some desired level. Aircraft controllability is a function of a number of factors, including control surfaces. These fundamental definitions are summarized in Table 12.1.

In an aircraft, there are two main groups of surfaces: (i) lifting surfaces and (ii) control surfaces. In a conventional aircraft, lifting surfaces primarily include the wing, horizontal tail, and vertical tail. As the name implies, lifting surfaces are to generate aerodynamic lift force. In contrast, control of an aircraft is applied through devices referred to as control surfaces. The control surfaces, however, are deflected by the pilot via a stick/yoke and pedal. In general, control surfaces may be broadly classified into two types: conventional and non-conventional. Conventional control surfaces are divided into two main groups: (i) primary control surfaces and (ii) secondary control surfaces (see Figure 12.1). The primary control surfaces (Figure 12.2) are in charge of control of the flight route and usually in a conventional aircraft are as follows: (i) aileron, (ii) elevator, and (iii) rudder.

Aircraft Design: A Systems Engineering Approach, First Edition. Mohammad H. Sadraey.

Table 12.1 Definition of fundamental terms

No.	Term	Definition
1	Trim, balance, and equilibrium	When the summation of all forces exerted on an aircraft and the summation of all moments about an aircraft center of gravity are zero, the aircraft is in "trim."
2	Control	A desired change in the aircraft trim condition from an initial trim point to a new trim point with a specified rate.
3	Stability	The tendency of an aircraft to oppose any input and return to the original trim point if disturbed by an undesired force or moment.
4	Static stability	The tendency of an aircraft to oppose any input if disturbed from the trim point.
5	Dynamic stability	The tendency of an aircraft to return to the original trim point if disturbed.

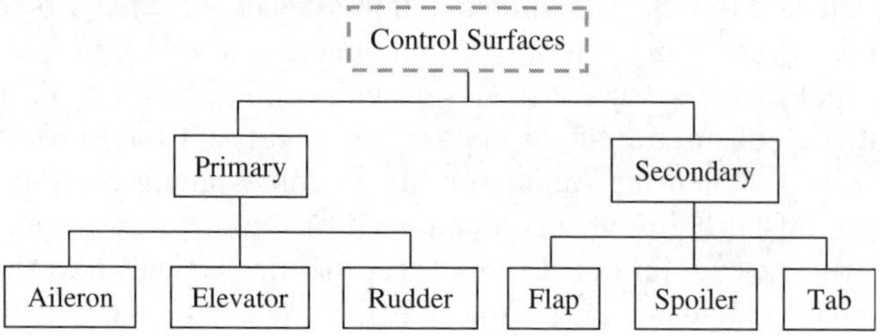

Figure 12.1 Classification of conventional control surfaces

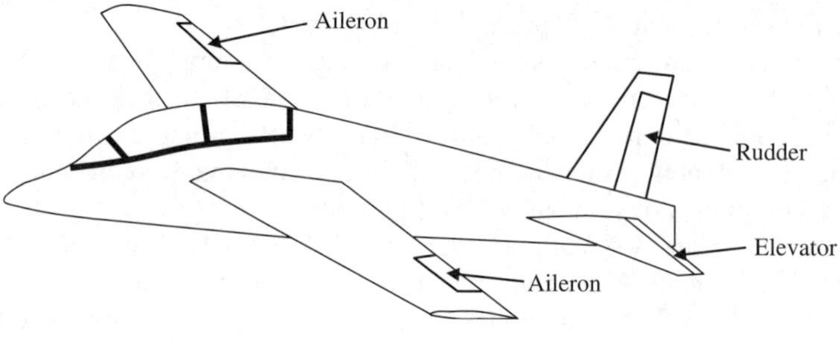

Figure 12.2 Primary control surfaces

In contrast, secondary surfaces are employed to reinforce primary control surfaces for minor or less important functions.

The primary control surfaces of aileron, elevator, and rudder are respectively utilized for lateral control, longitudinal control, and directional control. However, they also contribute largely to lateral trim, longitudinal trim, and directional trim of the aircraft. In the majority of aircraft configurations, lateral and directional motions are coupled; hence, the aileron also affects the directional motion and the rudder affects the lateral motion. Conventional primary control surfaces are like a plain flap, but their applications are different. When control surfaces are deflected, the cambers of their related lifting surfaces (wing, horizontal tail, or vertical tail) are changed. Thus, the deflection of a control surface varies the aerodynamic forces; and consequently, a resultant moment will influence the aircraft motion. The design of a high-lift device (e.g., flap) is addressed in Chapter 5, and the functions and applications of a tab are presented in Section 12.7.

To analyze the aircraft control, a coordinate-axis system must be defined. There are four coordinate systems: (i) earth-fixed, (ii) body-fixed, (iii) wind axis system, and (iv) stability axis system. Here, for the purpose of control, a body-fixed coordinate system is adopted where there are three orthogonal axes which follow the right-hand rule. The x-axis is along the fuselage (body) center line passing through the aircraft center of gravity, the y-axis is perpendicular to the x-axis and to the right (from a top view), and the z-axis is perpendicular to the xy plane (i.e., downward). Figure 12.3 illustrates the convention for positive directions of axes of aircraft. Positive roll is defined as a clockwise rotation about the x-axis as seen from the pilot seat (when in cruise (CR); right wing down, left wing up). Similarly, positive pitch is defined as a clockwise rotation about the y-axis as seen from the pilot seat (nose up). Finally, positive yaw is defined as a clockwise rotation about the z-axis as seen from the pilot seat (nose to right). Figure 12.4 and Table 12.2 demonstrate the convention for positive deflections of surfaces. These conventions are significant and are used to develop design techniques in this book.

An aircraft is capable of performing various maneuvers and motions. These may be broadly classified into three main groups: (i) longitudinal motion, (ii) lateral motion, and (iii) directional motion. In the majority of aircraft, longitudinal motion does not influence

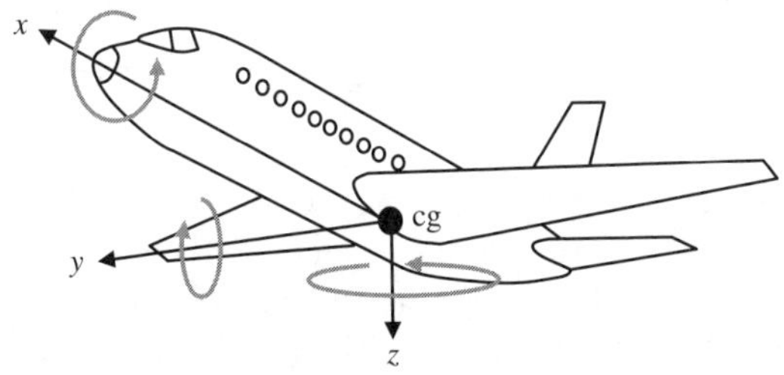

Figure 12.3 Axes and positive rotations convention

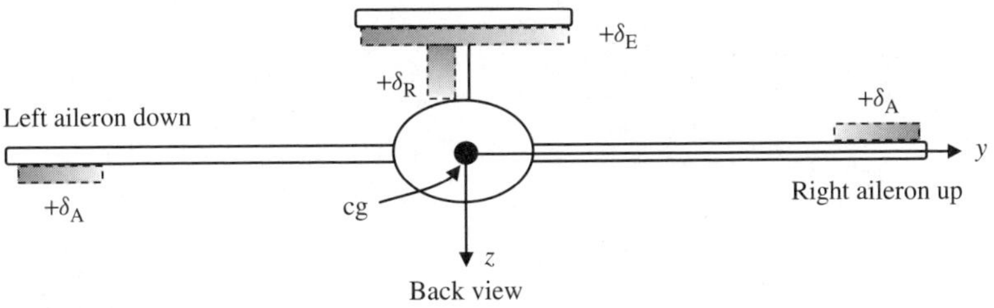

Figure 12.4 Convention for positive deflections of control surfaces

Table 12.2 Convention for positive control surface deflections

No.	Control surface	Symbol	Positive control surface deflection
1	Elevator	δ_E	Producing a negative pitching moment (down, $+\delta_E$; up, $-\delta_E$).
2	Aileron	δ_A	Generating a positive rolling moment. Left and right ailerons are considered ($\delta A_{\text{left-down}}, \delta A_{\text{right-down}}$); and $\delta_A = 0.5 \cdot (\delta_{\text{Aleft}} + \delta_{\text{Aright}})$.
3	Rudder	δ_R	Producing a positive side force and a negative yawing moment (left, $+\delta_R$; right, $-\delta_R$).

the lateral and directional motions. However, lateral and directional motions are often coupled; any lateral motion will often induce a directional motion, and any directional motion will often induce a lateral motion. The definition of these motions is as follows:

1. **Longitudinal motion**. Any aircraft motion in the xz plane is called longitudinal motion (e.g., pitch about the y-axis, plunging, climbing, cruising, pulling up, and descending). Lift, drag, and pitching moment have the major influence on this motion. The pitching motion is assumed to be a longitudinal motion.
2. **Lateral motion**. The aircraft rotation about the x-axis is called lateral motion (e.g., roll about the x-axis). Lift force and rolling moment have the major influence on this motion. The rolling motion is assumed to be a lateral motion.
3. **Directional motion**. The aircraft rotation about the z-axis and any motion along the y-axis is called directional motion (e.g., yaw about the z-axis, sideslipping, and skidding). Side-force and yawing moment have the major influence on this motion. The yawing motion is assumed to be a directional motion. A level turn is a combination of lateral and directional motions.

Secondary control surfaces (Figure 12.5) are in fact auxiliary control surfaces and are applied in special cases. These surfaces mainly include: (i) high-lift device (e.g., flap), (ii) tab, and (iii) spoiler. Usually a flap is used to increase the wing lift coefficient when the speed is low (i.e., take-off and landing). High-lift devices may be employed at the leading edge and trailing edge of the wing. The trailing edge high-lift device is often called a flap, while two groups of leading edge high-lift devices are slots and slats. To allow for an aileron to have a larger moment arm, the trailing edge flaps are in the inboard of the

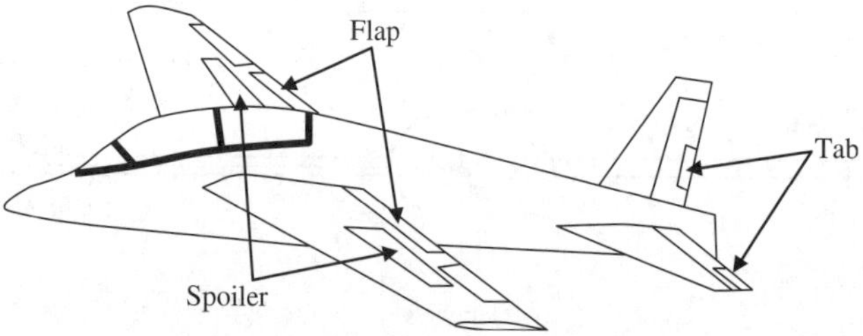

Figure 12.5 Secondary control surfaces

wing, and ailerons are in the outboard of the wing. This causes interference between the flap span and aileron span. In acrobatic aircraft where the significance of the aileron is more than that of the flap, the aileron is designed prior to the flap.

The design of a flap has been introduced in Chapter 5, but aileron design will be presented in Section 12.4. The spoiler essentially has two functions: (i) as a brake during landing and (ii) as an auxiliary device during roll. The third type of secondary control surface, the tab, has different types – such as trim tab, balance tab, geared tab, and flying tab – but its main role is to reduce the force necessary for control by the pilot. Spoilers are sometimes used as aileron substitutes, for roll control, especially when torsional aeroelasticity is critical. Spoilers are flat sheets (with no curve) on top of the wing used to decrease lift when deflected up. As the name implies, a spoiler degrades the lift, so they are not utilized at high speeds. In a number of high-speed aircraft, spoilers are used instead of ailerons. They are most effective roll controls at high speeds, and they make useful lift dampers to achieve maximum effect of wheel brakes on touchdown. Sailplanes and gliders employ spoilers to steepen the angles of glide by directly increasing the drag and reducing the lift-to-drag ratio.

A group of variables which are widely used in the design of control surfaces are control derivatives. The control derivatives are simply the rate of change of aerodynamic forces and moments (or their coefficients) with respect to a control surface deflection (e.g., elevator). Control derivatives represent the amount of change in an aerodynamic force or moment acting on an aircraft when there is a small change in the deflection of a control surface. The greater the control derivative, the more powerful is the corresponding control surface. Three most important non-dimensional control derivatives are $C_{\mathrm{l}_{\delta A}}$, $C_{\mathrm{m}_{\delta E}}$, and $C_{\mathrm{n}_{\delta R}}$. The unit of all non-dimensional control derivatives is 1/rad. The derivative $C_{\mathrm{l}_{\delta A}}$ is the rate of change of rolling moment coefficient with respect to a unit change in the aileron deflection (Equation (12.1)). The derivative $C_{\mathrm{m}_{\delta E}}$ is the rate of change of pitching moment coefficient with respect to a unit change in the elevator deflection (Equation (12.2)). The derivative $C_{\mathrm{n}_{\delta R}}$ is the rate of change of yawing moment coefficient with respect to a unit change in the rudder deflection (Equation (12.3)):

$$C_{\mathrm{l}_{\delta A}} = \frac{\partial C_{\mathrm{l}}}{\partial \delta_A} \tag{12.1}$$

$$C_{\mathrm{m}_{\delta E}} = \frac{\partial C_{\mathrm{m}}}{\partial \delta_E} \tag{12.2}$$

$$C_{\mathrm{n}_{\delta R}} = \frac{\partial C_{\mathrm{n}}}{\partial \delta_R} \tag{12.3}$$

After the aircraft main components (e.g., wing, tail, and landing gear) have been designed, the control power requirements may be expressed and interpreted in terms of the control derivatives. For instance, a rudder is designed to satisfy the requirements of $C_{\mathrm{n}_{\delta R}} < -0.4$ 1/rad for a fighter. Or an elevator is designed to satisfy the requirements of $C_{\mathrm{m}_{\delta E}} < -2$ 1/rad for a transport aircraft.

Figure 12.6 illustrates a flowchart that represents the control surfaces design process. In general, the design process begins with a trade-off study to establish a clear line between the stability and controllability requirements and ends with optimization. During the trade-off study, two extreme limits of flying qualities are examined and the border

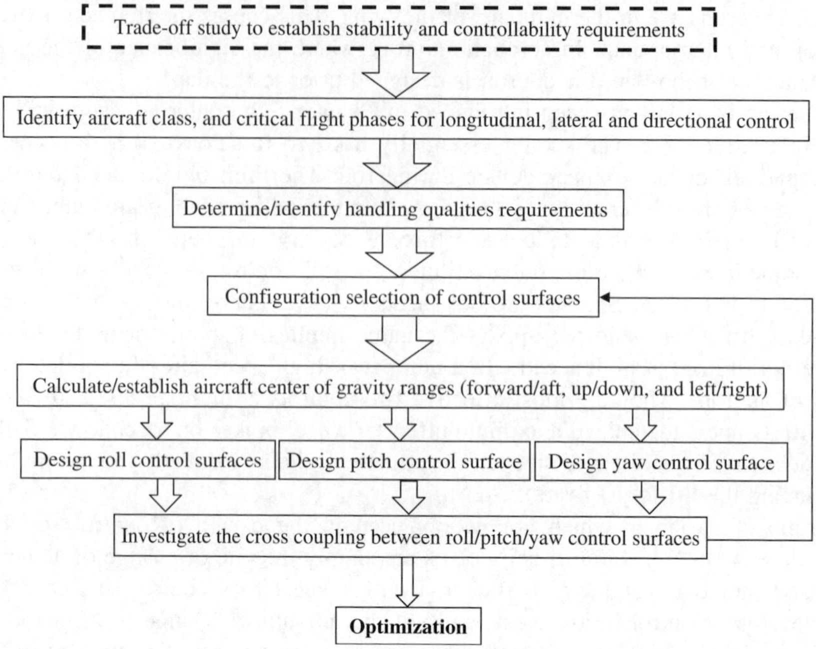

Figure 12.6 Control surfaces design process (US Air Force)

line between stability and controllability is drawn. For instance, a fighter can sacrifice stability to achieve higher controllability and maneuverability. Then, an automatic flight control system may be employed to augment the aircraft stability. In the case of a civil airliner, safety is the utmost goal so stability is clearly favored over controllability.

The results of this trade-off study will be primarily applied to establishing the most aft and the most forward allowable location of the aircraft center of gravity. Three control surfaces for roll control, pitch control, and yaw control are usually designed in parallel. Then the probable cross-coupling between the three surfaces is studied to ensure that each control surface is not negating the controllability features of the aircraft in other areas. If the cross-coupling analysis reveals an unsatisfactory effect on any control surfaces, one or more control surfaces must be redesigned to resolve the issue. Flight control systems, including control surfaces, should be designed with sufficient redundancy to achieve two orders of magnitude more reliability than some desired level. In FAR 23, one level of redundancy for a control system is required (i.e., power transmission line). The lines of power transmission (i.e., wire and pipe) should not be close to each other, should not be close to fuel tanks, and should not be close to hydraulic lines. In most Boeing aircraft, there are three separate hydraulic lines. If there is a leak in the hydraulic lines or if the engines become inoperative, there is an extra hydraulic system which is run independently. So, the Boeing 747 has four hydraulic systems. These design considerations provide a highly safe and reliable aircraft.

In this chapter you will find the design of the elevator, rudder, and aileron, but the detail design of the spoiler and tab is not considered. Most aircraft, of course, do not have a spoiler and the majority has a tab, however nearly all benefit from a flap. Apart

Table 12.3 Typical values for geometry of control surfaces

Control surface	Elevator	Aileron	Rudder
Control surface area/lifting surface area	$S_E/S_h = 0.15–0.4$	$S_A/S = 0.03–0.12$	$S_R/S_V = 0.15–0.35$
Control surface span/lifting surface span	$b_E/b_h = 0.8–1$	$b_A/b = 0.2–0.40$	$b_R/b_V = 0.7–1$
Control surface chord/lifting surface chord	$C_E/C_h = 0.2–0.4$	$C_A/C = 0.15–0.3$	$C_R/C_V = 0.15–0.4$
Control surface maximum deflection (negative)	−25 deg (up)	25 deg (up)	−30 deg (right)
Control surface maximum deflection (positive)	+20 deg (down)	20 deg (down)	+30 deg (left)

from conventional control surfaces (aileron, elevator, and rudder), there are other surfaces such as elevon, ruddervator, and flaperon that are seldom used. In this chapter, the design of conventional control surfaces is presented in detail, but the design of unconventional control surfaces is examined in brief. Table 12.3 presents typical values for the geometry of primary control surfaces.

This chapter is devoted to detail design of primary control surfaces. The configuration selection of control surfaces is addressed in Section 12.2. Fundamentals of handling qualities and the technique to evaluate them are described in Section 12.3. The aileron design process is presented in Section 12.4, the elevator design technique is examined in Section 12.5, and the rudder design procedure is introduced in Section 12.6. Section 12.7 is devoted to the concept of aerodynamic balance and mass balance of primary control surfaces. Three comprehensive and fully solved examples for the design of aileron, elevator, and rudder are presented in Section 12.8.

12.2 Configuration Selection of Control Surfaces

The first step in the design of control surfaces is to select the control surface configuration. The primary idea behind the design of flight control surfaces is to position them so that they function primarily as moment generators. They provide three types of rotational motion (roll, pitch, and yaw). A conventional configuration includes elevator, aileron, and rudder. Variations to this classical configuration lead to some variations in the arrangements of these control surfaces. Table 12.4 represents several control surface configuration options. Some types of control surface are tied to particular aircraft configurations; they must be selected for specific aircraft configurations. Table 12.4 also illustrates a few aircraft examples.

The control surface configuration selection is a function of aircraft configuration (e.g., wing, tail, and engine), cost, performance, controllability, power transmission, and operational requirements. The consequence of some aircraft configurations is to have a particular type of control surface. For instance, when a V-tail configuration is selected during the aircraft conceptual design phase, a ruddervator is the best candidate to control both yawing

Table 12.4 Control surface configuration options

No.	Control surface configuration	Aircraft configuration
1	Conventional (aileron, elevator, and rudder)	Conventional (or canard replacing elevator)
2	All-moving horizontal tail, rudder, and aileron	Horizontal tail and elevator combined
3	All-moving vertical tail, elevator, and aileron	Vertical tail and rudder combined
4	Flaperon, elevator, and rudder	Flap and aileron combined (e.g., X-29 and F-16 Falcon)
5	Taileron, rudder	All-moving horizontal tail (elevator) and aileron combined (e.g., F-16 Falcon)
6	Elevon, rudder (or equivalent)	Aileron and elevator combined (e.g., Dragon, F-117 Night Hawk, and Space Shuttle)
7	Ruddervator, aileron	V-tail (e.g., UAV Global Hawk and Predator)
8	Drag-rudder, elevator, and aileron	No vertical tail (e.g., DarkStar)
9	Canardvator, aileron	Elevator as part of canard, plus aileron
10	Four control surfaces	Cross (+ or ×) tail configuration (e.g., most missiles)
11	Aileron, elevator (or equivalent), and split rudder	No vertical tail. Aileron-like surfaces split into top and bottom sections (e.g., bomber B-2 Spirit)
12	Spoileron, elevator, and rudder	Spoiler and aileron combined (e.g., B-52)
13	Thrust vector control	Augmented or no control surfaces, VTOL UAV

VTOL: Vertical Take Off and Landing.

and pitching moments. Another example is when the designer decides to have a delta wing without aft tail. In such a case, an elevon is a great candidate as a means of control power to control the pitch rate and roll rate. The final decision on the control surface configuration will be the output of a trade-off study to balance and satisfy all design requirements in an optimum way. In general, unconventional control surfaces are more challenging to design, more complex to manufacture, and also harder to analyze. However, unconventional control surfaces are more efficient when a higher control power is required in a challenging design environment.

12.3 Handling Qualities

In designing primary control surfaces, several factors have an affect; one of the most important is the handling or flying qualities. Before entering into the principles and methods of design, we need to know more about the definition of handling qualities and their criteria. Handling qualities is defined as "those qualities of an aircraft which govern the ease and precision with which a pilot is able to perform his mission." Knowledge of these parameters enables a designer to imagine the nature of the aircraft's response to any command and/or disturbance. Handling qualities reflect the ease with which a pilot can carry out some particular mission with an aircraft that has a specific set of flying qualities. However, handling qualities depend not only upon the visual and motion cues available and the display of flight information in the cockpit, but also upon the control power of primary control surfaces.

Handling qualities involve the study and evaluation of the stability and control characteristics of an aircraft, as well as the pilot comfort level. They have a critical bearing on the safety of flight and on the ease of controlling the aircraft in steady flight, during maneuvers, and the response of the aircraft to an atmospheric disturbance. The importance of handling qualities is particularly marked when some aircraft exhibit such unwanted flight characteristics as pilot-induced oscillation or roll ratchet.

An aircraft system is a set of elements (including a human pilot), the relation between these elements, and the boundary around them. The science of ergonomics studies the interaction between people and machines. The implementation of ergonomics in system design will make the system work better by eliminating aspects of system functioning which are undesirable, such as inefficiency, fatigue, and user difficulties. References [1–5] introduce various aspects of ergonomics and its application to aviation and aircraft.

A human pilot is a variable and dynamic element closing the loop around the flight control systems. Therefore, handling qualities should be arranged to suit the pilot so that his adapted characteristic is best for the flight mission. The interaction between human pilot control input and the corresponding aircraft response must be such that the pilot can achieve the mission objectives with reasonable physical effort. Therefore, it is difficult to specify analytical performance criteria for the dynamic behavior/perception of a human pilot/passenger. However, vehicle-related handling qualities are usually characterized by a number of parameters, such as the time constant, damping ratio, and undamped natural frequency of the exponential/oscillatory response of the aircraft to an input.

Civil aviation standards authorities such as the FAA [6, 7] have addressed the handling qualities requirement only slightly. For instance, statements such as "The airplane must have adequate directional control during taxiing" (FAR 25.233) or "Land planes may have no uncontrollable longitudinal tendency" (FAR 25.231) are used to specify the requirements. Fortunately, military aviation standards authorities have defined detail specifications (MIL-STD-1797 [8]) (Military Standards), MIL-C-18244 (Control and Stabilization Systems), and MIL-F-87242 (Flight Controls) for military aircraft handling qualities. These specifications must be met by US military aircraft, but may be expanded/adapted to civil aircraft. MIL-F-878FC has now been superseded by MIL-STD-1797 [8], which contains additional information. Such specifications are basic in the design of control surfaces.

The control surfaces must be designed such that aircraft possess acceptable flying qualities anywhere inside the operational flight envelope, with allowable cg range and allowable aircraft weight (see Figure 11.6). The operational flight envelope defines the boundaries in terms of speed, altitude, and load factor within which the aircraft must be capable of operating in order to accomplish the desired mission. A typical operational flight envelope for a transport aircraft is shown in Figure 12.7.

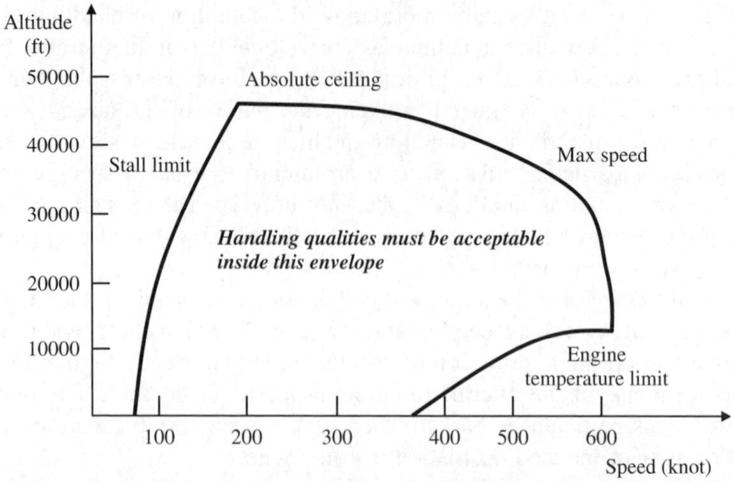

Figure 12.7 A typical operational flight envelope

12.3.1 Definitions

The handling quality specifications are frequently expressed with reference to aircraft classes, flight phases, and levels of acceptability. So, these terms are explained first, before discussing the specifications. In this book, it is essentially the recommendations of standards MIL-STD-1797 that are followed for fixed-wing aircraft.

12.3.1.1 Aircraft Classes

An aircraft is considered to belong to one of the four classes shown in Table 12.5. It is seen that classification is based on the weight of an aircraft as well as its maneuverability. The handling qualities of each class differ. According to MIL-F-8785C [9], for the purpose of handling qualities, aircraft are classified into four classes: I, I, II, and IV.

Table 12.5 Aircraft classes

Class	Aircraft characteristics
I	Small, light aircraft (maximum take-off mass less than 6000 kg) with low maneuverability
II	Aircraft of medium weight and low-to-medium maneuverability (maximum take-off mass between 6000 and 30 000 kg)
III	Large, heavy, and low-to-medium maneuverability aircraft (maximum take-off mass more than 30 000 kg)
IV	Highly maneuverable aircraft, no weight limit (e.g., acrobatic, missile, and fighter)

In Class I, small light aircraft such as (i) light utility, (ii) primary trainer, and (iii) light observation aircraft are included. GA aircraft may be considered as Class I air vehicles. Class II includes the following aircraft: (i) heavy utility/search and rescue, (ii) light or medium transport/cargo/tanker, (iii) early warning/electronic countermeasures/airborne command, control, or communications relay, (iv) anti-submarine, (v) assault transport, (vi) reconnaissance (RC), (vii) tactical bomber, (viii) heavy attack, and (ix) trainer for Class II.

Class III aircraft are as follows: (i) heavy transport/cargo/tanker, (ii) heavy bomber, (iii) patrol/early warning/electronic countermeasures/airborne command, control, or communications relay, and (iv) trainer for Class III. The following aircraft are in Class IV: (i) fighter/interceptor, (ii) attack, (iii) tactical RC, (iv) observation, and (v) trainer for Class IV. Civil transport aircraft may be considered as Class II/III air vehicles.

The handling qualities of each class differ, and their differences will be introduced later. In general, Class IV has the highest control requirements compared with the other three classes. The procuring activity will assign an aircraft to one of these classes, and the handling quality requirements for that class shall apply.

12.3.1.2 Flight Phases

The flight phase is another parameter which has a significant role in handling qualities. Flying quality requirements vary for the different phases of a mission. Figure 12.8 illustrates general operations in a typical flight mission. Take-off, climb, cruise, descent, and landing are the least operations necessary to have a conventional flight mission. Different aircraft in performing their missions may have the following phases and maneuvers: bank, spin, pull-out, chandelle, inverted flight, stall, S turn, aerial refueling, approach, patrol, loiter (LO), and so on. Some maneuvers need more pilot force and high accuracy, but some are very simple and easy. Whatever mission and aircraft is used to accomplish it, the mission is divisible into three phases of flight: phase A, phase B, and phase C.

Phase A includes non-terminal flight phases that require maneuvering, precision tracking, or precise flight-path control. Phase B involves those non-terminal flight phases that are normally accomplished using gradual maneuvers and without precision tracking, although accurate flight-path control may be required. Phase C involves terminal flight phases that are normally accomplished using gradual maneuvers and usually require accurate flight-path control. In general, phases A and B are non-terminal and phase C is terminal. The flight operations included in each category are tabulated in Table 12.6. Phase B is usually not critical, but phases A and C, depending on the mission of the aircraft, could be critical. In designing control surfaces, only the critical phase of flight is satisfied.

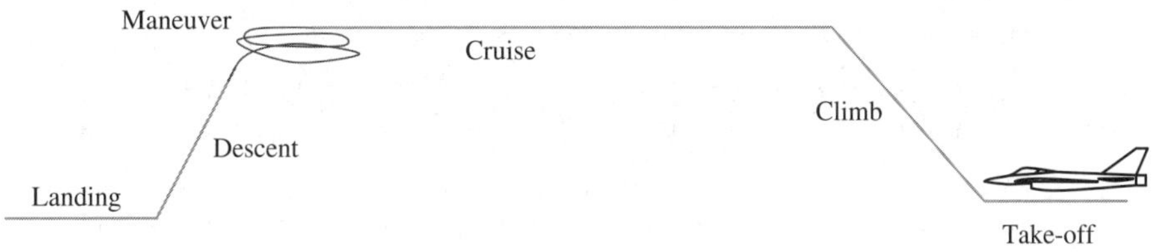

Figure 12.8 Main operations in a typical flight

Table 12.6 Flight phase categories [8]

Category	Examples of flight operation
A	(i) Air-to-air combat (CO); (ii) ground attack (GA); (iii) weapon delivery/launch (WD); (iv) aerial recovery (AR); (v) reconnaissance (RC); (vi) in-flight refueling (receiver) (RR); (vii) terrain following (TR); (viii) anti-submarine search (AS); (xi) close formation flying (FF); and (x) low-altitude parachute extraction system (LAPES) delivery.
B	(i) Climb (CL); (ii) cruise (CR); (iii) loiter (LO); (iv) in-flight refueling in which the aircraft acts as a tanker (RT); (v) descent (D); (vi) emergency descent (ED); (vii) emergency deceleration (DE); and (viii) aerial delivery (AD).
C	(i) Take-off (TO); (ii) catapult take-off (CT); (iii) powered approach (PA); (iv) wave-off/go-around (WO); and (v) landing (L).

Experience with aircraft operations demonstrates that certain flight phases require more stringent values of flying quality parameters than others. For instance, an air-to-air combat (CO) operation requires more dutch-roll damping than a cruising flight. Also, a given mission flight phase will generally have an aircraft normal state associated with it (e.g., flaps and gear down for landing approach and up for cruising flight).

12.3.1.3 Levels of Acceptability

The third point a control surface designer should know before considering the issue of handling qualities is levels of acceptability. The requirements for airworthiness and handling qualities are stated in terms of three distinct, specified values of control (or stability) parameters. Each value is a limiting condition necessary to satisfy one of the three levels of acceptability. These levels are related to the ability of the pilot to complete the missions for which the aircraft is intended. The definitions of these three levels are given in Table 12.7.

As defined in the specifications of Table 12.7, the level of acceptability relates to the ease of flight and flight safety. According to airworthiness standards, an aircraft with any level of acceptability from one to three is allowed to fly, but for the design of control surfaces, level 1 must be the objective. An aircraft with level 1 can only terminate flight

Table 12.7 Levels of acceptability

Level	Definition
1	Flying qualities clearly adequate for the mission flight phase.
2	Flying qualities adequate to accomplish the mission flight phase, but some increase in pilot workload or degradation in mission effectiveness, or both, exists.
3	Flying qualities such that the airplane can be controlled safely, but pilot workload is excessive or mission effectiveness is inadequate, or both. Category A flight phases can be terminated safely, and Category B and C flight phases can be completed.

Table 12.8 Levels of acceptability and pilot comfort

Level	Meaning	Pilot comfort level	Pilot status
1	Very comfortable	1 – 3	
2	Hardly comfortable	4 – 6	
3	Uncomfortable	7 – 10	

phase A safely and in other phases may be run out of control. When an aircraft is in level 1, there is no failure during phases of flight. When an aircraft has one failure per 1 000 000 flights, it will be considered to be at level 1. When an aircraft has one failure per 10 000 flights, it will be considered to be at level 2. If any aircraft has one failure per 100 flights, it is considered to be at level 3. An aircraft in level 3 is recommended to be retired to avoid an accident, because any time a system or component fails, an accident may occur. The control surfaces must be designed such that the level 1 of handling qualities is achieved.

The levels of acceptability may be determined on the basis of the pilot's opinion (personal feeling) of the flying characteristics of the aircraft. The Cooper–Harper rating scale [10] is a set of criteria used by test pilots and flight test engineers to evaluate the handling qualities of an aircraft during flight test operations. The scale ranges from 1 to 10, with 1 indicating the best handling characteristics and 10 the worst. The criterion is evaluative, and thus the scale is considered subjective. A high number in this scale demonstrates a deficiency in the design of control-related components. An interpretation of the handling characteristics in terms of pilot comfort level is illustrated in Table 12.8.

In a conventional aircraft, the lateral and directional motions are highly coupled, and the longitudinal motion often does not induce any lateral or directional motion (except at high angles of attack). For this reason, the handling qualities criteria are broadly divided into two groups: (i) longitudinal handling qualities and (ii) lateral-directional handling qualities. Basic criteria for longitudinal handling qualities are presented in Section 12.3.2, while Section 12.3.3 examines the lateral-directional handling qualities.

12.3.2 Longitudinal Handling Qualities

Any motion in the *xz* plane such as CR, CL, pitch, or plunging motion is referred to as a longitudinal motion. Motion variables such as forward speed (U), angle of attack (α), pitch angle (θ), and pitch rate (Q) are the most important parameters in longitudinal handling qualities analysis. Longitudinal handling qualities primarily rely on longitudinal control and longitudinal stability (i.e., the aircraft response to a desired pilot elevator deflection and to atmospheric disturbance). The more longitudinally stable an aircraft, the less longitudinally controllable and the less responsive the aircraft will be to elevator input.

If the aircraft designer desires the pilot to be comfortable when he/she applies commands to the aircraft, the controllability of the aircraft should be acceptable. Longitudinal handling qualities determine the acceptability of the aircraft response to pilot input and atmospheric disturbance. Therefore, the decision on longitudinal handling qualities is expressed in terms of longitudinal dynamic stability which can relate aircraft response to elevator deflection. Longitudinal handling qualities are criteria for the design of the elevator.

When an atmospheric disturbance hits the aircraft, or when the elevator is deflected, an aircraft response tends to contain two simultaneous and distinct modes: (i) long-period (often referred to as phugoid) mode and (ii) short-period mode. Both long-period (phugoid) motion and short-period motion are often of second-order response type, which involves oscillation. A second-order mode is a sinusoidal motion and may be modeled using a frequency (ω) and a damping ratio (ζ). In this section, handling qualities for phugoid response and short-period response are presented. In general, provided that the separation between the frequencies of the phugoid and short-period modes is small, handling qualities can arise. If the ratio between the frequency of phugoid mode and the frequency of short-period mode is less than 0.1 ($\omega_{ph}/\omega_{sp} < 0.1$), there may be some trouble with the longitudinal handling qualities. Longitudinal handling qualities are expressed in two main groups: (i) longitudinal control and (i) longitudinal stability; they are addressed in the following two sections.

12.3.2.1 Longitudinal Control

An aircraft must be longitudinally controllable, as well as maneuverable within the flight envelope (Figure 12.6). In a conventional aircraft, the longitudinal control is primarily applied though the deflection of the elevator (δ_E) and engine throttle setting (δ_T). There are two groups of requirements in the aircraft longitudinal controllability: (i) required pilot force and (ii) aircraft response to the pilot input. In order to deflect the elevator, the pilot must apply a force to the stick/yoke/wheel and hold it (in the case of an aircraft with a stick-fixed control system). In an aircraft with a stick-free control system, the pilot force is amplified through such devices as a tab or spring. Pilot force analysis is beyond the scope of this text; the interested reader is referred to such references as [11–16].

The aircraft response in the longitudinal control is frequently expressed in terms of pitch rate (q). However, the forward speed and angle of attack would be varied as well. The most critical flight condition for pitch control is when the aircraft is flying at a low speed. Two flight operations which feature a very low speed are take-off and landing. Take-off control is much harder than landing control due to safety considerations. A take-off operation is usually divided into three sections: (i) ground section, (ii) rotation or transition, and (iii) CL. The longitudinal control in a take-off is mainly applied during the rotation section, with the nose pitched up by rotating the aircraft about the main gear.

The longitudinal control handling quality requirements during take-off operation are stated as follows: in an aircraft with a tricycle landing gear, the pitch rate should have a value such that the take-off rotation does not take longer than a specified length of time. Since the take-off rotation dynamics is governed by Newton's second law, the take-off rotation time may readily be expressed in terms of the aircraft angular acceleration ($\ddot{\theta}$) about the main gear rotation point. For instance, in a transport aircraft, the acceptable value for the take-off rotation time is 3–5 seconds. The equivalent value for the angular rotation

Table 12.9 Take-off angular acceleration requirements

No.	Aircraft type	Rotation time during take-off (s)	Take-off pitch angular acceleration (deg/s^2)
1	Highly maneuverable (e.g., acrobatic GA and fighter)	0.2–0.7	12–20
2	Utility, semi-acrobatic GA	1–2	10–15
3	Normal general aviation	1–3	8–10
4	Small transport	2–4	6–8
5	Large transport	3–5	4–6
6	Remote control, model	1–2	10–15

rate to achieve such a requirement is 5–7 deg/s^2. This requirement must be satisfied when the aircraft center of gravity is located at the most forward location. Table 12.9 provides take-off angular acceleration requirements for various types of aircraft. This table has already been provided in Chapter 9, but is repeated here for convenience. These specifications are employed in the design of the elevator. The application of the requirement is examined in Section 12.5.

12.3.2.2 Longitudinal Stability

Longitudinal stability is defined as the tendency of an aircraft to return to its initial longitudinal trim point (e.g., steady-state angle of attack and forward speed) if disturbed by a longitudinal disturbance (e.g., vertical gust). There are static and dynamic longitudinal stability requirements. The static longitudinal stability is satisfied provided that the following condition is met:

$$C_{m_\alpha} \prec 0 \tag{12.4}$$

For the dynamic longitudinal stability, the aircraft response to a longitudinal disturbance is studied. The response of a longitudinally dynamically stable aircraft to a vertical gust

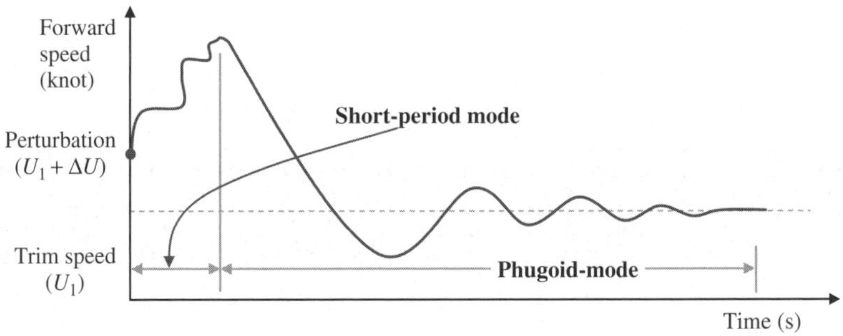

Figure 12.9 Typical response modes to a forward speed disturbance

Table 12.10 Phugoid mode requirement

Level of acceptability	Requirement
1	Damping ratio of phugoid mode (ζ_{ph}) ≥ 0.04
2	Damping ratio of phugoid mode (ζ_{ph}) ≥ 0.0
3	Time to double the amplitude at least 55 seconds

is an oscillatory one and comprises two modes: (i) short-period mode and (ii) long-period mode (Figure 12.9).

Phugoid Mode

In a dynamically longitudinally stable aircraft, the phugoid mode is a second-order response which the oscillation usually damps out in about a few minutes. The phugoid mode is characterized by a damping ratio and a frequency. Provided that the frequencies of the phugoid and the short-period modes of response are widely separated, either in stick-free or stick-fixed flight condition, the values of damping ratio shown in Table 12.10 must be achieved. In the design of the horizontal tail (including the elevator), the level 1 of acceptability must be considered.

Short-Period Mode

In a dynamically longitudinally stable aircraft, the short-period mode is a second-order oscillatory response which the oscillation usually damps out in about a few seconds. The short-period mode is also characterized by a damping ratio and a frequency. The longitudinal handling qualities related to the short-period response are mainly governed by the short-period damping ratio (ζ_{sp}). For an aircraft at any flight phase, Table 12.11 indicates specified values of damping ratio of the short-period mode at three levels and three flight phases.

At high speed, low values of the short-period damping ratio are less troublesome than at low speeds. A desired short-period natural frequency (ω_n) is often between 0.4 and 0.6. In the design of the horizontal tail and elevator, level 1 at the most critical flight phase must be achieved.

Table 12.11 Short-period mode damping ratio specification

Flight phase	Short-period damping ratio (ζ_s)					
	Level 1		Level 2		Level 3	
	Minimum	Maximum	Minimum	Maximum	Minimum	Maximum
A	0.35	1.3	0.25	2.0	0.15	No maximum
B	0.3	2.0	0.2	2.0	0.15	No maximum
C	0.35	1.3	0.25	2.0	0.15	No maximum

12.3.3 Lateral-Directional Handling Qualities

The design of rudder and aileron is primarily governed by lateral-directional handling qualities. The specification of handling qualities of lateral-directional motion is more involved than for longitudinal motion, and consequently requires more parameters. The lateral-directional handling qualities are divided into three groups: (i) lateral control, (ii) directional control, and (iii) lateral-directional stability. Lateral or roll control requirements govern the aircraft response to the aileron deflection, directional control requirements govern the aircraft response to the rudder deflection, and lateral-directional stability requirements address the aircraft transient response to the atmospheric disturbance. According to airworthiness standards, to turn an aircraft, the pilot must deflect the aileron and rudder simultaneously. A turn is usually a combination of lateral and directional motions.

In a conventional aircraft, lateral and directional motions are often coupled; any lateral motion induces a directional motion, and any directional motion induces a lateral motion. Hence, when the rudder is deflected to create a yaw moment, it simultaneously generates a rolling motion. This is due to the fact that the aerodynamic center of a vertical tail is usually above the aircraft cg. In contrast, when the aileron is deflected to create a rolling moment, it simultaneously generates directional motion. This is due to the sideslip motion as well as different wing drags on the left and right wing sections. In many aircraft, an interconnection between aileron and rudder is used to satisfy that part of the FARs which require negative aileron deflection to accompany positive rudder pedal force application.

When a lateral-directional disturbance hits an aircraft, the aircraft will often demonstrate three responses: rolling motion, spiral oscillation, and dutch-roll oscillation. The handling quality requirements for each mode are examined in this section. Here, some techniques and numbers will allow the designer to judge which aircraft possesses acceptable handling characteristics. This information gives the rudder and aileron designer the feeling of pilot satisfaction with lateral-directional handling quality characteristics.

12.3.3.1 Roll Control

Roll or lateral control requirements govern the aircraft response to the aileron deflection; thus, the requirements are employed in the design of the aileron. It is customary to specify the roll power in terms of the change of bank angle achieved in a given time in response to a step function in roll command. Thus, the aircraft must exhibit a minimum bank angle within a certain specified time in response to aileron deflection. The required bank angles and times are specified in Table 12.12 for various aircraft classes and different flight phases (according to MIL-F-8785C).

Roll performance in terms of a bank angle change ($\Delta\phi$) in a given time (t) is specified in Table 12.12(a) for Class I, (b) for Class II, (c) for Class III, and (d) for Class IV aircraft. The notation "60 deg in 1.3 seconds" in Table 12.12 indicates the maximum time it should take from an initial bank angle (say 0 deg) to reach a bank angle which is 60 deg different from the initial one, following full deflection of the aileron. It may also be interpreted as the maximum time it should take from a bank angle of -30 deg to $+30$ deg. For Class IV aircraft, for level 1, the yaw control should be free. For other aircraft and levels, it is permissible to use the yaw control to reduce any sideslip which tends to retard the roll rate. Such yaw control is not permitted to induce sideslip, which enhances the roll rate.

Table 12.12 Roll control requirements

(a) Time to achieve a specified bank angle change for Class I			
Level	Flight phase category		
	A	B	C
	Time to achieve a bank angle of 60° (s)	Time to achieve a bank angle of 45° (s)	Time to achieve a bank angle of 30° (s)
1	1.3	1.7	1.3
2	1.7	2.5	1.8
3	2.6	3.4	2.6

(b) Time to achieve a specified bank angle change for Class II					
Level	Runway	Flight phase category			
		A	B	C	C
		Time to achieve a bank angle of 45° (s)	Time to achieve a bank angle of 45° (s)	Time to achieve a bank angle of 30° (s)	Time to achieve a bank angle of 25° (s)
1	Land-based	1.4	1.9	1.8	–
	Carrier-based	1.4	1.9	2.5	–
2	Land-based	1.9	2.8	3.6	–
	Carrier-based	1.9	2.8	–	1.0
3	Land-based	2.8	3.8	–	1.5
	Carrier-based	2.8	3.8	–	2.0

(c) Time to achieve a 30° bank angle change for Class III				
Level	Speed range	Flight phase category		
		A (s)	B (s)	C (s)
1	Low	1.8	2.3	2.5
	Medium	1.5	2.0	2.5
	High	2.0	2.3	2.5
2	Low	2.4	3.9	4.0
	Medium	2.0	3.3	4.0
	High	2.5	3.9	4.0
3	All	3.0	5.0	6.0

Table 12.12 (*continued*)

(d) Time to achieve a specified bank angle change for Class IV						
Level	Speed range	Flight phase category				
		A			B	C
		30°	50°	90°	90° (s)	30° (s)
1	Very low	1.1 s	–	–	2.0	1.1
	Low	1.1 s	–	–	1.7	1.1
	Medium	–	–	1.3 s	1.7	1.1
	High	–	1.1 s	–	1.7	1.1
2	Very low	1.6 s	–	–	2.8	1.3
	Low	1.5 s	–	–	2.5	1.3
	Medium	–	–	1.7 s	2.5	1.3
	High	–	1.3 s	–	2.5	1.3
3	Very low	2.6 s	–	–	3.7	2.0
	Low	2.0 s	–	–	3.4	2.0
	Medium	–	–	2.6 s	3.4	2.0
	High	–	2.6 s	–	3.4	2.0

For a complete definition of the speed range, see Ref. [8]. The very-low-speed range represents speeds close to stall speed ($V_s \leq V < 1.3V_s$). The low-speed range represents take-off and approach speeds ($1.3V_s \leq V < 1.8V_s$). The medium-speed range represents speeds up to 70% of maximum level speed ($1.8V_s \leq V < 0.7V_{max}$). The high-speed range represents speeds from 70 to 100% of maximum level speed ($0.7V_{max} \leq V < V_{max}$). The design specification could be extracted based on the requirements in Table 12.12. The civil aircraft tend to have a lower roll control requirement and must be established by consultation with the customer.

For the case of a GA aircraft, Section 23.157 of Part 23 of FAR governs the rate of roll requirements as follows:

1. **Takeoff**. It must be possible, using a favorable combination of controls, to roll the airplane from a steady 30-deg banked turn through an angle of 60 deg, so as to reverse the direction of the turn within: (i) for an airplane of 6000 pounds or less maximum weight, 5 seconds from initiation of roll and (ii) for an airplane of over 6000 pounds maximum weight, (W+500)/1300 seconds, but not more than 10 seconds, where W is the weight in pounds. The requirement must be met when rolling the airplane in each direction with: (i) flaps in the takeoff position; (ii) landing gear retracted; (iii) for a single-engine airplane, at maximum takeoff power; and for a multi engine airplane with the critical engine inoperative and the propeller in the minimum drag position, and the other engines at maximum takeoff power; and (iv) the airplane trimmed at a speed equal to the greater of 1.2 V_{S1} or 1.1 V_{MC}, or as nearly as possible in trim for straight flight.
2. **Approach**. It must be possible, using a favorable combination of controls, to roll the airplane from a steady 30-deg banked turn through an angle of 60 deg, so as to reverse

the direction of the turn within: (i) for an airplane of 6000 pounds or less maximum weight, 4 seconds from initiation of roll and (ii) for an airplane of over 6000 pounds maximum weight, $(W + 2800)/2200$ seconds, but not more than 7 seconds, where W is the weight in pounds.

For the case of a civil transport aircraft, Section 5.147 of FAR Part 25 governs the lateral control requirements as follows:

> It must be possible to make 20° banked turns, with and against the inoperative engine, from steady flight at a speed equal to 1.3 V_{SR1}, with (i) the critical engine inoperative and its propeller (if applicable) in the minimum drag position; (ii) the remaining engines at maximum continuous power; (iii) the most unfavorable center of gravity; (iv) landing gear (a) retracted and (b) extended; (v) flaps in the most favorable CL position; and (vi) maximum takeoff weight.
>
> With the critical engine inoperative, roll response must allow normal maneuvers. Lateral control must be sufficient, at the speeds likely to be used with one engine inoperative, to provide a roll rate necessary for safety without excessive control forces or travel. Airplanes with four or more engines must be able to make 20° banked turns, with and against the inoperative engines, from steady flight at a speed equal to 1.3 V_{SR1}, with maximum continuous power, and with the airplane in the configuration prescribed by paragraph (b) of this section. With the engines operating, roll response must allow normal maneuvers (such as recovery from upsets produced by gusts and the initiation of evasive maneuvers). There must be enough excess lateral control in sideslips (up to sideslip angles that might be required in normal operation), to allow a limited amount of maneuvering and to correct for gusts.

For transport aircraft, it is suggested to also use the Class II and III military requirements as the base requirements.

The application of roll control requirements will be employed in the design of the aileron. In the aileron design process, level 1 of Table 12.12 must be considered. The application of these requirements is examined in Section 12.4. For instance, for level 1 flying qualities for a Class IV aircraft (e.g., a fighter) in an air-to-air CO (flight phase A), the minimum allowable time to achieve 90 deg (−45 deg to +45 deg) of bank angle in a roll is as short as 1.3 seconds. The time constant is defined as the time it takes for the response to reach 63% of the steady-state value.

12.3.3.2 Directional Control

In a conventional aircraft, directional control is usually maintained by the use of aerodynamic controls (e.g., rudder) alone at all airspeeds. There are a number of cases where directional control must be achievable within a specified set of limits and constraints. In this section, the most important ones are presented. Directional control characteristics enable the pilot to balance the yawing moments and control the yaw and sideslip. Sensitivity to the yaw control pedal forces shall be sufficiently high that directional control and force requirements can be met and satisfactory coordination can be achieved without unduly high pedal forces, yet sufficiently low that occasional improperly coordinated control inputs will not seriously degrade the flying qualities.

In a multi-engine aircraft, at all speeds above $1.4V_s$ with asymmetric loss of thrust from the most critical factor while the other engine(s) develop normal rated thrust, the airplane with yaw control pedals free may be balanced directionally in steady straight flight. The trim settings shall be those required for wings-level straight flight prior to the failure. When an aircraft in is directional trim with symmetric power/thrust, the trim change of propeller-driven airplanes with speed shall be such that wings-level straight flight can be maintained over a speed range of ±30% of the trim speed or ±100 knot equivalent airspeed, whichever is less (except where limited by boundaries of the service flight envelope) with the yaw control device (i.e., rudder). In the case of one engine inoperative (asymmetric thrust), it shall be possible to maintain a straight flight path throughout the operational flight envelope with yaw control device (e.g., rudder) not greater than 100 lb for levels 1 and 2 and not greater than 180 lb for level 3, without re-trimming.

Asymmetric loss of thrust may be caused by many factors including engine failure, inlet unstart, propeller failure, or propeller-drive failure. Following sudden asymmetric loss of thrust from any factor, the airplane shall be safely controllable in the cross-winds of Table 12.13 from the most unfavorable direction. An aircraft must be directionally controllable for the appropriate flight phases when any single failure or malfunction of the propulsive system, including inlet or exhaust, causes loss of thrust on one or more engines or propellers, considering also the effect of the failure or malfunction on all subsystems powered or driven by the failed propulsive system. Table 12.20 provides the maximum allowable cross-wind speed for several aircraft.

It must be possible to take off and land with normal pilot skill and technique in 90-deg cross-winds, from either side, of velocities up to those specified in Table 12.13. The rudder must be powerful enough to maintain directional trim in a cross-wind take-off/landing operation. For all airplanes except land-based airplanes equipped with cross-wind landing gear, or otherwise constructed to land in a large crabbed attitude, yaw and roll control power shall be adequate to develop at least 10 deg of sideslip in the power approach with yaw control pedal forces not exceeding some specified values as given in Ref. [8]. For level 1, roll control shall not exceed either 10 lb of force or 75% of control power available to the pilot. For levels 2 and 3, the roll control force shall not exceed 20 lb. Yaw and roll control power, in conjunction with other normal means of control, shall be adequate to maintain a straight path on the ground or other landing surface. This requirement applies in calm air and in cross-winds up to the values specified in Table 12.13 with cockpit control forces not exceeding the values specified in Ref. [8].

Table 12.13 Cross-wind velocity requirements

Level	Class	Cross-wind speed
1	I	20 knots
	II, III, and IV	30 knots
2	I	20 knots
	II, III, and IV	30 knots
3	I, II, III, and IV	One-half the value for levels 1 and 2

For the case of a GA aircraft, Section 23.147 of FAR Part 23 mandates the following lateral control requirement:

> For all airplanes, it must be shown that the airplane is safely controllable without the use of the primary lateral control system in any all-engine configuration(s) and at any speed or altitude within the approved operating envelope. For each multiengine airplane, it must be possible, while holding the wings level within 5 deg, to make sudden changes in heading safely in both directions. This ability must be shown at 1.4 V_{S1} with heading changes up to 15 deg, with: (i) the critical engine inoperative and its propeller in the minimum drag position; (ii) the remaining engines at maximum continuous power; (iii) the landing gear (a) retracted and (b) extended; and (iv) the flaps retracted. For each multiengine airplane, it must be possible to regain full control of the airplane without exceeding a bank angle of 45 deg, reaching a dangerous attitude, or encountering dangerous characteristics in the event of a sudden and complete failure of the critical engine, making allowance for a delay of 2 seconds in the initiation of recovery action appropriate to the situation, with the airplane initially in trim, in the following conditions: (i) maximum continuous power on each engine; (ii) the wing flaps retracted; and (iii) the landing gear retracted.

For the case of a civil transport aircraft, Section 25.147 of FAR Part 25 mandates the following directional control requirement:

> It must be possible, with the wings level, to yaw into the operative engine and to safely make a reasonably sudden change in heading of up to 15 deg in the direction of the critical inoperative engine. This must be shown at 1.3 V_{SR1} for heading changes up to 15 deg, and with: (i) the critical engine inoperative and its propeller in the minimum drag position; (ii) the power required for level flight at 1.3 V_{SR1}, but not more than maximum continuous power; (iii) the most unfavorable center of gravity; (iv) landing gear retracted; (v) flaps in the approach position; and (vi) maximum landing weight.

12.3.3.3 Lateral-Directional Stability

When a laterally directionally dynamically stable aircraft experiences a lateral-directional disturbance (i.e., a horizontal gust hits the vertical tail), the aircraft will oppose the disturbance, and eventually return to the initial trim point. There are static and dynamic lateral-directional stability requirements. The static lateral-directional stability is satisfied provided the following two conditions are met:

$$C_{l_\beta} \prec 0 \tag{12.5}$$

$$C_{n_\beta} \succ 0 \tag{12.6}$$

The dynamic lateral-directional stability is investigated through the response of an aircraft to a lateral-directional disturbance. In a conventional aircraft, the response of the aircraft involves a second-order oscillatory mode (often called dutch roll), plus two first-order modes (spiral and roll). This section examines some aspects of the lateral-directional handling qualities of an aircraft in terms of these three modes. For other lateral-directional handling qualities requirements, the interested reader is encouraged to consult Refs [1–5].

Table 12.14 Roll mode time constant specification (maximum value)

Flight phase	Aircraft class	T_R (s)		
		Level 1	Level 2	Level 3
A	I, IV	1.0	1.4	10
	II, III	1.4	3.0	10
B	All	1.4	3.0	10
C	I, IV	1.0	1.4	10
	II, III	1.4	3.0	10

Roll Subsidence Mode

The roll subsidence mode is a part of the aircraft response to any lateral-direction disturbance. In addition, when the trim bank angle (even in a level cruising flight where the bank angle is zero) is disturbed, a laterally directionally dynamically stable aircraft will return to its initial bank angle through a rolling motion mode. Furthermore, the roll mode indicates the rapidity of roll response to the application of roll control. As the name implies, the roll mode is a rolling motion about the x-axis that involves a change in bank angle. At any rate, the roll mode or roll subsidence mode is a first-order response which is characterized by a time constant. For an aircraft to feature acceptable lateral-directional handling qualities, the roll time constant (T_R) of the roll subsidence mode is required to be less than the specified values given in Table 12.14.

Spiral Mode

Another contributing mode of response to a lateral-directional disturbance is the spiral mode. A spiral mode is a first-order response which is characterized by a time constant. As the name implies, the spiral mode is a yawing motion about the z-axis which involves a change in yaw angle. In most conventional aircraft, the spiral mode is often unstable. Thus, there are no specific requirements for spiral stability in any aircraft. Therefore, the spiral mode is often allowed to be lightly unstable, but limits are placed on the minimum time for the mode to double the amplitude (i.e., allowable divergence of the mode).

Handling qualities specify an acceptable spiral mode by assuming that the aircraft is trimmed for straight and level flight, with no bank angle, no yaw rate, and with the cockpit controls free. The specification is given in terms of the time taken for the bank angle to double following an initial disturbance in the bank angle of up to 20 deg. The time taken must exceed the values given in Table 12.15. This table applies some requirements in the design of the aileron, which will be discussed in the next section.

Dutch-Roll Oscillation

The dutch-roll mode is a second-order mode of response to a lateral-directional disturbance and consists mainly of simultaneous sideslipping and yawing. The rolling motion is also present in most dutch rolls, however, it has a relatively negligible participation. Although the dutch-roll mode has little useful part to play in the lateral-directional dynamic stability of an aircraft, it does possess significant nuisance value. The dutch roll (i.e., second-order)

Table 12.15 Time to double amplitude in spiral mode

Aircraft class	Flight phase	Minimum time to double amplitude in spiral mode		
		Level 1 (s)	Level 2 (s)	Level 3 (s)
I and IV	A	12	8	4
	B and C	20	8	4
II and III	A, B, and C	20	8	4

Table 12.16 Dutch-roll mode handling qualities

Level	Flight phase	Aircraft class	Minimum ζ_d	Minimum $\zeta_d\omega_{n_d}$ (rad/s)	Minimum ω_{n_d} (rad/s)
1	A	I, IV	0.19	0.35	1.0
1		II, III	0.19	0.35	0.4
1	B	All	0.08	0.15	0.4
1	C	I, II, and IV	0.08	0.15	1.0
1		III	0.08	0.15	0.4
2	All	All	0.02	0.05	0.4
3	All	All	0.02	No limit	0.4

response is characterized by a damping ratio (ζ_d) and a frequency of oscillation (ω_d). The requirements of the important dutch-roll parameters, namely damping ratio, and the dutch-roll frequency are specified in Table 12.16. The frequency and damping ratio of the dutch-roll mode must exceed the values given in Table 12.16.

The lower limit on the dutch-roll damping ratio is the larger of the two values that come from the table, except that a value of 0.7 need not be exceeded for Class III. Furthermore, Class III aircraft may be exempted from some of the minimum frequency requirements, subject to specific approval. Aircraft with a large amount of roll–yaw coupling are subject to more stringent requirements. Most jet transport aircraft have dutch-roll problems, so they employ the yaw damper to augment lateral-directional dynamic stability. In the design of wing and vertical tail, and also lateral-directional control surfaces (i.e., rudder and aileron), the application of lateral-directional stability requirements will be employed.

12.4 Aileron Design

12.4.1 Introduction

The primary function of an aileron is the lateral (i.e., roll) control of an aircraft; however, it also affects the directional control. For this reason, the aileron and the rudder are usually designed concurrently. Lateral control is governed primarily through a roll rate (P). The aileron is structurally part of the wing and has two pieces, each located on the trailing edge of the outer portion of the wing left and right sections. Both ailerons are often used symmetrically, hence their geometries are identical. Aileron effectiveness is a measure of how good the deflected aileron is at producing the desired rolling moment. The generated rolling moment is a function of aileron size, aileron deflection, and distance from the aircraft fuselage center line. Unlike rudder and elevator which are displacement control, the aileron is rate control. Any change in the aileron geometry or deflection will change the roll rate, which subsequently varies constantly the roll angle.

The deflection of any control surface, including the aileron, involves a hinge moment. The hinge moments are the aerodynamic moments that must be overcome to deflect the control surfaces. The hinge moment governs the magnitude of the augmented pilot force required to move the corresponding actuator to deflect the control surface. To minimize the size and thus the cost of the actuation system, the ailerons should be designed so that the control forces are as low as possible.

In the design process of an aileron, four parameters need to be determined. They are: (i) aileron planform area (S_a), (ii) aileron chord/span (C_a/b_a), (iii) maximum up and down aileron deflection ($\pm\delta_{Amax}$), and (iv) location of inner edge of the aileron along the wing span (b_{ai}). Figure 12.10 shows the aileron geometry. As a general guidance, the typical values for these parameters are as follows: $S_a/S = 0.05–0.1$, $b_a/b = 0.2–0.3$, $C_a/C = 0.15–0.25$, $b_{ai}/b = 0.6–0.8$, and $\delta_{A_{max}} = \pm 30$ deg. Based on this, about 5–10% of the wing area is devoted to the aileron, the aileron-to-wing-chord ratio is about 15–25%, the aileron-to-wing-span ratio is about 20–30%, and the inboard aileron span is about 60–80% of the wing span. Table 12.17 illustrates the characteristics of the aileron of several aircraft.

Factors affecting the design of the aileron are: (i) the required hinge moment, (ii) the aileron effectiveness, (iii) the aerodynamic and mass balancing, (iv) the flap geometry, (v) the aircraft structure, and (vi) the cost. Aileron effectiveness is a measure of how effective the aileron deflection is in producing the desired rolling moment. Aileron effectiveness is a function of its size and its distance from the aircraft center of gravity. Hinge moments are also important because they are the aerodynamic moments that must be overcome to rotate the aileron. The hinge moments govern the magnitude of force required of the pilot

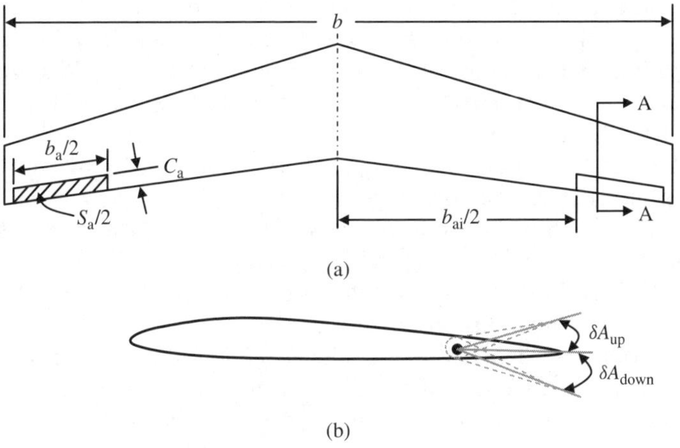

Figure 12.10 Geometry of aileron. (a) Top view of the wing and aileron; (b) Side view of the wing and aileron (Section A-A)

Table 12.17 Characteristics of aileron for several aircraft

No.	Aircraft	Type	m_{TO} (kg)	b (m)	C_A/C	Span ratio		δ_{Amax} (deg)	
						$b_i/b/2$	$b_o/b/2$	Up	Down
1	Cessna 182	Light GA	1 406	11	0.2	0.46	0.95	20	14
2	Cessna Citation III	Business jet	9 979	16.31	0.3	0.56	0.89	12.5	12.5
3	Air tractor AT-802	Agriculture	7 257	18	0.36	0.4	0.95	17	13
4	Gulfstream 200	Business jet	16 080	17.7	0.22	0.6	0.86	15	15
5	Fokker 100A	Airliner	44 450	28.08	0.24	0.6	0.94	25	20
6	Boeing 777-200	Airliner	247 200	60.9	0.22	0.32[a]	0.76[2]	30	10
7	Airbus 340-600	Airliner	368 000	63.45	0.3	0.64	0.92	25	20
8	Airbus A340-600	Airliner	368 000	63.45	0.25	0.67	0.92	25	25

[a] Inboard aileron.
[b] Outboard aileron.

to move the aileron. Therefore, great care must be used in designing the aileron so that the control forces are within acceptable limits for the pilot. Finally, aerodynamic and mass balancing deals with techniques to vary the hinge moments so that the stick force stays within an acceptable range. Handling qualities discussed in the previous section govern these factors. In this section, the principles of aileron design, design procedure, governing equations, constraints, and design steps as well as a fully solved example are presented.

12.4.2 Principles of Aileron Design

A basic item in the list of aircraft performance requirements is the maneuverability. Aircraft maneuverability is a function of engine thrust, aircraft mass moment of inertia, and control power. One of the primary control surfaces which causes the aircraft to be steered along its three-dimensional flight path (i.e., maneuver) to its specified destination is the aileron. Ailerons are like plain flaps placed at the outboard of the trailing edge of the wing. The right aileron and left aileron are deflected differentially and simultaneously to produce a rolling moment about the x-axis. Therefore, the main role of the aileron is in roll control; however, it will affect yaw control as well. Roll control is the fundamental basis for the design of an aileron.

Table 12.12 (lateral directional handling quality requirements) provides significant criteria to design the aileron. This table specifies the required time to bank an aircraft at a specified bank angle. Since the effectiveness of control surfaces is lowest at a slower speed, the roll control in take-off or landing operations is the flight phase at which the aileron is sized. Thus, in designing the aileron one must consider only level 1 and the most critical phases of flight that are usually in phase B.

Based on Newton's second law for rotational motion, the summation of all applied moments is equal to the time rate of change of angular momentum. If the mass and the geometry of the object (i.e., vehicle) are fixed, the law is reduced to a simpler version: the summation of all moments is equal to the mass moment of inertia time of the object about the axis, or rotation multiplied by the rate of change of angular velocity. In the case of a rolling motion, the summation of all rolling moments (including the aircraft

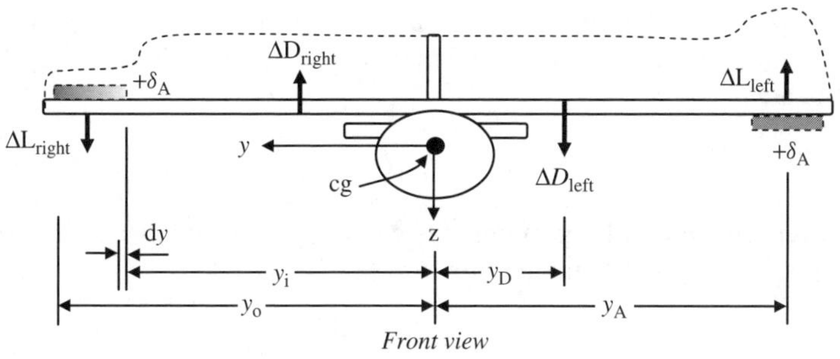

Figure 12.11 Incremental change in lift and drag in generating a rolling motion

aerodynamic moment) is equal to the aircraft mass moment of inertia about the x-axis multiplied by the time rate of change ($\partial/\partial t$) of the roll rate (P):

$$\sum L_{cg} = I_{xx}\frac{\partial P}{\partial t} \tag{12.7}$$

or

$$\dot{P} = \frac{\sum L_{cg}}{I_{xx}} \tag{12.8}$$

Generally speaking, there are two forces involved in generating the rolling moment: (i) an incremental change in wing lift due to a change in aileron angle and (ii) aircraft rolling drag force in the yz plane. Figure 12.11 illustrates the front view of an aircraft where incremental change in the lift due to aileron deflection (ΔL) and incremental drag due to the rolling speed are shown.

The aircraft in Figure 12.11 is planning to have a positive roll, so the right aileron is deflected up and the left aileron down (i.e., $+\delta_A$). The total aerodynamic rolling moment in a rolling motion is:

$$\sum M_{cg_x} = 2\Delta L \cdot y_A - \Delta D \cdot y_D \tag{12.9}$$

The factor 2 has been introduced in the moment due to lift to account for both left and right ailerons. The factor 2 is not considered for the rolling moment due to rolling drag calculations, since the average rolling drag will be computed later. The parameter y_A is the average distance between each aileron and the x-axis (i.e., aircraft center of gravity). The parameter y_D is the average distance between the rolling drag center and the x-axis (i.e., aircraft center of gravity). A typical location for this distance is about 40% of the wing semispan from the root chord.

In an aircraft with a short wing span and a large aileron (e.g., the fighter General Dynamics F-16 Fighting Falcon (Figure 4.6)), the drag does not influence the rolling speed considerably. However, in an aircraft with a long wing span and a small aileron (such as the bomber Boeing B-52 (Figures 8.6 and 9.4)), the rolling-induced drag force has a significant effect on the rolling speed. For instance, the B-52 takes about 10 seconds to have a bank angle of 45 deg at low speeds, while for a fighter such as the F-16 it takes only a fraction of a second for such a roll.

Owing to the fact that ailerons are located at some distance from the center of gravity of the aircraft, the incremental lift force generated by the ailerons deflected up/down creates a rolling moment:

$$L_A = 2\Delta L \cdot y_A \tag{12.10}$$

However, the aerodynamic rolling moment is generally modeled as a function of the wing area (S), wing span (b), and dynamic pressure ($\overline{q}$) as:

$$L_A = \overline{q} S C_l b \tag{12.11}$$

where C_l is the rolling moment coefficient and the dynamic pressure is:

$$\overline{q} = \frac{1}{2}\rho V_T^2 \tag{12.12}$$

where ρ is the air density and V_T is the aircraft true airspeed. The parameter C_l is a function of aircraft configuration, sideslip angle, rudder deflection, and aileron deflection. In a symmetric aircraft with no sideslip and no rudder deflection, this coefficient is linearly modeled as:

$$C_l = C_{l_{\delta A}} \delta_A \tag{12.13}$$

The parameter $C_{l_{\delta A}}$ is referred to as the aircraft rolling moment coefficient due to aileron deflection derivative and is also called the aileron roll control power. The aircraft rolling drag induced by the rolling speed may be modeled as:

$$D_R = \Delta D_{\text{left}} + \Delta D_{\text{right}} = \frac{1}{2}\rho V_R^2 S_{\text{tot}} C_{D_R} \tag{12.14}$$

where the aircraft average C_{D_R} is the aircraft drag coefficient in rolling motion. This coefficient is about 0.7–1.2, which includes the drag contribution of the fuselage. The parameter S_{tot} is the summation of the wing planform area, horizontal tail planform area, and vertical tail planform area:

$$S_{\text{tot}} = S_w + S_h + S_{vt} \tag{12.15}$$

The parameter V_R is the rolling linear speed in a rolling motion and is equal to the roll rate (P) multiplied by the average distance between the rolling drag center (see Figure 12.11) along the y-axis and the aircraft center of gravity:

$$V_R = P \cdot y_D \tag{12.16}$$

Since all three lifting surfaces (wing, horizontal tail, and vertical tail) contribute to the rolling drag, y_D is in fact the average of three average distances. The non-dimensional control derivative $C_{l_{\delta A}}$ is a measure of the roll control power of the aileron; it represents the change in rolling moment per unit change of aileron deflection. The larger the $C_{l_{\delta A}}$, the more effective the aileron is at creating a rolling moment. This control derivative may be calculated using the method introduced in Ref. [17]. However, an estimate of the roll control power for an aileron is presented in this section based on a simple *strip*

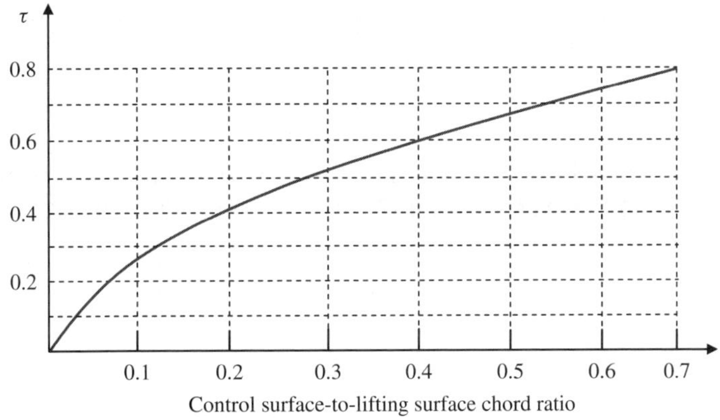

Figure 12.12 Control surface angle of attack effectiveness parameter

integration method. The aerodynamic rolling moment due to the lift distribution may be written in coefficient form as:

$$\Delta C_{\mathrm{l}} = \frac{\Delta L_{\mathrm{A}}}{\overline{q}Sb} = \frac{\overline{q}C_{L_{\mathrm{A}}}C_{\mathrm{a}}y_{\mathrm{A}}\,\mathrm{d}y}{\overline{q}Sb} = \frac{C_{L_{\mathrm{A}}}C_{\mathrm{a}}y_{\mathrm{A}}\,\mathrm{d}y}{Sb} \tag{12.17}$$

The section lift coefficient $C_{L_{\mathrm{A}}}$ on the sections containing the aileron may be written as:

$$C_{L_{\mathrm{A}}} = C_{L_\alpha}\alpha = C_{L_\alpha}\frac{\mathrm{d}\alpha}{\mathrm{d}\delta_{\mathrm{A}}}\delta_{\mathrm{A}} = C_{L_\alpha}\tau_{\mathrm{a}}\cdot\delta_{\mathrm{A}} \tag{12.18}$$

where τ_{a} is the aileron effectiveness parameter and is obtained from Figure 12.12, given the ratio between aileron chord and wing chord. Figure 12.12 is a general representative of the control surface effectiveness, it may be applied to the aileron (τ_{a}), elevator (τ_{e}), and rudder (τ_{r}). Thus, in Figure 12.12, the subscript of the parameter τ is dropped to indicate the generality.

Integrating over the region containing the aileron yields

$$C_{\mathrm{l}} = \frac{2C_{L_{\alpha\mathrm{w}}}\tau\delta_{\mathrm{A}}}{Sb}\int_{y_{\mathrm{i}}}^{y_{\mathrm{o}}} Cy\,\mathrm{d}y \tag{12.19}$$

where $C_{L_{\alpha\mathrm{w}}}$ has been corrected for three-dimensional flow and the factor 2 is added to account for the two ailerons. For the calculation in this technique, the wing sectional lift curve slope is assumed to be constant over the wing span. Therefore, the aileron sectional lift curve slope is equal to the wing sectional lift curve slope. The parameter y_{i} represents the inboard position of the aileron with respect to the fuselage center line, and y_{o} the outboard position of the aileron with respect to the fuselage center line (see Figure 12.11). The aileron roll control derivative can be obtained by taking the derivative with respect to δ_{A}:

$$C_{\mathrm{l}_{\delta\mathrm{A}}} = \frac{2C_{L_{\alpha\mathrm{w}}}\tau}{Sb}\int_{y_{\mathrm{i}}}^{y_{\mathrm{o}}} Cy\,\mathrm{d}y \tag{12.20}$$

The wing chord (C) as a function of y (along the span) for a tapered wing can be expressed by the following relationship:

$$C = C_r \left[1 + 2\left(\frac{\lambda - 1}{b}\right) y \right] \tag{12.21}$$

where C_r denotes the wing root chord and λ is the wing taper ratio. Substituting this relationship back into the expression for $C_{l_{\delta A}}$ (Equation (12.20)) yields:

$$C_{l_{\delta A}} = \frac{2C_{L_{\alpha w}}\tau}{Sb} \int_{y_i}^{y_o} C_r \left[1 + 2\left(\frac{\lambda - 1}{b}\right) y \right] y \, dy \tag{12.22}$$

or

$$C_{l_{\delta A}} = \frac{2C_{L_{\alpha w}}\tau C_r}{Sb} \left[\frac{y^2}{2} + \frac{2}{3}\left(\frac{\lambda - 1}{b}\right) y^3 \right]_{y_i}^{y_o} \tag{12.23}$$

This equation can be employed to estimate the roll control derivative $C_{l_{\delta A}}$ using the aileron geometry and estimating τ from Figure 12.12. Getting back to Equation (12.12), there are two pieces of aileron; one each at the left and right sections of the wing. These two pieces may have a similar magnitude of deflection or slightly different deflections, due to the adverse yaw. At any rate, only one value will enter into the calculation of the rolling moment. Thus, an average value of aileron deflection will be calculated as follows:

$$\delta_A = \frac{1}{2}\left[\left|\delta_{A_{left}}\right| + \left|\delta_{A_{right}}\right| \right] \tag{12.24}$$

The sign of this δ_A will later be determined based on the convention introduced earlier; a positive δ_A will generate a positive rolling moment. Substituting Equation (12.9) into Equation (12.7) yields:

$$L_A + \Delta D \cdot y_D = I_{xx} \dot{P} \tag{12.25}$$

As the name implies, $\dot{P}$ is the time rate of change of the roll rate:

$$\dot{P} = \frac{d}{dt} P \tag{12.26}$$

In contrast, the angular velocity about the x-axis (P) is defined as the time rate of change of the bank angle:

$$P = \frac{d}{dt} \Phi \tag{12.27}$$

Combining Equations (12.26) and (12.27) and removing dt from both sides results in:

$$\dot{P} \, d\Phi = P \, dP \tag{12.28}$$

Assuming that the aircraft is initially at a level cruising flight (i.e., $P_o = 0$, $\phi_o = 0$), both sides may be integrated as:

$$\int_0^{\phi} \dot{P} \, d\Phi = \int_0^{P_{ss}} P \, dP \tag{12.29}$$

Thus, the bank angle due to a rolling motion is obtained as:

$$\Phi = \int \frac{P}{\dot{P}} \mathrm{d}P \qquad (12.30)$$

where $\dot{P}$ is obtained from Equation (12.25). Thus:

$$\Phi = \int_0^{P_{ss}} \frac{I_{xx} P}{L_A + \Delta D \cdot y_D} \mathrm{d}P \qquad (12.31)$$

Both the aerodynamic rolling moment and the aircraft drag due to rolling motion are functions of roll rate. Plugging these two moments into Equation (12.31) yields:

$$\Phi_1 = \int_0^{P_{ss}} \frac{I_{xx} P}{\overline{q} S C_l b + \frac{1}{2}\rho \left(P \cdot y_D\right)^2 \left(S_w + S_h + S_{vt}\right) C_{D_R} \cdot y_D} \mathrm{d}P \qquad (12.32)$$

The aircraft rate of roll rate response to the aileron deflection has two distinct states: (i) a transient state and (ii) a steady state (see Figure 12.13). The integral limit for the roll rate (P) in Equation (12.32) is from an initial trim point of no roll rate (i.e., wing level and $P_o = 0$) to a steady-state value of roll rate (P_{ss}). Since the aileron is featured as a rate control, the deflection of the aileron will eventually result in a steady-state roll rate (Figure 12.13). Thus, unless the ailerons are returned to the initial zero deflection, the aircraft will not stop at a specific bank angle. Table 12.12 defines the roll rate requirements in terms of the desired bank angle (Φ_2) for the duration of t seconds. Equation (12.32) has a closed-form solution and can be solved to determine the bank angle (Φ_1) when the roll rate reaches its steady-state value.

When the aircraft has a steady-state (P_{ss}) roll rate, the new bank angle (Figure 12.14) after Δt seconds (i.e., $t_2 - t_{ss}$) is readily obtained by the following linear relationship:

$$\Phi_2 = P_{ss} \cdot \left(t_2 - t_{ss}\right) + \Phi_1 \qquad (12.33)$$

Due to the fact that the aircraft drag due to roll rate is not constant and is increased with an increase in the roll rate, the rolling motion is not linear. This implies that the variation

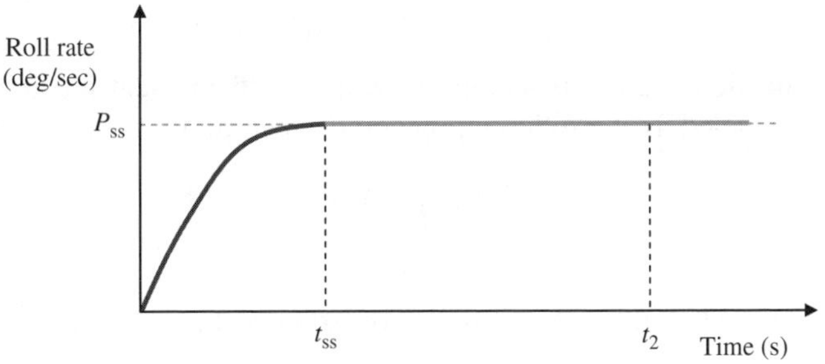

Figure 12.13 Aircraft roll rate response to an aileron deflection

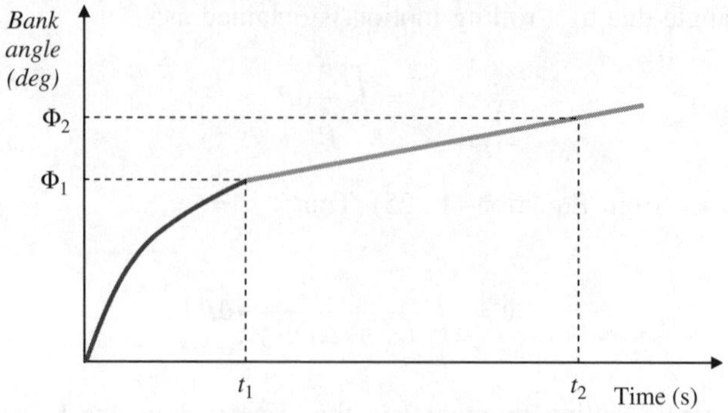

Figure 12.14 Aircraft bank angle response to an aileron deflection

of the roll rate is not linear, and there is an angular rotation about the x-axis. However, until the resisting moment against the rolling motion is equal to the aileron-generated aerodynamic rolling moment, the aircraft will experience an angular acceleration about the x-axis. Soon after the two rolling moments are equal, the aircraft will continue to roll with a constant roll rate (P_{ss}). The steady-state value for the roll rate (P_{ss}) is obtained by considering the fact that when the aircraft is rolling with a constant roll rate, the aileron-generated aerodynamic rolling moment is equal to the moment of aircraft drag in the rolling motion:

$$L_A = \Delta D_R \cdot y_D \tag{12.34}$$

Combining Equations (12.14)–(12.16), the aircraft drag due to the rolling motion is obtained as:

$$D_R = \frac{1}{2}\rho\left(P \cdot y_D\right)^2\left(S_w + S_h + S_{vt}\right)C_{D_R} \tag{12.35}$$

Inserting Equation (12.35) into Equation (12.34) yields:

$$L_A = \frac{1}{2}\rho\left(P \cdot y_D\right)^2\left(S_w + S_h + S_{vt}\right)C_{D_R} \cdot y_D \tag{12.36}$$

Solving for the steady-state roll rate (P_{ss}) results in:

$$P_{ss} = \sqrt{\frac{2 \cdot L_A}{\rho\left(S_w + S_h + S_{vt}\right)C_{D_R} \cdot y_D^3}} \tag{12.37}$$

In contrast, Equation (12.32) is simply a definite mathematical integration. This integration may be modeled as the following general integration problem:

$$y = k\int \frac{x\,dx}{x^2 + a^2} \tag{12.38}$$

According to Ref. [18], there is a closed-form solution to such integration as follows:

$$y = k\frac{1}{2}\ln\left(x^2 + a^2\right) \tag{12.39}$$

The parameters k and a are obtained by comparing Equation (12.38) with Equation (12.32):

$$k = \frac{2I_{xx}}{\rho y_D^3 \left(S_w + S_h + S_{vt}\right) C_{D_R}} \tag{12.40}$$

$$a^2 = \frac{V^2 S C_l b}{\left(S_w + S_h + S_{vt}\right) C_{D_R} y_D^3} \tag{12.41}$$

Hence, the solution to the integration in Equation (12.32) is determined as:

$$\Phi_1 = \left[\frac{I_{xx}}{\rho y_D^3 \left(S_w + S_h + S_{vt}\right) C_{D_R}} \ln\left(P^2 + \frac{V^2 S C_l b}{\left(S_w + S_h + S_{vt}\right) C_{D_R} y_D^3}\right)\right]_0^{P_{ss}} \tag{12.42}$$

Applying the limits (from 0 to P_{ss}) to the solution results in:

$$\Phi_1 = \frac{I_{xx}}{\rho y_D^3 \left(S_w + S_h + S_{vt}\right) C_{D_R}} \ln\left(P_{ss}^2\right) \tag{12.43}$$

Recall that we are looking to determine the aileron roll control power. In other words, we need to find how long it takes (t_2) to bank to a desired bank angle when the ailerons are deflected. This duration tends to have two parts: (i) the duration (t_{ss}) for the aircraft to reach the steady-state roll rate (P_{ss}) and (ii) the time (Δt_R) to roll linearly from Φ_{ss} to Φ_2 (see Figure 12.14):

$$t_2 = t_{ss} + \Delta t_R \tag{12.44}$$

where:

$$\Delta t_R = \frac{\Phi_2 - \Phi_1}{P_{ss}} \tag{12.45}$$

Comparing Figures 12.13 and 12.14 indicates that $t_1 = t_{ss}$. The time (t_{ss}) that it takes an aircraft to achieve a steady-state roll rate due to aileron deflection is a function of angular acceleration ($\dot{P}$). Based on classical dynamics, this accelerated roll can be expressed as:

$$\Phi_1 = \Phi_o + \frac{1}{2}\dot{P}\, t_{ss}^2 \tag{12.46}$$

It is assumed that the aircraft is initially at a wing-level flight condition (i.e., $\Phi_o = 0$). Hence:

$$t_1 = t_{ss} = \sqrt{\frac{2\Phi_1}{\dot{P}}} \tag{12.47}$$

where Φ_1 is determined from Equation (12.43). In addition, in an accelerated rolling motion, the relationship between the final roll rate (P_1) and the initial roll rate (P_o) is a function of the rate of roll rate ($\dot{P}$) and the final bank angle (Φ_1). Based on classical dynamics, an accelerated rolling motion can be expressed as:

$$P_1^2 - P_o^2 = 2\dot{P}\,\Phi_1 \tag{12.48}$$

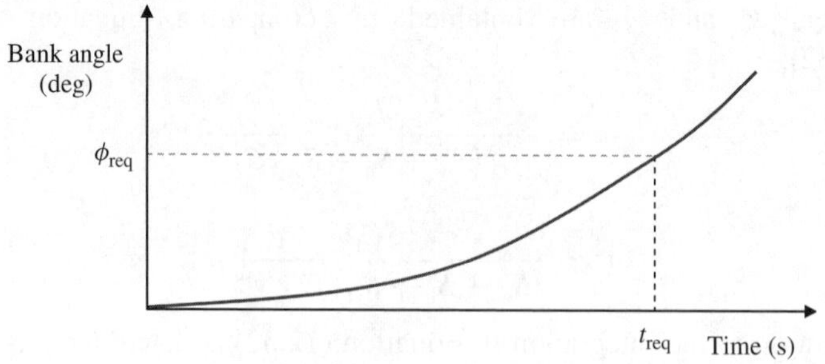

Figure 12.15 Bank angle versus time

It is assumed that the aircraft is initially at a wing-level flight condition (i.e., $P_o = 0$) and the new roll rate is the steady-state roll rate (i.e., $P_1 = P_{ss}$). Thus:

$$\dot{P} = \frac{P_{ss}^2}{2\Phi_1} \tag{12.49}$$

where P_{ss} is determined from Equation (12.45).

For GA and transport aircraft, the time to reach the steady-state rolling motion (t_1) is long (more than 10 seconds). Thus, the application of Equations (12.48) and (12.49) is often not needed for aileron design, since the roll requirement is within a few seconds. However, for a fighter aircraft and missile, the rolling motion (see Figure 12.15) is very fast (the time t_1 is within a few seconds), so the application of Equations (12.48) and (12.49) is usually needed for aileron design. For this reason, when the bank angle (Φ_1) corresponding to steady-state roll rate (P_{ss}) is beyond 90 deg, Equation (12.46) serves as a relationship for the required time to reach a desired bank angle. Therefore, the duration required (t_{req}) to achieve a desired bank angle (Φ_{des}) will be determined as follows:

$$t_2 = \sqrt{\frac{2\Phi_{des}}{\dot{P}}} \tag{12.50}$$

The equations and relationships introduced and developed in this section provide the necessary tools to design the aileron to satisfy the roll control requirements. Table 12.12 addresses the military aircraft roll control requirements; for a civil aircraft, it is suggested to adopt a similar list of requirements. To have the greatest roll control by an aileron to produce a rolling moment, consider the aileron outboard of the wing toward the wing tip. Therefore, a flap will be considered at the inboard of the wing. This approach will result in the smallest, lightest, and most economical aileron surfaces. The aileron design technique and the design procedure will be presented in Section 12.4.4.

12.4.3 Aileron Design Constraints

Any design problem in engineering is usually limited by various constraints and the aileron design is no exception. In this section, a number of constraints on the aileron design will be introduced.

12.4.3.1 Aileron Reversal

A number of aircraft, when flying near their maximum speed, are subject to an important aeroelastic phenomenon. No real structure is ideally rigid, and it has static and dynamic flexibility. Wings are usually produced from aerospace materials such as aluminum and composite materials and have structures which are flexible. This flexibility causes the wing to be unable to maintain its geometry and integrity, especially in high-speed flight operations. This phenomenon, which is referred to as aileron reversal, negatively influences the aileron effectiveness.

Consider the right section of a flexible wing with a downward-deflected aileron to create a negative rolling moment. At subsonic speeds, the increment in aerodynamic load due to aileron deflection has a centroid somewhere near the middle of the wing chord. At supersonic speeds, the control load acts mainly on the deflected aileron itself, and hence has its centroid even further to the rear. If this load centroid is behind the elastic axis of the wing structure, then a nose-down twist (α_{twist}) of the main wing surface (about the y-axis) results. The purpose of this deflection was to raise the right wing section. However, the wing twist reduces the wing angle of attack, and leads to a reduction of the lift on the right section of the wing (Figure 12.16). In extreme cases, the down-lift due to aeroelastic twist will exceed the commanded up-lift, so the net effect is reversed. This change in the lift direction will consequently generate a positive rolling moment.

This undesired rolling moment implies that the aileron has lost its effectiveness and the roll control derivative $C_{l_{\delta A}}$ has changed its sign. Such a phenomenon is referred to as aileron reversal. This phenomenon poses a significant constraint on the aileron design. In addition, the structural design of the wing must examine this aeroelasticity effect of the aileron deflection. The aileron reversal often occurs at high speeds. Most high-performance aircraft have an aileron reversal speed beyond which the ailerons lose their effectiveness. The F-14 fighter aircraft experiences aileron reversal at high speed.

Clearly, such aileron reversal is not acceptable within the flight envelope, and must be considered during the design process. A number of solutions for this problem are: (i) make the wing stiffer, (ii) limit the range of aileron deflections at high speed, (iii) employ two sets of ailerons – one set at the inboard wing section for high-speed flight and one set at the outboard wing section for high-speed flight, (iv) reduce the aileron chord, (v) use a spoiler for roll control, and (vi) move the ailerons toward the wing inboard section. The transport aircraft Boeing 747 has three different types of roll control device: inboard ailerons, outboard ailerons, and spoilers. The outboard ailerons are disabled except in low-speed flights when the flaps are also deflected. Spoilers are essentially flat plates of about 10–15% chord located just ahead of the flaps. When the spoilers are raised,

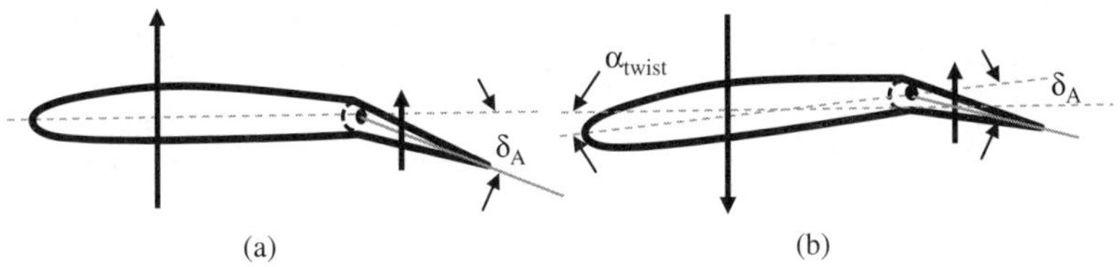

Figure 12.16 Aileron reversal. (a) An ideal and desired aileron; (b) An aileron with aileron reversal

they cause a flow separation and local loss of lift. Thus, to avoid roll reversal within the operational flight envelope, the wing structure must be designed with sufficient stiffness.

12.4.3.2 Adverse Yaw

When an airplane is banked to execute a turn, it is desired that the aircraft yaws and rolls simultaneously. Furthermore, it is beneficial to have the yawing and rolling moments in the same direction (i.e., both either positive or negative). For instance, when an aircraft is to turn to the right, it should be rolled (about the x-axis) clockwise and yawed (about the z-axis) clockwise. In such a turn, the pilot will be comfortable. Such a yawing moment is referred to as pro-verse yaw, and such a turn is a prerequisite for a coordinated turn. This yaw keeps the aircraft pointing into the relative wind. In contrast, if the aircraft yaws in a direction opposite to the desired turn direction (i.e., a positive roll, but a negative yaw), pilot will have an undesirable feeling and the aircraft turn is not coordinated. This yawing moment is referred to as adverse yaw. When a turn is not coordinated, the aircraft will either slip or skid.

To see why and how these turns may happen, see Figure 12.17 where the pilot is planning to turn to the right. For such a goal, the pilot must apply a positive aileron deflection (i.e., left-aileron down and right-aileron up). The lift distribution over the wing in a cruising flight is symmetric; that is, the right wing section lift and the left wing section lift are the same. When the left aileron is deflected down and the right aileron is deflected up, the lift distribution varies such that the right wing section lift is more than the left wing section lift. Such deflections create a clockwise rolling moment (Figure 12.17(a)), as desired.

However, the aileron deflection simultaneously alters the induced drag of the right and left wing differently. Recall that the wing drag has two components: zero-lift drag (D_o) and induced drag (D_i). The wing-induced drag is a function of the wing lift coefficient

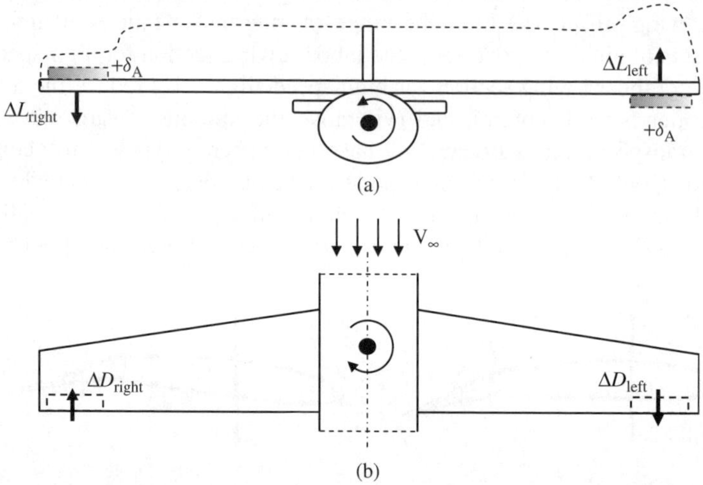

Figure 12.17 Adverse yaw due to wing drag. (a) Front view (positive roll); (b) Down view (negative yaw)

($C_{D_i} = K \cdot C_L^2$). Since the right wing section local lift coefficient is higher than the left wing section local lift coefficient, the right wing section drag is higher than the left wing section drag. The drag is an aerodynamic force and has an arm relative to the aircraft center of gravity. The drag direction is rearward, so this wing/drag couple generates a negative (see Figure 12.17(b)) yawing moment (i.e., adverse yaw). Thus, if the rudder is not deflected simultaneously with the aileron deflection, the direction of the aileron-generated rolling moment and the wing/drag-generated yawing moment would not be coordinated. Thus, when a pilot deflects a conventional aileron to make a turn, the aircraft will initially yaw in a direction opposite to that expected.

The phenomenon of adverse yaw imposes a constraint on the aileron design. To avoid such an undesirable yawing motion (i.e., adverse yaw), there are a number of solutions, four of which are as follows: (i) employ a simultaneous aileron/rudder deflection so as to eliminate the adverse yaw. This requires an interconnection between the aileron and the rudder. (ii) Differential ailerons; that is, the up-deflection of the aileron on the one side is greater than the down-deflection of the aileron on the other side. This causes an equal induced drag in the right and left wing during a turn. (iii) Employ a Frise aileron, in which the aileron hinge line is higher than the regular location. (iv) Employ a spoiler. Both the Frise aileron and the spoiler create a wing drag such that both wing section drags are balanced. Most Cessna aircraft use Frise ailerons, but most Piper aircraft employ differentially deflected ailerons. The critical condition for an adverse yaw occurs when the airplane is flying at slow speeds (i.e., high lift coefficient). This phenomenon means that the designer must consider the application of one or a combination of the above-mentioned techniques to eliminate adverse yaw.

12.4.3.3 Flap

The wing trailing edge in a conventional aircraft is the home for two control surfaces, one primary (i.e., aileron) and one secondary (i.e., trailing edge high-lift device such as flap). As the aileron and the flap are next to each other along the wing trailing edge, they impose a span limit on one another (Figure 12.16). The balance between the aileron span (b_a) and the flap span (b_f) is a function of the priority of roll control over take-off/landing performance. To improve the roll control power, the ailerons are to be placed on the outboard and the flap on the inboard part of the wing sections. The application of a high-lift device applies another constraint on the aileron design, which must be dealt with in the aircraft design process.

The spanwise extent of the aileron depends on the amount of span required for the trailing edge high-lift devices. In general, the outer limit of the flap is at the spanwise station where the aileron begins. The exact span needed for ailerons depends primarily on the roll control requirements. A low-speed aircraft usually utilizes about 40% of the total wing semispan for ailerons. This means that the flaps can start at the side of the fuselage and extend to the 60% semispan station. However, with the application of spoilers, the ailerons are generally reduced in size, and the flaps may extend to about 75% of the wing semispan. Furthermore, if a small inboard aileron is provided for gentle maneuvers, the effective span of the flaps is reduced.

If the take-off/landing performance is of higher importance in the priority list, try to devote a small span to the aileron so that a large span can be occupied by powerful flaps. This in turn means a lower stall speed and greater safety. In contrast, if the roll control has higher priority than the take-off/landing performance, the ailerons should be designed before the flaps are designed. Due to the importance of the roll control in a fighter aircraft, the span of the flaps must be selected to be as short as possible, so that the span of the aileron is long enough. Therefore, in a fighter aircraft, it is advised to design the aileron prior to designing the flap. In contrast, in the case of civil GA and transport aircraft, it is recommended to design the flap first, while in the case of a fighter aircraft, design the aileron first.

12.4.3.4 Wing Rear Spar

Another aileron design constraint in a conventional aircraft is applied by the wing rear spar. The aileron needs a hinge line to rotate about and to provide the aileron with sufficient freedom to operate. To have a lighter and less complicated wing structure, it is advised to consider the wing rear spar as the most forward limit for the aileron. This may limit the aileron chord but at the same time, improves the wing structural integrity. In addition, it is structurally better to have the same chord for aileron and flaps. This selection results in a lighter structure and allows the rear spar to hold both the flap and the aileron. Therefore, the aileron-to-wing attachment through the rear spar (see Figure 12.18) is considered as both a constraint and, at the same time, an attachment point.

12.4.3.5 Aileron Stall

When ailerons are deflected more than about 20–25 deg, flow separation tends to occur. Thus, the ailerons will lose their effectiveness. Furthermore, close to wing stall, even a small downward aileron deflection can produce flow separation and loss of roll control effectiveness. To prevent loss of roll control effectiveness, it is recommended to consider the aileron maximum deflection to be less than 25 deg (both up and down). Hence, the maximum aileron deflection is dictated by the aileron stall requirement. Table 12.19 provides a technique to determine the stall angle of a lifting surface (e.g., wing) when its control surface (e.g., aileron) is deflected.

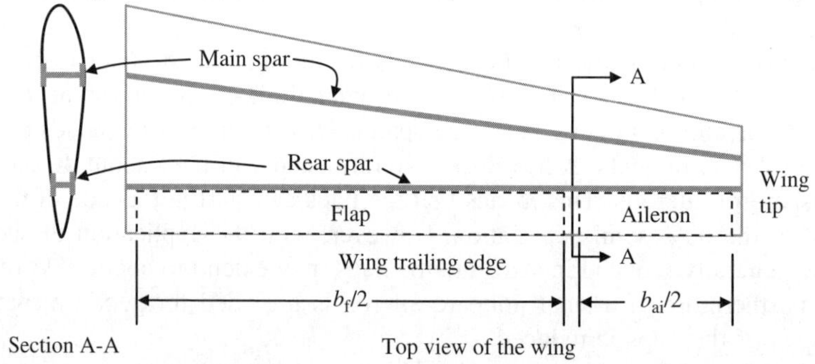

Figure 12.18 Flap, aileron, and rear spar

12.4.3.6 Wing Tip

Due to a spanwise component of airflow along the wing span, there is a tendency for the flow to leak around the wing tips. This flow establishes a circulatory motion that trails downstream of the wing. Thus, a trailing vortex is created at each wing tip. To consider the effects of vortex flow at the tip of the wing, the span of the ailerons must not run toward the wing tip. In other words, some distance must exist between the outer edge of the aileron and the tip of the wing (see Figure 12.16).

12.4.4 Steps in Aileron Design

In Sections 12.4.1–12.4.3, the aileron function, design criteria, parameters, governing rules and equations, formulation, and design requirements have been developed and presented. In addition, Section 12.3 introduces the roll control and lateral handling quality requirements for various aircraft and flight phases. In this section, the aileron design procedures in terms of design steps are introduced. It must be noted that there is no unique solution to satisfy the customer requirements in designing an aileron. Several aileron designs may satisfy the roll control requirements, but each will have unique advantages and disadvantages. Based on the systems engineering approach, the aileron detail design begins with identifying and defining design requirements and ends with optimization. The following are the aileron design steps for a conventional aircraft:

1. Layout design requirements (e.g., cost, control, structure, manufacturability, and operational).
2. Select roll control surface configuration.
3. Specify maneuverability and roll control requirements.
4. Identify the aircraft class and critical flight phase for roll control.
5. Identify the handling quality design requirements (Section 12.3) from resources such as aviation standards (e.g., Table 12.12). The design requirements primarily include the time (t_{req}) that it takes an aircraft to roll from an initial bank angle to a specified bank angle. The total desired bank angle is denoted as Φ_{des}.
6. Specify/select the inboard and outboard positions of the aileron as a function of wing span (i.e., b_{a_i}/b and b_{a_o}/b). If the flaps are already designed, identify the outboard position of the flap then consider the inboard location of the aileron to be next to the outboard position of the flap.
7. Specify/select the ratio between the aileron chord and the wing chord (i.e., C_a/C). An initial selection for the aileron leading edge may be considered as next to the wing rear spar.
8. Determine the aileron effectiveness parameter (τ_a) from Figure 12.12.
9. Calculate the aileron rolling moment coefficient derivative ($C_{l_{\delta A}}$). You may use references such as [17] or estimate the derivative by employing Equation (12.23).
10. Select the maximum aileron deflection ($\delta_{A_{max}}$). A typical value is about ±25 deg.
11. Calculate the aircraft rolling moment coefficient (C_l) when the aileron is deflected with the maximum deflection (Equation (12.13)). Both positive and negative deflections will serve the same.
12. Calculate the aircraft rolling moment (L_A) when the aileron is deflected with the maximum deflection (Equation (12.10)).

13. Determine the steady-state roll rate (P_{ss}) employing Equation (12.37).
14. Calculate the bank angle (Φ_1) at which the aircraft achieves the steady-state roll rate (Equation (12.43)).
15. Calculate the aircraft rate of roll rate ($\dot{P}$) that is produced by the aileron rolling moment until the aircraft reaches the steady-state roll rate (P_{ss}) by using Equation (12.49).
16. If the bank angle (Φ_1) calculated in step 14 is greater than the bank angle (Φ_{req}) of step 5, determine the time (t) that it takes the aircraft to achieve the desired bank angle using Equation (12.50). The desired bank angle is determined in step 5.
17. If the bank angle (Φ_1) calculated in step 14 is less than the bank angle (Φ_{req}) of step 5, determine the time (t_2) that it takes the aircraft to reach the desired bank angle (Φ_2 or Φ_{req}) using Equations (12.44) and (12.45).
18. Compare the roll time obtained in step 16 or 17 with the required roll time (t_{req}) expressed in step 5. In order for the aileron design to be acceptable, the roll time obtained in step 16 or 17 must be equal to or slightly longer than the roll time specified in step 5.
19. If the duration obtained in step 16 or 17 is equal to or longer than the duration (t_{req}) stated in step 5, the aileron design requirement has been met and move to step 23.
20. If the duration obtained in step 16 or 17 is shorter than the duration (t_{req}) stated in step 5, the aileron design has not met the requirement. The solution is to either increase the aileron size (aileron span or chord) or increase the aileron maximum deflection.
21. If the aileron geometry is changed, return to step 7. If the aileron maximum deflection is changed, return to step 10.
22. In case an increase in the geometry of the aileron does not resolve the problem, the entire wing must be redesigned or the aircraft configuration must be changed.
23. Check for aileron stall when deflected by its maximum deflection angle. If aileron stall occurs, the deflection must be reduced. Return to step 10.
24. Check the features of adverse yaw. Select a solution to prevent it.
25. Check the aileron reversal at high speed. If it occurs, either redesign the aileron or reinforce the wing structure.
26. Apply aerodynamic balance/mass balance, if necessary (Section 12.7).
27. Optimize the aileron design.
28. Calculate the aileron span, chord, area, and draw the final design for the aileron.

12.5 Elevator Design

12.5.1 Introduction

A fundamental requirement of a safe flight is longitudinal control; which is assumed to be the primary function of an elevator. An aircraft must be longitudinally controllable, as well as maneuverable within the flight envelope (Figure 12.7). In a conventional aircraft, the longitudinal control is primarily applied though the deflection of elevator (δ_E), and engine throttle setting (δ_T). Longitudinal control is governed through pitch rate (Q) and consequently angular acceleration ($\ddot{\theta}$) about the y-axis (or rate of pitch rate). Longitudinal control of an aircraft is achieved by providing an incremental lift force on horizontal tail.

Thus, elevator which is classified as a primary control surface is considered as a pitch control device.

The incremental tail lift can be generated by deflecting the entire tail or by deflecting elevator which is located at the tail trailing edge. Since the horizontal tail is located at some distance from the aircraft center of gravity, the incremental lift force creates a pitching moment about the aircraft cg. Pitch control can be achieved by changing the lift on either aft horizontal tail or canard.

There are two groups of requirements in the aircraft longitudinal controllability: (i) pilot force and (ii) aircraft response to the pilot input. In order to deflect the elevator, the pilot must apply a force to stick/yoke/wheel and hold it (in the case of an aircraft with a stick-fixed control system). In an aircraft with a stick-free control system, the pilot force is amplified through such devices as tab and spring. The pilot force analysis is out of scope of this text; the interested reader is referred to study references such as [11, 12].

In a conventional symmetric aircraft, the longitudinal control is not coupled with the lateral-directional control. Thus, the design of the elevator is almost entirely independent of the design of the aileron and the rudder. This issue simplifies the design of the elevator. In the design of the elevator, four parameters should be determined. They are: (i) elevator planform area (S_E), (ii) elevator chord (C_E), (iii) elevator span (b_E), and (iv) maximum elevator deflection ($\pm\delta_{E_{max}}$). As a general guidance, the typical values for these parameters are as follows: $S_E/S_h = 0.15 - 0.4$, $b_E/b_h = 0.8{-}1$, $C_E/C_h = 0.2{-}0.4$, $\delta_{E_{max\_up}} = -25\,\text{deg}$, and $\delta_{E_{max\_down}} = +20\,\text{deg}$. Figure 12.19 shows the geometry of the horizontal tail and elevator. As a convention, the up deflection of elevator is denoted negative, and down deflection as positive. Thus a, negative elevator deflection is creating a negative horizontal tail list while generating a positive (nose up) pitching moment.

Prior to the design of elevator, the wing and horizontal tail must be designed, as well as the most aft and most forward locations of aircraft center of gravity must be known. In this section, principals of elevator design, design procedure, governing equations, constraints, and design steps as well as a fully solved example are presented.

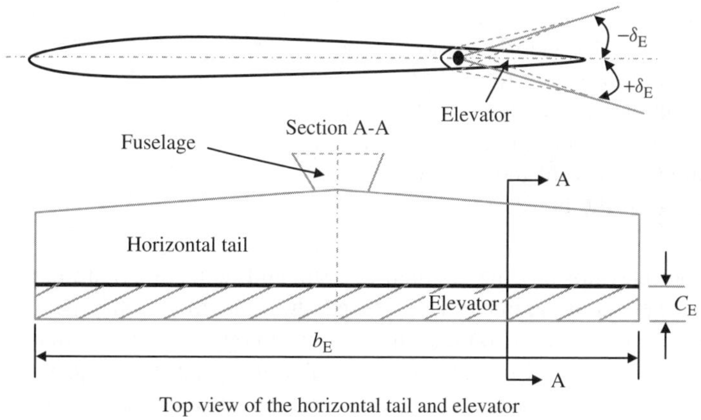

Figure 12.19 Horizontal tail and elevator geometry

12.5.2 Principles of Elevator Design

The elevator is a primary control surface placed on the trailing edge of the horizontal tail or canard. Longitudinal control and longitudinal trim are two main functions of the elevator, and it has a minor influence on the longitudinal stability. The elevator is flap-like and is deflected up and down. With this deflection, the camber of the airfoil of the tail is changed, and consequently the tail lift coefficient (C_{L_h}) is changed. The main objective of elevator deflection is to increase or decrease the tail plane lift and hence the tail plane pitching moment.

Factors affecting the design of an elevator are elevator effectiveness, elevator hinge moment, and elevator aerodynamic and mass balancing. The elevator effectiveness is a measure of how effective the elevator deflection is in producing the desired pitching moment. The elevator effectiveness is a function of elevator size and tail moment arm. Hinge moment is also important, because it is the aerodynamic moment that must be overcome to rotate the elevator. The hinge moment governs the magnitude of force required of the pilot to move the stick/yoke/wheel. Therefore, great care must be used in designing an elevator so that the stick force is within acceptable limits for the pilot. Aerodynamic and mass balancing (see Section 12.7) deal with the technique to vary the hinge moment so that the stick force stays within an acceptable range, and no aeroelastic phenomenon occurs.

The longitudinal control handling quality requirements during take-off are stated as follows: in an aircraft with a tricycle landing gear, the pitch rate should have a value such that the take-off rotation does not take longer than a specified length of time. Since the take-off rotation dynamics is governed by Newton's second law, the take-off rotation time may readily be expressed in terms of the aircraft angular acceleration ($\ddot{\theta}$) about the main gear rotation point. For instance, in a transport aircraft, the acceptable value for the take-off rotation time is 3–5 seconds. The equivalent value for the angular rotation rate to achieve such a requirement is 4–6 deg/s^2. This requirement must be satisfied when the aircraft center of gravity is located at the most forward location. Table 12.9 provides take-off angular acceleration requirements for various types of aircraft. These specifications are employed in the design of elevator.

In the elevator detail design process, the following parameters must be determined:

1. elevator chord-to-tail chord ratio (C_E/C_h);
2. elevator span-to-tail span ratio (b_E/b_h);
3. maximum up-elevator deflection ($-\delta_{E_{max}}$);
4. maximum down-elevator deflection ($+\delta_{E_{max}}$);
5. aerodynamic balance of the elevator;
6. mass balance of the elevator.

The first four elevator parameters (chord, span, and deflections) are interrelated. When the value of one elevator parameter is increased, the value of the other parameters could be decreased. In contrast, each parameter has a unique constraint. For instance, the elevator maximum deflection should be less than the value that causes flow separation or causes the horizontal tail to stall. In addition, ease of fabrication suggests having an elevator chord of span that is more convenient. Thus, for simplicity in design and manufacture, the elevator span is often selected to be equal to the horizontal tail span (i.e., $b_E/b_h = 1$).

When the elevator is deflected more than about 20–25 deg, flow separation over the tail tends to occur. Thus, the elevator will lose its effectiveness. Furthermore, close to horizontal tail stall, even a small downward elevator deflection can produce flow separation and loss of pitch control effectiveness. To prevent pitch control effectiveness, it is recommended to consider the elevator maximum deflection to be less than 25 deg (both up and down). Hence, the maximum elevator deflection is dictated by the elevator/tail stall requirement.

Provided that the elevator is designed to have full span (i.e., $b_E = b_h$), and the deflection is at its maximum allowable value, the elevator chord must be long enough to generate the desired change in the tail lift. However, as the elevator chord is increased, the tail becomes more prone to flow separation. If the required elevator chord is more than 50% of the horizontal tail chord (i.e., $C_E/C_h > 0.5$), an all-moving tail (i.e., $C_E = C_h$) is recommended. Fighter aircraft are often equipped with an all-moving horizontal tail to create the maximum amount of pitching moment in order to improve the pitch maneuverability. Most fighter aircraft have such a tail, since they are required to be highly maneuverable. Table 12.18 shows specifications of elevators for several aircraft.

The most critical flight condition for pitch control is when the aircraft is flying at a low speed due to the fact that the elevator is less effective. Two flight operations which feature a very low speed are take-off and landing. Take-off control is much harder than landing control due to safety considerations. A take-off operation is usually divided into three sections: (i) ground section, (ii) rotation or transition, and (iii) climb. The longitudinal control in a take-off is mainly applied during the rotation section, in which the nose is pitched up by rotating the aircraft about the main gear.

A fundamental criterion for elevator design is elevator effectiveness. The elevator effectiveness is representative of longitudinal control power and is frequently measured by three non-dimensional derivatives ($C_{m_{\delta E}}, C_{L_{\delta E}}, C_{L_{h_{\delta E}}}$) as follows:

1. The primary production of the elevator is an aircraft pitching moment to control the pitch rate. The non-dimensional derivative which represents the longitudinal control power derivative is the rate of change of aircraft pitching moment coefficient with respect to elevator deflection ($C_{m_{\delta E}}$). This is determined as:

$$C_{m_{\delta E}} = \frac{\partial C_m}{\partial \delta_E} = -C_{L_{\alpha_h}} \eta_h \cdot \overline{V}_H \cdot \frac{b_E}{b_h} \tau_e \tag{12.51}$$

   where $C_{L_{\alpha_h}}$ is the horizontal tail lift curve slope, $\overline{V}_H$ denotes the horizontal tail volume coefficient, and η_h is the horizontal tail dynamic pressure ratio. The parameter τ_e is the angle of attack effectiveness of the elevator, which is primarily a function of elevator-to-tail chord ratio (C_E/C_h). The latter variable (τ_e) is determined from Figure 12.12. The typical value for the derivative $C_{m_{\delta E}}$ is about −0.2 to −4 1/rad.
2. Another measure of elevator effectiveness is a parameter which represents the contribution of the elevator to aircraft lift ($C_{L_{\delta E}}$). This non-dimensional derivative is the rate of change of aircraft lift coefficient with respect to elevator deflection and is defined as follows:

$$C_{L_{\delta E}} = \frac{\partial C_L}{\partial \delta_E} = C_{L_{\alpha_h}} \eta_h \frac{S_h}{S} \frac{b_E}{b_h} \tau_e \tag{12.52}$$

where η_h is the horizontal tail dynamic pressure ratio and S_h is the horizontal tail planform area.

3. The third measure of elevator effectiveness is a non-dimensional derivative which represents the contribution of the elevator to tail lift ($C_{L_{h\delta E}}$). This derivative is the rate of change of tail lift coefficient with respect to elevator deflection and is defined as follows:

$$C_{L_{h\delta E}} = \frac{\partial C_{L_h}}{\partial \delta_E} = \frac{\partial C_{L_h}}{\partial \alpha_h}\frac{\partial \alpha_h}{\partial \delta_E} = C_{L_{\alpha_h}} \tau_e \tag{12.53}$$

The most significant elevator design requirement is the take-off rotation requirement. This design requirement is a function of aircraft mission and landing gear configuration.

Table 12.18 Specifications of elevators for several aircraft

No.	Aircraft	Type	m_{TO} (kg)	S_E/S_h	C_E/C_h	δ_{Emax} (deg)	
						Down	Up
1	Cessna 182	Light GA	1 406	0.38	0.44	22	25
2	Cessna Citation III	Business jet	9 979	0.37	0.37	15	15.5
3	Gulfstream 200	Business jet	16 080	0.28	0.31	20	27.5
4	AT-802	Agriculture	7 257	0.36	0.38	15	29
5	ATR 42-320	Regional airliner	18 600	0.35	0.33	16	26
6	Lockheed C-130 Hercules	Military cargo	70 305	0.232	0.35	15	40
7	Fokker F-28-4000	Transport	33 000	0.197	0.22	15	25
8	Fokker F-100B	Airliner	44 450	0.223	0.32	22	25
9	McDonnell Douglas DC-8	Transport	140 600	0.225	0.25	10	25
10	McDonnell Douglas DC-9-40	Transport	51 700	0.28	0.30	15	25
11	McDonnell Douglas DC-10-40	Transport	251 700	0.225	0.25	16.5	27
12	McDonnell Douglas MD-11	Transport	273 300	0.31	0.35	20	37.5
13	Boeing 727-100	Transport	76 820	0.23	0.25	16	26
14	Boeing 737-100	Transport	50 300	0.224	0.25	20	20
15	Boeing 777-200	Transport	247 200	0.30	0.32	25	30
16	Boeing 747-200	Transport	377 842	0.185	0.23	17	22
17	Airbus A-300B	Transport	165 000	0.295	0.30	17	30
18	Airbus 320	Transport	78 000	0.31	0.32	17	30
19	Airbus A340-600	Airliner	368 000	0.24	0.31	15	30
20	Lockheed L-1011 Tristar	Transport	231 000	0.215	0.23	0	25
21	Lockheed C-5A	Cargo	381 000	0.268	0.35	10	20

Table 12.19 Reduction in tail stall angle ($\Delta\alpha_{h_E}$, deg) when elevator is deflected

δ_E (deg)	Tail-to-elevator chord ratio C_E/C_h										
	0	0.1	0.2	0.3	0.4	0.5	0.6	0.7	0.8	0.9	1
0	0	0	0	0	0	0	0	0	0	0	0
±5	0	0.3	0.5	1.1	1.6	2.2	2.7	3.3	3.9	4.4	5
±10	0	0.6	1	2.1	3.2	4.4	5.5	6.6	7.7	8.9	10
±15	0	0.9	1.5	3.2	4.9	6.5	8.2	9.9	11.6	13.3	15
±20	0	1.2	2	4.2	6.5	8.7	11	13.2	15.5	17.7	20
±25	0	1.6	2.5	5.3	8.1	11	13.7	16.5	19.4	22.2	25
±30	0	1.9	3	6.4	9.7	13.1	16.5	19.9	23.2	26.6	30

Two popular configurations are: (i) nose gear or tricycle and (ii) tail gear. These two landing gear configurations tend to require different take-off rotations, as follows:

1. In aircraft with tricycle landing gear, the elevator must be powerful enough to rotate the aircraft about the main gear and lift the nose with specified angular pitch acceleration. This requirement shall be satisfied when the aircraft has 80% of take-off speed ($0.8V_{TO}$) and the aircraft center of gravity is at its most allowable forward position. This requirement is equivalent to a rotation at stall speed with a specified angular acceleration.
2. In aircraft with tail-gear configuration, the elevator must be such as to rotate the aircraft about the main gear and lift the tail with specified angular pitch acceleration. This requirement shall be satisfied when the aircraft has 50% of take-off speed ($0.5V_{TO}$) and the aircraft center of gravity is at its most allowable aft position.

The angular pitch acceleration requirement for various aircraft is given in Table 12.9. In a conventional aircraft, the take-off rotation when the aircraft cg is at its most forward location frequently requires the most negative elevator deflection (up). In contrast, the longitudinal trim when the aircraft cg is at its most aft location and the aircraft has the lowest allowable speed usually requires the most positive elevator deflection (down). The governing equations for take-off rotation operation and the technique to calculate pitch rate acceleration during rotation are developed in Section 12.5.3. The governing equations for longitudinal trim and the technique to calculate the desired elevator deflection are developed in Section 12.5.4.

There are generally crucial interrelations between elevator design, landing gear design, and aircraft weight distribution (i.e., aircraft cg positions). Any of these three components/parameters will impose a limit/constraint on the other two components/parameters. For instance, as the aircraft most allowable cg is pushed forward, the elevator is required to be larger. Furthermore, as the main gear in a tricycle configuration is moved rearward, the elevator needs to be more powerful. Hence, it is necessary that the elevator design group has a compromising attitude and a close relationship with the landing gear design team, and also with the aircraft weight distribution group. Sometimes, a slight change in the landing gear design may lead to a considerable improvement in the elevator design.

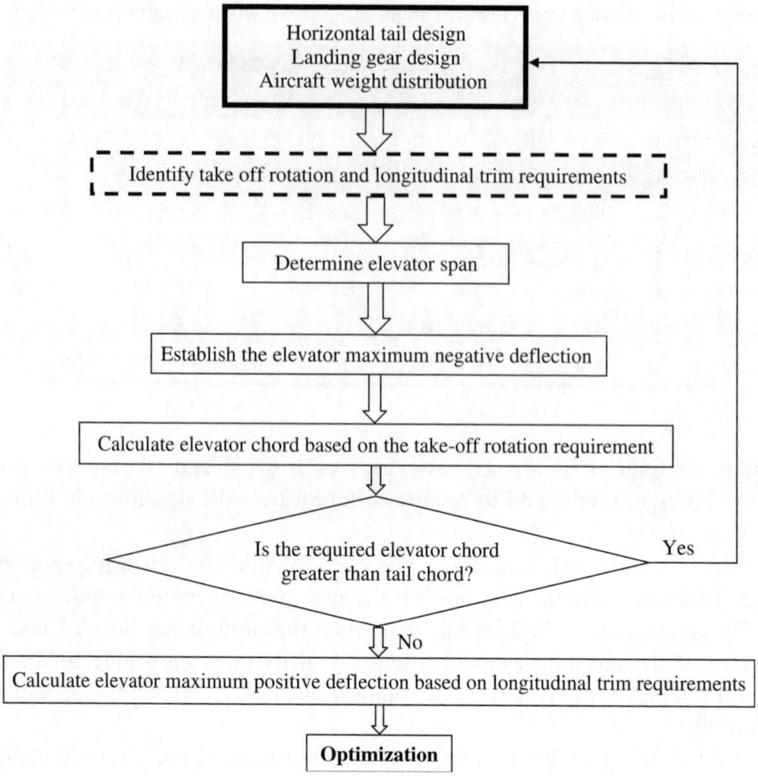

Figure 12.20 Elevator design flowchart

In the interest of minimizing the total cost/aircraft weight, changes must be incorporated as required, leading to a preferred design configuration. Therefore, the elevator and landing gear must be designed/evaluated/optimized simultaneously, and the aircraft cg must be positioned so as to provide the best design environment for both landing gear and elevator. An elevator design flowchart is presented in Figure 12.20.

12.5.3 Take-Off Rotation Requirement

For an aircraft with a landing gear configuration in which the main gear is behind the aircraft cg (e.g., tricycle landing gear), the take-off rotation requirement is employed to design the elevator. Most aircraft, to become airborne, must be rotated about the main gear to achieve the angle of attack required for lift-off. Exceptions to this are aircraft like the military bomber Boeing B-52. The take-off rotation requirement requires the elevator design to be such that the pitch angular acceleration ($\ddot{\theta}$) is greater than a desired value. In Chapter 9, the requirement is developed mathematically by focusing specifically on the relationship with landing gear design. In this section, the elevator design technique is established based on the technique developed in Chapter 9.

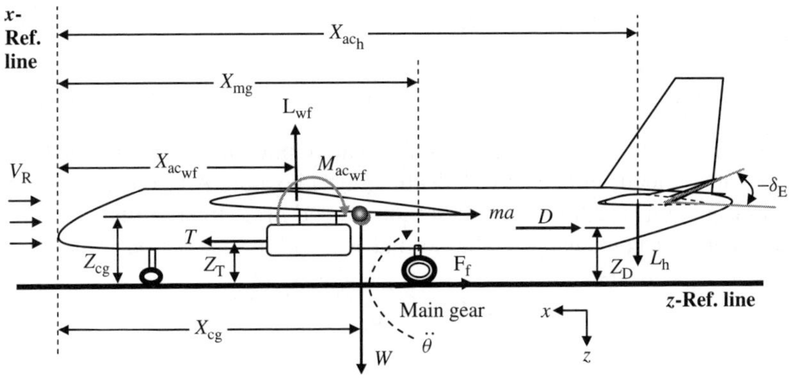

Figure 12.21 Forces and moments during take-off rotation

The angular acceleration about the main gear rotation point $\ddot{\theta}$ is a function of a number of parameters, including horizontal tail area, horizontal tail arm, aircraft weight, rotation speed, the distance between the main gear and the aircraft cg, and finally the elevator control power. Typical rotational accelerations are given in Table 12.9 for various types of aircraft. The rotation acceleration is the aircraft acceleration at the time the aircraft begins to rotate about the main gear. This speed must be slightly more than the stall speed (V_s):

$$V_R = 1.1 - 1.3V_s \tag{12.54}$$

However, for safety, the elevator is designed to rotate the aircraft with the desired acceleration at the stall speed (V_s).

In this section, an analysis of the elevator design to generate a given level of pitch angular acceleration about the main gear contact point is presented. Consider the aircraft with tricycle landing gear in Figure 12.21, which is at the onset of a rotation about the main gear in a take-off operation. The figure illustrates all forces and moments contributing to this moment of the take-off operation. The contributing forces include wing/fuselage lift (L_{wf}), horizontal tail lift (L_h), aircraft drag (D), friction force between tires and ground (F_f), aircraft weight (W), engine thrust (T), and acceleration force (ma). Note that the latter force (ma) is acting backward, due to Newton's third law, as a reaction to the acceleration. Furthermore, the contributing moments are the wing/fuselage aerodynamic pitching moment ($M_{o_{wf}}$) plus the moments of preceding forces about the rotation point. The distance between these forces is measured with respect to both the x reference line (i.e., fuselage nose) and the z reference line (i.e., ground), as shown in Figure 12.21.

For a conventional aircraft with a tricycle landing gear, the horizontal tail lift is negative during rotation. It is recommended to consider the ground effect in calculating lift and drag to achieve more accurate results. The ground friction coefficient, μ, depends on the type of terrain. Table 9.7 introduces the friction coefficients for different terrains.

There are three governing equations of motion that determine the aircraft equilibrium at the instant of rotation – two force equations and one moment equation:

$$\sum F_x = m\frac{dV}{dt} \Rightarrow T - D - F_f = ma \Rightarrow T - D - \mu N = ma \tag{12.55}$$

$$\sum F_z = 0 \Rightarrow L + N = W \Rightarrow L_{\mathrm{wf}} - L_{\mathrm{h}} + N = W \Rightarrow N = W - \left(L_{\mathrm{wf}} - L_{\mathrm{h}}\right) \tag{12.56}$$

$$\sum M_{\mathrm{cg}} = I_{yy_{\mathrm{mg}}} \ddot{\theta} \Rightarrow -M_W + M_D - M_T + M_{L_{\mathrm{wf}}} + M_{\mathrm{ac_{wf}}} + M_{L_{\mathrm{h}}} + M_a = I_{yy_{\mathrm{mg}}} \ddot{\theta} \tag{12.57}$$

Equation (12.57) indicates that the aircraft negative pitching moment must be overcome by an opposite moment created by deflecting the elevator. All contributing forces and moments in Equations (12.55)–(12.57) are introduced in Chapter 9; they are repeated and renumbered here for convenience. The normal force (N), the friction force (F_{f}), and the aircraft lift at take-off are:

$$N = W - L_{\mathrm{TO}} \tag{12.58}$$

$$F_{\mathrm{f}} = \mu N = \mu \left(W - L_{\mathrm{TO}}\right) \tag{12.59}$$

$$L_{\mathrm{TO}} = L_{\mathrm{wf}} + L_{\mathrm{h}} \tag{12.60}$$

The wing/fuselage lift (L_{wf}), horizontal tail lift (L_{h}), aerodynamic drag (D) forces, and wing/fuselage pitching moment about the wing/fuselage aerodynamic center are as follows. Recall that the horizontal tail lift is negative.

$$L_{\mathrm{h}} = \frac{1}{2}\rho V_{\mathrm{R}}^2 C_{L_{\mathrm{h}}} S_{\mathrm{h}} \tag{12.61}$$

$$L_{\mathrm{wf}} \cong \frac{1}{2}\rho V_{\mathrm{R}}^2 C_{L_{\mathrm{TO}}} S_{\mathrm{ref}} \tag{12.62}$$

$$D_{\mathrm{TO}} = \frac{1}{2}\rho V_{\mathrm{R}}^2 C_{D_{\mathrm{TO}}} S_{\mathrm{ref}} \tag{12.63}$$

$$M_{\mathrm{ac_{wf}}} = \frac{1}{2}\rho V_{\mathrm{R}}^2 C_{\mathrm{m_{ac_{wf}}}} S_{\mathrm{ref}} \overline{C} \tag{12.64}$$

where V_{R} denote the aircraft linear forward speed at the instant of rotation, S_{ref} represents the wing planform area, S_{h} is the horizontal tail planform area, ρ is the air density, and $\overline{C}$ is the wing mean aerodynamic chord. Furthermore, the four coefficients of C_D, $C_{L_{\mathrm{wf}}}$, $C_{L_{\mathrm{h}}}$, and $C_{\mathrm{m_{ac_{wf}}}}$ denote drag, wing/fuselage lift, horizontal lift, and wing/fuselage pitching moment coefficients respectively. In Equation (12.57), the clockwise rotation about the y-axis is assumed to be a positive rotation.

The contributing pitching moments in take-off rotation control are aircraft weight moment (M_W), aircraft drag moment (M_D), engine thrust moment (M_T), wing/fuselage lift moment ($M_{L_{\mathrm{wf}}}$), wing/fuselage aerodynamic pitching moment ($M_{\mathrm{ac_{wf}}}$), horizontal tail lift moment ($M_{L_{\mathrm{h}}}$), and linear acceleration moment (M_a). These moments are obtained as follows:

$$M_W = W\left(x_{\mathrm{mg}} - x_{\mathrm{cg}}\right) \tag{12.65}$$

$$M_D = D\left(z_D - z_{\mathrm{mg}}\right) \tag{12.66}$$

$$M_T = T\left(z_T - z_{\mathrm{mg}}\right) \tag{12.67}$$

$$M_{L_{\mathrm{wf}}} = L_{\mathrm{wf}}\left(x_{\mathrm{mg}} - x_{\mathrm{ac_{wf}}}\right) \tag{12.68}$$

$$M_{L_h} = L_h \left(x_{ac_h} - x_{mg}\right) \tag{12.69}$$

$$M_a = ma \left(z_{cg} - z_{mg}\right) \tag{12.70}$$

In Equations (12.65)–(12.70), the subscript "mg" denotes main gear, since the distances are measured from the main gear. The inclusion of the moment generated by the aircraft acceleration (Equation (12.70)) is due to the fact that based on Newton's third law, any action creates a reaction (ma). This reaction force produces a moment when its corresponding arm is taken into account. Substituting these moments into Equation (12.57) yields:

$$\sum M_{cg} = I_{yy}\ddot{\theta} \Rightarrow -W\left(x_{mg} - x_{cg}\right) + D\left(z_D - z_{mg}\right) - T\left(z_T - z_{mg}\right) + L_{wf}\left(x_{mg} - x_{ac_{wf}}\right)$$
$$+ M_{ac_{wf}} - L_h\left(x_{ac_h} - x_{mg}\right) + ma\left(z_{cg} - z_{mg}\right) = I_{yy_{mg}}\ddot{\theta} \tag{12.71}$$

where $I_{yy_{mg}}$ represents the aircraft mass moment of inertia about the y-axis at the main gear. In an aircraft with a tricycle landing gear, the tail lift moment, wing/fuselage moment, drag moment, and acceleration moment are all clockwise, while the weight moment, thrust moment, and wing/fuselage aerodynamic pitching moment are counterclockwise. These directions must be considered when assigning a sign to each. The role of the elevator in Equation (12.71) is to create a sufficient horizontal tail lift (L_h). The result is as follows:

$$L_h = \frac{\left[L_{wf}\left(x_{mg} - x_{ac_{wf}}\right) + M_{ac_{wf}} + ma\left(z_{cg} - z_{mg}\right) - W\left(x_{mg} - x_{cg}\right) + D\left(z_D - z_{mg}\right) - T\left(z_T - z_{mg}\right) - I_{yy_{mg}}\ddot{\theta}\right]}{x_{ac_h} - x_{mg}} \tag{12.72}$$

Then, this horizontal tail lift must be so as to satisfy the take-off rotation requirement. The elevator contribution to this lift is through the tail lift coefficient, which can be obtained by using Equation (12.61):

$$C_{L_h} = \frac{2L_h}{\rho V_R^2 S_h} \tag{12.73}$$

This tail lift coefficient is generally negative (about −1 to −1.5) and is a function of tail angle of attack (α_h), tail airfoil section [19] features, and tail planform parameters such as aspect ratio, sweep angle, and taper ratio. The horizontal tail lift coefficient is modeled as:

$$C_{L_h} = C_{L_{ho}} + C_{L_{\alpha_h}}\alpha_h + C_{L_{h_{\delta E}}}\delta_E \tag{12.74}$$

where $C_{L_{\alpha_h}}$ is the tail lift curve slope and $C_{L_{ho}}$ is the zero angle of attack tail lift coefficient. Most horizontal tails tend to use a symmetric airfoil section, so the parameter $C_{L_{ho}}$ is normally zero. Inclusion of this statement, and plugging Equation (12.53) into Equation (12.74) results in:

$$C_{L_h} = C_{L_{\alpha_h}}\alpha_h + C_{L_{\alpha_h}}\tau_e\delta_E = C_{L_{\alpha_h}}\left(\alpha_h + \tau_e\delta_E\right) \tag{12.75}$$

Recall that the tail angle of attack is already (see Chapter 6) defined as:

$$\alpha_h = \alpha + i_h - \varepsilon \tag{12.76}$$

where α is the aircraft angle of attack at the onset of rotation, i_h denotes the tail incidence angle, and ε represents the downwash angle which is determined through Equation (6.54). The aircraft angle of attack, when the aircraft is on the ground (i.e., onset of rotation), is usually zero.

The elevator designer can control the magnitude of the elevator control power by proper selection of the elevator geometry. Equation (12.75) enables the elevator designer to determine the elevator characteristics to satisfy the take-off rotation requirement. Knowing τ_e, one can use Figure 12.12 to estimate the elevator chord-to-tail chord ratio. This represents the minimum elevator area to satisfy the most crucial aircraft longitudinal control requirement. Note that the take-off rotation requirement dictates the maximum up-deflection of the elevator ($-\delta_{E_{max}}$). The maximum positive (down-) deflection of the elevator ($+\delta_{E_{max}}$) is dictated by the longitudinal trim requirement which will be examined in the next section. The elevator will normally generate its maximum negative pitching moment to maintain longitudinal trim when the aircraft is flying with the lowest velocity and the aircraft cg is at its most aft allowable location. However, the elevator will generally create its maximum positive pitching moment during take-off rotation when the aircraft cg is at its most forward allowable location and the aircraft has maximum take-off weight.

Satisfaction of the take-off rotation requirements frequently generates a conflict between design groups such as the landing gear design group, tail design group, weight and balance group, fuselage design group, propulsion system design group, and elevator design group. Each design group may focus on other design requirements and consider the rotation requirement at the end of the list. If that is the case, one design group will create a challenge for the other design groups which may not be resolved easily. The solution is to have all design groups debate the various solutions and adopt the least challenging one.

12.5.4 Longitudinal Trim Requirement

When all longitudinal moments and forces are in equilibrium, it is said that the aircraft is in longitudinal trim. In this section, we shall be concerned with longitudinal trim. The elevator plays a significant role in the aircraft longitudinal trim to fly at various trim conditions. To carry out the longitudinal trim analysis and to derive a relationship that represents the function of the elevator in longitudinal trim, consider the aircraft in Figure 12.22 which is cruising with a constant speed. The engine is located under the wing and the engine thrust (T) has an offset (z_T) from the aircraft center of gravity. The engine creates a positive pitching moment. It is assumed that the engine setting angle is zero.

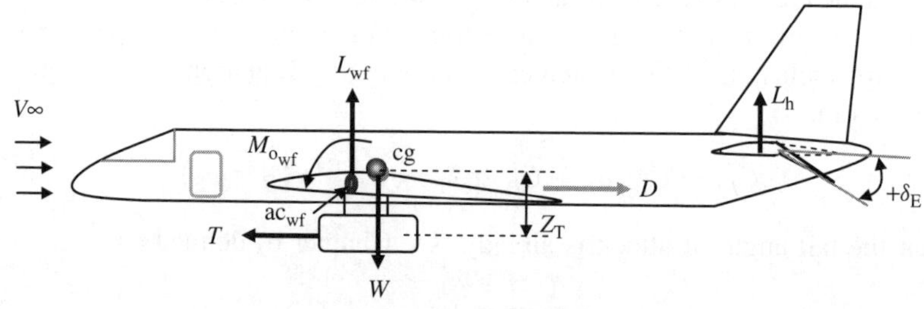

Figure 12.22 Longitudinal trim

The governing longitudinal trim equations are:

$$\sum F_z = 0 \Rightarrow L = W \tag{12.77}$$

$$\sum F_x = 0 \Rightarrow D = T \tag{12.78}$$

$$\sum M_{cg} = 0 \Rightarrow M_A + T \cdot z_T = 0 \tag{12.79}$$

It is also assumed that there is always sufficient thrust to balance the drag force, thus only two equations need to be expanded:

$$\overline{q} \cdot S \cdot C_L = W \tag{12.80}$$

$$\overline{q} \cdot S \cdot \overline{C} \cdot C_{\mathrm{m}} + T \cdot z_T = 0 \tag{12.81}$$

The aerodynamic forces and moments are functions of non-dimensional derivatives, so Equations (12.80) and (12.81) may be written as follows:

$$\overline{q} \cdot S \cdot \left(C_{L_{\mathrm{o}}} + C_{L_\alpha}\alpha + C_{L_{\delta \mathrm{E}}}\delta_{\mathrm{E}}\right) = W \tag{12.82}$$

$$\overline{q} \cdot S \cdot \overline{C} \cdot \left(C_{m_{\mathrm{o}}} + C_{\mathrm{m}_\alpha}\alpha + C_{\mathrm{m}_{\delta \mathrm{E}}}\delta_{\mathrm{E}}\right) + T \cdot z_T = 0 \tag{12.83}$$

The equations may be reformatted as:

$$C_{L_{\mathrm{o}}} + C_{L_\alpha}\alpha + C_{L_{\delta \mathrm{E}}}\delta_{\mathrm{E}} = \frac{W}{\overline{q} \cdot S} = C_{L_1} \tag{12.84}$$

$$C_{\mathrm{m_o}} + C_{\mathrm{m}_\alpha}\alpha + C_{\mathrm{m}_{\delta \mathrm{E}}}\delta_{\mathrm{E}} = -\frac{T \cdot z_T}{\overline{q} \cdot S \cdot \overline{C}} \tag{12.85}$$

where C_{L_1} is the steady-state aircraft lift coefficient at this cruising flight. It is useful to recast the equations in a matrix format:

$$\begin{bmatrix} C_{L_\alpha} & C_{L_{\delta \mathrm{E}}} \\ C_{\mathrm{m}_\alpha} & C_{\mathrm{m}_{\delta \mathrm{E}}} \end{bmatrix} \begin{bmatrix} \alpha \\ \delta_{\mathrm{E}} \end{bmatrix} = \begin{bmatrix} C_{L_1} - C_{L_{\mathrm{o}}} \\ -\dfrac{T \cdot z_T}{\overline{q} \cdot S \cdot \overline{C}} - C_{\mathrm{m_o}} \end{bmatrix} \tag{12.86}$$

This set of equations has two unknowns – aircraft angle of attack (α) and elevator deflection (δ_E). Solutions to this set of equations employing Cramer's rule are as follows:

$$\alpha = \frac{\begin{vmatrix} C_{L_1} - C_{L_{\mathrm{o}}} & C_{L_{\delta \mathrm{E}}} \\ -\dfrac{T \cdot z_T}{\overline{q} \cdot S \cdot \overline{C}} - C_{\mathrm{m_o}} & C_{\mathrm{m}_{\delta \mathrm{E}}} \end{vmatrix}}{\begin{vmatrix} C_{L_\alpha} & C_{L_{\delta \mathrm{E}}} \\ C_{\mathrm{m}_\alpha} & C_{\mathrm{m}_{\delta \mathrm{E}}} \end{vmatrix}} \tag{12.87}$$

$$\delta_{\mathrm{E}} = \frac{\begin{vmatrix} C_{L_\alpha} & C_{L_1} - C_{L_{\mathrm{o}}} \\ C_{\mathrm{m}_\alpha} & -\dfrac{T \cdot z_T}{\overline{q} \cdot S \cdot \overline{C}} - C_{\mathrm{m_o}} \end{vmatrix}}{\begin{vmatrix} C_{L_\alpha} & C_{L_{\delta \mathrm{E}}} \\ C_{\mathrm{m}_\alpha} & C_{\mathrm{m}_{\delta \mathrm{E}}} \end{vmatrix}} \tag{12.88}$$

or

$$\alpha = \frac{\left(C_{L_1} - C_{L_o}\right) C_{m_{\delta E}} + \left(\frac{T \cdot z_T}{\overline{q} \cdot S \cdot \overline{C}} + C_{m_o}\right) C_{L_{\delta E}}}{C_{L_\alpha} C_{m_{\delta E}} - C_{m_\alpha} C_{L_{\delta E}}} \tag{12.89}$$

$$\delta_E = -\frac{\left(\frac{T \cdot z_T}{\overline{q} \cdot S \cdot \overline{C}} + C_{m_o}\right) C_{L_\alpha} + \left(C_{L_1} - C_{L_o}\right) C_{m_\alpha}}{C_{L_\alpha} C_{m_{\delta E}} - C_{m_\alpha} C_{L_{\delta E}}} \tag{12.90}$$

where the aircraft static longitudinal stability derivative (C_{m_α}) is determined by Equation (6.67). The elevator deflection to maintain the aircraft longitudinal trim can be obtained directly from Equation (12.90). Note that if the thrust line is above the aircraft cg, the parameter z_T will be negative. The elevator angle must be large enough to maintain longitudinal trim at all flight conditions, particularly when the aircraft center of gravity is located at the most allowable aft position.

The elevator designer must synthesize an elevator in such a way that the longitudinal trim is not a limiting factor anywhere in the intended flight envelope and throughout the aircraft mission. In case the required elevator angle is more than about 30 deg, the designer needs to increase the elevator size or even the tail arm. This is to ensure that the elevator does not cause any flow separation over the tail during its application. Figure 12.23 shows the typical variations of elevator deflection versus aircraft speed to maintain aircraft longitudinal trim. As the figure illustrates, one of the objectives in the horizontal tail design is to require a zero elevator deflection during a cruising flight. Please note that as a convention, the up-deflection of the elevator is considered to be negative.

There is a constraint on the elevator design which must be considered and checked. The elevator deflection must not cause the horizontal tail to stall. The elevator deflection will decrease the tail stall angle. At the end of take-off rotation, the aircraft, wing, and tail angles of attack are all increased. The elevator may be thought of as a plain flap attached to the tail, so the tail stall angle depends upon the elevator chord and deflection. So, the elevator designer should check whether the tail stall occurs when the maximum elevator deflection is employed and the fuselage is lifted up. The recommendation is to keep the tail

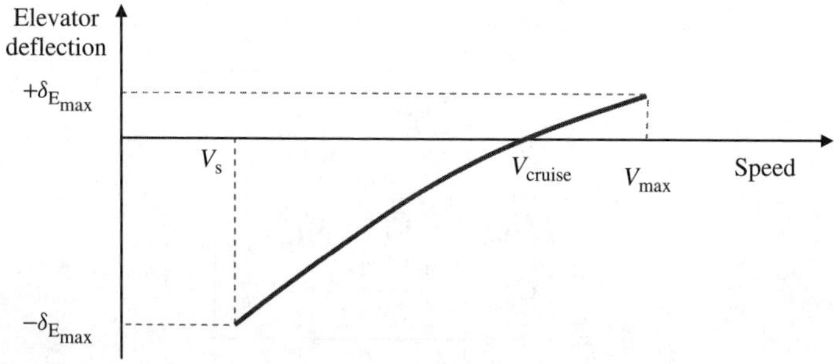

Figure 12.23 Typical variations of elevator deflection versus aircraft speed

within 2 deg of its stall angle of attack. Using Equation (12.76), the relationship between horizontal tail angle at take-off and fuselage take-off angle of attack (α_{TO}) is obtained:

$$\alpha_{h_{TO}} = \alpha_{TO}\left(1 - \frac{d\varepsilon}{d\alpha}\right) + i_h - \varepsilon_o \tag{12.91}$$

The fuselage take-off angle of attack may be assumed to be equal to the aircraft take-off angle of attack. This equation yields the maximum positive tail angle of attack, which must be less than the tail stall angle. In contrast, during a cruising flight with maximum speed, when the elevator is employed to maintain longitudinal trim, the maximum positive elevator deflection must be checked. The tail stall angle of attack during take-off rotation (α_{h_s}) is a function of a number of parameters including tail airfoil section, elevator deflection, and elevator chord and is determined by:

$$\alpha_{h_s} = \pm\left(\alpha_{h_{s:\delta E=0}} - \Delta\alpha_{h_E}\right) \tag{12.92}$$

where $\alpha_{h_{s:\delta E=0}}$ is the tail stall angle (typically about 14 deg) when an elevator is not employed. The parameter $\Delta\alpha_{h_E}$ is the magnitude of reduction in tail stall angle of attack due to elevator deflection, and must be determined using a wind-tunnel test or referring to aerodynamics references. Table 12.19 illustrates the empirical values (in degrees) for the parameter $\Delta\alpha_{h_E}$ as a function of elevator deflection and tail-to-elevator chord ratio.

When the elevator is designed, the generated horizontal tail lift coefficient needs to be calculated and compared with the desired tail lift coefficient. Tools such as computational fluid dynamics technique or lifting-line theory (Section 5.14) may be utilized for such calculation. One of the parameters in this evaluation process is the change in the tail zero-lift angle of attack due to elevator deflection ($\Delta\alpha_{o_E}$). The horizontal tail is a lifting surface and may be treated in the same way as the wing. Thus, the empirical Equation (5.39) is reformatted to approximate the parameter $\Delta\alpha_{o_E}$ as follows:

$$\Delta\alpha_{o_E} \approx -1.15 \cdot \frac{C_E}{C_h}\delta_E \tag{12.93}$$

The generated horizontal tail lift coefficient must be equal to the desired generated horizontal tail lift coefficient. The parameter $\Delta\alpha_{o_E}$ is employed in the application of lifting-line theory to approximate the tail lift coefficient as well as the tail lift distribution.

12.5.5 Elevator Design Procedure

In Sections 12.5.1–12.5.4 the elevator primary function, parameters, governing rules and equations, objectives, design criteria, and formulation, as well as design requirements, have been presented in detail. In addition, Figure 12.20 illustrates the design flowchart of the elevator. In this section, the elevator design procedure is introduced in terms of design steps. It must be noted that there is no unique solution to satisfy the customer requirements in designing an elevator. Several elevator designs may satisfy the requirements, but each will have unique advantages and disadvantages. It must be noted that there is a possibility that no elevator can satisfy the requirements due to the limits/constraints imposed by the tail design and landing gear design. In such a situation, the designer must return to redesign the tail and/or landing gear components.

Based on the systems engineering approach, the elevator detail design begins with identifying and defining design requirements and ends with optimization. The following are the elevator design steps for a conventional aircraft:

1. Layout the elevator design requirements (see Section 12.5.2).
2. Identify the take-off rotation acceleration requirement from Table 12.9.
3. Select the elevator span (see Table 12.3).
4. Establish the maximum elevator deflection to prevent flow separation (see Table 12.3).
5. Calculate the wing/fuselage lift (L_{wf}), aircraft drag (D), and wing/fuselage pitching moment about the wing/fuselage aerodynamic center using Equations (12.62)–(12.64).
6. Calculate the aircraft linear acceleration (a) during take-off rotation using Equation (12.55).
7. Calculate the contributing pitching moments during take-off rotation (i.e., aircraft weight moment (M_W), aircraft drag moment (M_D), engine thrust moment (M_T), wing/fuselage lift moment ($M_{L_{\mathrm{wf}}}$), wing/fuselage aerodynamic pitching moment ($M_{\mathrm{ac_{wf}}}$), and linear acceleration moment (M_a) using Equations (12.65)–(12.70). For this calculation, consider the most forward aircraft center of gravity.
8. Calculate the desired horizontal tail lift (L_{h}) during take-off rotation employing Equation (12.72). For this calculation, consider the most forward aircraft center of gravity.
9. Calculate the desired horizontal tail lift coefficient ($C_{L_{\mathrm{h}}}$) employing Equation (12.73).
10. Calculate the angle of attack effectiveness of the elevator (τ_{e}) employing Equation (12.75). In this calculation, the maximum negative elevator deflection (from step 4) is considered.
11. Determine the corresponding elevator-to-tail chord ratio ($C_{\mathrm{E}}/C_{\mathrm{h}}$) from Figure 12.12.
12. If the elevator-to-tail chord ratio ($C_{\mathrm{E}}/C_{\mathrm{h}}$) is more than 0.5, it is suggested to select an all-moving tail (i.e., $C_{\mathrm{E}}/C_{\mathrm{h}} = 1$).
13. If the angle of attack effectiveness of the elevator (τ_{e}) is greater than 1, there is no elevator which can satisfy the take-off rotation requirement by the current tail/landing gear specifications. In such a case, the horizontal tail and/or landing gear must be redesigned. Then, return to step 5.
14. Using an aerodynamic technique such as computational fluid dynamics or lifting-line theory (see Section 5.14), determine the horizontal tail lift distribution and horizontal tail lift coefficient when the elevator is deflected with its maximum negative angle (i.e., $-\delta_{\mathrm{E_{max}}}$).
15. Compare the produced horizontal tail lift coefficient of step 13 with the desired horizontal tail lift coefficient of step 9. These two numbers must be the same. If not, adjust the elevator chord or elevator span to vary the produced horizontal tail lift coefficient.
16. Calculate the elevator effectiveness derivatives ($C_{m_{\delta E}}, C_{L_{\delta E}}, C_{L_{h_{\delta E}}}$) from Equations (12.51)–(12.53). For these calculations, examine both the most aft and the most forward aircraft center of gravity.
17. Calculate the elevator deflection (δ_{E}) required to maintain longitudinal trim at various flight conditions using Equation (12.90). For these calculations, examine the most aft and most forward aircraft center of gravity, as well as various aircraft speeds.

18. Plot the variations of the elevator deflection versus airspeed and also versus altitude. For these calculations, consider both the most aft and most forward aircraft center of gravity.
19. Compare the maximum required down elevator deflection ($+\delta_{E_{max}}$) with the maximum deflection established in step 4. If the maximum required down elevator deflection of step 15 is greater than the maximum deflection established in step 4, there is no elevator which can satisfy the longitudinal trim requirements with the current tail/landing gear specification. In such a case, the horizontal tail and/or landing gear must be redesigned. Then, return to step 5.
20. Check whether or not the elevator deflection causes the horizontal tail to stall during take-off rotation by using Equation (12.92).
21. If tail stall will occur during take-off rotation, the elevator must be redesigned by reducing the elevator deflection and/or elevator chord. Return to step 3.
22. If tail stall will occur during take-off rotation, and neither of the two elevator parameters (i.e., elevator deflection and chord) may be reduced to prevent tail stall, other aircraft components such as horizontal tail, landing gear, or aircraft center of gravity must be redesigned/relocated.
23. Apply aerodynamic balance/mass balance, if necessary (Section 12.7).
24. Optimize the elevator.
25. Calculate elevator span, elevator chord, and elevator area then draw the top view and side view of the horizontal tail (including elevator) with dimensions.

12.6 Rudder Design

12.6.1 Introduction to Rudder Design

The rudder is a primary control surface and is responsible for the aircraft directional control. The rudder is a movable surface located on the trailing edge of the vertical tail. The rudder is the vertical counterpart to the elevator. When the rudder is rotated (i.e., deflected, δ_R), a lift force (i.e., side force, L_V) is created (Figure 12.24) by the rudder/vertical tail combination. Consequently, a yawing moment (N) about the aircraft center of gravity (about the aircraft z-axis) is generated. Thus, control of the yawing moment about the center of gravity is primarily provided by means of the rudder. The third unintended production of the rudder is a rolling moment. This is due to the fact

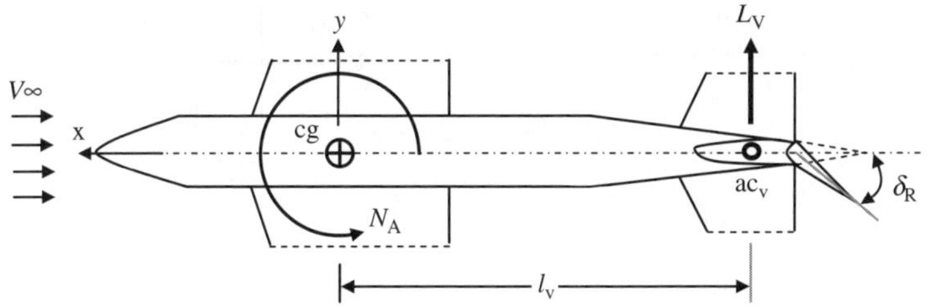

Figure 12.24 Directional control via rudder deflection (top view)

that the vertical tail (i.e., rudder) is usually placed above the aircraft cg. Two fundamental roles of the rudder are directional control and directional trim. Therefore, parameters of the rudder are determined by the directional trim and control requirements. The rudder control power must be sufficient to accomplish these two requirements in various flight conditions. The aircraft heading angle (ψ) is mainly determined through a directional control process.

There are interferences between rudder and aileron, and they are often applied simultaneously. Thus, the lateral and directional dynamics are frequently coupled. Thus, it is good practice to design the aileron and rudder concurrently. The rudder, similar to an elevator, is a displacement control device, while the aileron is a rate control device. The fundamentals of design of elevator and rudder are similar, but since their applications are different, the design of a rudder is generally more complicated. However, rudder deflections to the right and left are the same, but up and down elevator deflections are different.

In the design of the rudder, four parameters must be determined: (i) rudder area (S_R), (ii) rudder chord (C_R), (iii) rudder span (b_R), (iv) maximum rudder deflection ($\pm\delta_{R_{max}}$), and (v) location of inboard edge of the rudder (b_{Ri}). Figure 12.25 shows the vertical tail geometry and rudder parameters. Table 12.20 illustrates characteristics of the rudder for several aircraft. Table 12.3 shows typical values for the geometry of a rudder (ratio between rudder chord, span, and area to vertical tail chord, span, and area) from which one can select preliminary data.

The convention for the positive rudder deflection is defined as the deflection to the left (of the pilot). As Figure 12.24 demonstrates, a positive rudder deflection creates a positive side force (i.e., in the positive y direction) but results in a negative yawing moment (i.e., counterclockwise). In a symmetric aircraft with zero sideslip angle and zero aileron deflection, the yawing moment is determined by multiplying the vertical tail lift by the vertical tail arm:

$$N_A = l_v \cdot L_V \tag{12.94}$$

where l_v is the vertical tail arm and is the distance, along the x-axis, between the aircraft cg and the vertical tail aerodynamic center (ac_v). The vertical tail aerodynamic center is usually located at the quarter chord of the vertical tail mean aerodynamic chord.

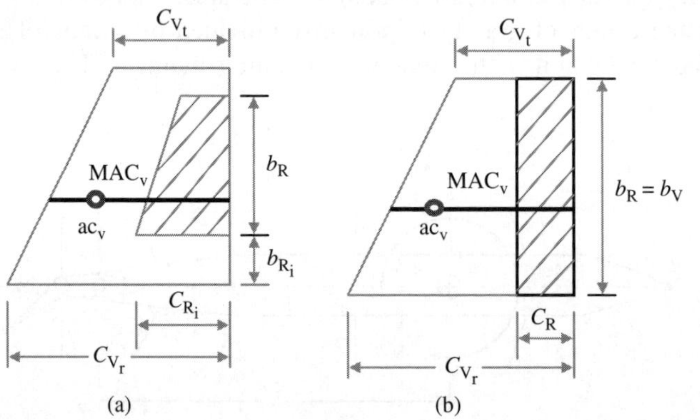

Figure 12.25 Vertical tail and rudder geometry. (a) A swept rudder; (b) A rectangular rudder

Table 12.20 Characteristics of rudder for several aircraft

No.	Aircraft	Type	m_{TO} (kg)	S_R/S_V	C_R/C_V	δ_{Rmax} (deg)	Max cross-wind speed (knot)
1	Cessna 182	Light GA	1 406	0.38	0.42	±24	–
2	Cessna 650	Business jet	9 979	0.26	0.27	±25	–
3	Gulfstream 200	Business jet	16 080	0.3	0.32	±20	–
4	Air tractor AT-802	Regional airliner	18 600	0.61	0.62	±24	–
5	Lockheed C-130E Hercules	Military cargo	70 305	0.239	0.25	±35	–
6	DC-8	Transport	140 600	0.269	35	±32.5	34
7	DC-10	Transport	251 700	0.145	38	±23/46[a]	30
8	Boeing 737-100	Transport	50 300	0.25	0.26	–	–
9	Boeing 777-200	Transport	247 200	0.26	0.28	±27.3	–
10	Boeing 747-200	Transport	377 842	0.173	0.22	±25	30
11	Lockheed C-5A	Cargo	381 000	0.191	0.2	–	43
12	Fokker 100A	Airliner	44 450	0.23	0.28	±20	30
13	Embraer ERJ145	Regional jet	22 000	0.29	0.31	±15	–
14	Airbus A340-600	Airliner	368 000	0.31	0.32	±31.6	–

[a] Tandem rudder.

The aircraft side force is primarily a function of dynamic pressure, vertical tail area (S_V), and in the direction of the vertical tail lift (L_V):

$$L_V = \overline{q} S_V C_{L_V} \tag{12.95}$$

where C_{L_V} is the vertical tail lift coefficient and is a function of vertical tail airfoil section, sideslip angle, and rudder deflection. The vertical tail lift coefficient is linearly modeled as:

$$C_{L_V} = C_{L_{V_o}} + C_{L_{V_\beta}} \beta + C_{L_{V_{\delta_R}}} \delta_R \tag{12.96}$$

The aircraft aerodynamic yawing moment is a function of dynamic pressure, wing area (S), and wing span (b), and is defined as:

$$N_A = \overline{q} S C_n b \tag{12.97}$$

where C_n is the yawing moment coefficient and is a function of aircraft configuration, sideslip angle, rudder deflection, and aileron deflection. The yawing moment coefficient is linearly modeled as:

$$C_n = C_{n_o} + C_{n_\beta} \beta + C_{n_{\delta A}} \delta_A + C_{n_{\delta R}} \delta_R \tag{12.98}$$

The parameter $C_{n\delta R}$ is referred to as the aircraft yawing moment coefficient due to rudder deflection derivative and is also called the rudder yaw control power. The rudder yaw control effectiveness is mainly measured by the rate of change of yawing moment with respect to rudder deflection angle. In a non-dimensional form:

$$C_{n\delta R} = \frac{\partial C_n}{\partial \delta_R} \tag{12.99}$$

The directional control derivative ($C_{n\delta R}$) depends strongly on the vertical tail size, vertical tail moment arm, and is determined by:

$$C_{n\delta R} = -C_{L\alpha_V} \overline{V}_V \eta_V \tau_r \frac{b_R}{b_V} \tag{12.100}$$

where $C_{L\alpha_V}$ denotes the vertical tail lift curve slope, $\overline{V}_V$ is the vertical tail volume coefficient, and η_V is the vertical tail dynamic pressure ratio (q_v/q_∞). The parameter τ_r is referred to as the rudder angle of attack effectiveness parameter and is a function of rudder chord-to-vertical tail chord ratio (C_R/C_V). It is determined through Figure 12.12. The contribution of the rudder size to the rudder control effectiveness is reflected by the rudder angle of attack effectiveness τ_r. The vertical tail volume coefficient is defined in Chapter 6 (Equation (6.72)), and is repeated here for convenience:

$$\overline{V}_V = \frac{l_V S_V}{bS} \tag{12.101}$$

Table 6.4 shows typical values for vertical tail volume coefficients. In large high-subsonic transport aircraft, directional control is provided by two in-tandem rudders; one is used for high-speed flights but both are employed in low-speed operations such as take-off and landing. For the purpose of reliability, rudders could be split into upper and lower halves, with independent signals and actuators plus redundant processors.

A rudder design flowchart is presented in Figure 12.26. As is observed, there are two checks which generate two feedback loops for the design procedure. In the first one, if the required rudder chord is greater than the vertical tail chord, the vertical tail must be redesigned or the aircraft cg must be relocated. In the second one, it is investigated whether the other rudder requirements are met, otherwise the designer has made a mistake in recognizing the critical role of the rudder. Both feedback loops demonstrate the iterative nature of the rudder design process. Various rudder design requirements are introduced in Section 12.6.2.

12.6.2 Fundamentals of Rudder Design

12.6.2.1 Rudder Design Requirements

The design requirements of the rudder are primarily driven by directional control and directional trim. Directional control is governed mainly through the yaw rate (R), while directional trim is often governed by the maximum rudder deflection (δ_{Rmax}). The FAA has a number of regulations for directional control, all of which must be addressed by a rudder designer.

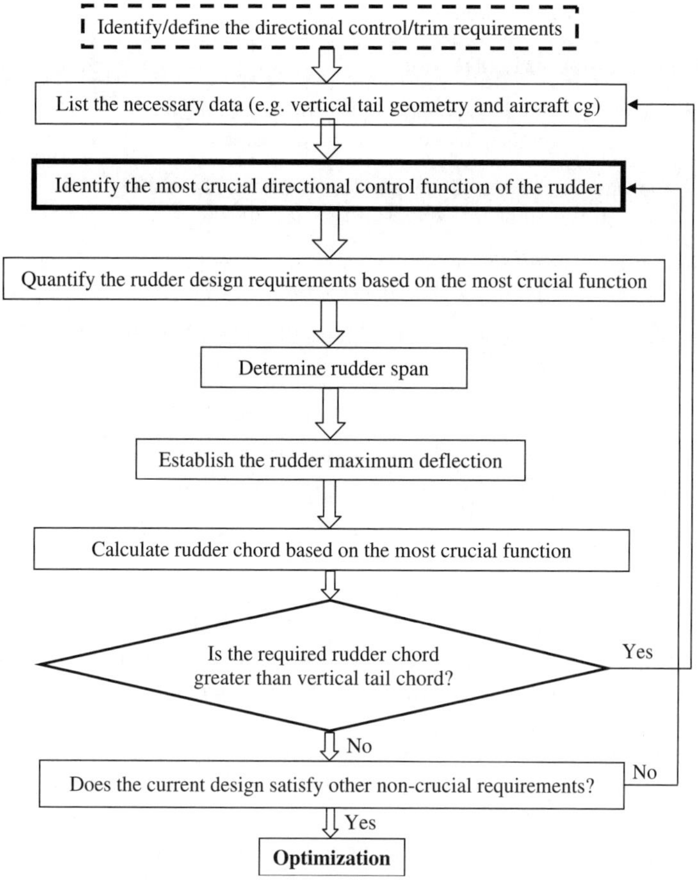

Figure 12.26 Rudder design flowchart

FAR Part 25 Section 25.147 requires the following:

> It must be possible, with the wings level, to yaw into the operative engine and to safely make a reasonably sudden change in heading of up to 15 deg in the direction of the critical inoperative engine. This must be shown at 1.3 V_S for heading changes up to 15 deg, and with (i) the critical engine inoperative and its propeller in the minimum drag position; (ii) the power required for level flight at 1.3 V_S, but not more than maximum continuous power; (iii) the most unfavorable center of gravity; (iv) landing gear retracted; (v) flaps in the approach position; and (vi) maximum landing weight.

There is a similar regulation for GA aircraft in FAR 23 and for military aircraft in MIL-STD. These requirements must be addressed in the design of the rudder.

The rudder plays different roles in different phases of flight for various aircraft. Six major functions of a rudder are: (i) cross-wind landing, (ii) directional control for balancing asymmetric thrust on multi-engine aircraft, (iii) turn coordination, (iv) spin recovery,

Table 12.21 Rudder design requirements

No.	Requirements	Brief description	Aircraft
1	Asymmetric thrust	When one engine fails, the aircraft must be able to overcome the asymmetric thrust.	Multi-engine aircraft
2	Cross-wind landing	An aircraft must maintain alignment with the runway during a cross-wind landing.	All
3	Spin recovery	An aircraft must be able to oppose the spin rotation and to recover from a spin.	Spinnable aircraft
4	Coordinated turn	The aircraft must be able to coordinate a turn.	All
5	Adverse yaw	The rudder must be able to overcome the adverse yaw that is produced by the ailerons.	All
6	Glide slope adjustment	The aircraft must be able to adjust the glide slope by increasing aircraft drag using a rudder deflection.	Glider aircraft

(v) adverse yaw, and (vi) glide slope adjustment for a glider. Table 12.21 tabulates these cases, which impose different requirements for various aircraft. In this section, these design requirements are introduced, formulated, and a technique to design the rudder to satisfy these requirements is developed.

Among these functions, one of them is usually the most critical depending upon the aircraft mission and configuration. From the six duties of rudder mentioned above, the first three are simple but the last three are more important. For instance, multi-engine aircraft often have directional trim in case of asymmetric thrust as the most critical case for a rudder. Single-engine aircraft often have maximum cross-wind landing as the most critical condition. In a spinnable aircraft, the spin recovery imposes the most critical rudder design requirement. The design of a rudder is performed with regard to the most critical role of the rudder. In some aircraft, spin recovery is critical, but in other aircraft, asymmetric power condition is critical. In normal unspinnable aircraft, cross-wind landing is often the most critical condition for a rudder, from which its design proceeds. Therefore, one of the first tasks of the rudder designer is to identify the most crucial case for a rudder to function within the aircraft flight envelope.

12.6.2.2 Asymmetric Thrust

In a multi-engine aircraft, when not all engines are located along the fuselage center line, directional trim must be achieved when the critical engine(s) fails (e.g., one or more engines are inoperative). The critical engine failure must represent the most critical mode of power plant failure with respect to controllability expected in service. In such a case, the operative engine(s) creates an undesirable yawing moment that must be nullified by the rudder. This design requirement is not applicable to a single-engine aircraft where the engine thrust is aligned with the fuselage center line (in fact, when the thrust line passes through the aircraft center of gravity). The same is true for a twin-engine aircraft when both engines are placed along the fuselage center line (such as the Voyager aircraft, where one prop-engine is located at the fuselage nose and the other one is at the fuselage rear section). The critical asymmetric power/thrust condition frequently occurs when all engines on one side of the aircraft fail at low speeds. The rudder must be powerful enough to overcome the yawing moment produced by the asymmetric thrust arrangement. Figure 12.27(a) shows a Boeing 777 with right engine inoperative.

Since one engine may suddenly fail during flight, the thrust of the other engine(s) may impose a yawing moment about the aircraft center of gravity, so it may disturb the aircraft directional trim, then deviate the direction of flight. In this condition, which is called the asymmetric power condition, the rudder deflection must produce a side force then a moment in order to counter the yawing moment of the working engine(s). The critical asymmetric power condition occurs for a multi-engine airplane when the engine(s) of one side fails at low flight speeds. FAR Section 25.149 requires a multi-engine transport aircraft to be directionally controllable at a critical speed referred to as the minimum controllable speed (V_{MC}). This speed may not exceed 1.13 stall speed at the most unfavorable cg location and the most critical take-off configuration. The rudder must be able to overcome the yawing moment produced by the asymmetric thrust arrangement. Furthermore, lateral control must be sufficient to roll the airplane, from an initial condition of steady flight through an angle of 20 deg in the direction necessary to initiate a turn away from the inoperative engine(s), in not more than 5 seconds.

In a single-engine aircraft, when the engine thrust line passes through the aircraft cg, this condition would not be conceivable and thus the rudder does not have such a role. However, the vertical tail and rudder are expected to offset the moment produced by the rotating propeller. To consider a safety margin, the author recommends the rudder designers to consider the minimum controllable speed (V_{MC}) to be 80% of the stall speed (i.e., $V_{MC} = 0.8V_s$). This suggested speed clearly requires that the aircraft be directionally trimmable during take-off ground roll (i.e., low altitude). The minimum controllable speed for a number of aircraft is as follows: Lockheed C-130 Hercules (Figure 5.4), 93.5 KEAS; Fokker 28, 71 KEAS; Airbus A-300B, 103 KEAS; Boeing 707-320B, 122 KEAS; and Boeing 747-200 (Figures 3.7, 3.12, and 9.4), 138 KEAS.

Suppose the aircraft pictured in Figure 12.27(c) lost power in its right engine (i.e., $T_R = 0$). Both engines are located a distance y_T from the fuselage center line. The resulting asymmetric thrust would produce a yawing moment about the aircraft cg equal to $T_L \cdot y_T$. In a steady-state trimmed flight, the thrust of the operative engine (i.e., T_L) must equal the aircraft drag; and the summation of the yawing moments must be zero. Therefore:

$$\sum N_{cg} = 0 \Rightarrow T_L y_T + L_V l_v = 0 \Rightarrow N_A = -T_L y_T \tag{12.102}$$

Inserting Equations (12.94), (12.97), and (12.98) into Equation (12.102) yields:

$$N_A = \bar{q}Sb\left(C_{n_o} + C_{n_\beta}\beta + C_{n_{\delta A}}\delta_A + C_{n_{\delta R}}\delta_R\right) \tag{12.103}$$

You may assume that the aircraft is symmetric about the xz plane (i.e., $C_{n_o} = 0$), the aileron is not deflected (i.e., $\delta_A = 0$), and there is no sideslip angle (i.e., $\beta = 0$). Therefore, the required rudder deflection to directional trim the aircraft in an asymmetric thrust condition is:

$$\delta_R = \frac{T_L y_T}{-\bar{q}SbC_{n_{\delta R}}} \tag{12.104a}$$

For other aircraft configurations such as an aircraft with three or more engines, a similar technique is employed to determine the required rudder deflection. If the aircraft possesses more than one engine on one side of the aircraft, the total yawing moment of the operative engines on one side is considered:

$$\delta_R = \frac{\sum_{i=1}^{n} T_{L_i} y_{T_i}}{-\overline{q} S b C_{n_{\delta R}}} \tag{12.104b}$$

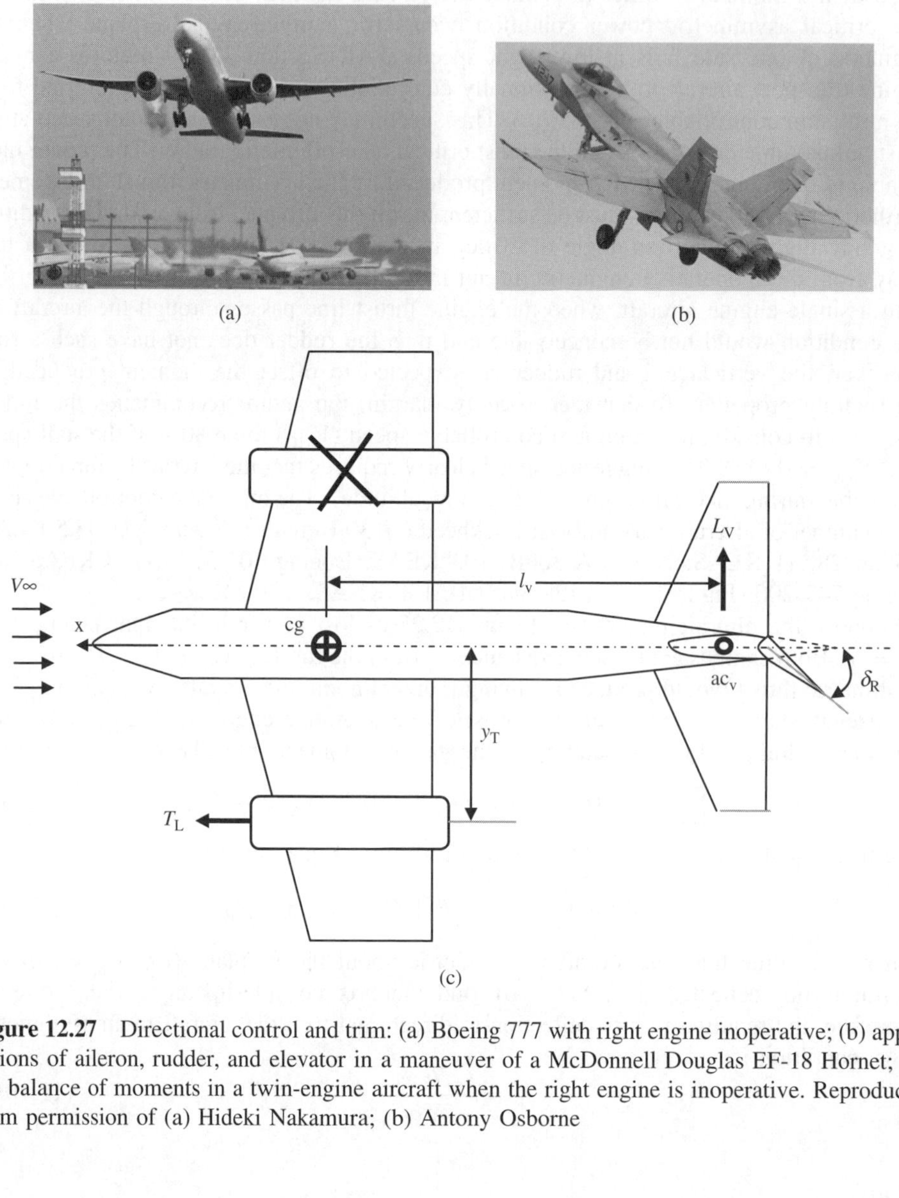

Figure 12.27 Directional control and trim: (a) Boeing 777 with right engine inoperative; (b) applications of aileron, rudder, and elevator in a maneuver of a McDonnell Douglas EF-18 Hornet; (c) the balance of moments in a twin-engine aircraft when the right engine is inoperative. Reproduced from permission of (a) Hideki Nakamura; (b) Antony Osborne

where n denotes the number of engine on one side of the aircraft. Given the aircraft geometry and engine thrust, one can calculate the rudder deflection to keep the aircraft directionally trimmed. The maximum rudder deflection is required when the aircraft has the lowest air speed, and the operative engine is generating its maximum thrust. In Figure 12.27, the rudder is deflected positively due to the right inoperative engine. In case the left engine is inoperative, a negative rudder deflection must be utilized and Equations (12.103) and (12.104) must be revised accordingly.

Example 12.1

A large transport aircraft with a maximum take-off mass of 65 000 kg is equipped with two turbofan engines each generating 116 kN of thrust. The lateral distance between the two engines is 12 m, and the maximum allowable rudder deflection is ±30 deg. Other characteristics of the aircraft are as follows:

$$C_{L_{\alpha V}} = 4.5\frac{1}{\text{rad}};\ S = 125\,\text{m}^2;\ b = 34\,\text{m};\ S_V = 26\ \text{m}^2;\ b_v = 7.6\ \text{m};\ b_R = b_v;$$

$$l_v = 18\ \text{m},\ C_R/C_V = 0.3;\ \eta_v = 0.97;\ V_s = 110\ \text{knot}$$

Is the rudder acceptable for maintaining directional trim in an asymmetric thrust flight condition?

Solution:

It is desired to directionally trim the aircraft in an asymmetric thrust flight condition when the aircraft minimum controllable speed is 80% of stall speed. So:

$$V_{MC} = 0.8\ \cdot V_S = 0.8 \cdot 110 = 88\,\text{knot} = 45.27\,\text{m/s}$$

The vertical tail volume coefficient is:

$$\overline{V}_V = \frac{l_V S_V}{bS} = \frac{18 \cdot 26}{34 \cdot 125} = 0.114 \qquad (12.101)$$

The rudder angle of attack effectiveness (τ_r) is a function of the rudder chord-to-vertical tail chord ratio (C_R/C_V). The parameter τ_r is given to be 0.3, so from Figure 12.12 the rudder angle of attack effectiveness is determined to be 0.52. The rudder control derivative is:

$$C_{n_{\delta R}} = -C_{L_{\alpha V}} \overline{V}_V \eta_V \tau_r \frac{b_R}{b_V} = -4.5 \cdot 0.114 \cdot 0.97 \cdot 0.52 \cdot 1 = -0.266\frac{1}{\text{rad}} \qquad (12.100)$$

The rudder deflection to balance the asymmetric thrust at sea level is:

$$\delta_R = \frac{T_L y_T}{-\overline{q} S b C_{n_{\delta R}}} = \frac{116\,000 \cdot \frac{12}{2}}{-\frac{1}{2} \cdot 1.225 \cdot (45.27)^2 \cdot 125 \cdot 34 \cdot (-0.266)} \qquad (12.104a)$$

or

$$\delta_R = 0.49 \text{ rad} = 28.06 \text{ deg}$$

The required rudder deflection is less than the maximum allowable rudder deflection (i.e., 28 < 30). Therefore, this rudder geometry is acceptable and can satisfy the asymmetric thrust balance requirement.

12.6.2.3 Cross-Wind Landing

One of the most important functions of a rudder in all types of aircraft is to maintain safe landing while a cross-wind is blowing. When a cross-wind blows during landing operation, and if the pilot does not react, the aircraft will exit out of the runway. The pilot is required to employ a special technique to maintain alignment with the runway during cross-wind landing. In general, the final approach under cross-wind conditions may be conducted in two ways: (i) with wings level (i.e., applying a drift correction in order to track the runway center line, this type of approach is called a crabbing) and (ii) with a steady sideslip (i.e., with the aircraft fuselage aligned with the runway center line, using a combination of into-wind aileron and opposite rudder to correct the drift). Most airlines recommend the first technique.

During the cross-wind landing, the rudder is applied to align the aircraft with the runway heading. The rudder must be powerful enough to permit the pilot to trim for the specified cross-winds. The reason why aircraft deviate toward cross-winds (and then change to the right direction of landing) is the directional (weathercock) stability of aircraft. In such a situation, the rudder produces a sideslip angle to maintain alignment with the runway. In this section, the first technique is addressed and governing equations are developed. When touching down with some crab angle on a dry runway, the aircraft automatically realigns with the direction of travel down the runway. However, if prevailing runway conditions and cross-wind components are considered inadequate for safe landing, the pilot may request the assignment of a more favorable runway.

According to airworthiness standards, aircraft must be able to land safely during a cross-wind with specified speed. For instance, according to CS-VLA Article 233, in every very light aircraft, landing may be carried out for 90-deg cross-winds of up to 10 knots. FAR Part 23 Section 233 requires that each GA aircraft must be able to carry out landing for 90-deg cross-winds of up to a wind velocity of 25 knots. There may be no uncontrollable ground-looping tendency in cross-wind landing. It is evident that the critical aircraft speed at cross-wind landing is the minimum speed ($1.1V_s$), which is a good criterion in the design of a rudder for single-engine GA aircraft. Operations in cross-wind conditions require strict adherence to applicable cross-wind limitations and operational recommendations. About 85% of cross-wind incidents and accidents occur at landing.

To evaluate the rudder power in a crabbed landing, consider the aircraft in Figure 12.28 which is approaching with a forward airspeed of U_1 along the runway. There is a cross-wind of V_W from the right that is creating a positive sideslip angle. The sideslip angle is defined as the angle between the flight direction and the relative wind:

$$\beta = \tan^{-1}\left(\frac{V_W}{U_1}\right) \tag{12.105}$$

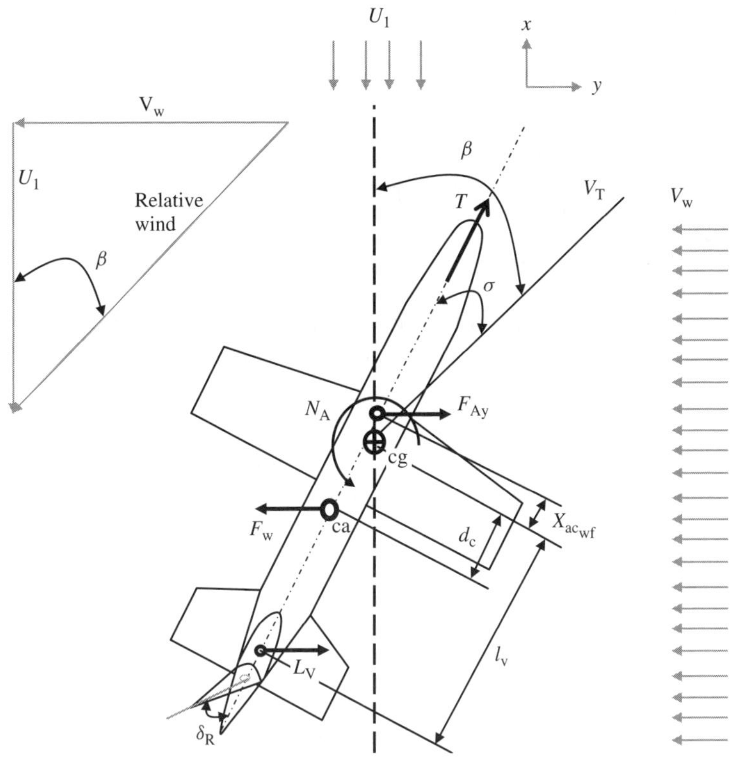

Figure 12.28 Forces and angles in cross-wind crabbed landing

The sideslip angle generates a yawing moment (N_A) and an aerodynamic side force (F_{Ay}) by the aircraft. The weathercock characteristic of the aircraft tends initially to rotate the aircraft about cg (the z-axis) and to yaw the aircraft toward the relative wind. The relative wind or aircraft total speed is the vector summation of the aircraft forward speed and wind speed:

$$V_T = \sqrt{U_1^2 + V_W^2} \tag{12.106}$$

In order to keep the aircraft landing direction along the runway, the rudder is employed to counteract the yawing moment created by the wind. The rudder produces a vertical tail lift along the y-axis (L_V), which consequently contributes to the aircraft yawing moment and aerodynamic side force. The application or rudder is to create a crab angle (σ) in order to prevent the aircraft from yawing to the relative wind and avoid drifting away from the runway. The rudder must be powerful enough to create the desired crab angle. The crab angle is defined as the angle between the fuselage center line and the runway (i.e., heading direction). Figure 12.28 shows all the forces and moments affecting the final approach operation while the aircraft is in a crabbed landing. The aircraft is in directional trim during a crabbed landing, so the following three force and moment equilibrium equations govern the flight condition:

$$\sum N_{cg} = 0 \Rightarrow N_A + F_w \cdot d_c \cos\sigma = 0 \tag{12.107}$$

$$\sum F_x = 0 \Rightarrow T\cos\sigma = D \tag{12.108}$$

$$\sum F_y = 0 \Rightarrow F_w = F_{A_y} \tag{12.109}$$

where d_c is the distance between the aircraft cg and the center of the projected side area of the aircraft, and l_v is the distance between the aircraft cg and the vertical tail aerodynamic center, T denotes engine thrust, and D is aircraft drag. The aircraft aerodynamic side force (F_{A_y}) and yawing moment (N_A) are determined as follows:

$$N_A = \overline{q}Sb\left(C_{n_o} + C_{n_\beta}(\beta - \sigma) + C_{n_{\delta R}}\delta_R\right) \tag{12.110}$$

$$F_{A_y} = \overline{q}S\left(C_{y_o} + C_{y_\beta}(\beta - \sigma) + C_{y_{\delta R}}\delta_R\right) \tag{12.111}$$

where S denotes the wing area and $\overline{q}$ is the dynamic pressure, which is a function of aircraft total speed:

$$\overline{q} = \frac{1}{2}\rho V_T^2 \tag{12.112}$$

The force of the cross-wind (F_w) acts on the center of the side area (ca) of the aircraft (see Figure 12.29). In a conventional aircraft, the center of the side area is always behind the aircraft center of gravity. This is mainly due to the fact that the vertical tail is located at the rear section of the fuselage. The cross-wind force is drag-like and shall be determined in a similar fashion. The force generated by the cross-wind (F_w) is a function of wind speed, aircraft side area, and side given by:

$$F_w = \frac{1}{2}\rho V_W^2 S_S C_{D_y} \tag{12.113}$$

where S_S represents the aircraft projected side area and C_{D_y} is the aircraft side drag coefficient. Typical values for the aircraft side drag coefficient for a conventional aircraft are 0.5–0.8. Only the aircraft aerodynamic side force (F_{A_y}) and yawing moment (N_A) are functions of the rudder deflections, so Equations (12.107) and (12.109) suffice for

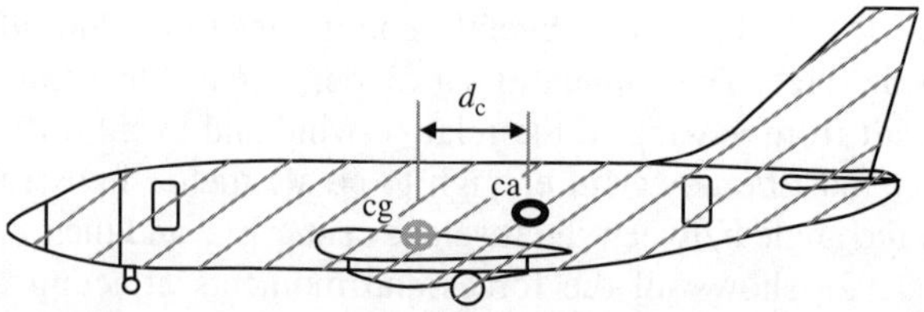

Figure 12.29 Aircraft projected side area and center of side area

the rudder design process. Inserting Equations (12.110)–(12.113) into Equations (12.107) and (12.109) results in:

$$\frac{1}{2}\rho V_{\rm T}^2 Sb\left(C_{\rm n_o} + C_{\rm n_\beta}(\beta - \sigma) + C_{\rm n_{\delta R}}\delta_{\rm R}\right) + F_{\rm w}\cdot d_{\rm c}\cos\sigma = 0 \tag{12.114}$$

$$\frac{1}{2}\rho V_{\rm W}^2 S_{\rm S} C_{D_{\rm y}} - \frac{1}{2}\rho V_{\rm T}^2 S\left(C_{\rm y_o} + C_{\rm y_\beta}(\beta - \sigma) + C_{\rm y_{\delta R}}\delta_{\rm R}\right) = 0 \tag{12.115}$$

In this set of equations, there are two unknowns: (i) rudder deflection ($\delta_{\rm R}$) and (ii) crab angle (σ). Solving these equations simultaneously yields two unknowns. Other flight parameters such as sideslip angle, engine thrust, aircraft drag, and wind force would be calculated separately. In designing the rudder, the designer should make certain that the rudder is powerful enough to allow for the aircraft to land safely in a cross-wind situation.

The four directional stability and control derivatives of $C_{\rm n_\beta}$, $C_{\rm n_{\delta R}}$, $C_{\rm y_\beta}$, and $C_{\rm y_{\delta R}}$ affect Equations (12.114) and (12.115). The directional control derivative $C_{\rm n_{\delta R}}$ was already introduced in Equation (12.100). The static stability derivative $C_{\rm n_\beta}$ is defined in Chapter 6 (Equation (6.73)), but is repeated here for convenience:

$$C_{\rm n_\beta} = K_{\rm f1} C_{L_{\alpha \rm V}}\left(1 - \frac{{\rm d}\sigma}{{\rm d}\beta}\right)\eta_{\rm V}\frac{l_{\rm Vt}S_{\rm V}}{bS} \tag{12.116}$$

The other two derivatives $C_{\rm y_\beta}$ and $C_{\rm y_{\delta R}}$ must be calculated using wind-tunnel testing or references such as [7]. These derivatives may be determined as following:

$$C_{\rm y_\beta} = \frac{\partial C_{\rm y}}{\partial \beta} \approx C_{\rm y_{\beta V}} = -K_{\rm f2} C_{L_{\alpha \rm V}}\left(1 - \frac{{\rm d}\sigma}{{\rm d}\beta}\right)\eta_{\rm V}\frac{S_{\rm V}}{S} \tag{12.117}$$

$$C_{\rm y_{\delta R}} = \frac{\partial C_{\rm y}}{\partial \delta_{\rm R}} = C_{L_{\alpha \rm V}}\eta_{\rm V}\tau_{\rm R}\frac{b_{\rm R}}{b_{\rm V}}\frac{S_{\rm V}}{S} \tag{12.118}$$

The parameter $K_{\rm f2}$ represents the contribution of the fuselage to the derivative $C_{\rm y_\beta}$ and depends strongly on the shape of the fuselage and its projected side area. The fuselage contribution to the derivative $C_{\rm y_\beta}$ tends to be positive. The typical value of $K_{\rm f2}$ for a conventional aircraft is about 1.3–1.4. The parameter ${\rm d}\sigma/{\rm d}\beta$ is referred to as the vertical tail sidewash gradient. Note that, in Figure 12.28, the rudder is deflected positively due to a positive sideslip angle. In case there is a cross-wind from the left (i.e., a negative sideslip angle), a negative rudder deflection must be utilized and Equations (12.105)–(12.118) must be revised accordingly.

In order to determine the center of the projected area of an aircraft (ca), the aircraft side view must be divided into several standard geometric shapes (segments) such as rectangle, triangle, and circle. By selecting a reference line (say fuselage nose), the distance between ca and the reference line (Figure 12.29) is obtained through the following mathematical relationship:

$$x_{\rm ca} = \frac{\sum_{i=1}^{n} A_i x_i}{\sum_{i=1}^{n} A_i} \tag{12.119}$$

where n represents the number of segments, A_i the projected side area of the ith segment, and x_i is the distance between the center of the projected side area of the ith segment and the reference line. The center of standard geometric shapes such as triangle and rectangle is known and may readily be obtained from standard mathematical handbooks such as Ref. [20]. Examples 12.2 and 12.6 will demonstrate the application of the technique.

Example 12.2

Problem statement: Consider a light transport aircraft with a take-off mass of 7400 kg, wing area of 32 m^2, and wing span of 8 m. The aircraft projected side area is 34 m^2 and the aircraft center of the projected side area is 1.8 m behind the aircraft cg. The aircraft approach speed is 82 knots and the maximum allowable rudder deflection is ±30 deg. Other aircraft characteristics, including two rudder-related derivatives, are as follows:

$$C_{n_\beta} = 0.1\frac{1}{\text{rad}};\ C_{n_{\delta_R}} = -0.08\frac{1}{\text{rad}};\ C_{y_\beta} = -0.6\frac{1}{\text{rad}};\ C_{y_{\delta R}} = 0.15\frac{1}{\text{rad}};$$
$$C_{D_y} = 0.6;\ C_{no} = 0;\ C_{y_o} = 0$$

Is the aircraft rudder powerful enough to allow for a safe crabbed landing when a perpendicular cross-wind of 30 knots is blowing? What about 25 knots?

Solution:

1. **Cross-wind of 30 knots**. Due to simplicity, a cross-wind from the right is assumed, generating a positive sideslip angle. The aircraft total speed, wind force, and sideslip angle are:

$$V_T = \sqrt{U_1^2 + V_W^2} = \sqrt{(82)^2 + (30)^2} = 87.316\,\text{knot} = 44.92\,\text{m/s} \quad (12.106)$$

$$F_W = \frac{1}{2}\rho V_W^2 S_S C_{D_y} = \frac{1}{2}\cdot 1.225\cdot(30\cdot 0.514)^2\cdot 34\cdot 0.6 = 2976\,\text{N} \quad (12.113)$$

$$\beta = \tan^{-1}\left(\frac{V_W}{U_1}\right) = \tan^{-1}\left(\frac{30}{82}\right) = 0.351\,\text{rad} = 20.1\,\text{deg} \quad (12.105)$$

Now, we have two equations and two unknowns:

$$\frac{1}{2}\rho V_T^2 Sb\left(C_{n_o} + C_{n_\beta}(\beta-\sigma) + C_{n_{\delta R}}\delta_R\right) + F_W\cdot d_c\cos\sigma = 0 \quad (12.114)$$

$$\frac{1}{2}\rho V_W^2 S_S C_{D_y} = \frac{1}{2}\rho V_T^2 S\left(C_{y_o} + C_{y_\beta}(\beta-\sigma) + C_{y_{\delta R}}\delta_R\right) \quad (12.115)$$

or

$$\frac{1}{2}\cdot 1.225\cdot(44.92\cdot 0.514)^2\cdot 32\cdot 8\left[0.1(0.351-\sigma) - 0.08\delta_R\right]$$
$$+ 2976\cdot 1.8\cos\sigma = 0 \quad (12.114)$$

$$2976 = \frac{1}{2} \cdot 1.225 \cdot (44.92 \cdot 0.514)^2 \cdot 32 \left(-0.6 \left(0.351 - \sigma\right) + 0.15\delta_R\right) \quad (12.115)$$

Simultaneous solution of these equations results in:

$$\delta_R = 0.64 \text{ rad} = 36.6 \text{ deg}$$

$$\sigma = 0.316 \text{ rad} = 18.12 \text{ deg}$$

The required rudder deflection (35.6 deg) exceeds the maximum allowable deflection (30 deg). Thus the aircraft is not able to handle a cross-wind of 30 knots.

2. **Cross-wind of 25 knots**. The same technique is employed, only the wind speed is 25 knots. The calculation results in a sideslip angle of +16.9 deg and a crab angle of 14.7 deg. In addition, the required rudder deflection is +29.64 deg, which is slightly less than the maximum allowable limit. Therefore, this aircraft is able to safely crab land with a cross-wind of 25 knots.

12.6.2.4 Spin Recovery

One of the most important roles of a rudder in the majority of airplanes is spin recovery. The most significant instrument to recover aircraft from a spin is a powerful rudder. Spin is a self-sustaining (auto-rotational) spiral motion of an airplane about the vertical (z) axis, during which the mean angle of attack of the wings is beyond the stall. Almost since man first flew, spinning has caused many fatal accidents. During the years 1965–1972, the US Navy lost an average of two aircraft per month and a total of 169 aircraft due to spin, the list of which is headed by 44 fighter aircraft F-4s (Phantom). This shows the crucial role of the rudder in a spin.

Spin is a high angle of attack/low airspeed situation; the airspeed will be hovering somewhere down in the stall area. Spin has two particular specifications: (i) fast rotation around the vertical axis and (ii) fully stalled wing. Spin usually starts after the wing stalls. One of the reasons aircraft enter a spin is that the inboard of the wing stalls before the outboard of the wing, in other words, the lift distribution over the wing is not elliptic. Spin is recovered by a procedure in which all the control surfaces (elevator, aileron, and rudder) contribute, particularly the rudder in an apparently unnatural way. The rudder is the most significant element in spin recovery to stop rotation. The primary control for spin recovery in many airplanes is a powerful rudder.

The rudder must be powerful enough to oppose the spin rotation in the first place. A spin follows departures in roll, yaw, and pitch from the condition of trim between the predominantly pro-spin moment due to the wings and the generally anti-spin moments due to other parts of the aircraft. If spin is not recovered, the aircraft will eventually crash. The criterion for rudder design in a spinnable aircraft may be spin recovery. Acrobatic and fighter airplanes are usually spinnable, but there are some airplanes (such as some transport aircraft) that are spin-proof or unspinnable.

In unspinnable aircraft, spin recovery is not a criterion for design of the rudder, that is, the rudder does not have to recover the aircraft from spin. According to airworthiness standards, in a spinnable aircraft, the rudder must have enough power to recover the spin in a limited time. For instance, EASA CS-VLA Article 221 requires that any very light aircraft must be able to recover from spin in a maximum period of one turn. FAR Part 23 Section 23.221 regulates spinning for normal category airplanes as follows:

> a single-engine, normal category airplane must be able to recover from a one-turn spin or a three-second spin, whichever takes longer, in not more than one additional turn after initiation of the first control action for recovery, or demonstrate compliance with the optional spin resistant requirements of this section.

Although the wing is fully stalled, left and right wing sections produce different lift. So, the aircraft begins to roll around the x-axis. Furthermore, since the drags of the right and left wing sections are different, the aircraft yaws toward the down-going wing (i.e., auto-rotation). In addition, since the aircraft is stalled, it loses lift and starts to dive, when in a normal spin entry the spin develops to an equilibrium state. Thus there is a mixture of stall, roll, yaw, and dive in a spin. Although the auto-rotation property of a wing (when a large part of it has an angle of attack beyond the stall) is the primary cause of a spin, this does not necessarily mean that a spin will occur.

There are damping moments provided by the fuselage, and the anti-spin moment from the vertical tail, which together counter the propelling moments from the wings. The result is that for a given combination of control settings there is one equilibrium rate of rotation at each angle of attack. A spin can only follow from auto-rotation if the equilibrium of pitching moments, and inertial moments, can be sustained. Aircraft which have such characteristics are called spinnable. But if equilibrium of the pitching moments and the inertial moments cannot be obtained simultaneously, the aircraft will recover itself.

The typical range of some spin parameters is as follows: angle of attack (α), 30–60 deg; rate of descent (ROD), 20–100 m/s; rate of spin (Ω), 20–40 rpm; helix angle (γ), 3–6 deg; and helix radius (R), half of the wing span. As the angle of attack increases the rate of rotation increases, and the helix radius decreases.

Basically, the rudder is not the only factor to feature an acceptable spin recovery. Two other significant factors are as follows: (i) aircraft mass distribution and aircraft moments of inertia, and (ii) fuselage side area and cross-section. It is very important that the inertia term be made anti-spin (negative for right spin) for recovery. When the magnitudes of pitch (I_{yy}) and roll (I_{xx}) inertia are close, the effect of the inertia term is small, and hence the rudder will be the primary control for spin recovery. But whenever the inertia term becomes significant, it has a considerable impact on the spin motion and thus the size of the rudder. The application of an aileron to aid recovery is generally not recommended, due to its nuisance impact. In some cases the use of ailerons, while stopping a spin, may suddenly cause a spin in the reverse direction.

Prior to the 1940s, due to placing fuel, stores, and engines on the wing, the changes in the inertia term were small (i.e., $I_{xx} - I_{yy} \approx 0$). Thus, the rudder was the only effective control to prevent spin and for spin recovery. However, today the changes in the inertia term are much more significant, because of the distribution of mass along the fuselage, so the pitch-to-roll inertia ratio (I_{yy}/I_{xx}) has a considerable value and the term $I_{xx} - I_{yy}$ has a large negative value. For this reason, although in modern aircraft the control surfaces are of similar size to those of older aircraft, their spin recovery characteristics are much less important.

When the aircraft mass concentration in the wing is greater than the mass concentration in the fuselage (e.g., sailplane), the aircraft moment of inertia will induce a pro-spin behavior. The net result is that the aircraft inertia produces an in-spin yawing moment which increases the outward sideslip. However, when the aircraft mass concentration in the wing is smaller than the mass concentration in the fuselage (e.g., fighter), the aircraft moment of inertia will induce an anti-spin behavior. Hence, the aircraft inertia creates an out-of-spin yawing moment which decreases the outward sideslip.

Damping provided by various parts of the aircraft, such as the fuselage and rudder, can counter the yawing moment of the wings during spin. So, provision of a large amount of damping in yaw for fuselage and vertical will be the two most effective means for prevention of a spin. The aerodynamic yawing moment due to rotation of the fuselage about the spin axes is largely dependent on the fuselage shape and its cross-section. In addition, the provision of strakes on the fuselage, in front of the tail, will increase the fuselage damping. Therefore, the aircraft designer can reduce the spin recovery load on the rudder by careful design of the fuselage and proper aircraft weight distribution.

When a steady-state spin is developed, the equilibrium of forces implies that the lift is equal to the centrifugal force and the aircraft weight is equal to the aircraft drag (see Figure 12.30). In order to stop the spin, a yawing moment is needed. In this situation, the three moments of inertia – rolling moment of inertia (I_{xx}), yawing moment of inertia (I_{yy}), and product of inertia (I_{zz}) – influence the recovery process. Newton's second law governs the aircraft rotation about the z-axis (i.e., yaw rate, R) in a spin recovery operation as follows:

$$N_{\mathrm{SR}} = \left(\frac{I_{xx}I_{zz} - I_{xz}^2}{I_{xx}}\right)_{\mathrm{w}} \dot{R}_{\mathrm{SR}} \tag{12.120}$$

The aircraft rolling moment of inertia (I_{xx}), the yawing moment of inertia (I_{yy}), and the product of inertia (I_{zz}) are generally calculated in the body-fixed axis system. Table 11.12

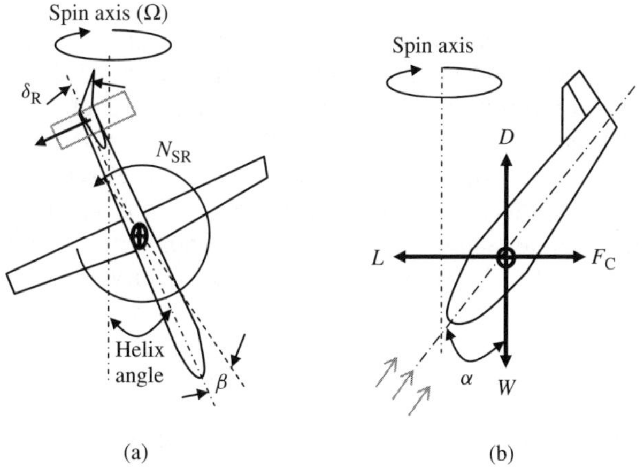

Figure 12.30 Forces and moments contributing to spin. (a) Rudder counteracts the yawing motion (top view); (b) Equilibrium of forces in spin (side view)

illustrates body-axis mass moments of inertia for several aircraft. During a spin where the aircraft possesses a high angle of attack, the aircraft is not yawing about the body z-axis but rather is in rotation about the wind z-axis. The subscript "w" in Equation (12.120) indicates that all three moments of inertia (I_{xx}, I_{zz}, and I_{xz}) must be computed in the wind-axis system. Thus, a transformation involving the aircraft angle of attack is necessary. This transformation is performed mathematically by the following matrix equation:

$$\begin{bmatrix} I_{xx_{\mathrm{w}}} \\ I_{zz_{w}} \\ I_{xz_{\mathrm{w}}} \end{bmatrix} = \begin{bmatrix} \cos^2\alpha & \sin^2\alpha & -\sin 2\alpha \\ \sin^2\alpha & \cos^2\alpha & \sin 2\alpha \\ \frac{1}{2}\sin 2\alpha & -\frac{1}{2}\sin 2\alpha & \cos 2\alpha \end{bmatrix} \begin{bmatrix} I_{xx_{\mathrm{B}}} \\ I_{zz_{\mathrm{B}}} \\ I_{xz_{\mathrm{B}}} \end{bmatrix} \tag{12.121}$$

It is interesting to note that the effect of a high angle of attack on I_{xz} is quite large. The desired rate of yaw rate (i.e., $\dot{R}$) is generally given by references such as FAR regulations. For instance, FAR Part 23 Section 23.221 requires that a GA aircraft must be able to recover from a one-turn spin in not more than 3 seconds. In addition, a typical value for the rate of spin (Ω) is about 20–40 rpm or 120–240 deg/s, where the higher rate is the most critical one. Hence, the rate of spin recovery ($\dot{R}_{\mathrm{SR}}$) is desired to be:

$$\dot{R}_{\mathrm{SR}} = \frac{\Omega}{t} = \frac{240\,\mathrm{deg/\,s}}{3\ \mathrm{s}} = 80\,\frac{\mathrm{deg}}{\mathrm{s}^2} = 1.396\,\frac{\mathrm{rad}}{\mathrm{s}^2} \approx 1.4\frac{\mathrm{rad}}{\mathrm{s}^2} \tag{12.122}$$

Therefore, it is suggested to design the rudder such that the aircraft is able to recover from a spin by the rate of yaw rate of 80 deg/s^2 or 1.4 rad/s^2. The desired aircraft counteracting yawing moment (N_{SR}), created by the rudder deflection, was already defined in Equation (12.97), where the aircraft airspeed is assumed to be equal to the stall speed. Hence,

$$N_{\mathrm{SR}} = \frac{1}{2}\rho V_{\mathrm{s}}^2 SbC_{\mathrm{n}_{\delta\mathrm{R}}}\delta_{\mathrm{R}} \tag{12.123}$$

Furthermore, the rudder control derivative $C_{\mathrm{n}_{\delta\mathrm{R}}}$ was already defined in Equation (12.100) for clean flight conditions. However during a spin, due to shielding of the rudder by the horizontal tail (see Figure 12.31), parts of the vertical tail and rudder are in the wake region. The shielding effect will negatively impact the effectiveness of the rudder and vertical tail. Thus, only unshielded areas of the vertical tail and rudder may contribute to the yawing moment generation and to the derivative $C_{\mathrm{n}_{\delta\mathrm{R}}}$. Therefore, Equations (12.100) and (12.101) should be modified to include the effective vertical tail area ($S_{\mathrm{V_e}}$), effective ruder area ($S_{\mathrm{R_e}}$), effective rudder span ($b_{\mathrm{R_e}}$), and effective rudder chord ($C_{\mathrm{R_e}}$). The effective rudder and vertical span, chord, and area are determined by measurement of the rudder and vertical tail areas beyond the wake region (see Figure 12.32). Thus, the rudder control derivative $C_{\mathrm{n}_{\delta\mathrm{R}}}$ during a spin is redefined as follows:

$$C_{\mathrm{n}_{\delta\mathrm{R}}} = -C_{L\alpha_{\mathrm{V}}}\overline{V}_{\mathrm{V_e}}\eta_{\mathrm{V}}\tau_{\mathrm{R}}\frac{b_{\mathrm{R_e}}}{b_{\mathrm{V}}} \tag{12.124}$$

Accordingly, the effective vertical tail volume ratio is given by:

$$\overline{V}_{\mathrm{V_e}} = \frac{l_{\mathrm{V}}S_{\mathrm{V_e}}}{bS} \tag{12.125}$$

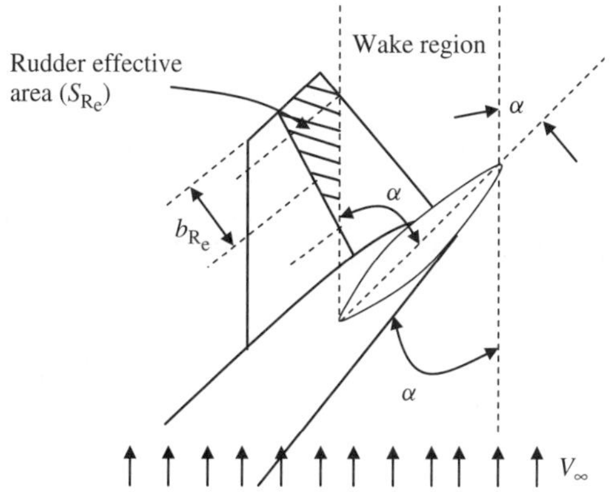

Figure 12.31 The influence of horizontal tail on the effectiveness of vertical tail and rudder

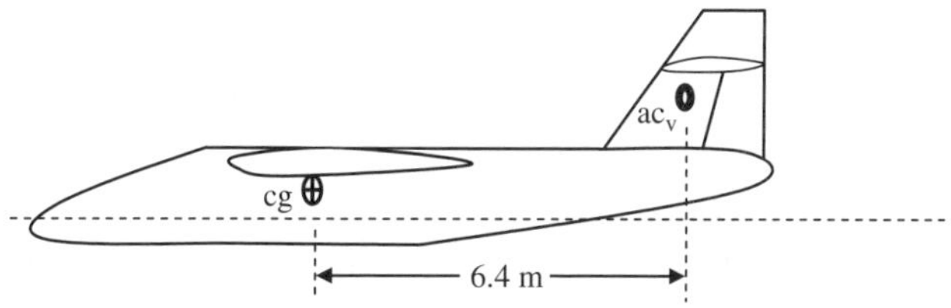

Figure 12.32 The aircraft in Example 12.3

The horizontal tail wake region is the area above a stalled horizontal tail where the tail has a high angle of attack α. Since the aircraft and tail angles of attack during spin vary and are functions of a number of parameters, a rule of thumb is established. As a rough rule, the horizontal tail wake region is considered to lie between two lines. The first line is drawn at the horizontal tail trailing edge with an orientation of 30 deg. The second line is drawn at the horizontal tail leading edge with an orientation of 60 deg. Therefore, the influence of the horizontal tail on the effectiveness of the vertical tail and rudder during a spin shall be applied using Equations (12.124) and (12.125).

Example 12.3

Consider the single-engine utility aircraft shown in Figure 12.32, with a maximum take-off mass of 1400 kg and a cruciform tail configuration:

$$C_{L_{\alpha V}} = 4.4\frac{1}{\text{rad}};\ S = 15\ \text{m}^2;\ b = 12\ \text{m};\ S_V = 2\ \text{m}^2;\ b_v = 2.3\ \text{m};\ b_R = 0.7b_v;$$

$$C_R/C_V = 0.4;\ \eta_v = 0.96;\ V_s = 55\ \text{knot};\ I_{xx_B} = 1150\ \text{kg m}^2;\ I_{zz_B} = 2400\ \text{kg m}^2;$$

$$I_{xz_B} = 120\ \text{kg m}^2$$

The maximum allowable rudder deflection is ±25 deg. Is this rudder able to satisfy the spin recovery requirement at 15 000 ft altitude? Assume the aircraft will spin at an angle of attack of 40 deg.

Solution:
The aircraft is spinning with a 40 deg angle of attack, so the moments of inertia must be transformed to the wind axes:

$$\begin{bmatrix} I_{xx_W} \\ I_{zz_W} \\ I_{xz_W} \end{bmatrix} = \begin{bmatrix} \cos^2\alpha & \sin^2\alpha & -\sin 2\alpha \\ \sin^2\alpha & \cos^2\alpha & \sin 2\alpha \\ \frac{1}{2}\sin 2\alpha & -\frac{1}{2}\sin 2\alpha & \cos 2\alpha \end{bmatrix} \begin{bmatrix} I_{xx_B} \\ I_{zz_B} \\ I_{xz_B} \end{bmatrix} \quad (12.121)$$

$$\begin{bmatrix} I_{xx_W} \\ I_{zz_W} \\ I_{xz_W} \end{bmatrix} = \begin{bmatrix} \cos^2(40) & \sin^2(40) & -\sin(80) \\ \sin^2(40) & \cos^2(40) & \sin(80) \\ \frac{1}{2}\sin(80) & -\frac{1}{2}\sin(80) & \cos(80) \end{bmatrix} \begin{bmatrix} 1150 \\ 2400 \\ 120 \end{bmatrix} \quad (12.121)$$

This transformation results in:

$$I_{xx_W} = 1548.3\,\text{kg m}^2;\ I_{zz_W} = 2001.7\,\text{kg m}^2;\ I_{xz_W} = -594.7\,\text{kg m}^2$$

It is desired to recover from a spin with a rate of 1.4 rad/s^2 (i.e., $\dot{R}_{SR} = 1.4\,\text{rad/s}^2$) when the aircraft spins with stall speed, so the required yawing moment to stop the spin with the desired rate will be:

$$N_{SR} = \left(\frac{I_{xx}I_{zz} - I_{xz}^2}{I_{xx}}\right)_W \dot{R}_{SR} = \left[\frac{1548.3 \cdot 2001.7 - (-594.7)^2}{1548.3}\right] \cdot 1.4 \quad (12.120)$$

or

$$N_{SR} = 2482.6\,\text{N m}$$

The horizontal tail will shield part of the vertical tail, but due to the cruciform configuration, no part of the rudder shall be in the wake region of the horizontal tail (i.e., $b_{Re} = b_R$). Hence, the effective vertical tail area is determined as follows:

$$C_V = \frac{S_V}{b_V} = \frac{2}{2.3} = 0.87\,\text{m} \quad (6.80)$$

$$S_{V_e} = S_V - (0.3 b_V C_V) = 2 - (0.3 \cdot 2.3 \cdot 0.87) = 1.4\,\text{m}^2$$

The effective vertical tail volume coefficient is:

$$\overline{V}_{V_e} = \frac{l_V S_{V_e}}{bS} = \frac{6.4 \cdot 1.4}{12 \cdot 15} = 0.05 \quad (12.125)$$

The rudder angle of attack effectiveness (τ_r) is a function of the rudder chord-to-vertical tail chord ratio (C_R/C_V). Given that the parameter C_R/C_V is 0.4, from Figure 12.12

the rudder angle of attack effectiveness (τ_r) is determined to be 0.6. The rudder control derivative is:

$$C_{n_{\delta R}} = -C_{L\alpha_V} \overline{V}_{V_e} \eta_V \tau_r \frac{b_{R_e}}{b_V} = -4.4 \cdot 0.05 \cdot 0.96 \cdot 0.6 \cdot 0.7 = -0.088 \frac{1}{\text{rad}} \quad (12.124)$$

The rudder deflection to balance the asymmetric thrust at sea level is:

$$N_{SR} = \frac{1}{2} \rho V_s^2 S b C_{n_{\delta R}} \delta_R \quad (12.123)$$

where the air density at 15 000 ft altitude is 0.768 kg/m^3. Thus,

$$\delta_R = \frac{2N_{SR}}{\rho V_s^2 S b C_{n_{\delta R}}} = \frac{2 \cdot 2482.6}{0.768 \cdot (55 \cdot 0.541)^2 \cdot 15 \cdot 12 \cdot (-0.088)} \quad (12.123)$$

or

$$\delta_R = -0.508 \text{ rad} = -29.11 \text{ deg}$$

The required rudder deflection is less than the maximum allowable rudder deflection (i.e., 29.11 < 30). Therefore, this rudder geometry is acceptable and can satisfy the spin recovery requirement.

12.6.2.5 Coordinated Turn

A simple and essential function of a rudder is carried out during a turning flight. A turn operation is basically performed by banking the aircraft using ailerons. The rudder's role in a turn is generally to coordinate the turn. A coordinated level turn is defined as a turn where the components of forces along the aircraft body-fixed y-axis sum to zero. In addition, it is desired that the aerodynamic side force (F_{A_y}) is equal to zero. This type of turn is desirable since it possesses a number of favorable features, such as: (i) no net lateral acceleration (i.e., no skidding/no slipping), (ii) constant turn radius, (iii) constant turn rate, (iv) even fuel distribution between left and right fuel tanks, and (v) passenger comfort. All of these are beneficial characteristics obtained by deflecting the rudder during a turning flight.

Simultaneous deflections of aileron and rudder will create a coordinated turn, although the aircraft may have a non-zero sideslip angle (β). The coordinated turn governing equations may be derived using Newton's second law on lateral-directional axes (Figure 12.33):

$$F_{A_{y_t}} = 0 = F_C - W \sin\phi \quad (12.126)$$

$$L_{A_t} = \left(I_{zz} - I_{yy}\right) R_1 Q_1 \quad (12.127)$$

$$N_{A_t} = I_{xz} R_1 Q_1 \quad (12.128)$$

where W represents the aircraft weight, ϕ the bank angle, R_1 the yaw rate, and Q_1 the pitch rate. Furthermore, F_C denotes centrifugal force, $F_{A_{y_t}}$ represents the aerodynamic

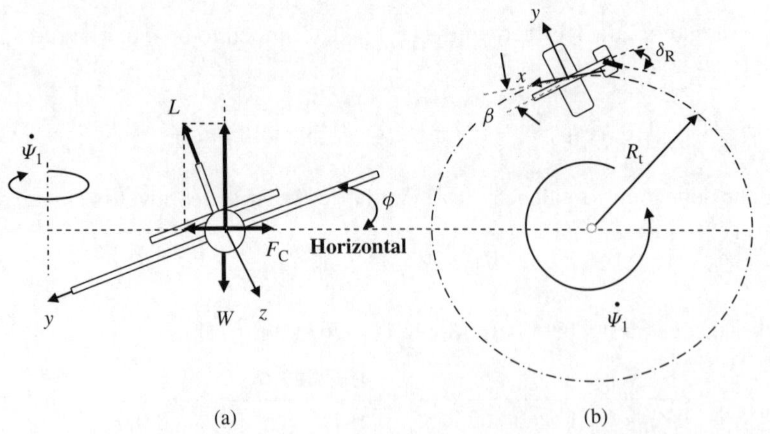

Figure 12.33 An aircraft in a turning flight. (a) Front view; (b) Top view

side force during turn, L_{A_t} denotes the aerodynamic rolling moment during turn, and N_{A_t} denotes the aerodynamic rolling moment during turn. These three lateral-directional forces and moments during a turn are given by:

$$F_C = m\frac{U_1^2}{R_t} \tag{12.129}$$

$$F_{A_{y_t}} = \frac{1}{2}\rho U_1^2 S\left(C_{y_\beta}\beta + C_{y_r}\frac{R_1 b}{2U_1} + C_{y_{\delta A}}\delta_A + C_{y_{\delta R}}\delta_R\right) \tag{12.130}$$

$$L_{A_t} = \frac{1}{2}\rho U_1^2 Sb\left(C_{l_\beta}\beta + C_{l_r}\frac{R_1 b}{2U_1} + C_{l\delta A}\delta_A + C_{l_{\delta R}}\delta_R\right) \tag{12.131}$$

$$N_{A_t} = \frac{1}{2}\rho U_1^2 Sb\left(C_{n_\beta}\beta + C_{n_r}\frac{R_1 b}{2U_1} + C_{n_{\delta A}}\delta_A + C_{n_{\delta R}}\delta_R\right) \tag{12.132}$$

where R_t is the turn radius, S the wing area, b the wing span, and U_1 the aircraft forward speed. The parameters $C_{y_\beta}; C_{y_r}; C_{y_{\delta A}}; C_{y\delta R}, C_{l_\beta}; C_{l_r}; C_{l\delta A}; C_{l_{\delta R}}$, and $C_{n_\beta}; C_{n_r}; C_{n_{\delta A}}; C_{n_{\delta R}}$ are all aircraft stability and control derivatives. The two variables of steady-state yaw rate (R_1) and pitch rate (Q_1) are determined as follows:

$$Q_1 = \dot{\psi}_1 \sin\phi = \frac{g\sin^2\phi}{U_1\cos\phi} \tag{12.133}$$

$$R_1 = \dot{\psi}_1 \cos\phi = \frac{g\sin\phi}{U_1} \tag{12.134}$$

where $\dot{\psi}_1$ is the rate of turn. The ratio between the lift (L) and the aircraft weight (W) is an important parameter in turning performance. It is called the load factor and represented by the symbol n:

$$n = \frac{L}{W} \tag{12.135}$$

Referring to Figure 12.33, the load factor will be equal to:

$$n = \frac{1}{\cos\phi} \tag{12.136}$$

It implies that as the aircraft bank angle (ϕ) increases, the load factor (n) will increase too. In order to determine the rudder deflection (δ_R) required to make the turn coordinated, Equations (12.130)–(12.132) must be solved simultaneously. The other two unknowns in this set of equations are usually the aileron deflection (δ_A) and the sideslip angle (β).

A special case for the rudder deflection during a turn is when the aircraft is experiencing an inoperative engine(s). The critical condition clearly would be at low speeds and the most unfavorable aircraft cg location. In such a turning flight, an asymmetric thrust yawing moment term will be added to Equation (12.132) as follows:

$$N_{A_t} + N_T = \frac{1}{2}\rho U_1^2 Sb\left(C_{n_\beta}\beta + C_{n_r}\frac{R_1 b}{2U_1} + C_{n_{\delta A}}\delta_A + C_{n_{\delta R}}\delta_R\right) \tag{12.137}$$

where N_T denotes the yawing moment produced by the operative (engines) about the aircraft cg (z-axis):

$$N_T = \sum_{i=1}^{n_e} T_i Y_i \tag{12.138}$$

It is important to realize that the amount of rudder deflection required to hold an engine-out condition can be significantly reduced by allowing the aircraft to *bank into the operating engine*(s). In the majority of aircraft, the aircraft is directionally uncontrollable if the aircraft banks into the in-operating engine(s) at low speeds (e.g., approach) since the required rudder deflection will be beyond the maximum allowable deflection.

12.6.2.6 Adverse Yaw

When an aircraft is banked to execute a turn operation, conventional ailerons tend to create a yawing moment that opposes the turn (i.e., adverse yaw). The yaw is generated by the difference between the induced drag (KC_L^2) of the downward-moving wing and the upward-moving wing. Recall that the aileron varies the wing lift locally when deflected. A down-deflected aileron will increase the local lift, while an up-deflected aileron will decrease the local lift. Hence, a downward-moving wing section has a decreased local lift and consequently a decreased induced drag. Similarly, an upward-moving wing section has an increased local lift and consequently an increased induced drag. Thus, a positive rolling moment (clockwise) will produce a negative yawing moment (counterclockwise) and a negative rolling moment (counterclockwise) will produce a positive yawing moment (clockwise). In this condition, the rudder must be able to overcome the adverse yaw so that a coordinated turn can be achieved.

The critical condition for the rudder to overcome the adverse yaw is when the airplane is flying slowly. Compared with other rudder functions (e.g., cross-wind landing and asymmetric thrust), eliminating an adverse yaw is not a critical role for a rudder. In contrast, there are techniques to eliminate adverse yaw without employing a rudder. Two convenient techniques to avoid adverse yaw are a Frise aileron and differential deflection.

Most Cessna aircraft are equipped with Frise ailerons, but most Piper aircraft employ differentially deflected ailerons.

12.6.2.7 Glide Slope Adjustment

Another role of a rudder is frequently applied during a gliding operation. One way to increase the glide slope and glide angle, in addition to pulling up the nose, is to use a rudder. When a glider, or an engine-powered aircraft with all engines inoperative, is approaching a runway, the only way to adjust the glide slope is to increase the drag while keeping the path. An effective technique to deliberately increase the aircraft drag is to increase the sideslip angle. During a glide, if the glide angle (γ) is less than a specific value, the glider or aircraft will pass over the runway and land on unsuitable ground. Under such a flight condition, the pilot will usually deflect the rudder to increase the aircraft drag in order to increase the glide angle, and then land safely. As the rudder deflection is increased, the glide slope gets steeper; thus a slow-rate glide is converted to a fast-rate glide. The rudder of a glider might be designed solely based on the glide slope requirements.

For a steady unaccelerated glide, the governing equations of motion are:

$$L - W\cos\gamma = 0 \tag{12.139}$$

$$D - W\sin\gamma = 0 \tag{12.140}$$

So the glide angle will be derived as:

$$\gamma = \tan^{-1}\left(\frac{1}{L/D}\right) = \tan^{-1}\left[2\sqrt{K \cdot C_{D_o}}\right] \tag{12.141}$$

For any sideslip angle (β), the drag force is increased and the forward speed is reduced. Thus, the aircraft drag during a glide involving a sideslip angle is obtained as:

$$D_{\text{GL}} = \frac{1}{2}\rho U_1^2 S C_{D_{\text{GL}}} \tag{12.142}$$

where the aircraft drag coefficient due to a sideslip angle is given by:

$$C_{D_{\text{GL}}} = C_{D_o} + KC_L^2 + C_{D_\beta}\beta \tag{12.143}$$

The derivative C_{D_β} is the rate of change of the aircraft drag with respect to the sideslip angle (β):

$$C_{D_\beta} = \frac{\partial C_D}{\partial \beta} \tag{12.144}$$

Reference [21] has introduced two non-dimensional coupling derivatives, one of which relates the sideslip angle to the aircraft drag coefficient (i.e., C_{D_β}). This derivative is given by:

$$C_{D_\beta} = [\text{sign}(\beta)]\frac{S_{\text{V}}}{S}\frac{2}{\pi} \tag{12.145}$$

The relationship between rudder deflection and sideslip angle is given by Equation (12.110). The rate of sink (ROS) in a gliding operation is obtained from the following equation:

$$ROS = U_1 \sin(\gamma) \tag{12.146}$$

The rudder deflection is adjusted to produce the desired ROS in a gliding flight.

12.6.3 Rudder Design Steps

In Sections 12.6.1 and 12.6.2 the rudder primary functions, parameters, governing rules and equations, design objectives, design criteria, and formulation, as well as design requirements, have been presented in detail. In addition, Figure 12.26 illustrates the design flowchart of the rudder. In this section, the rudder design procedure in terms of design steps is introduced. It must be noted that there is no unique solution to satisfy the customer requirements in designing a rudder. Several rudder designs may satisfy the requirements, but each will have unique advantages and disadvantages. It must be noted that there is a possibility that no rudder can satisfy the requirements due to the limits/constraints imposed by the vertical tail design and aircraft cg location. In such a situation, the designer must return to the vertical tail design and/or aircraft weight distribution and redesign/redistribute those components.

Based on the systems engineering approach, the rudder detail design process starts with identifying and defining the design requirements and ends with optimization. Since there are a number of directional control/trim requirements, a separate procedure is set for each rudder design requirement. If you can evaluate/identify the most crucial directional control function of the rudder (see Section 12.6.2), begin the design process with the requirements to satisfy the most crucial directional control function. In the case where it is very hard to recognize which of the directional control requirements are the most critical, follow the suggested ones as in Table 12.22.

Table 12.22 The most critical flight condition for a rudder

No.	Aircraft	The most critical flight condition
1	Glider/sailplane	Glide slope adjustment
2	Single-engine normal GA	Cross-wind landing
3	Single-engine utility/acrobatic GA	Spin recovery
4	Multi-engine normal GA	Asymmetric thrust
5	Multi-engine utility/acrobatic GA	Asymmetric thrust/spin recovery
6	Multi-engine transport (fuselage-installed engines)	Cross-wind landing
7	Multi-engine transport (wing-installed engines)	Asymmetric thrust/cross-wind landing
8	Military fighter	Directional maneuverability/spin recovery
9	Remote-controlled/model	Coordinated turn

12.6.3.1 Rudder Design Steps to Satisfy Asymmetric Thrust Requirements

The following are the rudder design steps for a conventional aircraft to satisfy asymmetric thrust requirements:

1. List the available/given data related to rudder design (e.g., vertical tail geometry, aircraft cg locations).
2. Identify the most unfavorable cg location and aircraft weight combination, the most unfavorable engines inoperative conditions, and the most unfavorable altitude for directional control. This will be set as the most critical condition.
3. Select the rudder span-to-vertical tail span ratio b_R/b_V (see Table 12.3).
4. Establish the maximum rudder deflection to prevent flow separation (see Table 12.3).
5. Determine/select the aircraft minimum controllable speed. FAR regulations provide some requirements for such speed. It is recommended to select a value equivalent to 80% of the stall speed to consider a safety factor.
6. Determine the required maximum yawing moment to directionally control/trim the aircraft in the most critical condition utilizing Equation (12.102).
7. Compute the rudder control derivative $C_{n\delta R}$ utilizing Equation (12.103) assuming that the maximum rudder deflection is employed.
8. Calculate the rudder angle of attack effectiveness (τ_R) employing Equation (12.100).
9. Determine the corresponding rudder-to-vertical tail chord ratio (C_R/C_V) from Figure 12.12.
10. If the rudder-to-vertical tail chord ratio (C_R/C_V) is more than 0.5, it is suggested to select an all-moving vertical tail (i.e., $C_R/C_V = 1$).
11. If the angle of attack effectiveness of the rudder (τ_r) is greater than 1, there is no rudder which can satisfy the most critical directional control/trim requirement with the current vertical tail/aircraft cg combination. In such a case the vertical tail must be redesigned and/or the aircraft cg must be relocated. Then, return to step 1.
12. Evaluate the rudder design to make certain that other rudder design requirements (e.g., cross-wind landing and spin recovery) are met. Otherwise, redesign the rudder based on the new most critical directional control requirement.
13. Check whether or not the rudder deflection causes the vertical tail to stall during directional control by using the technique introduced in Equation (12.92) and Table 12.19.
14. If vertical tail stall will occur during yawing motion, the rudder must be redesigned by reducing the rudder deflection and/or rudder chord. Return to step 4.
15. If vertical tail stall will occur during yawing motion, and neither of the two rudder parameters (i.e., rudder deflection and chord) may be reduced to prevent vertical tail stall, the other aircraft components such as vertical tail, engine location, or aircraft center of gravity must be redesigned/relocated.
16. Apply aerodynamic balance/mass balance, if necessary (Section 12.7).
17. Optimize the rudder.
18. Calculate the rudder span, rudder tip and root chords, and rudder area and then draw the top view and side view of the vertical tail (including rudder) with dimensions.

12.6.3.2 Rudder Design Steps to Satisfy Cross-Wind Landing Requirements

The following are the rudder design steps for a conventional aircraft to satisfy cross-wind landing requirements:

1. List the available/given data related to rudder design (e.g., vertical tail geometry, aircraft cg locations).
2. Identify the most unfavorable cg location and aircraft weight combination and the most unfavorable altitude for directional control. This will be set as the most critical condition.
3. Determine/select the maximum cross-wind speed (V_W) at which the aircraft must be able to land safely. FAR regulations provide some requirements for such wind speed.
4. Determine/select the aircraft approach speed. FAR regulations provide some requirements for such speed.
5. Determine the aircraft total airspeed (V_T) when a cross-wind is present using Equation (12.106). Assume the worst wind condition; that is, a perpendicular cross-wind to the runway.
6. Calculate the projected side area of the aircraft (S_S).
7. Determine the center of the projected side area of the aircraft (S_S) and its distance to the aircraft cg (d_c) using Equation (12.119).
8. Determine the aircraft side force produced by the cross-wind (F_w) using Equation (12.113).
9. Select the rudder span-to-vertical tail span ratio b_R/b_V (see Table 12.3).
10. Select the rudder-to-vertical tail chord ratio C_R/C_V (see Table 12.3).
11. Determine the aircraft sideslip angle (β) using Equation (12.105).
12. Calculate the aircraft sideslip derivatives C_{n_β} and C_{y_β} using Equations (12.116) and (12.117).
13. Calculate the rudder angle of attack effectiveness (τ_r) employing the equation from Figure 12.12.
14. Calculate the aircraft control derivatives $C_{y_{\delta R}}$ and $C_{n_{\delta R}}$ using Equations (12.118) and (12.100).
15. Compute the rudder control derivative by simultaneous solution of Equations (12.114) and (12.115). Another unknown variable is the sidewash angle (σ), which is beneficial in preparing the aircraft flight instruction manual.
16. If the rudder deflection is more than 30 deg, it is suggested to increase the rudder-to-vertical tail chord ratio up to an all-moving vertical tail (i.e., $C_R/C_V = 1$).
17. If the angle of attack effectiveness of the rudder (τ_R) is greater than 1, there is no rudder which can satisfy the most critical directional control/trim requirement with the current vertical tail/aircraft cg combination. In such a case, the vertical tail must be redesigned and/or the aircraft cg must be relocated. Then, return to step 1.
18. Evaluate the rudder design to make certain that other rudder design requirements (e.g., asymmetric thrust and spin recovery) are met. Otherwise, redesign the rudder based on the new most critical directional control requirement.

19. Check whether or not the rudder deflection causes the vertical tail to stall during the directional control by using the technique introduced in Equation (12.92) and Table 12.19.
20. If vertical tail stall will occur during yawing motion, the rudder must be redesigned by reducing the rudder deflection and/or rudder chord. Return to step 8.
21. If vertical tail stall will occur during yawing motion, and neither of the two rudder parameters (i.e., rudder deflection and chord) may be reduced to prevent vertical tail stall, the other aircraft components such as vertical tail, engine location, or aircraft center of gravity must be redesigned/relocated.
22. Apply aerodynamic balance/mass balance, if necessary (Section 12.7).
23. Optimize the rudder.
24. Calculate the rudder span, rudder tip and root chords, and rudder area and then draw the top view and side view of the vertical tail (including rudder) with dimensions.

12.6.3.3 Rudder Design Steps to Satisfy Spin Recovery Requirements

The following are the rudder design steps for a conventional aircraft to satisfy spin recovery requirements:

1. List the available/given data related to rudder design (e.g., vertical tail geometry, aircraft cg locations).
2. Identify the most unfavorable cg location and aircraft weight combination and the most unfavorable altitude for spin recovery. This will be set as the most critical condition.
3. Determine the aircraft angle of attack during a spin maneuver.
4. Calculate the aircraft mass moments of inertia I_{xx}, I_{zz}, and I_{xz} in the body-axis coordinate system. The technique to determine the aircraft mass moment of inertia is presented in Chapter 11 (Section 11.7).
5. Determine the aircraft mass moments of inertia I_{xx}, I_{zz}, and I_{xz} in the wind-axis coordinate system using Equation (12.120).
6. Determine the desirable rate of spin recovery. A typical value for the rate is given in Section 12.6.2.4.
7. Calculate the required yawing moment to stop the spin (N_{SR}) using Equation (12.120).
8. Compute the effective vertical tail area during spin operation. A graphical technique is depicted in Figure 12.31.
9. Calculate the effective vertical tail volume ratio using Equation (12.125).
10. Select the rudder span-to-vertical tail span ratio b_R/b_V (see Table 12.3).
11. Compute the effective rudder span during spin operation. A graphical technique is depicted in Figure 12.31.
12. Establish the maximum rudder deflection to prevent flow separation (see Table 12.3).
13. Compute the rudder control derivative $C_{n_{\delta R}}$ utilizing Equation (12.123) assuming that the maximum rudder deflection is employed.
14. Calculate the rudder angle of attack effectiveness (τ_R) employing Equation (12.124).
15. Determine the corresponding rudder-to-vertical tail chord ratio (C_R/C_V) from Figure 12.12.
16. If the rudder-to-vertical tail chord ratio (C_R/C_V) is more than 0.5, it is suggested to select an all-moving vertical tail (i.e., $C_R/C_V = 1$).

17. If the angle of attack effectiveness of the rudder (τ_R) is greater than 1, there is no rudder which can satisfy the most critical directional control/trim requirement with the current vertical tail/aircraft cg combination. In such a case, the vertical tail must be redesigned and/or the aircraft cg must be relocated. Then, return to step 1.
18. Evaluate the rudder design to make certain that other rudder design requirements (e.g., cross-wind landing and spin recovery) are met. Otherwise, redesign the rudder based on the new most critical directional control requirement.
19. Investigate whether or not the rudder deflection causes the vertical tail to stall during directional control by using the technique introduced in Equation (12.92) and Table 12.19.
20. If vertical tail stall will occur during yawing motion, the rudder must be redesigned by reducing the rudder deflection and/or rudder chord. Return to step 4.
21. If vertical tail stall will occur during yawing motion, and neither of the two rudder parameters (i.e., rudder deflection and chord) may be reduced to prevent vertical tail stall, the other aircraft components such as vertical tail, engine location, or aircraft center of gravity must be redesigned/relocated.
22. Apply aerodynamic balance and/or mass balance, if necessary (Section 12.7).
23. Optimize the rudder.
24. Calculate the rudder span, rudder tip and root chords, and rudder area and then draw the top view and side view of the vertical tail (including rudder) with dimensions.

12.7 Aerodynamic Balance and Mass Balance

Control surfaces (e.g., elevator, aileron, and rudder) are controlled by a human pilot in a manned aircraft. The pilot's link to the control surfaces is commonly by use of the elevator/aileron stick/yoke/wheel and rudder pedals. In addition, control surfaces interact with other aircraft structural elements to create aero-structural phenomena. The hinge moment created by a control surface must be such that the pilot is capable of handling the moment comfortably, and the effort required should be small enough to ensure that the pilot does not tire in prolonged application (say a few hours). Furthermore, the aerodynamic force produced by a control surface (if not managed properly) may interact with inertia and generate an undesirable structural phenomenon called flutter. These two aspects of control surfaces require more engineering attention and are governed by handling quality requirements (Section 12.3). Thus, when the area, span, chord, and maximum deflection of a control surface have been determined, the design of the control surface is not assumed complete. Two significant issues necessary to consider in the design of control surfaces are:

1. aerodynamic balance of control surfaces;
2. mass balance of control surfaces.

These two balances are frequently addressed for control surfaces such as elevator, aileron, and rudder. During this design operation a number of new control surface parameters are determined. Aerodynamic and mass balancing deals mainly with techniques to vary the aerodynamic force and hinge moment of a control surface so that the control force felt by the pilot stays within an acceptable range, and no undesired aeroelastic phenomenon occurs. Hence, two main objectives of aerodynamic and mass balance are:

1. reduction of required force felt by pilot;
2. avoiding flutter.

The hinge moment created by a control surface is defined similarly to other aircraft aerodynamic moments, as follows:

$$H = \frac{1}{2}\rho U_1^2 S_c C_c C_h \tag{12.147}$$

where S_c denotes the planform area of the control surface (e.g., S_A, S_E, and S_R), and C_c denotes the mean aerodynamic chord of the control surface (e.g., C_A, C_E, and C_R). The parameter C_h is the hinge moment coefficient and is given by:

$$C_h = C_{h_o} + C_{h_\alpha}\alpha_{LS} + C_{h_{\delta c}}\delta_c \tag{12.148}$$

where α_{LS} is the angle of attack of the lifting surface (e.g., tail and wing), δ_c is the control surface deflection (e.g., δ_A, δ_E, and δ_R), and δ_t is the tab deflection. The parameter C_{h_o} is the hinge moment coefficient for $\alpha_C = \delta_c = 0$, and is zero for a lifting surface symmetrical airfoil. The parameters C_{h_α} and $C_{h_{\delta c}}$ are two non-dimensional derivatives as follows:

$$C_{h_\alpha} = \frac{\partial C_h}{\partial \alpha_{LS}} \tag{12.149}$$

$$C_{h_{\delta c}} = \frac{\partial C_h}{\partial \delta_c} \tag{12.150}$$

These two derivatives are partial derivatives of the hinge moment coefficient (C_h) with respect to the lifting surface angle of attack (α_{LS}) and control surface deflection (δ_c) respectively. The derivation of these coefficients is beyond the scope of this text; Ref. [17] or Part VI of Ref. [22] may be used for their evaluation.

The hinge moment derivative C_{h_δ}, sometimes referred to as the control heaviness parameter, is the main parameter to reduce the hinge moment and control force. This derivative is a function of a number of variables, including the distance between *ac* of the control surface to the hinge, and the control surface nose curvature. The hinge moment derivative C_{h_α}, sometimes referred to as the control floating parameter, is the main parameter to produce the hinge moment and control force during an aircraft response to a gust. Both hinge moment derivatives C_{h_α} and $C_{h_{\delta c}}$ are usually negative; a typical value for C_{h_α} is about −0.1 1/rad and for $C_{h_{\delta c}}$ about −0.3 1/rad.

The pilot force applied to a stick/yoke/wheel/pedal is transmitted to a control surface via a power transmission system such as mechanical, hydraulic, or electric. The stick/yoke/pedal force (F_s) is related to the hinge moment through a factor referred to as the power transmission system gearing ratio:

$$F_s = G_C H \tag{12.151}$$

where G_C is the ratio between the linear/angular movement of the stick/wheel and the deflection of the control surface (see Figure 12.34):

$$G_C = \frac{\delta_c}{x_s} \quad \text{(for stick/pedal)} \tag{12.152a}$$

$$G_C = \frac{\delta_c}{\delta_y} \quad \text{(for yoke/wheel)} \tag{12.152b}$$

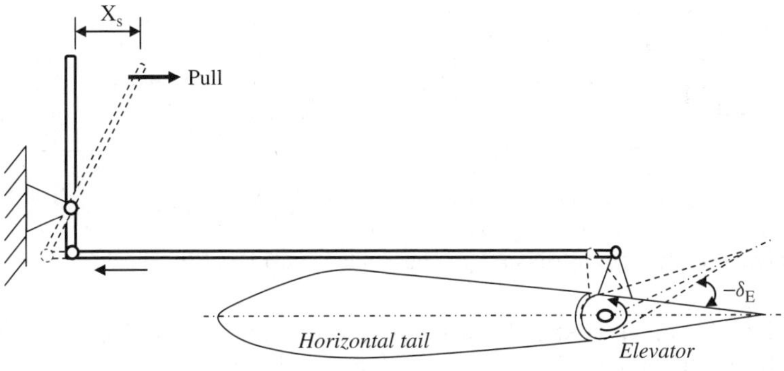

Figure 12.34 Stick movement, elevator deflection, and gearing ratio

where x_s is the linear movement of a stick or pedal and δ_y is the angular rotation of a yoke or wheel. In the following subsections, various techniques for aerodynamic and mass balance of control surfaces are addressed. For other techniques for aerodynamic balance, the interested reader is encouraged to consult Refs [12, 23, 24].

12.7.1 Aerodynamic Balance

In order to insure that the pilot is fully and comfortably capable of moving the stick/pedal and deflecting the control surfaces, the aerodynamic force and hinge moment of a control surface must be balanced or reduced. Obtaining linearly varying pilot forces for all flight conditions is an important handling quality (Section 12.3) requirement for both military and civil aircraft. There are various specific requirements in the FAR regulations for maximum and minimum forces permitted in various flight operations. For a small normal GA aircraft, a simple directly driven hinged surface (see Figure 12.34) is adequate.

Equations (12.147) and (12.151) demonstrate that, as the airplane mass, size, and airspeed grow, the forces required to move the control surfaces against the aerodynamic hinge moments become large. Aerodynamic balance of a control surface is simply a method of reducing hinge moment by techniques such as distribution of surface area about the hinge line, and employing devices such as tab and spring. The technique of aerodynamic balance is applied not only to manually controlled control surfaces, but also to power-assisted surfaces. In fact, in large transport aircraft, an artificial feel is often incorporated into the controls so that the pilot has a sense of feeling. An unbalanced control surface has values of C_{h_α} and $C_{h_{\delta c}}$ that are too large to give acceptable force. The application of aerodynamic balance will vary both derivatives C_{h_α} and $C_{h_{\delta c}}$, but by differing amounts. The designer must be careful not to overbalance the control surface, resulting in a positive $C_{h_{\delta c}}$.

The extent to which a control force is reduced is limited by non-linearities in the control force/control surface deflection relationship, and by the physical change in the external shape of the surface due to ice formation, and by the minimum allowable pilot control feel. A great deal of engineering effort is required to properly shape the leading edges of the tabs and to locate their hinge lines. These design features do not usually impact the control surface effectiveness, but they dominate the surface hinge moments, which in turn determine the forces felt by pilots. In case the control surface area ahead of the hinge line

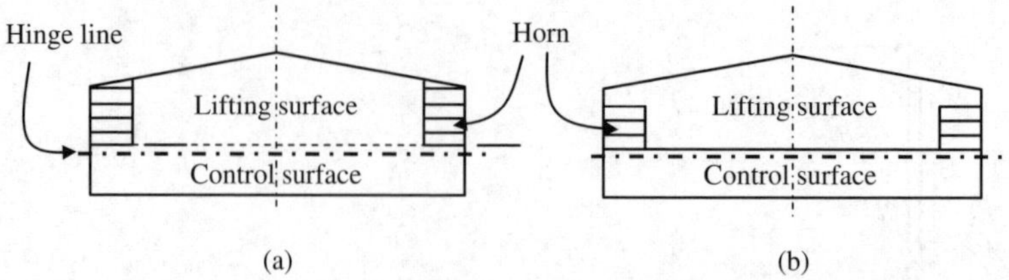

Figure 12.35 Horn balance. (a) Unshielded (full) horn; (b) Shielded horn

is increased, the resultant control lift acts on a small moment arm from the hinge. When a tab is utilized, the counteracting hinge moment will nullify/reduce the control surface hinge moment. There are many ways to aerodynamically balance a control surface, some of which are described here.

12.7.1.1 Horn Balance

A horn balance is the addition of an extra surface to the control surface ahead of the hinge line. A horn balance is a low-cost and simple technique and is used mostly on older and GA aircraft (Figure 12.35). There are two types of horn that both aerodynamically balance the control surfaces. The first type extends to the leading edge of the lifting surface, and is called an unshielded horn. The second type has part of the lifting surface ahead of it and is said to be a shielded horn. The effectiveness of a horn balance depends upon the area and moment of the horn ahead of the hinge, compared with the area and moment of the horn behind the hinge.

The control surface balance is defined as the ratio between the control surface area forward of the hinge line and that aft of the hinge line. The most popular application of the horn balance is the elevator, and next is the rudder. The horn balance is rarely used on an aileron. In a horn-equipped control surface, when the surface is deflected, the air that strikes the surface in front of the hinge line creates a pressure distribution, and hence a force, that helps deflect the surface even more. This counteracts the moment behind the hinge, tending to reduce the control surface deflection. A shielded horn is more favorable than an unshielded one; for instance, the formation of ice on an unshielded horn may create control problems. By careful design of the horn, the pilot stick force is significantly reduced. In addition, the horn has an icing problem so needs a de-icing device (e.g., electrically heated).

The horn balance of the rudder is depicted in Figure 8.6 (Marganski Swift S-1); in Figure 3.12 (Stampe-Vertongen SV-4C); in Figure 5.56 (Piper Super Cub); in Figure 6.8 (PZL-Mielec M-28B1R Bryza); in Figure 6.12 (Reims F337F Super Skymaster); in Figure 8.3 (Supermarine Spitfire); in Figure 9.4 (Piper PA-25-235 Pawnee); in Figure 5.44 (Pilatus PC-21); in Figure 6.27 (Gates Learjet 35A); and in Figure 10.5 (Embraer A-29B Super Tucano (EMB-314)).

12.7.1.2 Overhanging Balance

When the hinge line is moved aft (set back), closer to the control surface center of pressure, the surface is said to be aerodynamically balanced using nose overhang (Figure 12.36).

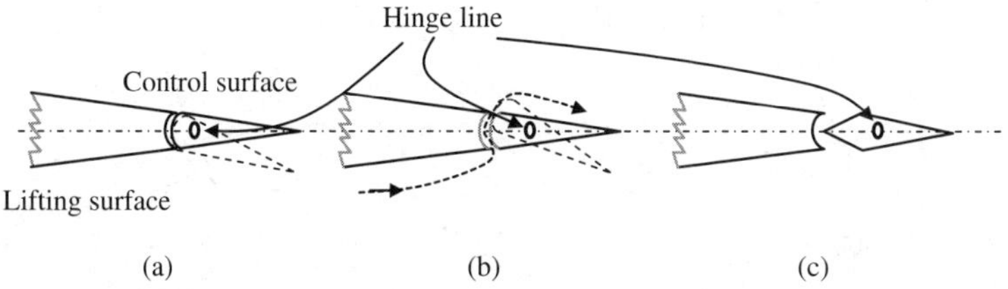

Figure 12.36 Control surfaces with overhang balance. (a) Normal unbalanced; (b) Set back round-nosed; (c) Set back sharp-nosed

This allows the leading edge of the control surface to emerge into the airstream when deflected. The shape of the leading edge of the control surface is essential in effectiveness of the overhang balance. The nose contour must be shaped so that the nose does not protrude into the airstream even at high control surface deflections. A highly curved protrusion, which would generate large suctions on the nose, would overbalance the surface, tending to move it to a higher deflection.

The leading edge of the control surface could be either blunted or sharpened. A blunt leading edge (either round or elliptic) will cause a suction flow around the radius and lightens the surface. However, sharpening the leading edge reduces the suction ahead of the hinge when the surface is deflected and heavies the surface. A blunt nose has a greater effect on the hinge moment than a sharp nose, but it may create non-linearities at small deflections. If the gap between the leading edge of the control surface and the trailing edge of the lifting surface is not sealed, the airflow leaks through the gap and tends to improve the balance.

12.7.1.3 Internal Sealed

Another type of aerodynamic balance, called internal sealed, utilizes an internal flexible airtight seal to feel the pressure difference between the top and bottom surfaces and to provide the balancing moment (see Figure 12.37). Each side of the seal is open to the atmosphere at the shroud trailing edges. The advantages are mainly that the seal prevents energy loss and flow from a high-pressure surface to a low-pressure one, and non-linearities do not occur. This results in a reduction of the induced drag by as much as about 5% of the total induced drag. The internal overhang effectiveness is comparable to the overhang balance. In terms of operation, the icing may sometimes block the normal movement of the seal, and even the normal deflection of the control surface, so a de-icing

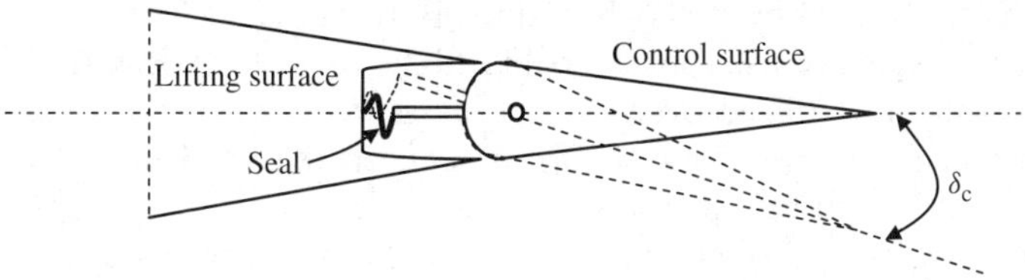

Figure 12.37 The internal sealed balance

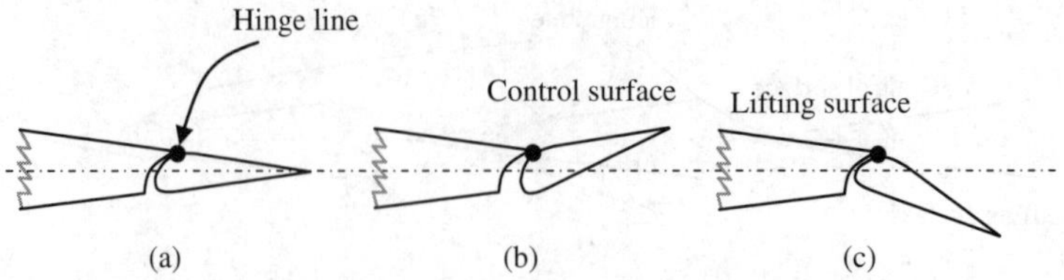

Figure 12.38 Control surface with frise balance. (a) Neutral; (b) Deflected up; (c) Deflected down

device must be provided. The effectiveness of the internal sealed balance is increased at high speeds.

12.7.1.4 Frise Balance

The idea of a Frise balance came originally from the Frise aileron. As discussed during aileron design (Section 12.4), the main purpose of the Frise aileron is to minimize the adverse yaw caused by aileron asymmetric drag. A Frise balance is applied by making the control surface have an unsymmetrical nose profile (Figure 12.38). When the Frise balanced control surface is deflected upward, it protrudes below the lifting surface's contour. At small angles of deflection the up-going aileron is overbalanced, and this helps to deflect the down-going aileron on the other side. The advantages of Frise ailerons are a large balance effect with a small setback hinge; and they are relatively easy to construct. The behavior of the control surface at low deflection angle is non-linear; rotating up overbalances the surface, while rotating down creates a heavy force. Therefore, the effectiveness of the Frise balance is a function of rigging point and hence careful rigging is needed for an effective balance.

12.7.1.5 Trim Tab

Control surfaces may be aerodynamically balanced without a nose treatment, by employing a tab at the trailing edge of the control surface. Tabs are secondary control surfaces placed at the trailing edges of the primary control surfaces. There are a number of tabs used in various aircraft in order to considerably reduce the hinge moment and control force. The most basic tab is a trim tab; as the name implies, it is used on elevators to longitudinally trim the aircraft in a cruising flight. Trim tabs are used to reduce the force the pilot applies to the stick to zero. Tab ensures that the pilot will not tire from holding the stick/yoke/wheel in a prolonged flight. Trailing edge tabs are employed as variable trimming devices, operated by stick/wheel directly from the cockpit.

Trim tabs are frequently used in reversible flight control systems (e.g., mechanical). However, trim tabs are utilized even in very large transport aircraft (such as KC-135) due to the fact that the loss of all engines is conceivable, and thus the pilot must be able to trim the jumbo aircraft with his/her body force. For instance, in the past history of flights of the Boeing 747, there have been at least three cases where all four engines became inoperative. One incident involved British Airways Flight 9, a scheduled flight from London Heathrow to Auckland. On June 24, 1982, a Boeing 747 aircraft flew into a cloud of volcanic ash thrown up by the eruption of Mount Galunggung, resulting in the failure of all four engines. The aircraft was diverted to Jakarta in the hope that enough engines could be restarted to allow it to land there. The aircraft was able to glide far enough to exit the ash cloud, and all engines were restarted, allowing the aircraft to land safely at the Halim Perdanakusuma Airport in Jakarta. This demonstrates the necessity of providing an alternate manual control of the control surfaces, even in a large transport aircraft with a hydraulic system.

To achieve a zero cockpit control force, the trim tab is deflected opposite to the elevator deflection. When a lifting surface is equipped with a tab, the hinge moment coefficient C_h is given by:

$$C_h = C_{h_o} + C_{h_\alpha}\alpha_{LS} + C_{h_{\delta c}}\delta_c + C_{h_{\delta t}}\delta_t \tag{12.153}$$

where δ_t represents the tab deflection and the parameter $C_{h_{\delta t}}$ is a non-dimensional derivative as follows:

$$C_{h_{\delta t}} = \frac{\partial C_h}{\partial \delta_t} \tag{12.154}$$

The effectiveness of the tab ($C_{h_{\delta t}}$) is a function of the tab geometry and tab hinge line. Figure 12.38 illustrates two types of trim tab arrangement. There are two hinges in a lifting surface (Figure 12.39(b)) whose control surface possesses a tab, one for control surface deflection and one for tab deflection. Trim tabs may be adjusted when the aircraft is on the ground (Figure 12.39(a)), or may be operated manually and set by the pilot during flight. Trim tabs are usually deflected by a device referred to as the trim wheel. The trim wheel and trim tab assist a pilot to longitudinally trim a large aircraft with his/her hand force (say 50 lbf), and to keep a large elevator at any deflection needed at any speed.

In large transport aircraft, an adjustable horizontal tail is generally used for the purpose of setting the yoke force equal to zero. The tab deflection is proportional to the control surface deflection. In general, the trim tab serves two functions: (i) it provides the ability to zero-out the stick/wheel/yoke force and (ii) it provides aircraft speed stability at the trim speed. The tab-to-control surface chord ratio is usually about 0.2–0.4. In an RC (Remote Controlled) model aircraft, home-built aircraft, and even in a small GA aircraft, a simple plate (i.e., tab) may be permanently added to the vertical tail to directionally balance the aircraft and to compensate for any imperfection in production. This makes

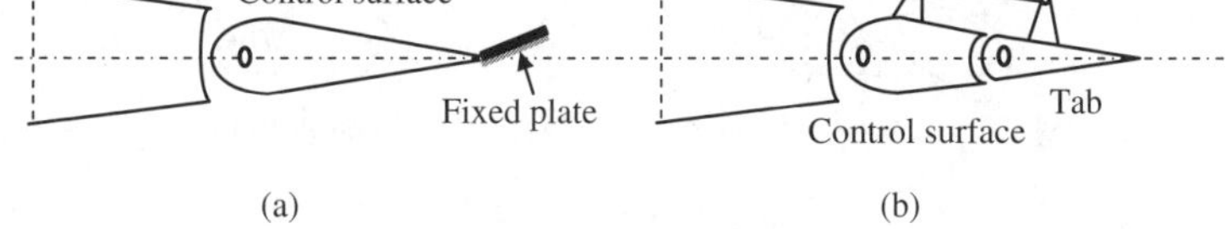

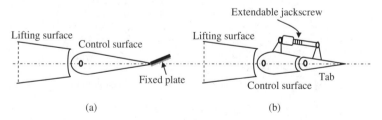

Figure 12.39 Trim tab. (a) Simple ground-adjustable trim tab; (b) Ground/flight-adjustable trim tab

the aircraft symmetric about the *xz* plane. The need for such a tab is not revealed until the first flight test. Large transport aircraft such as the Boeing 747 and Boeing 737 are equipped with a horizontal tail trim tab, as well as a vertical tail trim tab. Business jets such as the Cessna Citation 500 are equipped with vertical tail trim tab.

12.7.1.6 Balance Tab

Another useful tool in assisting the pilot to move the control surface (e.g., elevator) and reduce the amount of force that the pilot needs to apply to the stick/wheel is a balance tab. The balance tab (sometimes called a geared tab, geared balance tab, or link tab) is geared to the primary control surface in such a way that it moves in a given ratio to the control surface movement, and in the opposite direction. For example, if the pilot wishes to deflect the elevator down, the balance tab will deflect upward and the pressure distribution set up will create a force, and hence moment, to move the elevator down. Because they are placed at the trailing edge, balance tabs possess long moment arms and are very powerful in action. The geared tab lift is in a direction opposite to that of the basic control surface and thereby reduces the effectiveness of the surface. The tab creates a hinge moment about its own hinge, as well as a hinge moment about the hinge of the control surface, tending to increase the surface angle. The tab is connected directly to the lifting surface via a mechanical link; the length of the link may be adjusted. The tab neutral position (i.e., angle) with respect to the control surface can easily be adjusted to obtain the desired amount of balance.

A geared tab generally reduces the value of the derivative $C_{h_{\delta t}}$, with no considerable effect on the derivative C_{h_α} since the airfoil shape is not varied. A balance tab may be employed as a trim tab by making the tab follower link variable in length and by providing the pilot with the authority to vary the length of the link. This is often applied with the help of an electromechanical jackscrew in the tab arm.

A version of the balancing tab is called an anti-balance tab, where the tab moves in the same direction as the surface due to different connection to the lifting surface. Because of the opposite direction of the balance tab and the anti-balance tab, the balance tab exhibits a lagging behavior while the anti-balance tab exhibits a leading behavior. A lagging tab reduces the value of the derivative $C_{h_{\delta t}}$ (in a negative sense), while a leading tab increases the value of the derivative $C_{h_{\delta t}}$ (in a negative sense). By tailoring the value of the derivative $C_{h_{\delta t}}$, it is possible to achieve any desirable stick force versus speed gradient. Figure 12.40 illustrates the mechanical arrangements of a balance tab and an anti-balance tab.

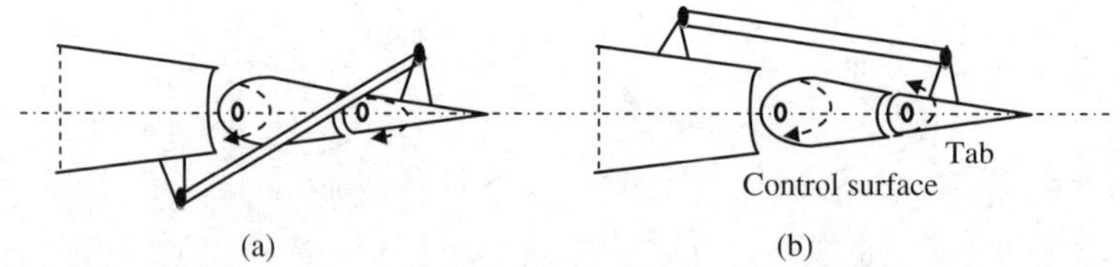

Figure 12.40 Balance tab. (a) Anti-balance tab; (b) Balance tab

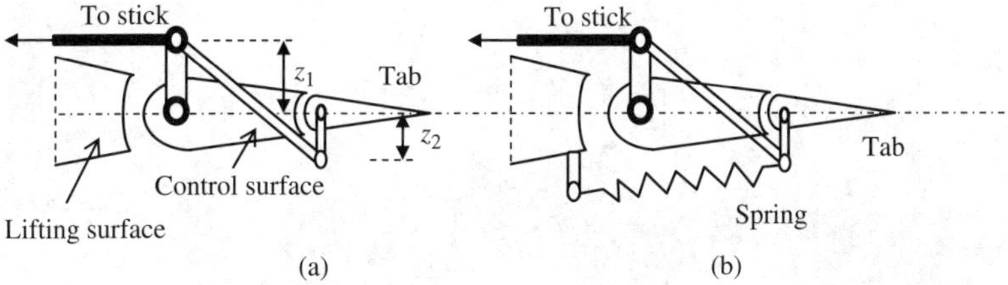

Figure 12.41 Servo tab and spring tab. (a) Servi tab; (b) Spring tab

12.7.1.7 Servo Tab

A servo tab is a tab in which the stick/wheel is connected directly to the tab, which is hinged to the control surface (Figure 12.41(a)). In contrast, in the cases of trim tab and balance tab, the stick/wheel is connected to the control surface. Thus, in the cases of trim tab and balance tab, when the pilot moves the stick/rotates the wheel, the control surface is deflected while in the case of a servo tab, when the pilot moves the stick/rotates the wheel, the tab is deflected. The deflection of the control surface is performed via a servo tab. In other words, the pilot controls the servo tab but the servo tab controls the control surface. The stick/wheel force depends on the hinge moment of both the control surface and the tab.

The servo tab effectiveness is a function of the ratio between the two arms of the tab and control surface (i.e., z_1/z_2). The servo tab is not employed in many modern aircraft, since its effectiveness at low speeds (particularly at stall) is not reliable. Similar to other tabs, the main function of the servo tab is to reduce the pilot force when the stick/wheel is moved. A very low stick/wheel force is achieved for a small arm ratio (z_1/z_2); however, the control is easily overbalanced. Large transport aircraft such as the Boeing KC-135 and Boeing 707 (Figure 12.42) are equipped with a horizontal tail trim tab, as well as a vertical tail trim tab. The military transport aircraft Lockheed C-130B (Figure 5.4) is equipped with a horizontal tail trim tab, as well as a servo tab.

12.7.1.8 Spring Tab

A spring tab is basically similar to a servo tab except a spring is added. The spring connects the tab to either a lifting surface or a control surface (Figure 12.41(b)). The addition of a spring to a servo tab will further reduce the stick/wheel force, so a spring tab may be assumed to be a variable servo tab. The effectiveness of the spring tab (C_{h_δ}) is a function of the ratio z_1/z_2 and the spring constant. The control force varies moderately with speed, since the effectiveness of the spring tab decreases with airspeed. Unlike the servo tab, the spring tab is not overbalanced even at stall speed, since the tab deflects only when there is a load on the control surface. One undesirable side-effect of the spring tab originates from the addition of a springy element. The response to pilot command is relatively slow, due to the floating action of the control surface. Spring tabs may be pre-loaded to avoid them coming into effect with a small pilot force. Figure 12.42 shows the aerodynamic balance of three aircraft (rudder tab of Boeing 707; aileron, elevator, and rudder tabs of ATR-72-600; and rudder tab of Mudry CAP-10B).

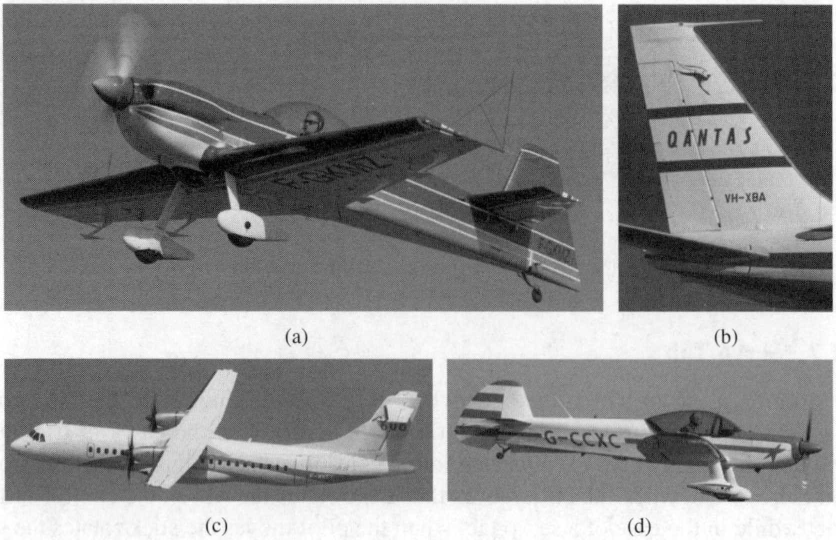

Figure 12.42 Aerodynamic balance and mass balance of various aircraft: (a) elevator and flap mass balance in CAP Aviation CAP-232; (b) rudder tab of Boeing 707; (c) aileron, elevator, and rudder tabs of ATR-72-600; (d) rudder tab of Mudry CAP-10B. Reproduced from permission of (a, d) Jenny Coffey; (b) A J Best; (c) Antony Osborne

12.7.2 Mass Balance

There are many aircraft, with reversible or irreversible flight control systems, that are equipped with a mechanism which allows the aircraft to fly with a condition usually referred to as control-free, stick-free, or yoke-free. Such a mechanism permits the control surface to be left in place (i.e., free to move) since the control surface's hinge moment has been set to zero by devices such as a tab. When an aircraft is operating control-free, the control surfaces will oscillate when a gust moves the control surface from the trimmed position. A freely oscillating control surface (e.g., elevator) may create an undesirable phenomenon called flutter. Flutter is a dynamic phenomenon and may lead to dynamic instability of the airplane that cannot be tolerated. Flutter is characterized as a high-frequency oscillation of the control surface caused by an interaction between the aerodynamic force (e.g., local lift) and weight of the control surface. This undesirable phenomenon can be prevented by stiffening the lifting surface's structure (wing, horizontal tail, and vertical tail) in bending and torsion. Another solution is to move the control surface center of gravity near to or in front of the hinge line.

One of the effective techniques to avoid flutter is mass balancing. Control surfaces (elevator, aileron, and rudder) are deflected around their hinge axes, close to their leading edges (about 5–10% of the chord). However, the center of gravity of a control surface is a little more aft (about 20–40% of the chord). This implies that the center of gravity and hinge axis of a control surface do not coincide (Figure 12.43(b)). Under such conditions, inertia will respond to any perturbation of the control surface deflection, which may lead

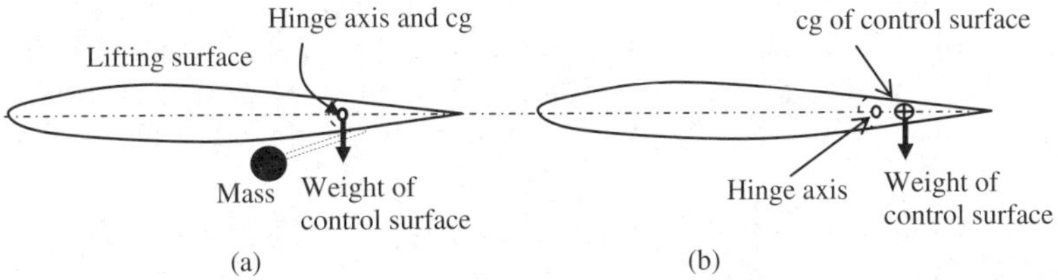

Figure 12.43 Mass balance of a control surface. (a) Mass balanced; (b) Unbalanced

to flutter. A counterbalancing mass needs to be set ahead of the surface to shift the center of gravity of a control surface forward to coincide with the hinge line (Figure 12.43(a)).

Many light aircraft and gliders which do not fly fast do not possess mass balanced control surfaces. Except for very low-speed aircraft, the flight control surfaces in an aircraft with a reversible flight control system are almost always mass balanced. The exact same effect of mass balance can be produced with the application of a mechanical spring. In an aircraft with a reversible flight control system, and with maximum speed above 100 knots, the mass balance is necessary. However, in an aircraft with irreversible flight control system, there is no need for mass balance. The transport aircraft Lockheed C-130 Hercules (Figure 5.4) and Lockheed C-5 Galaxy are mass balanced with uranium, so their control surfaces have a thick skin. Repainting an aircraft causes un-mass balancing, since the paint goes over the entire surface.

It is noticeable that the application of mass and spring balancing also improves the stick-free stability of an aircraft, as well as reducing the pilot force to deflect a control surface. In order to reduce the drag of the bare mass, it could be covered, faired, or aerodynamically shaped. Figure 12.42(a) shows the elevator and flap mass balance in CAP Aviation CAP-232.

12.8 Chapter Examples

In this section, three fully solved examples for the design of aileron, elevator, and rudder are provided. In these examples, the applications of the control surfaces design techniques and procedures are illustrated.

12.8.1 Aileron Design Example

Example 12.4

Problem statement: Design the roll control surface(s) for a land-based military transport aircraft to meet roll control MIL-STD requirements. The aircraft has a conventional configuration and the following geometry and weight characteristics:

$$m_{\mathrm{TO}} = 6500\,\mathrm{kg},\ S = 21\ \mathrm{m}^2,\ \mathrm{AR} = 8,\ \lambda = 0.7,\ S_{\mathrm{h}} = 5.3\ \mathrm{m}^2,\ S_{\mathrm{v}} = 4.2\ \mathrm{m}^2,$$

$$V_{\mathrm{s}} = 80\ \mathrm{knot},\ C_{L_{\alpha\mathrm{w}}} = 4.5\ 1/\mathrm{rad},\ I_{xx} = 28\,000\,\mathrm{kg\,m}^2$$

Furthermore, the control surface must be of low cost and manufacturable. The high-lift device has already been designed, and the outboard flap location is determined to be at 60% of the wing semispan. The wing rear spar is located at 75% of the wing chord.

Solution:

- **Step 1**. The problem statement specifies the maneuverability and roll control requirements to comply with MIL-STD.
- **Step 2**. Due to the aircraft configuration, simplicity of design, and desire for low cost, a conventional roll control surface configuration (i.e., aileron) is selected.
- **Step 3**. Hence, Table 12.12 will be the reference for the aileron design, which expresses the requirement as the time to achieve a specified bank angle change.
- **Step 4**. Based on Table 12.5, a land-based military transport aircraft with a mass of 6500 kg belongs to Class II. The critical flight phase for roll control is at the lowest speed. Thus, it is required that the aircraft must be roll controllable at approach flight condition. According to Table 12.6, the approach flight operation is considered as phase C. To design the aileron, a level of acceptability of 1 is considered. Therefore:

Class	Flight phase	Level of acceptability
II	C	1

- **Step 5**. The roll control handling qualities design requirement is identified from Table 12.12b, which states that the aircraft in Class II, flight phase C, for a level of acceptability of 1, is required to be able to achieve a bank angle of 30° in 1.8 seconds.
- **Step 6**. According to the problem statement, the outboard flap location is at 60% of the wing span. So the inboard and outboard positions of the aileron as a function of wing span (i.e., b_{a_i}/b and b_{a_o}/b) are tentatively selected to be at 70% and 95% of the wing span respectively.
- **Step 7**. The wing rear spar is located at 75% of the wing chord, so the ratio between aileron chord and wing chord (i.e., C_a/C) is tentatively selected to be 20%.
- **Step 8**. The aileron effectiveness parameter (τ_a) is determined from Figure 12.12. Since the aileron-to-wing chord ratio is 0.2, so the aileron effectiveness parameter will be 0.41.
- **Step 9**. The aileron rolling moment coefficient derivative ($C_{l_{\delta A}}$) is calculated employing Equation (12.22):

$$C_{l_{\delta A}} = \frac{2C_{L_{\alpha w}}\tau C_r}{Sb}\left[\frac{y^2}{2} + \frac{2}{3}\left(\frac{\lambda - 1}{b}\right)y^3\right]_{y_i}^{y_o} \tag{12.23}$$

We first need to determine the wing span, wing mean aerodynamic chord, and wing root chord:

$$\mathrm{AR} = \frac{b^2}{S} \Rightarrow b = \sqrt{S \cdot \mathrm{AR}} = \sqrt{21 \cdot 10} \Rightarrow b = 14.49\,\mathrm{m} \tag{5.19}$$

$$AR = \frac{b}{\overline{C}} \Rightarrow \overline{C} = \frac{b}{AR} = \frac{14.49}{10} \Rightarrow \overline{C} = 1.449\,\text{m} \tag{5.17}$$

$$\overline{C} = \frac{2}{3}C_r\left(\frac{1+\lambda+\lambda^2}{1+\lambda}\right) \Rightarrow 1.449 = \frac{2}{3}C_r\left(\frac{1+0.8+0.8^2}{1+0.8}\right) \Rightarrow C_r = 1.604\,\text{m} \tag{5.26}$$

The inboard and outboard positions of the aileron as a function of wing span are selected to be at 70% and 95% of the wing span respectively. Therefore:

$$y_i = 0.7\frac{b}{2} = 0.7 \cdot \frac{14.49}{2} = 5.072\,\text{m}$$

$$y_o = 0.95\frac{b}{2} = 0.95 \cdot \frac{14.49}{2} = 6.883\,\text{m}$$

Plugging the values for the parameters in Equation (12.23) gives the following:

$$C_{l_{\delta A}} = \frac{2 \cdot 4.5 \cdot 0.41 \cdot 1.604}{21 \cdot 14.49}$$

$$\left\{\left[\frac{6.883^2}{2} + \frac{2}{3}\left(\frac{0.8-1}{14.49}\right) \cdot 6.883^3\right] - \left[\frac{5.072^2}{2} + \frac{2}{3}\left(\frac{0.8-1}{14.49}\right) \cdot 5.072^3\right]\right\}$$

which yields:

$$C_{l_{\delta A}} = 0.176\frac{1}{\text{rad}}$$

- **Step 10**. A maximum aileron deflection ($\delta_{A_{max}}$) of ±20 deg is selected.
- **Step 11**. The aircraft rolling moment coefficient (C_l) when the aileron is deflected with the maximum deflection is:

$$C_l = C_{l_{\delta A}}\delta_A = 0.176 \cdot \frac{20}{57.3} = 0.061 \tag{12.13}$$

- **Step 12**. The aircraft rolling moment (L_A) when the aileron is deflected with the maximum deflection is calculated. The typical approach velocity is 1.1–1.3 times the stall speed, thus the aircraft is considered to approach with a speed of 1.3 V_s. In addition, the sea-level altitude is considered for the approach flight operation.

$$V_{app} = 1.3V_s = 1.3 \cdot 80 = 104\,\text{knot} = 53.5\,\text{m/s}$$

$$L_A = \frac{1}{2}\rho V_{app}^2 S C_l b = \frac{1}{2} \cdot 1.225 \cdot 53.5^2 \cdot 21 \cdot 0.061 \cdot 14.49 = 32692.6\,\text{Nm} \tag{12.11}$$

- **Step 13**. The steady-state roll rate (P_{ss}) is determined:

$$P_{ss} = \sqrt{\frac{2 \cdot L_A}{\rho\left(S_w + S_h + S_{vt}\right) C_{D_R} \cdot y_D^3}} \tag{12.37}$$

An average value of 0.9 is selected for the wing horizontal/tail vertical tail rolling drag coefficient. The drag moment arm is assumed to be at 40% of the wing span, so:

$$y_D = 0.4\frac{b}{2} = 0.4 \cdot \frac{14.49}{2} = 2.898\ \frac{\text{m}}{\text{s}}$$

$$P_{\text{ss}} = \sqrt{\frac{32692.6}{1.225\,(21 + 5.3 + 4.2) \cdot 0.9 \cdot (2.898)^3}} = 8.937\ \frac{\text{rad}}{\text{s}} \quad (12.37)$$

- **Step 14**. Calculate the bank angle (Φ_1) at which the aircraft achieves the steady-state roll rate:

$$\Phi_1 = \frac{I_{xx}}{\rho y_D^3 \left(S_{\text{w}} + S_{\text{h}} + S_{vt}\right) C_{D_{\text{R}}}} \ln\left(P_{\text{ss}}^2\right)$$

$$= \frac{28\,000}{1.225 \cdot (2.898)^3\,(21 + 5.3 + 4.2) \cdot 0.9} \ln\left(8.937^2\right) \quad (12.43)$$

$$\Phi_1 = 149.82\,\text{rad} = 8584.14\,\text{deg}$$

- **Step 15**. Calculate the aircraft rate of roll rate ($\dot{P}$) that is produced by the aileron rolling moment until the aircraft reaches the steady-state roll rate (P_{ss}):

$$\dot{P} = \frac{P_{\text{ss}}^2}{2\Phi_1} = \frac{8.937^2}{2 \cdot 149.82} = 0.267\ \frac{\text{rad}}{\text{s}^2} \quad (12.49)$$

- **Steps 16 and 17**. The bank angle (Φ_1) calculated in step 14 is compared with the bank angle (Φ_{req}) of step 5. Since the bank angle calculated in step 14 (i.e., 8584 deg) is greater than the bank angle (Φ_{req}) of step 5 (i.e., 30 deg), the time it takes the aircraft to achieve the bank angle of 30 deg is determined:

$$t_2 = \sqrt{\frac{2\Phi_{\text{des}}}{\dot{P}}} = \sqrt{\frac{2 \cdot 30}{0.267}} = 1.982\ \text{s} \quad (12.47)$$

- **Step 18**. The roll time obtained in step 16 or 17 is compared with the required roll time (t_{req}) expressed in step 5. The roll time obtained in step 16 or 17 to achieve the bank angle of 30 deg (i.e., 1.982 seconds) is longer than the roll time expressed in step 5 (i.e., 1.8 seconds). Hence the current aileron design does not satisfy the requirements and must be redesigned.
- **Steps 19 and 20**. The duration obtained in step 16 or 17 is shorter than the duration (t_{req}) stated in step 5, so the aileron design has not met the requirement. The solution is either to increase the aileron size (aileron span or chord) or to increase the aileron maximum deflection. Due to aileron stall concerns, and the location of the rear spar, the maximum aileron deflection and the aileron chord-to-wing chord ratio are not altered. The flap outboard location is at 60% of the wing span, thus the safest solution is to increase the aileron span.

- **Step 21**. By trial and error, it is determined that the aileron inboard span at 61% of the wing span will satisfy the roll control requirements. The calculation is as follows:

$$y_{\mathrm{i}} = 0.61\frac{b}{2} = 0.61 \cdot \frac{14.49}{2} = 4.42\,\mathrm{m}$$

$$y_{\mathrm{o}} = 0.95\frac{b}{2} = 0.95 \cdot \frac{14.49}{2} = 6.883\,\mathrm{m}$$

$$C_{l_{\delta A}} = \frac{2 \cdot 4.5 \cdot 0.41 \cdot 1.604}{21 \cdot 14.49}$$

$$\left\{\left[\frac{6.883^2}{2} + \frac{2}{3}\left(\frac{0.8-1}{14.49}\right) \cdot 6.883^3\right] - \left[\frac{4.42^2}{2} + \frac{2}{3}\left(\frac{0.8-1}{14.49}\right) \cdot 4.42^3\right]\right\}$$

$$C_{l_{\delta A}} = 0.228\ \frac{1}{\mathrm{rad}}$$

$$C_l = C_{l_{\delta A}}\delta_A = 0.228 \cdot \frac{20}{57.3} = 0.08 \tag{12.13}$$

$$L_A = \frac{1}{2}\rho V_{\mathrm{app}}^2 S C_l b = \frac{1}{2} \cdot 1.225 \cdot 53.5^2 \cdot 21 \cdot 0.08 \cdot 14.49 = 42\,429.6\,\mathrm{Nm} \tag{12.11}$$

$$P_{\mathrm{ss}} = \sqrt{\frac{2 \cdot L_A}{\rho\left(S_{\mathrm{w}} + S_{\mathrm{h}} + S_{\mathrm{vt}}\right) C_{D_R} \cdot y_D^3}} \tag{12.37}$$

$$P_{\mathrm{ss}} = \sqrt{\frac{42429.6}{1.225\,(21 + 5.3 + 4.2) \cdot 0.9 \cdot (2.898)^3}} = 10.181\ \frac{\mathrm{rad}}{\mathrm{s}} \tag{12.37}$$

$$\Phi_1 = \frac{I_{xx}}{\rho y_D^3\left(S_{\mathrm{w}} + S_{\mathrm{h}} + S_{\mathrm{vt}}\right) C_{D_R}} \ln\left(P_{\mathrm{ss}}^2\right)$$

$$= \frac{28{,}000}{1.225 \cdot (2.898)^3\,(21 + 5.3 + 4.2) \cdot 0.9} \ln\left(10.181^2\right)$$

$$\Phi_1 = 158.74\,\mathrm{rad} = 9095\,\mathrm{deg}$$

$$\dot{P} = \frac{P_{\mathrm{ss}}^2}{2\Phi_1} = \frac{10.181^2}{2 \cdot 158.74} = 0.327\ \frac{\mathrm{rad}}{\mathrm{s}^2} \tag{12.49}$$

$$t_2 = \sqrt{\frac{2\Phi_{\mathrm{des}}}{\dot{P}}} = \sqrt{\frac{2 \cdot 30}{0.327}} = 1.791 \tag{12.50}$$

The variations of bank angle versus time are plotted in Figure 12.44.

- **Step 26**. Aerodynamic balance/mass balance (beyond the scope of this example).
- **Step 27**. Optimization (beyond the scope of this example).

- **Step 28**. Geometry.

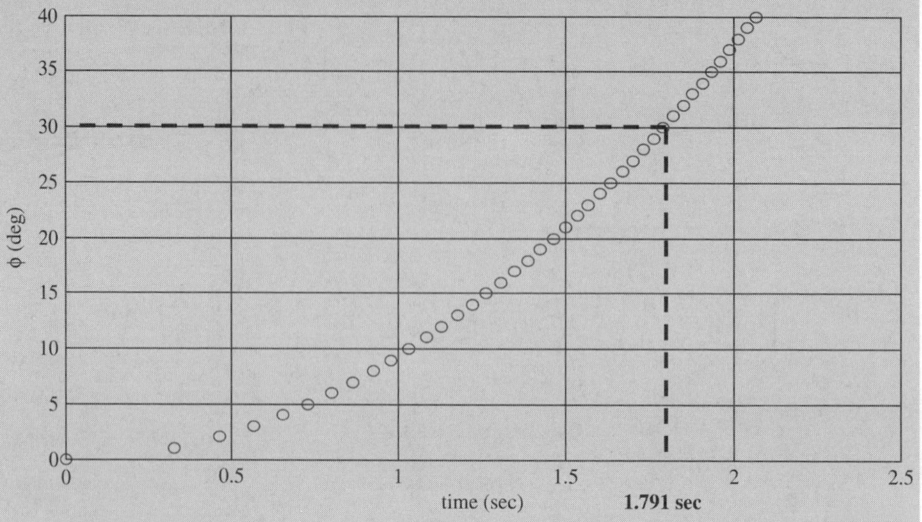

Figure 12.44 Variations of bank angle versus time for aircraft in Example 12.4

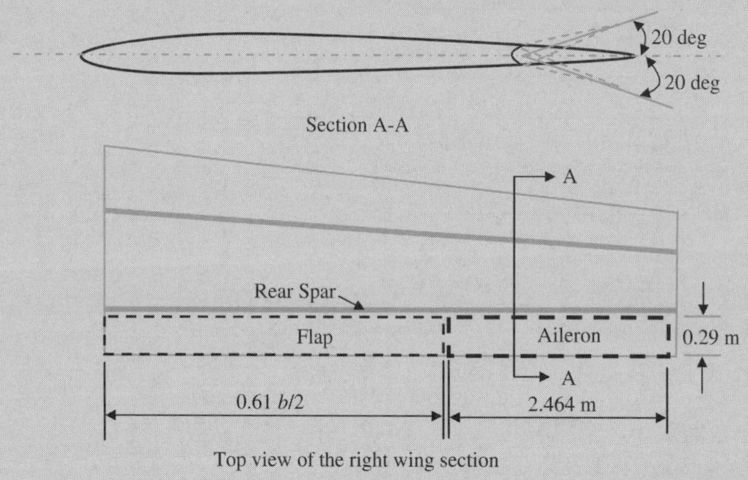

Figure 12.45 Wing and aileron of Example 12.4

The geometry of each aileron is as follows:

$$b_A = y_{o_A} - y_{i_A} = 6.883 - 4.42 = 2.464\,\text{m}$$

$$C_A = 0.2C_w = 0.2 \cdot 1.449 = 0.29\,\text{cm}$$

The overall planform area of both left and right ailerons is:

$$A_A = 2b_A C_A = 2 \cdot 2.464 \cdot 0.29 = 1.428\,\text{m}^2$$

The top and side views of the right wing section including the aileron are shown in Figure 12.45. It must be noted that the aileron chord was assumed to be 20% of the wing chord throughout the aileron span. However, the wing is tapered, so in fact the aileron needs to be tapered too. To make the application and the fabrication of the aileron easier, a constant chord aileron is selected; hence, a change in the aileron chord must be applied in the later design stage to correct for the assumption.

12.8.2 Elevator Design Example

Example 12.5

Problem statement: Figure 12.46 illustrates the geometry of a high-wing twin-jet engine light utility aircraft which is equipped with a tricycle landing gear. Design an elevator for this aircraft which has the following characteristics:

$$m_{TO} = 20\,000\,\text{kg}, V_s = 85\,\text{KEAS}, I_{yy} = 150\,000\,\text{kg m}^2, T_{max} = 2 \cdot 28\,\text{kN},$$

$$L_f = 23\,\text{m}, V_c = 360\,\text{KTAS (at 25 000 ft)}, C_{L_o} = 0.24, C_{D_{o_C}} = 0.024,$$

$$C_{D_{o_{TO}}} = 0.038, C_{L_\alpha} = 5.7\,1/\text{rad}$$

Wing:

$$S = 70\text{ m}^2, \text{AR} = 8, C_{L_{\alpha_{wf}}} = C_{L_{\alpha_w}} = 5.7\text{ 1/rad}, e = 0.88,$$

$$\lambda = 1, C_{L_{flap_{TO}}} = 0.5, \quad C_{m_{ac_{wf}}} = 0.05, i_w = 2\text{ deg}, h_o = 0.25, \alpha_{s_{TO}} = 12\text{ deg}$$

Horizontal tail:

$$S_h = 16\text{ m}^2, b_h = 9\text{ m}, C_{L_{\alpha_h}} = 4.3\text{ 1/rad}, i_h = -1\text{ deg}, \lambda_h = 1, \eta_h = 0.96,$$

$$\alpha_{h_s} = 14\text{ deg, airfoil section: NACA 0009}, \alpha_{twist} = 0$$

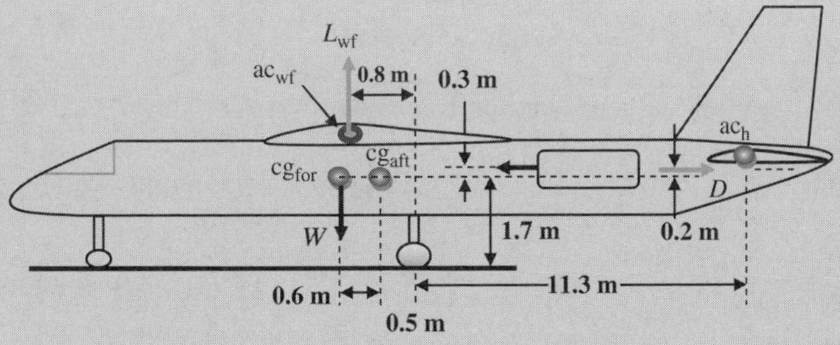

Figure 12.46 Aircraft geometry of Example 12.5

Solution:

- **Step 1**. The elevator design requirements are identified as follows:
  - Take-off rotation (longitudinal control) requirement. It is assumed that the airport is located at sea-level altitude.
  - Longitudinal trim requirements (within flight envelope).
  - Low cost.
  - Manufacturability.
- **Step 2**. Based on Table 12.9, the take-off pitch angular acceleration for this type of aircraft must be between 10 and 15 deg/s^2. A value of 12 deg/s^2 for the take-off pitch angular acceleration is tentatively selected.
- **Step 3**. Table 12.3 suggests a value of 0.8–1 for the elevator span-to-tail span ratio. A value of 1 is tentatively selected.
- **Step 4**. Table 12.3 suggests a value of −25 deg for the elevator maximum deflection. A value of −25 deg is tentatively selected.
- **Step 5**. Calculation of the wing/fuselage lift (L_{wf}), aircraft drag (D), and wing/fuselage pitching moment about the wing/fuselage aerodynamic center. The air density at sea level is 1.225 kg/m^3, and at 25 000 ft is 0.549 kg/m^3. To obtain the wing mean aerodynamic chord, the following calculations are made:

$$b = \sqrt{S \cdot \text{AR}} = \sqrt{70 \cdot 8} = 23.66\,\text{m} \tag{5.19}$$

$$\overline{C} = \frac{S}{b} = \frac{70}{23.66} = 2.96\,\text{m} \tag{5.18}$$

To find the aircraft drag, we have:

$$K = \frac{1}{\pi \cdot e \cdot \text{AR}} = \frac{1}{3.14 \cdot 0.88 \cdot 8} = 0.045 \tag{5.22}$$

$$C_{L_C} = \frac{2W}{\rho V_C^2 S} = \frac{2 \cdot 20\,000 \cdot 9.81}{0.549 \cdot (360 \cdot 0.5144)^2 \cdot 70} = 0.297 \tag{5.1}$$

$$C_{L_{\text{TO}}} = C_{L_C} + \Delta C_{L_{\text{flap}}} = 0.297 + 0.5 = 0.797 \tag{4.69c}$$

$$C_{D_{\text{TO}}} = C_{D_{o_{\text{TO}}}} + KC_{L_{\text{TO}}}^2 = 0.038 + 0.045 \cdot 0.797^2 = 0.067 \tag{4.68}$$

$$V_{\text{R}} = V_{\text{S}} = 85\,\text{knot} = 43.73\,\text{m/s} \tag{12.54}$$

Thus, the longitudinal aerodynamic forces and moment are:

$$D_{\text{TO}} = \frac{1}{2}\rho_o V_{\text{R}}^2 S C_{D_{\text{TO}}} = \frac{1}{2} \cdot 1.225 \cdot (43.73)^2 \cdot 70 \cdot 0.067 = 5472\,\text{N} \tag{12.63}$$

$$L_{\text{TO}} \approx L_{\text{wf}} = \frac{1}{2}\rho_o V_{\text{R}}^2 S_{\text{ref}} C_{L_{\text{Tof}}} = \frac{1}{2} \cdot 1.225 \cdot (43.73)^2 \cdot 70 \cdot 0.797$$

$$= 65\,371\,\text{N} \tag{12.62}$$

$$M_{\mathrm{ac_{wf}}} = \frac{1}{2}\rho_{\mathrm{o}} V_{\mathrm{R}}^{2} C_{\mathrm{m_{ac_{wf}}}} S_{\mathrm{ref}} \overline{C} = \frac{1}{2} \cdot 1.225 \cdot (43.73)^{2} \cdot (0.05) \cdot 70 \cdot 2.96$$
$$= 12\,125\,\mathrm{Nm} \quad (12.64)$$

- **Step 6**. Calculation of aircraft linear acceleration (a) during take-off rotation using Equation (12.55). The runway is assumed to be concrete, so from Table 9.7 (Chapter 9), a ground friction of 0.04 is selected:

$$F_{\mathrm{f}} = \mu \left(W - L_{\mathrm{TO}}\right) = 0.04 \ \left(20\,000 \cdot 9.81 - 65\,371\right) = 5230.5\,\mathrm{N} \quad (12.59)$$

Aircraft linear acceleration at the time of take-off rotation:

$$a = \frac{T - D_{\mathrm{TO}} - F_{\mathrm{R}}}{m} = \frac{2 \cdot 28\,000 - 5472 - 5230.5}{20\,000} \Rightarrow a = 2.265\,\frac{\mathrm{m}}{\mathrm{s}^{2}} \quad (12.55)$$

- **Step 7**. Calculation of the contributing pitching moments in the take-off rotation. The clockwise rotation about the y-axis is considered to be a positive direction. The most forward aircraft center of gravity is considered.

$$M_{W} = W\left(x_{\mathrm{mg}} - x_{\mathrm{cg}}\right) = -20\,000 \cdot 9.81 \cdot 1.1 = -215\,746\,\mathrm{Nm} \quad (12.65)$$

$$M_{D} = D\left(z_{D} - z_{\mathrm{mg}}\right) = 5472 \cdot 1.9 = 10\,397\,\mathrm{Nm} \quad (12.66)$$

$$M_{T} = T\left(z_{T} - z_{\mathrm{mg}}\right) = -2 \cdot 28\,000 \cdot (1.7 + 0.3) = -112\,000\,\mathrm{Nm} \quad (12.67)$$

$$M_{L_{\mathrm{wf}}} = L_{\mathrm{wf}}\left(x_{\mathrm{mg}} - x_{\mathrm{ac_{wf}}}\right) = 65\,371 \cdot 0.8 = 52\,297\,\mathrm{Nm} \quad (12.68)$$

$$M_{a} = ma\left(z_{\mathrm{cg}} - z_{\mathrm{mg}}\right) = 20\,000 \cdot 2.265 \cdot 1.7 = 77005.5\,\mathrm{Nm} \quad (12.70)$$

- **Step 8**. Calculate the desired horizontal tail lift (L_{h}) during take-off rotation employing Equation (12.72). For this calculation, consider the most forward aircraft center of gravity.

$$L_{\mathrm{h}} = \frac{\left[L_{\mathrm{wf}}\left(x_{\mathrm{mg}} - x_{\mathrm{ac_{wf}}}\right) + M_{\mathrm{ac_{wf}}} + ma\left(z_{\mathrm{cg}} - z_{\mathrm{mg}}\right) + W\left(x_{\mathrm{mg}} - x_{\mathrm{cg}}\right) + D\left(z_{D} - z_{\mathrm{mg}}\right) + T\left(z_{T} - z_{\mathrm{mg}}\right) - I_{yy\mathrm{mg}}\ddot{\theta}\right]}{x_{\mathrm{ac_{h}}} - x_{\mathrm{mg}}}$$

$$L_{\mathrm{h}} = \frac{\left[52\,297 + 12\,125 + 77005.5 - 215\,746 + 10\,397 - 112\,000 - \left(150\,000 \cdot \frac{12}{57.3}\right)\right]}{11.3} \quad (12.72)$$

or

$$L_{\mathrm{h}} = -18\,348\,\mathrm{N}$$

- **Step 9**. Calculation of the desired horizontal tail lift coefficient (C_{L_h}):

$$C_{L_h} = \frac{2L_h}{\rho_o V_R^2 S_h} = \frac{2 \cdot (-18\,348)}{1.225 \cdot 43.73^2 \cdot 16} \Rightarrow C_{L_h} = -0.979 \tag{12.73}$$

- **Step 10**. Calculation of the angle of attack effectiveness of the elevator (τ_e). In this calculation, the maximum elevator deflection is considered.

$$C_{L_h} = C_{L_{\alpha h}} \left(\alpha_h + \tau_e \delta_E\right) \tag{12.75}$$

The tail angle of attack is already defined as:

$$\alpha_h = \alpha + i_h - \varepsilon \tag{12.76}$$

where the downwash effect is determined as follows (Equation (6.54)):

$$\varepsilon_o = \frac{2C_{L_w}}{\pi \cdot AR} = \frac{2C_{L_{TO}}}{\pi \cdot AR} = \frac{2 \cdot 0.797}{3.14 \cdot 8} = 0.063\,\text{rad} = 3.63\,\text{deg} \tag{6.55}$$

$$\frac{\partial \varepsilon}{\partial \alpha} = \frac{2C_{L_{\alpha w}}}{\pi \cdot AR} = \frac{2 \cdot 5.7}{\pi \cdot 8} = 0.454\,\text{deg}/\text{deg} \tag{6.56}$$

The wing angle of attack (α_w) at take-off may be assumed to be equal to the wing incidence (i_w). Thus:

$$\varepsilon = \varepsilon_o + \frac{\partial \varepsilon}{\partial \alpha}\alpha_w = 0.063 + 0.454 \cdot \frac{2}{57.3} = 0.079\,\text{rad} = 4.54\,\text{deg} \tag{6.54}$$

Hence, the horizontal tail angle of attack at the instance of take-off rotation is:

$$\alpha_h = \alpha + i_h - \varepsilon = 2 - 1 - 4.54 = -3.54\,\text{deg} \tag{12.76}$$

The angle of attack effectiveness of the elevator from Equation (12.75) is:

$$\tau_e = \frac{\alpha_h + \left(C_{L_h}/C_{L_{\alpha h}}\right)}{\delta_{E_{max}}} = \frac{\frac{-3.54}{57.3} + \frac{-0.979}{4.3}}{-25/57.3} \Rightarrow \tau_e = 0.664 \tag{12.75}$$

- **Step 11**. The corresponding elevator-to-tail chord ratio (C_E/C_h) for a τ_e of 0.664 (from Figure 12.12) is determined to be 0.49. Thus:

$$\frac{C_E}{C_h} = 0.49$$

In other words, the elevator-to-tail chord ratio is determined to be 49%.

- **Steps 12 and 13**. **checked**.
- **Step 14**. Calculation of horizontal tail lift coefficient using lifting-line theory, when the elevator is deflected with its maximum negative angle (i.e., $-\delta_{E_{max}}$). The change in the tail lift coefficient when the elevator is deflected is:

$$\Delta\alpha_{o_E} \approx -1.15 \cdot \frac{C_E}{C_h}\delta_E = -1.15 \cdot 0.49 \cdot (-25) = 14.088\,\text{deg} \tag{12.93}$$

To apply the lifting-line theory, the MATLAB program in Chapter 8.3 is used and a few parameters such as tail area, tail span, and $\Delta\alpha_{o_E}$ are changed. When the program is executed, the tail lift distribution in Figure 12.47 is produced.

The corresponding tail lift coefficient is −1.14, which is slightly greater than the desired tail lift coefficient of −0.979. Thus the elevator is acceptable.

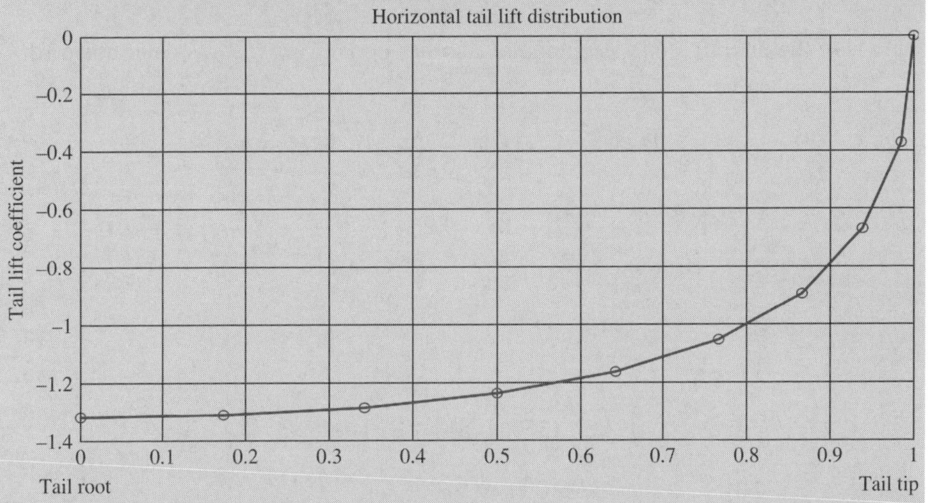

Figure 12.47 Tail lift distribution when elevator is deflected −25 deg

- **Step 15**. **checked**.
- **Step 16**. Calculation of the elevator effectiveness derivatives ($C_{m_{\delta E}}, C_{L_{\delta E}}, C_{L_{h_{\delta E}}}$). For these calculations, the *most aft* aircraft center of gravity and the *most forward* aircraft center of gravity are considered. For the case of the most aft cg, we have the following:

$$\overline{V}_H = \frac{l_h \cdot S_h}{S\overline{C}} = \frac{(11.3 + 0.5) \cdot 16}{70 \cdot 2.96} = 0.912 \tag{6.24}$$

$$C_{m_{\delta E}} = -C_{L_{\alpha h}}\eta_h \cdot \overline{V}_H \cdot \frac{b_E}{b_h}\tau_e = -4.3 \cdot 0.96 \cdot 0.912 \cdot 1 \cdot 0.664 = -2.5\,\frac{1}{\text{rad}} \tag{12.51}$$

$$C_{L_{\delta E}} = C_{L_{\alpha h}}\eta_h\frac{S_h}{S} \cdot \frac{b_E}{b_h}\tau_e = 4.3 \cdot 0.96 \cdot \frac{16}{70} \cdot 1 \cdot 0.664 = 0.626\,\frac{1}{\text{rad}} \tag{12.52}$$

$$C_{L_{h_{\delta E}}} = C_{L_{\alpha h}}\tau_e = 4.3 \cdot 0.664 = 2.85\,\frac{1}{\text{rad}} \tag{12.53}$$

The case of the most forward aircraft cg is addressed in Step 18.

- **Step 17**. Calculation of the elevator deflection (δ_E) required to maintain longitudinal trim at various flight conditions. The elevator deflection when the aircraft cg is located at its most aft position and the aircraft is flying with its maximum speed is calculated as follows. For this case, the distance between the tail ac and the aircraft cg is equal to 11.8 m (i.e., 11.3 + 0.5).

$$\delta_E = -\frac{\left(\frac{T \cdot z_T}{\overline{q} \cdot S \cdot \overline{C}} + C_{m_o}\right) C_{L_\alpha} + \left(C_{L_1} - C_{L_o}\right) C_{m_\alpha}}{C_{L_\alpha} C_{m_{\delta E}} - C_{m_\alpha} C_{L_{\delta E}}} \qquad (12.90)$$

where the aircraft static longitudinal stability derivative (C_{m_α}) is determined as follows:

$$C_{m_\alpha} = C_{L_{\alpha wf}}\left(h - h_o\right) - C_{L_{\alpha_h}} \eta_h \frac{S_h}{S}\left(\frac{l_h}{\overline{C}}\right)\left(1 - \frac{d\varepsilon}{d\alpha}\right) \qquad (6.67)$$

$$C_{m_\alpha} = 5.7\left(\frac{0.8 - 0.5}{2.96}\right) - 4.3 \cdot 0.96 \cdot \frac{16}{70}\left(\frac{11.3 + 0.5}{2.96}\right)(1 - 0.454)$$

$$= -1.479 \frac{1}{\text{rad}} \qquad (6.67)$$

Thus:

$$\overline{q} = \frac{1}{2}\rho V^2 = \frac{1}{2} \cdot 1.225 \cdot (360 \cdot 0.514)^2 = 21008 \text{ Pa}$$

$$C_{L_1} = \frac{2W}{\rho V_C^2 S} = \frac{2 \cdot 20\,000 \cdot 9.81}{1.225 \cdot (360 \cdot 0.5144)^2 \cdot 70} = 0.133 \qquad (5.1)$$

$$\delta_E = -\frac{\left(\frac{56\,000 \cdot (-0.3)}{21\,008 \cdot 70 \cdot 2.96} + 0.05\right) \cdot 5.7 + (0.133 - 0.24) \cdot (-1.479)}{5.7 \cdot (-2.5) - (-1.479) \cdot (0.626)} \qquad (12.90)$$

or

$$\delta_E = 0.033 \text{ rad} = +1.888 \text{ deg}$$

- **Step 18**. Plot the variations of the elevator deflection versus airspeed and also versus altitude. For these calculations, the *most aft* aircraft center of gravity and the *most forward* aircraft center of gravity are considered. The following Matlab program was written to calculate and plot the variations of elevator deflection to maintain longitudinal trim at various flight conditions:

```
clc
clear all
Vmax = 185; % m/s
Sw=70; % m^2
Sh = 16; % m^2
```

```
Cbar= 2.96; % m
Vs = 44; %m/sec
Tmax= 56000; %N
rho = 1.225; % kg/m^3
Cmo = 0.05;
zT = -0.3; %m
CLa = 5.2; %1/rad
CLah = 4.3; % 1/rad
CLa_wf = CLa;
g = 9.81; %m/s^2
m = 20000; % kg
CLo = 0.24;
taw = 0.664;
etha_h = 0.96;
lh = 11.3; % m from main landing gear
de_da = 0.454;
CLdE=-CLah*etha_h*Sh*taw/Sw;
% Most aft cg
xcg = 0.5; % m from main landing gear
h_to_ho = 0.3/Cbar; % m
l_h1 = lh+xcg; %m
VH1 = (l_h1*Sh)/(Sw*Cbar);
CmdE1 = -CLah*etha_h*VH1*taw;
Cma1 = CLa_wf*h_to_ho-CLah*etha_h*Sh*(l_h1/Cbar)*(1-de_da)/Sw;
% Most forward cg
xcg = 1.1; % m from main landing gear
h_to_ho = -0.3/Cbar; % m
l_h2 = lh+xcg; % m
VH2 = (l_h2*Sh)/(Sw*Cbar);
CmdE2 = -CLah*etha_h*VH2*taw;
Cma2 = CLa_wf*h_to_ho-CLah*etha_h*Sh*(l_h2/Cbar)*(1-de_da)/Sw;
i =1;
for U1=Vs:Vmax;
qbar=0.5*rho*U1^2;
CL1= (m*g)/(qbar*Sw);
f1=((Tmax*zT)/(qbar*Sw*Cbar))+Cmo;
dE1(i)=-((f1*CLa)+(CL1-CLo)*Cma1)z(CLa*CmdE1-Cma1*CLdE);
dE2(i)=-((f1*CLa)+(CL1-CLo)*Cma2)/(CLa*CmdE2-Cma2*CLdE);
V(i)=U1;
i=i+1;
end
plot(V/0.5144,dE1*57.3,'o',V/0.5144,dE2*57.3,'*')
grid
xlabel ('Speed (knot)')
ylabel ('\delta_E (deg)')
legend('Most aft cg','Most forward cg')
```

The results are plotted in Figures 12.48 and 12.49. Figure 12.49 shows the variations of elevator deflection with respect to aircraft speed at sea level to maintain

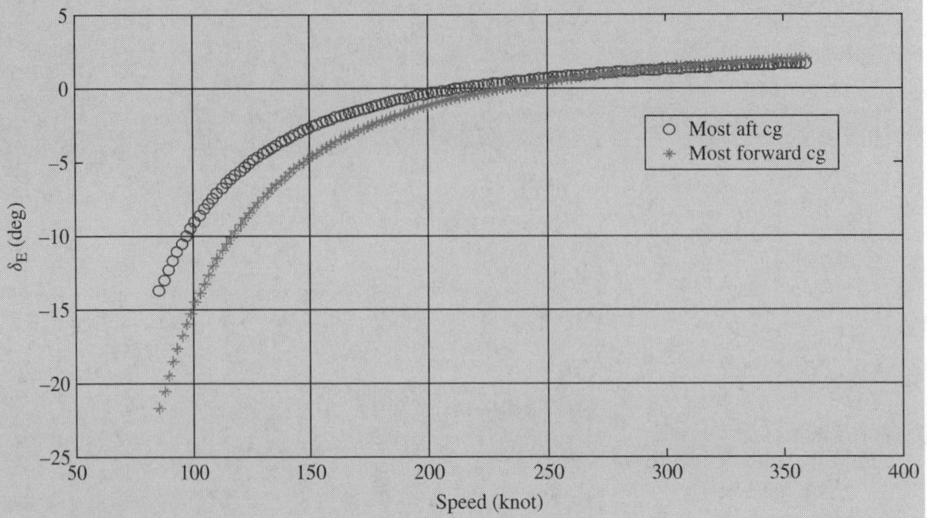

Figure 12.48 Variations of elevator deflection with respect to aircraft speed at sea level

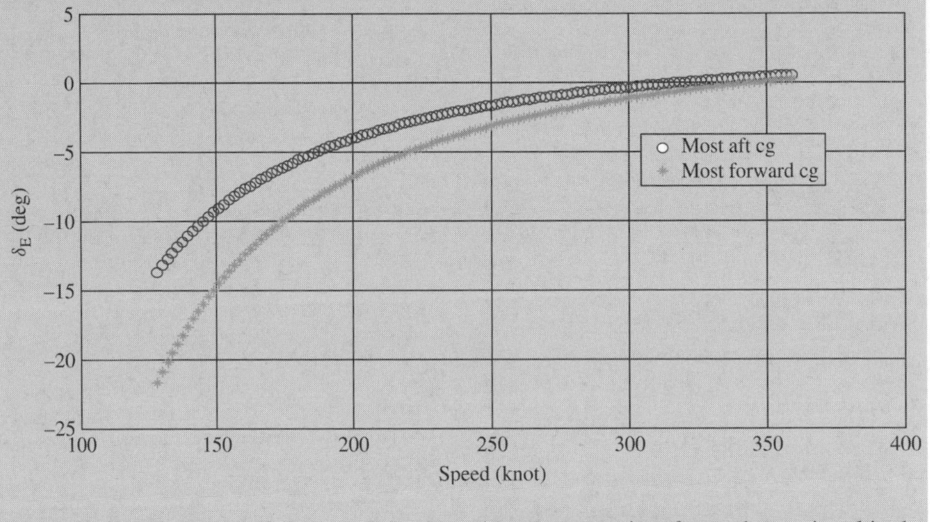

Figure 12.49 Variations of elevator deflection with respect to aircraft speed at cruise altitude

longitudinal trim in a cruising flight. However, Figure 12.49 illustrates the variations of elevator deflection with respect to aircraft speed at cruise altitude (i.e., 25 000 ft). Note that the above Matlab program was updated to include the cruise altitude air density in order to produce Figure 12.49. A figure such as Figure 12.48, which demonstrates the variations of elevator deflection with respect to aircraft speed in order to maintain longitudinal trim in a cruising flight, is often referred to as trim curve.

The results of steps 4–11 indicate that the maximum negative (up) elevator deflection is −25 deg, while the results of step 16 (Figures 12.48 and 12.49) demonstrate that the maximum positive (down) elevator deflection is +1.96 deg. Therefore:

$$\delta_{E_{max\_up}} = -25\,\text{deg}$$

$$\delta_{E_{max\_down}} = +1.96\,\text{deg}$$

- **Step 19**. **checked**.
- **Step 20**. We need to check to make certain that the elevator deflection is not causing the horizontal tail to stall during take-off rotation. It is assumed that the fuselage during take-off rotation is lifted up to 2 deg below the wing stall angle:

$$\alpha_{TO} = \alpha_{s_{TO}} - 2 = 12 - 2 = 10\,\text{deg}$$

Hence, the horizontal tail take-off angle is:

$$\alpha_{h_{TO}} = \alpha_{TO}\left(1 - \frac{d\varepsilon}{d\alpha}\right) + i_h - \varepsilon_o = 10 \cdot (1 - 0.454) - 1 - 3.636 = 0.828\,\text{deg} \tag{12.91}$$

The tail stall angle of attack during take-off rotation (α_{h_s}) is:

$$\alpha_{h_s} = \pm\left(\alpha_{h_{s:\delta E=0}} - \Delta\alpha_{h_E}\right) \tag{12.92}$$

where $\alpha_{h_{s:\delta E=0}}$ is the tail stall angle when the elevator is not employed and is given to be 14 deg. The parameter $\Delta\alpha_{h_E}$ is the magnitude of reduction in tail stall angle of attack due to elevator deflection and is determined using Table 12.19.

With an elevator deflection of 25 deg and an elevator chord ratio of 0.49, Table 12.19 shows a value of 10.71 deg for the parameter $\Delta\alpha_{h_E}$. Thus:

$$\alpha_{h_s} = \alpha_{h_{s:\delta E=0}} - \Delta\alpha_{h_E} = 14 - 10.71 = 3.29\,\text{deg} \tag{12.92}$$

Since the tail angle of attack at the end of rotation (i.e., 0.828 deg) is less than the tail stall angle when the elevator is deflected (i.e., 3.29 deg), the horizontal tail does not stall during take-off rotation. Therefore, that elevator is acceptable and passes all tests.

- **Steps 21 and 22**. **checked**.
- **Step 23**. Aerodynamic balance/mass balance (beyond the scope of this example).
- **Step 24**. Optimization (beyond the scope of this example).
- **Step 25**. Finally, the elevator geometry is as follows:

$$\frac{b_E}{b_h} = 1 \Rightarrow b_E = b_h = 9\,\text{m}$$

$$S_h = b_h\overline{C}_h \Rightarrow \overline{C}_h = \frac{S_h}{b_h} = \frac{16}{9} = 1.788\,\text{m} \tag{6.66}$$

$$\frac{C_E}{C_h} = 0.49 \Rightarrow C_E = 0.49C_h = 0.49 \cdot 1.788 = 0.871\,\text{m}$$

$$S_E = b_E C_E = 9 \cdot 0.871 = 7.84\,\text{m}^2$$

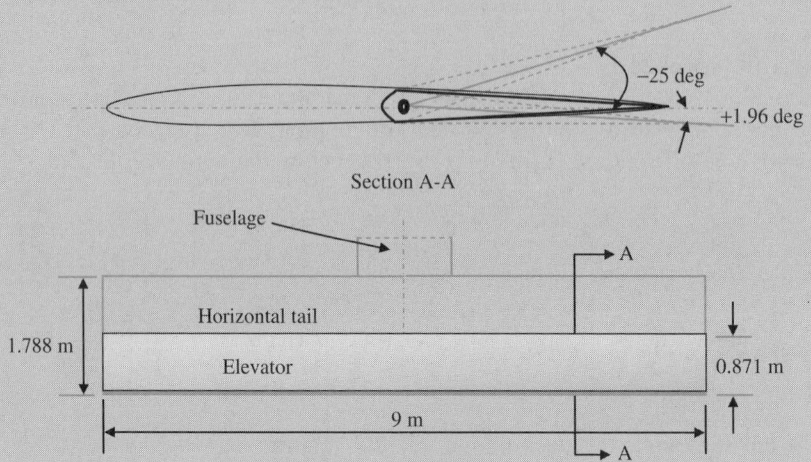

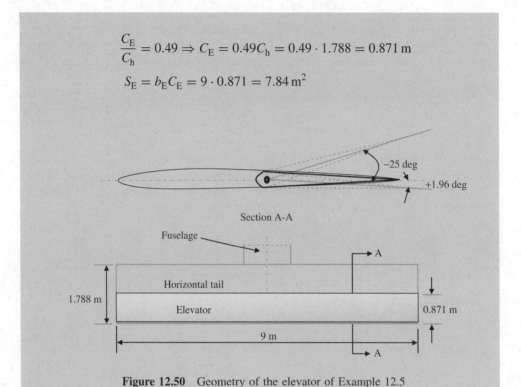

Figure 12.50 Geometry of the elevator of Example 12.5

Figure 12.50 depicts the horizontal tail and elevator geometry.

12.8.3 Rudder Design Example

Example 12.6

Problem statement: A large transport aircraft with a maximum take-off mass of 260 000 kg is equipped with four turbofan engines each generating 140 kN of thrust. The distance between the most aft and the most forward cg is 1.5 m. The top view and side view of the aircraft are shown in Figure 12.51; the fuselage has a cylindrical shape. Other characteristics of the aircraft are as follows:

$$S = 365\,\text{m}^2;\ b = 60\,\text{m};\ S_V = 50\,\text{m}^2;\ V_s = 120\,\text{knot};\ C_{L_{\alpha V}} = 4.5\frac{1}{\text{rad}};\ \eta_v = 0.97;$$

$$L_f = 63\,\text{m};\ D_f = 5.5\,\text{m};\ \frac{d\sigma}{d\beta} = 0;\ C_{n_o} = 0;\ C_{y_o} = 0$$

The aircraft is not spinnable, and is required to be able to land safely when there is a cross-wind of 40 knots. Design a rudder for this aircraft.

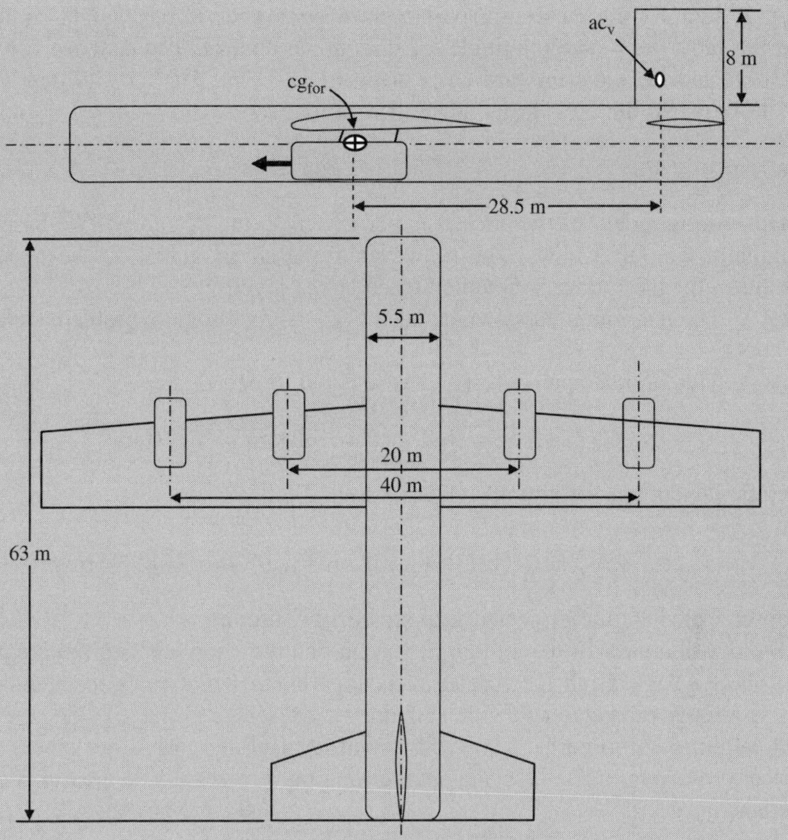

Figure 12.51 Top view and side view of the aircraft in Example 12.6

Solution:

- **Step 0**. The first step is to layout the design requirements and to identify the most critical one. Since the aircraft is not spinnable, so the design requirements are: (i) asymmetric thrust directional trim, (ii) cross-wind landing, (iii) coordinated turn, and (iv) adverse yaw. According to Table 12.22, the most critical rudder design requirements for a multi-engine wing-installed engines transport aircraft are either asymmetric thrust or cross-wind landing. At this moment, it is not evident which is the most crucial. Since the cross-wind is 40 knots (a relatively high value), it is assumed that cross-wind landing is the most critical design requirement. Nonetheless, the asymmetric thrust will also be investigated. Therefore, the steps of rudder design based on cross-wind landing (Section 12.6.3.2) will be followed.
- **Step 1**. List the available/given data related to rudder design.

  The vertical tail is rectangular with a planform area of 50 m^2. From Figure 12.51, the vertical tail span is 8 m and the distance between the most forward cg and the vertical tail aerodynamic center is 27 m.

- **Step 2**. Identify the most unfavorable cg location and aircraft weight combination, and the most unfavorable altitude for directional control. The distance between the most aft and the most forward cg is given to be 1.5 m. The most critical condition would be when the cg is at its most aft location, so:

$$l_{\mathrm{v}} = 28.5 - 1.5 = 27\ \mathrm{m}$$

Other weight data of the aircraft is not given, so the design will be based on the maximum take-off weight. The runway altitude is not given, so the most critical condition for the rudder is assumed to be at sea level.

- **Step 3**. The maximum cross-wind speed (V_{W}) is given by the problem statement as 40 knots.
- **Step 4**. The aircraft approach speed is selected to be $1.1V_{\mathrm{s}}$:

$$U_1 = 1.1\,V_{\mathrm{s}} = 1.1 \cdot 120 = 132\,\mathrm{knot} = 67.91\,\mathrm{m/s}$$

- **Step 5**. Determine the aircraft total airspeed (V_{T}):

$$V_{\mathrm{T}} = \sqrt{U_1^2 + V_{\mathrm{W}}^2} = \sqrt{(132)^2 + (40)^2} = 137.93\,\mathrm{knot} = 70.95\,\mathrm{m/s} \qquad (12.106)$$

- **Step 6**. Calculate the projected side area of the aircraft.

From a side view of the aircraft, it is seen that the wing and engines are projected into the fuselage. Both the fuselage and the vertical tail have a rectangular shape, so the aircraft projected side area is mainly the fuselage projected side area plus the vertical tail planform area. The geometry of the landing gear is not given, so 2% is added to the sum of the fuselage and vertical tail projected side area to account for the landing gear:

$$S_{\mathrm{S}} = 1.02\left[S_{\mathrm{f}} + S_{\mathrm{V}}\right] = 1.02\left[L_{\mathrm{f}}D_{\mathrm{f}} + S_{\mathrm{V}}\right] = 1.02 \cdot (63 \cdot 5.5 + 50) = 404.4\,\mathrm{m}^2$$

- **Step 7**. Determine the center of the projected side area of the aircraft (S_{S}) and its distance to the aircraft cg (d_{c}). Both the fuselage and the vertical tail have a rectangular shape, so the center of the fuselage is at the midpoint of the fuselage length from the nose (63/2 = 31.5 m). The vertical tail chord is:

$$S_{\mathrm{V}} = b_{\mathrm{V}} \cdot \overline{C}_{\mathrm{V}} \Rightarrow C_{\mathrm{V}} = \frac{S_{\mathrm{V}}}{b_{\mathrm{V}}} = \frac{50}{8} = 6.25\,\mathrm{m} \qquad (6.80)$$

The center of the vertical tail area is at the midpoint of the vertical tail chord from its leading edge (6.25/2 = 3.125 m). The overall center of the projected side area of the aircraft from the fuselage nose is determined as follows:

$$x_{\mathrm{ca}} = \frac{\sum_{i=1}^{n} A_i x_i}{\sum_{i=1}^{n} A_i} = \frac{\left(L_{\mathrm{f}}D_{\mathrm{f}}\right) x_{\mathrm{f}} + S_{\mathrm{V}} x_{\mathrm{V}}}{L_{\mathrm{f}}D_{\mathrm{f}} + S_{\mathrm{V}}}$$

$$= \frac{(63 \cdot 5.5) \cdot \dfrac{63}{2} + 50 \cdot \left[(63 - 6.25) + \dfrac{6.25}{2} \right]}{(63 \cdot 5.5) + 50} \tag{12.119}$$

or

$$x_{ca} = 35.078 \text{ m}$$

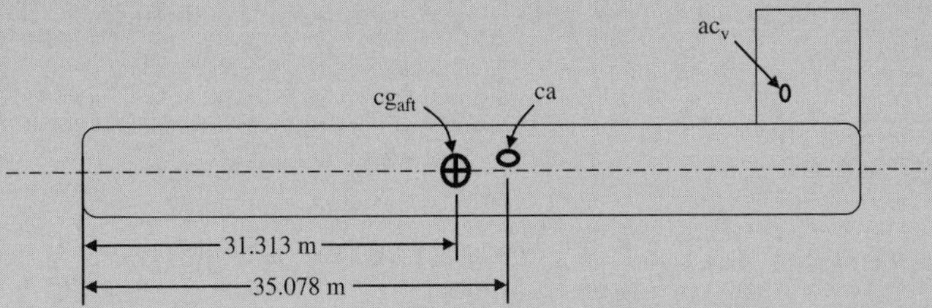

Figure 12.52 Center of projected side area

The vertical tail aerodynamic center is assumed to be at its quarter chord, so the distance between the center of gravity of the aircraft and the fuselage nose is (see Figure 12.52):

$$x_{cg} = L_f - l_V - 0.75C_V = 63 - 27 - 0.75 \cdot 6.25 = 31.313 \, \text{m}$$

Thus, the distance between the center of the projected side area of the aircraft and the aircraft cg (see Figure 12.52) is:

$$d_c = x_{ca} - x_{cg} = 35.078 - 31.313 = 3.766 \, \text{m}$$

Therefore the center of the projected side area of the aircraft is 3.766 m behind the aircraft cg.

- **Step 8**. Determine the aircraft side force produced by the cross-wind (F_w). For simplicity, a cross-wind from the right is assumed, generating a positive sideslip angle. Based on the side view of the aircraft (the fuselage has a cylindrical shape), a side drag coefficient of 0.6 is selected (Section 12.6.2.3). The sea-level air density is 1.225 kg/m^3.

$$F_w = \frac{1}{2} \rho V_W^2 S_S C_{D_y} = \frac{1}{2} \cdot 1.225 \cdot (40 \cdot 0.514)^2 \cdot 404.4 \cdot 0.6 = 62\,936 \, \text{N} \tag{12.113}$$

- **Step 9**. The rudder span-to-vertical tail span ratio b_R/b_V (from Table 12.3) is tentatively selected to be 1.
- **Step 10**. The rudder-to-vertical tail chord ratio (C_R/C_V) is tentatively selected to be 0.3 (Table 12.3).

- **Step 11**. Determine the aircraft sideslip angle (β):

$$\beta = \tan^{-1}\left(\frac{V_W}{U_1}\right) = \tan^{-1}\left(\frac{40}{132}\right) = 0.294\,\text{rad} = 16.86\,\text{deg} \qquad (12.105)$$

- **Step 12**. Calculate the aircraft sideslip derivatives C_{n_β} and C_{y_β}.

From Section 6.8.1, the parameter K_{f1} is selected to be 0.75; and from Section 12.6.2.3, the parameter K_{f2} is selected to be 1.35. Thus:

$$C_{n_\beta} = K_{f1} C_{L_{\alpha V}} \left(1 - \frac{d\sigma}{d\beta}\right) \eta_V \frac{l_{Vt} S_V}{bS} = 0.75 \cdot 4.5 \cdot (1 - 0) \cdot 0.96 \cdot \frac{27 \cdot 50}{60 \cdot 365} \qquad (12.116)$$

or

$$C_{n_\beta} = 0.2 \frac{1}{\text{rad}}$$

$$C_{y_\beta} = -K_{f2} C_{L_{\alpha V}} \left(1 - \frac{d\sigma}{d\beta}\right) \eta_V \frac{S_V}{S} = -1.35 \cdot 4.5 \cdot (1 - 0) \cdot 0.96 \cdot \frac{50}{365} \qquad (12.117)$$

or

$$C_{y_\beta} = -0.8 \frac{1}{\text{rad}}$$

- **Step 13**. A value of 0.51 for the rudder angle of attack effectiveness (τ_r) is extracted from Figure 12.12 for a control surface chord/lifting surface chord of 0.3.
- **Step 14**. Calculate the aircraft control derivatives $C_{y_{\delta R}}$ and $C_{n_{\delta R}}$:

$$C_{y_{\delta R}} = C_{L_{\alpha V}} \eta_V \tau_r \frac{b_R}{b_V} \frac{S_V}{S} = 4.5 \cdot 0.96 \cdot 0.51 \cdot 1 \cdot \frac{50}{365} = 0.302 \frac{1}{\text{rad}} \qquad (12.118)$$

$$C_{n_{\delta R}} = -C_{L_{\alpha V}} V_V \eta_V \tau_r \frac{b_R}{b_V} = -4.5 \cdot 0.062 \cdot 0.96 \cdot 0.51 \cdot 1 = -0.136 \frac{1}{\text{rad}} \qquad (12.100)$$

- **Step 15**. Compute the rudder deflection.

By simultaneous solution of Equations (12.114) and (12.115), the two unknowns of rudder deflection and sidewash angle are calculated:

$$\frac{1}{2}\rho V_T^2 S b \left(C_{n_o} + C_{n_\beta}(\beta - \sigma) + C_{n_{\delta R}} \delta_R\right) + F_w \cdot d_c \cos\sigma = 0 \qquad (12.114)$$

$$\frac{1}{2}\rho V_W^2 S_S C_{D_y} = \frac{1}{2}\rho V_T^2 S \left(C_{y_o} + C_{y_\beta}(\beta - \sigma) + C_{y_{\delta R}} \delta_R\right) \qquad (12.115)$$

or

$$\frac{1}{2} \cdot 1.225 \cdot (70.95)^2 \cdot 365 \cdot 60 \cdot \left[0.2 \cdot (0.294 - \sigma) - 0.136\delta_R\right]$$
$$+62936 \cdot 3.766 \cos\sigma = 0$$

$$62936 = \frac{1}{2} \cdot 1.225 \cdot (70.95)^2 \cdot 365 \left(-0.799 \cdot (0.294 - \sigma) + 0.302\delta_R\right)$$

Simultaneous solution of these equations results in:

$$\delta_R = 0.458\,\text{rad} = 26.2\,\text{deg}$$

$$\sigma = 0.191\,\text{rad} = 10.95\,\text{deg}$$

The required rudder deflection (26.2 deg) is less than the maximum allowable deflection (30 deg). Thus, the aircraft is able to handle a cross-wind of 40 knots and land safely.

- **Steps 16 and 17**. **checked**.
- **Step 18**. Now that the vertical tail is designed based on the cross-wind landing requirements, it is time to evaluate the rudder design to ensure the other rudder design requirements (e.g., asymmetric thrust and spin recovery) are met. Since the aircraft is not spinnable, the only major requirement is asymmetric thrust directional control. Since the rudder geometry and derivatives are known, we start from step 5 of the procedure in Section 12.6.3.1.
- **Step: 18-5**. Determine/select the aircraft minimum controllable speed (V_{MC}).

  For safety, it is decided to select a value equivalent to 80% of the stall speed:

$$V_{MC} = 0.8 \cdot V_s = 0.8 \cdot 120 = 96\,\text{knot} = 49.39\,\text{m/s}$$

- **Step: 18-6**. Determine the required maximum yawing moment to directionally control/trim the aircraft in the most critical condition. It is desired to directionally trim the aircraft in an asymmetric thrust flight condition when the aircraft is flying with minimum controllable speed.

  The aircraft has four engines, so the most crucial flight condition is when two engines ($n = 2$) on one side suddenly lose power and become inoperative. The distance between two internal engines is 20 m, and between two external engines is 40 m. Each engine is generating 140 kN of thrust. Hence, the rudder deflection to balance the asymmetric thrust at sea level is:

$$\delta_R = \frac{\sum_{i=1}^{n2} T_{L_i} y_{T_i}}{-\bar{q} S b C_{n_{\delta R}}} = \frac{\left(140\,000 \cdot \frac{10}{2}\right) + \left(140\,000 \cdot \frac{20}{2}\right)}{-\frac{1}{2} \cdot 1.225 \cdot (49.39)^2 \cdot 365 \cdot 60 \cdot (-0.136)} \tag{12.104b}$$

or

$$\delta_R = 0.945\,\text{rad} = 54.16\,\text{deg}$$

The required rudder deflection exceeds the maximum allowable rudder deflection (i.e., 54.16 > 30). Therefore, this rudder geometry is not acceptable to satisfy the asymmetric thrust balance requirement when the aircraft is flying with a speed equal to 80% of the stall speed.

At this point, there are two options open to the designer: (i) redesign the rudder and (ii) redefine the minimum controllable speed. Due to the fact that the aircraft may have a minimum controllable speed slightly greater than the stall speed, the second option is chosen. The maximum allowable rudder deflection of 30 deg is employed. Thus, the new minimum controllable speed is calculated as follows:

$$V_{MC} = \sqrt{\frac{\sum_{i=1}^{n2} T_{L_i} y_{T_i}}{-\frac{1}{2}\rho S b C_{n_{\delta R}} \delta_R}} = \sqrt{\frac{\left(140\,000 \cdot \frac{20}{2}\right) + \left(140\,000 \cdot \frac{40}{2}\right)}{-\frac{1}{2} \cdot 1.225 \cdot 365 \cdot 60 \cdot (-0.136) \cdot \frac{30}{57.3}}} \tag{12.104b}$$

or

$$V_{MC} = 66.356\,\text{m/s} = 129\,\text{knot}$$

This minimum controllable speed is about 7.5% greater than the aircraft stall speed:

$$\frac{V_{MC}}{V_s} = \frac{129}{120} = 1.075$$

- **Steps 19–23**. The rest of the example is left to the interested reader to continue.
- **Step 24**. Calculate the rudder span, rudder chord, and rudder area:

$$\frac{C_R}{C_V} = 0.3 \Rightarrow C_R = 0.3C_V = 0.3 \cdot 6.25 = 1.875\,\text{m}$$

$$\frac{b_R}{b_V} = 1 \Rightarrow b_R = b_V = 8\,\text{m}$$

$$S_R = b_R C_R = 8 \cdot 1.875 = 15\,\text{m}^2$$

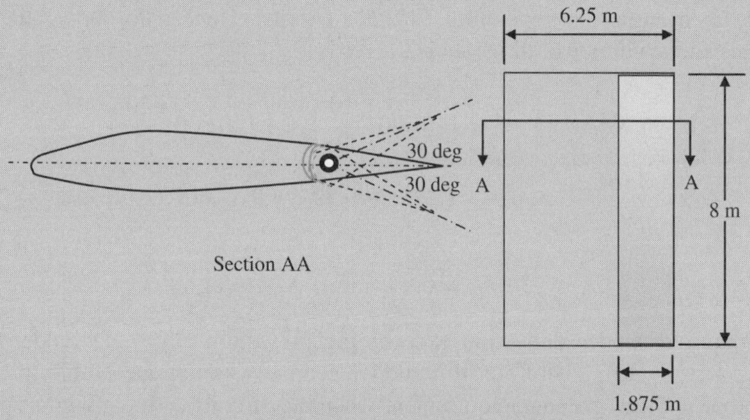

Figure 12.53 Geometry for the rudder of Example 12.6

The geometry of the rudder is depicted in Figure 12.53.

Problems

1. Identify the characteristics of the ailerons (C_A/C, b_A/b, S_A/S, and $b_{A_i}/b/2$) for the following aircraft: Cessna 182; Boeing 767; Airbus 340; F-16; Panavia Tornado; and Mooney M20M.

   You may use the manufacturers' websites, or references such as [25, 26], via measurement from the three-view of the aircraft.

2. Identify the characteristics of the elevator (C_E/C_h, b_E/b_h, and S_E/S_h) for the following aircraft: Dassault Rafale; Grob G 850 Strato 2C; Dassault Falcon 2000; PC-12; Boeing 787; F-22 Raptor; and Cessna 750.

   You may use the manufacturers' websites, or references such as [25, 26], via measurement from the three-view of the aircraft.

3. Identify the characteristics of the rudder (C_R/C_V, b_R/b_V, and S_R/S_V) for the following aircraft: Lockheed C-130 Hercules; Airbus 380; Global Hawk; Eurofighter; Boeing 737-500; Starkraft SK-700; and F/A-18 Hornet.

   You may use the manufacturers' websites, or references such as [25, 26], via measurement from the three-view of the aircraft.

4. A business jet transport aircraft with a maximum take-off mass of 20 000 kg is equipped with three engines each generating 21 kN of thrust. One engine is placed along the fuselage center line at the rear fuselage, but the other two engines are located beside the rear fuselage with a lateral distance of 3.8 m. Other characteristics of the aircraft are as follows:

$$S = 50\,\text{m}^2;\ b = 20\,\text{m};\ S_V = 14\,\text{m}^2;\ b_v = 4.2\,\text{m};\ b_R = 0.6\,b_v;\ l_v = 7.5\,\text{m};$$

$$C_{L_{\alpha V}} = 4.4\frac{1}{\text{rad}};\ C_R/C_V = 0.25;\ \eta_v = 0.97;\ V_s = 106\,\text{knot};\ \delta_{R_{max}} = \pm 30\,\text{deg}$$

   Is the rudder capable of maintaining directional trim in an asymmetric thrust flight condition?

5. A twin-turboprop cargo aircraft with a maximum take-off mass of 16 000 kg is equipped with two engines each generating 1400 kW of power. Two engines are located on the wing with a lateral distance of 7.1 m. Other characteristics of the aircraft are as follows:

$$S = 60\ \text{m}^2;\ b = 26\ \text{m};\ S_V = 17\ \text{m}^2;\ b_v = 5.1\ \text{m};\ b_R = b_v;\ l_v = 10.4\ \text{m};$$

$$C_{L_{\alpha V}} = 4.4\ \frac{1}{\text{rad}};\ C_R/C_V = 0.4;\ \eta_v = 0.95;\ V_s = 92\ \text{knot};\ \delta_{R_{max}} = \pm 30\ \text{deg}$$

   Is the rudder capable of maintaining directional trim in an asymmetric thrust flight condition?

6. A short-range jet transport aircraft with a maximum take-off mass of 44 000 kg is equipped with four engines each generating 31.1 kN of thrust. All engines are located under the wing. The lateral distance between two internal engines is 8.2 m,

and between two external engines is 13.2 m. Other characteristics of the aircraft are as follows:

$$S = 77\ \text{m}^2;\ b = 26.2\ \text{m};\ S_V = 23\ \text{m}^2;\ b_v = 4.9\ \text{m};\ b_R = 0.9b_v;\ l_v = 15.6\ \text{m};$$

$$C_{L\alpha V} = 4.2\frac{1}{\text{rad}};\ C_R/C_V = 0.32;\ \eta_v = 0.94;\ V_s = 97\ \text{knot};\ \delta_{R_{max}} = \pm 30\ \text{deg}$$

Is the rudder powerful enough to maintain the directional trim in an asymmetric thrust flight condition?

7. Consider a light transport aircraft with a take-off mass of 9000 kg, wing area of 34 m^2, and wing span of 9 m. The aircraft projected side area is 31 m^2 and the aircraft center of projected side area is 2.2 m behind the aircraft cg. The aircraft approach speed is 85 knot and the maximum allowable rudder deflection is ±30 deg. Other aircraft characteristics, including related stability and control derivatives, are as follows:

$$C_{n_\beta} = 0.07\frac{1}{\text{rad}};\ C_{n_{\delta R}} = -0.1\frac{1}{\text{rad}};\ C_{y_\beta} = -0.45\frac{1}{\text{rad}};\ C_{y_{\delta R}} = 0.18\frac{1}{\text{rad}};$$

$$C_{D_y} = 0.5;\ C_{n_o} = 0;\ C_{y_o} = 0$$

Is the aircraft rudder powerful enough to allow for a safe crabbed landing when a perpendicular cross-wind of 40 knots is blowing? What about 25 knots?

8. Consider a light GA aircraft with a take-off mass of 2000 kg, wing area of 24 m^2, and wing span of 14 m. The aircraft projected side area is 26 m^2 and the aircraft center of projected side area is 1.3 m behind the aircraft cg. The aircraft approach speed is 55 knot and the maximum allowable rudder deflection is ±30 deg. Other aircraft characteristics, including related stability and control derivatives, are as follows:

$$C_{n_\beta} = 0.09\frac{1}{\text{rad}};\ C_{n_{\delta R}} = -0.2\frac{1}{\text{rad}};\ C_{y_\beta} = -0.4\frac{1}{\text{rad}};\ C_{y_{\delta R}} = 0.19\frac{1}{\text{rad}};$$

$$C_{D_y} = 0.7;\ C_{n_o} = 0;\ C_{y_o} = 0$$

Is the aircraft rudder powerful enough to allow for a safe crabbed landing when a perpendicular cross-wind of 25 knots is blowing?

9. A twin-turboprop cargo aircraft has a take-off mass of 22 000 kg, wing area of 61 m^2, and wing span of 27 m. The aircraft projected side area is 92 m^2 and the aircraft center of projected side area is 3.4 m behind the aircraft cg. The aircraft approach speed is 96 knot and the maximum allowable rudder deflection is ±25 deg. Other aircraft characteristics, including two rudder-related derivatives, are as follows:

$$C_{n_\beta} = 0.18\frac{1}{\text{rad}};\ C_{n_{\delta R}} = -0.11\frac{1}{\text{rad}};\ C_{y_\beta} = -1.1\frac{1}{\text{rad}};\ C_{y_{\delta R}} = 0.17\frac{1}{\text{rad}};$$

$$C_{D_y} = 0.64;\ C_{n_o} = 0;\ C_{y_o} = 0$$

Calculate the maximum cross-wind speed at which the aircraft can safely crab land at 5000 ft altitude.

10. A large jet transport aircraft has a take-off mass of 260 000 kg, wing area of 360 m^2, and wing span of 60 m. The aircraft projected side area is 400 m^2 and the aircraft center of projected side area is 4.2 m behind the aircraft cg. The aircraft approach speed is 115 knot and the maximum allowable rudder deflection is ±30 deg. Other aircraft characteristics, including two rudder-related derivatives, are as follows:

$$C_{n_\beta} = 0.12\,\frac{1}{\text{rad}};\; C_{n_{\delta R}} = -0.09\,\frac{1}{\text{rad}};\; C_{y_\beta} = -1.3\,\frac{1}{\text{rad}};\; C_{y_{\delta R}} = 0.14\,\frac{1}{\text{rad}};$$

$$C_{D_y} = 0.51;\; C_{n_o} = 0;\; C_{y_o} = 0$$

Calculate the maximum cross-wind speed at which the aircraft can safely crab land at 5000 ft altitude.

11. A single-engine acrobatic aircraft has a maximum take-off mass of 2100 kg and a wing area of 18 m^2. The aircraft stall speed is 60 KEAS. Other aircraft characteristics are as follows:

$$b = 12\text{ m};\; S_V = 4.2\text{ m}^2;\; b_v = 3.1\text{ m};\; b_R = b_v;\; C_R/C_V = 0.45;\; \eta_v = 0.95;$$

$$\delta_{R_{max}} = 25\text{ deg};\; C_{L_{\alpha V}} = 4.4\,\frac{1}{\text{rad}};\; I_{xx_B} = 12\,000\text{ kg m}^2;\; I_{zz_B} = 15\,000\text{ kg m}^2;$$

$$I_{xz_B} = 0$$

Is this rudder able to satisfy the spin recovery requirement at 10 000 ft altitude? Assume the aircraft will spin at an angle of attack of 50 deg.

12. A fighter aircraft has a maximum take-off mass of 16 000 kg and a wing area of 52 m^2. The aircraft stall speed is 86 KEAS. Other aircraft characteristics are as follows:

$$b = 11\text{ m};\; S_V = 9.4\text{ m}^2;\; b_v = 2.9\text{ m};\; b_R = b_v;\; C_R/C_V = 0.28;\; \eta_v = 0.98;$$

$$C_{L_{\alpha V}} = 4.1\,\frac{1}{\text{rad}};\; \delta_{R_{max}} = 30\text{ deg};\; I_{xx_B} = 35\,000\text{ kg m}^2;\; I_{zz_B} = 180\,000\text{ kg m}^2;$$

$$I_{xz_B} = 2200\text{ kg m}^2$$

Is this rudder powerful enough to satisfy the spin recovery requirement at 20 000 ft altitude? Assume the aircraft will spin at an angle of attack of 60 deg.

13. A jet trainer aircraft has a maximum take-off mass of 1800 kg and a wing area of 13 m^2. The aircraft stall speed is 75 KEAS. Other aircraft characteristics are as follows:

$$b = 8\text{ m};\; S_V = 3.8\text{ m}^2;\; b_v = 2.1\text{ m};\; b_R = b_v;\; C_R/C_V = 0.4;\; \eta_v = 0.95;$$

$$\delta_{R_{max}} = 30\text{ deg};\; C_{L_{\alpha V}} = 4.6\,\frac{1}{\text{rad}};\; I_{xx_B} = 1100\text{ kg m}^2;\; I_{zz_B} = 7500\text{ kg m}^2;$$

$$I_{xz_B} = 300\text{ kg m}^2$$

Is this rudder powerful enough to satisfy the spin recovery requirement at 15 000 ft altitude? Assume the aircraft will spin at an angle of attack of 70 deg.

14. A land-based trainer aircraft has a maximum take-off mass of 1750 kg and a wing area of 14 m^2. The high-lift device has already been designed and the outboard flap location is determined to be at 70% of the wing semispan. The wing rear spar is located at 75% of the wing chord. The aircraft has a conventional configuration and the following geometry and weight characteristics:

$$AR = 6,\ \lambda = 0.75,\ S_h = 4.8\,\text{m}^2,\ S_v = 4.1\ \text{m}^2,\ V_s = 74\ \text{knot},\ C_{L_{\alpha w}} = 4.8\ 1/\text{rad},$$

$$I_{xx} = 1200\,\text{kg m}^2$$

Design the ailerons to meet the MIL-STD roll control requirements.

15. A ground attack jet aircraft has a maximum take-off mass of 5000 kg and a wing area of 16 m^2. The high-lift device has outboard location at 65% of the wing semispan, and the wing rear spar is located at 80% of the wing chord. The aircraft has a conventional configuration and the following geometry and weight characteristics:

$$AR = 7,\ \lambda = 0.8,\ S_h = 4.8\,\text{m}^2,\ S_v = 4.1\ \text{m}^2,\ V_s = 74\ \text{knot},\ C_{L_{\alpha w}} = 4.8\ 1/\text{rad},$$

$$I_{xx} = 1200\,\text{kg m}^2$$

Design the ailerons to meet the MIL-STD roll control requirements.

16. A twin-turbofan airliner aircraft has a maximum take-off mass of 45 000 kg and a wing area of 93 m^2. The stall speed is 98 knot. The aircraft has a T-tail configuration and the following geometry and weight characteristics:

$$AR = 8.4,\ \lambda = 0.3,\ S_h = 22\,\text{m}^2,\ S_v = 12.3\ \text{m}^2,\ C_{L_{\alpha w}} = 4.8\ 1/\text{rad},$$

$$I_{xx} = 510\,000\,\text{kg m}^2$$

The aircraft is expected to satisfy the roll control requirement of a military transport aircraft. Design the ailerons.

17. Figure 12.54 illustrates the geometry of a high-wing twin-jet engine long-range transport aircraft equipped with tricycle landing gear. The aircraft has the following characteristics:

$$m_{TO} = 41\,000\,\text{kg},\ V_s = 105\ \text{KEAS},\ I_{yy} = 1\,700\,000\,\text{kgm}^2,\ T_{max} = 2 \cdot 65\,\text{kN},$$

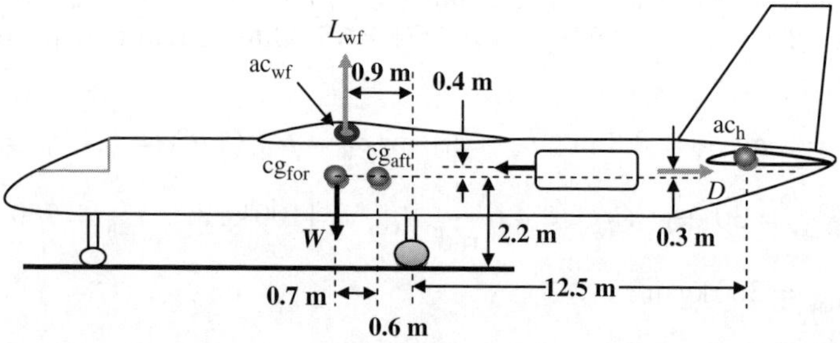

Figure 12.54 Aircraft geometry for Problem 17

$L_f = 30\,\text{m}, V_c = 490\,\text{KTAS (at 30 000 ft)}, C_{L_o} = 0.2, C_{D_{oC}} = 0.021,$

$C_{D_{oTO}} = 0.042, C_{L_\alpha} = 5.3\ 1/\text{rad}$

Wing : $S = 85\ \text{m}^2$, $\text{AR} = 10$, $C_{L_{\alpha_{wf}}} = C_{L_{\alpha_w}} = 5.41/\text{rad}$, $e = 0.93$, $\lambda = 0.4$,

$\Delta C_{L_{flapTO}} = 0.7$, $C_{m_{ac_{wf}}} = -0.03$, $i_w = -1\ \text{deg}$, $h_o = 0.23$, $\alpha_{s_{TO}} = 13\ \text{deg}$

Horizontal tail : $S_h = 22\ \text{m}^2$, $b_h = 7\ \text{m}$, $C_{L_{\alpha_h}} = 4.1\ 1/\text{rad}$, $i_h = -1.5\ \text{deg}$,

$\lambda_h = 0.6$, $\eta_h = 0.96$, $\alpha_{h_s} = 15\ \text{deg}$

Airfoil section : NACA 0012, $\alpha_{twist} = 0$

Design an elevator for this aircraft.

18. Figure 12.55 illustrates a low-wing twin-engine turboprop commuter aircraft with tricycle landing gear. The aircraft has the following characteristics:

$m_{TO} = 7500\,\text{kg}, V_s = 89\,\text{KEAS}, I_{yy} = 130\,000\ \text{kg m}^2, P_{max} = 2 \cdot 746\,\text{kW},$

$L_f = 18\,\text{m}, V_C = 246\,\text{KTAS (at 20 000 ft)}, C_{L_o} = 0.15, C_{D_{oC}} = 0.024,$

$C_{D_{oTO}} = 0.045$, $C_{L_\alpha} = 5.3\ 1/\text{rad}$

Wing: $S = 30\ \text{m}^2$, $b = 18\ \text{m}$, $C_{L_{\alpha_{wf}}} = 5.4\ 1/\text{rad}$, $e = 0.93$, $\lambda = 0.5$,

$\Delta C_{L_{flapTO}} = 0.96$, $C_{m_{ac_{wf}}} = -0.03$, $i_w = 1\ \text{deg}$, $h_o = 0.22$, $\alpha_{s_{TO}} = 13\ \text{deg}$

Horizontal tail: $S_h = 5.7\ \text{m}^2$, $b_h = 4.6\ \text{m}$, $C_{L_{\alpha_h}} = 4.1\ 1/\text{rad}$, $i_h = -1.5\ \text{deg}$,

$\lambda_h = 0.6$, $\eta_h = 0.96$, $\alpha_{h_s} = 15\ \text{deg}$

Airfoil section: NACA 0012, $\alpha_{twist} = 0$

Design an elevator for this aircraft.

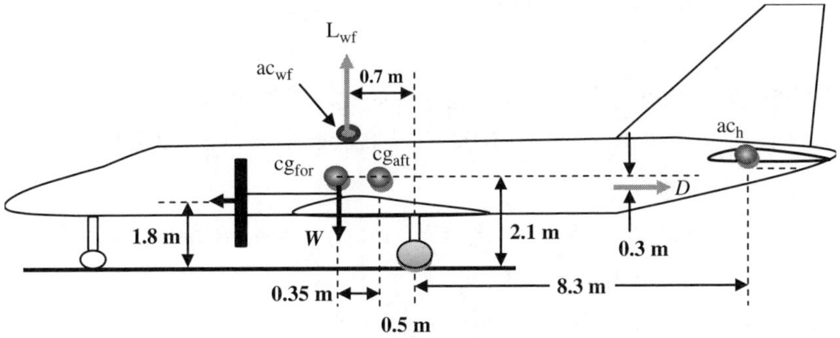

Figure 12.55 Aircraft geometry for Problem 18

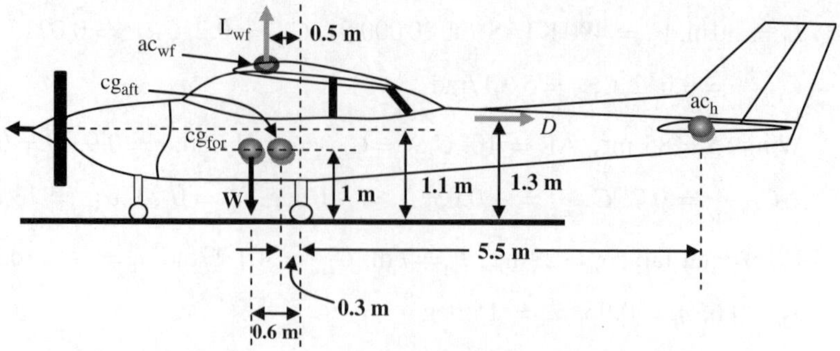

Figure 12.56 Aircraft geometry for Problem 19

19. Figure 12.56 illustrates a high-wing single-engine turboprop GA aircraft with tricycle landing gear. The aircraft has the following characteristics:

$m_{TO} = 3500\,kg$, $V_s = 60$ KEAS, $I_{yy} = 5000\,kg\,m^2$, $P_{max} = 447\,kW$, $L_f = 11$ m, $V_C = 160$ KTAS (at 15 000 ft), $C_{L_o} = 0.1$, $C_{D_{oC}} = 0.029$, $C_{D_{oTO}} = 0.05$, $C_{L_\alpha} = 5.3$ 1/rad

Wing : $S = 25\,m^2$, $b = 16\,m$, $C_{L_{\alpha_{wh}}} = C_{L_{\alpha w}} = 5.61/rad$, $e = 0.9$, $\lambda = 0.8$, $\Delta C_{L_{flapTO}} = 0.5$, $C_{m_{ac_{wf}}} = -0.05$, $i_w = 2\,deg$, $h_o = 0.21$, $\alpha_{s_{TO}} = 13\,deg$

Horizontal tail: $S_h = 6.5\ m^2$, $b_h = 5.2$ m, $C_{L_{\alpha_h}} = 4.4$ 1/rad, $i_h = -2$ deg, $\lambda_h = 0.7$, $\eta_h = 0.9$, $\alpha_{h_s} = 14\,deg$

Airfoil section: NACA 0009, $\alpha_{twist} = 0$

Design an elevator for this aircraft.

20. Figure 12.57 illustrates a jet fighter aircraft with tricycle landing gear. The aircraft has the following characteristics:

$m_{TO} = 12500\,kg$, $V_s = 180$ KEAS, $I_{yy} = 110\,000\,kg\,m^2$, $T_{max} = 131$ kN, $L_f = 15$ m, $M_C = 2$ (at 50 000 ft), $C_{L_o} = 0.1$, $C_{D_{oC}} = 0.029$, $C_{D_{oTO}} = 0.034$, $C_{L_\alpha} = 5.3$ 1/rad

Wing: $S = 28\ m^2$, AR = 3.2, $C_{L_{\alpha_{wh}}} = C_{L_{\alpha w}} = 5.6$ 1/rad, $e = 0.9$, $\lambda = 0.3$, $\Delta C_{L_{flapTO}} = 0.5$, $C_{m_{ac_{wf}}} = -0.02$, $i_w = 0$ deg, $h_o = 0.23$ (at subsonic speeds), $\alpha_{s_{TO}} = 14$ deg

Horizontal tail: $S_h = 6.5\ m^2$, $b_h = 5.6$ m, $C_{L_{\alpha_h}} = 4.4$ 1/rad, $i_h = -2$ deg, $\lambda_h = 0.7$, $\eta_h = 0.9$, $\alpha_{h_s} = 14$ deg

Airfoil section: NACA 64A-204, $\alpha_{twist} = 0$

Design an elevator for this aircraft.

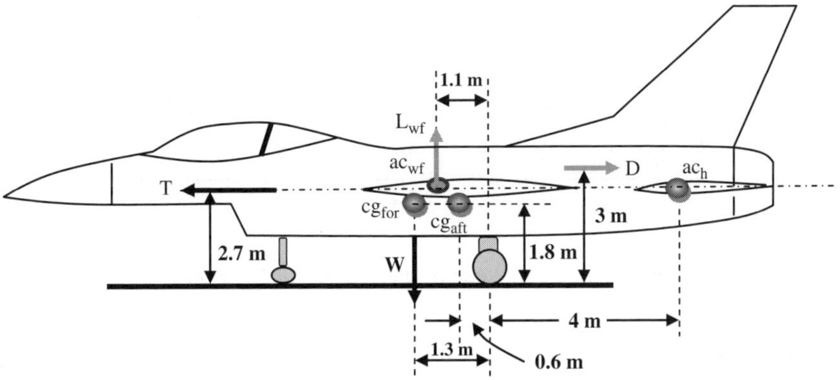

Figure 12.57 Aircraft geometry for Problem 20

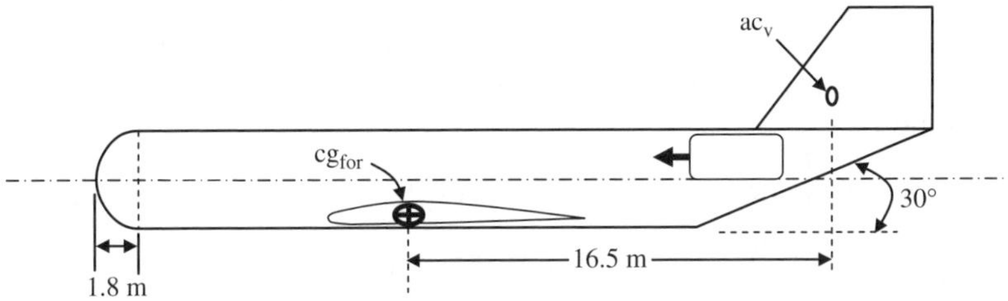

Figure 12.58 Side view of the aircraft for Problem 21

21. A twin-turbofan engine transport aircraft has a maximum take-off mass of 63 000 kg. Each engine placed at the rear fuselage generates 82.3 kN of thrust. The distance between the most aft and the most forward cg is 0.8 m, and between the two engines is 6 m. The side view of the aircraft is shown in Figure 12.58, and the fuselage has a cylindrical shape with a semispherical nose. Other characteristics of the aircraft are as follows:

$$S = 115\,\text{m}^2,\ b = 33\,\text{m}, S_\text{V} = 35\,\text{m}^2\ ,\ \text{AR} = 1.2,\ \lambda_\text{V} = 0.8,\ V_\text{s} = 110\ \text{knot},$$
$$L_\text{f} = 41\,\text{m},\ D_\text{f} = 3.6\,\text{m},\ C_{L_{\alpha\text{V}}} = 4.5\ \text{1/rad},\ \eta_\text{v} = 0.96,\ \text{d}\sigma/\text{d}\beta = -0.06,\ C_{\text{n}_\text{o}} = 0,$$
$$C_{y_\text{o}} = 0$$

The aircraft is not spinnable, and is required to be able to land safely when there is a cross-wind of 35 knots. Design a rudder for this aircraft.

22. A jet trainer aircraft has a maximum take-off mass of 1700 kg and a wing area of 13 m^2. The side view of the aircraft is shown in Figure 12.59, and the fuselage has a cylindrical shape. Other characteristics of the aircraft are as follows:

$$\text{AR} = 6,\ S_\text{V} = 4.1\ \text{m}^2, \text{AR}_\text{V} = 1.2,\ \lambda_\text{V} = 0.6,\ V_\text{s} = 70\ \text{knot},\ L_\text{f} = 9\ \text{m},$$
$$D_\text{f} = 1.8\ \text{m},\ \eta_\text{V} = 0.96,\ C_{L_{\alpha\text{V}}} = 4.5\ \text{1/rad},\ \text{d}\sigma\big/\text{d}\beta = -0.1, C_{\text{n}_\text{o}} = 0,\ C_{\text{y}_\text{o}} = 0,$$
$$I_{xx_\text{B}} = 1200\,\text{kg}\,\text{m}^2,\ I_{zz_\text{B}} = 7000\,\text{kg}\,\text{m}^2,\ I_{xz_\text{B}} = 300\,\text{kg}\,\text{m}^2$$

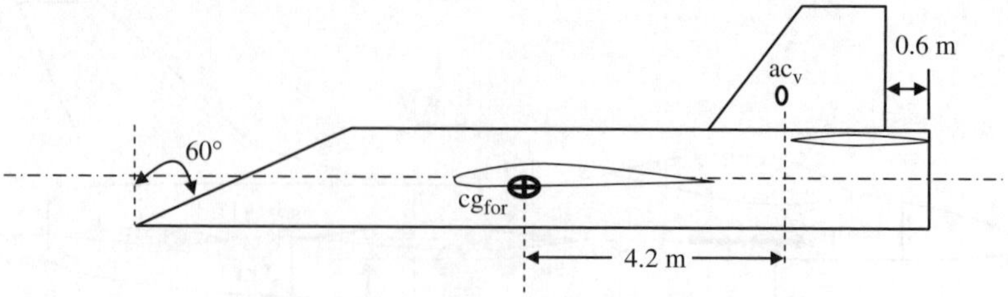

Figure 12.59 Side view of the aircraft for Problem 22

The aircraft is spinnable at 60 deg angle of attack. The aircraft is required to be able to land safely when there is a cross-wind of 25 knots, as well as able to recover from a spin at 5000 ft altitude in a maximum period of one turn. Design a rudder for this aircraft.

23. A single-engine acrobatic aircraft has a maximum take-off mass of 2300 kg and a wing area of 20 m^2. The aircraft stall speed is 61 KEAS. Other aircraft characteristics are as follows:

$$b = 13\ \text{m},\ S_V = 4.5\ \text{m}^2,\ b_v = 3.3\ \text{m},\ \eta_v = 0.95,\ C_{L_{\alpha V}} = 4.4\ 1/\text{rad},$$

$$I_{xx_B} = 13\,000\ \text{kg m}^2,\ I_{zz_B} = 16\,000\ \text{kg m}^2,\ I_{xz_B} = 0$$

Design a rudder such that the aircraft can recover from a spin at an angle of attack of 50 deg at 5000 ft altitude in a maximum period of one turn.

24. A jet fighter aircraft has a maximum take-off mass of 15 000 kg and a wing area of 45 m^2. The aircraft has the following characteristics:

$$\text{AR} = 3.4,\ S_V = 12\ \text{m}^2,\ \text{AR}_v = 1.3\ \text{m},\ \eta_v = 0.95,\ C_{L_{\alpha V}} = 4.4\ 1/\text{rad},$$

$$I_{xx_B} = 30\ 000\ \text{kg m}^2, I_{zz_B} = 200\,000\ \text{kg m}^2, I_{xz_B} = 2600\ \text{kg m}^2$$

Design a rudder such that the aircraft can recover from a spin at an angle of attack of 70 deg at 20 000 ft altitude in a maximum period of one turn.

References

[1] Hawkins, F.H. (1998) *Human Factors in Flight*, 2nd edn, Ashgate.
[2] Bridger, R.S. (2008) *Introduction to Ergonomics*, 3rd edn, CRC Press.
[3] Kroemer, K.H.E., Kroemer, H.B., and Kroemer-Elbert, K.E. (2000) *Ergonomics: How to Design for Ease and Efficiency*, 2nd edn, Prentice Hall.
[4] Salyendy, G. (2006) *Handbook of Human Factors and Ergonomics*, 3rd edn, Wiley-VCH Verlag GmbH.
[5] Vink, P. (2011) *Aircraft Interior Comfort and Design; Ergonomics Design Management: Theory and Applications*, CRC Press.
[6] Federal Aviation Regulations (2011) Part 23 , Airworthiness Standards: Normal, Utility, Aerobatic, and Commuter Category Airplanes, Federal Aviation Administration, Department of Transportation, Washington.
[7] Federal Aviation Regulations (2011) Part 25, Airworthiness Standards: Transport Category Airplanes, Federal Aviation Administration, Department of Transportation, Washington.

[8] MIL-STD-1797 (1997). *Flying Qualities of Piloted Aircraft*, Department of Defense, Washington, DC.
[9] MIL-F-8785C (1980). *Military Specification: Flying Qualities of Piloted Airplanes*, Department of Defense, Washington, DC.
[10] Harper, R.P. and Cooper, G.E. (1986) Handling qualities and pilot evaluation. *Journal of Guidance, Control, and Dynamics*, **9** (5) 515–529.
[11] Stevens, B.L. and Lewis, F.L. (2003) *Aircraft Control and Simulation*, 2nd edn, Wiley-VCH Verlag GmbH.
[12] Roskam, J. (2007) *Airplane Flight Dynamics and Automatic Flight Control*, DAR Corporation.
[13] Mclean, D. (1990) *Automatic Flight Control Systems*, Prentice-Hall.
[14] Nelson, R. (1989) *Flight Stability and Automatic Control*, McGraw Hill.
[15] McCormick, B.W. (1979) *Aerodynamics, Aeronautics and Flight Mechanics*, Wiley-VCH Verlag GmbH.
[16] Etkin, B. and Reid, L.D. (1996) *Dynamics of Flight-Stability and Control*, 3rd edn, Wiley-VCH Verlag GmbH.
[17] Hoak, D.E., Ellison, D.E. and Fink, R.D. (1978) USAF Stability and Control DATCOM, Flight Control Division, Air Force Flight Dynamics Laboratory, Wright-Patterson AFB, Ohio.
[18] Spiegel, M.R. and Liu, J. (1999) *Mathematical Handbook of Formulas and Tables*, 2nd edn, Schaum's Outlines, McGraw-Hill.
[19] Abbott, I.H. and Von Doenhoff, A.F. (1959) *Theory of Wing Sections*, Dover, New York.
[20] Spiegel, M.R. and Liu, J. (1999) *Schaum's Outline Series in Mathematical Handbook of Formulas and Tables*, McGraw-Hill.
[21] Sadraey, M. and Colgren, R. (2006) Derivations of major coupling derivatives, and the state space formulation of the coupled equations of motion. 6th AIAA Aviation Technology, Integration and Operations Conference (ATIO), Wichita, KS, September 25–27, AIAA-2006-7790.
[22] Roskam, J. (2003) *Airplane Design*, DAR Corporation.
[23] Torenbeek, E. (1996) *Synthesis of Subsonic Airplane Design*, Delft University Press.
[24] Stinton, D. (2001) *The Design of the Aeroplane*, AIAA.
[25] Jackson, P. *Jane's All the World's Aircraft*, Jane's Information Group, 1996–2011.
[26] Bourke, J. (2012) RC groups, http://www.rcgroups.com/forums/showthread.php?t=557457.
[27] Joint Aviation Requirements CS-23 (2007) Normal, Utility, Aerobatic, and Commuter Category Aeroplanes, European Aviation Safety Agency.
[28] Joint Aviation Requirements CS-25 (2007) Large Aeroplanes, European Aviation Safety Agency.

Appendices

Appendix A

Standard Atmosphere, SI Units

Altitude (m)	*T* (K)	*P* (N/m^2)	ρ (kg/m^3)
0	288.15	101 325	1.225
1000	281.65	89 876	1.1117
2000	275.15	79 501	1.007
3000	268.67	70 121	0.9093
4000	262.18	61 660	0.8193
5000	255.69	54 048	0.7364
6000	249.20	47 217	0.6601
7000	242.71	41 105	0.590
8000	236.23	35 651	0.526
9000	229.74	30 800	0.467
10 000	223.26	26 500	0.413
11 000	216.78	22 700	0.365
12 000	216.66	19 399	0.312
13 000	216.66	16 579	0.267
14 000	216.66	14 170	0.228
15 000	216.66	12 112	0.195
16 000	216.66	10 353	0.166
17 000	216.66	8850	0.142
18 000	216.66	7565	0.122
19 000	216.66	6467	0.104
20 000	216.66	5529	0.089
21 000	216.66	4727	0.076
22 000	216.66	4042	0.065
23 000	216.66	3456	0.056
24 000	216.66	2955	0.047
25 000	216.66	2527	0.041

$\rho_o = 1.225\,\text{kg/m}^3$, $T_o = 15°\text{C} = 288.15\,\text{K}$, $P_o = 101\,325\,\text{N/m}^2$, $a_o = 340.29\,\text{m/s}$, $\mu_o = 1.785 \times 10^{-5}\,\text{kg/m/s}$.

Aircraft Design: A Systems Engineering Approach, First Edition. Mohammad H. Sadraey.

Appendix B

Standard Atmosphere, British Units

Altitude (ft)	T (°R)	P (lb/ft^2)	ρ (slug/ft^3)	Altitude (ft)	T (°R)	P (lb/ft^2)	ρ (slug/ft^3)
0	518.7	2116.2	0.002378	31 000	408.3	601.6	0.000858
1000	515.12	2040.9	0.002308	32 000	404.7	574.6	0.000827
2000	511.5	1967.7	0.002241	33 000	401.2	548.5	0.000796
3000	508	1896.7	0.002175	34 000	397.6	523.5	0.000767
4000	504.43	1827.7	0.002111	35 000	394.1	499.3	0.000738
5000	500.86	1761	0.002048	36 000	390.5	476.1	0.000710
6000	497.3	1696	0.001987	37 000	390	453.9	0.000678
7000	493.7	1633.1	0.001897	38 000	390	432.6	0.000646
8000	490.2	1572.1	0.001868	39 000	390	412.4	0.000616
9000	486.6	1513	0.001811	40 000	390	393.1	0.000587
10 000	483	1455.6	0.001755	41 000	390	374.7	0.00056
11 000	479.5	1400	0.001701	42 000	390	357.2	0.000533
12 000	475.9	1346.2	0.001648	43 000	390	340.5	0.000509
13 000	472.4	1294.1	0.001596	44 000	390	324.6	0.000485
14 000	468.8	1243.6	0.001545	45 000	390	309.5	0.000462
15 000	465.2	1195	0.001496	46 000	390	295	0.00044
16 000	461.7	1147.5	0.001448	47 000	390	281.2	0.00042
17 000	458.1	1101.7	0.001401	48 000	390	268.1	0.0004
18 000	454.5	1057.5	0.001355	49 000	390	255.5	0.000381
19 000	451	1014.7	0.001311	50 000	390	243.6	0.000364
20 000	447.4	973.3	0.001267	51 000	390	232.2	0.000347
21 000	443.9	933.3	0.001225	52 000	390	221.4	0.00033
22 000	440.3	894.6	0.001184	53 000	390	211	0.000315
23 000	436.8	857.3	0.001143	54 000	390	201.2	0.0003
24 000	433.2	821.2	0.001104	55 000	390	191.8	0.000286
25 000	429.6	786.3	0.001066	56 000	390	182.8	0.000273
26 000	426.1	752.7	0.00103	57 000	390	174.3	0.00026
27 000	422.5	720.3	0.000993	58 000	390	166.2	0.000248
28 000	419	689	0.000958	59 000	390	158.4	0.000236
29 000	415.4	658.8	0.000923	60 000	390	151	0.000225
30 000	411.9	629.7	0.00089	61 000	390	144	0.000215

$\rho_o = 0.002378$ slug/ft^3, $T_o = 518.7$°R, $P_o = 2116.2$ lb/ft^2 $= 14.7$ psi, $a_o = 1116.4$ ft/s, $\mu_o = 1.199 \times 10^{-4}$ lb/ft s.

Index

Aircraft Design: A Systems Engineering Approach, First Edition. Mohammad H. Sadraey.